*Solid State Chemistry
and its Applications*

Solid State Chemistry and its Applications

ANTHONY R. WEST,

Department of Chemistry
University of Aberdeen

JOHN WILEY & SONS

Chichester · New York · Brisbane · Toronto · Singapore

Library of Congress Cataloging in Publication Data:

West, Anthony R.
  Solid State Chemistry and its Applications
  Includes index.
  1. Solid state chemistry. I. Title.
0D47.W47   1984   540'.421   83-21607

ISBN 0 471 90377 9 (U.S.)

British Library Cataloguing in Publication Data:

West, Anthony R.
  Solid state chemistry and its applications.
  1. Solid state chemistry
  I. Title
  541'.0421   QD478

ISBN 0 471 90377 9

Photosetting by Thomson Press (India) Limited, New Delhi.
Printed in Great Britain by Courier International Ltd, Tiptree, Essex.

Contents

Preface

Solid state chemistry is a subject that is very relevant to modern technology. Nevertheless, it has been rather neglected in the teaching of inorganic and physical chemistry. Perhaps, in part, this arises because although there are many specialized works on selected topics in solid state chemistry, there are few if any, that give a general coverage of the subject ranging from synthesis, characterization and structural considerations to properties and applications. This book was written in an attempt to fill the gap and perhaps even define what is meant by solid state chemistry. It was not written with any special course in mind. In retrospect, it may serve as a moderately detailed and non-mathematical introduction to the subject at either undergraduate or postgraduate level. The coverage is broad and cuts across many specialities; inevitably there are inaccuracies in the treatment and areas where the coverage is inadequate. Choice of subject matter was difficult, especially for the later chapters and I am conscious that certain areas, such as catalysis have been excluded. However, I hope that the aim of achieving a uniform coverage will offset any such deficiencies. Many people read and commented upon various sections, for which I thank them. To my wife, I owe an enormous debt for her understanding and encouragement over the seven-year period during which the manuscript was prepared. She also helped greatly with the practicalities by typing a large part of the manuscript. Many of the diagrams (the good ones!) were drawn by Mr Steve Black and Mrs Judee Kerr.

A. R. WEST

Aberdeen, March 1984

Chapter 1

What is Solid State Chemistry?

This book sets out to introduce what, in the author's view, constitutes solid state chemistry. There are many areas of overlap and interest in the solid state sciences: solid state physics, solid state chemistry, materials science, ceramics and ceramic engineering, mineralogy and metallurgy. Only in recent years, however, has solid state chemistry begun to be recognized as a separate branch of physical science although, arguably, it is the most central of the solid state sciences. What, then, is solid state chemistry?

Solid state chemistry is concerned with the synthesis, structure, properties and applications of solid materials. The materials are usually inorganic, but not exclusively so. Thus, metals may be including during consideration of crystal structures and related areas such as crystal defects, solid solutions, phase transitions and phase diagrams. Organic solids are also included in cases (a) where they exhibit interesting physical properties such as high electrical conductivity and (b) where the reactions of organic solids are subject to topochemical control, i.e. where they are influenced by their molecular packing arrangement. Minerals are included since they are examples of inorganic solids that occur in nature. The materials of interest are usually crystalline, but not always so; thus, some aspects of glasses are very relevant to solid state chemistry.

The large majority of inorganic solids are non-molecular and their structure is determined by the manner in which the atoms or ions are packed together in three dimensions. The variety and complexity of structure types which they show is central to an appreciation of solid state chemistry. This includes not only the description and classification of crystal structures, with a working knowledge of space groups, but also an evaluation of the factors that influence and control crystal structures. By contrast, in molecular substances the structure and properties are those of the individual molecules. The fact that many of them are solids at room temperature is incidental. Solid state chemistry is therefore concerned mainly with non-molecular materials.

An additional and important structural aspect is the defect structure of solids. All solids contain defects of some kind and often these have a great influence on properties such as electrical conductivity, mechanical strength and chemical reactivity. Closely related to this is the subject of solid solutions. These allow the compositions of solids to be varied significantly within the framework of the same

1

overall crystal structure. By varying the composition in this way, it is possible to systematically control or modify many properties.

The methods that are used to synthesize solids are obviously very important. Several methods are available including solid state reaction, vapour phase transport, precipitation and electrochemical methods. The solid may be prepared in a variety of forms, including single crystal, powder, solid piece, etc. Many of the synthetic methods that are used are unique to solid state chemistry and are not encountered in other branches of chemistry.

The physical techniques and methods that are used to analyse and study solids are also often quite different from those that are used in other branches of chemistry. Thus, there is much less emphasis on spectroscopic methods and, instead, more emphasis on a variety of diffraction (principally X-ray diffraction) and microscopic (principally electron microscopic) techniques. This is not to say that spectroscopic methods are unimportant in solid state chemistry: they are indeed very important in certain specialized areas. However, spectroscopic methods cannot be used to great effect either for the characterization of solids or for general structure determination. This contrasts sharply with the situation in molecular inorganic and organic chemistry where, for many years, a combination of spectroscopic methods has provided the principal means of elucidating molecular structures.

X-ray diffraction has two principal uses in solid state chemistry. First, almost all crystal structures are solved by single crystal X-ray diffraction. Exceptions arise when either electron or neutron diffraction is used or when powdered samples are used for X-ray diffraction. Second, all crystalline solids have their own characteristic X-ray powder diffraction pattern which may be used as a means of 'fingerprinting'. However, powder diffraction has many other applications besides straightforward phase identification. These include: study of polymorphism, phase transitions and solid solutions, determination of accurate unit cell parameters, particle size measurement, phase diagram determination. As far as analytical techniques go, powder X-ray diffraction is unusual, if not unique, since it is a technique for *phase* analysis. This contrasts with other analytical techniques, including various of the spectroscopies, which give an *elemental* analysis.

The conditions used for the synthesis of compounds, their stability once prepared and any reactions that they may undergo, are represented by the appropriate equilibrium phase diagram. The ability to interpret phase diagrams is vitally important in solid state chemistry. Unfortunately, phase diagrams of solid systems, especially complex ones, is an area that is often neglected in chemistry.

The importance of a knowledge of the bonding forces that hold solids together is self-evident. However, quantitative calculations of, for example, bond energies are difficult to make because most inorganic solids have bonds that are a blend of ionic and covalent. Thus, the treatment of ideal, extreme bond types—ionic bonds by lattice energy calculations and covalent bonds by molecular orbital theory—is possible and refined calculations may be made, for instance, of the

NaCl structure or the H_2 molecule. The number of substances that may be treated in this way is very limited, however. Most so-called ionic substances have a degree of covalent bonding, in which the outer electron clouds on the anions are polarized towards the extra positive charge on the cations. Also, most so-called covalent substances have bonds which exhibit some polarity, arising from the difference in electronegativity of the elements present. Many solids can be treated according to band (or zone) theory. This may be applied qualitatively to rationalizing the structure, spectroscopic behaviour, conductivity, etc., of a wide range of solids. It is particularly invaluable in understanding the conductivity of metals and semiconductors. Vast tracts of solid state physics are concerned with the detailed electronic structure of solids, analysed according to band theory; most of this lies outside the sphere of solid state chemistry.

There is great diversity in the range of properties and applications of solids and an analysis of structure—property relations is fundamental to solid state chemistry. The structure of solids on at least three levels must be considered: the crystal structure of ideally perfect solids; the defect structure of solids, including the structure of surfaces; and the texture of polycrystalline solids. Sometimes a particular property is controlled entirely by one level of structure, e.g. the hardness of diamond relative to graphite is a consequence of its crystal structure. More often, however, the property depends on an interplay between the different levels of structure. For instance, in semiconductors such as silicon and gallium arsenide it appears to be essential that they have the diamond structure. However, the actual level of conductivity depends on the defect or dopant concentration. It is also important that single crystal material be used since the presence of grain boundaries, in a polycrystalline sample, would have deleterious effects. The structure at all levels is important in these semiconductors, therefore. In other properties, such as certain magnetic behaviour and mechanical strength, grain boundaries have a beneficial effect and supplement the behaviour observed with single crystal materials.

The study of structure–property relations is a very fruitful area and one with immense possibilities for the development of new materials or materials with unusual combinations of properties. Thus one can envisage the potential of 'crystal engineering' whereby it is increasingly possible to design new materials which have a specific structure and properties.

Chapter 2

Preparative Methods

Various methods are available for preparing solids, the method adopted depending to a certain extent on the form of the desired product. Thus, crystalline solids may take the form of:

(a) a single crystal that is as pure and as free from defects as possible.
(b) a single crystal whose structure has been modified by the creation of defects, usually as a consequence of introducing specific impurities,
(c) a powder, i.e. a large number of small crystals,
(d) a polycrystalline solid piece, e.g. a pellet or a ceramic tube, in which a large number of crystals are present, in various orientations,
(e) a thin film.

In addition, an important class of solid materials is non-crystalline, amorphous or glassy. Non-crystalline solids may also be prepared in various forms, e.g. as tubes, pellets or thin films.

The above classes of solids each have their own special preparative methods, many of which are not encountered in 'wet' chemical laboratories. Some of these methods are discussed in this chapter.

2.1 Solid state reaction

2.1.1 General principles

Probably the most widely used method for the preparation of polycrystalline solids (i.e. powders) is the direct reaction, in the solid state, of a mixture of solid starting materials. Solids do not usually react together at room temperature over normal timescales and it is necessary to heat them to much higher temperatures, often 1000 to 1500 °C, in order for reaction to occur at an appreciable rate. This shows that both thermodynamic and kinetic factors are important in solid state reactions; thermodynamic considerations show whether or not a particular reaction should occur by considering the changes in free energy that are involved; kinetic factors determine the rate at which the reaction occurs.

In discussing the factors that influence reactions of solids, it is useful to refer to a specific reaction. In what follows, let us consider the reaction of MgO and Al_2O_3, in a 1:1 molar ratio, to form spinel $MgAl_2O_4$.

Reaction conditons. Although, from thermodynamic considerations MgO and Al_2O_3 should react to form $MgAl_2O_4$, in practice the reaction rate is extremely slow at normal temperatures. Only above ~ 1200 °C does reaction begin to occur to any significant extent and for complete reaction of a powder mixture to occur it may be necessary to heat it for several days at, for example, 1500 °C.

Structural considerations. $MgAl_2O_4$ has a crystal structure that shows similarities and differences to those of both MgO and Al_2O_3; both MgO and spinel have a cubic close packed array of oxide ions, in contrast to Al_2O_3 which has a distorted hexagonal close packed array of oxide ions; on the other hand, the Al^{3+} ions occupy octahedral sites in both Al_2O_3 and spinel, whereas the Mg^{2+} ions are octahedral in MgO but tetrahedral in $MgAl_2O_4$.

6

Why are solid state reactions difficult? In order to appreciate why solid state reactions take place with great difficulty and only at high temperatures, consider the reaction of two crystals of MgO and Al_2O_3 which are in intimate contact across one shared face (Fig. 2.1a). After an appropriate heat treatment, the crystals have partially reacted to form a layer of $MgAl_2O_4$ at the interface, as shown in Fig. 2.1(b). The first stage of reaction is the formation of $MgAl_2O_4$ nuclei. This nucleation is rather difficult because of (a) the considerable differences in structure between reactants and product and (b) the large amount of structural reorganization that is involved in forming the product: bonds must

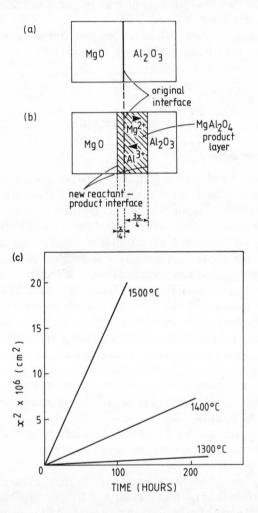

Fig. 2.1 (a, b) Schematic reaction, by interdiffusion of cations, of single crystals of MgO and Al_2O_3 to give $MgAl_2O_4$. (c) Thickness, x, of $NiAl_2O_4$ product layer as a function of temperature and time formed by reaction of NiO and Al_2O_3. (From Pettit, Randklev and Felton, 1966)

be broken and reformed and atoms must migrate, perhaps over considerable distances (on an atomic scale). Ions such as Mg^{2+} in MgO and Al^{3+} in Al_2O_3 are normally regarded as being trapped on their appropriate lattice sites and it is difficult for them to hop into empty, adjacent sites. Only at very high temperatures do such ions have sufficient thermal energy to enable them occasionally to jump out of their normal lattice sites and diffuse through the crystals. In summary, therefore, nucleation of $MgAl_2O_4$ probably involves some reorganization of the oxide ions at the site of the potential nucleus together with the interchange of Mg^{2+} and Al^{3+} ions across the interface between the MgO and Al_2O_3 crystals.

Although nucleation is a difficult process, the subsequent stage, involving growth of the product layer, may well be even more so. In order for further reaction to occur and the $MgAl_2O_4$ layer to grow thicker, counter diffusion of Mg^{2+} and Al^{3+} ions must occur right through the existing $MgAl_2O_4$ product layer (Fig. 2.1b) to the new reaction interfaces. At this stage, there are two reaction interfaces: that between MgO and $MgAl_2O_4$ and that between $MgAl_2O_4$ and Al_2O_3. Let us assume that the rate limiting step for further reaction is the diffusion of Mg^{2+} and Al^{3+} ions to and from these interfaces. Since the diffusion rates are slow, even at high temperatures further reaction takes place only slowly and at a decreasing rate as the spinel layer grows thicker.

Reaction rates. A detailed study of the kinetics of a related reaction, that between polycrystalline pellets of NiO and Al_2O_3 to form $NiAl_2O_4$ spinel, has shown that interdiffusion of cations through the spinel product layer is indeed the rate controlling step. In the simple case of lattice diffusion through a planar layer, diffusion is governed by a parabolic rate law of the form:

$$\frac{dx}{dt} = k \cdot x^{-1}$$

or

$$x = (k't)^{1/2} \tag{2.1}$$

where x is the amount of reaction (here equal to the thickness of the growing, spinel layer), t is time, and k, k' are rate constants. It was shown that the rate of formation of $NiAl_2O_4$ fitted equation (2.1) by measuring the thickness, x, of the spinel product layer, as in Fig. 2.1(b), as a function of time. Results are shown for three temperatures in Fig. 2.1(c): x^2 is plotted against time and straight line plots are obtained of slopes equal to k'. The reaction occurs much more quickly with increasing temperature, as would be expected. The activation energy for the reaction could be obtained from an Arrhenius plot of log k' versus T^{-1}.

Wagner reaction mechanism. The mechanism of reaction between MgO and Al_2O_3 described above, involving the counterdiffusion of Mg^{2+} and Al^{3+} ions through the product layer followed by further reaction at two reactant–product interfaces, is known as the Wagner mechanism. In order to maintain charge balance, for every three Mg^{2+} ions which diffuse to the right-hand interface

(Fig. 2.1b) two Al^{3+} ions must diffuse to the left-hand interface. The reactions that occur at the two interfaces may be written, ideally, as:

(a) Interface $MgO/MgAl_2O_4$:

$$2Al^{3+} - 3Mg^{2+} + 4MgO \rightarrow MgAl_2O_4$$

(b) Interface $MgAl_2O_4/Al_2O_3$:

$$3Mg^{2+} - 2Al^{3+} + 4Al_2O_3 \rightarrow 3MgAl_2O_4$$

Overall reaction:

$$4MgO + 4Al_2O_3 \rightarrow 4MgAl_2O_4$$

It can be seen that reaction (b) gives three times as much spinel product as reaction (a) and, hence, the right-hand interface should grow or move at three times the rate of the left-hand interface. This mechanism has been tested experimentally for a similar reaction, that between MgO and Fe_2O_3 to form $MgFe_2O_4$ spinel; it was found that the two interfaces moved in the ratio of 1:2.7, close to the ideal value of 1:3. In reactions such as the above, in which the interfaces between reactant and product can be clearly seen, perhaps due to a difference in colour, the movement of the interfaces may be used as a marker to monitor the progress of the reaction. This effect of using a marker is known as the *Kirkendall effect*.

The above discussion shows that three of the important factors that influence the rate of reaction between solids are (a) the area of contact between the reacting solids and hence their surface areas, (b) the rate of nucleation of the product phase and (c) the rates of diffusion of ions through the various phases and especially through the product phase. Clearly, it is necessary to maximize all of these factors in order to reduce the time taken for solids to react together. Let us consider each in a little more detail.

Surface area of solids. The surface area of a given amount of solid varies enormously, depending on whether it is in the form of a fine powder, coarse powder or a single crystal; i.e. surface area depends on particle size. This may be shown by some simple calculations, using MgO as an example. MgO has a density of $3.58 \, g \, cm^{-3}$ and hence a single crystal of MgO, in the form of a perfect cube and of mass 3.58 g, would have a volume of $1 \, cm^3$. The crystal would have six faces, each of area $1 \, cm^2$, and hence the total surface area of the crystal would be $6 \, cm^2$ or $6 \times 10^{-4} \, m^2$.

Suppose the crystal is now crushed into a fine powder in which each particle is also a perfect cube but of edge dimension $10 \, \mu m$ ($10^{-3} \, cm$). This is a typical particle size for a sample that has been ground with an agate mortar and pestle or ball-milled for, say, one hour. Note, however, that in practice a distribution of particle sizes would result. The powder now contains $(10^3)^3 = 10^9$ crystallites, each of surface area $6 \times 10^{-6} \, cm^2$. Hence, the total surface area of the powder is $6 \times 10^3 \, cm^2$ or $6 \times 10^{-1} \, m^2$.

Finally, consider the same mass of MgO but in the form of a very fine powder with cubic crystallites of edge $100 \text{Å} (10^{-6} \text{cm})$. Very long grinding times would be needed to achieve such a small crystal size, but by alternative methods, e.g. precipitation from solution or decomposition of salts, it may be readily achieved. The original mass of MgO would now contain 10^{18} crystallites of total surface area $10^{18} \times (6 \times 10^{-16}) = 600 \text{m}^2$.

In conclusion, therefore, the surface area of MgO, or any solid, increases greatly with decreasing particle size. In the form of a single crystal, a few grams of solid has a surface area about equal to the area of a large postage stamp. In the form of a finely divided powder, its surface area is about equal to the area of a hundred metre running track!

The surface area of reacting solids has a great influence on reaction rates since the total area of contact between the grains of the reacting solids depends approximately on the total surface area of the grains. Although it is theoretically feasible, in certain cases, for all the surfaces of the reacting solids to be in intimate contact, this is unlikely to happen in practice and usually the contact area is considerably less than the total surface area. The area of contact may be increased somewhat by pressing the reacting powder into a pellet, but even at relatively high pressures, e.g. 10^5 p.s.i., the pellets are usually porous and the crystal contacts are not maximized: typically, cold-pressed pellets are 20 to 40 per cent porous. A further increase in contact area and reduction in pellet porosity may be achieved by compressing the pellet at high temperatures, i.e. by *hot pressing*. The combined effects of temperature and pressure may cause the particles to fit together better, but the densification process is usually slow and may require several hours.

Although the surface area of solids largely controls the area of contact between reacting grains in a mixture, it does not appear directly in the equation for the rate of reaction, such as equation (2.1). However, it is included indirectly since there is an inverse correlation between the thickness of the product layer, x, and the area of contact. For a given mass of reactants, the thickness x of the product layer which corresponds to a certain degree of reaction, say 50 per cent, decreases with decreasing particle size. In this way, particle size and surface area affect the value of x.

Reactivity of solids—nucleation and diffusion rates. In the reaction of two solids to form a product, two stages may usually be identified: nucleation of the product and its subsequent growth. Different factors are important in these two stages.

Nucleation is facilitated if there is a structural similarity between the product and one or both of the reactants because this reduces the amount of structural reorganization that is necessary for nucleation to occur. For instance, in the reaction of MgO and Al_2O_3 to form spinel, the spinel product has a similar oxide ion arrangement to that in MgO. Spinel nuclei may therefore form at or on the surface of the MgO crystals such that the oxide arrangement is essentially continuous across the MgO–spinel interface. In such a process, the nucleating phase (spinel) makes use of a matching or partial matching of its own structure to that of an existing phase (MgO) in the reaction mixture and the nucleation step is easier as a consequence.

Topotactic and epitactic reactions. For reactions in which nucleation makes use of a structural similarity between the nucleus and an existing phase, there is usually a clear orientational relationship between the structures of the reactant and product. Two types of oriented reactions or transformations have been identified: *topotactic reactions* and *epitactic reactions.* Both require a structural relationship between the two phases, but in epitactic reactions this is restricted to the actual interface between the two crystals. Hence the two structures may (hypothetically) have a common arrangement of oxide ions at the interface but the oxide arrangement to either side develops in a different way. Epitactic reactions therefore require only a two-dimensional structural similarity at the crystal interface.

Topotactic reactions are more specific than epitactic ones because they require not only the structural similarity at the interface but also that this similarity should continue into the bulk of both crystalline phases. This could occur for the topotactic growth of spinel on MgO if the cubic close packed oxide ion arrangement continued across the interface. Topotactic and epitactic reactions are of common occurrence since their nucleation step is usually easier than that for reactions in which there is no structural or orientational relationship between reactant and product.

Oriented nucleation processes, such as described above, are subject to one further constraint: not only must the two phases have a similar structural arrangement at the interface but also the dimensions of the two structures, interatomic separations, etc., should be similar. If the two structures have quite different interatomic distances (e.g. compare the oxygen–oxygen separation in MgO and BaO, both of which have the rock salt structure) then they cannot be matched over a large area of contact. It has been estimated that a difference in interfacial lattice parameters of about 15 per cent between nucleus and substrate is the most that can be tolerated if oriented nucleation is to occur.

Surface structure and reactivity. The ease of nucleation of product phases also depends on the actual surface structure of the reacting phases. Direct information on surface structure is difficult to obtain, although in recent years several new techniques for the study of solid surfaces have been developed. However, from a consideration of crystal structures it can be shown that in most crystals the structure cannot be the same over the entire crystal surface. An exception would be the surface structure of a material such as MgO, with the rock salt structure, in which the crystals formed perfect cubes, as in Fig. 2.2(c). Each cube face has Miller indices (Chapter 5) of the type (100) and hence the same structure, in which anions and cations alternate at the corners of a square grid (Fig. 2.3a). An example in which two different surface structures appear in the same crystal is shown in Fig. 2.2(b). This is a cube (of, for example, MgO) in which all corners are absent, to reveal {111} faces. In this orientation, the {111} planes of MgO contain alternate

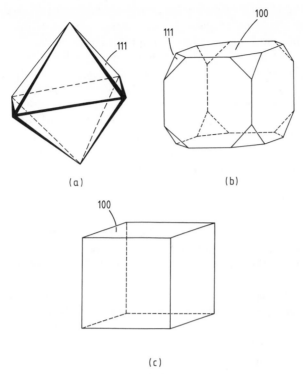

Fig. 2.2 Crystals of cubic symmetry in the form of (a) an octahedron, (b) a cubooctahedron and (c) a cube

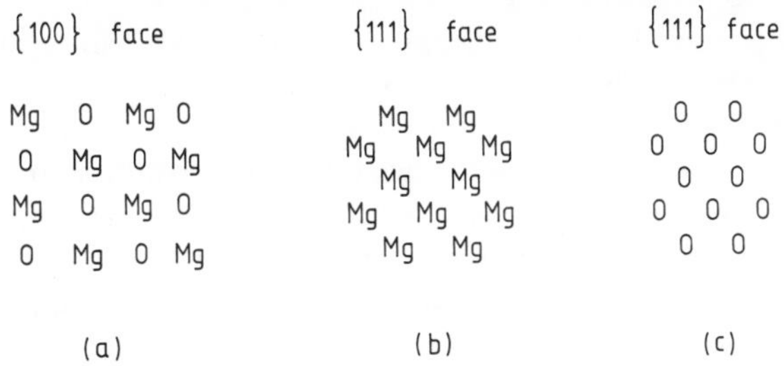

$\{100\}$ face $\{111\}$ face $\{111\}$ face

```
Mg  O  Mg  O          Mg    Mg            O   O
                    Mg   Mg   Mg        O   O   O
 O  Mg  O  Mg          Mg    Mg            O   O
Mg  O  Mg  O        Mg   Mg   Mg        O   O   O
 O  Mg  O  Mg          Mg    Mg            O   O
```

     (a)   (b)   (c)

Fig. 2.3 Surface structures of a MgO crystal displaying (a) (100) and (b, c) (111) faces

layers of Mg^{2+} and O^{2-} ions (Fig. 2.3b and c); hence a $\{111\}$ surface will be either a complete layer of Mg^{2+} ions or a complete layer of O^{2-} ions. This is clearly very different to the structure of the $\{100\}$ surfaces.

Since different surfaces have different structures, their reactivity is likely to differ considerably. Evidence that this is so comes from the morphology of crystals: a well-known example is the case of NaCl crystallizing from aqueous solution. The normal morphology is as cubes presenting $\{100\}$ faces (Fig. 2.2c), but if urea is added to the solution the crystals that form are octahedra (Fig. 2.2a) in which each face is a {111} face. The urea appears not to be incorporated into the crystals but must act to influence the reactivity of the various surfaces.

A general guideline to crystal growth and morphology is that *those surfaces that are most prominent in a crystal are those that grow most slowly*. This can be understood from Fig. 2.2. Let us start with a crystal (b) that has both {100} and {111} surfaces (i.e. a cube with its corners sliced off). If only the {111} faces are permitted to grow, this has the effect of completing the corners of the cube to give (c). If, however, only the {100} faces can grow, an octahedron (a) results. Clearly, therefore, urea must act to reduce the growth of {111} faces of NaCl compared to the growth of the {100} faces. The latter then 'grow out of' the crystal so that, eventually, only the {111} faces are left.

It may be concluded, therefore, that, in general, different parts of the surface of a crystal have a different structure and hence different reactivity. It is difficult, however, to give general rules about which surfaces are the most reactive.

Reactivity of solids also depends greatly on the presence of crystal defects. The occurrence and influence of defects inside crystals has received considerable attention and is reasonably well understood (Chapter 9). Less well understood but at least as important is the defective structure at or close to surfaces. Probably the best-known example of the importance of surface defects is the influence of emergent screw dislocations at crystal surfaces (Chapter 9) on crystal growth mechanisms. An example is shown schematically in Fig. 2.4, in which the crystal surface is in the form of spiral growth steps. This arises when the original surface of the nucleus or crystal is not exactly planar but is somewhat distorted in the region of the emergent dislocation. The distorted region provides convenient sites for atoms, ions or molecules to attach themselves and hence for the crystal to grow larger. The process is able to continue indefinitely, in principle at least, because the new crystal surface continues to exhibit the dislocation and growth steps. Dislocations are discussed further in Chapter 9.

These latter two examples, of crystal morphology and crystal growth mechanisms, are typical of crystals grown from either solution or vapour phases. Undoubtedly, however, the factors that control them are important in the reaction of solids in general, including solid state reactions.

Equation (2.1) relates the rate of solid state reaction to the diffusion of ions

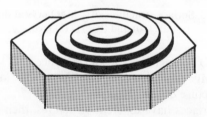

Fig. 2.4 Schematic crystal surface showing spiral growth steps caused by the presence of a screw dislocation

through the bulk of the crystals and, especially, through the product phase. Diffusion of ions is enhanced greatly by the presence of crystal defects, especially vacancies and interstitials and probably, also, dislocation pipes. The defects may be intrinsic to the crystals or may be governed by the presence of impurities. A further discussion of defects and diffusion is given in Chapter 9.

Further difficulties in studying solid state reactions. The reaction discussed above involving single crystals of MgO and Al_2O_3 has in fact been tested experimentally. While the results obtained were in general agreement with the scheme outlined here, additional complicating factors were found to be present. These are worth mentioning since they indicate the complexity of, and difficulty in studying, solid state reactions.

(a) At the temperature of the reaction, $\sim 1500\,^\circ$C, spinel exists as a range of solid solutions on the join $MgO-Al_2O_3$. The solid solution compositions cover the range between 50 and ~ 60 mol % Al_2O_3, i.e. the formula varies from $MgAl_2O_4$ to $\sim Mg_{0.73}Al_{2.18}O_4$. This means that, in the initial stages of reaction at least, the spinel product may have a variable composition. Thus, the spinel produced at the MgO–spinel interface will be the most magnesium-rich spinel, $MgAl_2O_4$, but that at the Al_2O_3–spinel interface will be the most magnesium-deficient spinel, $Mg_{0.73}Al_{2.18}O_4$. Consequences of this are (i) that the spinel product exhibits compositional inhomogeneity and (ii) that with continued reaction, the two interfaces do not move at relative rates given by the simple ratio $1:3$.

(b) During the later stages of reaction and/or during cooling of partially reacted crystals, the spinel product breaks away from the MgO parent crystal, presumably due to volume differences of the two phases. This makes the study of solid state diffusion rates and crystal orientation effects difficult.

(c) In cases where the reacting crystals are not in good contact there is evidence that an additional mass transfer mechanism, involving the gas phase, is important. Vapour phase transport is discussed further in Section 2.7.7.

2.1.2 Experimental procedure

Solid state reactions are so rarely discussed in textbooks that it is worth while to give details of the experimental procedures involved. A typical procedure is outlined in the following, using the synthesis of spinel as an example.

Reagents. For the synthesis of $MgAl_2O_4$, MgO and Al_2O_3 would be the obvious starting materials to use. These should be dried thoroughly prior to weighing, especially MgO which is hygroscopic, by heating at high temperature, e.g. 200 to 800 °C, for a few hours. Fine grained materials should be used if possible in order to maximize surface areas and hence reaction rates. Alternatively, and probably better, $MgCO_3$ (or some other oxysalt of magnesium) could be used as the source of MgO since it is less hygroscopic than MgO and on decomposition at high temperature, 600 to 900 °C, a fine grained MgO with high surface area and reactivity should result. It is not necessary to carry out the decomposition of the $MgCO_3$ as a separate step, prior to mixing with the Al_2O_3, but it is better carried out in situ, as the first step in the reaction with Al_2O_3 on heating. Other more reactive sources of Al_2O_3 could also be used, e.g. $Al(OH)_3$ or polymorphs of Al_2O_3 other than the normal, stable, α form, corundum. However, the composition of such materials (e.g. their water content) should be known accurately if they are to be used for quantitative work.

Mixing. After the reactants have been weighed out in the required amounts, they are mixed together. For manual mixing of small quantities ($\lesssim 20$ g total), this may be done with an agate mortar and pestle. Agate is preferable to, for example, porcelain since it is hard, unlikely to contaminate the mixture and, with its smooth, non-porous surface, is easy to clean afterwards. Homogenization of the mixture is aided greatly by adding sufficient of a volatile organic liquid—acetone or alcohol are suitable—to form a paste. During the process of grinding and mixing, the organic liquid gradually volatilizes and after 10 to 15 minutes it has usually evaporated completely. If too much liquid is added such that the solids can swirl in the liquid, then sedimentation may occur (which cancels the effect of mixing!) and much larger times of mixing are needed before all the liquid evaporates.

For quantities much larger than ~ 20 g, manual mixing is very tedious since normal-sized mortars cannot cope with the large quantities of material; mixing must then be done in batches. In such cases, mechanical mixing may be better, using a ball mill, and may take several hours.

Container material. For the subsequent reaction at high temperatures, it is necessary to choose a suitable container material which is chemically inert to the reactants under the heating conditions used. The noble metals, platinum and gold, are usually suitable, although expensive. Of the two, platinum is usually better since it has a higher melting point, ~ 1700 °C, than gold, 1063 °C, and is considerably harder; gold-based alloys such as Au–Pd are sometimes used as they are harder than pure gold. The containers may be crucibles or boats made from foil; crucibles should be reusable many times whereas foil boats are thin and fragile and have very limited lifetimes. For low temperature reactions other metals may be used, e.g. nickel below 600 to 700 °C. Various inert, refractory inorganic materials are also used for containers, such as crucibles of Al_2O_3, stabilized ZrO_2 and SiO_2. These are prone to attack at high temperatures, however, especially by alkali oxides.

Heat treatment. The heating programme to be used depends very much on the form and reactivity of the reactants. MgO and Al_2O_3 are both inert, refractory materials and are unlikely to react together to any significant degree below 1200 to 1300 °C. A mixture of MgO and Al_2O_3 may therefore be placed in a furnace directly, at, for example, 1400 to 1600 °C. If one or more of the reactants is an oxysalt, e.g. $MgCO_3$, the first stage of the reaction must be the decomposition of the oxysalt and the mixture should be heated first at an appropriate temperature for a few hours so that decomposition occurs in a controlled manner. If this stage is omitted and the mixture is heated directly at a higher temperature, the decomposition may occur very vigorously and may cause the sample to froth or spit out of the container. Na_2CO_3 is a particularly bad example of this if it is heated initially much above 700 °C. Reaction to give the final product usually requires several hours, or even days, depending on the reaction temperature. Reaction is often greatly facilitated by cooling and grinding the sample periodically. This is because during heating, sintering and grain growth of both reactant and product phases (Chapter 20) usually occur in addition to the main reaction, causing a reduction in the surface area of the mixture. The effect of grinding is to maintain a high surface area, as well as to bring fresh surfaces into contact. The reaction rate may also be speeded up by pelleting the samples prior to heating, thereby increasing the area of contact between the grains.

In order to speed up reaction rates, it would perhaps seem that the obvious approach is to raise the temperature as high as possible. This is indeed possible for mixtures of MgO and Al_2O_3 which may be safely heated to ~ 2000 °C (but in a suitable container). In many other systems, however, additional problems occur at high temperatures, especially the volatilization of one or more components of the reacting mixture. Alkali oxides, for instance, volatilize at high temperatures. It is not possible to give rules about 'safe' temperatures, below which volatilization does not occur, since volatilization rates depend considerably on composition. For instance, using Li_2CO_3 as the source of Li_2O in reactions, melting begins to occur at ~ 700 °C and Li_2CO_3 evaporates quickly at higher temperatures. When mixed with other components, however, the evaporation rate may be greatly reduced; e.g. mixtures of Li_2CO_3 and Al_2O_3 in the molar ratio 1:5, which react to form $LiAl_5O_8$, may be heated gradually to ~ 1200 °C before loss of lithia becomes a serious problem.

Care may also be needed, in control of either temperature or atmosphere, if the components have variable oxidation states and a certain oxidation state is desired in the product phase. Thus, in the synthesis of phases containing Fe^{2+}, it is necessary to maintain a reducing atmosphere in contact with the sample in order to prevent oxidation to Fe^{3+} from occurring.

Analysis. The products of solid state reactions are usually in the form of a powder or a sintered, polycrystalline piece. Large single crystals are not usually obtained by this method although crystals up to ~ 0.1 mm in size may sometimes be obtained. The main technique for analysing the product is usually X-ray powder diffraction since this technique tells us which crystalline phases are present. Each crystalline material has its own characteristic X-ray powder pattern which serves as a 'finger print' for its identification. The X-ray powder method serves to indicate whether the reaction is complete, therefore, by showing that the original reactants have disappeared and by showing that unwanted side-products or intermediates have not formed. If the product is a new phase its X-ray powder pattern will obviously not have been reported previously.

The X-ray powder method does not give a chemical analysis of the product. Hence, if there are doubts over its chemical composition (e.g. if alkali volatilization is suspected to have occurred), additional analytical techniques may be needed (e.g. X-ray fluorescence or atomic absorption analysis) in order to determine the overall chemical composition. A more detailed survey of the methods used to analyse and obtain chemical and structural information about solids is given in Chapter 3.

2.1.3 Coprecipitation as a precursor to solid state reaction

In normal solid state reactions, the reactants are mixed together manually or mechanically and the subsequent reaction rate depends to a large degree on the particle size of the reactants, the degree of homogenization achieved on mixing and the intimacy of contact between the grains, as well as the obvious effect of temperature. By using coprecipitation procedures, it is sometimes possible to achieve a high degree of homogenization together with a small particle size and thereby speed up the reaction rate. This may be illustrated by the synthesis of $ZnFe_2O_4$ spinel. In one method, oxalates of zinc and iron are used as the reactants; these are dissolved in water in the ratio of 1:1, the solutions then being mixed and heated to evaporate the water. Oxalates of zinc and iron are gradually precipitated together and the resulting fine powder is a solid solution that contains the cations mixed together essentially on an atomic scale. The precipitated solids are filtered off and calcined (i.e. heated) in the usual way, but because of the high degree of homogenization, much lower reaction temperatures are sufficient for reaction to occur (e.g. $\sim 1000\,^{\circ}C$ for the formation of $ZnFe_2O_4$). The overall reaction may be written:

$$Fe_2((COO)_2)_3 + Zn(COO)_2 \rightarrow ZnFe_2O_4 + 4CO + 4CO_2$$

The method has also been successfully used for the preparation of other spinels: $CoFe_2O_4$, $MnFe_2O_4$ and $NiFe_2O_4$. The method does not work well in cases where (a) the two reactants have very different solubilities in water, (b) the reactants do not precipitate at the same rate or (c) supersaturated solutions commonly occur. It is often not suitable for the preparation of high purity, accurately stoichiometric phases, therefore.

2.1.4 Other precursor methods

Variations of the above coprecipitation method have been devised in which various stoichiometric mixed salts are precipitated as intermediates. For instance, ferrite spinels such as $NiFe_2O_4$ have been prepared via a basic double acetate pyridinate of nickel and iron, of stoichiometry $Ni_3Fe_6(CH_3COO)_{17}O_3OH \cdot 12C_5H_5N$. The Ni:Fe ratio is accurately 1:2 in this

Table 2.1 *Synthesis of stoichiometric chromites.* (From Whipple and Wold, 1962).

Chromite	Precursor	Ignition, °C
$MgCr_2O_4$	$(NH_4)_2Mg(CrO_4)_2 \cdot 6H_2O$	1100–1200
$NiCr_2O_4$	$(NH_4)_2Ni(CrO_4)_2 \cdot 6H_2O$	1100
$MnCr_2O_4$	$MnCr_2O_7 \cdot 4C_5H_5N$	1100
$CoCr_2O_4$	$CoCr_2O_7 \cdot 4C_5H_5N$	1200
$CuCr_2O_4$	$(NH_4)_2Cu(CrO_4)_2 \cdot 2NH_3$	700–800
$ZnCr_2O_4$	$(NH_4)_2Zn(CrO_4)_2 \cdot 2NH_3$	1400
$FeCr_2O_4$	$NH_4Fe(CrO_4)_2$	1150

compound and purification may be carried out by recrystallization from pyridine. The pyridinate is itself prepared from a basic double acetate hydrate phase. In order to prepare $NiFe_2O_4$, the pyridinate crystals are slowly heated to 200 to 300 °C in order to burn off the organic material, followed by heating in air at ~ 1000 °C for two to three days.

Similar methods have been used to synthesize chromite spinels, MCr_2O_4, where M = Mg, Zn, Cu, Mn, Fe, Co, Ni. Thus, manganese chromite, $MnCr_2O_4$, was prepared from precipitated $MnCr_2O_7 \cdot 4C_5H_5N$, which was gradually heated to 1100 °C. During heating, hexavalent chromium in the dichromate was reduced to trivalent chromium; the mixture was finally heated in a hydrogen-rich atmosphere at 1100 °C to ensure that all the manganese was present in the divalent state. The precursors used to synthesize the other chromites are given in Table 2.1. By careful control of the experimental conditions, these precursor methods are capable of yielding phases of accurate stoichiometry. This is important since several chromites and ferrites are valuable magnetic materials whose properties may be sensitive to purity and stoichiometry, Chapter 16.

In principle, coprecipitation and other precursor methods should be applicable to a wide variety of solid state syntheses. However, each synthesis requires its own special conditions, precursor reactions, etc., and it is not possible to give a universal set of conditions which are applicable to all such syntheses.

2.1.5 Kinetics of solid state reactions

The above discussions have indicated some of the factors that influence reactions of solids. Many factors are usually involved in a particular reaction and consequently the analysis of kinetic data may be difficult.

In heterogeneous reactions between, say, two solids, the actual reaction to form the product occurs at an interface. In kinetic studies of reaction rates, it is desired to know which is the slow, rate controlling step in a reaction; there are at least three clear possibilities for this:

(a) transport of matter to the reaction interface,
(b) reaction at the interface,
(c) transfer of matter away from the reaction interface.

The approaches used to analyse rate data are rather different from those used for gas phase reactions. In the latter, the concept of reaction order is very useful; in this, the rate of change of concentration, c, of one species depends on the nth power of concentration, i.e.

$$\frac{dc}{dt} = -kc^n \tag{2.2}$$

where n is the reaction order and k is a constant. From the value of n, important insight into reaction mechanisms may be obtained, such as the number of molecules involved in a particular reaction step.

For most solid state reactions, however, it is usually incorrect and misleading to think in terms of reaction order, since the reactions do not involve molecules. The data may still be represented *empirically* in this way and it is sometimes found that n is not a simple integer but may be fractional.

It is not proposed to cover further the kinetics of solid state reactions here; the interested reader is referred to the book by Schmalzried (1971).

2.2 Crystallization of solutions, melts, glasses and gels

These crystallization methods have one feature which makes them particularly useful for preparations: the starting material for crystallization, i.e. aqueous solution, melt, glass or gel, is usually homogeneous, single phase and amorphous. This may greatly facilitate formation of the crystalline product, since long range diffusion of ions may not be necessary and the product may form at much lower temperatures than by solid state reaction. The method can also sometimes yield metastable phases which are difficult or impossible to prepare by other means. The kinetic and thermodynamic features of certain of these reactions, especially the crystallization of supercooled melts and glasses, are discussed in Chapter 18.

2.2.1 Solutions and gels: zeolite synthesis

By crystallization from aqueous solutions, crystalline phases may be prepared at low temperatures, e.g. 25 to 100 °C, whereas much higher temperatures would be required for a normal solid state synthesis. The method is suitable for mixed salts—carbonates, sulphates, etc.—and, particularly, for the preparation of hydrate phases. As a synthetic route to new structures, it is valuable for those structures which do not have high temperature stability. An important application is in the synthesis of molecular sieve zeolites; these are hydrated aluminosilicate framework structures which contain large channels and cavities (e.g. channels of 4 to 6 Å diameter) such that a variety of organic and inorganic substances may enter the zeolite structure. The starting materials for zeolite synthesis may be aqueous solutions of silicate and aluminate anions which are mixed together with alkali. A gel forms by a process of copolymerization of the silicate and aluminate anions and on subsequent heat treatment, perhaps under a high water vapour pressure (hydrothermal treatment), crystals of zeolite are produced. This may be represented, schematically, for a sodium aluminosilicate zeolite, by the following reactions:

$$NaAl(OH)_4(aq) + Na_2SiO_3(aq) + NaOH(aq)$$

$$\downarrow \; 25\,°C$$

$$(Na_a(AlO_2)_b(SiO_2)_c \cdot NaOH \cdot H_2O)\, gel$$

$$\downarrow \; 25 \text{ to } 175\,°C$$

$$Na_x(AlO_2)_x(SiO_2)_y \cdot mH_2O$$

Zeolite crystals

Zeolites are an important group of materials since they have many industrial applications, especially in catalysis. Zeolite synthesis is an active area of research since it is desired to synthesize novel structures with different capacities for ion exchange or molecular sieve action. Some general conditions for zeolite synthesis (after Breck) are:

(a) Reactive starting materials should be used such as freshly coprecipitated gels or amorphous solids.
(b) Relatively high pH, introduced in the form of an alkali metal hydroxide or other strong base, is important.
(c) Low temperature hydrothermal conditions with concurrent low autogeneous pressure at saturated water vapour pressure are needed.
(d) A high degree of supersaturation of the components of the gel is desirable leading to the nucleation of a large number of crystals.

2.2.2 Melts

The crystallization of melts is very similar to the crystallization of solutions, the only difference, usually, being the temperature: melts are high temperature liquids whereas aqueous solutions are liquids at low temperatures. Thus, on melting together the solid starting materials, complete homogenization occurs and recrystallization takes place on subsequent cooling of the melt. Crystallization of melts (as well as solutions) is a valuable method for growing single crystals since, in the presence of the liquid phase and in the absence of too many crystal nuclei, large crystals readily grow.

In order to use and appreciate melt crystallization methods, a knowledge of the relevant phase diagram is necessary, since the phase diagram provides a diagrammatic representation of the conditions (of temperature and composition) for crystal growth. For the crystallization of congruently melting crystals from liquid of the same composition, the only information necessary is the melting point. Often, however, especially in the 'flux growth' method, the liquid and the crystallizing phase have different compositions. This may be illustrated for a simple eutectic system, A–B (Fig. 2.5). Let us suppose that it is desired to grow crystals of compound A, of melting point T_A. The effect of adding a small amount, e.g. 10 to 20 per cent, of a second material, B, is to drastically reduce the melting

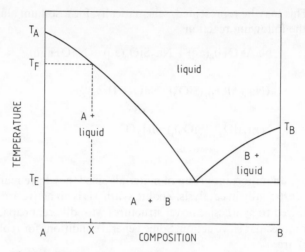

Fig. 2.5 Binary eutectic phase diagram showing use of flux
B to grow crystals of A

temperature and to cause melting to occur over a range of temperatures. For instance, if the composition of the original A–B mixture is given by X, melting of the mixture begins to occur at temperature T_E and is not complete until temperature T_F. Hence, over the range of temperatures between T_E and T_F, the partially melted mixture contains crystals of A and a liquid whose composition changes with temperature. In order to obtain crystals of A, the mixture of composition X may be cooled slowly through the temperature range T_F to T_E; alternatively, the mixture may be heated isothermally at some temperature

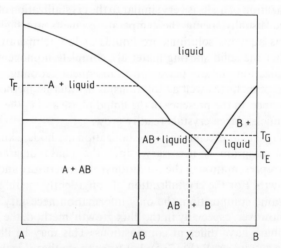

Fig. 2.6 Binary phase diagram containing incongruently melting compound AB and showing method for growing crystals of AB

between T_E and T_F. On subsequent cooling to room temperature, the remaining liquid will probably crystallize to give a fine grained eutectic structure containing small crystals of A and B; embedded in this eutectic structure are the larger crystals of A which grew as the 'primary phase' at high temperature. The crystals of A may be large since they were allowed to grow, in the presence of liquid, over a considerable period of time, and in this respect the method is analogous to crystal growth from aqueous solution.

Similar principles apply to more complex phase diagrams and an important example is in the crystallization of incongruently melting phases, shown in Fig. 2.6. If it is desired to prepare or grow crystals of the incongruently melting compound AB, this is unlikely to be achieved readily by simply cooling liquid of composition AB since, on reducing the temperature below T_F, crystals of A rather than AB begin to form. However, by cooling a liquid of composition X, crystals of AB form, in the presence of liquid, over the temperature range T_G to T_E. Phase diagrams, their construction, interpretation and uses, are discussed in more detail in Chapter 11.

2.2.3 Glasses

It is sometimes useful to prepare crystalline materials via glassy intermediates; an important application is in the manufacture of glass-ceramic materials. An example is lithium disilicate, $Li_2Si_2O_5$, which is an important component of several commercial glass-ceramics; crystalline $Li_2Si_2O_5$ melts at 1032 °C. A mixture of Li_2O and SiO_2, in a molar ratio of 1:2 may be melted at, for example, 1100 °C for a few hours in a platinum crucible to produce a homogeneous melt. On cooling rapidly or quenching to room temperature, the melt viscosity increases and a homogeneous, transparent glass results. The glass transition temperature is ~ 450 °C and on reheating at 500 to 700 °C crystallization of $Li_2Si_2O_5$ occurs readily in a few hours. The method may clearly only be used to prepare crystalline compounds which are also capable of glass formation. Since the number and range of glass-forming systems is limited, this restricts the applicability of the method considerably. Glass and glass-ceramics are discussed further in Chapter 18.

2.3 Vapour phase transport methods

An interesting and potentially valuable preparative method developed in recent years, notably by Schäfer (1971), is the method of chemical transport via the vapour phase. The method may be used for the synthesis of new compounds, for the growth of single crystals or for the purification of a compound. Basically, the method consists of a tube, usually of silica glass, which contains the reactant(s), A, at one end and which is sealed, either under vacuum or more usually with an atmosphere of a gaseous transporting agent, B (Fig. 2.7a). The tube is placed inside a furnace such that a temperature gradient exists inside the tube; typically, there is a temperature change of 50 °C along the length of the tube.

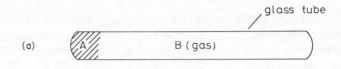

Equilibrium : A (s) + B (g) ⇌ A B (g)

Fig. 2.7 Simple vapour phase transport experiment for the growth of single crystals of substance A

Materials A and B react together to form gaseous AB which subsequently decomposes at the other end of the tube to redeposit crystals of A (Fig. 2.7b).

The process shown in Fig. 2.7 is for the simple case of transport of material A along the length of the tube. The method depends on the existence of a reversible equilibrium between reactant A, transporting agent B and gaseous product AB. If the formation of AB is endothermic, then, preferentially, the reaction to form AB occurs at temperature T_2 (hot end) and gaseous AB is transported to the cooler end of the tube where it decomposes to deposit crystals of A at temperature T_1. Conversely, if the formation of AB is exothermic, the reactant is arranged to be at the cold end of the tube and the product is formed at the hot end.

The equilibrium equation shown in Fig. 2.7 has an associated equilibrium constant, K, whose magnitude is usually temperature dependent. Hence, the equilibrium concentration of AB changes with temperature and is different at temperatures T_2 and T_1; this causes a concentration gradient of gaseous AB to build up which provides the driving force for transport by gaseous diffusion. In order for an efficient vapour phase transport reaction to occur, the equilibrium positions of the transport reaction at the two temperatures should not be too extreme, i.e. they should not lie too far either to the left or to the right.

An example of an endothermic reaction which can be used for chemical transport is the formation of gaseous PtO_2 at 1200 °C or higher:

$$Pt(s) + O_2 \rightleftharpoons PtO_2(g)$$

The PtO_2 diffuses to lower temperatures where it may deposit well-formed crystals of platinum metal. In furnaces that contain heating elements of platinum, it is common to find platinum crystals deposited on cooler parts of the furnace wall: these form by a process of vapour phase transport.

The *van Arkel method* for the purification of certain metals makes use of an exothermic reaction between metal and iodine to form a gaseous iodide, e.g.:

$$Cr + I_2 \rightleftharpoons CrI_2(g)$$

Since the formation of CrI_2 is exothermic, chromium metal is redeposited at a higher temperature. Other metals which may be purified by this method include Ti, Hf, V, Nb, Cu, Ta, Fe and Th. By this method, the metals may be extracted from their carbides, nitrides and oxides.

An elegant variation of this method is to transport *two* substances in opposite directions along the temperature gradient; this is possible if one reaction is exothermic and the other endothermic. For instance, WO_2 and W may be separated by this means using I_2 and H_2O as the gas phase: W is deposited at 1000 °C and WO_2 at 800 °C. The reactions involved are:

$$WO_2(s) + I_2(g) \underset{800\,°C}{\overset{1000\,°C}{\rightleftharpoons}} WO_2I_2(g)$$

and

$$W(s) + 2H_2O(g) + 3I_2(g) \underset{1000\,°C}{\overset{800\,°C}{\rightleftharpoons}} WO_2I_2(g) + 4HI(g)$$

Another example is the possible separation of Cu and Cu_2O using HCl as the transporting agent. The reactions involved are:

$$Cu_2O(s) + 2HCl(g) \underset{900\,C}{\overset{500\,°C}{\rightleftharpoons}} 2CuCl(g) + H_2O(g)$$

and

$$Cu(s) + HCl(g) \underset{500\,°C}{\overset{600\,°C}{\rightleftharpoons}} CuCl(g) + \tfrac{1}{2}H_2(g)$$

Since CuCl forms exothermically from Cu_2O and endothermically from Cu, the Cu_2O is redeposited at a higher temperature and Cu at a lower temperature.

The above examples are simple cases of transport in which the reactant and product phases are essentially the same. For use as a preparative method, the transport reaction is coupled with a subsequent reaction, represented schematically as:

At T_2: $A(s) + B(g) \rightleftharpoons AB(g)$

At T_1: $AB(g) + C(s) \rightleftharpoons AC(s) + B(g)$

Overall: $A(s) + C(s) \rightleftharpoons AC(s)$

There are many examples of the use of this method for the preparation of binary, ternary and even quaternary compounds. Some examples (after Schäfer, 1971) are:

(a) Preparation of Ca_2SnO_4. CaO and SnO_2 react slowly according to the equation:

$$2CaO + SnO_2 \rightarrow Ca_2SnO_4$$

In the presence of gaseous CO, the reaction may be greatly speeded up since SnO_2 converts to gaseous SnO and can be chemically transported:

$$SnO_2(s) + CO \rightleftharpoons SnO(g) + CO_2$$

The gaseous SnO subsequently reacts with CaO and CO_2.

(b) Preparation of nickel chromite, $NiCr_2O_4$. Reaction of NiO and Cr_2O_3 is normally slow but is speeded up in the presence of oxygen since gaseous CrO_3 forms which migrates to the NiO:

$$Cr_2O_3(s) + \tfrac{3}{2}O_2 \rightleftharpoons 2CrO_3(g)$$
$$2CrO_3(g) + NiO(s) \rightarrow NiCr_2O_4(s) + \tfrac{3}{2}O_2$$

(c) Preparation of niobium silicide, Nb_5Si_3. Metallic niobium and silica, SiO_2, do not react together if heated under vacuum at, for example, $1100\,^\circ C$, but in the presence of traces of H_2 gaseous SiO forms which migrates to the niobium:

$$SiO_2(s) + H_2 \rightleftharpoons SiO(g) + H_2O$$
$$3SiO(g) + 8Nb \rightarrow Nb_5Si_3 + 3NbO$$

Alternatively, in the presence of I_2, gaseous NbI_4 is formed and is transported to the SiO_2:

$$Nb(s) + 2I_2 \rightleftharpoons NbI_4(g)$$
$$11NbI_4 + 3SiO_2 \rightarrow Nb_5Si_3 + 22I_2 + 6NbO$$

(d) Preparation of aluminium sulphide, Al_2S_3. Aluminium and sulphur react together only slowly at, for example, $800\,^\circ C$, since the liquid Al becomes coated with a skin of Al_2S_3 which acts as a diffusion barrier to further reaction. In the presence of I_2, however, and a temperature gradient of $100\,^\circ C$, Al_2S_3 separates as large colourless crystals at the cool, $700\,^\circ C$, end. This is because the reaction product, Al_2S_3, is chemically transportable via the formation of gaseous AlI_3:

$$2Al + 3S \rightarrow Al_2S_3$$
$$Al_2S_3(s) + 3I_2 \rightleftharpoons 2AlI_3(g) + \tfrac{3}{2}S_2(g)$$

(e) Preparation of Cu_3TaSe_4. The elements, Cu, Ta and Se, are heated at $800\,^\circ C$ in the presence of gaseous iodine; transport in a temperature gradient occurs to give crystals of Cu_3TaSe_4 at $750\,^\circ C$, presumably by means of a gaseous complex iodide intermediate.

(f) Preparation of zinc tungstate, $ZnWO_4$. The oxides ZnO and WO_3 are heated at $1060\,^\circ C$ in the presence of Cl_2 gas. Intermediate volatile chlorides form which are transported and crystals of $ZnWO_4$ are deposited at $980\,^\circ C$.

These examples illustrate the importance of the gas phase as a transporting agent and its subsequent influence on reaction rates. Gases react much more quickly than do solids because mobilities are increased. In addition, the gaseous phase is often important in normal 'solid state' reactions under isothermal conditions, where it may act as a rapid means of transporting matter from one crystal to another.

2.4 Modification of existing structures by ion exchange and intercalation reactions

In these reactions, the structure and composition of an existing crystal is modified so as to yield new compounds. This may be done by *intercalation*—inserting extra atoms or ions into the structure—or by *ion exchange*—replacing ions in the structure with other ions from, for example, an aqueous solution or molten salt. Clearly, in order for a crystal to undergo one of these reactions it must possess certain structural characteristics, the most important of which is a degree of structural openness which permits foreign atoms or ions to readily diffuse into or out of the crystal. Some examples are as follows. In the case of graphite and TiS_2 intercalation compounds, the host crystals possess a layer or lamellar structure and the layers are pushed apart as foreign atoms or ions penetrate the interlayer space; the layers move back together if the intercalated atoms subsequently diffuse out of the crystal. In the β-alumina structure, open layers also exist, which permit the easy migration of Na^+ ions; by immersing Na β-alumina in a suitable molten salt, the Na^+ ions may be replaced by a wide variety of other cations. The zeolite molecular sieves possess three-dimensional, alumino-silicate framework structures which contain networks of interconnecting cages and tunnels. These normally contain hydrated cations which may be readily exchanged by other cations using a suitable ion exchange procedure. Apart from the value of reactions such as these in the preparation of new compounds, some of them are of considerable commercial importance and potential. Thus, by virtue of their ion exchange capabilities, zeolites have many applications, including the use as water softeners; TiS_2 and other transition metal dichalcogenides may intercalate Li^+ ions to give, for example, $LiTiS_2$, and these have potential uses as reversible electrodes in solid state batteries. Some examples are considered below in more detail.

2.4.1 Graphite intercalation compounds

Graphite is the classic example of a host crystal that is capable of intercalating a wide variety of atoms, ions and molecules, and as such it has been extensively studied. Graphite has a planar ring structure, shown in Fig. 2.8(a), and it is possible to intercalate, for example, alkali cations, halide anions, ammonia and amines, oxysalts and metal halides between the carbon layers.

Some typical reactions and conditions for the formation of graphite intercalation compounds are as follows:

$$\text{Graphite} \xrightarrow[25\,°C]{HF/F_2} \text{graphite fluoride}, C_{3.6}F \text{ to } C_{4.0}F$$
$$\text{(black)}$$

$$\text{Graphite} \xrightarrow[450\,°C]{HF/F_2} \text{graphite fluoride}, CF_{0.68} \text{ to } CF$$
$$\text{(white)}$$

$$\text{Graphite} + \text{K(melt or vapour)} \rightarrow C_8K$$
$$\text{(bronze)}$$

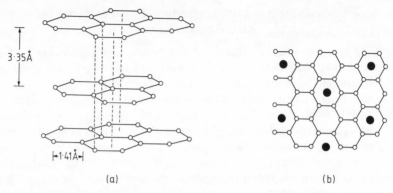

Fig. 2.8 Structures of (a) graphite, in oblique projection showing the two-layer stacking sequence and (b) graphite potassium, C_8K. In (b) the graphite layers are superposed in projection but the potassium atoms are not and various stacking sequences are possible. The structure in (b) is typical of many donor and acceptor complexes of graphite

$$C_8K \xrightarrow[\text{vacuum}]{\text{partial}} \underset{\text{(steel blue)}}{C_{24}K} \rightarrow C_{36}K \rightarrow C_{48}K \rightarrow C_{60}K$$

$$\text{Graphite} + H_2SO_4(\text{conc.}) \rightarrow C_{24}{}^+(HSO_4)^- \cdot 2H_2SO_4 + H_2$$

$$\text{Graphite} + FeCl_3 \rightarrow \text{graphite}/FeCl_3 \text{ intercalate}$$

$$\text{Graphite} + Br_2 \rightarrow C_8Br$$

Most of these reactions are reversible; thus C_8K forms on exposure of graphite to molten potassium and the potassium may be subsequently removed under vacuum. The reactions are readily reversible because the structure and planarity of the carbon layers are essentially unaffected by intercalation.

Let us consider the structure of graphite and one of these compounds in more detail. In graphite, the individual carbon layers (Fig. 2.8a), are not close packed layers since the carbon coordination number is only three. The layers are stacked together with a two-layer repeat unit which may be represented... ABABA..., although this is somewhat different from the normal hexagonal stacking sequence. Thus, some carbon atoms are directly over carbon atoms in the layer below whereas others are over the space in the middle of the rings. The carbon atoms may be regarded as sp^2 hybridized, with an additional p orbital, containing a single electron, perpendicular to the plane of the rings. These p orbitals overlap with similar p orbitals on adjacent carbon atoms, resulting in an infinite, two-dimensional, delocalized π system, rather like an extended benzene molecule. The carbon–carbon distance, 1.41 Å, is similar to that in benzene and is intermediate between normal, single and double bond values. Graphite has a high thermal and electrical conductivity as a consequence of the delocalized π electrons which, in band theory terminology, may be regarded as forming the conduction band of graphite. Bonding within the graphite layers is strong, with an average C—C bond strength of 1.5, but adjacent layers are held together by much weaker van der Waals bonds, as evidenced by the relatively large interlayer spacing of 3.35 Å.

This weak bonding permits suitable foreign atoms to intercalate between the layers and push them apart to, for example, 5.5 Å in C_4F, ~ 6.6 Å in CF and 5.41 Å in C_8K.

The structures of the intercalation compounds are often not known with certainty; the probable structure of one—graphite potassium, C_8K—is shown in Fig. 2.8(b). The relative positions of the carbon layers are different to those in the pure graphite structure in that they now form an ...AAA... stacking sequence. Potassium ions are sandwiched between pairs of carbon rings, giving K^+ a coordination number of twelve. If all such sites were occupied, the stoichiometry C_2M would result, but in C_8K only one quarter are occupied, in an ordered fashion. The electronic structure of graphite is modified on intercalation of potassium atoms since partial electron transfer from potassium to graphite occurs, resulting in a polar structure which may nominally be represented as $C_8^-K^+$. In C_8K, graphite is therefore behaving as an electron acceptor. In other compounds, such as graphite–halogen compounds, it acts as an electron donor to the halogen, e.g. in graphite bromide with the nominal ionic formula $C_8^+Br^-$. In both C_8K and C_8Br, the conduction band is still only partly occupied and these compounds are good conductors of electricity.

2.4.2 Transition metal dichalcogenide and other intercalation compounds

The disulphides of transition metals in groups 4, 5 and 6 have layered structures and are capable of forming intercalation compounds. Probably the most well studied is TiS_2; it has a sandwich structure in which layers of titanium atoms occupy octahedral sites between layers of sulphur atoms. Adjacent sandwiches are held together by weak van der Waals bonds and can be prized apart fairly readily, thereby allowing foreign species to be inserted, as shown schematically in Fig. 2.9.

A simple intercalation reaction, and one with considerable applications in high density batteries, is the reaction between lithium and TiS_2 to form Li_xTiS_2, where $0 < x < 1$. Lithium ions enter the space between TiS_2 sandwiches causing an expansion of about 10 per cent (for $x = 1$) perpendicular to the plane; at the same

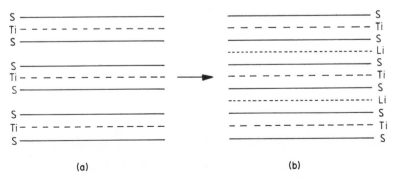

Fig. 2.9 Schematic layered structure of (a) TiS_2 and (b) $LiTiS_2$

time, reduction of Ti^{4+} to Ti^{3+} occurs. The reaction is usually carried out in an organic solvent by using n-butyl lithium as the source of lithium:

$$xC_4H_9Li + TiS_2 \xrightarrow[\text{inert atmosphere}]{\text{hexane}} Li_xTiS_2 + \frac{x}{2}C_8H_{18}$$

The reaction takes place readily at room temperature and the Li_xTiS_2 solid product may be filtered off and washed with hexane. n-Butyl lithium is a very useful lithiating agent since it is relatively mild; agents which have a higher lithium activity, e.g. metallic Li, may cause further reduction of the chalogenide to occur beyond the intercalation stage, to give lower chalcogenides or even metallic titanium.

Li_xTiS_2 may also be prepared by an electrochemical method, shown in Fig. 2.10. TiS_2 powder forms the cathode by bonding on to a metal grid with a polymer, e.g. Teflon. This is immersed in a Li^+-containing electrolyte, e.g. $LiClO_4$ dissolved in a polar non-aqueous solvent, dioxolane. A sheet of lithium metal is also dipped into the electrolyte and serves as the anode. On closing the external circuit between the anode and cathode, an electron current flows in the external circuit and Li^+ ions are transported from the anode, through the electrolyte, to the cathode where they intercalate the TiS_2. The rate of intercalation and its extent (reaction may be stopped at a particular x value) may be controlled by applying a small reverse potential across the cell. Prior to reaction, the cell shown in Fig. 2.10 has an open circuit voltage of ~ 2.5 V and this gives the Li/TiS_2

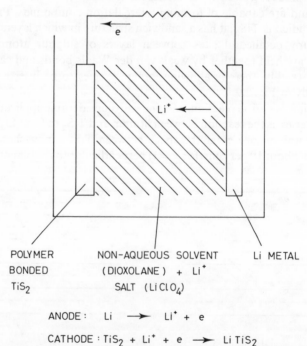

POLYMER   NON-AQUEOUS SOLVENT   Li METAL
BONDED   (DIOXOLANE) + Li^+
TiS_2 SALT ($LiClO_4$)

ANODE: $Li \longrightarrow Li^+ + e$

CATHODE: $TiS_2 + Li^+ + e \longrightarrow LiTiS_2$

Fig. 2.10 Electrochemical intercalation of lithium by TiS_2

couple considerable potential for energy storage (see Chapter 13). A battery constructed in this way has three to four times the energy stored in a lead–acid battery of the same weight.

The two methods described for the synthesis of Li_xTiS_2 are typical of a large variety of possible intercalation reactions. By such methods it is possible to intercalate other alkali cations, Cu^+, solvated ions, Ag^+, H^+, NH_3, organic amines (i.e. molecules that can donate electrons and therefore act as Lewis bases) and metallocenes. Also, a large number of layered and tunnel compounds are capable of intercalation. As well as graphite and the disulphides and diselenides of many transition metals, the list includes Ta_2S_2C, $NiPS_3$, $FeOCl$, V_2O_5, MoO_3, TiO_2, MnO_2, WO_3, etc. A particularly well-known example, which is not discussed further here, is the formation of tungsten bronzes, e.g. Na_xWO_3, by reaction of WO_3 and sodium metal.

2.4.3 Ion exchange reactions

In structures which contain, for instance, an anion array that has open layers or interconnected channels, it may be possible to replace some of the cations by ion exchange and thereby synthesize new compounds. This has been done in the β-alumina family of compounds, whose layered structure is shown schematically in Fig. 2.11. The structure consists of 'spinel blocks', approximately 9 Å thick, that are separated from each other by relatively open 'conduction planes'. These conduction planes contain (a) bridging oxygen atoms that act as props between the spinel blocks and (b) Na^+ ions which are mobile and impart to β-alumina its solid electrolyte properties. By immersing β-alumina in suitable molten salts at, for example, 300 °C, it is possible to ion exchange Na^+ for a large variety of cations such as Li^+, K^+, Rb^+, Ag^+, Cu^+, Tl^+, NH_4^+, In^+, Ga^+, NO^+ and H_3O^+. More recently, it has been found that divalent cations such as Ca^{2+} may enter the conduction planes by immersing in molten salts at 600 to 800 °C (to maintain electroneutrality, each Ca^{2+} replaces two Na^+ ions). The extent of the ion exchange under equilibrium conditions depends on the melt composition, as

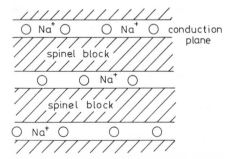

Fig. 2.11 Schematic layered structure of β-alumina of approximate composition $Na_2O \cdot 8Al_2O_3$

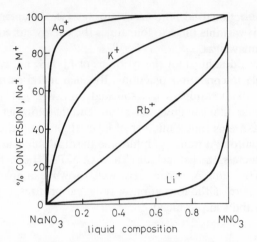

Fig. 2.12 Equilibria between β-alumina and binary nitrate melts at 300 to 350 °C. (From Yao and Kummer, 1967)

shown in Fig. 2.12 for the equilibria between β-alumina crystals and binary nitrate melts containing $NaNO_3$ and another metal nitrate at 300 to 350 °C. Thus, it is particularly easy to exchange Na^+ for Ag^+. Also, it is possible to effect a controlled partial ion exchange by controlling the composition of the nitrate melt. Ion exchanged β-aluminas may also be prepared by electrochemical methods, similar to those described in the preceding section.

Ion exchange reactions have been studied in considerable detail in β-alumina, but the same principles could probably be applied to a wide variety of crystals, especially alkali-containing crystals. Ion exchange reactions are constrained by both kinetic and thermodynamic factors. Kinetic factors are influenced largely by the mobility of ions: at elevated temperatures, e.g. ~ 300 °C, alkali ions are often quite mobile in crystals and are capable of ion exchange on immersion in a suitable melt. Thus, $Ag_2Si_2O_5$, which has a silicate sheet structure, has been prepared by immersing $Na_2Si_2O_5$ crystals in molten $AgNO_3$ at 280 °C. However, it is considerably more difficult to replace or introduce ions with a valency greater than unity since such ions form stronger bonds, whether ionic or covalent, and tend to be immobile. Much higher temperatures may therefore be necessary to effect their ion exchange.

Whereas the rate of ion exchange depends on kinetic factors, the extent of ion exchange, if indeed it occurs at all, depends on thermodynamic factors, as shown in Fig. 2.12 for β-alumina. Ion exchange involves equilibria of ions between the crystal and the melt and, in particular, depends on the activities of the two cations involved in the crystal and in the melt.

2.4.4 Synthesis of new metastable phases by 'Chimie Douce'

The precursor methods described in Sections 2.1.3 and 2.1.4 have the advantage that reaction takes place at much lower temperatures than when using normal solid state reaction procedures. Another use of precursor methods is in the synthesis of new, metastable phases which cannot be prepared by other routes. Although these phases are thermodynamically metastable they are often kinetically stable to quite high temperatures. Usually there is a close structural relationship between the precursor phase and final product. Often the new phases therefore have unusual structures and interesting properties. The French have coined the delightful term 'chimie douce' for this preparative method. Two examples of its use are as follows.

A new polymorph of titanium dioxide, $TiO_2(B)$, has been prepared by Tornaux and coworkers from the precursor phase $K_2Ti_4O_9$ which was first prepared by normal solid state reaction of KNO_3 (source of K_2O) and TiO_2 at 1000 °C for two days. This was then hydrolysed with nitric acid at room temperature to yield a solid product of stoichiometry $H_2Ti_4O_9 \cdot H_2O$. The final stage was to heat this material at 500 °C where it lost water to give the new $TiO_2(B)$ polymorph. The crystal structure of $TiO_2(B)$ is closely related to that of $K_2Ti_4O_9$: hydrolysis and dehydration leads to the effective removal of K_2O from $K_2Ti_4O_9$, leaving the remainder of the structure intact. The structure of $TiO_2(B)$ is built of TiO_6 octahedra, but these are linked up in a different way to the structures of the other TiO_2 polymorphs, rutile, anatase and brookite.

A new form of tungsten trioxide, hexagonal WO_3, has been prepared by Figlarz and coworkers via a tungstic acid gel precursor. Solutions of sodium tungstate, Na_2WO_4, and hydrochloric acid were mixed and the resulting gel treated hydrothermally (Section 2.7.8) at 120 °C to give a crystalline hydrate phase, $3WO_3 \cdot H_2O$. This was then dehydrated by gradually heating in air to 420 °C to give the new material. Hexagonal WO_3 has a more open structure than

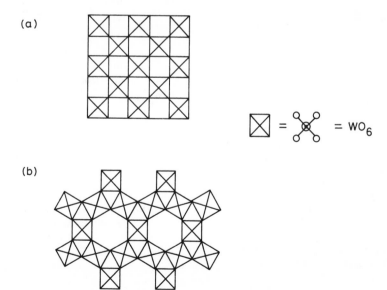

Fig. 2.13 The structures of (a) cubic ReO_3 and (b) hexagonal WO_3. The normal, monoclinic WO_3 structure is a slightly distorted version of (a)

the normal monoclinic form which has a distorted ReO_3 structure. Both forms are built of corner-sharing WO_6 octahedra, as shown schematically in Fig. 2.13.

2.5 Electrochemical reduction methods

A wide variety of crystalline solids have been prepared, some for the first time, others as well-formed single crystals using electrochemical reduction methods. The most common method is to electrolyse a molten mixture of the components, from which the resulting products usually crystallize out. In this method one of the species present undergoes reduction; examples are the reduction of transition metal ions to lower valence states and the reduction of oxyanions, such as phosphate to phosphide, carbonate to carbide and borate to boride. The conditions required to synthesize a certain compound, e.g. the correct melt composition to use, have to be found largely by trial and error and the reactions involved are often not well understood; the reduction potentials of the various ions involved are often not known at the temperature of the melt. The method has, however, considerable potential for further development and applications.

In this method, the melt usually contains a mixture of (a) a metal oxide, (b) an alkali borate (or phosphate, carbonate, etc.) and (c) an alkali halide. The mixture is contained in a crucible which is chemically inert to reaction with the melt and

Table 2.2 *Some electrochemical reduction reactions.* (From Wold and Bellavance, 1972, p. 279)

Melt constituents	Product of electrolysis	Temperature ($^\circ$C)	Comments
$CaTiO_3, CaCl_2$	$CaTi_2O_4$	850	Product was new titanium (III) compound
Na_2WO_4, WO_3	Na_xWO_3		Formation of tungsten bronze
Na_2MoO_3, MoO_3	MoO_2	675	Large crystals of MoO_2 produced
NaOH, Ni electrodes	$NaNiO_2$		Nickel obtained from cathode
$Na_2B_4O_7, NaF,$ V_2O_5, Fe_2O_3	FeV_2O_4	850	Vanadium spinel product
$Na_2B_4O_7, NaF, WO_3,$ Na_2SO_4	WS_2	800	
$NaPO_3, Fe_2O_3, NaF$	FeP	925	Product purified by vapour phase transport (with I_2)
Na_2CrO_4, Na_2SiF_6	Cr_3Si		
$Na_2Ge_2O_5, NaF, NiO$ $Li_2B_4O_7, LiF,$	Ni_2Ge		
Ta_2O_5	TaB_2	950	

the products of electrolysis; it may be made of alumina (for oxide synthesis) or graphite (for synthesis of carbides, sulphides, etc.). Electrodes, which should also be inert, may be of platinum or graphite. A selection of compounds prepared by this method, together with the melt constituents, is given in Table 2.2. Further examples of the use of electrochemical methods are discussed in Section 2.4.2. (intercalation compounds).

2.6 Preparation of thin films

Thin films, both crystalline and amorphous, are very important in modern technology. For instance, they may be used to form protective coatings on materials and they play a key role in the miniaturization of components in electronic devices. They also sometimes have special properties which derive from their thinness and, in particular, from their very large surface area to volume ratio: the structures and properties of the surfaces of solids are often quite different from the interior. Various methods are used to prepare thin films; they fall into two main groups, chemical and physical methods. A brief survey of each is given.

2.6.1 Chemical and electrochemical methods

Cathodic deposition. This is a standard method of electroplating; two metal electrodes are dipped into an electrolyte solution and on application of an external field across the electrodes, metal ions from the solution are deposited on the cathode as a thin film. In order to maintain charge balance, the anodic metal gradually dissolves in the electrolyte.

Electroless deposition. This is similar to cathodic deposition except that it takes place in the absence of an applied external field. It is commonly used for the deposition of nickel films. Both this and the cathodic deposition method suffer from the disadvantage that their use is limited mainly to the deposition of metallic films on to substrates that are electronically conducting, i.e. metals.

Anodic oxidation. This is an electrolytic method for producing oxide films on the surfaces of metals such as Al, Ta, Nb, Ti and Zr. These metals form the anode that dips into a liquid electrolyte such as a salt or acid solution. Oxide ions are attracted to the anode to form a thin layer of, for example, Al_2O_3; on increasing the field strength, more oxide ions diffuse through the oxide layer to the metal surface and, hence, the oxide layer grows thicker. An equilibrium thickness is usually reached which depends on the magnitude of the applied field. Anodic layers may also be formed by exposure of the metal to a glow discharge.

Thermal Oxidation. Many substances oxidize in air, especially at high temperatures, and in some cases the product is an inert film which inhibits further oxidation. Thus, metallic aluminium forms an oxide film that is 30 to 40 Å thick at room temperature but whose thickness increases with temperature. The method is not restricted to oxide films; thus some metals, on exposure to, for example, ammonia at elevated temperatures, become coated with a metal nitride film.

Chemical vapour deposition. This method makes use of some of the principles involved in vapour phase transport reactions, outlined in Section 2.2.3. It is used to obtain very pure films of crystalline semiconductors including III–V compounds. The films are formed by decomposition of gaseous molecules. The decomposition may be achieved by, for example, pyrolysis (i.e. heating), photolysis (i.e. irradiation with IR or UV light) or chemical reaction, as shown in the following examples:

$$GeH_4 \xrightarrow{\text{heat or } h\nu} Ge$$

$$Si(CH_2CH_3)_4 \xrightarrow[\text{air}]{\text{heat}} SiO_2$$

$$SiCl_4 + 2H_2 \rightarrow Si + 4HCl$$

$$SiH_4 \rightarrow Si + 2H_2$$

Alternatively, use may be made of equilibria between various species, in which the position of equilibrium varies with temperature; e.g.:

$$2SiI_2 \rightleftharpoons SiI_4 + Si$$

This reaction goes increasingly to the right-hand side with decreasing temperature and hence silicon may deposit on a cold substrate.

Vapour deposited films tend to match the structure of the substrate, if possible, resulting in epitaxial growth mechanisms. Epitaxially grown films may also form by crystallization or precipitation from a liquid phase onto the substrate. To do this, the compound to be deposited is dissolved in a low melting metal such as indium or lead and on subsequent cooling the compound precipitates out.

2.6.2 Physical methods

Cathode sputtering. The apparatus used for this technique is outlined in Fig. 2.14(a). Basically, it consists of a bell jar which contains a reduced pressure — 10^{-1} to 10^{-2} torr—of an inert gas, argon or xenon. This gas is subjected to a potential drop of several kilovolts. A glow discharge is thereby created in the inert gas from which positive ions are accelerated towards the cathode (target). These high energy ions remove material from the cathode which then condenses on the surroundings, including the substrates to be coated, which are placed in a suitable position relative to the cathode. The mechanism of sputtering, or removal of material from the cathode, appears to involve the transfer of momentum from the gaseous ions to the cathode in such a way that atoms or ions are then ejected from the cathode. Modern cathode sputtering equipment has various refinements, including the means to prevent permanent contamination of the substrate by inert gas atoms or ions.

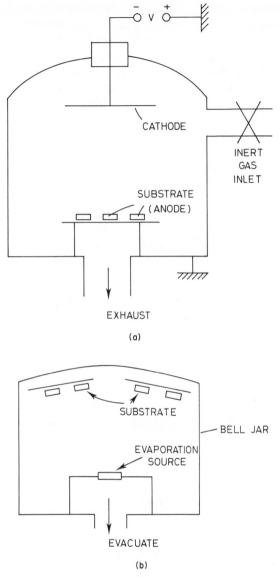

Fig. 2.14 (a) Cathode sputtering equipment for thin film deposition. (b) Vacuum evaporation equipment for thin film deposition

Vacuum evaporation. This appears to be the most widely used and versatile method for the preparation of thin films. The apparatus is outlined in Fig. 2.14(b). The system operates under a high vacuum, 10^{-6} torr or better. Material from the evaporation source is converted into the gaseous phase by, for example, heating or electron bombardment. The gaseous material then deposits on the substrate and its surrounding as a film. Various substrate materials are used, depending on the subsequent application of the film that is to be deposited. For electronic applications it acts both as a mechanical support and an electrical insulator. Typical substrates used are ceramics (Al_2O_3), glass, alkali halides, silicon, germanium, etc. A wide variety of source materials for evaporation may be used, including metals, alloys, semiconductors, insulators and inorganic salts. These are placed in containers made of, for example, tungsten, tantalum or molybdenum, which withstand very high temperatures and are chemically unreactive towards the material being evaporated.

It is often important to thoroughly clean the substrate surface prior to evaporation and this may be carried out by a sequence of steps involving: ultrasonic cleaning in, for example, a detergent solution, degreasing in, for example, alcohol, degassing under vacuum and, finally, ion bombardment to remove the surface layers from the substrate. Such cleaning may be necessary in order to obtain good adhesion of the film to the substrate or it may simply be desirable for reasons of the purity required.

2.7 Growth of single crystals

Crystals may be grown from vapour, liquid or solid phases although, usually, only the first two give crystals of sufficient size to be used in applications or for property measurements. A brief summary only is given of some of the various methods.

2.7.1 Czochralski method

This is basically a method for the growth of a single crystal from a melt of the same composition. A seed crystal is placed in contact with (on the surface of) the

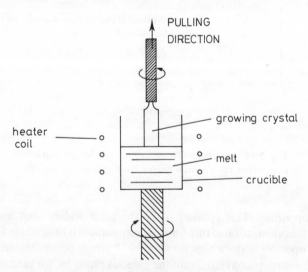

Fig. 2.15 Czochralski method for crystal growth

melt, whose temperature is maintained slightly above its melting point. As the seed is gradually pulled out of the melt (Fig. 2.15), the melt solidifies on the surface of the seed to give a rod-shaped crystal in the same crystallographic orientation as the original seed. The melt and growing crystal are usually rotated counterclockwise during pulling in order to maintain a constant temperature, melt uniformity, etc. The method is widely used for the growth of crystals of semiconducting materials, Si, Ge, GaAs, etc.; usually an inert gas atmosphere at high pressure is used in order to prevent loss of As, P, etc. It has also been used to produce laser generator materials such as $Ca(NbO_3)_2$ doped with neodymium.

2.7.2 Bridgman and Stockbarger methods

These methods are also based on the solidification of a stoichiometric melt but, in these, oriented solidification of the melt is achieved by effectively passing the melt through a temperature gradient such that crystallization occurs at the cooler end. This is achieved in the Stockbarger method by arranging for a relative displacement of the melt and a temperature gradient (Fig. 2.16a). In the Bridgman method, the melt is inside a temperature gradient furnace and the furnace is gradually cooled so that solidification begins at the cooler end (Fig. 2.16b). In both methods, it is again advantageous to use a seed crystal and atmospheric control may be necessary.

2.7.3 Zone melting

This is related to the Stockbarger method but the thermal profile through the furnace is such that only a small part of the charge is melted at any one time (Fig. 2.16c). Initially that part of the charge in contact with the seed crystal is melted. As the boat is pulled through the furnace, oriented solidification onto the seed occurs and, at the same time, more of the charge melts. This also forms a well-known method for the purification of solids, the *zone-refining technique*. It makes use of the principle that impurities usually concentrate in the liquid rather than in the solid phase. Impurities are therefore 'swept out' of the crystal by the moving molten zone. The method has been used for the purification and crystal growth of high melting metals such as tungsten.

2.7.4 Precipitation from solution or melt: flux method

In contrast to the above methods in which melts solidify to give crystals that have the same composition as the melt, precipitation methods involve the growth of crystals from a solvent of different composition to the crystals. The solvent may be one of the constituents of the desired crystal, e.g. crystallization of salt hydrate crystals using water as the solvent, or the solvent may be an entirely separate liquid element or compound in which the crystals of interest are partially soluble, e.g. SiO_2 and various high melting silicates may be precipitated from low melting borate or halide melts. In these cases, the solvent melts are sometimes referred to

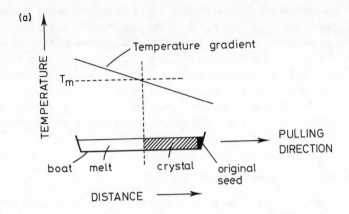

(a)

TEMPERATURE →

Temperature gradient

T_m

boat   melt   crystal   original seed

PULLING DIRECTION

DISTANCE →

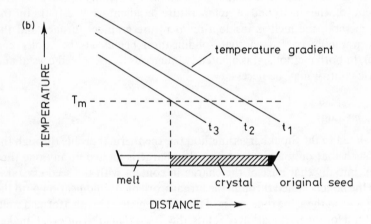

(b)

TEMPERATURE →

temperature gradient

T_m

t_3 t_2 t_1

melt   crystal   original seed

DISTANCE →

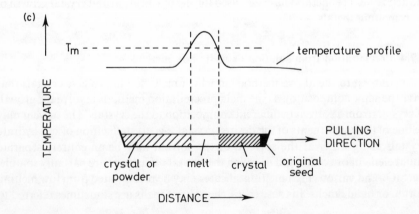

(c)

TEMPERATURE →

T_m

temperature profile

crystal or powder   melt   crystal   original seed

PULLING DIRECTION

DISTANCE →

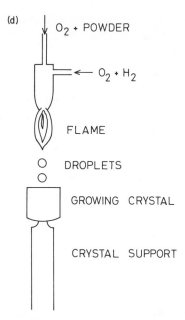

(d)

O_2 + POWDER

← O_2 + H_2

FLAME

○ DROPLETS
○

GROWING CRYSTAL

CRYSTAL SUPPORT

Fig. 2.16 (a) Stockbarger method: T_m = crystal melting point. (b) Bridgman method: times t_1, t_2 and t_3 are shown. (c) Zone melting method. (d) Verneuil method

as *fluxes* since they effectively reduce the melting point of the crystals by a considerable amount. The method has recently been used to grow crystals of β- and β''-alumina solid electrolytes using a borate flux. Further details of flux growth are given in Section 2.2.2, where the importance of having information on the relevant phase diagram is stressed.

2.7.5 Epitaxial growth of thin layers

Single crystals in the form of thin layers are often required for application in electronic devices and special methods may be needed for their preparation. The use of a seed crystal in Czochralski and other methods implies an orientation relationship between the growing crystal and the seed. The crystal may grow epitaxially (two-dimensional structural relationship between seed and crystal) or topotaxially (three-dimensional relationship) on the seed. In the epitaxial growth of thin layers, oriented growth of the crystal occurs on the surface of a substrate. The substrate may be a crystal of the same or similar composition or it may be an entirely different material whose lattice parameters, at its surface, match those of the growing crystal, to within a few per cent. In this way, thin layers of GaAs have been grown from the vapour phase by deposition onto various substrates that include Al_2O_3, spinel $MgAl_2O_4$, Ge and ThO_2. Thin film crystals may also be prepared by epitaxial growth from liquids, in which case a knowledge of the relevant phase diagram is necessary. In these methods, attention to the surface condition of the substrate is extremely important in order to ensure that it is as free from defects and impurities as possible.

2.7.6 Verneuil flame fusion method

This method was first used in 1904 for growing crystals of high melting oxides, including artificial gemstones such as ruby and sapphire. The starting material, in the form of a fine powder, is passed through an oxyhydrogen flame or some other high temperature torch or furnace (Fig. 2.16d). After melting has taken place in the flame, the droplets fall on to the surface of the growing crystal or seed where they solidify, leading to crystal growth. The method has recently been used to prepare single crystals of CaO with a melting point $\sim 2600\,^{\circ}\text{C}$ by using a plasma torch to melt the CaO powder.

2.7.7 Vapour phase transport

See Section 2.3.

2.7.8 Hydrothermal methods

See Section 2.8.1.

2.7.9 Comparison of different methods

For each crystalline substance, there are optimum conditions for satisfactory crystal growth. In preparing to grow crystals of a new material, much preliminary work, perhaps lasting several months, is needed in order to find suitable growth conditions. The quality of the resulting crystals therefore depends very much on the experimenters' skill. In addition, the various techniques that are available have certain intrinsic advantages and disadvantages; these are summarized in Table 2.3.

Table 2.3 *Comparison of crystal growth methods*

Method	Advantages	Disadvantages
Melt growth (Czochralski, Bridgman–Stockbarger, Verneuil)	Rapid growth rates, giving large crystals; simple apparatus	Crystal quality may be poor with inhomogeneities and large defect concentrations
Solution growth (water crystallization, flux growth, hydrothermal method)	Isothermal conditions with slow growth rates give quality crystals of low defect concentration	Slow growth rates; problems of contamination by container or flux

2.8 High pressure and hydrothermal methods

High pressure and hydrothermal methods are finding increasing applications in materials science and solid state chemistry. They are technologically important, both as an important method of crystal growth and for the synthesis of new materials with useful properties. They are also interesting scientifically since the use of high pressures provides an additional parameter or lever for obtaining fundamental information on the structures, behaviour and properties of solids.

In most high pressure methods, the sample is effectively squeezed between the jaws of opposed rams or anvils. Hydrothermal methods differ in that water under pressure is present inside the reaction vessel. For convenience, hydrothermal methods and their applications are outlined first, followed by other (dry) high pressure methods.

2.8.1 Hydrothermal methods

Hydrothermal methods utilize water under pressure and at temperatures above its normal boiling point as a means of speeding up the reactions between solids. The water performs two roles. The water—as liquid or vapour—serves as the pressure transmitting medium. In addition, some or all of the reactants are partially soluble in the water under pressure and this enables reactions to take place in, or with the aid of, liquid and/or vapour phases. Under these conditions, reactions may occur that, in the absence of water, would occur only at much higher temperatures. The method is therefore particularly suited for the synthesis of phases that are unstable at higher temperatures. It is also a useful technique for growth of single crystals; by arranging for a suitable temperature gradient to be present in the reaction vessel, dissolution of the starting material may occur at the hot end and reprecipitation at the cooler end.

Since hydrothermal reactions must be carried out in closed vessels, the pressure–temperature relations of water at constant volume are important. These are shown in Fig. 2.17. The *critical temperature* of water is 374 °C. Below 374 °C, two fluid phases, liquid and vapour, can coexist. Above 374 °C only one fluid phase, *supercritical water*, ever exists. Curve AB represents the *saturated steam curve*. At pressures below this curve liquid water is absent and the vapour phase is not saturated with respect to steam; on the curve the vapour is composed of saturated steam which is in equilibrium with liquid water; above the curve, liquid water is effectively under compression and the vapour phase is absent.

The dashed curves in Fig. 2.17 may be used to calculate the pressure that is developed inside a vessel after it has been partially filled with water, closed and heated to a certain temperature. Thus, curve BC corresponds to a vessel that is initially 30 per cent full of water: at, for example, 600 °C, a pressure of 800 bar is generated inside the closed vessel. Although Fig. 2.17 applies strictly to pure water, the curves are modified little, provided the solubility of solids present in the reaction vessel is small.

The design of hydrothermal equipment is basically a tube, usually of steel,

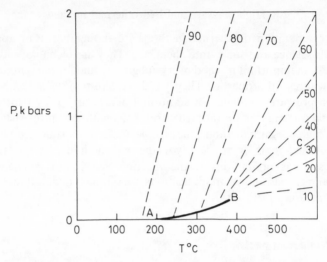

Fig. 2.17 Pressure–temperature relations for water at constant volume. Dashed curves represent pressures developed inside a closed vessel; numbers represent the percentage degree of filling of the vessel by water at ordinary *P, T*. (After Kennedy, 1950)

closed at one end. The other end has a screw cap with a gasket of soft copper to provide a seal. Alternatively, 'the 'bomb' may be connected directly to an independent pressure source, such as a hydraulic ram; this is known as the 'cold seal' method. The reaction mixture and an appropriate amount of water are placed inside the bomb which is then sealed and placed inside an oven at the required temperature.

Applications (*a*) *synthesis of new phases : calcium silicate hydrates.* Hydrothermal methods have been used successfully for the synthesis of many materials. A good example is the family of calcium silicate hydrates, many of which are important components of set cement and concrete. Typically, lime, CaO and quartz, SiO_2, are heated with water at temperatures in the range 150 to 500 °C and pressures of 0.1 to 2.0 kbar. Each calcium silicate hydrate has, for its synthesis, optimum preferred conditions of: composition of starting mix, temperature, pressure and time. For example, xonotlite, $Ca_6Si_6O_{17}(OH)_2$, may be prepared by heating equimolar mixtures of CaO and SiO_2 at saturated steam pressures in the range 150 to 350 °C. By varying the experimental conditions, H.F.W. Taylor and others have been able to unravel the chemistry of this complex family of solids. More details are given in Chapter 19.

(*b*) *Growth of single crystals.* For the growth of single crystals by hydrothermal methods it is often necessary to add a *mineralizer*. A mineralizer is any compound added to the aqueous solution that speeds up its crystallization. It usually operates by increasing the solubility of the solute through the formation of soluble species that would not usually be present in the water. For instance, the

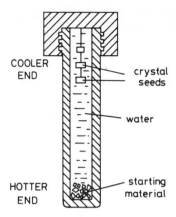

COOLER END

crystal seeds

water

HOTTER END

starting material

Fig. 2.18 Schematic hydrothermal bomb used for crystal growth

solubility of quartz in water at 400 °C and 2 kbar is too small to permit the recrystallization of quartz, in a temperature gradient, within a reasonable space of time. On addition of NaOH as a mineralizer, however, large quartz crystals may be readily grown (Fig. 2.18). Using the following conditions, crystals of kilogram size have been grown: quartz and 1.0 M NaOH solution are held at 400 °C and 1.7 kbar; at this temperature some of the quartz dissolves. A temperature gradient is arranged to exist in the reaction vessel and at 360 °C the solution is supersaturated with respect to quartz which precipitates onto a seed crystal. In summary, therefore, quartz dissolves in the hottest part of the reaction vessel, is transported through-out the vessel via convection currents and is precipitated in cooler parts of the vessel where its solubility in water is lower. Quartz single crystals are used in many devices in radar and sonar, as piezoelectric transducers, as monochromators in X-ray diffraction, etc. Annual world production of quartz single crystals, using hydrothermal and other methods, is currently a staggering 600 tons!

Using similar methods, many substances have been prepared as high quality single crystals, e.g. corundum (Al_2O_3) and ruby (Al_2O_3 doped with Cr^{3+}).

2.8.2 Dry high pressure methods

Techniques are now available for generating static pressures of several hundred kilobars at both ambient and high temperatures; using shock wave methods the attainable P, T range may be extended still further. Experimental techniques are not detailed here since the techniques are rather specialized. They vary from a simple 'opposed anvil' arrangement, whereby the sample is effectively squeezed between two pistons or rams, one of which is fixed and the other connected to a hydraulic jack, to more complex designs involving three or four anvils or rams.

Applications: Synthesis of unusual crystal structures. Phases synthesized at high pressures tend to have higher densities than phases synthesized at atmospheric

Table 2.4 *High pressure polymorphism of some simple solids*

Solid	Normal structure and coordination number	Typical transformation conditions P(kbar) T(°C)		High pressure structure and coordination number
C	Graphite, 3	130	3000	Diamond, 4
CdS	Wurtzite 4:4	30	20	Rock salt 6:6
KCl	Rock salt 6:6	20	20	CsCl, 8:8
SiO_2	Quartz 4:2	120	1200	Rutile, 6:3
Li_2MoO_4	Phenacite 4:4:3	10	400	Spinel, 6:4:4
$NaAlO_2$	Ordered Wurtzite 4:4:4	40	400	Ordered Rock salt 6:6:6

pressure and this sometimes gives rise to unusually high coordination numbers (see also Chapter 12). For example, the coordination of silicon in SiO_2 and the silicates is, with very few exceptions, tetrahedral. An exception is one of the high pressure polymorphs of SiO_2, called stishovite, which is formed at pressures above 100 to 120 kbar. Stishovite has the rutile structure and therefore contains octahedrally coordinated silicon. Some other examples of increased coordination number in high pressure polymorphs are given in Table 2.4.

By use of high pressures it is possible to stabilize ions in unusual oxidation states, such as Cr^{4+}, Cr^{5+}, Cu^{3+}, Ni^{3+} and Fe^{4+}. Thus chromium normally occurs as Cr^{3+} and Cr^{6+} with octahedral and tetrahedral coordination, respectively. However, various perovskite phases containing octahedrally coordinated Cr^{4+} have been prepared at high pressures, such as $PbCrO_3$, $CaCrO_3$, $SrCrO_3$ and $BaCrO_3$.

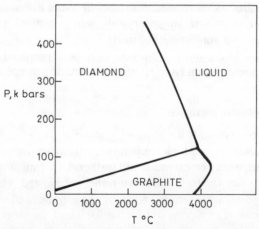

Fig. 2.19 The P, T phase diagram for carbon. (After Bundy, 1963)

Possibly the main industrial application of high pressure methods is in the synthesis of diamonds from graphite. Suitable thermodynamic conditions for the transformation are given by reference to the P, T phase diagram for carbon, given in Fig. 2.19, although the transformation rates are slow, even at pressures and temperatures that are well within the diamond phase field.

Questions

2.1   Explain why reactions in the solid state take place only slowly in most cases. What can be done to speed up reaction rates?

2.2   How would you attempt to prepare the following by solid state reaction, given a free choice of reagents: (a) Na_3PO_4 (m.p. $> 1500\,°C$); (b) $NaAlO_2$ (m.p. $> 1500\,°C$); (c) $Y_3Fe_5O_{12}$ (YIG, ferrimagnetic garnet, Chapter 16); (d) $RbAg_4I_5$ (solid electrolyte, Chapter 13); (e) Ca_3SiO_5 (cement, Chapter 19); (f) $Pb(Zr_{0.5}Ti_{0.5})O_3$ (PZT ferroelectric, Chapter 15); (g) $Y_2Si_3O_3N_4$ (sialon ceramic, Chapter 20)?

2.3   What kind of factors would you have to take into consideration in planning to make a kinetic study of a solid state reaction between, for example, powders of MgO and Al_2O_3 to form $MgAl_2O_4$ spinel? How would you try to analyse the results?

2.4   How would you attempt to make an intercalation compound of graphite and fluorine? Details of how this has been successfully prepared recently are given in *New Scientist* (1983) July 21, p 191.

2.5   From a consideration of the phase diagram for carbon (Fig. 2.19), how would you try to synthesize diamond?

2.6   With the aid of Fig. 2.12, how would you replace 50 per cent of the Na^+ ions in β-alumina by: (a) Ag^+, (b) K^+, (c) Rb^+? How would you determine the degree of ion exchange?

2.7   Calculate the activation energy for the reaction between NiO and Al_2O_3 to form $NiAl_2O_4$ using the data given in Fig. 2.1c. Check your answer with that given in the literature.

2.8   Furnaces that contain platinum-based heating elements are often found, after a period of use at high temperatures, to have a deposit of Pt crystals in cooler, outer regions of the furnace, well away from the Pt heating elements. How do you account for this?

2.9   Under what conditions can vapour phase transport methods be used to purify metals?

2.10 How would you attempt to prepare the following given a free choice of normal laboratory reagents and any equipment that you desired. (a) a single crystal rod of very high purity Si; (b) a single crystal film of Si, approx. 1 μm thick; (c) a coating of boron on a substrate of metallic iron and subsequent conversion to a surface layer of iron boride; (d) the stishovite polymorph of SiO_2, with the rutile crystal structure; (e) polycrystalline $Li_{0.7}Ti^{3+}_{0.7}Ti^{4+}_{0.3}S_2$; (f) large single crystals of quartz?

References

R. M. Barrer (1982). *Hydrothermal Chemistry of Zeolites*, Academic Press.

V. V. Boldyrev, M. Bulens and B. Delmon (1979). *The Control of the Reactivity of Solids*, Elsevier.

D. W. Breck (1974). Zeolite Molecular Sieves, Structure, Chemistry and Use, Wiley.

P. P. Budnikov and A. M. Ginstling (1968). *Principles of Solid State Chemistry*, Applied Science.

F. B. Bundy (1963). *J. Chem. Phys.*, **38**, 631.

P. G. Dickens and M. S. Whittingham (1968). The tungsten bronzes and related compounds. *Quart. Revs.*, **22**, 30–44

B. Gerand, G. Nowogrocki, J. Guenot and M. Figlarz (1979). *J. Solid State Chem.*, **29**, 429.

P. Hagenmuller (1972). *Preparative Methods in Solid State Chemistry*, Academic Press.

J. M. Honig and C. N. R. Rao (1981). *Preparation and Characterisation of Materials*, Academic Press.

G. C. Kennedy (1950). *Am. J. Sci.*, **248**, 540.

W. D. Kingery, H. K. Bowen and D. R. Uhlmann (1976). *Introduction to Ceramics*, Wiley.

R. Marchand, L. Brohan and M. Tournaux (1980). *Materials Res. Bull.*, **15**, 1129.

S. Mroczkowski (1980). Needs and opportunities in crystal growth, *J. Chem. Ed.*, **57**, 537.

F. S. Pettit, E. H. Randklev and E. J. Felten (1966). *J. Amer. Ceram. Soc.*, **49**, 199.

R. Roy (1977). *J. Amer. Ceramic Soc.*, **60**, 350.

H. Schäfer (1971). Preparative solid state chemistry: the present position, *Angew. Chem. Int'l Ed.*, **10**, 43–50.

H. Schmalzried (1971). *Solid State Reactions*, Verlag Chemie.

Whipple and A. Wold (1962). *J. Inorganic Nucl. Chem.*, **24**, 23–27.

A. Wold (1980). The preparation and characterisation of materials, *J. Chem. Ed.*, **57**, 531.

A. Wold and D. Bellavance (1972). In *Preparative Methods in Solid State Chemistry*, (Ed. P. Hagenmuller), Academic Press.

Y. F. Yao and J. T. Kummer (1967). *J. Inorg. Nucl. Chem.*, **29**, 2453.

Chapter 3

Characterization of Inorganic Solids : Applications of Physical Techniques

3.1 General approach

The simplest and most obvious first question to ask about an inorganic substance is 'What is it?'. The methods that are used to answer this come into two main categories, depending on whether the substance is molecular or non-molecular. If the substance is molecular, whether it be solid, liquid or gaseous, identification is usually carried out by some combination of spectroscopic methods and chemical analysis. If the substance is non-molecular and crystalline, identification is usually carried out by X-ray powder diffraction supplemented, where necessary, by chemical analysis. Each crystalline solid has its own characteristic X-ray powder pattern which may be used as a 'fingerprint' for its identification. The powder patterns of most known inorganic solids are included in an updated version of the Powder Diffraction File; by using an appropriate search procedure, unknowns can usually be identified rapidly and unambiguously.

Once the substance has been identified, the next stage is to determine its structure, if this is not known already. For molecular materials, details of the molecular geometry may be obtained from further spectroscopic measurements. Alternatively, if the substance is crystalline, X-ray crystallography may be used, in which case information is also obtained on the way in which the molecules pack together in the crystalline state. For molecular substances, this usually completes the story as far as identification and structure determination are concerned; attention may then be focused on other matters such as properties or chemical reactivity.

For non-molecular substances, however, the word 'structure' takes on a whole new meaning. In order for a solid to be well characterized, one needs to know about:

(a) the form of the solid, whether it is single crystal or polycrystalline and, if the latter, what is the number, size, shape and distribution of the crystalline particles;
(b) the crystal structure;

Table 3.1 *Characterization of solids and techniques available*

	Bond type	Electronic structure	Elemental analysis	Polycrystalline texture	Surface structure	Crystal defects	Local structure, CN, etc.	Crystal structure	Unit cell, space group	Amorphous or crystalline	Phase identification
X-ray diffraction	(√)			(√)		(√)	(√)	√	√	√	√
Electron diffraction and microscopy			√	√	√	√		√	√	√	√
Neutron diffraction						√		√	√	(√)	(√)
Optical microscopy				√	(√)	(√)				√	√
IR spectroscopy	(√)		(√)		√		√				√
UV, visual spectroscopy	√	√	(√)				(√)	√			
NMR, ESR spectroscopy	(√)	(√)	(√)				(√)	√			(√)
Electron spectroscopy—ESCA, XPS, UPS, AES, EELS	√	√	√		√		(√)	√			
X-ray spectroscopy—XRF AEFS, EXAFS	(√)	(√)	√		√		(√)	√			
Mössbauer spectroscopy	(√)		(√)				√				

(c) the crystal defects that are present—their nature, number and distribution;

(d) the impurities that are present and whether they are distributed at random or are concentrated into small regions;

(e) the surface structure, including any compositional inhomogeneities or absorbed surface layers.

No single technique is capable of providing a complete characterization of a solid. Rather, a variety of techniques are used, in combination. Sometimes, however, one is interested in only one structural aspect, in which case a single technique may provide the required information.

There are three main categories of physical technique which may be used to characterize solids; these are diffraction, microscopic and spectroscopic techniques. In addition, other techniques such as thermal analysis, magnetic measurements and physical property measurements may give valuable information in certain cases. In Table 3.1 a listing of some of the techniques is given together with the structural information that each is capable of providing.

The principal purpose of this chapter is to amplify the information that is contained in Table 3.1. It is assumed that the reader has at least a passing familiarity with the different techniques. Only limited explanations of each are given and instead attention is focused on the applications to solids and the kind of information that is obtained. Some techniques, especially X-ray diffraction, are discussed in more detail in later chapters. Thermal analysis and physical property measurements—conductivity, magnetism, etc.—are not included in this chapter but are also discussed in later chapters.

3.2 Survey of techniques and their applications to solids

3.2.1 Diffraction techniques

Because of its central rôle in solid state chemistry, the subject of X-ray diffraction is treated in considerable depth in later Chapters. It is relevant, however, for the purpose of providing an overall perspective, to summarise here the main features of the various X-ray techniques, and their uses.

3.2.1.1 X-ray powder diffraction

An X-ray powder diffraction pattern is a set of lines or peaks, each of different intensity and position (d-spacing or Bragg angle, θ), on either a strip of photographic film or on a length of chart paper (Fig. 3.1). For a given substance the line positions are essentially fixed and are characteristic of that substance. The intensities may vary somewhat from sample to sample, depending on the method of sample preparation and the instrumental conditions. For identification purposes, principal note is taken of line positions together with a semi-quantitative consideration of intensities. Some applications of X-ray powder diffraction are now given.

50

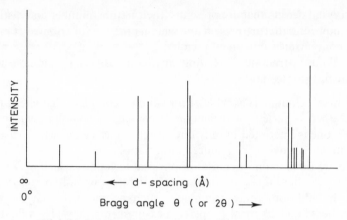

Fig. 3.1 Schematic X-ray powder diffraction pattern

Phase identification. Each crystalline substance has its own characteristic powder diffraction pattern which may be used for its identification. Standard patterns are given in the Powder Diffraction File (known as the JCPDS File or, formerly, as the ASTM File). The inorganic section of the File now contains over 35000 entries and is increasing at the rate of about 2000 per year. The file also contains a section for organic compounds. Substances are indexed on one of two methods. The Hanawalt index uses the eight most intense lines and the Fink index uses the first eight lines, of longest *d*-spacing, in the powder pattern.

Mixtures of substances may be identified provided, of course, that the patterns of the component phases are available for comparison. For many applications, and especially where the number of possible compounds is limited, it is preferable to keep one's own set of standard patterns, rather than resorting to the File with every new sample. This is particularly so when using Guinier films to identify samples. The powder X-ray method may be used as a rough check on purity provided the impurities are present as a separate crystalline phase(s). The lower limit of detection of impurity phases in routine work is usually in the range 1 to 5 per cent. Under favourable conditions such as when looking for a specific impurity, the detection limit may be decreased considerably by either increasing the sensitivity, if a diffractometer is used, or by increasing the time of exposure, if a film method is used.

Quantitative phase analysis. The amount of a particular crystalline phase in a mixture may be determined by quantitative X-ray powder diffraction. The procedure is straightforward but somewhat tedious and prone to errors. It is necessary to add an internal standard, which is a well-crystallized phase such as α-Al_2O_3, to the sample in a closely controlled amount (e.g. 10 per cent by weight). A line in the powder pattern of the phase of interest is selected and its intensity is compared with that of a suitable internal standard line. The amount of the phase present can be determined by interpolation from a previously constructed calibration graph of intensity against composition.

Determination of accurate unit cell parameters. The positions (*d*-spacings) of the lines in a powder pattern are governed by the values of the unit cell parameters (*a, b, c, α, β, γ*). Unit cell lattice parameters are normally determined by single crystal methods but the values obtained are often accurate to only two or three significant figures. More accurate cell parameters may be obtained from the powder pattern, provided the various lines have been assigned Miller indices *hkl* and their positions have been measured accurately. Using a least squares minimization procedure, unit cell parameters accurate to four or five significant figures may usually be obtained. Accurate unit cell parameters are particularly useful (a) for enabling complex powder patterns to be indexed, (b) for studying the effects of composition on cell parameters and (c) for measuring thermal expansion coefficients.

Solid solution lattice parameters. The lattice parameters of solid solution series often show a small but detectable variation with composition. This provides a useful means of characterizing solid solutions and, in principle, lattice parameters may be used as an indicator of composition. If the composition dependence is linear then the so-called Vegard's Law is obeyed. Deviations from Vegard's Law often occur in metallic solid solutions; the reasons for the deviations are not well understood. In non-metallic solid solutions, deviations from Vegard's Law are much less common and, when they occur, can usually be ascribed to a structural feature in the solid solutions. For instance, one cause of a positive deviation from Vegard's Law is the occurrence of incipient immiscibility in an apparently homogenous and random solid solution series.

Crystal structure determination. Crystal structures are solved by analysing the intensities of diffracted X-ray beams. Normally single crystal samples are used but powders may be used in cases where (a) single crystals are not available and (b) the structure is fairly simple and only a limited number of atomic coordinates must be determined in order to solve the structure. As an example of the latter, suppose that one has prepared a new phase which appears to have the spinel structure. Powder diffraction may be used to confirm its structure and to decide whether it is a normal or an inverse spinel.

Particle size measurement. X-ray powder diffraction may be used to measure the average crystal size in a powdered sample, provided the average diameter is less than about 2000 Å. The lines in a powder diffraction pattern are of finite breadth but if the particles are very small the lines are broader than usual. The broadening increases with decreasing particle size. The limit is reached with particle diameters in the range roughly 20 to 100 Å; then the lines are so broad that they effectively 'disappear' into the background radiation.

For particles that are markedly non-spherical, it may be possible to estimate the shape since different lines in the powder pattern are broadened to differing degrees.

Short range order in non-crystalline solids. Crystalline solids give diffraction patterns that have a number of sharp lines (Fig. 3.1). Non-crystalline solids—glasses, gels—give diffraction patterns that have a small number of very broad humps (see Fig. 18.6b silica glass). From these humps, information on local structure may be obtained. The results are usually presented as a radial distribution function (RDF) (see Fig. 18.7). This shows the probability of finding an atom as a function of distance from a reference atom. Information is thereby obtained on coordination environments (first and outer spheres) and bond distances. Such information is useful for testing the validity of structural models for non-crystalline solids. It has been applied to simple glasses based on, for example, SiO_2, B_2O_3, etc., and to amorphous carbon. While the method has undoubtedly been important in providing structural information on such materials and in testing theories of glass structure, there are severe limitations to its use. Data analysis is not straightforward and with complex multi-component glasses little useful information can be obtained. This is unfortunate since new glassy materials are finding various applications, e.g. neodymium laser glass, amorphous semiconductors, glassy metals and ionically conducting glasses. Structural information on these materials is badly needed.

Crystal defects and disorder. Certain types of defect and disorder that occur in crystalline solids may be detected by a variety of diffraction effects. The measurement of particle size from X-ray line broadening has already been mentioned. Another possible source of line broadening is strain within the crystals. This may be present in plastically deformed (i.e. work hardened) metals.

The thermal motion of atoms, which is inevitably present in all substances above absolute zero, causes a reduction in peak intensities and an increase in the level of background radiation. This is particularly noticeable at high temperatures and as the melting point of the sample is approached. Account of such thermal motions is taken in structure determinations by introducing the so-called temperature factor.

The technique of small angle X-ray scattering (SAXS) is used for detecting inhomogeneities on the scale of 10 to 1000 Å. Diffracted radiation is measured at angles just off the incident beam (i.e. at a small number of degrees 2θ or at the left end of Fig. 3.1) using a Krattky camera. This diffracted radiation appears as a shoulder on the intense beam of undiffracted radiation centred at $0°\,2\theta$. In favourable cases, inhomogeneities such as two-phase structures associated with immiscibility in solid or liquid solutions may be studied.

3.2.1.2 *High temperature X-ray powder diffraction*

Thermal expansion coefficients. The thermal expansion of, for example, metals is conventionally measured by dilatometry using rod-shaped specimens. An alternative and rather unconventional method is to use high temperature X-ray powder diffraction (HTXR). By this means, the change in unit cell parameters with temperature is measured and from this the thermal expansion coefficients may be calculated. For cubic materials, the results obtained by dilatometry and HTXR should agree well. Exceptions may arise if the crystal structure changes significantly with temparature and, especially, if a significant number of atom or ion vacancies is produced at high temperature. Vacancies arise when atoms leave their sites and effectively migrate to the surface of the specimen. In such cases, the coefficients determined dilatometrically may exceed the X-ray values.

The X-ray method is distinctly advantageous for non-cubic materials. The expansion of all three unit cell edges may be determined, together with any change(s) in the cell angles. By dilatometry, however, only a single average coefficient is measured (unless single crystal samples are available and the expansion is measured in different directions).

Thermal expansion data are important in several areas apart from the obvious ones concerning metals. The list includes ceramics and glass-ceramics that are resistant to thermal shock, insulating ceramic substrates for use in printed circuit boards and any situation where, for example, a ceramic to ceramic or ceramic to metal seal is necessary.

Polymorphism and phase transitions. High temperature X-ray powder diffraction is a valuable technique for obtaining structural information on polymorphs and phases that exist only at high temperatures. It is particularly useful for studying high temperature structures that cannot be preserved to room temperature by quenching. Other high temperature structures may be quenched to room temperature where they are kinetically stable and may therefore be studied at leisure by normal techniques. An example of a high temperature polymorph that cannot be quenched to room temperature is β-quartz; the stable room temperature polymorph of SiO_2 is α-quartz but this transforms to β-quartz on heating above 573 °C. When β-quartz is cooled, it reverts rapidly to α-quartz. The only way to obtain structural information on β-quartz is by X-ray diffraction at high temperatures. An example of a high temperature polymorph that may be preserved at room temperature, where it is kinetically stable, is α-$CaSiO_3$, pseudo-wollastonite. This phase is thermodynamically stable only above 1125 °C but it may be readily preserved to room temperature. The polymorph that is stable at room temperature is β-$CaSiO_3$, wollastonite.

Both film-based and diffractometer instruments are available for high temperature powder work. The diffractometer is preferable for accurate measurement of d-spacings or intensities. Film methods have the advantage that a continuous X-ray record, with changing time or temperature, may be obtained; this may give valuable information on the mechanisms of phase transitions. X-ray instruments are also available for studying samples under other non-ambient conditions, e.g. at low temperatures or at high pressure.

3.2.1.3 *Single crystal X-ray diffraction*

There are several single crystal X-ray diffraction techniques. Most use diffraction cameras and the results take the form of patterns of spots on

54

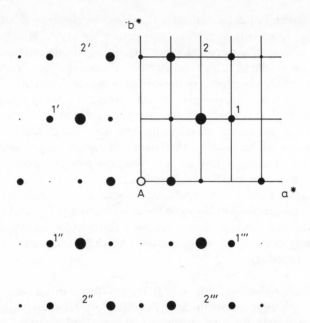

Fig. 3.2 Schematic single crystal X-ray precession photograph through a section of the reciprocal lattice. Relative intensities are indicated by the size of the spots

photographic films. A schematic diagram of a photograph taken with a precession camera is shown in Fig. 3.2. The centre, A, of the film corresponds to the position of the straightthrough, undiffracted beam of radiation. The spots are caused by diffracted X-ray beams and fall on the corners of an imaginary grid. In this diagram the grid is rectangular and has been sketched in for the upper right-hand quadrant. Other grid shapes also occur, e.g. square and parallelogram (angles $\neq 90°$). The grid size and shape forms part of what is known as the *reciprocal lattice* and is related inversely to the size and shape of the unit cell of the crystal. The two axes of the grid are marked for convenience as **a*** and **b***. From the distance apart of spots in these directions, the values of the unit cell dimensions **a** and **b** may be calculated. The spots in this particular examples are arranged with a symmetrical distribution of intensity. For instance, spots 1, 1′, 1″, 1‴ all have the same intensity. The distribution of intensities, and in particular whether or not the pattern is symmetric, is governed by the symmetry of the unit cell.

Single crystal X-ray diffraction methods have the following applications. A further discussion is given in Section 5.4.2.

Determination of unit cell and space group. The unit cell type of a crystal, i.e. cubic, tetragonal, monoclinic, etc., may be determined from single crystal X-ray photographs such as Fig. 3.2. Basically, one looks for a symmetrical arrangement of spots which may then be correlated with the symmetry of the unit cell. For instance, if the left-hand side of the photograph is a mirror image of the right-hand side (as in Fig. 3.2) then the vertical axis, b^*, is an axis of symmetry. In Fig. 3.2, a^* is also an axis of symmetry since the top and bottom halves of the photograph are mirror-symmetric. Usually more than one photograph, corresponding to more than one section through the reciprocal lattice, is needed in order to determine the unit cell unequivocally. For a section such as shown in Fig. 3.2 it is possible, for instance, that the associated crystal symmetry is *orthorhombic*: the two axes a^* and b^* are both symmetry axes, are of unequal length and are separated by an angle of 90°. However, information on the third axis, c^*, would be needed in order to confirm this.

Once the unit cell has been determined, the next stage is to decide what is the space group. This is done essentially by looking for patterns of absent spots in the X-ray photographs. For instance, alternate spots in a row may be absent or perhaps entire rows are absent. The latter will be apparent only when comparing photographs through different sections of the reciprocal lattice. These absent spots are called *systematic absences*. From the patterns of systematic absences, it is possible to determine the *lattice type*—face centred, body centred, etc.—and whether or not the crystal possesses elements of *space symmetry*, i.e. *screw axes* and/or *glide planes*. In Fig. 3.2, some spots are shown as absent, e.g. at 2, 2′, 2″, 2‴, but these appear not to be systematic absences. Using information on systematic absences (if any), possible space groups for the crystal may be deduced. It is often the case that more than one space group has the same set of systematic absences; the correct space group for the crystal will then usually emerge only when the crystal structure has been determined.

Crystal structure determination. The raw material for solving crystal structures is diffracted X-ray intensity data. A single crystal may give up to two or three thousand possible diffracted beams and it is necessary to measure the intensity of a significant proportion of these. This may be done using film methods by measuring, for instance, the relative intensity of spots in photographs such as Fig. 3.2. It is more common nowadays, however, to use a single crystal automatic diffractometer. This is an instrument which scans systematically around the crystal. It measures the intensity of the diffracted beam at each intersection of the imaginary grid (reciprocal lattice) shown in Fig. 3.2.

The intensity data are processed with the aid of a set of complex computer programs. The final result is a three-dimensional electron density map which effectively shows the positions of all the atoms in the unit cell.

The rôle that crystal structure determination by X-ray diffraction has played in inorganic and solid state chemistry cannot be overemphasized. Virtually all our knowledge of crystal structures has been obtained by this method over a period of some seventy years. This knowledge has been fundamental to understanding crystalline materials, their structures, properties and, hence, applications.

Electron distribution, atom size and bonding. The end result of a crystal structure determination is a map showing the positions of atoms in the unit cell. From this, structural information such as coordination numbers and bond distances may be readily deduced. With high accuracy measurements on relatively simple structures, it is possible to go one stage further and investigate the distribution of valence electrons in the structure. This gives information on (a) the sizes of the atoms or ions and (b) the bond type, whether ionic or covalent.

An electron density map through a section of the LiF structure is shown in Fig. 8.1. It shows approximately circular regions of high electron density. These regions give the positons of the atoms or ions. The total number of electrons associated with each atom or ion may be estimated semi-quantitatively by computing the volume under each peak. In the case or LiF, the electron density drops almost to zero along the line connecting the atom centres; this provides direct evidence that the bonding is predominantly ionic. For other structures, however, the electron density does not drop to zero between the atom centres; this shows that the bonding is covalent, at least partially.

An estimate of atom or ion size may be made by looking for the position of the electron density minimum along the line connecting the atom centres, as shown in Fig. 8.2 for LiF. This has led to the rather startling result that cations are considerably larger and anions considerably smaller than has been assumed for many years. For instance, well-established tables of ionic radii, such as the Pauling set, lead one to expect that in, for example, MgO, the oxide ion (radius 1.45 Å) is much larger than the magnesium ion (radius 0.65 Å). Accurate electron density maps show, however, that the ions are much more nearly equal in size (oxide, 1.09 Å, magnesium, 1.02 Å). Much of our thinking about crystal structures is based on Pauling-type radii and it will be some years before the significance of the new X-ray results is fully assimilated and assessed.

Crystal defects and disorder. Information on crystal defects and/or disorder may sometimes be obtained by single crystal photographic methods. Random point defects are not readily detected but significant disorder in one or two dimensions usually gives a measurable diffraction effect. If one-dimensional disorder is present, such as associated with stacking faults or twinning, then certain of the spots in photographs such as Fig. 3.2 appear instead as streaks. If two-dimensional disorder is present, as in a fibrous structure in which the fibres are parallel but are randomly orientated about the fibre axis, then the diffraction spots are smeared out into discs.

3.2.1.4 *Electron diffraction*

For the single crystal X-ray diffraction studies described above it is necessary to have crystals that are at least 0.05 mm in diameter. Otherwise, the intensities of the diffracted beams are too weak to be detected clearly. This is because the efficiency with which X-rays are diffracted is very low. Often, however, crystals as large as 0.05 mm are simply not available or cannot be prepared. In such cases, electron diffraction may be used. This technique makes use of the wave properties of electrons and because the scattering efficiency of electrons is high, small samples may be used. The results take the form of patterns of spots on photographic films, rather like the precession photographs shown in Fig. 3.2.

One disadvantage of electron diffraction is that secondary diffraction commonly occurs. Because the scattering efficiency of electrons is high, the diffracted beams are strong. Secondary diffraction occurs when these diffracted beams effectively become the incident beam and are diffracted by another set of lattice planes. There are two undesirable consequences of secondary diffraction. First, under certain circumstances, extra spots may appear in the diffraction pattern; care is therefore needed in the interpretation of diffraction patterns. Second, the intensities of diffracted beams are unreliable and cannot be used quantitatively for crystal structure determination.

In spite of these disadvantages, electron diffraction is very useful and complements the various X-ray techniques. Thus with X-rays, the scattering efficiency is small, secondary diffraction is rarely a problem and intensities are reliable, but (relatively) large samples are needed. On the other hand, with electrons, the scattering efficiency is high and although intensities are unreliable very small samples can be studied. Some applications of electron diffraction are as follows. Other applications are discussed in Section 3.2.2.

Unit cell and space group determination. The technique is very useful for obtaining unit cell and space group information, in a similar way to that outlined above for single crystal X-ray methods. For crystals smaller than 0.01 to 0.02 mm in diameter, it is, in fact, the only reliable method for obtaining such information. The other methods that are sometimes used, such as graphical or computer-assisted indexing of X-ray powder patterns, give results that are not always 100 per cent reliable. In spite of its utility, electron diffraction is very much underused in solid state chemistry for unit cell and space group determination.

Phase indentification. Electron diffraction is unsuitable as a routine method of phase identification in relatively large (e.g. 10 mg or more) samples and should only be used if X-ray powder methods are unavailable. It is useful, however, (a) when only very small quantities are available, (b) for thin film samples and (c) for detecting small amounts of impurity phases. In all these cases, there would be insufficient material to show up in X-ray diffraction.

3.2.1.5 *Neutron diffraction*

Neutron diffraction is a very expensive technique. In order to get a sufficiently intense source of neutrons, a nuclear reactor is needed. Few laboratories have their own neutron facility and, instead, experiments are carried out at central laboratories which provide a user service (e.g. at Grenoble, France). In spite of its high cost, neutron diffraction is a valuable technique and can provide information, especially on magnetic materials, that is not attainable with other techniques. Clearly, it is never used when alternative techniques, such as X-ray diffraction, can solve a particular problem.

Crystal structure determination. The size of sample for neutron diffraction work is relatively large, ~ 1 cm^3. Since crystals of this size are often not available, crystallographic studies are usually carried out on polycrystalline samples. A powder neutron diffraction pattern looks very much like an X-ray one (Fig. 3.1).

There are several characteristic differences between neutron and X-ray diffraction. First, the neutron source gives a continuous spectrum of radiation without the intense characteristic peaks that are present in X-ray spectra (see Fig. 5.2). In order to have monochromatic neutrons, it is necessary to select a particular wavelength for use and by some means filter out the remainder. Most of the available neutron energy is wasted, therefore, and the beam that is used is weak and not particularly monochromatic.

A recent advance which offers exciting prospects is the method of 'pulsed source with time of flight analysis'. This uses the entire neutron spectrum (variable wavelength, λ) with a fixed diffraction angle, θ. By using a pulsed source, the diffracted radiation is separated according to its time of flight and, hence, according to its wavelength. The fundamental law of diffraction is Bragg's Law, $n\lambda = 2d \sin \theta$. In the time of flight method, λ and d (the d-spacing) are the variables at fixed θ. This compares with conventional diffraction techniques in which d and θ are the variables at fixed λ. The pulsed method gives much more rapid data collection. It has the additional advantage that it may be used for studies of short time relaxation phenomena, especially in experiments where samples are subjected to pulsed magnetic fields.

A second difference between neutron and X-ray diffraction is that the scattering powers of atoms towards neutrons are quite different to those towards X-rays. In the latter, scattering power is a simple function of atomic number, and light atoms such as hydrogen diffract X-rays only very weakly. With neutrons, the atomic nuclei, rather than the extra-nuclear electrons, are responsible for the scattering and, in fact, hydrogen is a strong scatterer of neutrons. There is no simple dependence of neutron scattering power on atomic number and, additionally, some atoms cause a change of phase of π (or $\lambda/2$) in the diffracted neutron beam.

Neutron diffraction may be used for crystallographic work in cases where X-ray diffraction is inadequate. It has been much used to locate light atoms, especially hydrogen in hydrides, hydrates and organic structures. Usually, the main part of the structure is solved by X-ray methods and neutron diffraction is subsequently used to provide the finishing touches and locate the light atoms. Neutron diffraction is also used to distinguish between atoms that have similar X-ray scattering powers, such as manganese and iron or cobalt and nickel. The neutron scattering powers of these atoms are different and, for instance, superlattice phenomena, associated with Mn/Fe ordering in alloys, are readily observed by neutron diffraction.

Magnetic structure analysis. Magnetic properties depend on the presence of unpaired electrons, especially in d or f orbitals. Since neutrons possess a magnetic dipole moment they interact with unpaired electrons, giving rise to an additional scattering effect. This forms the basis of a powerful technique for studying the magnetic structure of materials. A simple example of magnetic structure and order is shown by NiO. Nickel oxide has the face centred cubic rock salt structure, as revealed by X-ray diffraction. However, when examined by neutron diffraction, extra peaks are observed which indicate the presence of a superstructure. This arises because the unpaired d electrons (in the e_g orbitals) are arranged so as to be antiparallel in alternate layers of nickel atoms. Neutrons detect this ordering of the spins whereas X-rays do not. Neutron diffraction has been used to study ferro-, ferri- and antiferromagnetic materials and, in fact, provided the first direct proof, in MnO, of the existence of antiferromagnetic phenomena. Magnetic phenomena depend on the interactions between electrons on different atoms and neutron diffraction provides an insight into the nature and strength of these exchange interactions. Further details are given in Chapter 16.

Inelastic scattering, soft modes and phase transitions. 'Slow' neutrons possess kinetic energy that is comparable to the thermal energy levels in a solid. Such neutrons are inelastically scattered by phonons (i.e. vibrational modes) in the solid. From an analysis of the energy of the scattered neutrons, information on phonons and interatomic forces is obtained. For magnetic materials, further information on their electron exchange energy is obtained.

Displacive phase transitions are believed to be associated with the instability of a lattice vibration. A certain type of vibrational mode in the low temperature structure effectively collapses at the critical temperature. Such modes are called soft modes. They may be studied by IR and Raman spectroscopy, provided the vibrations involved are spectroscopically active, and also by neutron scattering. The latter technique is useful since it is not limited by the spectroscopic selection rules; by measuring the inelastic scattering about a number of points in the reciprocal lattice of the crystal, the atomic displacements that are responsible for the soft mode and the phase transition may be determined. For instance, the mechanism of the displacive transition in quartz, SiO_2, at $573\,°C$ has been analysed by recording the neutron spectra at several temperatures below and above $573\,°C$.

3.2.2 Microscopic techniques

As a first step in examining a solid, it is usually well worth while to have a look at it under magnification. This may involve no more than a brief look with a polarizing microscope or it may involve a more in-depth study using one or more instruments. Materials that visually appear to be somewhat similar, such as fine powders of white sand or table salt, may look quite different under the microscope. Thus, the crystals in these two particular examples have a different morphology (i.e. shape) and their optical properties, as studied in plane polarized light, are quite distinct.

Various kinds of microscope are available and they can be divided into two groups: optical and electron. With optical microscopes, particles down to a few micrometres ($1\,\mu\mathrm{m} = 10^4\,\text{Å} = 10^{-3}\,\mathrm{mm}$) in diameter may be seen under high magnification. The lower limit is reached when the particle size approaches the wavelength of visible light, 0.4 to 0.7 μm. For submicrometre-sized particles it is essential to use electron microscopy. With this technique it is possible to image features that have diameters as small as a few angstroms. Both optical and electron microscopy use two types of instrument in that the sample can be viewed in transmission (i.e. the beam of light or electrons passes through the sample) or in reflection (the beam of light or electrons is reflected off the sample surface).

3.2.2.1 *Optical microscopy*

Optical microscopes are of two basic types, as indicated above. The *polarizing* or *petrographic microscope* is a transmission instrument. It is widely used by geologists and mineralogists and can be used profitably in solid state chemistry. Samples are usually in the form of either fine powders or thin slices cut off a solid piece. The *metallurgical* or *reflected light microscope* is suitable for looking at the surfaces of materials, especially opaque ones. It is much used in metallurgy, mineralogy and ceramics.

Samples for the polarizing microscope are often fine powders with particle sizes in the range 10 to 100 μm. On this scale, substances may often be transparent whereas they would be opaque in bulk form. The sample is immersed in a liquid whose refractive index is close to that of the sample. If this is not done and the solid is examined in air, much of the light is scattered off the surface of the sample rather than transmitted through it. It is then difficult, if not impossible, to measure the various optical properties of the solid.

The basic components of a polarizing microscope are indicated in Fig. 3.3. The source may give either white or monochromatic light. This shines on the polarizer and only that component whose vibration direction is parallel to that of the polarizer is permitted to pass through. The resulting plane polarized light passes through lenses, apertures and accessories (not shown) and on to the sample which is mounted on the microscope stage. Light transmitted by the sample and

EYEPIECE LENS

↑

ANALYSER (in or out)

↑

OBJECTIVE LENS (interchangeable, to vary magnification)

↑

SAMPLE

↑

POLARIZER

↑

LIGHT SOURCE (white or monochromatic light)

Fig. 3.3 Basic components of a polarizing microscope

immersion liquid is picked up by the objective lens. The instrument usually has several of these, of different magnification, which are readily interchangeable. The analyser may be placed in or out of the path of light. It is similar to the polarizer but is oriented so that its vibration direction is at 90° to that of the polarizer. When the analyser is 'in', only light vibrating in the correct direction is permitted to pass through and on to the eyepiece. When the analyser is 'out', the microscope behaves as a simple magnifying microscope. Various other attachments (not shown) may also be present between the sample and the eyepiece.

In practice, samples are usually examined with the analyser alternately 'out' and 'in'. With the analyser 'out', the instrument can be focused and a first examination made, noting for instance the size and shape of the particles. With the analyser 'in', the sample is viewed between crossed polars, and one can immediately tell whether the sample is *isotropic* (dark) or *anisotropic* (bright or coloured). By rotating the sample and stage (in some instruments, the polars rotate and the stage is fixed), the *extinction directions* can be seen if the crystals are anisotropic; from the nature of the extinction, information on the quality of the crystals may be obtained. If it is desired to measure the refractive index (or indices) of the sample, the analyser is again taken out and various immersion liquids, of different refractive index, are tried until one is found in which the sample is effectively invisible. The refractive index of the sample then matches that of the liquid. It is possible to do this systematically; using the *Becke line method*, it is possible to tell whether the refractive index of the sample is larger or smaller than that of the immersion liquid.

The above measurements are simple and fairly rapid. Often the information obtained is all that is required from microscopy. If necessary, further measurements can be made. The variation of refractive index with crystal orientation may be studied using the concept of the *optical indicatrix*. Anisotropic crystals may be classed into *uniaxial* or *biaxial*, depending on whether they have one or two *optic axes*. When viewed down an optic axis, anisotropic crystals appear to be isotropic, i.e. they are dark between crossed polars. If a convergent beam of light is shone onto the sample, a *conoscopic examination* may be made and interference figures are seen; from these, further information on uniaxial and biaxial crystals is obtained.

The reflected light microscope is similar to the transmission one except that the source and objective lens are on the same side of the sample. It is used to examine solid lumps of opaque material such as metals, minerals and ceramics. The amount of information that can be obtained depends very much on the care and skill with which the sample is prepared. It is best to have a flat polished surface which has been treated chemically to cause preferential etching of part of the sample. Some phases etch away more quickly than others and this gives relief to the initially flat surface; other phases take on a coloured hue after etching. The information that can be obtained by this technique primarily concerns the texture of the solid, i.e. the phases present, their identification, the number, size and distribution of particles.

Some uses of optical microscopy are now given.

Crystal morphology and symmetry. Before the advent of X-ray diffraction ($\sim$ 1910), morphological data provided an important means of classifying crystals. The method used and still uses a goniometric stage on which a crystal is mounted. The crystal can be rotated on the stage and effectively viewed from all angles under the microscope. From the shape of the crystal, information on the internal symmetry of the crystal structure may be obtained. With well-formed crystals, the number and disposition of the crystal faces can be determined, which may indicate the class (i.e. point group) of the crystal. Nowadays, goniometry is little used in solid state science.

It is possible, however, to obtain quite a lot of useful information from a cursory examination of powdered samples using the polarizing microscope. For instance:

(a) The crystal fragments often have some kind of characteristic shape, especially if the crystals cleave easily and in a particular orientation.

(b) By examining between crossed polars, it is possible to classify substances into isotropic and anisotropic. Isotropic crystalline substances are limited to those of cubic symmetry although amorphous solids such as glasses and gels are also isotropic. Isotropic substances appear 'dark' between crossed polars since plane polarized light passes through them without modification. Anisotropic substances cause a partial rotation of the plane of polarization of light as it passes through them. The emergent beam has a component that is vibrating parallel to the vibration direction of the analyser and hence anisotropic substances appear 'light'. Anisotropic substances include all non-cubic crystalline solids. The vast majority of crystalline solids come into this category. It is often possible to distinguish between uniaxial crystals (of hexagonal, trigonal or tetragonal symmetry) and biaxial crystals (orthorhombic, monoclinic and triclinic symmetry). In uniaxial crystals, the optic axis is parallel to the unique symmetry axis (six-, three- or fourfold, respectively) and when viewed down this axis such crystals appear to be isotropic. In a powdered sample it is quite common to find crystals in this orientation, especially if they exhibit preferred orientation. For instance, crystals of hexagonal symmetry often exist as thin hexagonal plates and these naturally tend to lie on their flat faces. Although biaxial crystals have two optic axes, these are not usually parallel to any pronounced edge or feature of the crystal. The chance of finding biaxial crystals oriented with an optic axis parallel to the light beam is quite small.

(c) When anisotropic substances are viewed between crossed polars they usually appear 'light'. On rotating the sample on the microscope stage, the sample becomes 'dark' in a certain position. This is known as *extinction*. On rotating a sample, extinction occurs every 90°; at the 45° position, the sample exhibits maximum brightness. If the sample has parallel extinction, i.e. the extinction occurs when a pronounced feature of the crystal, such as an edge, is parallel to the direction of vibration of the polarized light, then it is likely that the crystal possesses some symmetry, i.e. it is not triclinic but is of monoclinic symmetry, or higher.

By using steps (a) to (c), outlined above, it is often possible to make a good guess at the symmetry and the unit cell of a crystalline substance. This can be a useful prelude to making more time-consuming X-ray diffraction studies. If crystals have a recognizable morphology, use can be made of this to save time in orienting crystals for X-ray work or in measuring some physical property of the crystal as a function of crystal orientation.

Phase identification, purity and homogeneity. Crystalline substances may be identified according to their optical properties, refractive indices, optic axes, etc. Much use is made of this in mineralogy. Standard tables are available with which optical data for an unidentified mineral may be compared. The method is little used outside mineralogy for the identification of complete unknowns. However, where only a limited number of substances are possible, microscopy can be a very powerful method of identification. For instance, in the synthesis of new compounds or phase diagram studies on a specific system, only a limited number of phases are likely to appear. Provided the general appearance and optical properties of most or all are known, then optical microscopy may provide an extremely rapid method of phase analysis. In the space of a few minutes, a sample can be prepared for examination and analysed in the microscope. Often, this gives sufficient information to answer a particular question regarding phase content or purity and may avoid the necessity of making more time-consuming analyses by other techniques.

The purity of a sample may be checked rapidly provided the impurities form a separate crystalline or amorphous phase. Low impurity levels are easily detected, especially if the optical properties of the impurity are markedly different from those of the major phase.

The quality and homogeneity of glasses may be checked readily with a polarizing microscope. For instance, in the preparation of silicate glasses using sand (SiO_2) as one of the raw materials, it may take a long time before all the SiO_2 crystals dissolve during the melting operation. The presence of undissolved silica may not be noticed from a visual inspection of a glass sample, whereas it would be obvious immediately under the microscope. The compositional homogeneity of glass, once all the crystalline raw materials have melted, may be checked by measuring the refractive index of a random selection of fragments of crushed glass. The method works because the refractive index of glass is usually composition dependent. Although glass is isotropic, sometimes it shows stress birefringence. Glasses that have not been annealed properly or are in a highly stressed condition, for some reason or other, appear light and apparently anisotropic between crossed polars.

The quality of single crystals may be assessed with a polarizing microscope. Good quality crystals should show a sharp extinction; i.e. on rotating the crystal relative to the polars, extinction should occur simultaneously throughout the entire crystal. Crystal aggregates may show an extinction that is wavy or irregular. If the crystal divides into strips which extinguish alternately as the crystal rotates, then the crystal is likely to be twinned. Other optical effects are also associated with twinning such as striation and crosshatching. Twinning may occur either (a) as a result of the crystal growth mechanism or (b) as a consequence of a phase transition from a high symmetry to a lower symmetry phase, as in paraelectric–ferroelectric or paramagnetic–ferromagnetic transitions. Optical microscopy is a powerful method for studying the latter, especially if a hot stage is fitted so that the sample temperature can be varied under the microscope.

Crystal defects—grain boundaries and dislocations. Reflected light microscopy on polished and etched materials can provide much information on internal crystal surfaces (e.g. grain boundaries) and line defects (dislocations). Such defects are always present, even in good quality single crystals. They can be detected because, for example, in the region of an emergent dislocation at the crystal surface, the crystal structure is in a stressed conditon; if the surface is treated with a suitable chemical reagent etching may occur preferentially at such stressed sites, resulting in etch pits. Dislocation densities may be determined by counting the number of etch pits per unit area; this has been applied to, for example, metals and alkali halide crystals. Grain boundaries also etch away preferentially and low angle grain boundaries are seen as rows of dislocation etch pits.

3.2.2.2 *Electron microscopy*

Electron microscopy is an extremely versatile technique capable of providing structural information over a wide range of magnification. At one extreme, scanning electron microscopy (SEM) complements optical microscopy for studying the texture, topography and surface features of powders or solid pieces; features up to tens of micrometres in size can be seen and, because of the depth of focus of SEM instruments, the resulting pictures have a definite three-dimensional quality. At the other extreme, high resolution electron microscopy (HREM) is capable, under favourable circumstances, of giving information on an atomic scale, by direct lattice imaging. Resolution of ~ 2 Å has been achieved, which means that it is now becoming increasingly possible to 'see' individual atoms. However, lest any one should think that HREM is on the verge of solving all remaining problems concerning the structure of materials, it must be emphasized that there are still formidable obstacles to be overcome before this goal is achieved; there is no immediate prospect of redundancy for more conventional crystallographers!

Electron microscopes are of either transmission or reflection design. For examination in transmission, samples should usually be thinner than ~ 2000 Å. This is because electrons interact strongly with matter and are completely absorbed by thick particles. Sample preparation may be difficult, especially if it is not possible to prepare thin foils. Thinning techniques, such as ion bombardment, are used, but not always satisfactorily, especially with, for

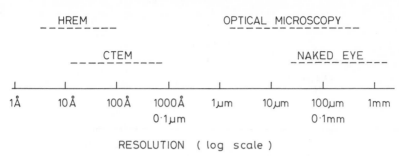

RESOLUTION (log scale)

Fig. 3.4 Working ranges of various techniques used for viewing solids. CTEM = conventional transmission electron microscopy; HREM = high resolution electron microscopy; SEM = scanning electron microscopy

example, polycrystalline ceramics. There is also a danger that ion bombardment may lead to structural modification of the solid in question or that different parts of the material may be etched away preferentially in the ion beam. One possible solution is to use higher voltage instruments, e.g. 1 MV. Thicker samples may then be used since the beam is more penetrating; in addition, the amount of background scatter is reduced and higher resolution may be obtained. Alternatively, if the solid to be examined can be crushed into a fine powder then at least some of the resulting particles should be thin enough to be viewed in transmission.

With reflection instruments, sample thickness is no longer a problem and special methods of sample preparation are not required. It is usually necessary only to coat the sample with a thin layer of metal, especially if the sample is a poor electrical conductor, in order to prevent the build-up of charge on the surface of the sample. The main reflection instrument is the SEM. It covers the magnification range between the lower resolution limit of optical microscopy ($\sim 1\,\mu$m) and the upper practical working limit of transmission electron microscopy (TEM) ($\sim 0.1\,\mu$m) although, in fact, SEM can be used to study structure over a much wider range, from $\sim 10^{-2}$ to $\sim 10^2\,\mu$m. The approximate working ranges of different kinds of microscope are summarized in Fig. 3.4.

Some SEM and TEM instruments have the very valuable additional feature of providing an elemental analysis of sample composition. There are various names for the technique including electron probe microanalysis (EPMA), electron microscopy with microanalysis (EMMA) and analytical electron microscopy (AEM). The main mode of operation makes use of the fact that when a sample is placed in the microscope and bombarded with high energy electrons, many things can happen (Fig. 3.5), including the generation of X-rays. These X-rays are characteristic emission spectra of the elements present in the sample. By scanning either the wavelength (wavelength dispersive, WD) or the energy (energy dispersive, ED) of the emitted X-rays (4) it is possible to identify the elements present. If a suitable calibration procedure has been adopted, a quantitative

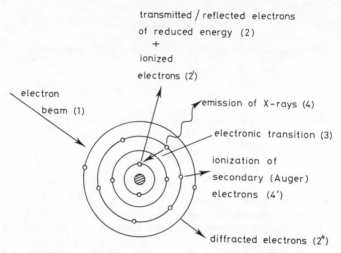

transmitted / reflected electrons
of reduced energy (2)
+
ionized
electrons (2')

electron
beam (1)

emission of X-rays (4)

electronic transition (3)

ionization of
secondary (Auger)
electrons (4')

diffracted electrons (2")

Fig. 3.5 Some of the processes that occur on bombarding a sample with electrons (in, for example, an electron microscope)

elemental analysis may be made. At present, only elements heavier than and including sodium can be determined; lighter elements do not give suitable X-ray spectra. For lighter elements, however, there are alternative techniques, Auger spectroscopy and electron energy loss spectroscopy (EELS) (2).

Generation of low energy Auger electrons is another process (4') that occurs on electron bombardment of the sample in the microscope (Fig. 3.5). The energy of Auger electrons is characteristic of the atom from which they are emitted. Auger emission occurs when electrons, and not X-rays, are ejected from the sample as an energy release mechanism. EELS detects the transmitted/reflected electrons (2) which were responsible for the initial ionization of electrons (2'). Since energy is required to ionize the atoms (2') the EELS electrons are of reduced energy

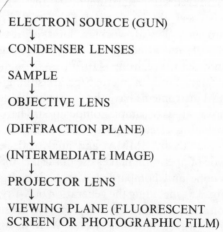

ELECTRON SOURCE (GUN)
↓
CONDENSER LENSES
↓
SAMPLE
↓
OBJECTIVE LENS
↓
(DIFFRACTION PLANE)
↓
(INTERMEDIATE IMAGE)
↓
PROJECTOR LENS
↓
VIEWING PLANE (FLUORESCENT
SCREEN OR PHOTOGRAPHIC FILM)

Fig. 3.6 Basic components of a transmission electron microscope

compared to the energy of the incident electron beam (1). EELS and Auger techniques are discussed further in Section 3.2.3.

The basic components of a TEM instrument are listed in Fig. 3.6. Electrons emitted from a tungsten filament (electron gun) are accelerated through a high voltage (50 to 100 kV). Their wavelength is related to the accelerating voltage, V, by

$$\lambda = h(2meV)^{-1/2} \tag{3.1}$$

where m and e are the mass and charge of the electron. At high voltage, as the velocity of the electron approaches the velocity of light, m is increased by relativistic effects. The electron wavelengths are much smaller than the X-ray wavelengths used in diffraction experiments, e.g. $\lambda \sim 0.04\,\text{Å}$ at 90 kV accelerating voltage. Consequently, the Bragg angles for diffraction are small and the diffracted beams are concentrated into a narrow cone centred on the undiffracted beam (Section 3.2.1).

In order to use electrons, instead of light, in a microscope it is necessary to be able to focus them. As yet, no one has found a substance that can act as a lens to focus electrons but, fortunately, electrons may be focused by an electric or magnetic field. Electron microscopes contain several electromagnetic lenses. The condenser lenses are used to control the size and angular spread of the electron beam that is incident on the sample. Transmitted electrons then pass through a sequence of lenses—objective, intermediate and projector—and form a magnified image of the sample on a fluorescent viewing screen. Photographs may be taken if so desired. By changing the relative position of the viewing screen, the diffraction pattern, rather than the image, of the specimen is seen and can be photographed. The region of the sample that is chosen for imaging may be controlled by an aperture placed in the intermediate image plane. This is particularly important in examining polycrystalline materials that contain more than one phase. The technique is then known as selected area electron diffraction. The quality of photographs, especially where the particles of interest are difficult to pick out from the background, may be improved by dark field imaging. In this, only the diffracted beams from the particle of interest are allowed to recombine to form the image.

In the scanning electron microscope, electrons from the electron gun are focused to a small spot, 50 to 100 Å in diameter, on the surface of the sample. The electron beam is scanned systematically over the sample, rather like the spot on a television screen. Both X-rays and secondary electrons are emitted by the sample; the former are used for chemical analysis and the latter are used to build up an image of the sample surface which is displayed on a screen. A limitation with SEM instruments is that the lower limit of resolution is $\sim 100\,\text{Å}$. A recent advance is the development of the scanning transmission electron microscope (STEM). This combines the scanning feature of the SEM with the intrinsically higher resolution obtainable with TEM. Some applications of electron microscopes that primarily involve electron diffraction have been given in Section 3.2.1.4. Other uses are as follows.

Particle size and shape, texture, surface detail. Electron microscopy, especially SEM, is invaluable for surveying materials under high magnification and providing information on particle sizes and shapes. The results complement those obtained from optical microscopy by providing information on submicrometre-sized particles.

Crystal defects. Using the dark field technique on, for example, thin foils with TEM, crystal defects such as dislocations, stacking faults, antiphase boundaries and twin boundaries may be seen directly. The domain structure of ferromagnetic and ferroelectric materials may be observed. With HREM, it is now possible to see detail on an atomic scale although, as yet, clear images of atoms are not usually obtained. Variations in local structure such as site occupancies and vacancies can be observed directly; this, therefore, complements the results obtained by conventional X-ray crystallography which yield an average structure.

Precipitation and phase transitions. Many glasses exhibit a liquid immiscibility texture on a submicrometre scale and this can be seen by electron microscopy. Precipitation phenomena in crystalline systems are widespread and often technologically important. Hardening processes in alloys often involve precipitation; e.g. the formation of cementite, Fe_3C, precipitates in the hardening of steels. Similar processes are increasingly being studied in inorganic (i.e. non-metallic) and ceramic systems. Many minerals have undergone precipitation (or exsolution) reactions over geological timescales. In all such cases, EM can be used to study the nature of the precipitates, i.e. their crystal structures (by electron diffraction), their crystallographic orientation relative to the parent structure and the overall texture of the solid.

Chemical analysis. The use of EMMA and related techniques enables the composition of small regions of solid, $\sim 1000\,\text{Å}$ in diameter, to be determined. This may be applied to small individual particles or to larger masses, in which case compositional variations across the solid may be determined. This is becoming extremely valuable (a) as a straightforward means of elemental analysis and (b) to study inhomogeneous materials such as complex mineral aggregates, cements and concretes.

3.2.3 Spectroscopic techniques

There are many different spectroscopic techniques but all work on the same basic principle. This is that, under certain conditions, materials are capable of absorbing or emitting energy. The energy can take various forms. Usually it is electromagnetic radiation but it also can be sound waves, particles of matter, etc. The experimental results, or spectra, take the form of a plot of intensity of absorption or emission (y axis) as a function of energy (x axis). The energy axis is often expressed in terms of either frequency, f, or wavelength, λ, of the appropriate radiation. The various terms are interrelated by the classic equation

$$E = hf = hc\lambda^{-1} \tag{3.2}$$

where h is Planck's constant (6.6×10^{-34} J sec), c is the velocity of light (2.998×10^{10} cm sec^{-1}), f is frequency (in hertz, cycles per second) and λ is the wavelength (in centimetres). Units of λ^{-1} are cm^{-1} or wavenumbers and E is expressed in joules. Chemists usually prefer joules per mole for units of E, in which case the above equation is multiplied by Avogadro's number, N. Some useful interconversions between the different energy units, obtained by substituting for the constants into equation (3.2) are

$$
\begin{aligned}
E &= 3.991 \times 10^{-10} f \simeq 4 \times 10^{-10} f \, \text{J mol}^{-1} && (f \text{ in sec}^{-1}) \\
E &= 11.97\lambda^{-1} \simeq 12\lambda^{-1} \, \text{J mol}^{-1} && (\lambda \text{ in cm}) \\
E &= 3 \times 10^{10} \lambda^{-1} \, \text{sec}^{-1} && (\lambda \text{ in cm}) \\
E(\text{eV}) &\simeq 96 E(k \, \text{J mol}^{-1})
\end{aligned}
\tag{3.3}
$$

The electromagnetic spectrum covers an enormous span of frequency, wavelength and, therefore, energy. The different spectroscopic techniques operate over different, limited frequency ranges within this broad spectrum, depending on the processes and magnitudes of the energy changes that are involved (Fig. 3.7). At the low frequency, long wavelength end, the associated energy changes are small, < 1 J mol^{-1}, but may be sufficient to cause the reversal of spins of either nuclei or electrons in an applied magnetic field. Thus, the nuclear magnetic resonance (NMR) technique operates in the radiowave region at, for example, 100 MHz (10^8 Hz) and detects changes in nuclear spin state.

At higher frequencies and shorter wavelengths, the amount of associated energy increases and, for instance, the vibrational motions of atoms in molecules or solids may be altered by the absorption or emission of infrared (IR) radiation. At still higher frequencies, electronic transitions within atoms may occur from one energy level to another. For electronic transitions involving outer (valence) shells, the associated energy usually lies in the visible and ultraviolet regions; however, for inner shell transitions, much larger energies are involved and fall in the X-ray region.

Although many of the spectroscopic techniques were initially developed for and applied to molecular materials, often liquids and gases, they are finding many applications in solid state studies. Spectroscopic measurements on solids complement well the results obtained from (X-ray) diffraction since spectroscopy gives information on local order whereas diffraction is concerned primarily with long range order. Spectroscopic techniques may be used to determine coordination numbers and site symmetries; variations in local order can be detected, as can impurities and imperfections; amorphous materials, such as glasses and gels, can be studied just as easily as crystalline materials. By contrast, the long range periodic structures of crystals can be determined only by diffraction techniques (and occasionally by HREM); however, this results in an

70

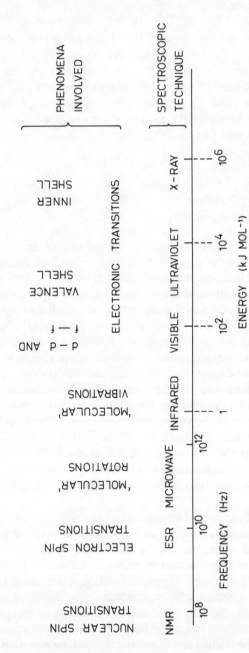

Fig. 3.7 Principal regions of the electromagnetic spectrum and the associated spectroscopic techniques

average picture of the local structure in which information on defects, impurities and subtle variations in local order may effectively be lost. Also, diffraction techniques are of limited use for studying amorphous materials.

Some of the principal spectroscopic techniques and their uses in solid state chemistry are as follows.

3.2.3.1 *Vibrational spectroscopy: IR and Raman*

Atoms in solids vibrate at frequencies of approximately 10^{12} to 10^{13} Hz. Vibrational modes, involving pairs or groups of bonded atoms, can be excited to higher energy states by absorption of radiation of appropriate frequency. IR spectra and closely related Raman spectra are plots of intensity of absorption (IR) or scattering (Raman) as a function of frequency or wavenumber. In the IR technique, the frequency of the incident radiation is varied and the quantity of radiation absorbed or transmitted by sample is obtained. In the Raman technique, the sample is illuminated with monochromatic light, usually generated by a laser. Two types of scattered light are produced by the sample. Rayleigh scatter emerges with exactly the same energy and wavelength as the incident light. Raman scatter, which is usually much less intense than Rayleigh scatter, emerges at either longer or shorter wavelength than the incident light. Photons of light from the laser, of frequency v_0, induce transitions in the sample and the photons gain or lose energy as a consequence. For a vibrational transition of frequency v_1, associated Raman lines of frequency $v_0 \pm v_1$ appear in the scattered beam. This scattered light is detected in a direction perpendicular to the incident beam.

IR and Raman spectra of solids are usually complex with a large number of peaks, each corresponding to a particular vibrational transition. A complete assignment of all the peaks to specific vibrational modes is possible with molecular materials and, in favourable cases, is possible with non-molecular solids. The IR and Raman spectra of a particular solid are usually quite different since the two techniques are governed by different selection rules. The number of peaks that are observed with either technique tends to be considerably less than the total number of vibrational modes and different modes may be active in the two techniques. For instance, in order for a particular mode to be IR active, the associated dipole moment must vary during the vibrational cycle. Consequently, centrosymmetric vibrational modes are IR inactive. The principal selection rule for a vibrational mode to be Raman active is that the nuclear motions involved must produce a change in polarizability.

IR and Raman spectra are much used for the straightforward identification of specific functional groups, especially in organic molecules. In inorganic solids, covalently bonded linkages such as hydroxyl groups, trapped water and oxyanions—carbonate, nitrate, sulphate, etc.—give rise to intense IR and Raman peaks. Some examples are shown in Fig. 3.8. The three spectra are quite different in the region 300 to 1500 cm^{-1}, largely due to the different oxyanions present in each sample. Spectrum (c) contains an additional absorption doublet at ~ 3500 cm^{-1} caused by the H_2O molecules in the gypsum. Peaks that occur in

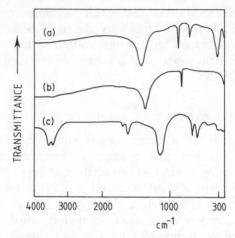

Fig. 3.8 Infared absorption spectra of
(a) calcite, CaCO$_3$, (b) NaNO$_3$, (c) gypsum,
CaSO$_4$·2H$_2$O. (Data taken from P. A.
Estep-Barnes, p. 532 in Zussman, 1977)

the region ~ 3000 to 3500 cm^{-1} are usually characteristic of OH groups in some
form: the frequencies of the peak maxima depend on the O—H bond strength
and information may be obtained on, for instance, the location of the OH group,
whether it belongs to a water molecule and whether or not hydrogen bonding is
present.

Peaks associated with the vibrational modes of covalently bonded groups,
such as oxyanions, usually occur at relatively high frequencies, above
~ 300 cm^{-1} (Fig. 3.8). At lower frequencies in the far infrared region, lattice
vibrations give rise to peaks; e.g. alkali halides give broad lattice absorption
bands in the region 100 to 300 cm^{-1}. The peak positions depend inversely on the
mass of the anions and cations, as shown in the following sequences (positions in
cm^{-1}): LiF(307), NaF(246), KF(190), RbF(156), CsF(127); and LiCl(191),
NaCl(164), KCl(141), RbCl(118), CsCl(99).

Since inorganic solids tend to give characteristic vibrational spectra, these may
be used for identification purposes; by setting up a file of reference spectra it
should be possible to identify unknowns. This has already been done, with
considerable success, for organic molecules, but has not yet been adequately done
for inorganic solids. Perhaps, with the comprehensive coverage and success
achieved by the JCPDS X-ray powder diffraction file, there is less demand for a
second method of fingerprinting inorganic solid materials.

An interesting example of the use of laser Raman spectroscopy for fingerprinting crystalline solids is in the distinction between two of the polymorphs of silica,
quartz and cristobalite (Fig. 3.9). In the region 100 to 500 cm^{-1}, each polymorph
gives three sharp peaks, but at different positions in the two cases. Use has been
made of these spectral differences in the analysis of ash from the recent Mount St
Helens volcanic eruption in the United States.

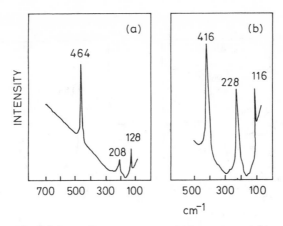

Fig. 3.9 Laser Raman spectra of (a) quartz and (b) cristobalite. (From Farwell and Gage, 1981)

Apart from possible uses in routine sample identification, vibrational spectra may be used to characterize solids and provide some structural information. For these purposes, a deeper understanding of the spectra is necessary and, in particular, the assignment of peaks to specific vibrational modes is desirable. Methods are available which can be used to carry out this assignment; they will not be discussed here since they are rather complicated and as yet have been used only for simple crystal structures.

An interesting use of laser Raman spectroscopy has been in the study of small variations in the compositions of the crystalline ferroelectrics, $LiNbO_3$ and $LiTaO_3$. At high temperatues, $\sim 1150\,°C$, $LiNbO_3$ (ideal composition: 50 mol% Li_2O 50 mol% Nb_2O_5) forms a short range of solid solutions whose Nb_2O_5 content ranges from ~ 49 to ~ 55 mol% (Fig. 3.10a). At lower temperatures, $\sim 600\,°C$, the extent of the solid solutions is much more limited, ~ 49.5 to 51 mol% Nb_2O_5. The Raman spectra of the solid solutions were found to be very sensitive to composition (Fig. 3.10b) and, in particular, peaks broadened significantly with increasing Nb_2O_5 content. From measurements of peak breadth, a correlation with the crystal composition was obtained and hence the variation of the solid solution extent with temperature could be determined. Such information is important since $LiNbO_3$ is a valuable ferroelectric material, whose Curie temperature varies with composition and hence with the conditions of crystal growth.

The above example used variation in the breadths of Raman lines to monitor solid compositions. In other favourable cases, shifts in peak positions occur and can be used to characterize and monitor solid solution composition. An example is the solid solution series between ZrO_2 and HfO_2. The Raman spectra are complex and show approximately 15 peaks in the range 100 to 700 cm^{-1}. With varying solid solution composition, some of the peak positions change smoothly and provide a sensitive indicator of composition. Data are shown for two peaks in Fig. 3.10(c). The peak positions vary by 30–40 wavenumbers across the solid

74

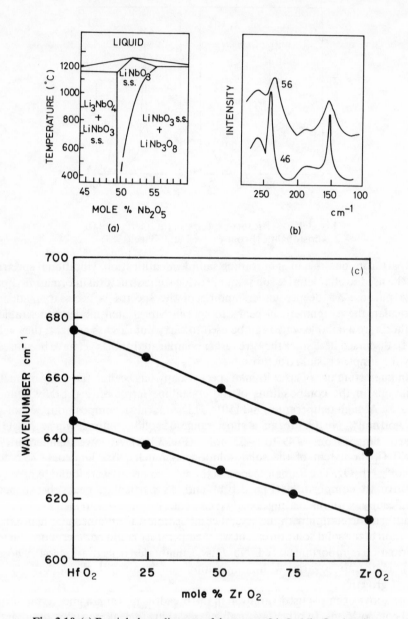

Fig. 3.10 (a) Partial phase diagram of the system $Li_2O-Nb_2O_5$ showing the stability range of $LiNbO_3$ solid solutions. (b) Laser Raman spectra of two different compositions showing that the peak breadths vary with solid solution composition. (From Scott and Burns, 1972). 3.10(c) Relation between wavenumber and position of Raman bands as a function of HfO_2-ZrO_2 solid solution composition.

solution. This variation is much larger than the half width of the peaks (i.e. the peak width at half the peak intensity) which was ~ 8 wavenumbers. In this particular example, Raman spectra are a more sensitive indicator of solid solution composition than X-ray powder diffraction patterns since the X-ray powder patterns of ZrO_2 and HfO_2 are quite similar.

It should be possible to use IR and Raman techniques to study other similar problems in solid state chemistry. This could be particularly useful since often, X-ray diffraction techniques, especially lattice parameter measurements, are insensitive to small variations in stoichiometry.

3.2.3.2 *Visible and ultraviolet spectroscopy*

Transitions of electrons between outermost energy levels are associated with energy changes in the range $\sim 10^4$ to $10^5 \, cm^{-1}$ or $\sim 10^2$ to $10^3 \, kJ \, mol^{-1}$. These energies span the range from the near infrared through the visible to the ultraviolet (Fig. 3.7) and are therefore often associated with colour. Various types of electronic transition occur and may be detected spectroscopically; some are shown schematically in Fig. 3.11. The two atoms A and B are neighbouring atoms in some kind of solid structure; they may be, for instance, an anion and a cation in an ionic crystal. The inner electron shells are localized on the individual atoms. The outermost shells may overlap to form delocalized bands of energy levels. Four basic types of transition are indicated in Fig. 3.11:

(i) Promotion of an electron from a localized orbital on one atom to a higher energy but still localized orbital on the same atom. The spectroscopic absorption band associated with this transition is sometimes known as an *exciton band*. Transitions in category (i) include (a) d–d and f–f transitions in transition metal compounds, (b) outer shell transitions in heavy metal compounds, e.g. $6s$–$6p$ in lead (II) compounds, (c) transitions associated with

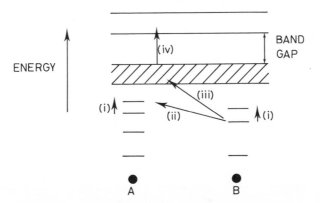

Fig. 3.11 Possible electronic transitions in a solid. These involve electrons in localized orbitals and/or delocalized bands

defects such as trapped electrons or holes, e.g. colour centres (F, H, etc.) in alkali halides and (d) transitions involving, for example, silver atoms in photochromic glasses: colloidal silver is precipitated initially on photo-irradiation and subsequent electronic transitions occur within the reduced silver atoms.

(ii) Promotion of an electron from a localized orbital on one atom to a higher energy but still localized orbital on an adjacent atom. The associated absorption bands are known as *charge transfer spectra*. The transitions are usually 'allowed transitions' according to the spectroscopic selection rules and hence the absorption bands are intense. Charge transfer processes are, for example, responsible for the intense yellow colour of chromates; an electron is transferred from an oxygen atom in a $(CrO_4)^{2-}$ tetrahedral complex anion to the central chromium atom. In mixed valence transition metal compounds, e.g. in magnetite, Fe_3O_4, charge transfer processes also occur.

(iii) Promotion of an electron from a localized orbital on one atom to a delocalized energy band, the conduction band, which is characteristic of the entire solid. In many solids the energy required to cause a transition such as this is very high but in others, especially those containing heavy elements, the transition occurs in the visible/ultraviolet region and the materials are photoconductive, e.g. some chalcogenide glasses are photoconductive.

(iv) Promotion of an electron from one energy band (the valence band) to another band of higher energy (the conduction band). The magnitude of the band gap in semiconductors (Si, Ge, etc.) may be determined spectroscopically; a typical semiconductor has a band gap of 1 eV, 96 kJ mol^{-1}, which lies between the visible and UV regions.

The appearance of a typical UV and visible absorption spectrum is shown schematically in Fig. 3.12. It contains two principal features. Above a certain energy or frequency, intense absorption occurs. Since the transmittance of the

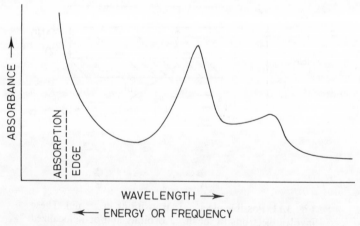

Fig. 3.12 Schematic typical UV/visible absorption spectrum

sample drops to essentially zero at the *absorption edge*, this places a high frequency limit on the spectral range that can be investigated. If it is desired to go to frequencies above the absorption edge, then reflectance techniques must be used. Transitions of types (ii) and (iii) in particular are responsible for the appearance of the absorption edge. Its position varies considerably between different materials. In electronically insulating ionic solids it may occur in the ultraviolet, but in photoconducting and semiconducting materials it may occur in the visible or even in the near infrared spectral regions.

The second feature is the appearance of broad absorption peaks or bands at frequencies below that of the absorption cut-off. These are generally associated with type (i) transitions.

Visible and UV spectroscopy has a variety of applications associated with the local structure of materials. This is because the positions of the absorption bands are sensitive to coordination environment and bond character. Some specific applications to solids are now given.

Structural studies on glass. Information on the local structure of amorphous materials may be obtained by UV and visible spectroscopy. Often it is necessary to add a small quantity of a spectroscopically active species to the raw materials. This may be a transition metal compound, in which case the d–d spectra of the transition metal ion are recorded. Alternatively, heavy metal cations such as Tl^+, Pb^{2+} may be added. From the nature of the spectra, coordination numbers can usually be deduced and, hence, information can be obtained on the availability of such sites in the glass structure. The spectra of ions such as Pb^{2+} are sensitive to the degree of covalent character in the bonds between a Pb^{2+} ion and the neighbouring anions. This provides a means for studying the basicity of glasses. Basicity is associated with electron-donating ability and, for instance, non-bridging oxide ions which contain a net unit negative charge are highly basic in a silicate glass structure. By contrast, bridging oxide ions, which link two silicon atoms, contain very little excess negative charge and are not basic. Pb^{2+} ions may therefore be added to a glass in order to probe the basicity of the sites available to it (see Duffy and Ingram, 1976).

Redox equilibria in glasses may be studied. One of the most important is the Fe^{2+}–Fe^{3+} couple since the Fe^{3+} ion is responsible for the green/brown colour of many glasses. For many applications, such as glasses for use in optical fibre communications, the presence of these ions and their associated spectra is highly undesirable.

Study of laser materials. Laser materials often contain a transition metal ion as the active species, e.g. the ruby laser is essentially Al_2O_3 doped with a small amount of Cr^{3+}; the neodymium glass laser consists of a glass doped with Nd^{3+} ions. Information on the Cr^{3+}, Nd^{3+} ions and their coordination environment in the host structure may be obtained from the UV and visible spectra. Laser action results from a population inversion in which a large number of electrons are promoted into a higher energy level. These subsequently drop back into a lower level and, at the same time, emit the laser beam. It is therefore very important to have a detailed knowledge of the available energy levels and possible electronic transitions.

3.2.3.3 *Nuclear magnetic resonance (NMR) spectroscopy*

NMR spectroscopy has made an enormous impact on the determination of molecular structure over the last two or three decades, but until very recently has yielded only limited structural information about solids. Molecular substances, especially organic molecules in the liquid state, give NMR spectra which at high resolution are composed of a number of sharp peaks. From the positions and relative intensities of the peaks it is often possible to tell which atoms are bonded together, coordination numbers, next nearest neighbours, etc. In the solid state, by contrast, broad featureless peaks are observed from which little direct structural information can be obtained. Strenuous efforts have been made to sharpen these broad peaks that are observed in solids and recently, considerable success has been achieved. In one technique, the 'magic angle spinning' (MAS) technique, the sample is rotated at a high velocity at a critical angle of 54.74° to the applied magnetic field.

NMR spectroscopy is a technique that involves the magnetic spin energy of atomic nuclei. For elements that have a non-zero nuclear spin, such as ^1H, ^2H, ^6Li, ^7Li, ^{13}C and ^{29}Si, but not, for example, ^{12}C, ^{16}O, or ^{28}Si, an applied magnetic field will influence the energy of the nuclei. The magnetic energy levels split into two groups, depending on whether the nuclear spins are aligned parallel or antiparallel with the applied magnetic field. The magnitude of the energy difference between parallel and antiparallel spin states is small, ~ 0.01 J mol^{-1} for an applied magnetic field of 10^4 G (1 T). This amount of energy is associated with the radiofrequency region of the electromagnetic spectrum. NMR spectrometers operating at, for example, 50 MHz can therefore induce nuclear spin transitions. The magnitude of the energy change and the associated frequency of absorption depends not only on the particular element involved but also on its chemical environment. Thus, in organic molecules, hydrogen atoms bonded to different types of carbon atom or different functional groups may be distinguished since they absorb at slightly different frequencies. Note that NMR instruments are usually operated at fixed frequency, e.g. 220 MHz, and it is the magnitude of the energy difference between the parallel and antiparallel spin states that is varied, by varying the applied magnetic field strength.

Applications :(a) structural studies. Conventional NMR measurements on solids give broad, featureless bands which are of little use for structural work. By using refinements such as the MAS technique, the broad bands collapse to reveal a fine structure. Lippmaa and others (1980) have applied this technique to crystalline silicates and found that the ^{29}Si NMR spectra give peaks whose positions depend on the nature of the silicate anion. In particular, the spectra can distinguish between isolated SiO_4 tetrahedra and SiO_4 tetrahedra linked by sharing

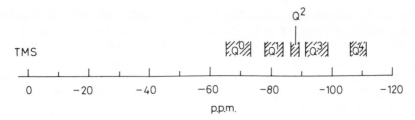

Fig. 3.13 Positions of ^{29}Si NMR peaks in silicates as a function of the degree of condensation, Q, of the silicate anion. (From E. Lippmaa *et al.*, 1980)

common corners (oxygen atoms) to one, two, three or four other tetrahedra. In order to facilitate the discussion, it is customary to assign to each silicon (or its SiO_4 tetrahedron) a 'Q value' which represents the number of adjacent SiO_4 tetrahedra to which it is directly bonded. Q values range from zero (as in orthosilicates such as Mg_2SiO_4 with isolated tetrahedra) to four (as in three-dimensional framework structures such as SiO_2, in which all four corners are shared). The positions of the ^{29}Si NMR peaks, i.e. the *chemical shifts* relative to the internal standard (tetramethylsilane, $(CH_3)_4Si$, or TMS), depend approximately on the Q value, as indicated in Fig. 3.13. For each Q value, a range of chemical shifts is observed, depending on other features of the crystal structure.

The ^{29}Si NMR spectrum of a calcium silicate, xonotlite, is shown in Fig. 3.14. The silicate anion of xonotlite is an infinite double chain or ladder with a cross link or rung at every third tetrahedron in each chain, as shown schematically in the diagram. Two types of silicon are present, therefore—Q^2 and Q^3—and in the relative amounts of 2:1. The NMR peaks appear in the appropriate positions for Q^2 and Q^3 silicon atoms and their intensities are in the ratio of 2:1, as expected.

NMR spectroscopy can also be used to probe the fine structure of alumino-silicate crystal structures. In these, the aluminium atoms can play two possible

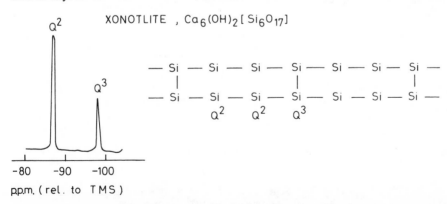

Fig. 3.14 The ^{29}Si NMR spectrum of xonotlite (from Lippmaa *et al.* 1980). Also shown in a highly schematic manner is the structure of the silicate double chain anion and the Q values of the silicon atoms involved. Oxygen atoms are omitted

rôles. Either they occupy octahedral sites and do not really form part of the aluminosilicate framework of linked tetrahedra. Or, more often, they occupy tetrahedral sites, similar to silicon, and do form part of the aluminosilicate framework. In the latter case, the chemical shift of a particular silicon atom depends on how many aluminium atoms are in its second coordination sphere (Fig. 3.15). For instance, a Q^4 silicon atom in an aluminosilicate framework is surrounded by four other tetrahedral atoms and, of these, any number between zero and four may be aluminium atoms (i.e. AlO_4 tetrahedra). It is found that the chemical shift increases with decreasing number of aluminium neighbours, from -84 p.p.m. with four aluminium neighbours, as in nepheline, $KNa_3(AlSiO_4)_4$, to $\sim$ -108 p.p.m. with no aluminium neighbours (as in SiO_2). The crystal structure of natrolite, $Na_2(Al_2Si_3O_{10})\cdot 2H_2O$, contains two types of Q^4 silicon atoms with three and two neighbouring aluminium atoms, respectively. This is reflected in the ^{29}Si NMR spectrum which shows two peaks at -87.7 and -95.4 p.p.m. (Fig. 3.15c). In some aluminosilicates, the Al, Si distributions are disordered over the tetrahedral sites, which is indicated by a broadening of the NMR lines. An extreme example is sanidine, $K(AlSi_3O_8)$, in which the Al, Si positions appear to

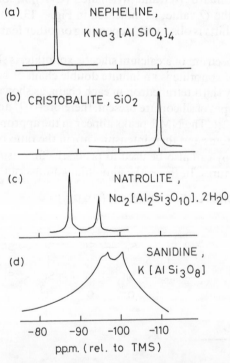

Fig. 3.15 Schematic ^{29}Si NMR spectra of silicates containing different Q^4 silicon atoms. The number of aluminium atoms in the second coordination sphere are (a) 4, (b) 0, (c) 3 and 2, (d) 2 and 1. (From Lippmaa et al., 1980)

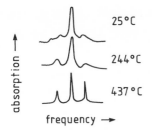

absorption →

frequency →

25°C

244°C

437°C

Fig. 3.16 Temperature-dependent ^{7}Li NMR spectra of β-eucryptite, LiAlSiO$_4$. (From Brinkman *et al.*, 1981)

be completely disordered; its NMR signal gives a very broad line which is ~ 15 to 20 p.p.m. wide at half-height.

In summary, NMR spectroscopy appears to be a potentially powerful technique for studying the local structure of silicates and other inorganic solids. It has been used for such work only since ~ 1980 and, therefore, considerable future developments are hoped for. The information that is obtained complements very well the 'average' structural information that is obtained by X-ray diffraction. Thus the nature of the Al, Si distribution in silicate structures has long been a thorny problem for mineralogists and crystallographers and only in favourable cases have X-ray measurements fully resolved the structure. (The X-ray scattering powers of aluminium and silicon are very similar and hence it is difficult to distinguish between them.)

(*b*) *Atomic migration in solids.* Atomic migration processes can be studied in cases where 'motional narrowing' of the normally broad solid state spectra occur. An example, given in Fig. 3.16, is the ^{7}Li NMR spectrum of crystalline eucryptite, β-LiAlSiO$_4$, at three different temperatures. At room temperature, a complex spectrum with several, overlapping broad peaks is observed. As the temperature is raised, the peaks become increasingly narrower. This narrowing is attributed to increasing movement of the Li$^+$ ions with increasing temperature. At room temperature, they are effectively trapped in their sites in the aluminosilicate framework and β-eucryptite has a very low Li$^+$ ion conductivity. With increasing temperature, the Li$^+$ ions begin to hop from site to site, and above ~ 400 °C, eucryptite is a moderately good solid electrolyte. Similar motional narrowing effects have been observed with other solid electrolytes in, for example, the ^{23}Na NMR spectrum of β-alumina and the ^{19}F spectrum of fluoride ion conductors such as PbF$_2$. The peak breadths may be measured as a function of temperature and, from this, the activation energy for conduction obtained.

The mechanism of Li$^+$ ion conduction in Li$_3$N has been deduced very elegantly by ^{7}Li NMR. Li$_3$N contains two crystallographically distinct Li$^+$ ions, Li(1) and Li(2), but only the NMR peaks associated with Li(2) showed significant motional narrowing with increasing temperature. The Li(2)$^+$ ions are therefore the major current carriers in Li$_3$N. The crystal structure of Li$_3$N is a layered structure in which layers, of constitution 'Li$_2$N' and containing the Li(2)$^+$ ions, are propped apart by Li(1)$^+$ ions. The 'Li$_2$N' layers are shown in projection in Fig. 3.17(a); the Li(2)$^+$ ions form a hexagonal network with nitride ions in the

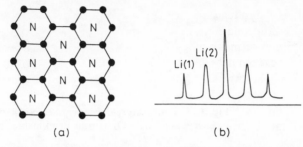

Fig. 3.17 (a) Structure of the 'Li$_2$N' layers in Li$_3$N; solid circles are Li(2)$^+$ ions. (b) The ^{7}Li NMR spectrum of Li$_3$N. (From Rabenau, 1982)

centres of the rings. In three dimensions, these 'Li$_2$N' layers are stacked directly over each other, with Li(1)$^+$ ions sandwiched between the nitride ions of adjacent layers. The ^{7}Li NMR spectrum of Li$_3$N is shown in Fig. 3.17(b). It contains a central line and two pairs of satellites; the inner pair has twice the intensity (integrated area) of the outer pair and is therefore assigned to the Li(2)$^+$ ions, which are twice as numerous as the Li(1)$^+$ ions. With increasing temperature, the inner satellites were observed to narrow much more significantly than the outer satellites, thereby confirming that the Li(2)$^+$ ions are the major current carriers. The Li(2)$^+$ ions are able to move in Li$_3$N largely by virtue of the fact that a small number, 1 to 2 per cent of the Li(2) sites are vacant.

3.2.3.4 *Electron spin resonance (ESR) spectroscopy*

The ESR technique is closely related to NMR; it detects changes in electron spin configuration. ESR depends on the presence of permanent magnetic dipoles, i.e. unpaired electrons, in the sample, such as those which occur in many transition metal ions. The reversal of spin of these unpaired electrons in an applied magnetic field is recorded. The magnitude of the energy change is again small, ~ 1.0 J mol^{-1}, although somewhat larger than the corresponding NMR energy changes (Fig. 3.7). ESR spectrometers therefore operate at microwave frequencies, e.g. at 2.8×10^{10} Hz (28 GHz), with an applied magnetic field of, for example, 3000 G. In practice, the spectra are obtained by varying the magnetic field at constant frequency. Absorption of energy associated with the spin transition occurs at the resonant condition:

$$\Delta E = hf = g\beta_e H \tag{3.4}$$

where β_e is a constant, the Bohr magneton ($\beta_e = eh/4\pi mc = 9.723 \times 10^{-12}$ J G^{-1}), and H is the strength of the applied magnetic field. The factor g, the gyromagnetic ratio, has a value of 2.0023 for a free electron but varies significantly for paramagnetic ions in the solid state. The value of g depends on the particular paramagnetic ion, its oxidation state and coordination number. It is therefore responsible for the position of an absorption peak and is analogous to the 'chemical shift' of NMR spectra.

ESR spectra of solids often show broad absorption peaks (as do NMR spectra) and, therefore, certain conditions must be met in order to get sharp peaks from which useful information may be obtained. One source of line broadening is spin–spin interactions between neighbouring unpaired electrons. This is overcome by having only a low concentration of unpaired electrons, e.g. 0.1 to 1 per cent of a paramagnetic transition metal ion dissolved in a diamagnetic host structure. A second source of line broadening is the occurrence of low lying excited states near to the ground state energy of the paramagnetic ion. This leads to frequent electron transitions, short relaxation times and broad peaks. To overcome this, the spectra are recorded at low temperatures, often at the liquid helium temperature, 4.2 K. The interpretation of ESR spectra is greatly simplified if the

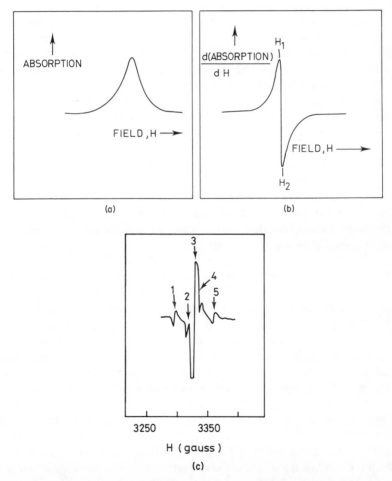

Fig. 3.18 (a) Schematic ESR absorption peak, (b) the first derivative of the absorption and (c) the ESR spectrum of CrO_4^{3-} in Ca_2PO_4Cl at 77 K. (From Greenblatt, 1980)

paramagnetic species has only one unpaired electron, as in, for example, d^1 transition metal ions, V^{4+}, Cr^{5+}, etc.

ESR spectra are usually presented as the first derivative of the absorption (Fig. 3.18b), rather than as the absorption itself, a. ESR spectra often comprise a set of closely spaced peaks. The 'hyperfine splitting' which causes multiple peaks arises because, in addition to the applied external magnetic field, there are internal nuclear magnetic fields present associated with the transition metal ion itself or with surrounding ligands. For instance, ^{53}Cr has a nuclear spin moment, I, of $\frac{3}{2}$ and the spectrum of $^{53}Cr^{5+}$ (d^1 ion) will be split into $2I + 1 = 4$ hyperfine lines. An example is shown in Fig. 3.18(c) for the ESR spectrum of Cr^{5+}, as CrO_4^{3-}, dissolved in apatite, Ca_2PO_4Cl, at 77K. Naturally occurring chromium is a mixture of isotopes, the main one being ^{52}Cr which has a nuclear spin moment of zero. Unpaired electrons on the $^{52}Cr^{5+}$ ions are responsible for the very intense central line, 3, in the spectrum. The four small, equally spaced lines 1, 2, 4 and 5 are associated with unpaired electrons on the small percentage of $^{53}Cr^{5+}$ ions which have $I = \frac{3}{2}$. In some cases, additional fine structure, 'superhyperfine splitting' may occur due to interactions of an unpaired electron with the nuclear moment of neighbouring ions.

From ESR spectra such as Fig. 3.18, one can (with appropriate experience!) obtain information on the paramagnetic ion and its immediate environment in the host structure. Specifically, one may determine:

(a) the oxidation state, electronic configuration and coordination number of the paramagnetic ion,
(b) the ground state d orbital configuration of the paramagnetic ion and any structural distortions arising from, for example, a Jahn–Teller effect,
(c) the extent, if any, of covalency in the bonds between the paramagnetic ion and its surrounding anions or ligands.

Since the paramagnetic ion is present only in small amounts, it is assumed that its site symmetry is identical to that of the host ion that it substitutionally replaces. For instance, it has been shown that in chromium-doped apatites, $M_5(PO_4)_3X : M = Ca, Sr, Ba$; $X = Cl, F$, the CrO_4^{3-} tetrahedron is distorted in such a way as to be squashed or compressed along one of the fourfold inversion axes of the tetrahedron. It is very likely, therefore, that the PO_4 tetrahedra have a similar distortion. On the other hand, in $BaSO_4$, the SO_4 tetrahedra are distorted by stretching along a four-fold inversion axis. This was shown by doping $BaSO_4$ with MnO_4^{2-} containing Mn^{6+} (also a d^1 ion) and recording the ESR spectra.

A technique related to ESR and which is useful for observing the hyperfine and superhyperfine splittings is electron nuclear double resonance (ENDOR). Basically, it is a combination of NMR and ESR spectroscopy. In the ENDOR technique, the NMR frequencies of nuclei adjacent to a paramagnetic centre are scanned. The resulting NMR spectrum shows fine structure due to interactions with the paramagnetic ion. By scanning in turn each of the nuclei (that have a non-zero nuclear spin moment) in the region of the paramagnetic species, a detailed plan of the local atomic structure is obtained. The method has been applied successfully to the study of crystal defects, such as colour centres (trapped electrons in alkali halide crystals; see Chapter 9), radiation damage effects and doping in, for example, phosphors and semiconductors.

3.2.3.5 *X-ray spectroscopies: XRF, AEFS, EXAFS*

The X-ray region of the electromagnetic spectrum is probably the most generally useful region for structural studies, analysis and characterization of solids. There are three main ways in which X-rays are used: for diffraction, emission and absorption, Fig. 3.19a.

Diffraction techniques utilize a monochromatic beam of X-rays which is diffracted by the sample; the techniques are used for both crystal structure determination and phase identification and are covered in Section 3.2.1 and Chapter 5.

Emission techniques utilize the characteristic X-ray emission spectra of elements, which are generated by, for instance, bombardment with high energy electrons. The spectra are used in the chemical analysis of both bulk samples (X-ray fluoresence) and submicroscopic particles (analytical electron microscopy, EPMA, etc.). They also have limited uses in determination of local structure, coordination numbers, etc.

Absorption techniques measure the absorption of X-rays by samples, especially at energies in the region of absorption edges. They are powerful techniques for studying local structure but are not accessible to most people since a synchrotron radiation source is needed.

Emission techniques. On bombardment of matter with, for instance, high energy electrons or X-rays, inner shell electrons may be ejected from atoms. Outer shell electrons then drop into the vacancies in the inner levels and the excess energy is released in the form of electromagnetic radiation, often X-radiation (Figs 3.5 and 3.19). Each element gives a characteristic X-ray emission spectrum (see Fig. 5.2) composed of a set of sharp peaks. The spectra are different for each element since the peak positions depend on the difference in energy between electron levels, e.g. $2p$ and $1s$, which, in turn, depend on atomic number (Moseley's Law). X-ray emission spectra may, therefore, be used for elemental analysis, both qualitatively by looking for peaks at certain positions or wavelengths and quantitatively by measuring peak intensities and comparing them against a calibration chart. X-ray fluorescence (XRF) does just this: a solid sample is bombarded with high energy X-rays and the resulting emission spectrum recorded. From the spectral peak positions the elements present can be identified and from their intensities a quantitative analysis made. Analytical electron microscopy and related techniques such as electron probe microanalysis (Section 3.2.2.2) operate in the same way. The sample is bombarded with high energy electrons; the diffracted electrons are used to record both electron micrographs and electron diffraction patterns;

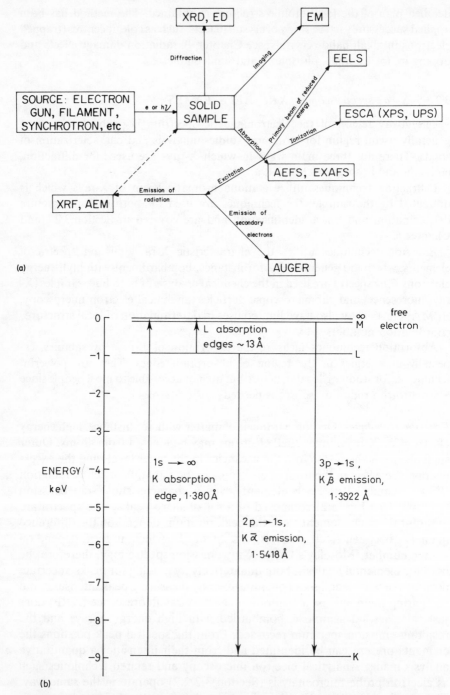

Fig. 3.19 (a) Flowsheet showing relations between various diffraction, microscopic and spectroscopic techniques associated with electrons and X-rays. (b) Electronic transitions responsible for emission and absorption X-ray spectra. Wavelength values are for copper

the emitted X-rays are used for elemental analysis. Analytical electron micro-scopy is rapidly becoming an invaluable technique for the chemical analysis of solids, in particular because very small particles or regions of a solid may be characterized. The heterogeneous nature of many solids, such as cements, steels and catalysts, may therefore be studied.

X-ray emission spectra also have uses in determining local structure such as coordination numbers and bond distances. This is because the peak positions vary slightly, depending on the local environment of the atoms in question. As an example, the Al $K\beta$ emission bands (corresponding to the $3p \rightarrow 1s$ transition) for three widely different materials, metallic aluminium, α-Al_2O_3 and sanidine, $KAlSi_3O_8$, are shown in Fig. 3.20. The peaks clearly occur in different positions and, in the case of Al_2O_3 and sanidine, appear to be poorly resolved doublets. By comparing the spectra for a wide range of aluminium-containing oxide com-pounds, it has been found that approximate correlations exist between (a) peak position and coordination number of the aluminium (it is usually 4 or 6) and (b) peak position and Al—O bond distance for a given coordination number. Similar studies on silicates (in which silicon is almost always tetrahedrally coordinated) have shown a correlation between the Si $K\beta$ peak position and Si—O bond length; since the Si—O bond length varies according to whether the oxygen bridges two silicate tetrahedra or is non-bridging, this may be used to study polymerization of silicate anions in, for example glasses, gels and crystals.

For elements that can exist in more than one valence state or oxidation state, a correlation usually exists between peak position and oxidation state. Thus, XRF may be used to distinguish the different oxidation states of, for example, sulphur ($-$ II to $+$ VI) in a variety of sulphur-containing compounds.

Absorption techniques. Atoms give characteristic X-ray absorption spectra as well as characteristic emission spectra. These arise from the various ionization and intershell transition that are possible, as shown in Fig. 3.19a. In order to ionize a

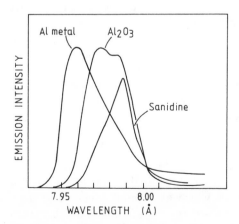

Fig. 3.20 Al $K\beta$ emission spectra of three aluminium-containing materials. (From White and Gibbs, 1969)

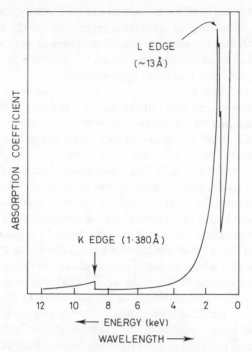

Fig. 3.21 Variation of X-ray absorption coefficient with wavelength for copper metal. (From Stern, 1976)

1s electron in copper metal an energy of almost 9 keV, corresponding to a wavelength of 1.380 Å, is needed. Much less energy (∼ 1 keV) is needed to ionize L shell (2s, 2p) electrons. The X-ray absorption spectrum of copper is shown in Fig. 3.21. It takes the form of a smooth curve which increases rapidly at low X-ray energies and superposed on which are K, L, etc., absorption edges. The K absorption edge represents the minimum energy that is required to ionize a 1s electron in copper. The absorption coefficient therefore undergoes an abrupt increase as the energy is increased to this value. The L absorption edge (in fact, three closely spaced L edges are usually seen) represents ionization of 2s, 2p electrons. At intermediate energies of the incident X-ray beam, e.g. at 4 keV, electrons may be ionized from L or outer shells and they leave the atom with a net kinetic energy, E, given by

$$E = hv - E_0$$

where hv is the energy of the incident X-ray photon (4 keV in this case) and E_0 is the critical ionization energy required to free the electron from the atom.

The wavelengths at which absorption edges occur depend on the relative separation of the atomic energy levels in atoms which, in turn, depend on atomic number (Moseley's Law). They are, therefore, characteristic for each element and may be used for identification purposes in a similar manner to emission spectra.

This section is, however, concerned primarily with two other techniques which monitor variations in the absorption spectra, especially in the region of the absorption edge. These techniques are absorption edge fine structure (AEFS) and extended X-ray absorption fine structure (EXAFS).

Although X-ray absorption techniques have been in use since the 1930s, a recent development involving the use of synchrotrons and storage rings as a source of X-rays has given them a new impetus. This is because synchrotron radiation, produced when charged particles, electrons, are accelerated in a magnetic field, is a very intense spectrum of continuous X-ray wavelengths. Using it, much higher sensitivity is achieved and more information can be obtained from the absorption spectra. The number of laboratories that have a particle accelerator suitable for producing synchrotron radiation is, of course, very limited; for instance, there is only one in the United Kingdom, at Daresbury. Others are at Orsay, France, and Stanford, United States. Such laboratories therefore provide a central user service.

AEFS. In the region of an absorption edge, fine structure associated with inner shell transitions is often seen, e.g. the K edge of copper may show additional peaks due to transitions $1s \rightarrow 3d$ (but not in Cu^{+1} compounds), $1s \rightarrow 4s$, $1s \rightarrow 4p$. The exact peak positions depend on details of oxidation state, site symmetry, surrounding ligands and the nature of the bonding. The absorption spectra may therefore be used to probe local structure. Examples are shown in Fig. 3.22, on an expanded energy or wavelength scale (compared to Fig. 3.21), for two copper compounds, CuCl and $CuCl_2 \cdot 2H_2O$. For each, the K absorption edge, with superposed fine structure peaks ($1s \rightarrow 4p$ etc.) is seen. The entire spectrum is displaced to higher energies in $CuCl_2 \cdot 2H_2O$, reflecting the higher oxidation state

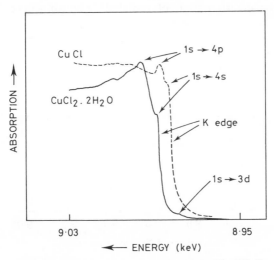

Fig. 3.22 AEFS spectra of CuCl and $CuCl_2 \cdot 2H_2O$.
(From Chan, Hu and Gamble, 1978)

of copper (+ 2) in CuCl$_2$·2H$_2$O compared with that (+ 1) in CuCl and, therefore, the increased difficulty in ionizing K shell electrons.

EXAFS. Whereas the AEFS technique examines at high resolution the details of the fine structure in the region of an absorption edge, the EXAFS technique examines the variation of absorption with energy (or wavelength) over a much wider range, extending out from the absorption edge to higher energies by up to ~ 1 keV. The absorption usually shows a ripple, known also as the Kronig fine structure (Fig. 3.23) from which, with suitable data processing, information on local structure and, especially, bond distances may be obtained. Explanations of the origin of the ripple will not be attempted here. Suffice it to say that it is related to the wave properties of the electron; the ionized photoelectrons interact with neighbouring atoms in the solid which then act as secondary sources of scattering for the photoelectrons. Interference between adjacent scattered waves may occur and this influences the probability of absorption of an incident X-ray photon occurring. The degree of interference depends on the wavelength of the photoelectron (and hence on the wavelength of the incident X-ray photons) and the local structure, including interatomic distances, in the region of the emitting atom. EXAFS is therefore a kind of in situ electron diffraction in which the source of the electron is the actual atom which participates in the X-ray absorption event. Using Fourier transform techniques, it is possible to analyse the ripple pattern and obtain something that looks like a radial distribution function (RDF), as shown in Fig. 18.7.

EXAFS is a technique for determining local structure and is equally suitable for non-crystalline as well as crystalline materials. It is particularly valuable for studying disordered and amorphous materials such as glasses, gels and amorphous metals since structural information on them is generally hard to obtain. For the determination of radial distribution curves in amorphous

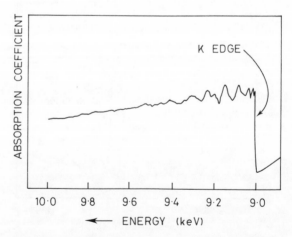

Fig. 3.23 EXAFS spectrum of copper metal. (From E. A. Stern, 1976)

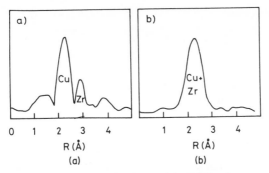

Fig. 3.24 EXAFS-derived partial RDSs for an amorphous $Cu_{46}Zr_{54}$ alloy: (a) Zr K edge, (b) Cu K edge. (From Gurman, 1982)

materials (i.e. graphs showing the probability of finding an atom as a function of distance from a central atom), EXAFS may in future be used in preference to conventional diffraction techniques. This is because EXAFS has one great advantage: by tuning in to the absorption edge of each element present in the material in turn, a partial RDF for each element may be constructed. By contrast, conventional diffraction techniques give only a single averaged RDF for all the elements present.

An example shown in Fig. 3.24 is for the alloy $Cu_{46}Zr_{54}$. The RDFs are Fourier transforms derived from (a) the zirconium K edge at 18 keV and (b) the copper K edge at 9 keV. The positions of the peaks in Fig. 3.24 are related to, but not directly equal to, interatomic distances. From the RDFs, it was shown that each zirconium atom is surrounded by an average of 4.6 Cu atoms at 2.74 Å and 5.1 Zr atoms at 3.14 Å: copper–copper distances are 2.47 Å.

Studies similar to these on metallic glasses enable structural models for the glasses to be tested, for instance (a) whether the dense, random packing of spheres is an appropriate model or (b) whether chemical ordering effects occur whereby there is a preference for a certain type of neighbouring atom around a particular atom. Information on bonding may be obtained from the interatomic distances; for instance, for a given pair of atoms, metallic bonds are expected to be somewhat longer than covalent bonds. Departures from non-random structure in metallic alloys may be observed, e.g. the beginning of clustering, phase separation or Guinier–Preston zone formation.

EXAFS has been used to study silicate and other oxide glasses. The results (for GeO_2 glass, at least) support a random network model of glass, rather than a microcrystallite model (Chapter 18). Coordination numbers can be determined. Thus in GeO_2 glass, the germanium is tetrahedrally coordinated to oxygen (as in silica glass) but on addition of the network modifying oxide, Li_2O, some of the germanium atoms change to octahedral coordination. In sodium silicate glasses the Na^+ ions occupy five coordinate sites as they also do in some crystalline sodium silicates.

3.2.3.6 *Electron spectroscopies: ESCA, XPS, UPS, AES, EELS*

Electron spectroscopy techniques measure the kinetic energy of electrons that are emitted from matter as a consequence of bombarding it with ionizing radiation or high energy particles. Various processes take place when atoms are exposed to ionizing radiation (see Fig. 3.5 and 3.19). The simplest is the direct ionization of an electron from either a valence or an inner shell. The kinetic energy, E, of the ionized electron is equal to the difference between the energy, hv, of the incident radiation and the binding energy or ionization potential, E_b, of the electron, i.e. $E = hv - E_b$. For a given atom, a range of E_b values is possible, corresponding to the ionization of electrons from different inner and outer valence shells, and these E_b values are characteristic for each element. Measurement of E, and therefore E_b, provides a means of identification of atoms and forms the basis of the ESCA (electron spectroscopy for chemical analysis) technique developed by Siegbahn and coworkers in Uppsala (1967). The ionizing radiation that is used in ESCA is usually either X-rays (Mg $K\alpha$, 1254 eV or Al $K\alpha$, 1487 eV monochromatic radiation) or ultraviolent light (He discharge, 21.4 and 40.8 eV for the $2p \rightarrow 1s$ transitions in He and He$^+$, respectively) and the techniques are then also known as XPS (X-ray photoelectron spectroscopy) and UPS (ultraviolet photoelectron spectroscopy), respectively. The main difference between XPS and UPS is in the electron shells that are accessible for ionization. Thus, inner shell electrons may be ionized in XPS but only outer electrons—in valence shells, molecular orbitals or energy bands—may be ejected in UPS.

A related technique is Auger electron spectroscopy (AES). In this, the electrons that are ejected and detected are not the primary ionized electrons, as in ESCA, but are produced by secondary processes involving the decay of ionized atoms from excited states to lower energy states (Fig. 3.25). The ESCA process may be visualized as:

$$\text{Atom A} \xrightarrow{\text{radiation}} A^{+*} + e^- \quad \text{(ESCA)}$$

where A^{+*} refers to an ionized atom which is in an excited state. This excited state condition arises either if the electron is ejected from an inner shell, leaving a vacancy, or if other electrons in the atom have been promoted to higher, normally

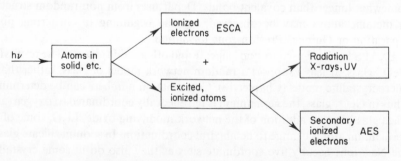

Fig. 3.25 Origins of ESCA and Auger spectra

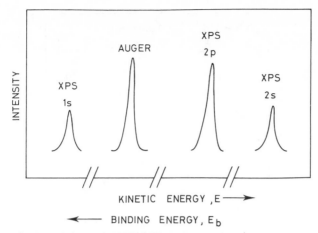

Fig. 3.26 Schematic XPS, AES spectrum of Na^+ in a sodium-containing solid. The Auger peak arises from an initial $1s$ vacancy which is then filled by a $2p$ electron causing ionization of another $2p$ (Auger) electron

empty levels, during irradiation. The excited atom decays when electrons drop into vacancies in lower energy levels. Energy is consequently released by one of two methods:

$$A^{+*} \rightarrow A^+ + h\nu \qquad \text{(X-rays, UV)}$$

or

$$A^{+*} \rightarrow A^{++} + e^- \qquad \text{(Auger electrons)}$$

The energy may be emitted as electromagnetic radiation: this is, in fact, the normal method by which X-rays are produced, although for lighter atoms, UV photons are generated instead. Alternatively, the energy may be transferred to another (outer shell) electron in the same atom which is then ejected. Such secondary ionized electrons are known as Auger electrons.

AES spectra are usually observed at the same time as ESCA spectra. A schematic example is shown in Fig. 3.26. Often, however, AES spectra are complex (but not in Fig. 3.26) and difficult to interpret. AES is not, as yet, a widely used technique although this may change in the near future.

In the last decade or so, ESCA has proved to be a powerful technique for determining energy levels in atoms and molecules. In solid state sciences, it is particularly useful as a technique for studying surfaces because the electrons that are produced in ESCA are not very energetic (usually their energy is much less than 1 keV) and are rapidly absorbed by solid matter. Consequently, they cannot escape from solids unless they are ejected within ~ 20 to $50\,\text{Å}$ (2 to 5 nm) of the surface. Information on bulk structure of solids is also obtained, provided the surface layer is representative of the bulk.

Applications :(a) chemical shifts and local structure. Some limited success has been achieved in using XPS to probe the local structure of solids. This has been

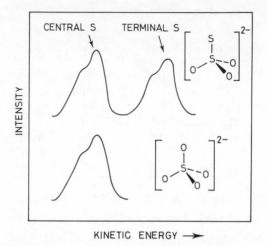

Fig. 3.27 Schematic XPS $2p$ spectra of sodium thiosulphate and sodium sulphate. Note that each peak is a doublet depending on the spin orbit states, $\frac{1}{2}$ and $\frac{3}{2}$, of the $2p$ electron

possible because the binding energies of electrons in a particular atom may show a small variation, depending on the immediate environment of the atom and its charge or oxidation state. The idea is to measure the 'chemical shift' of an atom, relative to a standard, and thereby obtain information on local structure (as, for example, NMR). Examples of chemical shift effects are shown in Figs 3.27 and 3.28. In sodium thiosulphate, $Na_2S_2O_3$, the two types of sulphur atom may be distinguished (Fig. 3.27). Peaks are of equal height, indicating equal numbers of each. Assignment of the higher kinetic energy peaks to the terminal S atom is made on the basis that this atom carried more negative charge than the central S

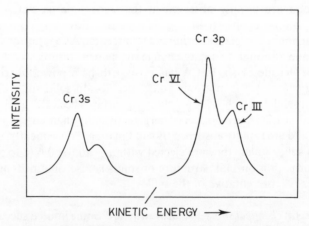

Fig. 3.28 XPS spectrum of Cr $3s$, $3p$ electrons in KCr_3O_8

atom and is, therefore, easier to ionize. Hence E_b is less and $(hv - E_b)$ greater for the terminal S atom than for the central S atom. By comparison, sodium sulphate, Na_2SO_4, shows a single sulphur $2p$ peak at the same energy as that for the central sulphur atom in $Na_2S_2O_3$.

The compound KCr_3O_8 is a mixed valence compound better written as $KCr^{III}(Cr^{VI}O_4)_2$. Its XPS spectrum shows doubled peaks for chromium both for $3s$ and $3p$ electrons (Fig. 3.28). The intensities are in the ratio of 2:1 and the peaks are assigned to the Cr^{VI} and Cr^{III} oxidation states. This fits with the formula and also with the expectation that E_b is greater for Cr^{VI} than for Cr^{III}.

Although the two examples above show clearly the influence of local structure effects in the ESCA spectra, these examples are by no means typical. In many other cases, the chemical shifts associated with different oxidation states or local environments may be quite small and the spectra insensitive to local structure. The promise of ESCA as a local structure probe, applicable to virtually all elements of the periodic table, is therefore as yet only partly fulfilled.

(b) *Bonding and band structure.* ESCA has been used to investigate the band structure of solids, especially in metals and semiconductors. An interesting example is shown in Fig. 3.29 for a series of sodium tungsten bronzes. These have the ReO_3 structure but contain variable amounts of intercalated sodium. W^{VI} is d^0 and hence, ideally, WO_3 should have no $5d$ electrons. On addition of sodium to WO_3, to form Na_xWO_3, electrons from sodium enter the tungsten $5d$ band

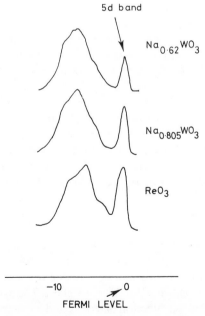

5d band

$Na_{0.62}WO_3$

$Na_{0.805}WO_3$

ReO_3

−10　　　　0

FERMI LEVEL

Fig. 3.29 XPS spectra for tungsten bronzes. (From Campagna *et al.*, 1975)

(leaving Na^+). This shows up in the XPS spectrum as a peak representing ionization from the $5d$ band and whose intensity increases with increasing sodium content. The spectrum of ReO_3 is shown for comparison. Re^{VI} is d^1 and hence its spectrum is similar to that which might be expected for $Na_{1.0}WO_3$.

(c) *Surface studies*. The principal use and potential of ESCA in solid state sciences is probably for studying surfaces, mainly because it is one of the few techniques that is surface selective. It may be used purely as an analytical method for detecting which elements are present at a surface and in particular for detecting layers of absorbed molecules on, for example, the surfaces of metals. This is clearly of great relevance to catalysis. Coatings may be analysed; for instance, a GeO_2 layer on the surface of germanium metal may be detected, since a chemical shift effect is seen. Sometimes, surface states in the electron band structure of a metal can be detected. These appear as electron orbitals localized at or near the surface, rather than delocalized throughout the bulk of the metal. In order to detect them, very clean surfaces are needed.

(d) *Electron energy loss spectroscopy*. EELS or ELS is a technique associated with analytical electron microscopy (Section 3.2.2.2). It can be used for elemental analysis, for light elements such as carbon and nitrogen and for studying energy levels in surface layers of solids. In the EELS technique, a monochromatic incident beam of electrons is used which causes ionization of inner shell electrons in atoms of the sample (see Fig. 3.5). These ionizations are a necessary precursor to the generation of X-rays. The electrons that are responsible for these ionizations suffer an energy loss as a consequence. An EELS spectrum is a plot of intensity of these electrons against energy loss. An intense peak occurs at zero energy loss. This corresponds to electrons that are either scattered elastically or do not interact with the sample. The other peaks in the EELS spectrum correspond to the electrons responsible for inner shell ionizations. The peaks are usually weak and broad and the spectra increase in complexity with increasing atomic number. EELS is a particularly useful technique for analysing light atoms and therefore complements X-ray fluorescence which is more useful with heavier elements. A derivative of EELS is extended energy loss fine structure spectroscopy (EXELFS), which is concerned with fine structure in the EELS spectrum. EXELFS is the electron analogue of the X-ray technique EXAFS.

3.2.3.7 *Mössbauer spectroscopy*

Mössbauer or γ-ray spectroscopy is akin to NMR spectroscopy in that it is concerned with transitions that take place inside atomic nuclei. The incident radiation that is used is a highly monochromatic beam of γ-rays whose energy may be varied by making use of the Doppler effect. The absorption of γ-rays by the sample is monitored as a function of energy and a spectrum is obtained which usually consists of a number of poorly resolved peaks. From an appropriate analysis of the spectrum, information on local structure—oxidation states, coordination numbers and bond character—may be obtained.

The γ-rays that are used in Mössbauer spectroscopy are produced by decay of radioactive elements such as $^{57}Fe^*_{26}$ or $^{119}Sn^*_{50}$. The γ-rays appear in the electromagnetic spectrum (Fig. 3.7) to the high energy (right-hand) side of X-rays. The γ-emission is associated with a change in population of energy levels in the nuclei responsible, rather than with a change in atomic mass or number. Under certain conditions of 'recoilless emission', all of the energy change in the nuclei is transmitted to the emitted γ-rays and this gives rise to a highly monochromatic beam of radiation. This radiation may then be absorbed by a sample that contains similar atoms to those responsible for the emission. However, the nuclear energy levels of the absorbing atoms do vary somewhat, depending on the oxidation state of the element, its coordination number, etc., and hence some means of modulating either the energy of the incident radiation or the energy levels within the nuclei of the sample is required. In practice, the energy of the γ-rays is modified by making use of the Doppler effect. With the sample placed in a fixed position, the γ-ray source is moved at a constant velocity either towards or away from the sample. This has the effect of either increasing or decreasing the energy of the γ-rays incident upon the sample. In this way, the energy of the γ-rays may be varied and the γ-ray absorption spectrum of the sample determined. Only a limited number of isotopes can be induced to emit γ-rays suitable for Mössbauer work. The most widely used isotopes are ^{57}Fe and ^{119}Sn and, hence, Mössbauer spectroscopy has been most used on iron- and tin-containing substances. Other elements and isotopes which have been profitably studied include ^{129}I, ^{99}Ru and ^{121}Sb. A schematic illustration of the technique is given in Fig. 3.30.

Several types of information may be obtained from Mössbauer spectra. In the

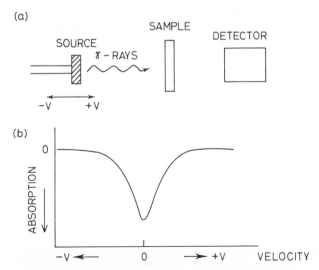

Fig. 3.30 (a) The Mössbauer effect. The energy of the γ-rays is modified when emitted from a moving source (Doppler effect). (b) Typical single line spectrum obtained when source and sample are identical

simplest case, in which both the emitter and sample are identical, the resonant absorption peak occurs when the source is stationary (Fig. 3.30b). When the emitter and sample are not identical, however, the absorption peak is shifted. This chemical shift, δ, arises because the nuclear energy levels in the atoms concerned have been modified by changes in the extra-nuclear density distribution in the atoms. In particular, chemical shifts may be correlated with the density at the nucleus of the outer shell s electrons. Chemical shifts are controlled principally by oxidation state, coordination number and the type of bonding. A summary given in Fig. 3.31 of the chemical shifts observed in various iron-containing ionic solids shows the influence of both charge and coordination number on the chemical shift. Use may be made of this for diagnostic purpose, e.g. to determine the nature of iron in inorganic compounds, minerals, etc.

For those nuclei that have a nuclear spin quantum number $I > \frac{1}{2}$, the distribution of positive charge inside the nucleus is non-spherical and a quadrupole moment, Q, results. The net effect of this is to cause splitting of the nuclear energy levels and hence splitting of the peaks in the Mössbauer spectrum. For both ^{57}Fe and ^{119}Sn, the peaks are split into doublets. The separation of the doublets, known as the quadrupole splitting, Δ, is, like the chemical shift, δ, sensitive to local structure and oxidation state.

A second type of line splitting which is important in the study of magnetic interactions is the magnetic hyperfine Zeeman splitting. This arises when a nucleus, of spin I, is placed in a magnetic field; each nuclear energy level splits into $2I + 1$ sublevels. The magnetic field usually arises from magnetic exchange effects in ferro-, antiferro- and paramagnetic samples or it can be an externally applied field.

Applications. The principle use of Mössbauer spectroscopy in solid state inorganic chemistry is for the determination of local structure. As well as the determination of oxidation state and coordination number, which has already been mentioned, the electronic configuration of, for example, iron, whether it has

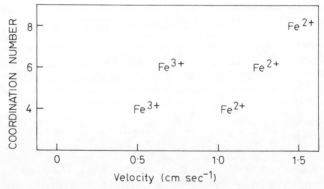

Fig. 3.31 Chemical shifts in iron-containing compounds. (From Bancroft, 1973)

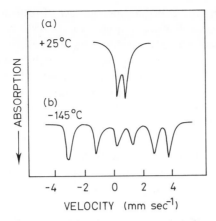

Fig. 3.32 Mössbauer spectrum of $KFeS_2$ (a) above and (b) below the Neel temperature. (From Greenwood, 1967)

a high or low spin, can be determined together with any site distortion away from, for example, octahedral or tetrahedral symmetry. This has been quite useful in mineralogy because of the common occurrence of iron in the earth's crust in a variety of minerals. It has also been a valuable aid to structure determinations since X-ray crystallography cannot normally distinguish between Fe^{2+} and Fe^{3+} ions. In favourable cases, the purity and quality of a crystalline substance may be checked. Thus extra peaks may appear due to the impurities, or the spectrum may show, for example, peak broadening, indicative of poor crystal quality.

The study of hyperfine splitting and, especially, its temperature dependence provides information on magnetic ordering. For instance, $KFeS_2$ is antiferromagnetic below 245 K and hyperfine splitting results in a spectrum containing six peaks. Above 245 K only quadrupole coupling occurs and the spectrum reduces to a doublet (Fig. 3.32).

Questions

3.1 Under what conditions may (i) optical microscopy, (ii) powder X-ray diffraction, be used to determine the purity of a solid material?

3.2 Compare the methods that may be used to identify and characterize crystalline samples of (a) iodoform, CHI_3, and (b) CdI_2.

3.3 How would lightly crushed samples of the following differ when viewed in a polarising microscope (a) SiO_2 glass, (b) quartz crystal, (c) NaCl, (d) Cu metal?

3.4 Calculate the wavelength of the electrons accelerated through 50 kV in an electron microscope.

3.5 Explain the following abbreviations: EXAFS, EELS, XRF, ESCA, MASNMR, SEM, TEM, AEM.

3.6 What technique(s) might you use to determine the following: (a) the oxidation state and coordination number of iron in brown/green bottle glass; (b) the coordination number of Cr^{3+} in a crystal of ruby; (c) the nature of a surface layer on a piece of aluminium metal; (d) the nature of some microscopic, crystalline inclusions in a polycrystalline, silicon carbide ceramic piece; (e) the Ca:Si ratio in particles of hydrated cement; (f) the location of the hydrogen atoms in a powdered sample of palladium hydride; (g) whether or not MnO possessed a magnetically ordered superstructure.

3.7 What are the approximate energy values, in units of (a) eV, (b) kJ mole^{-1}, (c) cm^{-1}, (d) Hz, for (i) changes of electron spin orientation (ii) d–d transitions in NiO, (iii) the K absorption edge in Cu metal, (iv) lattice vibrations in KCl.

3.8 Which microscopic technique(s) would you use to study (i) the texture of a piece of metal, (ii) the homogeneity of a sample of powdered glass, (iii) defects such as dislocations, stacking faults and twinning, (iv) the possible contamination of a sample of salt by washing soda?

3.9 Explain why it is generally much easier to characterize organic materials than non-molecular inorganic materials.

3.10 Suppose that you are trying to synthesize a new zeolite using methods outlined in Chapter 2 and that you end up with a white solid. How would you determine if it was a new zeolite, its structure, composition and purity.

References

G. E. Bacon (1962). *Neutron Diffraction*, Oxford.

G. M. Bancroft (1973). *Mössbauer Spectroscopy*, McGraw-Hill.

A. Bianconi, L. Incoccia and S. Stipcich (Eds) (1983). *EXAFS and Near Edge Structure*, Springer-Verlag.

L. S. Birks (1969). *X-ray Spectrochemical Analysis*, 2nd ed., Interscience.

D. Briggs (1977). *Handbook of X-ray and Ultraviolet Photoelectron Spectroscopy*, Heyden.

D. Brinkman et al. (1981). *Solid State Ionics*, 5, 434.

M. Campagna et al. (1975). *Rev. Lett. Phys.*, 34, 738.

S. I. Chan, V. W. Hu and R. C. Gamble (1978). *J. Mol. Str.*, 45, 239.

J. A. Duffy and M. D. Ingram (1976). An interpretation of glass chemistry in terms of the optical basicity concept. *J. Non-Cryst. Solids*, 21, 373.

LeRoy Eyring (1980). The application of high-resolution electron microscopy to problems in solid state chemistry, *J. Chem. Ed.*, 57, 565.

S. O. Farwell and D. R. Gage (1981). *Anal. Chem.*, 53, 1529A.

M. Greenblatt (1980). *J. Chem. Ed.*, 57, 546.

N. N. Greenwood (1967). The Mössbauer spectra of chemical compounds, *Chem. Brit.*, 3, 56.

S. J. German (1982). EXAFS studies in materials science, *J. Mat. Sci.*, 17, 1541.

N. H. Hartshorne and A. Stuart (1971). *Practical Optical Crystallography*, Arnold.

P. W. Hawkes (1972). *Electron Optics and Electron Microscopy*, Taylor and Francis, London.

J. M. Honig and C. N. R. Rao (Eds) (1981). *Preparation and Characterisation of Materials*, Academic Press.

Ron Jenkins (1974). *An Introduction to X-ray Spectrometry*, Heyden.

M. A. Krebs and R. A. Condrate, Sr., *J. Amer. Ceram. Soc.* (1982) C144.

E. Lippmaa *et al.* (1980). *J. Amer. Chem. Soc.*, **102**, 4889.

W. Wayne Meinke (1973). Characterisation of solids—chemical composition, in *Treatise on Solid State Chemistry* (Ed. N. B. Hannay), Vol. I, Plenum Press.

R. E. Newnham and Rustum Roy (1975). Structural characterisation of solids, in *Treatise on Solid State Chemistry* (Ed. N. B. Hannay), Vol. 2, p. 437, Plenum Press.

A. Rabenau (1982). *Solid State Ionics*, **6**, 277.

M. W. Roberts (1981). Photoelectron spectroscopy and surface chemistry, *Chem. Brit.*, **1981**, 510.

B. A. Scott and G. Burns (1972). *J. Amer. Ceram. Soc.*, **55** (5), 225.

K. Siegbahn *et al.* (1967). *ESCA : Atomic, Molecular and Solid State Structure Studied by Means of Electron Spectroscopy*, Almquist and Wicksells, Uppsala.

E. A. Stern (1976). *Sci. Amer.*, **234** (4), 96.

E. W. White and G. V. Gibbs (1969). *Amer. Miner.*, **54**, 931.

J. Zussman (Ed.) (1977). *Physical Methods in Determinative Mineralogy*, Academic Press.

Chapter 4

Thermal Analysis

Thermal analysis may be defined as the measurement of physical and chemical properties of materials as a function of temperature. In practice, however, the term thermal analysis is used to cover certain specific properties only. These are enthalpy, heat capacity, mass and coefficient of thermal expansion. Measurement of the coefficient of thermal expansion of metal bars is a simple example of thermal analysis. Another example is the measurement of the change in weight of oxysalts or hydrates as they decompose on heating. With modern equipment, a wide range of materials may be studied. Uses of thermal analysis in solid state science are many and varied and include the study of solid state reactions, thermal decompositions and phase transitions and the determination of phase diagrams. Most kinds of solid are 'thermally active' in one way or another and may be profitably studied by thermal analysis.

The two main thermal analysis techniques are *thermogravimetric analysis* (TGA), which automatically records the change in weight of a sample as a function of either temperature or time, and *differential thermal analysis* (DTA), which measures the difference in temperature, ΔT, between a sample and an inert reference material as a function of temperature; DTA therefore detects changes in heat content. A technique that is closely related to DTA is *differential scanning calorimetry* (DSC). In DSC, the equipment is designed to allow a quantitative measure of the enthalpy changes that occur in a sample as a function of either temperature or time. A fourth thermal analysis technique is *dilatometry*, in which the change in linear dimension of a sample as a function of temperature is recorded. Dilatometry has long been used to measure coefficients of thermal expansion of metals; recently, it has acquired a new name, *thermomechanical analysis* (TMA), and has been applied to more diverse materials and problems, e.g. to the quality control of polymers.

With modern, automatic thermal analysis equipment it is possible to do TGA, DTA and DSC using the same instrument; with some models, TGA and DTA may be carried out simultaneously. Thermal analysis equipment is necessarily rather complicated and expensive, in order that a wide variety of thermal events and properties may be studied, both rapidly and with high sensitivity and accuracy. The basic principles of operation of each technique are simple, however.

In this chapter, the basic principles of TGA, DTA and DSC are described, together with applications of each; instrumental descriptions are omitted.

4.1 Thermogravimetric analysis (TGA)

Thermogravimetry is a technique for measuring the change in weight of a substance as a function of temperature or time. The results usually appear as a continuous chart record; a schematic, typical, single step decomposition reaction is shown in Fig. 4.1. The sample, usually a few milligrams in weight, is heated at a constant rate, typically in the range 1 to $20\,°C\,min^{-1}$, and has a constant weight W_i, until it begins to decompose at temperature T_i. Under conditions of dynamic heating, decomposition usually takes place over a range of temperatures, T_i to T_f, and a second constant weight plateau is then observed above T_f, which corresponds to the weight of the residue W_f. The weights W_i, W_f and the difference in weight ΔW are fundamental properties of the sample and can be used for quantitative calculations of compositional changes, etc. By contrast, the temperatures T_i and T_f depend on variables such as heating rate, the nature of the solid (e.g. its particle size) and the atmosphere above the sample. The effect of atmosphere can be dramatic, as shown in Fig. 4.2 for the decomposition of $CaCO_3$: in vacuum, the decomposition is complete by $\sim 500\,°C$, but in CO_2 at

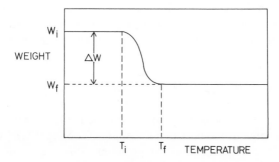

Fig. 4.1 Schematic thermogram for a single step decomposition reaction

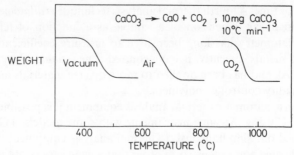

Fig. 4.2 Decomposition of CaCO₃ in different atmospheres

one atmosphere pressure, decomposition does not even commence until above 900 °C. T_i and T_f pertain to the particular experimental conditions, therefore, and do not necessarily represent equilibrium decomposition temperatures.

Applications of TGA are discussed later, together with those of DTA, since it is often advantageous to use both techniques to study a particular problem.

<h3 style="text-align:center">4.2 Differential thermal analysis (DTA) and differential
scanning calorimetry (DSC)</h3>

Differential thermal analysis is a technique in which the temperature of a sample is compared with that of an inert reference material during a programmed change of temperature. The temperature of sample and reference should be the same until some thermal event, such as melting, decomposition or change in crystal structure, occurs in the sample, in which case the sample temperature either lags behind (if the change is endothermic) or leads (if the change is exothermic) the reference temperature.

The reason for having both a sample and a reference is shown in Fig. 4.3. In (a), a sample is shown heating at a constant rate and its temperature, T_s, is monitored continuously with a thermocouple. The temperature of the sample as a function of time is shown in Fig. 4.3(b); the plot is linear until an endothermic event occurs in the sample, e.g. melting at temperature T_c. The sample temperature remains constant at T_c until the event is completed; it then increases rapidly to catch up with the temperature required by the programmer. The thermal event in the sample at T_c therefore appears as a rather broad deviation from the sloping baseline in (b). Such a plot is insensitive to small heat effects since the time taken to complete such processes may be short and hence the deviation from the baseline is small. Further, any spurious variations in the baseline, caused by, for example, fluctuations in the heating rate, would appear as apparent thermal events. Because of its insensitivity, this technique has only limited applications; its main use historically has been in the 'method of cooling curves' which was used to determine phase diagrams: the sample temperature was recorded on cooling rather than on heating and since the heat effects associated with solidification and crystallization are usually large, they could be detected by this method.

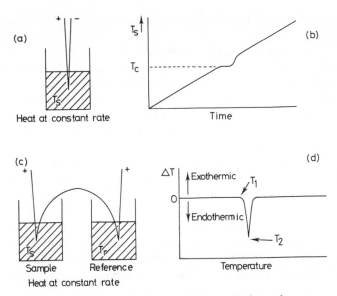

Fig. 4.3 The DTA method. Graph (b) results from the set-up shown in (a) and graph (d), a typical DTA trace, results from the arrangement shown in (c)

In Fig. 4.3(c), the arrangement normally used in DTA is shown. Sample and reference are placed side by side in a heating block which is either heated or cooled at a constant rate; identical thermocouples are placed in each and are connected 'back to back'. When the sample and reference are at the same temperature, the net output of this pair of thermocouples is zero. When a thermal event occurs in the sample, a temperature difference, ΔT, exists between the sample and reference which is detected by the net voltage of the thermocouples. A third thermocouple (not shown) is used to monitor the temperature of the heating block and the results are presented as ΔT against temperature (Fig. 4.3d). A horizontal baseline, corresponding to $\Delta T = 0$, occurs and superposed on this is a sharp peak due to the thermal event in the sample. The temperature of the peak is taken either as the temperature at which deviation from the baseline begins, T_1, or as the peak temperature, T_2. While it is probably more correct to use T_1, it is often not clear where the peak begins and, therefore, it is more common to use T_2. The size of the ΔT peak may be amplified so that events with very small enthalpy changes may be detected. Figure 4.3(d) is clearly, therefore, a much more sensitive and accurate way of presenting data than Fig. 4.3(b) and is the normal method for presenting DTA results.

Commercial DTA instruments are available which enable the temperature range -190 to $1600\,°C$ to be covered. Sample sizes are usually small, a few milligrams, because then there is less trouble with thermal gradients within the sample which could lead to reduced sensitivity and accuracy. Heating and cooling rates are usually in the range 1 to $50\,°C\,min^{-1}$. With the slower rates, the sensitivity is reduced because ΔT for a particular event decreases with decreasing heating rate.

DTA cells are usually designed for maximum sensitivity to thermal changes, but this is often at the expense of losing a calorimetric response; thus peak areas or peak heights are only qualitatively related to the magnitude of the enthalpy changes occurring. It is possible to calibrate DTA equipment so that quantitative enthalpy values from the peak areas can be obtained, but the calibration is usually tedious. If calorimetric data are required it is usually better and easier to use differential scanning calorimetry (DSC).

DSC is very similar to DTA. A sample and an inert reference are also used in DSC but the cell is designed differently. In some DSC cells, the sample and reference are maintained at the same temperature during the heating programme and the extra heat input to the sample (or to the reference if the sample undergoes an exothermic change) required in order to maintain this balance is measured. Enthalpy changes are therefore measured directly. In other DSC cells, the difference in temperature between the sample and reference is measured, as in DTA, but by careful attention to cell design the response of the cell is calorimetric.

4.3 Applications of DTA (DSC) and TGA

4.3.1 General comments

Uses of thermal analysis in solid state science are many and varied. Generally, DTA is more versatile than TGA: TGA detects effects which involve weight

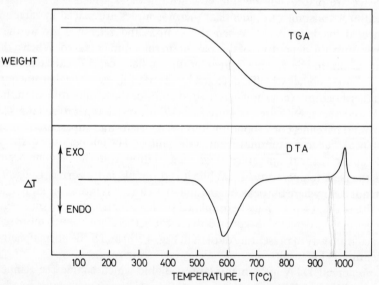

Fig. 4.4 Schematic TGA and DTA curves for kaolin minerals. Curves vary depending on the sample structure and composition, e.g. the TGA weight loss and associated DTA endotherm can occur anywhere in the range 450 to 750 °C

changes only. DTA also detects such effects, but, in addition, detects other effects, such as polymorphic transitions, which do not involve changes in weight. For many problems, it is advantageous to use both DTA and TGA because the DTA events can then be classified into those which do and those which do not involve weight change. An example is the decomposition of kaolin, $Al_4(Si_4O_{10})(OH)_8$ (Fig. 4.4). By TGA, a change in weight occurs at ~ 500 to $600\,°C$, which corresponds to dehydration of the sample; this dehydration also shows up on DTA as an endothermic event. A second DTA effect occurs at 950 to $980\,°C$, which has no counterpart in the TGA trace: it corresponds to a recrystallization reaction in the dehydrated kaolin. This latter DTA effect is exothermic, which is unusual; it means that the structure that is obtained between ~ 600 and $950\,°C$ is metastable and the DTA exotherm signals a decrease in enthalpy of the sample and therefore a change to a more stable structure (Chapter 12). The details of the structural changes that occur at this transition have still not been fully resolved.

Another useful ploy is to follow the thermal changes on cooling as well as on heating. This enables a separation of reversible changes, such as melting/solidification, from irreversible changes, such as most decomposition reactions. A schematic DTA sequence illustrating reversible and irreversible changes is shown in Fig. 4.5. Starting with a hydrated material, dehydration is the first event that occurs on heating and appears as an endotherm. The dehydrated material undergoes a polymorphic transition, which is also endothermic, at some higher temperature. Finally, the sample melts, giving a third endotherm. On cooling, the melt crystallizes, as shown by an exothermic peak, and the polymorphic change also occurs, exothermically, on cooling, but rehydration does not occur. The diagram shows two reversible and one irreversible process. It should be clear that

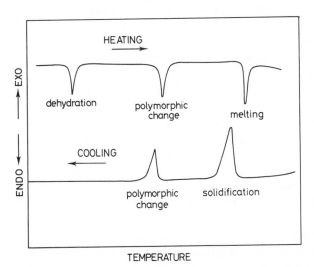

Fig. 4.5 Some schematic reversible and irreversible changes

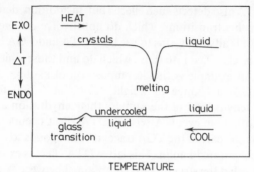

Fig. 4.6 Schematic DTA curves showing melting of crystals on heating and large hysteresis on cooling, which gives rise to glass formation

if a particular process, on heating, is endothermic, then the reverse process, on cooling, must be exothermic.

On studying reversible processes, which are observed on both heating and cooling, it is common to observe *hysteresis*; for instance, the exotherm that appears on cooling may be displaced to occur at lower temperatures than the corresponding endotherm which appears on heating. Ideally, the two processes should occur at the same temperature but hystereses ranging from a few degrees to several hundred degrees are commonplace. The two reversible changes in Fig. 4.5 are shown with small but definite hystereses.

Hysteresis depends not only on the nature of the material and the structural changes involved—difficult transitions involving, for example, the breaking of strong bonds are likely to exhibit much hysteresis—but also on the experimental conditions, such as the rates of heating and cooling. Hysteresis occurs particularly on cooling at relatively fast rates; in some cases, if the cooling rate is sufficiently fast, the change can be suppressed completely. The change is then effectively irreversible under those particular experimental conditions. An example of this which is of great industrial importance is associated with glass formation, as shown schematically in Fig. 4.6. Beginning with a crystalline substance, e.g. silica, an endotherm appears when the substance melts. On cooling, the liquid does not recrystallize but becomes supercooled; as the temperature drops so the viscosity of the supercooled liquid increases until eventually it becomes a glass. Hence, the crystallization has been supressed entirely; in other words, hysteresis is so great that crystallization has not occurred. In the case of SiO_2, the liquid is very viscous even above the melting point, $\sim 1700\,^{\circ}C$, and crystallization is slow, even at slow cooling rates.

4.3.2 Specific applications

4.3.2.1 *Glasses*

The previous section finished with a discussion on hysteresis in glass-forming systems. An important use of DTA and DSC in glasses is to measure the glass

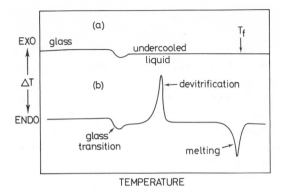

Fig. 4.7 DTA on heating of (a) a glass that does not devitrify and shows no thermal event apart from the glass transition and (b) a glass that devitrifies above T_g

transition temperature, T_g. This appears not as a clear peak but rather as a broad anomaly in the baseline of the DTA curve, as shown in Figs 4.6 and 4.7; T_g represents the temperature at which the glass transforms from a rigid solid to a supercooled, albeit very viscous, liquid. The glass transition is an important property of a glass since it represents the upper temperature limit at which the glass can really be used and also provides a convenient and readily measurable parameter for studying glasses. For glasses which kinetically are very stable, such as silica glass, the glass transition at T_g is the only thermal event observed on DTA since crystallization is usually too sluggish to occur (Fig. 4.7a). For other glasses, however, crystallization or devitrification may occur at some temperature above T_g and below the melting point, T_f. Devitrification appears as an exotherm and is followed by an endotherm at a higher temperature that corresponds to the melting of these same crystals (Fig. 4.7b). Examples of glasses which readily devitrify are metallic glasses, which may be prepared as thin films by rapidly quenching certain liquid alloy compositions. Other types of important glass-forming materials are amorphous polymers and amorphous chalcogenide semiconductors (Chapter 18).

4.3.2.2 *Polymorphic phase transitions and control of properties.*

Polymorphic phase transitions may be studied easily and accurately by DTA; since many physical or chemical properties of a particular sample may be modified or changed completely as a consequence of a phase transition, their study is extremely important. For example, it may be desired to prevent a transition from occurring in a particular material or to modify the temperature at which the transition occurs. Instead of designing or looking for completely new materials it is often better to modify the properties of existing materials by the formation of solid solutions with certain additives. Phase transition temperatures often vary greatly with solid solution composition and hence DTA may provide a sensitive monitor of both the property and the composition of the material.

Examples are:

(a) Ferroelectric $BaTiO_3$ has a Curie temperature of $\sim 120\,°C$ which may be determined by DTA; substitution of other ions for Ba^{2+} or Ti^{4+} causes the Curie temperature to vary.
(b) In cements, the β-polymorph of Ca_2SiO_4 has superior cementitious properties to the γ-polymorph. On cooling cement clinker as it comes out of the cement kiln, it is therefore desirable that the high temperature α'-polymorph should transform to β rather than to γ and various additives are used to ensure that this happens. The effect of different additives on the $\alpha'-\beta$ and $\alpha'-\gamma$ transitions can be studied by DTA.
(c) In refractories, transitions such as $\alpha \rightleftharpoons \beta$ quartz or quartz $\rightleftharpoons$ cristobalite have a deleterious effect on silica refractories because volume changes associated with each transition reduce the mechanical strength of the refractory. These transitions, which should be prevented from occurring if possible, may be monitored by DTA.

4.3.2.3 *Characterization of materials*

The occurrence of DTA effects can be used as a means of characterizing or analysing materials. If a substance is of completely unknown identity then it is unlikely that it can be identified from DTA alone, but DTA can be useful in picking out differences in a group of materials, e.g. the kaolin minerals mentioned above. It can also be used as a guide to purity in certain cases, e.g. the $\alpha-\gamma$ transition in iron is very sensitive to impurities: on addition of $0.02\,wt\%$ carbon, the transition temperature is reduced from 910 to $723\,°C$. Melting points are also often greatly affected by the presence of impurities, especially if the impurity gives rise to a low-melting eutectic. TGA may be used similarly to determine purity, by comparing the weight loss for decomposition of a particular substance with the value expected theoretically for decomposition of the pure compound.

4.3.2.4 *Phase diagram determination*

DTA is a powerful method for the determination of phase diagrams, especially when used in conjunction with other techniques, such as X-ray diffraction for the identification of the crystalline phases present. Its use is illustrated in Fig. 4.8(b) for two compositions in the simple binary eutectic system shown in Fig. 4.8(a). On heating composition A, melting begins to occur at the eutectic temperature, T_2, and gives rise to an endothermic peak. However, this is superposed on a much broader endothermic peak which terminates approximately at temperature T_1, and which is due to the continued melting that occurs over the temperature range T_2 to T_1. For this composition, an estimate of both solidus, T_2, and liquidus, T_1, temperatures may therefore be made. Composition B corresponds to the eutectic composition. On heating, this transforms completely to liquid at the eutectic temperature, T_2, and gives a single, large endotherm, peaking at T_2, on DTA.

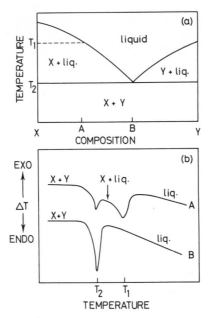

Fig. 4.8 Use of DTA for phase diagram determination; (a) a simple binary eutectic system; (b) schematic DTA traces for two compositions, A and B, on heating.

Thus, if DTA traces for a range of mixtures between X and Y could be compared, all should give an endotherm at T_2, whose magnitude depended on the degree of melting that occurred at T_2 and hence on the closeness of the sample composition to the eutectic composition, B. In addition, all compositions, apart from B, should give a broad endotherm peaking at some temperature above T_2; this is due to the completion of melting at the liquidus. The temperature of this peak should vary with composition.

Polymorphic phase transitions also appear on phase diagrams, at subsolidus temperatures, and these may be determined very readily by DTA, especially if solid solutions form and the transition temperature is composition dependent. Examples are given in Chapters 11 and 12.

A comment on the quality of baselines obtained on DTA traces may be made here. Ideally, horizontal baselines, as shown in Figs 4.3 to 4.7, are very much the exception in real systems. Baselines are often gently sloping, either up or down, and their slopes may change with temperature; also the baseline may be different on either side of a peak, especially if the peak represents a major event such as melting. Often peaks are preceded by a premonitory drift in the baseline which makes it very difficult to determine the onset temperature of a peak. Such premonitory phenomena may be associated with an increase in the concentration of crystal defects, i.e. increasing disorder, as the transition is approached and it is

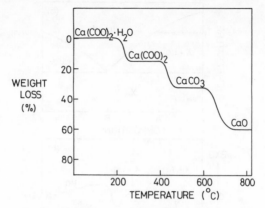

Fig. 4.9 Schematic, stepwise decomposition of calcium oxalate hydrate by TGA

difficult to separate, in the DTA trace, these premonitory phenomena from the actual onset of the transition.

4.3.2.5 *Decomposition pathways*

In multistage decomposition processes, TGA, either alone or in conjunction with DTA, may be used to separate and determine the individual steps. A well-known example, the decomposition of calcium oxalate monohydrate, is illustrated in Fig. 4.9, from which it can be seen that decomposition occurs in three stages giving, as intermediates, anhydrous calcium oxalate and calcium carbonate. Many other examples of similar multistage decompositions could be given in hydrates, hydroxides, oxysalts and minerals.

4.3.2.6 *Kinetics*

TGA and DTA may be used for various kinds of kinetic study. An accurate and rapid TGA method is to study decomposition reactions isothermally; the TGA furnace is arranged at a pre-set temperature and the sample introduced directly at this temperature. After allowing 2 to 3 minutes for the sample to equilibrate at the furnace temperature, the decomposition of the sample with time can be followed. The process may be repeated at other temperatures and the results analysed to determine reaction mechanisms, activation energies, etc.

A method which has considerable potential but which has some inherent difficulties in data analysis is a kinetic study based on a single, dynamic heating cycle of either TGA or DTA. The method is clearly very rapid but analysis of the results can be difficult since two variables, time and temperature, are effectively considered simultaneously. Provided the kinetic law for the process is inde-

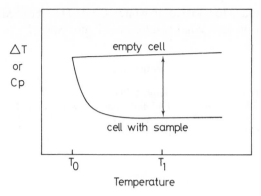

Fig. 4.10 Schematic measurement of heat capacity at a temperature T_1 well above the starting temperature T_0

pendent of temperature it is possible to obtain meaningful and reliable results, although the errors are probably larger than in isothermal measurements.

4.3.2.7 *Enthalpy and heat capacity measurements*

It has already been mentioned that DTA may be used to determine, semi-quantitatively, the enthalpies of transitions or reactions, provided that a suitable calibration of the instrument is made. For a given set of instrumental and experimental conditions, it is possible to obtain enthalpy values from the area of DTA peaks.

With DSC cells, or with DTA cells that have been designed for calorimetric response, such measurements may be made rather more accurately and, in addition, the heat capacity of substances or phases may be measured as a function of temperature. For this, a comparison is made between the baseline curve for an empty cell and that for a cell containing the sample, as shown schematically in Fig. 4.10. Details are not given since the calculation depends on the design of the instrument used and several designs are available commercially.

Questions

4.1 What kind of DTA and TGA traces would you expect to obtain on heating samples of the following until they became liquid: (i) beach sand, (ii) window glass, (iii) salt, (iv) washing soda, (v) epsom salts, (vi) metallic Ni, (vii) ferroelectric $BaTiO_3$, (viii) a clay mineral?

4.2 Which of the following would you expect to give a reversible DTA effect with or without hysteresis (i) melting of salt, (ii) decomposition of $CaCO_3$, (iii) melting of beach sand, (iv) oxidation of metallic Mg, (v) decomposition of $Ca(OH)_2$?

4.3 Salt is often added to icy roads in winter. Could DTA be used to quantify the effects of salt on ice? What results would you expect?

References

T. Daniels (1973). *Thermal Analysis*, Kogan Page.
R. C. Mackenzie (1970). *Differential Thermal Analysis*, Vols I and II, Academic Press.
M. I. Pope and M. D. Judd (1977). *Differential Thermal Analysis*, Heyden.
W. W. Wendlandt (1974). *Thermal Methods of Analysis*, Wiley.

Chapter 5

X-ray Diffraction

116

Undoubtedly the most important and useful technique in solid state chemistry, X-ray diffraction, has been in use since the early part of this century for the fingerprint characterization of crystalline materials and for the determination of their crystal structures. This chapter deals with the basic principles of diffraction and describes in some detail the powder method and its applications. A brief description of single crystal methods and their applications is given. The methods used for solving crystal structures are largely omitted but sections are included that together with Chapter 6 may help guide the reader through the crystallographic literature. It is important to be able to comprehend papers that report new crystal structures without necessarily going into the mathematics of the methods that are used to solve them.

5.1 X-rays and their generation

X-rays are electromagnetic radiation of wavelength $\sim 1\text{Å}$ (10^{-10}m). They occur in that part of the electromagnetic spectrum between γ-rays and the

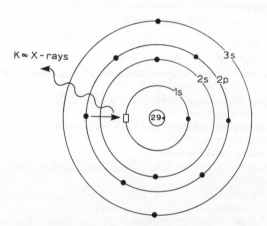

Fig. 5.1 Generation of Cu $K\alpha$ X-rays. A $2p$ electron falls into the empty $1s$ level ($\square$) and the excess energy is released as X-rays

ultraviolet. X-rays are produced when high energy charged particles, e.g. electrons accelerated through 30000 V, collide with matter. The electrons are slowed down or stopped by the collision and some of their lost energy is converted into electromagnetic radiation. Such processes give 'white radiation', X-rays which have wavelengths ranging upwards from a certain lower limiting value. This lower wavelength limit corresponds to the X-rays of highest energy and occurs when all the kinetic energy of the incident particles is converted into X-rays. It may be calculated from the formula, $\lambda_{min}(\text{Å}) = 12400/V$, where V is the accelerating voltage.

The X-rays which are used in almost all diffraction experiments are produced by a different process that leads to *monochromatic X-rays*. A beam of electrons, again accelerated through, say, 30 kV is allowed to strike a metal *target*, often copper. The incident electrons have sufficient energy to ionize some of the copper $1s$ (K shell) electrons (Fig. 5.1). An electron in an outer orbital ($2p$ or $3p$) immediately drops down to occupy the vacant $1s$ level and the energy released in the transition appears as X-radiation. The transition energies have fixed values and so a spectrum of characteristic X-rays results (Fig. 5.2). For copper, the $2p \rightarrow 1s$ transition, called $K\alpha$, has a wavelength of 1.5418 Å and the $3p \rightarrow 1s$ transition, $K\beta$, 1.3922 Å. The $K\alpha$ transition occurs much more frequently than the $K\beta$ transition and it is this more intense $K\alpha$ radiation which results that is used in diffraction experiments. In fact, the $K\alpha$ transition is a doublet, $K\alpha_1 = 1.54051$ Å and $K\alpha_2 = 1.54433$ Å, because the transition has a slightly different energy for the two possible spin states of the $2p$ electron which makes the transition, relative to the spin of the vacant $1s$ orbital. In some X-ray experiments, diffraction by the $K\alpha_1$ and $K\alpha_2$ radiations is not resolved and a single line or spot is observed instead of a doublet (e.g. in powder diffractometry at low angle). In other experiments, separate diffraction peaks may be observed; if desired, this can be overcome by removing the weaker $K\alpha_2$ beam from the incident radiation (Sections 5.6.1 and 5.6.2).

The wavelengths of the $K\alpha$ lines of the target metals commonly used for X-ray generation are given in Table 5.1. The wavelengths are related to the atomic number, Z, of the metal, by *Moseley's Law*:

$$f^{1/2} = \left(\frac{c}{\lambda}\right)^{1/2} \propto Z \tag{5.1}$$

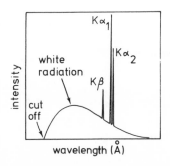

Fig. 5.2 X-ray emission spectrum of a metal, e.g. copper

Table 5.1 *X-ray wavelengths of commonly used target materials*

Target	$K\alpha_1$	$K\alpha_2$	$K\bar{\alpha}$*	Filter
Cr	2.2896	2.2935	2.2909	V
Fe	1.9360	1.9399	1.9373	Mn
Cu	1.5405	1.5443	1.5418	Ni
Mo	0.7093	0.7135	0.7107	Nb.
Ag	0.5594	0.5638	0.5608	Pd

* $\bar{\alpha}$ is the intensity-weighted average of α_1 and α_2.

where f is the frequency of the $K\alpha$ line. Hence the wavelength of the $K\alpha$ line decreases with increasing atomic number.

The X-ray emission spectrum of an element such as copper (Fig. 5.2) has two main features. The intense, monochromatic peaks, caused by electronic transitions within the atoms, have wavelengths that are characteristic of the element i.e. copper. These monochromatic peaks are superposed on a background of 'white' radiation, mentioned earlier, which is produced by the general interaction of high velocity electrons with matter. In order to generate the characteristic monochromatic radiation, the voltage used to accelerate the electrons needs to be sufficiently high ($\gtrsim 10$ kV) so that ionization of the copper $1s$ electrons may occur.

In the generation of X-rays (Fig. 5.3), the electron beam, provided by a heated tungsten filament, is accelerated towards an anode by a potential difference of ~ 30 kV. The electrons strike the target, a piece of copper fixed to the anode, and a spectrum of X-rays, such as shown in Fig. 5.2, is emitted. The chamber, known as the *X-ray tube*, is evacuated in order to avoid collisions between air particles and either the incident electrons or emitted X-rays. The X-rays leave the tube through 'windows' made of beryllium. The absorption of X-rays on passing through materials depends on the atomic weight of the elements present in the material. Beryllium with an atomic number of 4 is, therefore, one of the most suitable window materials. For the same reason, lead is a very effective material for shielding X-ray equipment and absorbing stray radiation. While an X-ray tube is in operation, continuous cooling of the anode is necessary. Only a small fraction of the energy of the incident electron beam is converted into X-rays. Most of the

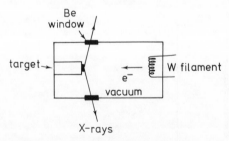

Fig. 5.3 Schematic design of a filament X-ray tube

energy is converted into heat and the anode would soon melt if it were not cooled.

For most diffraction experiments, a monochromatic beam of X-rays is desired and not a continuous spectrum. In the spectrum of X-rays emitted by copper (or any metal), the $K\alpha$ line(s) is the most intense and it is desired to filter out all the other wavelengths, leaving the $K\alpha$ line for diffraction experiments. For copper radiation, a sheet of nickel foil is very effective in carrying out this separation. The energy required to ionize $1s$ electrons of nickel corresponds to a wavelength of 1.488 Å, which lies between the values for the $K\alpha$ and $K\beta$ lines of the copper emission spectrum. Cu $K\beta$ radiation, therefore, has sufficient energy to ionize $1s$ electrons of nickel whereas Cu $K\alpha$ radiation does not. Nickel foil is effective in absorbing the Cu $K\beta$ radiation and most of the white radiation, leaving a monochromatic, reasonably clean beam of $K\alpha$ radiation. A lighter element, such as iron, would absorb $K\alpha$ radiation as well as $K\beta$, because its *absorption edge* is displaced to higher wavelengths. On the other hand, a heavier element, such as zinc, would transmit both $K\alpha$ and $K\beta$ radiations while still absorbing much of the higher energy white radiation. The atomic number of the element in the filter generally is one or two less than that of the target material, (Table 5.1). An alternative method of obtaining monochromatic X-rays uses a single crystal monochromator and is discussed in Section 5.6.2.

5.2 Diffraction

5.2.1 An optical grating and diffraction of light

As an aid to understanding the diffraction of X-rays by crystals, let us consider the diffraction of light by an optical grating. This gives a 1-dimensional analogue of the three-dimensional process that occurs in crystals.

An optical grating may consist of a piece of glass on which have been ruled a large number of accurately parallel and closely spaced lines. The separation of the lines should be a little larger than the wavelength of light, say 10000 Å. The grating is shown in projection as a row of points in Fig. 5.4. Consider what happens to a beam of light which hits the grating perpendicular to the plane of the grating. A piece of glass without the lines would simply transmit the light, but in the grating the lines act as secondary point (or, rather, line) sources of light and re-

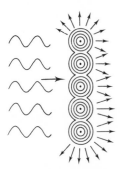

Fig. 5.4 Lines on an optical grating act as secondary sources of light

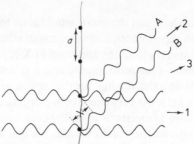

Fig. 5.5 Constructive interference in directions 1 and 2

radiate light in all directions. Interference then occurs between the waves originating from each line source. In certain directions, adjacent beams are in phase with each other and *constructive interference* occurs to give a resultant diffracted beam in that direction. Two such directions are shown in Fig. 5.5. In direction 1, parallel to the incident beam, the diffracted beams are obviously in phase. In direction 2, the beams are also in phase although beam B is now exactly one wavelength behind beam A. At directions between 1 and 2, beam B lags behind beam A by a fraction of one wavelength, and *destructive interference* occurs. For a certain direction, 3, B is exactly half a wavelength behind A and complete destructive interference or *cancellation* occurs. For other directions between 1 and 2, the destructive interference is only partial. Thus, directions 1 and 2 have maximum light intensity and this falls off gradually to zero as the angle changes to direction 3. In the optical grating, however, there are not just two parallel diffracted beams A and B but several hundred or thousand, one for each line on the grating. This causes the resultant diffracted beams to sharpen enormously after interference so that intense beams occur in directions 1 and 2 with virtually no intensity over the whole range of directions between 1 and 2.

The directions, such as 2, in which constructive interference occurs are governed by the wavelength of the light, λ, and the separation, a, of the lines on the grating. Consider the diffracted beams 1 and 2 (Fig. 5.6) which are at an angle, ϕ, to the direction of the incident beam. If 1 and 2 are in phase with each other, the

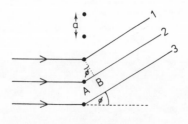

Fig. 5.6 Diffraction of light by an optical grating

distance AB must equal a whole number of wavelengths; i.e.

$$AB = \lambda, 2\lambda, \ldots, n\lambda$$

But

$$AB = a \sin \phi$$

Therefore,

$$a \sin \phi = n\lambda \qquad (5.2)$$

This equation gives the conditions under which constructive interference occurs and relates the spacing of the grating to the light wavelength and the diffraction order, n. Hence, depending on the value of $a \sin \phi$, one or more diffraction orders, corresponding to $n = 1$, 2, etc., may be observed.

We can now understand why the separation of the lines on the grating must be of the same order of magnitude as, but somewhat larger than, the wavelength of light. The condition for the first order diffracted beam to occur is $a \sin \phi = \lambda$. The maximum value of $\sin \phi$ is 1, corresponding to $\phi = 90°$ but realistically, in order to observe first order diffraction, $\sin \phi < 1$ and, therefore, $a > \lambda$. If $a < \lambda$, only the zero order direct beam is observable.

If, on the other hand, $a \gg \lambda$, individual diffraction orders ($n = 1, 2, 3, \ldots$, etc.) are so close together as to be unresolved and, effectively, a diffraction continuum results. This is because, for large values of a, $\sin \phi$ and, hence, ϕ must be very small. Therefore $\phi_{n=1} \gtrsim 0$ and the first order beam is not distinguishable from the primary beam. Visible light has wavelengths in the range 4000 to 7000 Å and so, in order to observe well-separated spectra, grating spacings are usually 10000 to 20000 Å.

The other condition to be observed in the construction of an optical grating is that the lines should be accurately parallel. If this were not so, ϕ would vary over the grating and the diffraction spectra would be blurred or irregular and of poor quality generally.

5.2.2 Crystals and diffraction of X-rays

By analogy with the diffraction of light by an optical grating, crystals, with their regularly repeating structures, should be capable of diffracting radiation that has a wavelength similar to the interatomic separation, ~ 1 Å. Three types of radiation are used for crystal diffraction studies: X-rays, electrons and neutrons. Of these, X-rays are by far the most useful but electron and neutron diffraction both have important specific applications and are discussed in Chapter 3.

The X-ray wavelength commonly employed is the characteristic $K\alpha$ radiation, $\lambda = 1.5418$ Å, emitted by copper. When crystals diffract X-rays, it is the atoms or ions which act as secondary point sources and scatter the X-rays; in the optical grating, it is the lines scratched or ruled on the glass surface which cause scattering.

Historically, two approaches have been used to treat diffraction by crystals. These are as follows.

5.2.2.1 *The Laue equations*

Diffraction from a hypothetical 1-dimensional crystal, constituting a row of atoms, may be treated in the same way as diffraction of light by an optical grating because, in projection, the grating is a row of points. An equation is obtained which relates the separation, a, of the atoms in the row, the X-ray wavelength, λ, and the diffraction angle, ϕ; i.e.

$$a \sin \phi = n\lambda$$

A real crystal is a three-dimensional arrangement of atoms for which three *Laue equations* may be written:

$$a_1 \sin \phi_1 = n\lambda$$
$$a_2 \sin \phi_2 = n\lambda$$
$$a_3 \sin \phi_3 = n\lambda$$

Each equation corresponds to the diffraction condition for rows of atoms in one particular direction and three directions or axes are needed in order to represent the atomic arrangement in the crystal. For a diffracted beam to occur, these three equations must all be satisfied simultaneously.

The Laue equations provide a rigorous and mathematically correct way to describe diffraction by crystals. The drawback is that they are cumbersome to use. The alternative theory of diffraction, based on Bragg's Law, is much simpler and is used almost universally in solid state chemistry. No further discussion of the Laue equations is given in this book.

5.2.2.2 *Bragg's Law*

The Bragg approach to diffraction is to regard crystals as built up in layers or planes such that each acts as a semi-transparent mirror. Some of the X-rays are reflected off a plane with the angle of reflection equal to the angle of incidence, but the rest are transmitted to be subsequently reflected by succeeding planes.

The derivation of Bragg's Law is shown in Fig. 5.7. Two X-ray beams, 1 and 2, are reflected from adjacent planes, A and B, within the crystal and we wish to

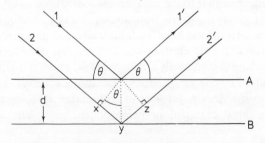

Fig. 5.7 Derivation of Bragg's Law for X-ray diffraction

know under what conditions the reflected beams 1′ and 2′ are in phase. Beam 22′ has to travel the extra distance xyz as compared to beam 11′, and for 1′ and 2′ to be in phase, distance xyz must equal a whole number of wavelengths. The perpendicular distance between pairs of adjacent planes, the *d-spacing*, *d*, and the angle of incidence, or *Bragg angle*, θ, are related to the distance xy by

$$xy = yz = d \sin \theta$$

Thus

$$xyz = 2d \sin \theta$$

But

$$xyz = n\lambda$$

Therefore

$$2d \sin \theta = n\lambda \qquad Bragg's\ Law \qquad (5.3)$$

When Bragg's Law is satisfied, the reflected beams are in phase and interfere constructively. At angles of incidence other than the Bragg angle, reflected beams are out of phase and destructive interference or cancellation occurs. In real crystals, which contain thousands of planes and not just the two shown in Fig. 5.7, Bragg's Law imposes a stringent condition on the angles at which reflection may occur. If the incident angle is incorrect by more than a few tenths of a degree, cancellation of the reflected beams is usually complete.

For a given set of planes, several solutions of Bragg's Law are usually possible, for $n = 1, 2, 3$, etc. It is customary, however, to set n equal to 1 and for situations where, say, $n = 2$, the d-spacing is instead halved by doubling up the number of planes in the set; hence n is kept equal to 1. (Note that $2\lambda = 2d \sin \theta$ is equivalent to $\lambda = 2(d/2) \sin \theta$.)

It is difficult to give an explanation of the nature of the semi-transparent layers or planes that is immediately convincing. This is because they are a concept rather than a physical reality. Crystal structures, with their regularly repeating patterns, may be referred to a three-dimensional grid and the repeating unit of the grid, the *unit cell* (see the next section), can be found. The grid may be divided up into sets of planes in various orientations and it is these planes which are considered in the derivation of Bragg's Law. In some cases, with simple crystal structures, the planes do correspond to layers of atoms, but this is not generally the case.

Some of the assumptions upon which Bragg's Law is based may seem to be rather dubious. For instance, it is known that diffraction occurs as a result of interaction between X-rays and atoms. Further, the atoms do not *reflect* X-rays but scatter or diffract them in all directions. Nevertheless, the highly simplified treatment that is used in deriving Bragg's Law gives exactly the same answers as are obtained by a rigorous mathematical treatment. We therefore happily use terms such as reflexion (often deliberately spelt incorrectly!) and bear in mind that we are fortunate to have such a simple and picturesque, albeit inaccurate, way to describe what in reality is a very complicated process. Further discussion of Bragg's Law and diffraction must wait until we have considered some basic rules and definitions about the symmetry and structures of crystals.

5.3 Definitions

5.3.1 Unit cells and crystal systems

Crystals are built up of a regular arrangement of atoms in three dimensions; this arrangement can be represented by a repeat unit or motif called the *unit cell*. The unit cell is defined as *the smallest repeating unit which shows the full symmetry of the crystal structure*. Let us see exactly what this means, first of all in two dimensions. A section through the NaCl structure is shown in Fig. 5.8(a). Possible repeating units are given in Fig. 5.8(b) to (e). In each, the repeat unit is a square and adjacent squares share edges and corners. Adjacent squares are identical, as they must be by definition; thus, all the squares in (b) have Cl⁻ ions at their corners and centres. The repeat units in (b), (c) and (d) are all of the same size and, in fact, differ only in their relative position. This brings us to an important point. The choice of origin of the repeating unit is to a certain extent a matter of personal taste, even though the size and shape or orientation of the cell are fixed. The repeat unit of NaCl is usually chosen as either (b) or (c) rather than (d) because it is easier to draw the unit and visualize the structure as a whole if the unit contains atoms or ions at special positions such as corners, edge centres, etc. Another guideline is that the origin is usually chosen so that the symmetry of the structure is evident (Section 5.3.3).

In the hypothetical case that two-dimensional crystals of NaCl could be formed, the repeat unit shown in Fig. 5.8(e) or its equivalent, with chlorine at the corners and sodium in the middle, would be the correct unit. Comparing (e) and, for example, (b), both are square and show the symmetry of the structure; as (e) is half the size of (b), (e) would be preferred according to the definition of the unit cell

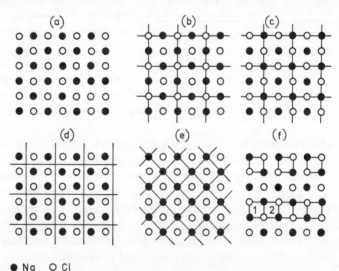

● Na   ○ Cl

Fig. 5.8 (a) Section through the NaCl structure, showing (b) to (e) possible repeat units and (f) incorrect units

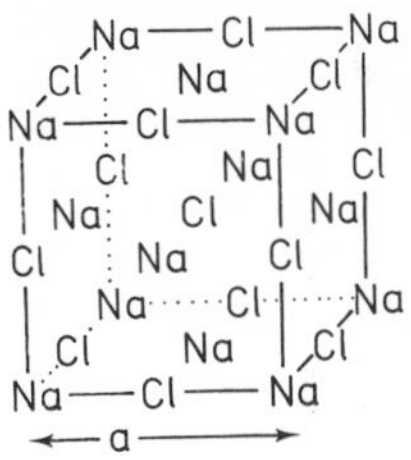

Fig. 5.9 Cubic unit cell of NaCl, $a = b = c$

given above. In three dimensions, however, the unit cell of NaCl is based on (b) rather than (e) because only (b) shows the cubic symmetry of the structure (Section 5.3.3).

In Fig. 5.8(f) are shown two examples of what is *not* a repeat unit. The top part of the diagram contains isolated squares whose area is one quarter of the squares in (b). It is true that each square is identical but it is not permissible to isolate unit cells or areas from each other, as appears to happen here. The bottom part of the diagram contains units that are not identical; thus square 1 has a sodium in its top right corner whereas 2 has a chlorine in this position.

The unit cell of NaCl in three dimensions in shown in Fig. 5.9; it contains Na^+ ions at the corners and face centre positions with Cl^- ions at the edge centres and body centre. Each face of the unit cell looks like the unit area shown in Fig. 5.8(c). As in the two-dimensional case, the choice of origin is somewhat arbitrary and an equally valid unit cell could be chosen in which the Na^+ and Cl^- ions were interchanged. The unit cell of NaCl is cubic. The three edges of the cell, a, b and c are equal in length. The three angles of the cell α (between b and c), β (between a and c) and γ (between a and b) are all equal to 90°. A cubic unit cell also possesses certain symmetry elements, and these symmetry elements together with the shape define the cubic unit cell.

The seven *crystal systems* listed in Table 5.2 are the seven independent unit cell shapes that are possible in three-dimensional crystal structures. Each crystal system is governed by the presence or absence of symmetry in the structure and the essential symmetry for each is given in the third column. Let us next deal with symmetry because it is of fundamental importance in solid state chemistry and, especially, in crystallography.

5.3.2 Symmetry, point symmetry and point groups

Symmetry is most easily defined by the use of examples. Consider the silicate tetrahedron shown in Fig. 5.10(a). If it is rotated about an axis passing along one of the Si—O bonds, say the vertical one, then every 120° the tetrahedron finds itself in an identical position. Effectively, the three basal oxygens change position with each other every 120°. During a complete 360° rotation, the tetrahedron passes through three such identical positions. The fact that different (i.e. > 1) identical orientations are possible means that the SiO_4 tetrahedron possesses symmetry. The axis about which the tetrahedron may be rotated is called a *symmetry element* and the process of rotation a *symmetry operation*.

Table 5.2 *The seven crystal systems*

Crystal system	Unit cell shape[†]	Essential symmetry	Space lattices
Cubic	$a = b = c,\ \alpha = \beta = \gamma = 90°$	Four threefold axes	P, F, I
Tetragonal	$a = b \neq c,\ \alpha = \beta = \gamma = 90°$	One fourfold axis	P, I
Orthorhombic	$a \neq b \neq c,\ \alpha = \beta = \gamma = 90°$	Three twofold axes or mirror planes	P, F, I, A(B or C)
Hexagonal	$a = b \neq c,\ \alpha = \beta = 90°, \gamma = 120°$	One sixfold axis	P
Trigonal (a)	$a = b \neq c,\ \alpha = \beta = 90°, \gamma = 120°$	One threefold axis	P
(b)	$a = b = c,\ \alpha = \beta = \gamma \neq 90°$	One threefold axis	R
Monoclinic*	$a \neq b \neq c,\ \alpha = \gamma = 90°, \beta \neq 90°$	One twofold axis or mirror plane	P, C
Triclinic	$a \neq b \neq c,\ \alpha \neq \beta + \gamma \neq 90°$	None	P

* Two settings of the monoclinic cell are used in the literature. The one given here, which is most commonly used, and the other $a \neq b \neq c,\ \alpha = \beta = 90°, \gamma \neq 90°$.

[†] The symbol $\neq$ means not necessarily equal to. Sometimes, crystals possess *pseudo-symmetry* in which, say, the unit cell is geometrically cubic but does not possess the essential symmetry elements for cubic symmetry, and the symmetry is lower, perhaps tetragonal.

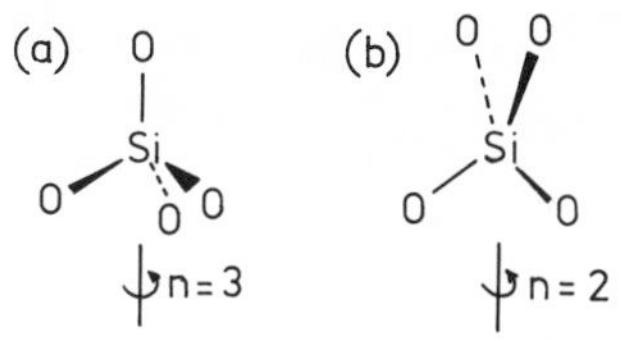

Fig. 5.10 (a) Threefold and (b) twofold rotation axes

The symmetry elements that are important in crystallography are listed in Table 5.3. There are two nomenclatures for labelling symmetry elements, the Hermann–Mauguin system used in crystallography and the Schönflies system used in spectroscopy. It would be ideal if there was only one system which everybody used, but this is unlikely to come about since (a) both systems are very well established and (b) crystallographers require elements of space symmetry that spectroscopists do not, and vice versa, (c) spectroscopists use a more extensive range of point symmetry elements than crystallographers.

The symmetry element described above for the silicate tetrahedron is a *rotation axis*, with symbol n. Rotation about this axis by $360/n$ degrees gives an identical orientation and the operation is repeated n times before the original configuration is regained. In this case, $n = 3$ and the axis is a *threefold rotation axis*. The SiO_4 tetrahedron possesses four threefold rotation axes, one in the direction of each Si—O bond.

When viewed from another angle, SiO_4 tetrahedra possess twofold rotation axes (Fig. 5.10b) which pass through the central silicon and bisect the O—Si—O bonds. Rotation by 180° leads to an indistinguishable orientation of the tetrahedron. The tetrahedron possesses three of these twofold axes. (According to the Schönflies system, there are six twofold axes; each axis is counted twice, to include the two possible directions of each axis, i.e. up or down). The *identity operation* corresponds to $n = 1$ (rotation by 360°). It is apparently trivial, being equivalent to doing nothing, but it is important in the application of group theory to point symmetry.

Table 5.3 *Symmetry elements*

	Symmetry element	Notation	
		Hermann–Mauguin (crystallography)	Schönflies (spectroscopy)
Point symmetry	Mirror plane	m	σ_v, σ_h
	Rotation axis	$n(= 2, 3, 4, 6)$	$C_n(C_2, C_3, \text{etc.})$
	Inversion axis	$\bar{n}(= \bar{1}, \bar{2}, \text{etc.})$	—
	Alternating axis (rotoreflection)	—	$S_n(S_1, S_2, \text{etc.})$
	Centre of symmetry	$\bar{1}$	i
Space symmetry	Glide plane	n, d, a, b, c	—
	Screw axis	$2_1, 3_1, \text{etc.}$	—

128

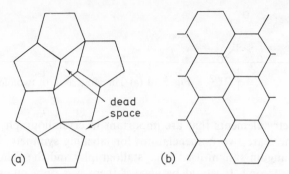

Fig. 5.11 (a) The impossibility of forming a close packed array of pentagons. (b) A close packed layer of hexagons.

Crystals may display rotational symmetries of 1, 2, 3, 4 and 6. Others, such as $n = 5, 7$, are never observed. This is not to say that molecules which have pentagonal symmetry cannot exist in the crystalline state. They can, of course, but their fivefold symmetry cannot be exhibited by the crystal as a whole. This is shown in Fig. 5.11(a), where a fruitless attempt has been made to pack together pentagons to form a complete layer; for hexagons with sixfold rotation axes (b) a close packed layer is easily produced.

A *mirror plane, m*, exists when two halves of a molecule or ion can be interconverted by carrying out the imaginary process of reflection across the mirror plane. The silicate tetrahedron possesses three mirror planes, one of which is shown in Fig. 5.12(a). The silicon and two oxygens lie on the mirror plane and are unaffected by the process of reflection. The other two oxygens are interchanged on reflection.

The *centre of symmetry*, $\bar{1}$, exists when any part of a molecule or ion can be reflected through this centre of symmetry, which is a point, and an identical arrangement found on the other side. An $Si_2O_7^{6-}$ group which has a linear Si–O–

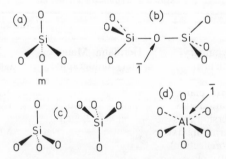

Fig. 5.12 Symmetry elements: (a) mirror plane, (b) centre of symmetry, (c) absence of centre of symmetry in a tetrahedron, (d) centre of symmetry in an octahedron

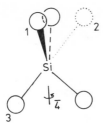

Fig. 5.13 Fourfold inversion axis

Si bridge (it is not usually linear) and with the two SiO_4 tetrahedra in the staggered conformation has a centre of symmetry at the bridging oxygen (Fig. 5.12b). If a line is drawn from any oxygen, through the centre of symmetry and extended an equal distance on the other side, it terminates at another oxygen. A single tetrahedron (e.g. SiO_4) does not have a centre of symmetry (located at Si); the two orientations of the tetrahedra in Fig. 5.12(c) are not identical. On the other hand an octahedron, e.g. AlO_6, is centrosymmetric (Fig. 5.12d).

The *inversion axis*, $\bar{n}$, is a combined symmetry operation involving rotation (according to n) and inversion through the centre. A $\bar{4}$ (fourfold inversion) axis is shown in Fig. 5.13. The first stage involves rotation by $360/4 = 90°$ and takes, for example, oxygen 1 to position 2. This is followed by inversion through the centre, at Si, and leads to the position of oxygen 3. Oxygens 1 and 3 are therefore related by a $\bar{4}$ axis. Possible inversion axes in crystals are limited to $\bar{1}, \bar{2}, \bar{3}, \bar{4}$ and $\bar{6}$ for the same reason that only certain pure rotation axes are allowed. The onefold inversion axis is simply equivalent to the centre of symmetry and the twofold inversion axis is the same as a mirror plane perpendicular to that axis.

The symmetry elements discussed so far are elements of *point symmetry*. For all of them, at least one point stays unchanged during the symmetry operation, i.e. an atom lying on a centre of symmetry, a rotation axis or a mirror plane does not move during the respective symmetry operations. Finite-sized molecules can only possess point symmetry elements, whereas crystals may have extra symmetries that include translation steps as part of the symmetry operation. Many molecules and crystals possess more than one element of point symmetry, but the number of combinations of symmetry elements that may occur in crystals is limited to 32. These are known as the crystallographic *point groups*.

5.3.3 Symmetry: the choice of unit cell and crystal system

The geometric shapes of the various crystal systems (unit cells) are listed in Table 5.2. These shapes do not *define* the unit cell; they are merely a consequence of the presence of certain symmetry elements. A cubic unit cell is defined as one having four three-fold symmetry axes (Fig. 5.14) and it is an automatic consequence of this condition that $a = b = c$ and $\alpha = \beta = \gamma = 90°$. The *essential* symmetry elements by which each crystal system is defined are listed in Table 5.2. In most crystal systems, other symmetry elements are also present. For instance, cubic crystals have many others, including three four-fold axes passing through the centres of each pair of opposite cube faces (Fig. 5.14).

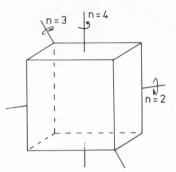

Fig. 5.14 Two-, three- and fourfold axes of a cube

The *tetragonal* unit cell is characterized by a single fourfold axis and is exemplified by the structure of CaC_2. This is related to the NaCl structure but, because the carbide ion is cigar-shaped rather than spherical, one of the cell axes becomes longer than the other two (Fig. 5.15a). A similar tetragonal cell may be drawn for NaCl by replacing Na for Ca and Cl for C_2; it occupies half the volume of the true, cubic unit cell (Fig. 5.15b). The choice of a tetragonal unit cell for NaCl is rejected because it does not show the full cubic symmetry of the crystal (see Fig. 5.8e and the discussion with it).

The *trigonal* system is characterized by a single threefold axis. Its shape can be derived from that of a cube by stretching or compressing the cube along one of its body diagonals (Fig. 5.16). The threefold axis parallel to this body diagonal is retained but those along the other body diagonal directions are all destroyed. All three cell edges remain the same length and all three angles stay the same but are not equal to $90°$. It is possible to describe such a trigonal cell for NaCl with $\alpha = \beta = \gamma = 60°$ where Na^+ ions are at the corners and a Cl^- ion is in the body centre, but this is again unacceptable because NaCl has symmetry higher than trigonal. $NaNO_3$ has a structure that may be regarded as a trigonal distortion of the NaCl structure: instead of spherical Cl^- ions, it has triangular nitrate groups, and the presence of these effectively causes a compression along one body diagonal (or

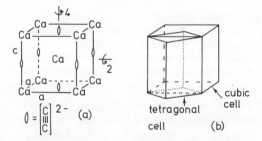

Fig. 5.15 (a) Tetragonal unit cell of CaC_2; (b) relation between tetragonal and cubic cells for NaCl

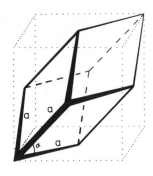

Fig. 5.16 Derivation of a trigonal unit cell from a cubic cell

rather an expansion in the plane perpendicular to the diagonal). All fourfold symmetry axes and all but one of the threefold axes are destroyed.

The trigonal crystal system is one of the most difficult to work with. The trigonal cell can be defined with either rhombohedral axes (as above) or with hexagonal axes (Table 5.2). There has been a controversy among crystallographers lasting for many years over the exact status of the trigonal crystal system. Some claim that it should not be regarded as a separate system but should be included as a subsystem of the hexagonal system. The majority accord it with independent status, however, and it has been treated as such in Table 5.2.

The *hexagonal* crystal system is discussed in some detail in Chapter 7, where a drawing of the hexagonal unit cell is given in Fig. 7.6.

The *orthorhombic* unit cell may be regarded as something like a shoebox in which the angles are all 90° but the sides are of unequal length. It usually possesses several mirror planes of symmetry and several twofold axes; the minimum requirement for orthorhombic symmetry is the presence of three mutually perpendicular mirror planes or twofold axes.

The *monoclinic* unit cell may be regarded as derived from our orthorhombic shoebox by a shearing action in which the top face is partially sheared relative to the bottom face and in a direction parallel to one of the box edges. As a consequence, one of the angles departs from 90° and most of the symmetry is lost, apart from a mirror plane and/or a single twofold axis.

The *triclinic* system possesses no symmetry at all, which is reflected in the shape of the unit cell.

5.3.4 Space symmetry and space groups

The symmetry of finite-sized molecules is limited to the elements of point symmetry whereas crystals, with their infinite repeating structures (for all practical purposes), may also possess elements of *space symmetry*. This involves a combination of point symmetry elements—rotation or mirror plane reflection—with incremental translational steps through the structure.

The *screw axis* combines translation and rotation; the atoms or ions in a crystal which possesses screw axes appear to lie on spirals about these axes. A schematic example is shown in Fig. 5.17(a). The symbol for a screw axis, X_Y,

132

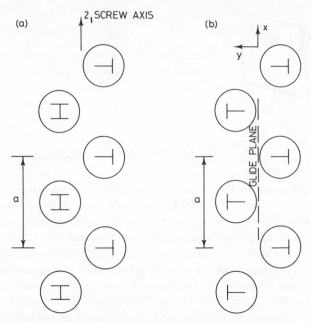

Fig. 5.17 Arrangements of coins with heads (H) and tails (T), illustrating (a) a 2_1 screw axis parallel to a and (b) an a glide plane perpendicular to b

indicates a translation by the fraction Y/X of the unit cell edge in the direction of the screw axis, together with a simultaneous rotation by $\frac{1}{X} \times 360°$ about the screw axis. Thus, a 4_2 axis parallel to a involves translation by $a/2$ and rotation by $90°$, and this process is repeated twice for every unit cell.

The *glide plane* combines translation and reflection and is shown schematically in Fig. 5.17(b). Translation may be parallel to any of the unit cell axes (a, b, c), to a face diagonal (n) or to a body diagonal (d). The a, b, c and n glide planes all have a translation step of half the unit cell in that direction; by definition, the d glide has a translation which is a quarter that of the body diagonal. For the axial glide planes a, b and c, it is important to know both the direction of translation and the reflection plane, e.g. an a glide may be perpendicular to b (i.e. in the ac plane) or perpendicular to c.

A fundamental characteristic of crystals and their structures is the *space group*. This is a set of symbols which summarizes information about the crystal system, lattice type and elements of point and space symmetry. Combination of the fourteen Bravais lattices with possible point and space symmetry elements gives rise to a total of 230 possible space groups. Chapter 6 is devoted to space groups and the relation between space groups and crystal structure.

5.3.5 Lattice, Bravais lattice

It is very useful to be able to represent the manner of repetition of atoms, ions or molecules in a crystal by an array of points, the array being called a *lattice* and

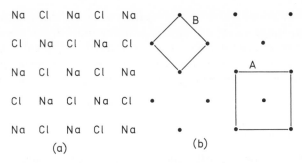

Fig. 5.18 (a) Representation of the NaCl structure in two dimensions by (b) an array of lattice points

the points *lattice points*. The section of the NaCl structure (Fig. 5.18a) may be represented by an array of points (b); each point represents one Na^+ and one Cl^- but whether the point is located at Na^+, at Cl^- or in between is irrelevant. The unit cell may be constructed by linking up the lattice points; two ways of doing this, A and B, are shown in (b). A cell such as B which contains lattice points only at the corners is *Primitive*, P, whereas a cell such as A which contains additional lattice points is *centred*. Several types of centred lattice are possible. The *face centred lattice*, F, contains additional lattice points in the centre of each face (Fig. 5.19a); NaCl is face centred cubic. A *side centred lattice* contains extra lattice points on only one pair of opposite faces, e.g. C-centred (Fig. 5.19b), whereas a *body centred lattice*, I, has an extra lattice point at the body centre (Fig. 5.19c). α-iron is *body centred cubic* because it has a cubic unit cell with iron atoms at the corner and body centre positions. CsCl is also cubic with cesium atoms at corners and chlorine atoms at the body centre (or vice versa), but it is primitive. This is because, in order for a lattice to be body centred, the atom or group of atoms which are located at or near the corner must be identical to those at or near the body centre position.

The combination of crystal system and lattice type gives the *Bravais lattice* of a structure. There are fourteen possible Bravais lattices. They can be deduced from Table 5.2 by taking the different allowed combinations of crystal system and space lattice, e.g. primitive monoclinic, C-centred monoclinic and primitive triclinic are three of the fourteen possible Bravais lattices. The lattice type plus

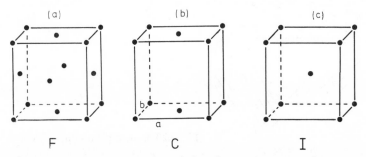

Fig. 5.19 (a) Face centred, (b) side centred and (c) body centred lattices

134

unit cell combinations which are absent from the table either (a) would violate symmetry requirements, e.g. a C-centred lattice cannot be cubic because it would not have the necessary threefold axes, or (b) may be represented by a smaller, alternative cell, e.g. a face centred tetragonal cell can be redrawn as a body centred tetragonal cell; the symmetry is still tetragonal but the volume is halved (Fig. 5.15b).

5.3.6 Lattice planes, Miller indices and directions

The concept of lattice planes (Section 5.2.2.2) is apparently straightforward but is a source of considerable confusion because there are two separate ideas which can easily become mixed. Any close packed structure, such as metal structures, ionic structures—NaCl, CaF_2, etc.—may, in certain orientations, be regarded as being built up of layers or planes of atoms stacked to form a three-dimensional structure. These layers are often related in a simple manner to the unit cell of the crystal such that, for example, a unit cell face may coincide with a layer of atoms. The reverse is not necessarily true, however, especially in more complex structures, and, for example, unit cell faces or simple sections through the unit cell often do not coincide with layers of atoms in the crystal. *Lattice planes*, which are a concept introduced with Bragg's Law, are defined purely from the shape and dimensions of the unit cell. Lattice planes are entirely imaginary and simply provide a reference grid to which the atoms in the crystal structure may be referred. Sometimes, a given set of lattice planes coincides with layers of atoms, but not usually.

Consider the two-dimensional array of lattice points shown in Fig. 5.20. This array of points may be divided up into many different sets of rows and for each set there is a characteristic perpendicular distance, d, between pairs of adjacent rows. In three dimensions, these rows become planes and adjacent planes are separated by the *interplanar d-spacing, d*. Bragg's Law treats X-ray as being diffracted from these various sets of lattice planes and the Bragg diffraction angle, θ, for each set is related to the d-spacing by Bragg's Law (Section 5.2.2.2).

Lattice planes are labelled by assigning three numbers known as *Miller indices* to each set. The derivation of Miller indices is illustrated in Fig. 5.21. The origin of the unit cell is at point O and two planes are shown which are parallel

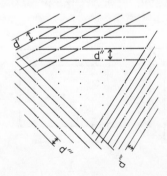

Fig. 5.20 Lattice planes (in projection)

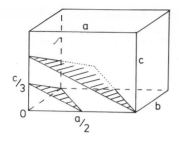

Fig. 5.21 Derivation of Miller indices

and pass obliquely through the unit cell. A third plane in this set must, by definition, pass through the origin. Each of these planes continues out to the surface of the crystal and in so doing cuts through many more unit cells; also, there are many more planes in this set parallel to the two shown, but which do not pass through this particular unit cell. In order to assign Miller indices to a set of planes, first consider that plane which is adjacent to the one that passes through the origin. Second, find the intersection of this plane on the three axes of the cell and write these intersections as fractions of the cell edges. The plane in question cuts the x axis at $a/2$, the y axis at b and the z axis at $c/3$; the fractional intersections are $\frac{1}{2}$, 1, $\frac{1}{3}$. Third, take reciprocals of these fractions; this gives (213). These three integers are the Miller indices of the plane and all other planes that

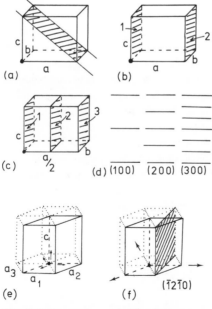

Fig. 5.22 Examples of Miller indices: (a) (101), (b) (100), (c) (200), (d) (h00), (e) labelling of axes in a hexagonal cell, (f) $(\bar{1}2\bar{1}0)$ plane, origin at solid circle, positive a directions indicated by arrows

136

are parallel to it and are separated from adjacent planes by the same d-spacing.

Some other examples are shown in Fig. 5.22. In (a), the shaded plane cuts x, y and z at $1a$, ∞b and $1c$, i.e. the plane is parallel to b. Taking reciprocals of 1, ∞ and 1 gives us (101) for the Miller indices. A Miller index of 0 means, therefore, that the plane is parallel to that axis. In Fig. 5.22(b), the planes of interest comprise opposite faces of the unit cell. We cannot directly determine the indices of plane (1) as it passes through the origin. Plane (2) has intercepts of $1a$, ∞b and ∞c and Miller indices of (100). Figure 5.22(c) is similar to (b) but there are now twice as many planes as in (b). To find the Miller indices, consider plane (2) which is the one that is closest to the origin but without passing through it. Its intercepts are $\frac{1}{2}$, ∞ and ∞ and the Miller indices are (200). A Miller index of 2 therefore indicates that the plane cuts the relevant axis at half the cell edge. This illustrates an important point. After taking reciprocals, do not divide through by the highest common factor. A common source of error is to regard, say, the (200) set of planes as those planes interleaved between the (100) planes, to give the sequence (100), (200), (100), (200), (100), ... The correct labelling is shown in Fig. 5.22(d). If extra planes are interleaved between adjacent (100) planes then *all* planes are labelled as (200).

The general symbol for Miller indices is (hkl). It is not necessary to use commas to separate the three letters or numbers and the indices are enclosed in curved brackets. The symbol $\{\ \}$ is used to indicate sets of planes that are equivalent; for example, the sets (100), (010) and (001) are equivalent in cubic crystals and may be represented collectively as $\{100\}$.

Miller indices of planes in hexagonal crystals are an exception in that four indices are often used $(hkil)$. The value of the i index is derived from the reciprocal of the fractional intercept of the plane in question on the a_3 axis (Fig. 5.22e) in exactly the same way that the other indices are derived. The i index is, to a certain extent, a piece of redundant information because the relation $h + k + i = 0$ always holds, as in, for example, $(10\bar{1}1)$, $(\bar{2}110)$, $(1\bar{2}11)$, etc. A bar over an index means that the opposite or negative direction of the corresponding axis was used in defining the indices (see Section 5.3.10). In Fig. 5.22 (f), the $(\bar{1}2\bar{1}0)$ plane is shown as an example. Sometimes hexagonal indices are written with the third index as a dot, e.g. $(\bar{1}2.0)$, and in other cases the i index is omitted completely, as in $(\bar{1}20)$.

Directions in crystals and lattices are labelled by first drawing a line that passes

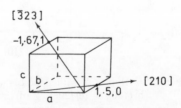

Fig. 5.23 Indices of directions [210] and [$\bar{3}$23]

through the origin and parallel to the direction in question. Let the line pass through a point with general fractional coordinates x, y, z; the line also passes through $2x, 2y, 2z$; $3x, 3y, 3z$, etc. These coordinates, written in square brackets $[x, y, z]$, are the indices of the direction; x, y and z are arranged to be the set of smallest possible integers, by division or multiplication throughout by a common factor. Thus $[\frac{1}{2}\frac{1}{2}0]$, $[110]$, $[330]$ all describe the same direction, but conventionally $[110]$ is used. For cubic systems, an $[hkl]$ direction is always perpendicular to the (hkl) plane of the same indices, but this is only sometimes true in non-cubic systems. Sets of directions which, by symmetry, are equivalent, e.g. cubic $[100]$, $[010]$, etc., are written with the general symbol $\langle 100 \rangle$. Some examples of directions and their indices are shown in Fig. 5.23.

5.3.7 d-spacing formulae

We have already defined the d-spacing of a set of planes as the perpendicular distance between any pair of adjacent planes in the set and it is this d value that appears in Bragg's Law. For a cubic unit cell, the (100) planes simply have a d-spacing of a, the value of the cell edge (Fig. 5.22b). For (200) in a cubic cell, $d = a/2$, etc. For orthogonal crystals (i.e. $\alpha = \beta = \gamma = 90°$), the d-spacing for any set of planes is given by the formula

$$\frac{1}{d_{hkl}^2} = \frac{h^2}{a^2} + \frac{k^2}{b^2} + \frac{l^2}{c^2} \tag{5.4}$$

The equation simplifies for tetragonal crystals, in which $a = b$, and still further for cubic crystals with $a = b = c$; i.e. for cubic crystals,

$$\frac{1}{d^2} = \frac{h^2 + k^2 + l^2}{a^2} \tag{5.5}$$

As a check, for cubic (200), $h = 2$, $k = l = 0$ and $1/d^2 = 4/a^2$. Therefore,

$$d = \frac{a}{2}$$

Monoclinic and especially, triclinic crystals have much more complicated d-spacing formulae because each angle that is not equal to $90°$ becomes an additional variable. The relevant formulae for d-spacings and unit cell volumes are given in Appendix 6.

5.3.8 Lattice planes and d-spacings—how many are possible?

The number of possible sets of lattice planes and their corresponding d-spacings is usually large, but finite for two reasons. First, the wavelength of the X-ray beam that is used places a lower limit on the d-spacings that may be experimentally observed: from Bragg's Law, $n\lambda = 2d \sin \theta$ and $d = n\lambda/2 \sin \theta$. The maximum value of $\sin \theta$ is 1 (when $2\theta = 180°$) and this places a lower limit on the observed d-spacings of $d = \lambda/2$ (for $n = 1$). For Cu $K\alpha$, $\lambda/2 \simeq 0.77$ Å and, if it is

Table 5.4 *Calculated d-spacings for an or-*
thorhombic cell, for $a = 3.0$, $b = 4.0$, $c = 5.0 \text{Å}$

hkl	d(Å)
001	5.00
010	4.00
011	3.12
100	3.00
101	2.57
110	2.40
111	2.16

desired to measure d-spacings that are smaller than this, the target metal in the X-ray tube must be changed for one which gives a shorter wavelength radiation, e.g. molybdenum (see Table 5.1).

Second, the number of sets of planes is limited because the Miller indices, used in calculating d-spacing values can have only integral values (i.e. $h, k, l =$ $0, 1, 2, \ldots$). There is an inverse relation between d-spacing and the magnitude of the Miller indices. Thus, the largest d-spacings which may be observed correspond to Miller indices such as (100), (010), (001), (110), etc. If the unit cell dimensions are known, it is possible to calculate all the possible d-spacings by substituting values of h, k and l into the appropriate d-spacing formula. For example, consider an orthorhombic substance that has cell parameters $a = 3.0$, $b = 4.0$, $c = 5.0 \text{Å}$. The d-spacings are given by equation (5.4):

$$\frac{1}{d_{hkl}^2} = \frac{h^2}{9} + \frac{k^2}{16} + \frac{l^2}{25}$$

Limiting ourselves to h, k and l values of 0 and 1, the various possible hkl combinations and their calculated d-spacings are given in Table 5.4. The hkl values are listed in order of decreasing d-spacing; hence 011 appears above 100. Obviously, the list could be extended for larger h, k, l values and, for example, could be terminated when a certain minimum d-spacing had been reached. If h, k, l values of 2 were included then, for example, 002 would appear in the list with a d-spacing of 2.5Å.

5.3.9 Systematically absent reflections

In Section 5.3.8, the factors that govern the *maximum* number of possible lattice planes and d-spacings are discussed. In principle, each of these sets of planes will diffract (or reflect) X-rays, but in many cases the resultant intensity is zero. These *absent reflections* may be divided into two groups: those that are absent due to some quirk in the structure and those that are absent due to the symmetry or type of lattice possessed by the structure. The latter are known as *systematic absences*. Systematic absences arise if either the lattice type is non-primitive (I, F, etc.) or if elements of space symmetry (screw axes, glide planes) are present.

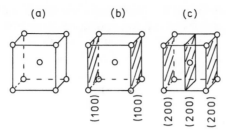

Fig. 5.24 (a) Body centred cubic α-Fe,
(b) (100) planes, (c) (200) planes

As an example of absences due to lattice type, consider α-Fe. (Fig. 5.24a), which is body centred cubic. Reflection from the (100) planes (Fig. 5.24b) has zero intensity and is systematically absent. This is because, at the Bragg angle for these planes, the body centre atoms which lie midway between adjacent (100) planes diffract X-rays exactly 180° out of phase relative to the corner atoms which lie on the (100) planes. Averaged over the whole crystal, there are equal numbers of corner and body centre atoms and the beams diffracted by each cancel completely. In contrast, a strong 200 reflection is observed because all the atoms lie on (200) planes (Fig. 5.24c) and there are no atoms lying between (200) planes to cause destructive interference. It is easy to show, by similar arguments, that the 110 reflection is observed whereas 111 is systematically absent in α-Fe. For each non-primitive lattice type there is a simple characteristic formula for systematic absences (Table 5.5). For a body centred cell reflections for which $(h + k + l)$ is odd are absent, e.g. reflections such as 100, 111, 320, etc., are systematically absent.

Systematic absences caused by the presence of space symmetry elements are rather complicated and difficult to describe and will be dealt with only briefly. Absences arise because, in certain orientations of a crystal that possesses space symmetry elements, the dimensions of one or more of the unit cell edges appear to be reduced (often by half). For the screw axis shown in Fig. 5.17(a), the true cell repeat, a, is indicated; however, when the $(h00)$ planes only are considered the cell repeat appears to be $a/2$. Because the $(h00)$ planes are perpendicular to x, any

Table 5.5 *Systematic Absences due to lattice type*

Lattice type	Rule for reflection to be observed*
Primitive, P	None
Body centred, I	$hkl; h + k + l = 2n$
Face centred, F	$hkl; h, k, l$ either all odd or all even
Side centred, e.g. C	$hkl; h + k = 2n$
Rhombohedral, R	$hkl; -h + k + l = 3n$
	or $(h - k + l = 3n)$

* If space symmetry elements are present, additional rules limiting the observable reflections may apply.

difference in rotational orientation about x does not affect diffraction from the $(h00)$ planes. In other words, the position of atoms in planes perpendicular to x does not affect the intensity of $(h00)$ reflections since these intensities depend only on the atomic positions along the x axis. Thus, by considering diffraction from the $(h00)$ planes alone, it is not possible to distinguish between a simple translation of $a/2$ and a translation of $a/2$ combined with a rotation of $180°$ about x. For a 2_1 screw axis parallel to x, reflections such as $(100), (300), \ldots, (h00)$: $h = 2n + 1$ are systematically absent.

It is difficult to show pictorially how the presence of glide planes leads to certain systematic absences. Suffice it to say that it does and that there are several types of glide plane, depending on the magnitude and direction of the translation step and on the orientation of the reflection plane. For example, a glide plane which has a translation of $b/2$ and is reflected across the bc plane is characterized by the absence of reflections of the type $0kl : k = 2n + 1$; i.e. for the $0kl$ planes the length of the b cell edge appears to be halved. Further discussion of screw axes and glide planes is given in Chapter 6.

5.3.10 Multiplicities

For cubic materials, lattice planes such as $(013), (031), (103), (130)$, etc., all have the same d-spacing, as can be shown readily from the d-spacing formula for cubic crystals (equation 5.5). In a powder X-ray pattern, the variable coordinate is d-spacing, or Bragg angle θ, and, therefore, reflections which have the same d-spacing will be superposed. The multiplicity of a powder line is the number of lines, one from each set of planes, that are superposed to give the observed line. Multiplicities may be calculated readily if the crystal symmetry is known: the object is to find the maximum possible number of (hkl) combinations which are equivalent, taking both positive and negative values of h, k and l. The maximum multiplicity possible is 48 and occurs for cubic symmetry and h, k, l reflections where $h \neq k \neq l \neq 0$; i.e.

hkl	$hk\bar{l}$	hlk	$hl\bar{k}$	lkh	$lk\bar{h}$
$hk\bar{l}$	$\bar{h}k\bar{l}$	$hl\bar{k}$	$\bar{h}l\bar{k}$	$lk\bar{h}$	$\bar{l}k\bar{h}$
$h\bar{k}l$	$\bar{h}\bar{k}l$	$h\bar{l}k$	$\bar{h}\bar{l}k$	$l\bar{k}h$	$\bar{l}\bar{k}h$
$\bar{h}kl$	$\bar{h}k\bar{l}$	$\bar{h}lk$	$\bar{h}l\bar{k}$	$\bar{l}kh$	$\bar{l}k\bar{h}$
lhk	$l\bar{h}\bar{k}$	klh	$kl\bar{h}$	khl	$kh\bar{l}$
$lh\bar{k}$	$\bar{l}h\bar{k}$	$kl\bar{h}$	$\bar{k}l\bar{h}$	$kh\bar{l}$	$\bar{k}h\bar{l}$
$l\bar{h}k$	$\bar{l}\bar{h}k$	$k\bar{l}h$	$\bar{k}\bar{l}h$	$k\bar{h}l$	$\bar{k}\bar{h}l$
$\bar{l}hk$	$\bar{l}h\bar{k}$	$\bar{k}lh$	$\bar{k}l\bar{h}$	$\bar{k}hl$	$\bar{k}h\bar{l}$

For orthorhombic crystals, h, k and l cannot be interchanged, as $a \neq b \neq c$, but negative and positive permutations are still possible to give a general multiplicity

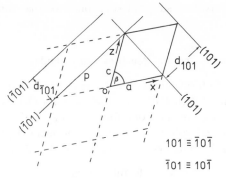

Fig. 5.25 Negative and positive Miller indices

of eight; i.e.

$$
\begin{array}{cccc}
hkl & hk\bar{l} & h\bar{k}l & \bar{h}kl \\
h\overline{kl} & \bar{h}k\bar{l} & \bar{h}\bar{k}l & \overline{hkl}
\end{array}
$$

The number of possible permutations for any symmetry is reduced when h, k or $l = 0$. For example, the cubic 100 powder line has a multiplicity of only six, i.e. 100, $\bar{1}00$, 010, $0\bar{1}0$, $00\bar{1}$ and $00\bar{1}$. Orthorhombic 100 has a multiplicity of two, i.e. 100 and $\bar{1}00$.

Negative Miller indices indicate that the opposite or negative directions of the relevant axes must be used in defining the Miller indices. This is seen in Fig. 5.25 which shows a monoclinic cell, as heavy lines, in projection down b. The origin of the cell, O, and the chosen positive directions for x and z are marked. For illustration consider the $\{101\}$ planes. These planes are all parallel to y and, therefore, perpendicular to the plane of the paper. Two planes of the sets (101) and ($\bar{1}01$) are drawn as light lines. In defining the ($\bar{1}01$) planes, the plane p, which is the plane adjacent to the one (not shown) that passes through the origin, cuts the x axis at -1 and the z axis at 1. It should be apparent from the scale of the drawing that $d_{101} \neq d_{\bar{1}01}$ unless $\beta = 90°$, as in, say, orthorhombic crystals.

The ($\bar{1}0\bar{1}$) planes (not shown) are exactly the same as the (101) planes and may simply be regarded as the (101) planes 'looked at from the opposite direction'. It is important, however, to give the (101) and ($\bar{1}0\bar{1}$) planes separate status, especially in single crystal diffraction studies. Similarly, the planes ($\bar{1}01$) and ($10\bar{1}$) are the same but are treated separately in diffraction. For general planes, (hkl), this equality or 'counting twice' of d-spacings is for the pairs of planes (hkl) and ($\overline{hkl}$). All powder lines therefore have a minimum multiplicity of two. In single crystal diffraction studies (hkl) and ($\overline{hkl}$) may be observed as separate reflections. The intensity of corresponding (hkl) and ($\overline{hkl}$) reflections is usually the same but under certain circumstances, such as when *anomalous dispersion* is present, the intensities are not equal. It is often stated that 'diffraction patterns in reciprocal space possess a centre of symmetry'. This grandiose statement means that hkl and

$\overline{hkl}$ reflections are equivalent, subject to the condition that anomalous dispersion is not present.

5.3.11 Unit cell contents, crystal densities and formulae

The unit cell, by definition, must contain at least one formula unit, whether it be an atom, ion pair, molecule, etc. In centred cells, and sometimes in primitive cells, the unit cell contains more than one formula unit. A simple relation may be derived between the cell volume, the number of formula units in the cell, the formula weight and the bulk crystal density. The density is given by

$$D = \frac{\text{mass}}{\text{volume}} = \frac{\text{formula weight}}{\text{molar volume}}$$

$$= \frac{\text{FW}}{\text{volume of formula unit} \times N}$$

where N is Avogrado's number. If the unit cell, of volume V, contains Z formula units, then

$$V = \text{volume of one formula unit} \times Z$$

Therefore,

$$D = \frac{\text{FW} \times Z}{V \times N} \tag{5.6}$$

V is usually expressed as Å^3 and must be multiplied by 10^{-24} to give densities in the normal units of grams per cubic centimetre. Substituting for Avogrado's number, the above formula reduces to

$$D = \frac{\text{FW} \times Z \times 1.66}{V} \tag{5.7}$$

If V is in Å^3 in equation (5.7), the units of D are in grams per cubic centimetre.
This simple formula has a number of uses, as shown by the following examples:

(a) It can be used to check that a given set of crystal data are consistent and that, for example, an erroneous formula weight has not been assumed.
(b) It can be used to determine any of the four variables if the other three are known. This is most common for Z (which must be a whole number) but is also used to determine FW and D.
(c) By comparison of D_{obs} (the experimental density of a material) and $D_{\text{X-ray}}$ (the density calculated from the above formula), information may be obtained on: the presence of crystal defects such as vacancies as opposed to interstitials, the mechanisms of solid solution formation, the porosity of ceramic pieces.

Considerable confusion often arises in determining the value of the contents, Z, of a unit cell. This is because atoms or ions that lie on corners, edges or faces of the unit cell are also shared between adjacent cells, and this must be taken into consideration.

For example, α-Fe (Fig. 5.24a) has $Z = 2$. The corner iron atoms, of which there are eight, are each shared between eight neighbouring unit cells. Effectively, each contributes only $\frac{1}{8}$ to the particular cell in question, giving $8 \times \frac{1}{8} = 1$ net iron atom for the corners. The body centre iron atom lies entirely inside the unit cell and counts as one. Hence $Z = 2$.

For NaCl (Fig. 5.9), $Z = 4$, i.e. $4(Na^+ Cl^-)$. The corner Na^+ ions again count as one. The face centre Na^+ ions, of which there are six, count as $\frac{1}{2}$ each, giving a total of $1 + 3 = 4\,Na^+$ ions. The Cl^- ions at the edge centres, of which there are twelve, count as $\frac{1}{4}$ each, which, together with Cl^- at the body centre, give a total of $12 \times \frac{1}{4} + 1 = 4\,Cl^-$ ions.

5.4 The X-ray diffraction experiment

When reduced to basic essentials, the X-ray diffraction experiment, Fig. 5.26, requires an X-ray source, the sample under investigation and a detector to pick up the diffracted X-rays. Within this broad framework, there are three variables which govern the different X-ray techniques:

(a) radiation—monochromatic or of variable λ;
(b) sample—single crystal, powder or a solid piece;
(c) detector—radiation counter or photographic film.

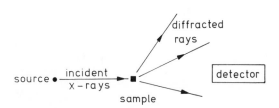

Fig. 5.26 The X-ray diffraction experiment

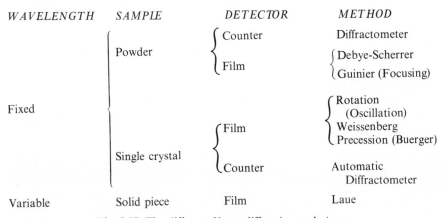

WAVELENGTH	SAMPLE	DETECTOR	METHOD
		Counter	Diffractometer
	Powder	Film	Debye-Scherrer
			Guinier (Focusing)
Fixed		Film	Rotation (Oscillation)
			Weissenberg
			Precession (Buerger)
	Single crystal	Counter	Automatic Diffractometer
Variable	Solid piece	Film	Laue

Fig. 5.27 The different X-ray diffraction techniques

These are summarized for the most important techniques in Fig. 5.27. With the exception of the Laue method, which is used almost exclusively by metallurgists and is not discussed here, monochromatic radiation is nearly always used. An outline of powder and single crystal methods is given in this section with a more comprehensive account of powder methods being given in Section 5.6.

5.4.1 The powder method—principles and uses

The principles of the powder method are shown in Fig. 5.28. A monochromatic beam of X-rays strikes a finely powdered sample that, ideally, has crystals randomly arranged in every possible orientation. In such a powder sample, the various lattice planes are also present in every possible orientation. For each set of planes, therefore, at least some crystals must be oriented at the Bragg angle, θ, to the incident beam and thus, diffraction occurs for these crystals and planes. The diffracted beams may be detected either by surrounding the sample with a strip of photographic film (Debye–Scherrer and Guinier focusing methods) or by using a movable detector, such as a Geiger counter, connected to a chart recorder (diffractometer).

The original powder method, the *Debye–Scherrer method*, is little used

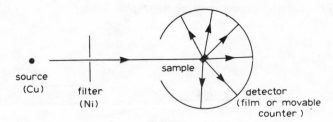

Fig. 5.28 The powder method

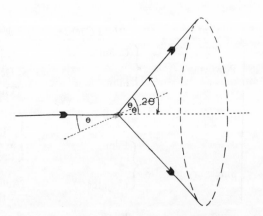

Fig. 5.29 The formation of a cone of diffracted radiation in the powder method

nowadays, but since it is simple it is instructive to consider its mode of operation. For any set of lattice planes, the diffracted radiation forms the surface of a cone, as shown in Fig. 5.29. The only requirement for diffraction is that the planes be at angle θ to the incident beam; no restriction is placed on the angular orientation of the planes about the axis of the incident beam. In a finely powdered sample, crystals are present at every possible angular position about the incident beam and the diffracted beams that result appear to be emitted from the sample as cones of radiation. (Each cone is in fact a large number of closely spaced diffracted beams.) If the Bragg angle is θ, then the angle between diffracted and undiffracted beams is 2θ and the angle of the cone is 4θ. Each set of planes gives its own cone of radiation. The cones are detected by a thin strip of film wrapped around the sample (Fig. 5.28); each cone intersects the film as two short arcs (Fig. 5.30), which are symmetrical about the two holes in the film (these allow entry and exit of incident and undiffracted beams). In a well-powdered sample, each arc appears as a continuous line, but in coarser samples the arcs may be spotty due to the relatively small number of crystals present.

To obtain d-spacings from the Debye–Scherrer film, the separation, S, between pairs of corresponding arcs is measured. If the camera (film) radius, R, is known, then

$$\frac{S}{2\pi R} = \frac{4\theta}{360} \qquad (5.8)$$

from which 2θ and therefore d may be obtained for each pair of arcs. The disadvantages of this method are that exposure times are long (6 to 24 hours) and that closely spaced arcs are not well resolved. This is because, although the incident beam enters the camera through a pinhole slit and collimator tube, the beam is somewhat divergent and the spread increases in the diffracted beams. If, in an effort to increase the resolution, a finer collimator is used, the resulting diffracted beams have much less intensity and longer exposure times are needed. Apart from considerations of the extra time involved, the amount of background radiation detected by the film (as fogging) increases with exposure time and, consequently, weak lines may be lost altogether in the background.

In modern film methods (*Guinier focusing methods*) a *convergent*, intense incident beam is used with the result that excellent resolution of lines is obtained and exposure times are much reduced (10 min to 1 hr). A convergent beam is obtained by placing a bent single crystal of quartz or graphite between the X-ray source and the sample. The orientation of this bent crystal is adjusted so that it

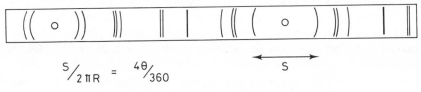

$$\frac{S}{2\pi R} = \frac{4\theta}{360}$$

Fig. 5.30 Schematic Debye–Scherrer photograph

diffracts the incident beam and converts it from a divergent beam into a convergent one. The beam then strikes the sample and the diffracted beams are arranged to focus at the surface of the film. Further details are given in Section 5.6. A typical schematic, Guinier film is shown in Fig. 3.1.

The other modern powder technique is *diffractometry*, which gives a series of peaks on a strip of chart paper. A convergent incident beam is again used (Section 5.6) to give fairly good resolution of peaks. Both peak positions and intensities (peak heights) are readily obtained from the chart to make this a very useful and rapid method of phase analysis.

The most important use of the powder method is in the qualitative identification of crystalline phases or compounds. While most chemical methods of analysis give information about the *elements* present in a sample, powder diffraction is very different and perhaps unique in that it tells which *crystalline compounds* or *phases* are present but gives no direct information about their chemical constitution.

Each crystalline phase has a characteristic powder pattern which can be used as a fingerprint for identification purposes. The two variables in a powder pattern are peak position, i.e. *d*-spacing, which can be measured very accurately if necessary, and intensity, which can be measured either qualitatively or quantitatively. It is rare but not unknown that two materials have identical powder patterns. More often, two materials have one or two lines with common *d*-spacings, but on comparing the whole patterns, which may contain between ~ 5 and 100 observed lines, the two are found to be quite different. In more extreme cases, two substances may happen to have the same unit cell parameters and, therefore, the same *d*-spacings, but since different elements are probably present in the two, their intensities are quite different. The normal practice in using powder patterns for identification purposes is to pay most attention to the *d*-spacings but, at the same time, check that the intensities are roughly correct.

For the identification of unknown crystalline materials, an invaluable reference source is the *Powder Diffraction File* (Joint Committee on Powder Diffraction Standards, Swarthmore, USA), previously known as the ASTM file, which contains the powder patterns of about 35000 materials; new entries are added at the current rate of ~ 2000 p.a. In the search indices, materials are classified either according to their most intense peaks or according to the first eight lines in the powder pattern in order of decreasing *d*-spacing. Identification of an unknown is usually possible within 30 min of obtaining its measured powder pattern. Problems arise if the material is not included in the file (obviously!) or if the material is not pure but contains lines from more than one phase.

For many types of work, the materials being analysed may not be completely unknown but may be restricted to a range of possible phases. It is then much easier to have to hand a standard pattern of all the phases likely to be encountered. Comparison of the unknown with the standard patterns leads to identification in a matter of minutes. Guinier films are admirably suited to this type of work since they are small and can be easily compared by simply lining up the back stop mark (corresponding to $0°\,2\theta$) on each.

Table 5.6 *Some uses of the powder method*

Characterization of materials by X-ray 'fingerprints'
Qualitative phase analysis (presence or absence of phases)
Quantitative phase analysis
Refinement of unit cell parameters
Study of solid solution formation
Determination of crystal size
Study of crystal distortion by stress
Measurement of thermal expansion coefficients (HTXR)*
Determination of high temperature phase diagrams (HTRX)*
Study of phase transformations
Crystal structure determination
Study of the reactions of solids

* HTXR = high temperature X-ray diffraction.

The powder method has many important secondary uses, especially in the general area of applied crystallography. These are discussed in Section 5.6 and in various places throughout this book; a summary is given in Table 5.6.

5.4.2 Single crystal methods—principles and uses

The main uses of single crystal methods are, as indicated in Section 3.2.1.3, to determine unit cells and space groups and, if there is sufficient interest, to measure the intensities of reflections and carry out a full crystal structure determination. Monochromatic X-rays are generally used, an optimum crystal size is ~ 0.2 mm diameter and the detector may be either film or counter. Only film methods are discussed here. The sole purpose of the counter method, using a single crystal diffractometer, is to collect intensity data. As such it is an extremely valuable instrument but it is not really a technique which a solid state chemist can use for himself as part of his armoury of experimental techniques and, therefore, is not discussed further.

There are three main film techniques: the rotation or oscillation method, the Weissenberg method and the precession method. A brief outline of each is given next but details of how the instruments work are omitted. The latter two methods are rather complicated and anyone wishing to understand them is recommended to get direct instruction and practical experience in their use.

In the *rotation method*, a crystal is mounted by sticking it onto, for example, a

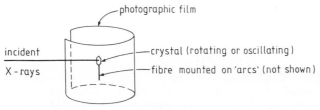

Fig. 5.31 Schematic single crystal rotation method

glass fibre. The crystal is *set* so that one of its unit cell axes is vertical and it is made to oscillate or rotate about this axis (Fig. 5.31). A horizontal beam of X-rays strikes the crystal and the diffracted beams are detected by a piece of film bent round the crystal in the form of a cylinder. After developing the film, parallel rows of spots are seen (Fig. 5.32), each spot corresponding to diffraction from one particular set of planes in the crystal. There are two main uses of these photographs. From oscillation photographs (with the crystal oscillated through, for example, 15° about the vertical axis), the presence of symmetry in the distribution and intensity of spots in the photograph is looked for. In particular, if the top half of the photograph is a mirror image of the bottom half, as it is in Fig. 5.32, then the vertical axis, about which the crystal is set, is an axis of symmetry. Not all unit cell axes are symmetry axes. For instance, in an orthorhombic structure all three axes are symmetry axes but in a monoclinic structure, only the unique *b* axis is a symmetry axis. The second use is for determining the magnitude of the unit cell dimension in the vertical direction. Although this is straightforward, a special chart is needed in order to measure the distance between adjacent rows of spots. This distance in *reciprocal space* is inversely related to the magnitude of the unit cell dimension in *real space*. Thus, if the rows are closely spaced the unit cell dimension is large, and vice versa.

Let us consider briefly how the *zero layer* in Fig. 5.32 arises. The zero layer is the row that passes through the centre of the film. Suppose the crystal has orthorhombic symmetry and is set with *c* vertical; *a* and *b* are therefore horizontal. From the definition of Miller indices, we know that planes of the type (*hk0*) have one direction in common in that they are all parallel to *c*. Since the incident X-ray beam is horizontal and therefore perpendicular to *c*, it follows that all beams diffracted from the (*hk0*) planes are also horizontal. These beams radiate horizontally from the crystal and are detected on film as the zero layer row of spots.

The next row of spots, the *first layer*, corresponds to reflections from the (*hk1*) set of planes, and so on. The position of the spots on the zero layer row depends

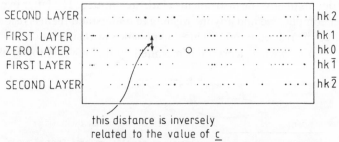

Fig. 5.32 Schematic oscillation photograph obtained in the rotation method. It is assumed that the crystal *c* axis is vertical. Further details are not given since a reference chart is needed in order to measure up the photograph

on the d-spacings of the planes in question and therefore on the values of h and k. Planes that have large d-spacings diffract X-rays at low Bragg angles and these are the spots closest to the hole in the centre of the film (this hole is the exit for undiffracted X-rays and corresponds to a Bragg angle of zero).

In studying a new crystal, one usually starts with the rotation camera. Once the crystal is set with one axis vertical, it is possible to tell from the photographs the value of that unit cell dimension. To obtain the remaining cell dimensions (two sides, three angles), the individual rows of spots have to be analysed. If the crystal is of high symmetry and does not give too many spots on the rotation photograph, it may be possible to obtain the remaining information by a suitable graphical analysis of the photograph. It is more common, however (if one has the equipment), to put the crystal onto either a Weissenberg or a precession camera. Essentially, both of these cameras take a single row of spots (as observed in the rotation photograph) and separate the spots into two dimensions.

In the *Weissenberg method*, the crystal is also surrounded cylindrically by film but metal screens are placed between the crystal and the film such that only one row of spots, or layer line, is allowed through. During exposure, the crystal undergoes a slow oscillation and at the same time the film translates up and

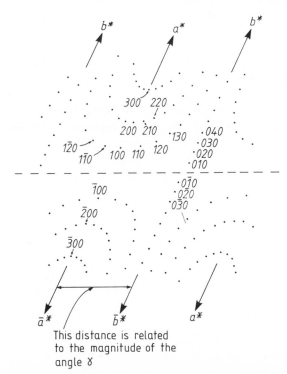

Fig. 5.33 Schematic zero layer ($hk0$) Weissenberg photograph. All spots are shown with equal intensity. In practice, intensities vary

down. The coupling of the film and crystal motions is rather complicated and will not be considered here. A schematic, simplified Weissenberg photograph is shown in Fig. 5.33. The pattern of spots is distorted because of the manner of the coupling between the two motions. From the photograph, one first looks for the two axes; these are labelled as a^* and b^* if the crystal is set about c; note that each axis appears on the photograph several times. (The starring of axes is associated with the reciprocal lattice; it is discussed later and in Appendix 7.) The axial spots lie on straight lines but all the others are on curves. The separation of the spots along the a^* and b^* axes is related inversely to the value of the cell parameter for those axes and so the second and third unit cell dimensions may be obtained from the photographs. The distance between two axial rows of spots is related to the angle between these two axes in the unit cell, i.e. to the angle γ for axes a^* and b^*. Because of the distorted nature of Weissenberg photographs it is almost essential, in measuring them up, to superpose the film over a suitably scaled reference grid (not shown). It then becomes a straightforward process to obtain the relevant unit cell information. Some of the spots in Fig. 5.33 have been assigned Miller indices, hkl. The method used for this assignment is similar to that used for precession photographs, discussed next.

The *precession method* gives photographs that are much easier to interpret and measure and also are more attractive aesthetically (see Fig. 3.2 and Section 3.2.1.3). Each layer line of a rotation photograph is converted, in a precession photograph, into a two-dimensional network of spots. One first inspects the photograph for the presence of symmetry in the distribution and intensity of the spots and thereby locates the two axes a^* and b^* (sometimes it is possible to choose more than one set of valid axes, especially if there is not much symmetry evident from the photograph). Next, it is important (in Weissenberg films also) to assign Miller indices to the spots. To see how this is done, let us suppose that our

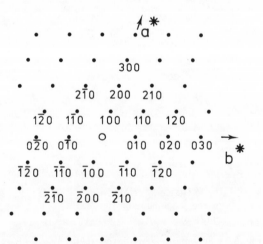

Fig. 5.34 Indexing of a precession photograph,
$hk0$ layer

photograph is a zero layer photograph for a crystal set about c (Fig. 5.34) and for which the angle between a^* and b^* is not 90°. All the spots, therefore, have Miller indices that belong to the set ($hk0$). The spots are shown of the same size and intensity but in practice this is most unlikely to occur. The spots in the row that forms the a^* axis have indices ($h00$) which are labelled, working out from the centre of the film, with increasing h values. Similarly, the row of spots that forms the b^* axis is the ($0k0$) set of reflections and is labelled accordingly. Negative directions of a^* and b^* are indicated by the bar over the appropriate Miller indices. We have now labelled the two axes and so all remaining spots may be indexed, as shown.

A fundamental feature of precession photographs is that distances on them, e.g. between pairs of spots, are inversely proportional to the corresponding distances in the actual crystal. Consider the $h00$ reflections. We know that $d_{100} = 2d_{200} = 3d_{300}$, etc. On the precession photograph, the reverse is true. The distance of the 100 spot from the origin (the centre of the film) is exactly half the distance of the 200 spot from the origin, and so on. This brings us to the distinction between *real space*, in which the crystal exists, and *reciprocal space*, which describes the directions followed by diffracted beams and hence the patterns of spots on single crystal photographs. The starring of axes, a^* and b^*, is done to distinguish the axes of the *reciprocal lattice* from those of their real space counterparts, a and b. If the unit cell in real space is orthogonal ($\alpha = \beta = \gamma = 90°$), then real and reciprocal axes are parallel because, for example, $\alpha^* = 180° - \alpha = 90° = \alpha$, etc. This is not necessarily true for non-orthogonal crystals, however, if $\alpha^* \neq \alpha$, etc. A derivation of the reciprocal lattice is given in Appendix 7.

The beauty of the precession method is that, unlike the Weissenberg method, its photographs give an undistorted picture of the reciprocal lattice. This makes their interpretation very easy and logical. A slight disadvantage is that the theory behind the operation of the camera is rather complicated.

The use of single crystal methods in determining the nature of the unit cell and its dimensions has been mentioned several times. An extension of this is the determination of lattice type and possible space group(s) by inspecting the Weissenberg or precession photographs for systematically absent reflections. Once Miller indices have been assigned to spots, it is easy to see if certain groups of spots are absent (remember, centred lattices, screw axes and glide planes all cause systematic absences). For example, the arrangement of spots in Fig. 5.35(a) shows that the a^* row has every other spot absent in the $hk0$ layer; let us assume that there are no systematic absences in the $hk1$ layer (not shown). From this, we can say that there are no general $hk1$ absences and no general $hk0$ absences, but for $h00$ we have the condition that, for reflection to occur, $h = 2n$ (because 100, 300, 500, etc., are absent). This means that parallel to a in the crystal, there is a 2_1 screw axis.

In Fig. 5.35(b) and (c) are shown schematic zero layer and first layer photographs for a body centred lattice. From an examination of (b) in isolation one would perhaps choose 1 and 2 for axes of the reciprocal lattice but it would then be difficult to subsequently interpret (c) because in (c) no spots lie on the axes,

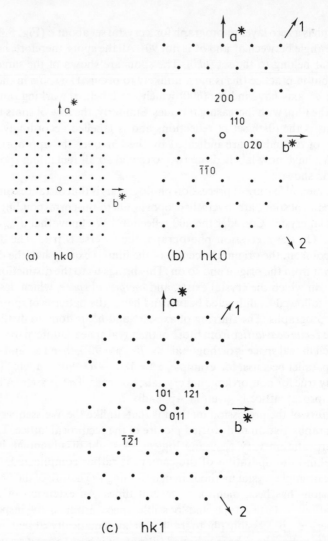

Fig. 5.35 Reciprocal lattices showing (a) 2_1 screw axis parallel to a, (b) and (c) body centring

1 and 2. The correct axes are marked a^* and b^*; in (b) alternate spots are systematically absent which leads to the condition for reflection; $hk0: h + k = 2n$ (e.g. 200, 400, 110, 130, etc., are present). In (c) also, alternate spots are absent but now the condition for reflection is $hk1: h + k = 2n + 1$ (e.g. 101, 011, 211, etc., are present). If we assume that layers $hk2$, $hk4$, etc., have the same pattern of absences as $hk0$ and that $hk3$, $hk5$, etc., are similar to $hk1$, then we arrive at the general condition for reflection $hkl: h + k + l = 2n$, which corresponds to a body centred lattice type, Table 5.5. The reader may now like to work out how a face centred lattice appears in reciprocal space. The answer is given in Appendix 7.

Once the crystal system, lattice type and systematic absences have been determined, a short list of possible space groups—sometimes one, but, more usually, two or three—for the crystal can be found by comparing the absences with those for the different space groups (given in *International Tables for X-Ray Crystallography*, Vol. 1). For example, suppose our crystal has a primitive orthorhombic unit cell and its only systematic absences indicate a 2_1 screw axis parallel to one of the unit cell axes. From an inspection of all possible space groups we find that our crystal must belong to the space group $P222_1$ (Chapter 6).

5.5 Intensities

Intensities of X-ray reflections are important for two main reasons. First, quantitative measurements of intensity are necessary in order to determine unknown crystal structures. Second, qualitative or semi-quantitative intensity data are needed in using the powder fingerprint method to characterize materials and especially in using the powder diffraction file to identify unknowns. Although this book is not concerned with the methods of crystal structure determination, it is considered important that the factors which control the intensity of X-ray reflections be understood. The topic falls into two parts: the intensity scattered by individual atoms and the resultant intensity scattered from the large number of atoms that are arranged periodically in a crystal.

5.5.1 Scattering of X-rays by an atom

Atoms diffract or scatter X-rays because an incident X-ray beam, which can be described as an electromagnetic wave with an oscillating electric field, sets each electron of an atom into vibration. A vibrating charge such as an electron emits radiation and this radiation is in phase or *coherent with* the incident X-ray beam. The electrons of an atom therefore act as secondary point sources of X-rays. Coherent scattering may be likened to an elastic collision between the wave and the electron: the wave is deflected by the electron without loss of energy and, therefore, without change of wavelength. The intensity of the radiation scattered coherently by 'point source' electrons has been treated theoretically and is given by the Thomson equation: *Point source*

$$I_p \alpha \tfrac{1}{2}(1 + \cos^2 2\theta) \qquad (5.9)$$

where I_p is the scattered intensity at any point, P, and 2θ is the angle between the directions of the incident beam and the diffracted beam that passes through P. From this equation it can be seen that the scattered beams are most intense when parallel or antiparallel to the incident beam and are weakest when at 90° to the incident beam. The Thomson equation is also known as the *polarization factor* and is one of the standard angular correction factors that must be applied during the processing of intensity data (for use in structure determination).

At this point, it is worth mentioning that X-rays can interact with electrons in a different way to give *Compton scattering* (Section 5.6.8). Compton scattering is rather like an elastic collision in that the X-rays lose some of their energy on impact and so the scattered X-rays are of longer wavelength than the incident X-rays. They are also no longer in phase with the incident X-rays; nor are they in phase with each other. A close similarity exists between Compton scattering and the generation of white radiation in an X-ray tube; both are examples of incoherent scattering that are sources of background radiation in X-ray diffraction experiments. As Compton scattering is caused by interaction between X-rays and the more loosely held outer valence electrons, it is an important effect with the lighter elements and can have a particularly deleterious effect on the powder patterns of organic materials such as polymers.

The X-rays that are scattered by an atom are the resultant of the waves scattered by each electron in the atom. The electrons may be regarded as particles that occupy different positions in an atom and interference occurs between their scattered waves. For scattering in the direction of the incident beam (Fig. 5.36a) beams 1′ and 2′, all electrons scatter in phase irrespective of their position. The scattered intensity is, then, the sum of the individual intensities. The *scattering factor*, or *form factor*, f, of an atom is proportional to its atomic number, Z, or, more strictly, to the number of electrons possessed by that atom.

For scattering at some angle 2θ to the direction of the incident beam, a phase difference, corresponding to the distance XY, exists between beams 1″ and 2″. This phase difference is usually rather less than one wavelength (i.e. XY < 1.5418 Å for Cu $K\alpha$ X-rays) because distances between electrons within an atom are short. As a result, only partial destructive interference occurs between 1″ and 2″. The net effect of interference between the beams scattered by all the electrons in the atom is to cause a gradual decrease in scattered intensity with increasing angle, 2θ. For example, the scattering power of copper is proportional to 29 (i.e. Z) at $2\theta = 0°$, to 14 at 90° and to 11.5 at 120°. It should also be apparent from Fig. 5.36(a) that for a given angle, 2θ, the net intensity decreases with decreasing X-ray wavelength. The form factors of atoms are given in *International Tables for X-ray Crystallography*, Vol. 3 (1952). They are tabulated against ($\sin\theta/\lambda$) to include the effect of both angle and X-ray wavelength; examples are shown in Fig. 5.36(b).

Two consequences of the dependence of form factors on $\sin\theta/\lambda$ and atomic number are as follows. First, the powder patterns of most materials contain only weak lines at high angles (above ~ 60 to $70°\, 2\theta$). Although several factors contribute to powder intensities, the main reason for this effect is that the atoms scatter only weakly at high angles (Fig. 5.36b). Second, in crystal structure determinations using X-rays, it is difficult to locate light atoms because their diffracted radiation is so weak. Thus hydrogen atoms cannot usually be located unless all the other elements present are also extremely light (e.g. in boron hydride crystals). Atoms that have as many electrons as oxygen can usually be located easily unless a very heavy atom such as uranium is present. Structures that are particularly difficult to solve are those in which a considerable number or all of the atoms present have similar atomic number, e.g. large organic molecules with carbon, nitrogen and oxygen atoms. In such cases, a common ploy is to make a

(a)

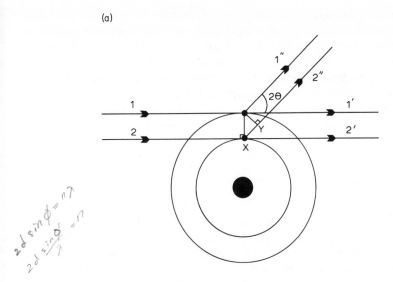

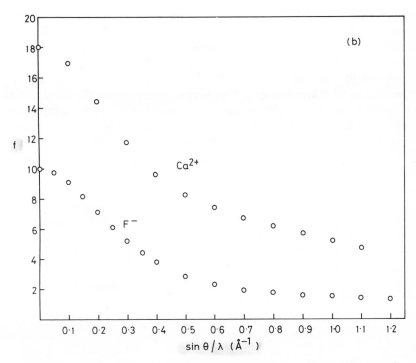

Fig. 5.36 (a) Scattering of X-rays by electrons in an atom. (b) Form factors of Ca²⁺ and F⁻

derivative of the compound of interest which contains a heavy metal atom. The heavy atoms can be detected very readily and because they determine the phase of diffracted beams, a lead is given towards placing the remaining atoms. Because of their similar atomic numbers, aluminium and silicon are very difficult to distinguish, which causes problems in determinations of aluminosilicate structures. One of the advantages of using neutrons instead of (or as well as) X-rays for crystallographic work is that the neutron form factors are not a simple function of atomic number. Light atoms, e.g. hydrogen and lithium are often strong neutron scatterers (Section 3.2.1.5).

5.5.2 Scattering of X-rays by a crystal

Each atom in a material acts as a secondary point source of X-rays. If the material is non-crystalline, beams are scattered by the atoms in all directions, but in crystalline materials the scattered beams interfere destructively in most possible directions. In other directions, interference is constructive or only partially destructive, resulting in the X-ray beams that are detected in diffraction experiments. The amount of interference and hence the resultant intensity depend on the relative phases of the beams scattered by each atom and therefore on the atomic positions in the crystal structure.

In Section 5.3.9, the phenomenon of systematically absent reflections is treated. Examples are given where both lattice centring and the presence of elements of space symmetry cause sets of reflections to be absent. The treatment is now extended to consider some special examples of partially destructive interference followed by consideration of general expressions for intensities.

Consider the rock salt structure. It is face centred cubic and, therefore, only those reflections may be observed for which hkl are either all odd or all even (Table 5.5). From this rule, for instance, 110 is systematically absent but 111 may be observed. Both of these planes are shown for NaCl in Fig. 5.37. In Fig. 5.37(a), (110) planes have Na$^+$ and Cl$^-$ ions lying on the planes and equal numbers of the same ions lying midway between the planes. Complete cancellation of the 110

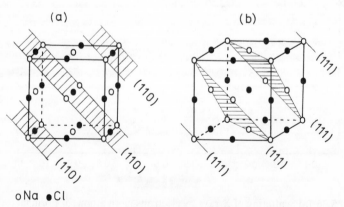

oNa •Cl

Fig. 5.37 (a) (110) and (b) (111) planes in NaCl

Table 5.7 *X-ray powder diffraction patterns for potassium halides.* (Data from Joint Committee on Powder Diffraction Standards, Swarthmore)

(hkl)	KF, $a = 5.347$Å		KCl, $a = 6.2931$Å		KI, $a = 7.0655$Å	
	d(Å)	I	d(Å)	I	d(Å)	I
111	3.087	29	—		4.08	42
200	2.671	100	3.146	100	3.53	100
220	1.890	63	2.224	59	2.498	70
311	1.612	10	—	—	2.131	29
222	1.542	17	1.816	23	2.039	27
400	1.337	8	1.573	8	1.767	15

reflection therefore occurs. In Fig. 5.37(b), (111) planes have Na^+ ions lying on the planes and Cl^- ions midway between the planes. The Na^+ and Cl^- ions scatter exactly 180° out of phase with each other for these planes, but since they have different scattering powers the destructive interference that occurs is only partial. The intensity of the 111 reflection in materials that have the rock salt structure is, therefore, related to the difference in atomic number of anion and cation. For the potassium halides, the 111 intensity is zero for KCl, since K^+ and Cl^- are isoelectronic, and its intensity should increase in the order

$$KCl < KF < KBr < KI$$

Some data which confirm this are given in Table 5.7.

Similar effects may be found in other simple crystal structures. In primitive cubic CsCl, if the difference between cesium and chlorine is ignored the atomic positions are the same as in body centred α-Fe (Fig. 5.24). The 100 reflection is systematically absent in α-Fe but is an observed reflection with CsCl because the scattering powers of Cs^+ and Cl^- are different, i.e. $f_{Cs^+} \neq f_{Cl^-}$.

5.5.3 Intensities—general formulae and a model calculation for CaF_2

Each atom in a crystal scatters X-rays by an amount related to the scattering power, f, of that atom. In summing the individual waves to give the resultant diffracted beam, both the *amplitude* and *phase* of each wave are important. If we know the atomic positions in the structure, the amplitude and phase appropriate to each atom in the unit cell may be calculated and the summation carried out by various mathematical methods, therefore simulating what happens during diffraction. Let us consider first the relative phases of different atoms in the unit cell. In Fig. 5.38(a) are drawn two (100) planes of a crystal that has an orthogonal (i.e. $\alpha = \beta = \gamma = 90°$) unit cell. The atoms A, B, C, A' lie on the a axis (perpendicular to (100) planes) with A and A' at the origin of adjacent unit cells. For the 100 reflection, A and A' scatter in phase because their phase difference is exactly one wavelength, 2π radians (Bragg's Law). Atom B, situated halfway between adjacent (100) planes, has a fractional x coordinate (relative to A) of $\frac{1}{2}$. The phase difference between (waves diffracted from) A and B is $\frac{1}{2} \cdot 2\pi = \pi$, i.e. atoms

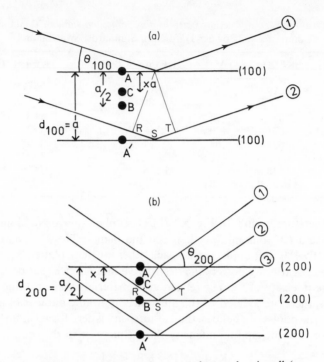

Fig. 5.38 (a) (100) planes for an orthogonal unit cell ($\alpha = \beta = \gamma = 90°$). Atoms A, B, C, A′ lie on the a cell edge. (b) (200) planes for the same unit cell as in (a)

A and B are exactly out of phase. Atom C has a general fractional coordinate x (at distance xa from A) and, therefore, a phase relative to A of $2\pi x$.

Consider, now, the 200 reflection for the same unit cell (Fig. 5.38b). Atoms A and B have a phase difference of 2π for the 200 reflection and scatter in phase, whereas their phase difference is π for the 100 reflection (in order to obey Bragg's Law, if d is halved, $\sin\theta$ must double; thus $\theta_{200} \gg \theta_{100}$). Comparing the Bragg diffraction conditions for the (100) and (200) planes, the effect of halving d is to double the relative phase difference between pairs of atoms such as A and B; therefore, A and C have a phase difference of $(2x \cdot 2\pi)$ for the (200) reflection.

For the general case of an $h00$ reflection, the d-spacing between adjacent ($h00$) planes is $(1/h)a$ (for an orthogonal cell); the phase difference, δ, between A and C is given by

$$\delta = 2\pi hx \tag{5.10}$$

The phase difference between atoms depends, therefore, on two factors: the Miller indices of the reflection that is being considered and the fractional coordinates of the atoms in the unit cell. The above reasoning may be extended readily to a general three-dimensional situation. For reflection from the set of planes with indices (hkl), the phase difference, δ, between atoms at the origin and a position with fractional coordinates (x, y, z) is given by

$$\delta = 2\pi(hx + ky + lz) \qquad (5.11)$$

This is an important formula and is applicable to all unit cell shapes. Let us use it on a simple structure, γ-Fe, which is face centred cubic with atoms at the corner and face centre positions, i.e. with fractional coordinates:

$$(0,0,0); \qquad (\tfrac{1}{2},\tfrac{1}{2},0); \qquad (\tfrac{1}{2},0,\tfrac{1}{2}); \qquad (0,\tfrac{1}{2},\tfrac{1}{2})$$

These coordinates may be substituted into the formula for δ to give four phases:

$$0, \qquad \pi(h + k), \qquad \pi(h + l), \qquad \pi(k + l)$$

How do these vary with the Miller indices? If h, k and l are either all even or all odd, the phases are in multiples of 2π and, therefore, are in phase with each other.

If, however, one, say h, is odd and the other two, k and l, are even, the four phases reduce to

$$0, \qquad (2n + 1)\pi, \qquad (2n + 1)\pi, \qquad 2n\pi$$

The first and last are π out of phase with the middle two and complete cancellation occurs. The γ-Fe structure is a simple example of a face centred cubic lattice in which the iron atoms correspond to lattice points and, in fact, we have just proved the condition for systematic absences due to face centring (Table 5.5). The reader may like to prove the condition for systematic absences in a body centred cubic structure, e.g. by working out the phases of the atoms for the structure of α-Fe.

The second major factor that affects intensities is the amplitude of the individual waves scattered by each atom as given by the scattering power, f. From Section 5.5.1, f is proportional to atomic number, Z, and decreases with increasing Bragg angle, θ.

We now wish to generalize the treatment to consider any atom in the unit cell. For atom j, the diffracted wave of amplitude f_j and phase δ_j may be represented by a sine wave of the form

$$F_j = f_j \sin(\omega t - \delta_j) \qquad (5.12)$$

The waves diffracted from each atom in the cell have the same angular frequency, ω, but may differ in f and δ. The resultant intensity is obtained from the summation of the individual sine waves. Mathematically, addition of waves may be carried out by various methods, including vector addition and by the use of complex numbers. In complex notation, wave j may be written as

$$F_j = f_j(\cos\delta_j + i\sin\delta_j) \qquad (5.13)$$

or as

$$F_j = f_j e^{i\delta_j} \qquad (5.14)$$

The intensity of a wave is proportional to the square of its amplitude; i.e.

$$I \propto f_j^2 \qquad (5.15)$$

and is obtained by multiplying the equation for the wave by its complex conjugate; i.e.

$$I \propto (f_j e^{i\delta_j})(f_j e^{-i\delta_j})$$

and, therefore

$$I \propto f_j^2$$

Alternatively,

$$[f_j(\cos \delta_j + i \sin \delta_j)][f_j(\cos \delta_j - i \sin \delta_j)] = f_j^2(\cos^2 \delta_j + \sin^2 \delta_j) = f_j^2$$

Substituting the expression for δ, the equation of a diffracted wave becomes

$$
\begin{aligned}
F_j &= f_j \exp 2\pi i(hx_j + ky_j + lz_j) \\
&= f_j[\cos 2\pi(hx_j + ky_j + lz_j) + i \sin 2\pi(hx_j + ky_j + lz_j)]
\end{aligned}
\tag{5.16}
$$

When written in these forms, the summation over the j atoms in the unit cell may be carried out readily, to give the *structure factor* or *structure amplitude*, F_{hkl}, for the *hkl* reflection; i.e.

$$F_{hkl} = \sum_{j=1-n} (f_j e^{i\delta_j})$$

or

$$F_{hkl} = \sum_j f_j(\cos \delta_j + i \sin \delta_j) \tag{5.17}$$

The intensity of the diffracted beam I_{hkl} is proportional to $|F_{hkl}|^2$ and is obtained from

$$
\begin{aligned}
I_{hkl} \propto |F_{hkl}|^2 &= \left[\sum_j f_j(\cos \delta_j + i \sin \delta_j) \right]\left[\sum_j f_j(\cos \delta_j - i \sin \delta_j) \right] \\
&= \sum_j (f_j \cos \delta_j)^2 + \sum_j (f_j \sin \delta_j)^2
\end{aligned}
\tag{5.18}
$$

This latter is a very important formula in crystallography because by using it the intensity of any *hkl* reflection may be calculated from a knowledge of the atomic coordinates in the unit cell. Let us see one example of its use. Calcium fluoride, CaF_2, has the fluorite structure with atomic coordinates in the face centred cubic unit cell:

$$
\begin{array}{ll}
\text{Ca} & (0,0,0) \quad (\tfrac{1}{2},\tfrac{1}{2},0) \quad (\tfrac{1}{2},0,\tfrac{1}{2}) \quad (0,\tfrac{1}{2},\tfrac{1}{2}) \\
\text{F} & (\tfrac{1}{4},\tfrac{1}{4},\tfrac{1}{4}) \quad (\tfrac{1}{4},\tfrac{1}{4},\tfrac{3}{4}) \quad (\tfrac{1}{4},\tfrac{3}{4},\tfrac{1}{4}) \quad (\tfrac{3}{4},\tfrac{1}{4},\tfrac{1}{4}) \\
& (\tfrac{3}{4},\tfrac{3}{4},\tfrac{1}{4}) \quad (\tfrac{3}{4},\tfrac{1}{4},\tfrac{3}{4}) \quad (\tfrac{1}{4},\tfrac{3}{4},\tfrac{3}{4}) \quad (\tfrac{3}{4},\tfrac{3}{4},\tfrac{3}{4})
\end{array}
$$

Substitution of these coordinates into the structure factor equation, (5.18), yields

$$
\begin{aligned}
F_{hkl} &= f_{Ca}[\cos 2\pi(0) + \cos \pi(h + k) + \cos \pi(h + l) + \cos \pi(k + l)] \\
&\quad + if_{Ca}[\sin 2\pi(0) + \sin \pi(h + k) + \sin \pi(h + l)]
\end{aligned}
$$

$$+ \sin \pi(k+l)] + f_F[\cos \pi/2(h+k+l)$$
$$+ \cos \pi/2(h+k+3l) + \cos \pi/2(h+3k+l)$$
$$+ \cos \pi/2(3h+k+l) + \cos \pi/2(3h+3k+l)$$
$$+ \cos \pi/2(3h+k+3l) + \cos \pi/2(h+3k+3l)$$
$$+ \cos \pi/2(3h+3k+3l)] + if_F[\sin \pi/2(h+k+l)$$
$$+ \sin \pi/2(h+k+3l) + \sin \pi/2(h+3k+l) + \sin \pi/2(3h+k+l)$$
$$+ \sin \pi/2(3h+3k+l) + \sin \pi/2(3h+k+3l)$$
$$+ \sin \pi/2(h+3k+3l) + \sin \pi/2(3h+3k+3l)]$$

Since the fluorite structure is face centred cubic, h, k and l must be either all odd or all even for an observed reflection; for any other combination, $F = 0$ (try it!). Consider the reflection 202:

$$F_{202} = f_{Ca}(\cos 0 + \cos 2\pi + \cos 4\pi + \cos 2\pi)$$
$$+ if_{Ca}(\sin 0 + \sin 2\pi + \sin 4\pi + \sin 2\pi)$$
$$+ f_F(\cos 2\pi + \cos 4\pi + \cos 2\pi + \cos 4\pi + \cos 4\pi + \cos 6\pi$$
$$+ \cos 4\pi + \cos 6\pi) + if_F(\sin 2\pi + \sin 4\pi + \sin 2\pi + \sin 4\pi + \sin 4\pi$$
$$+ \sin 6\pi + \sin 4\pi + \sin 6\pi)$$

That is,

$$F_{202} = f_{Ca}(1 + 1 + 1 + 1) + if_{Ca}(0 + 0 + 0 + 0)$$
$$+ f_F(1 + 1 + 1 + 1 + 1 + 1 + 1 + 1)$$
$$+ if_F(0 + 0 + 0 + 0 + 0 + 0 + 0 + 0)$$

or

$$F_{202} = 4f_{Ca} + 8f_F$$

The 202 reflection in CaF_2 has a d-spacing of 1.929 Å ($a = 5.464$ Å). Therefore

$$\theta_{202} = 23.6° \quad \text{and} \quad \sin \theta/\lambda = 0.259 \quad \text{for } \lambda = 1.5418 \text{ Å(Cu } K\alpha)$$

Form factors for calcium and fluorine are given in Fig. 5.36(b); for $\sin \theta/\lambda = 0.259$, by interpolation,

$$f_{Ca} = 12.65 \quad \text{and} \quad f_F = 5.8$$

Therefore,

$$F_{202} = 97$$

This calculation may be made for a series of hkl reflections and the results, after scaling, may be compared with the observed values (Table 5.8). In solving unknown crystal structures, the objective is always to obtain a model structure for which the calculated structure factors, F_{hkl}^{calc}, are in good agreement with those obtained from the experimental intensities, i.e. F_{hkl}^{obs}.

An important feature which simplifies the above calculations is that all the sine terms are zero. This is because the origin of the unit cell is also a centre of symmetry. For each atom at position (x, y, z) there is a centrosymmetrically

Table 5.8 *Structure factor calculations for CaF$_2$*

$d(\text{Å})$	hkl	I	Multiplicity	$I/(\text{multiplicity} \times L_p)$	F^{obs}	F^{calc}	F^{obs} scaled	$\|\,|F^{obs}| - |F^{calc}|\,\|$
3.143	111	100	8	0.409	0.640	67	90	23
1.929	202	57	12	0.476	0.690	97	97	0
1.647	311	16	24	0.098	0.313	47	44	3
1.366	400	5	6	0.193	0.439	75	62	13
1.254	331	4	24	0.047	0.217	39	31	8

$\sum F^{obs}$ scaled $= 324$

$\sum |F^{obs} - F^{calc}| = 47$

$R = \dfrac{\sum |\Delta F|}{\sum F^{obs}} = \dfrac{47}{324} = 0.15$

related atom at $(-x, -y, -z)$ [e.g. F at $(\frac{1}{4},\frac{1}{4},\frac{1}{4})$ and $(-\frac{1}{4}-\frac{1}{4}-\frac{1}{4})$, i.e. $(1-\frac{1}{4}, 1-\frac{1}{4}, 1-\frac{1}{4})$ or $(\frac{3}{4},\frac{3}{4},\frac{3}{4})$] and since $\sin(-\delta) = -\sin\delta$, the summation of the sine terms over the unit cell contents is zero. If, on the other hand, one of the F atoms was taken as the origin of the cell, the sine terms would be non-zero because, F, with its immediate coordination environment of 4Ca arranged tetrahedrally, does not lie on a centre of symmetry. Many structures, of course, belong to non-centric space groups, in which case the complete calculation of F using both cosine and sine terms cannot be avoided.

A further discussion of Table 5.8 is deferred to Section 5.5.5.

5.5.4 Factors that affect intensities

Intensities depend on several factors and not only on the structure factor discussed above. The main factors are:

(a) Polarization factor—angular dependence of intensity scattered by electrons (Section 5.5.1).

(b) Structure factor—dependence on the position of atoms in the unit cell and their scattering power (Section 5.5.3).

(c) Lorentz factor—a geometric factor that depends on the particular type of instrument used and varies with θ. Usually lumped with (a) to give the L_p factor (given in International Tables for X-ray Crystallography, Vol. 2, 266–90).

(d) Multiplicities—the number of reflections that contribute to an observed powder line (Section 5.3.10).

(e) Temperature factor—thermal vibrations of atoms cause a decrease in the intensities of diffracted beams and an increase in background scatter (Section 5.6.8).

(f) Absorption factor—absorption of X-rays by the sample and depends on the form of the sample and geometry of the instrument. Ideally, for single crystal work, crystals should be spherical so as to have the same absorption factor in all directions.

(g) Preferred orientation—occurs if the samples used in powder diffraction do not have a completely random arrangement of crystal orientations (Sections 5.4.1 and 5.6.1).

(h) Extinction—crystals that are nearly perfect have a reduced diffracting power, unimportant in powders.

These factors need to be considered quantitatively only if one is interested in carrying out work related to crystal structure determinations. For usage of (a) single crystal methods to determine unit cells and (b) powder methods to fingerprint materials, it is normal practice to use the raw intensity data without applying any of these correction factors.

5.5.5 R-factors and structure determination

In Section 5.5.3, it was shown how the structure factor, F_{hkl}^{calc}, may be calculated for any hkl reflection from a knowledge of the coordinates of the atoms in the unit cell. The values of F_{hkl}^{calc} for the first five lines in the powder pattern of CaF_2 are given in Table 5.8, column 7. The experimental intensities are given in column 3 and the intensities after correction for the L_p factor and multiplicities (Section 5.5.4(a), (c) and (d) in column 5. The observed structure factor, F_{hkl}^{obs}, is related to the corrected intensities by the relation: $F_{hkl}^{obs} = \sqrt{I_{corr}}$, and these values are given in column 6. In order to be able to compare the values of F_{hkl}^{obs} and F_{hkl}^{calc}, they must be scaled such that $\sum F_{hkl}^{obs} = \sum F_{hkl}^{calc}$. Multiplication of each F_{hkl}^{obs} value by 141 gives the scaled values in column 8. The measure of agreement between the individual, scaled F_{hkl}^{obs} and F_{hkl}^{calc} values is given by the *residual factor* or *R-factor*, defined as follows:

$$R = \frac{\sum ||F^{obs}| - |F^{calc}||}{\sum |F^{obs}|} \qquad (5.19)$$

Values of the numerator are listed in column 9 and an R-factor of 0.15 (or 15 per cent after multiplying by 100) is obtained.

In solving unknown crystal structures, one is guided, among other things, by the value of R; the lower it is, the more likely is the structure to be correct. The calculation given for CaF_2 is rather artificial since only five reflections were used (one normally uses hundreds or thousands of reflections), but it serves as an illustration. It is not possible to give hard and fast rules about the relation between the magnitude of R and the likely correctness of the structure, but, usually, when R is less than 0.1 to 0.2, the proposed structure is essentially correct. A structure which has been solved fully using good quality intensity data has R typically in the range 0.02 to 0.06.

5.5.6 Electron density maps

An electron density map is a plot of the variation of electron density throughout the unit cell. During the processes of solving an unknown structure it is often useful to construct electron density maps (Fourier maps) in order to try and locate atoms. As the structure refinement proceeds, the quality of the electron density map usually improves: the background electron density decreases and, at the same time, more peaks due to individual atoms become resolved. We are concerned here, not with the methods of structure refinement but only with the results, and the final electron density map obtained at the end of a structure determination is an important piece of information. Electron density maps usually take the form of sections through the structure at regular intervals; by superposing these, a three-dimensional picture of the electron density distribution may be obtained. In Fig. 5.39 is shown the electron density distribution for a section through a very simple structure, NaCl. The section is parallel to one face of the unit cell and passes through the centres of the Na^+, Cl^- ions. It has the following features.

An electron density map resembles a geographical contour map. The contours represent lines of constant electron density throughout the structure. Peaks of electron density maxima may be distinguished clearly and these correspond to

Fig. 5.39 Electron density map for NaCl

atoms; the coordinates of the atoms in the unit cell are given by the coordinates of the peak maxima. The assignment of peaks to particular atoms is made from the relative heights of the electron density peaks: the peak height is proportional to the number of electrons possessed by that atom, which apart from very light atoms is approximately equal to the atomic number of that atom. In Fig. 5.39, two types of peak of relative heights 100 and 50, are seen; these are assigned to chlorine and sodium, respectively. (The atomic numbers of chlorine and sodium are 17 and 11; for ions, the number of electrons are 18 (Cl^-) and 10(Na^+). The experimental maxima are therefore in fair agreement with the expected values.)

Electron density maps also show that our mental picture of atoms as spheres is essentially correct, at least on a time average. The electron density drops to almost zero at some point along the line connecting pairs of adjacent atoms in Fig. 5.39 and this supports the model of ionic bonding in NaCl. In other structures which have covalent bonding, there is residual electron density between atoms on the electron density map. However, in other than very simple structures, such as the alkali halides, there is one serious difficulty in using an electron density map to determine quantitatively the distribution of valence electrons. In most structure refinements, both the position and thermal vibration factors of atoms are allowed to vary in order to achieve the best agreement between measured and calculated structure factors and intensities. The final parameters may represent a compromise, therefore, and the electron density map, which is greatly influenced by the thermal vibration factors, is not necessarily a true representation of the distribution of valence electrons. In refining more simple structures, the atomic coordinates are usually known exactly; this gives rather more accuracy to the thermal vibration factors (or temperature factors) and hence to the electron density map.

5.6 Modern X-ray powder techniques and their applications

5.6.1 Powder diffractometers

The most commonly used powder X-ray instrument is the powder diffracto-meter. Its mode of operation is outlined in Section 5.4.1. It has a proportional, scintillation or Geiger counter as the detector which is connected to a chart recorder or sometimes to a means of digital output. In normal use, the counter is set to scan over a range of 2θ values at a constant angular velocity (it is common practice to refer to the angle 2θ between diffracted and undiffracted beams, rather than to the Bragg angle, θ). Usually, the range 10 to 80° 2θ is sufficient to cover the most useful part of the powder pattern. A typical diffractometer trace is shown in Fig. 18.6a for SiO_2. The scale is linear in 2θ and the d-spacings of the peaks may be calculated from Bragg's Law or obtained from standard tables of d versus 2θ. The scanning speed of the counter is usually $2°\ 2\theta\ min^{-1}$ and, therefore, about 30 min are needed to obtain a trace. Intensities are taken as peak heights, unless very accurate work is being done, in which case areas may be measured; the most intense peak is given intensity of 100 and the rest are scaled accordingly.

If very accurate d-spacings or intensities are desired, slower scanning speeds (e.g. $\frac{1}{8}°\ 2\theta\ min^{-1}$) are used. To obtain accurate d-spacings, an internal standard (a pure material, such as KCl, whose d-spacings are known accurately) is mixed in with the sample. A correction factor, which may vary with 2θ, is obtained from the discrepancy between observed and true d-spacings of the standard and is then applied to the pattern that is being measured. Accurate intensities are obtained from peak areas by cutting out the peaks and weighing them, by measuring their area with a device such as a planimeter or by using an automatic counter fitted to the diffractometer.

Samples for diffractometry take various forms: they include thin layers of the fine powder sprinkled onto a glass slide smeared with vaseline and thin flakes pressed onto a glass slide. Different people prefer different methods of sample preparation and the objective is always to obtain a sample which contains a random arrangement of crystal orientations. If the crystal arrangement is not random, then *preferred orientation* exists and can introduce errors, sometimes very large, into the measured intensities. Preferred orientation is a serious problem for materials that crystallize in a characteristic, very non-spherical shape, e.g. clay minerals which usually occur as thin plates or some cubic materials which crystallize as cubes and, on crushing, break up into smaller cubes. In a powder aggregate of such materials, the crystals tend to sit on their faces, resulting in a far from random average orientation.

The big disadvantage of early Debye–Scherrer cameras is that incident and

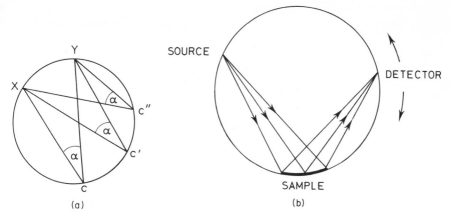

Fig. 5.40 (a) Theorem of a circle used to focus X-rays. (b) Arrangement of sample, source and detector on the circumference of a circle

diffracted beams are, inevitably, somewhat divergent and of low intensity. In diffractometers and modern focusing cameras, a convergent X-ray beam is used; this gives a dramatic improvement in resolution and, because much more intense beams may be used, exposure times are greatly reduced. It is not possible to focus or converge X-rays using the X-ray equivalent of an optical lens; instead, use is made of certain geometric properties of the circle in order to obtain a convergent X-ray beam. These properties are illustrated in Fig. 5.40(a). The arc XY forms part of a circle and all angles subtended on the circumference of this circle by the arc XY are equal, i.e. $XCY = XC'Y = XC''Y = \alpha$. Suppose that X is a source of X-rays and XC, XC′ represent the extremities of a divergent X-ray beam emitted from X. If the beam is diffracted by a sample which covers the arc between C and C′ such that the diffracting planes are tangential to the circle, then the diffracted beam, represented by CY and C′Y, will focus to a point at Y. The principle of the focusing method is therefore to arrange that the source of X-rays, the sample and the detector all lie on the circumference of a circle (Fig. 5.40b).

The focusing geometry of the diffractometer is shown schematically in Fig. 5.41. The *focusing circle* is dashed and has the source, S, sample and the receiving slit of the detector at F, all on its circumference. The focusing circle is not of constant size but decreases in radius as the Bragg angle θ increases (movement of F around the diffractometer circle with increasing θ is indicated by the arrow). The importance of having a flat sample surface can be seen (ideally it should be curved and change its radius of curvature with scanning angle, but this is not practicable); if the surface is uneven, or deviates much from the circumference of the circle, then the focusing action is lost.

The solid circle in Fig. 5.41 is the *diffractometer circle* and is of constant size. The sample is at its centre and the detector, F, scans around its circumference. In order to preserve the focusing action with changing 2θ, the surface of the sample must stay tangential to the focusing circle. This is achieved by coupling the

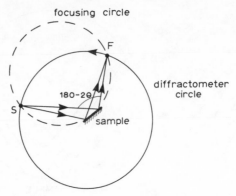

Fig. 5.41 Focusing geometry of the diffractometer

sample to the detector so that as the detector scans at angular velocity $2\theta\,\mathrm{min}^{-1}$, the sample rotates in the same direction at $\theta\,\mathrm{min}^{-1}$.

5.6.2 Focusing (Guinier) cameras

The same focusing principle that is basic to the construction of diffractometers is also used in focusing cameras, although several different arrangements are found in commercial instruments. An addition feature of focusing cameras is the inclusion of a *crystal monochromator* which serves two functions: to give highly monochromatic radiation and to produce an intense, convergent X-ray beam. There are several sources of background scattering in diffraction experiments (Section 5.6.8), one of which is the presence of radiation of wavelength different from that of the $K\alpha$ radiation. $K\alpha$ radiation may be separated from the rest by the use of filters or, better, by a crystal monochromator.

A crystal monochromator consists of a large single crystal of, for example, quartz, oriented such that one set of planes which diffracts strongly ($10\bar{1}1$ for quartz) is at the Bragg angle to the incident beam. This Bragg angle is calculated for $\lambda_{K\alpha_1}$ and so only the $K\alpha_1$ rays are diffracted, giving monochromatic radiation. (In fact, overtone reflections may occur because the $(20\bar{2}2)$ planes diffract X-rays of wavelength $\frac{1}{2}\lambda_{K\alpha}$ at the same Bragg angle. It is an easy matter to ensure that these overtone reflections have weak or negligible intensity.) If a flat crystal monochromator were used, much of the $K\alpha$ radiation would be lost since the X-ray beam emitted from a source is naturally divergent; only a small amount of the $K\alpha$ component would therefore be at the correct Bragg angle to the monochromator. To improve the efficiency, the crystal monochromator is bent, in which case a divergent X-ray beam may be used which is diffracted by the crystal monochromator to give a beam that is intense, monochromatic and convergent.

The arrangement of a focusing or Guinier camera which uses a crystal monochromator M and also makes use of the theorem of the circle described above is shown in Fig. 5.42. The convergent beam of monochromatic radiation

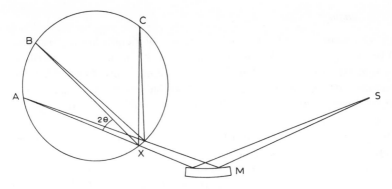

Fig. 5.42 Crystal monochromator M, source S and sample X, in a focusing camera

passes through the sample at X. Radiation that is not diffracted comes to a focus at A, where a beam stop is placed in front of the film to prevent its blackening. Various beams diffracted by the sample focus at B, C, etc. We know from the theorem of the circle that A, B, C and X must lie on the circumference of a circle. The film is placed in a cassette which is in the form of a short cylinder and lies on the circle ABC. The scale of the film is linear in 2θ, as is the chart output from a diffractometer. A schematic film is as shown in Fig. 3.1 except that instead of peaks of different height, lines of different intensity or different degrees of blackness are seen. Film dimensions are $\sim 1 \times 15$ cm which makes them very convenient to handle. The line at $0°\,2\theta$ or ∞ d-spacing corresponds to the undiffracted beam at A in Fig. 5.42. This is the reference position on the film. The mark is made by removing the beam stop for a fraction of a second while the X-rays are switched on. If required, a scale may be printed onto the film and the positions of the lines, relative to A, may be measured with a travelling microscope or, better, by microdensitometry; 2θ values and d-spacings may then be computed or obtained from tables.

The Guinier method is capable of giving accurate d-spacings, if desired, and the results are comparable to those obtained by diffractometry using very slow scanning speeds. Intensities of the lines on the films are either estimated visually or may be measured quantitatively using microdensitometry. Sample sizes are very small, 1 mg or less, and necessary exposure times vary between 5 min and 1 hr, depending on factors such as the crystallinity of the sample and the presence or absence of heavy elements which absorb X-rays.

5.6.3 Measurement of powder patterns and comparison of diffractometry with film methods

A powder pattern has three main features that may be measured quantitatively. In decreasing order of relative importance, these are (a) d-spacings, (b) intensities and (c) line profiles.

5.6.3.1 *d-spacings*

For routine measurements and for purposes of identification, no special care need be taken over sample preparation nor in the measurement of the film or diffractometer trace. For identification of completely unknown materials, diffractometry is probably quicker and somewhat easier. It is important to have approximate intensity values as well as reasonably accurate *d*-spacings and both may be determined directly from the diffractometer chart. Further, with film methods, there is a delay time of 1 to 2 hours during which the film is developed and prepared ready for examination.

The great advantage of film methods over diffractometry occurs when the powder patterns of different samples are to be compared. It is almost impossible to compare several 1 metre lengths of chart paper by trying to match or superpose the different traces. Yet with small Guinier films, several may be compared directly on a viewing screen. For work in specific areas, e.g. clay minerals, one can soon build up a file of standard films of all the phases likely to be encountered. Identification of unknowns and, probably more important, mixtures of unknowns then becomes straightforward and rapid.

For accurate measurement of *d*-spacings, diffractometry is normally regarded as the best method and most of the patterns in the powder diffraction file have been obtained by diffractometry. An internal standard of accurately known *d*-spacings must be added to the sample in order to eliminate instrumental error. A slow scanning speed, e.g. $\frac{1}{8}° 2\theta \, min^{-1}$, is used so that the scale of the trace may be greatly expanded, and if possible only high angle reflections are used. A powder pattern obtained by either diffractometry or film methods is squashed up at its low angle end and hence accurate *d*-spacings are best measured in the back-reflection region using high angle reflections. Care must be taken that the diffractometer is well adjusted and that a smooth sample surface is presented to the incident beam so as to give good focusing action. The disadvantage of this method is that with the slow scanning speeds several hours may be needed to record a significant part of the powder pattern.

The superiority of diffractometry for accurate measurement of *d*-spacings is now being challenged by advances in focusing camera techniques. Both the position and intensity of lines may be determined from a microdensitometric scan of a Guinier film. The resulting *d*-spacings may be as accurate as those obtained by diffractometry and the process is much quicker: only a short exposure time is required to take the photograph, irrespective of the subsequent use that is made of the film.

5.6.3.2 *Intensities*

It is by no means a straightforward exercise to obtain reliable powder X-ray intensity data. Sample preparation is very important as it may be difficult if not impossible to avoid preferred orientation of crystals within the powder specimen. Powders should be ground down, preferably to size of 1 to 10 μm, and it may be

Table 5.9 *Comparison of diffractometers and focusing cameras*

Feature	Diffractometer	Focusing camera
Exposure time	30 min usually	10 min–1 hr
Accuracy of 2θ values	Good–very good	Good–very good
Intensities	Very good	Poor–fair
Peak shape	Very good	Poor–fair
Comparison of different samples	Poor (clumsy)	Excellent
Resolution of closely spaced lines	Good–excellent	Excellent
Amount of sample required	0.05–2 g	~ 1 mg
Storage and retrieval of results	Clumsy, unless done by computer	Easy
Approximate cost of equipment (excluding generator)	£15000	£5000

For comparison purposes the following were used: 1020 diffractometer; Hägg focusing camera (both Philips).

worthwhile to sieve samples prior to X-ray diffraction. One or two large (e.g. 1 mm diameter) crystals in an otherwise fine powder can cause havoc with intensity measurements.

Intensities are normally measured by diffractometry, as peak heights or peak areas at slow scanning speeds. It is difficult to obtain quantitative intensities from films unless a microdensitometer is also available.

5.6.3.3 *Peak shape (line profiles)*

For certain specialized applications, the shape of peaks may yield valuable information. Peaks have a finite breadth, for reasons to be discussed later, but extra broadening may occur if (a) stresses are present in the crystals, e.g. in metal pieces that have been cold worked (Section 5.6.6), or (b) the size of the crystals is less than about 2000Å diameter (Section 5.6.5). The standard method for measuring peak profiles is diffractometry.

The relative merits of diffractometers and focusing cameras are summarized in Table 5.9.

5.6.4 High temperature powder diffraction

Several commercial instruments are available for recording powder patterns at high temperatures. Some are diffractometers fitted with a small furnace around the sample. The powder pattern is recorded in exactly the same way as for room temperature operations. Inert, refractory construction materials such as tungsten and iridium are used and very high temperatures, e.g. 2000 °C, are attainable.

Film methods are also available and a particularly elegant one is the Guinier–

172

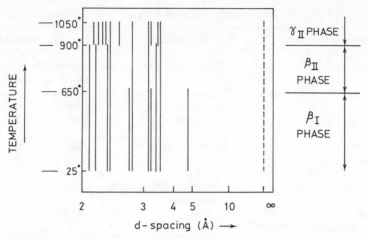

Fig. 5.43 Schematic high temperature Guinier powder photograph of
Li_2ZnSiO_4

Lenne camera which operates up to $\sim 1200\,°C$. A thin powdered specimen, mounted on a fine platinum gauze is suspended in the middle of a small furnace. A quartz crystal monochromator is used to provide a convergent X-ray beam and the diffraction pattern is recorded on film by the focusing method. The sample and furnace may be programmed to heat or cool at a certain rate and a continuous X-ray photograph is taken. The film is rectangular and bent to lie on the focusing circle (cylinder) of the camera. It can be translated at a constant velocity such that only a narrow strip of film, e.g. 5 mm wide, is exposed to the diffracted X-ray beams at any one time. A schematic photograph showing the polymorphic changes that occur on heating Li_2ZnSiO_4 is shown in Fig. 5.43. The horizontal axis is 2θ or d-spacing, as usual, and the vertical axis is temperature. One advantage of this particular camera is that because a continuous record is obtained, it is possible to follow phase transformations directly. This is often more useful than having two separate patterns that had been recorded before and after the changes involved took place.

Let us briefly consider the changes that appear on the photograph. With increasing temperature, the sequence $\beta_I \xrightarrow{650\,°C} \beta_{II} \xrightarrow{900\,°C} \gamma_{II}$ Li_2ZnSiO_4 is observed. For the $\beta_I \rightarrow \beta_{II}$ transition, some of the lines in the β_I powder pattern simply disappear to give the powder pattern of β_{II}. Such a transition is characteristic of order–disorder phenomena (in this case, the ordering is probably orientational ordering of MO_4 tetrahedra) in which the low temperature phase is an ordered superstructure of the disordered high temperature phase. The $\beta_{II} \rightarrow \gamma_{II}$ transition is rather different in that some of the β_{II} lines disappear and new lines appear in the γ_{II} powder pattern. This indicates that some major structural reorganization occurs during the transition. The discontinuity in some of the lines indicates that a change in volume accompanies the $\beta_{II} \rightarrow \gamma_{II}$ transition and allows us to classify the transition thermodynamically as first order (see Chapter 12).

High temperature powder diffraction is invaluable for identifying and studying phases that exist only at high temperatures. Many phases, e.g. β-quartz stable above 573 °C, undergo a phase change during cooling (to α-quartz) and no matter how fast the cooling rate, the transition cannot be suppressed. The only way to study such phases is, therefore, at high temperatures.

A more technical application of high temperature powder diffraction is in the measurement of coefficients of thermal expansion, data which for some materials may be very difficult to obtain using conventional dilatometry. For non-cubic crystals, the expansion is usually anisotropic and the different axial expansion coefficients may be determined readily by the X-ray method. Knowledge of expansion coefficients is very important for materials that are used in high temperature environments or which experience large temperature changes during use, e.g. some metals and ceramics.

5.6.5 Effect of crystal size on the powder pattern—particle size measurement

If the average crystal size in a powder is below a certain limit ($\sim 2000\,\text{Å}$ diameter), additional broadening of diffracted X-ray beams occurs. From measurement of this extra broadening an average particle size may be obtained. In the absence of extra broadening due to small particle size, powder lines or peaks have a finite breadth for several reasons: the radiation is not absolutely monochromatic, the $K\alpha$ line has an intrinsic breadth due to the Heisenberg uncertainty principle and the focusing geometry of the instrument may not be perfect for a variety of reasons. In order to understand why small particle size leads to line broadening it is necessary to consider the conditions under which diffraction may occur if the incident angle is slightly different from the Bragg angle, θ_B. A qualitative explanation is as follows.

The Bragg angle represents the condition under which each plane in a crystal diffracts exactly one wavelength later than the preceding plane. All diffracted beams are therefore in phase and constructive interference occurs. For an incident beam at a slightly greater angle, θ_1 (Fig. 5.44), there is a phase lag of slightly greater than one wavelength, $\lambda + \delta\lambda$, for rays diffracted from subsequent planes. By the time the $(j + 1)$th plane is reached, let the cumulative, incremental

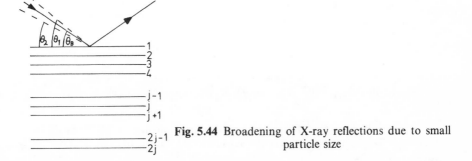

Fig. 5.44 Broadening of X-ray reflections due to small particle size

phase lag $\sum \delta\lambda$, be equal to half a wavelength; i.e.

$$j\delta\lambda = \lambda/2$$

Planes 1 and $(j + 1)$ are exactly π out of phase for radiation that is incident and diffracted at θ_1 and, therefore, cancel each other. If the crystal contains a total of $2j$ planes, then the net diffracted intensity at θ_1 is equal to zero because the rays diffracted from planes $1 \to j$ exactly cancel the rays diffracted from planes $(j + 1) \to 2j$. The angular range θ_B to θ_1 is the range over which the intensity of the diffracted beam falls from a maximum, at θ_B, to zero, at θ_1. A similar lower limiting angle, θ_2, occurs for which rays diffracted from adjacent planes have a phase difference of $\lambda - \delta\lambda$.

The magnitude of the angular range θ_1 to θ_2, and hence the breadth of the diffraction peak, is governed by the number of planes $2j$, and hence the crystal thickness. If the number of planes is very large, no significant broadening occurs because $\delta\lambda$ and therefore $(\theta_2 - \theta_1)$ is negligibly small. The commonly accepted formula for particle size broadening is the Scherrer formula:

$$t = \frac{0.9\lambda}{B\cos\theta_B} \tag{5.20}$$

where t is the thickness of the crystal (in angstroms), λ the X-ray wavelength and

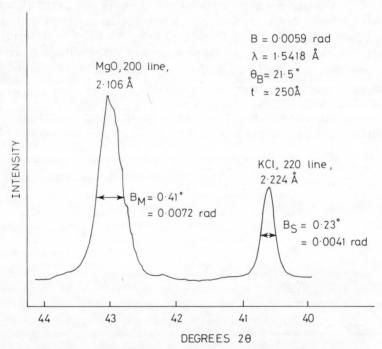

Fig. 5.45 Part of a diffractometer trace for a mixture of MgO (small particle size) and KCl (internal standard)

θ_B the Bragg angle. The line broadening, B, is measured from the extra peak width at half the peak height and is obtained from the Warren formula:

$$B^2 = B_M^2 - B_S^2 \tag{5.21}$$

where B_M is the measured peak width in radians at half peak height (Fig. 5.45) and B_S is the corresponding width of a peak of a standard material, mixed in with the sample, whose particle size is considerably greater than 2000 Å and which has a diffraction peak near to the relevant peak of the sample.

With good experimental techniques, broadening of high angle lines may be detected for crystal thickness up to ~ 2000 Å (e.g. for a crystal containing 2000 planes of d-spacing 1 Å). For a thickness of 50 to 500 Å the broadening is very easy to detect and measure (Fig. 5.45). The lower limit of detection occurs when the peaks become so broad that they disappear into the background radiation. For very small particle size, it is best to use low angle peaks if possible because, for a given crystal thickness, the broadening increases with angle. In extreme cases, peaks may be observable at low angles while high angle peaks cannot be distinguished from the background.

5.6.6 Effect of stress on a powder pattern

Crystals that are under stress may exhibit anomalous powder patterns. The whole powder pattern may be shifted to lower d-spacings if the crystals are under a uniform compressive stress such that a contraction of the unit cell occurs. If the stress is non-uniform, different crystals or different parts of the same crystal may be deformed to differing degrees and the powder lines become broadened. Commonly, both effects occur and lines may be both displaced and broadened.

Stresses may be (a) caused by the application of an external pressure or (b) generated internally as a consequence of a chemical reaction taking place inside the crystals. An example of (a) is the work hardening of metals in which residual distortions are present in the crystals after treatment. Examples of (b) are more varied and include coherent precipitation of supersaturated solid solutions (age hardening of metals and ceramics) and the occurrence of some phase transitions during cooling (if there is a change in volume or shape of the crystals and they are embedded in a solid matrix, then the rigid environment of the matrix may prevent the transition from occurring to completion).

5.6.7 Refinement of unit cell parameters and indexing of powder patterns

Unit cell parameters are often determined with the aid of single crystal X-ray photographs. The symmetry or unit cell type is first obtained from an inspection of the photographs. Positions of selected spots are then measured as accurately as possible to give values for the unit cell parameters. There are several intrinsic limitations to the accuracy of the values thus obtained—lack of internal standard, shrinkage of film, etc.—and accuracy of axial parameters is usually to between 0.05 and 0.2 per cent. Angles (for monoclinic and triclinic unit cells) can usually be measured to about 1°*. For many materials and applications, such values are sufficiently accurate, e.g. if determination of the unit cell is merely one step in solving a crystal structure. It is very often the case, however, especially in solid state chemistry, that there is intrinsic interest in the powder patterns of crystalline phases and more accurate cell dimensions are desired. It is usually essential to assign Miller indices to the powder lines and often a line cannot be indexed unambiguously if the cell parameters are known only approximately. On the other hand, accurate cell parameters may be obtained from a least squares refinement on the d-spacings of at least several high angle powder lines whose indexing is known for certain. A circular situation may exist in which the determination of accurate lattice parameters and indexing of the powder pattern are intimately related and one is not possible without knowledge of the other. Usually, with patience, the problem can be solved by an iterative method especially if at least a few low angle lines may be indexed unambiguously. Least squares refinement of their d-spacings lead to more accurate cell parameters. The theoretical d-spacings calculated for these new cell parameters then enable a few more lines to be indexed with certainty and the least squares cycle is repeated. By this method, axial parameters may be obtained routinely accurate to 0.002 per cent and angles accurate to $\sim 0.1°$.

For powder patterns which are particularly difficult to index, i.e. for which there are two or more plausible sets of (hkl) values for some or all of the lines, single crystal photographs may be additionally useful. The intensities of the various candidate (hkl) reflections may be estimated qualitatively from the indexed single crystal photographs and the strongest of these almost certainly corresponds to the reflection that gives the powder line. This is because spots that are weak on single crystal photographs are not usually observed in powder photographs unless the latter are grossly overexposed.

Various computer programs are available with which one can supposedly index powder patterns without the necessity for prior knowledge of the crystal system. Great care must be exercised in using these programs since they can give information on the unit cell and cell parameters which may be incorrect or misleading. If one has an independent check that a particular unit cell, which has been used in the program to index the powder lines, is indeed the correct one, then these programs are useful; if not, the results are, at best, only tentative.

Problems may arise particularly for crystals which have neither cubic nor triclinic crystal systems. For cubic materials, the d-spacings are controlled by only one parameter, the cubic cell edge a, and it is a straight forward matter to index cubic powder patterns. For triclinic crystals, there is no single unit cell which is the correct one and the choice of unit cell is of no great consequence. For all other crystal systems, however, the powder patterns are controlled by two or more variables and computer fitting of d-spacing data with speculative unit cell parameters is not necessarily reliable.

*Greater accuracy is obtained using modern single crystal diffractometers, typically 0.03% for axes and $\pm 0.05°$ for angles.

5.6.8 Sources of background radiation—fluorescence

The quality of an X-ray powder diffraction pattern is governed to a considerable extent by the level of background radiation which is present. In mild cases it may be difficult to pick out the weaker reflections but in serious cases, as when *fluorescence* occurs, the intensity of all the diffracted beams may be reduced considerably at the same time that a large increase in background scattering occurs.

Several sources of background scattering, with their remedies, are as follows:

(a) Collisions between air molecules and diffracted X-ray beams. For high quality Guinier focusing films it is worth while to evacuate the box containing the sample and film.

(b) The presence of white radiation in the incident beam. This is best eliminated by using a single crystal monochromator.

(c) Fluorescence. This occurs when the sample acts as a secondary source of X-rays. If the fluorescent radiation is weak it may be absorbed by placing a filter between the sample and detector, e.g. a strip of nickel foil placed over the film. If it is strong, it is best to change the wavelength of the primary beam, e.g. by replacing an X-ray tube containing a copper target by one with an iron or molybdenum target.

Fluorescence occurs when the radiation in the primary beam (i.e. emitted by the copper target) knocks out inner shell electrons within atoms of the sample. Electrons in the outer shells drop down to occupy empty levels in the inner shells and, in so doing, they emit their excess energy in the form of X-rays. The sample is therefore acting as a secondary source of X-rays. The amount of fluorescent radiation produced in this way depends on the atomic number of the atoms in the sample relative to that of the target material and is best seen by example. Cu $K\alpha$ radiation, of wavelength 1.5418 Å, is generated by the electronic transition $2p \rightarrow 1s$ (Fig. 5.1). In order to create a vacant $1s$ level in the first place, the $1s \rightarrow \infty$ ionization potential is needed and this energy difference corresponds to a wavelength for copper of 1.3804 Å, Fig. 3.19. Incident X-rays of wavelength somewhat less than or equal to 1.3804 Å may therefore ionize a Cu/$1s$ electron. Similarly, Cu $K\alpha$ X-rays may ionize electrons in atoms whose ionization energy has a value corresponding to $\lambda > 1.5418$ Å. Ionization potentials of $1s$ electrons in nickel, cobalt and iron correspond to 1.4880, 1.6081 and 1.7433 Å and therefore Cu $K\alpha$ radiation may ionize $1s$ electrons in cobalt and iron but not in nickel. Samples containing cobalt and iron fluoresce strongly in Cu $K\alpha$ radiation. Lighter atoms also fluoresce, but less strongly, since fluorescence is strongest when the incident radiation has a wavelength that is only slightly shorter than the absorption edge ($\equiv IP$) of the atoms.

(d) Compton scattering. When an X-ray beam strikes a sample two types of scattered X-rays are produced. In the first, the incident beam sets the electrons of the atoms in to vibration; X-rays of the same wavelength as the incident beam are re-emitted and are the characteristic diffracted radiation with which we are familiar. This is called *coherent, unmodified radiation*. The incident X-ray beam may also interact inelastically with the outer, more loosely bound electrons of the sample atoms. Some of the energy of the X-rays is inevitably lost and the resulting scattered X-rays, *modified Compton X-rays*, have slightly longer wavelength than the incident beam. Compton scattering contributes to the general background scatter and is particularly serious for the lighter elements. For this reason, organic and organic-based (e.g. polymeric) materials may give poor quality powder patterns, due to the combined effects of reduced diffracted intensity and increased background intensity. The intensity of Compton scattering increases with increasing angle (in contrast to diffracted radiation which decreases in intensity at higher angles; see Section 5.5) and so it is common to see powder patterns of, for example, polymers which have well-defined strong lines at high d-spacings but only a general background scattering at lower d-spacings. There is no real remedy for Compton scattering.

(e) Crystal imperfections and temperature diffuse scattering. Any kind of imperfection in the crystals of the sample causes a certain amount of diffuse scattering at angles other than the various Bragg angles. It cannot be avoided. The ideal powder pattern is obtained for crystals with perfect three-dimensional regularity, free from strain, imperfections, surface effects and at absolute zero. The effect of small particle size and non-uniform stress on powder peaks has been mentioned earlier. Atomic vibrations are an important source of diffuse scattering and these increase as the melting point is approached. It is particularly noticeable in high temperature powder diffraction results that the intensities of powder lines get progressively weaker as the temperature is raised and at the same time the background scattering increases. A useful guideline is that for a given material and set of conditions the total diffracted intensity is constant. If peak intensities are reduced, the intensity must reappear elsewhere—probably in the background.

5.6.9 A powder pattern is a crystal's 'fingerprint'

The powder X-ray diffraction method is very important and useful in qualitative phase analysis because every crystalline material has its own characteristic powder pattern; indeed, the method is often called the powder fingerprint method. There are two main factors which determine powder patterns: (a) the size and shape of the unit cell and (b) the atomic number and position of the various atoms in the cell. Thus, two materials may have the same crystal structure but almost certainly they have quite distinct powder patterns. For example, KF, KCl and KI all have the rock salt structure and should show the same set of lines in their powder patterns, but, as can be seen from Table 5.7, both the positions and intensities of the lines are different in each. The positions or d-spacings of the lines are shifted because the unit cells are of different size and, therefore, the a parameter in the d-spacing formula varies. Intensities are different because different anions with different atomic numbers and therefore different scattering powers are present in the three materials, even though the atomic coordinates are the same for each (i.e. cations at corner and face centre positions, etc.). KCl is a rather extreme example because the intensities of 111 and 311 reflections are too small to measure, but it serves to illustrate the importance of scattering power of the atoms present. Intensities are discussed in more detail in Section 5.5.

The powder pattern has two characteristic features, therefore: the d-spacings of the lines and their intensity. Of the two, the d-spacing is far more useful and capable of precise measurement. The d-spacings should be reproducible from sample to sample unless impurities are present to form a solid solution or the material is in some stressed, disordered or metastable condition. On the other hand, intensities are more difficult to measure quantitatively and often vary from sample to sample. Intensities can usually be measured only semi-quantitatively and may show variation of, say, 20 per cent from sample to sample (much more if preferred crystal orientation is present). Thus, the differences in tabulated intensities for, say, the 220 reflection of the three materials in Table 5.7 are probably not too significant.

The likelihood of two materials having the same cell parameters and d-spacings decreases considerably with decreasing crystal symmetry. Thus, cubic materials have only one variable, a, and there is a fair chance of finding two materials with the same a value. On the other hand, triclinic powder patterns have six variables, a, b, c, α, β and γ, and so accidental coincidences are far less likely. Problems of identification, if they occur at all, are most likely to be experienced with high symmetry, especially cubic, materials.

5.6.10 Structure determination from powder patterns

Although structure determination is normally carried out using single crystal X-ray data, there are instances where powder data can be used and may even be advantageous. The structures of metals and alloys have generally been solved from powder data. They are often cubic, hexagonal or tetragonal and it is a straightforward exercise to index their powder patterns and calculate the cell dimensions. Many or all of the atoms in the unit cell lie on special positions such as the origin, face centres, body centres, etc., and so the number of positional parameters which is variable and must be determined is either small or zero. The proposed structure can then be confirmed by comparing the intensities calculated from the model with those observed experimentally; there may be only 5 to 10 lines observed in the powder pattern but this is sufficient for the purpose.

Structure determination of non-metallic materials by the powder method is rather more difficult because these materials often have unit cells of low symmetry and/or there are a considerable number of positional variables for the atoms in the unit cell. Occasionally, if a suitable single crystal of the new phase cannot be obtained and if the crystal system and unit cell parameters may be determined by electron diffraction, the powder intensities may be used for a structure determination (electron diffraction intensities are very unreliable, even when compared with powder X-ray intensities). The limit to the complexity of structure which can be tested is governed by the number of powder lines whose intensities can be measured and whose indexing is known for certain. Solving a structure is akin to solving a set of simultaneous equations; there must not be more variables than equations if the equations are to be solved.

Although the powder method is not often used for the determination of completely unknown structures it does have many secondary uses related to crystal structures and can be a relatively quick method of obtaining results. For example, suppose that one has prepared a new phase which apparently has a perovskite structure. Computer programs are available which generate a calculated powder pattern from a given set of atomic coordinates. Into the program are fed the coordinates of the proposed structure together with the scattering factor data for the atoms in the structure. The program then calculates the intensities of all possible reflections and these may be compared with the observed powder intensities of the new phase in order to judge whether or not the postulated structure is correct.

In the above example, no refinement in atomic coordinates of a proposed structure is carried out; all that is done is to test a proposed model for correctness. Other cases exist where a limited part of a structure may be in doubt and powder data may be used to decide which of the various possibilities is correct. Good examples are spinels. These may be normal, inverse or intermediate; the different forms are distinguished by the way in which some or all of the cations arrange themselves over the available sites. For a particular material the powder pattern may be calculated for different cation arrangements and that which gives best agreement between observed and calculated intensities is probably correct. For problems such as this there is really no need for single crystal data and solution of the problem can be obtained within a matter of days instead of the weeks or months that are needed to carry out a single crystal analysis.

It has already been mentioned that powder data may be used for the study of materials which cannot be prepared in suitable-sized single crystal form. A related use is in the crystallographic study of materials at high temperatures. High temperature single crystal methods are not often used whereas high temperature powder diffractometry is an established technique. The structural transition at $\sim 270\,°C$ in cristobalite, SiO_2, has been followed by powder diffractometry. The structure of the high temperature form was solved—it has a disordered structure with a choice of six positions for each oxygen in a non-linear Si–O–Si bridge—and the changes that occur during the low–high transition were deduced.

Powder X-ray methods are invaluable for obtaining structural information about solid solutions—substitutional, interstitial or otherwise—and order-disorder phenomena in, for example, alloys. A further discussion is given in Chapter 10.

5.6.11 Powder patterns from single crystals—the Gandolfi camera

There are several instances where it may be desirable to have the powder pattern appropriate to a single crystal sample. To achieve this, a modified Debye–Scherrer camera, known as the Gandolfi camera, is used. In this camera, the crystal sample is stuck onto a fibre and mounted on a rotation device. This device causes the crystal to rotate, simultaneously and at different angular velocities, about two axes. Over a period of time, the crystal presents itself to the X-ray beam in a very large number of orientations. A completely random time-averaged orientation of the crystal is not achieved but nevertheless a sufficiently large number of orientations is presented that a recognizable powder pattern may be obtained after an exposure time of 1 to 2 days.

The Gandolfi camera is used in mineralogy in cases where a specimen is rare or unusual and should be preserved intact rather than crushed up in order to take a conventional X-ray powder pattern. In other cases only one or two grains of a particular crystalline phase may be available and there is insufficient material for a conventional powder pattern.

5.6.12 Powder patterns calculated from crystal structure data

Crystal structures are solved using intensity data, usually from single crystals but occasionally from powders (Section 5.6.10). It is sometimes useful to carry out the reverse exercise and, for a known crystal structure, calculate the corresponding powder pattern. This involves calculation of the d-spacings at which lines appear and their corresponding intensities. Calculation of d-spacings is straightforward and requires knowledge only of the unit cell dimensions and the appropriate d-spacing formula (see Section 5.3.7 and Appendix 6). Calculation of intensities is a little more complicated and requires a knowledge of the crystal structure and, in particular, the coordinates of all the atoms in the unit cell. The first step is to calculate the structure factors, F_{hkl}^{calc}, for all possible reflections of interest. The procedure for doing this is given in Section 5.5.3. The second step is to convert the structure factors to intensities (equation 5.18) and correct for the L_p factors and multiplicities.

Calculated powder patterns have several uses. They can be used:

(a) to provide reference intensity data for materials which are subject to preferred orientation, e.g. crystals with a platy texture,
(b) to show that a single crystal, whose structure has been solved, is representative of a bulk sample from which the crystal was taken,
(c) if observed and calculated powder patterns match well, to confirm that a new crystalline phase has the crystal structure which has been postulated for it,
(d) to generate powder patterns for hypothetical crystal structures and
(e) to assist in indexing complex powder patterns, i.e. in deciding the appropriate hkl values for a particular powder line.

182

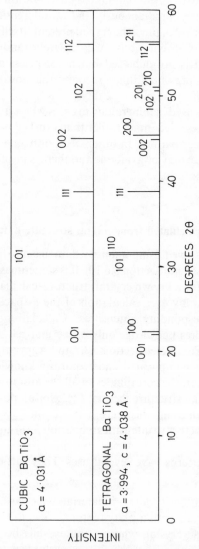

Fig. 5.46 Powder patterns of cubic and tetragonal BaTiO₃ showing the influence of crystal symmetry and multiplicities on the number of lines that are observed.

5.6.13 Influence of crystal symmetry and multiplicities on powder patterns

A general observation about the relative complexity of powder patterns is that the number of lines which appear increases with decreasing crystal symmetry. Thus, simple cubic substances give only a few lines whereas triclinic materials may give up to a hundred. This is explained in Section 5.3.10 as being due to the effect of multiplicity; cubic materials do have a large number of lines, just as many as a material with a similar sized, triclinic unit cell, but in cubic materials, many of the lines overlap and the number of distinct lines that may be seen is greatly reduced. A simple example of this is shown in Fig. 5.46; schematic powder patterns are given for two polymorphs of the perovskite phase, $BaTiO_3$. One polymorph is cubic. The other is tetragonal but the distortion from a cubic-shaped unit cell is not large; the c axis is about 1 per cent longer than a. As can be seen in Fig. 5.46, the tetragonal distortion leads to an increase in the number of observed powder lines. Thus 001 and 100 appear as separate lines because they have different d-spacings, whereas they overlap in the cubic polymorph; similarly 110 and 101 ($\equiv 011$) appear as separate lines in the tetragonal polymorph but overlap in the cubic form. Not all lines in the cubic form separate into doublets in

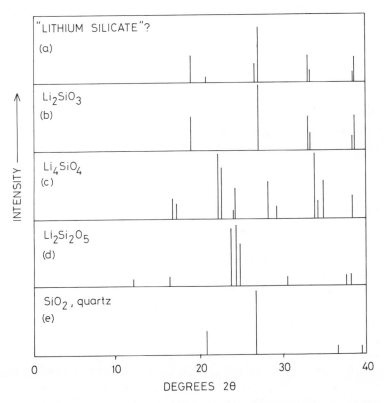

Fig. 5.47 Powder patterns of (a) a bottle labelled 'lithium silicate' and (b) to (e) standard lithium silicate and silica phases

the tetragonal form, however. Thus, 111 stays as a single line but 102 separates into three lines, 102, 201 and 210.

5.6.14 Powder patterns of mixtures of phases

Mixtures of crystalline phases may be analysed very easily and effectively using Guinier films. An illustration of the method is given in Fig. 5.47. Some time ago, the author purchased a bottle of what was advertised as Li_4SiO_4. The bottle arrived labelled 'lithium silicate'; part of its X-ray powder pattern is shown schematically in (a) together with patterns of standard lithium silicate phases and silica in (b) to (e). Comparison of the films showed that the bottle contained predominantly Li_2SiO_3 and a small amount of quartz, but none of the expected phase, Li_4SiO_4!

Questions

5.1 Using the $K\alpha_1$ data of Table 5.1, verify graphically Moseley's Law. What wavelength do you expect for $Co\,K\alpha_1$ radiation?

5.2 What symmetry elements do the following tetrahedral-shaped molecules possess: (a) CH_3Cl, (b) CH_2Cl_2, (c) CH_2ClBr?

5.3 Show that the following Bravais lattices are equivalent:
   (a) C-tetragonal and P-tetragonal
   (b) F-tetragonal and I-tetragonal
   (c) B-monoclinic and P-monoclinic (b unique axis)
   (d) C-monoclinic and I-monoclinic (b unique axis)

5.4 What is the probable lattice type of crystalline substances that give the following observed reflections?
   (a) 110, 200, 103, 202, 211
   (b) 111, 200, 113, 220, 222
   (c) 100, 110, 111, 200, 210
   (d) 001, 110, 200, 111, 201

5.5 Calculate the 2θ and d values for the first five lines in the X-ray powder pattern, Cu $K\alpha$ radiation, of a primitive cubic substance with $a = 5.0\,\text{Å}$. What is the multiplicity of each line?

5.6 At 20 °C, Fe is body centred cubic, $Z = 2$, $a = 2.866\,\text{Å}$. At 950 °C, Fe is face centred cubic, $Z = 4$, $a = 3.656\,\text{Å}$. At 1425 °C, Fe is again body centred cubic, $Z = 2$, $a = 2.940\,\text{Å}$. At each temperature, calculate (a) the density of iron, (b) the metallic radius of iron atoms.

5.7 The value of n in Bragg's Law is always set equal to 1. What happens to the higher order diffraction peaks?

5.8 Silver oxide, Ag_2O, has a cubic unit cell, $Z = 2$, with atomic coordinates Ag: $\frac{1}{4}\frac{1}{4}\frac{1}{4}, \frac{3}{4}\frac{3}{4}\frac{1}{4}, \frac{3}{4}\frac{1}{4}\frac{3}{4}, \frac{1}{4}\frac{3}{4}\frac{3}{4}$; O: 000, $\frac{1}{2}\frac{1}{2}\frac{1}{2}$. What are the atomic coordinates if the unit cell is displaced so that an Ag atom is at the origin? What is the lattice type of Ag_2O? What is the coordination number of Ag and O? Does the structure possess a centre of symmetry?

5.9 A cubic alkali halide has its first six lines with d-spacing 4.08, 3.53, 2.50, 2.13, 2.04 and 1.77 Å. Assign Miller indices to the lines and calculate the value of the unit cell dimension. The alkali halide has density $3.126\,\text{g cm}^{-3}$. Identify the alkali halide.

5.10 An imaginary orthorhombic crystal has two atoms of the same kind per unit cell located at 000 and $\frac{1}{2}\frac{1}{2}0$. Derive a simplified structure factor equation for this. Hence, show that, for a C-centred lattice, the condition for reflection is hkl: $h + k = 2n$.

5.11 Derive a simplified structure factor master equation for the perovskite structure of $SrTiO_3$. Atomic coordinates are Sr: $\frac{1}{2}\frac{1}{2}\frac{1}{2}$; Ti: 000; O: $\frac{1}{2}00$, $0\frac{1}{2}0$, $00\frac{1}{2}$.

5.12 The 111 reflection in the powder pattern of KCl has zero intensity but in the powder pattern of KF it is fairly strong. Explain.

5.13 The alloy gold–copper has a face centred cubic unit cell at high temperatures in which the Au, Cu atoms are distributed at random over the available corner and face centre sites. At lower temperatures, ordering occurs: Cu atoms are located preferentially on the corner sites and one pair of face centre sites; Au atoms are located on the other two pairs of face centre sites. What effects would you expect this ordering process to have on the X-ray powder pattern?

5.14 The X-ray powder pattern of orthorhombic Li_2PdO_2 includes the following lines: 4.68 Å (002), 3.47 Å (101), 2.084 Å (112). Calculate the values of the unit cell parameters. The density is $4.87\,\text{g cm}^{-3}$; what are the cell contents?

5.15 An ammonium halide, NH_4X, has the CsCl structure at room temperature, $a = 4.059$ Å, and transforms to the NaCl structure at 138 °C, $a = 6.867$ Å.
   (a) The density of the room temperature polymorph is $2.431\ \text{g cm}^{-3}$. Identify the substance.
   (b) Calculate the d-spacings of the first four lines in the powder pattern of each polymorph.
   (c) Calculate the percentage difference in molar volume between the two polymorphs, ignoring thermal expansion effects.
   (d) Assuming an effective radius of 1.50 Å for the spherical NH_4^+ ion and that anions and cations are in contact, calculate the radius of the anion in each structure. Are the anions in contact in the two structures?

5.16 'Each crystalline solid gives a characteristic X-ray powder diffraction pattern which may be used as a fingerprint for its identification.' Discuss the reasons for the validity of this statement and indicate why two solids with similar structures, e.g. NaCl and NaF, may be distinguished by their powder patterns.

5.17 Show by means of qualitative sketches the essential differences between the X-ray powder diffraction patterns of (a) a 1:1 mechanical mixture of powders of NaCl and AgCl and (b) a sample of (a) that has been heated to produce a homogeneous solid solution.

5.18 A sample of aluminium hydroxide was shown by chemical analysis to contain a few per cent of Fe^{3+} ions as impurity. What effect, if any, would the Fe^{3+} ions have on the powder pattern if it was present (a) as a separate iron hydroxide phase and (b) substituting for Al^{3+} in the crystal structure of $Al(OH)_3$.

References

L. V. Azaroff (1968). *Elements of X-ray Crystallography*, McGraw-Hill.

C. S. Barrett and T. B. Massalski (1966). *Structure of Metals*, 3rd ed., McGraw-Hill.

F. Donald Bloss (1971). *Crystallography and Crystal Chemistry*, Holt, Rinehart and Winston.

M. J. Buerger (1960). *Crystal Structure Analysis*, Wiley.

C. W. Bunn (1961). *Chemical Crystallography, An Introduction to Optical and X-ray Methods*, Clarendon Press.

B. D. Cullity (1978). *Elements of X-ray Diffraction*, Addison Wesley.

L. S. Dent Glasser (1977). *Crystallography and its Applications*, Van Nostrand Reinhold.

N. F. M. Henry, H. Lipson and W. A. Wooster (1960). *The Interpretation of X-ray Diffraction Photographs*, Macmillan.

N. F. M. Henry and K. Lonsdale (Eds) (1952). *International Tables for X-ray Crystallography*, Vol. 1, Kynoch Press.

R. Jenkins and J. L. DeVries (1970). *Worked examples in X-ray Analysis*, Springer-Verlag.

H. P. Klug and L. E. Alexander (1974). *X-ray Diffraction Procedures for Polycrystalline and Amorphous Materials*, 2nd ed., Wiley.

M. F. C. Ladd and R. A. Palmer (1978). *Structure Determination by X-ray Crystallography*, Plenum Press.

E. W. Nuffield (1966). *X-ray Diffraction Methods*, Wiley.

H. S. Peiser, H. P. Rooksby and A. J. C. Wilson (1960). *X-ray Diffraction by Polycrystalline Materials*, Chapman and Hall.

D. E. Sands (1969). *Introduction to Crystallography*, W. A. Benjamin.

G. H. Stout and L. H. Jensen (1968). *X-ray Structure Determination : A Practical Guide*, Macmillan.

B. K. Vainshtein (1981). *Modern Crystallography*, Springer-Verlag.

B. E. Warren (1969). *X-ray Diffraction*, Addison Wesley.

E. J. W. Whittaker (1981). *Crystallography*, Pergamon.

A. J. C. Wilson (1970). *Elements of X-ray Crystallography*, Addison Wesley.

M. M. Woolfson (1970). *An Introduction to X-ray Crystallography*, Cambridge University Press.

J. Wormald (1973). *Diffraction Methods*, Clarendon Press.

Chapter 6

Point Groups, Space Groups and Crystal Structure

A working knowledge of space groups is valuable for anyone interested in crystal chemistry, even though one does not wish to become involved in the actual determination of crystal structures. Without such a knowledge, one is entirely dependent for information about crystal structures on the availability of three-dimensional crystal models and good quality drawings in the literature. Once the basic principles of space groups are grasped, however, it is possible to make drawings of a structure for oneself, from different orientations if required, or even to construct one's own three-dimensional models. All that is needed is a listing of the atomic coordinates in the structures and details of the relevant space group.

This chapter is written for the non-crystallographer; short-cuts are made and many of the complications and subtleties of space groups are avoided. The main objective is to show the relation between space groups and sets of atomic coordinates, on the one hand, and three-dimensional crystal structures, on the other. As a background, it is assumed that the reader is familiar with unit cells, crystal systems, Bravais lattices and the elements of point and space symmetry. Summaries of these are given in Section 5.3 and Tables 5.2, 5.3. We begin with a discussion of point groups. Although point groups are not absolutely essential to the stated objective of this chapter it is well worth while to make their acquaintance. Point groups are much simpler than space groups because the elements of translational symmetry are absent from point groups. They therefore provide a relatively painless method of introduction to the subject while introducing the necessary concepts at an early stage.

6.1 Point groups

The elements of point symmetry which may be observed in crystals are the rotation axes 1, 2, 3, 4 and 6, the inversion axes $\bar{1}$, $\bar{2}$, $\bar{3}$, $\bar{4}$ and $\bar{6}$, and the mirror plane, m (which is equivalent to $\bar{2}$). These symmetry elements may occur either alone or in various possible combinations with each other to give a total of thirty-two possible crystallographic point groups. The method of drawing and labelling point groups which is used here is the same as that recommended by *International*

Table 6.1 *Point symmetry elements*

Symmetry element	Written symbol	Graphical symbol
	1	None
Rotation	2	❘
axes	3	▲
	4	◆
	6	⬢
	$\bar{1}$	None*
Inversion	$\bar{2}(\equiv m)$	——†
	$\bar{3}(\equiv 3 + \bar{1})$	▲
axes	$\bar{4}$	◈
	$\bar{6}(= 3/m)$	⬡
Mirror plane	m	——

* The inversion axis, $\bar{1}$, equivalent to a centre of symmetry, is represented as ○ in space groups but does not have a formal graphical representation in point groups, even though it is present in many point groups.
† The inversion axis $\bar{2}$ does not have a separate graphical symbol other than that of the mirror plane equivalent to it.

Table 6.2 *The thirty-two point groups*

Crystal system	Point groups
Triclinic	1, $\bar{1}$
Monoclinic	2, *m*, 2/*m*
Orthorhombic	222, *mm*2, *mmm*
Tetragonal	4, $\bar{4}$, 4/*m*, 422, 4*mm*, $\bar{4}$2*m*, 4/*mmm*
Trigonal	3, $\bar{3}$, 32, 3*m*, $\bar{3}$*m*
Hexagonal	6, $\bar{6}$, 6/*m*, 622, 6*mm*, $\bar{6}$*m*2, 6/*mmm*
Cubic	23, *m*3, 432, $\bar{4}$3*m*, *m*3*m*

Tables for X-ray Crystallography, Vol. 1. The symbols for the different point symmetry elements are given in Table 6.1. The thirty-two point groups, classified according to their crystal system, are listed in Table 6.2 and Appendix 5.

6.1.1 Representation of point groups and selected examples

Point groups are represented graphically as *stereograms*. Stereograms are used a lot in, for example, geology to represent direction in crystals and to show the relative orientation of crystal faces. For present purposes, all we need to know is that point groups are represented by a circle (a sphere in projection), usually with one of the axes perpendicular to the plane of the circle and passing through its

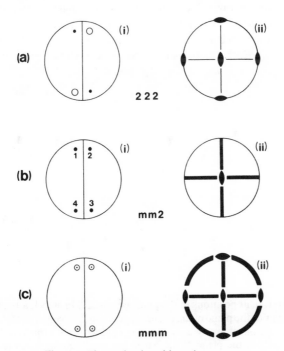

Fig. 6.1 The orthorhombic point groups

190

centre. The three orthorhombic point groups are drawn out in Fig. 6.1. For each, two diagrams are used. The right-hand one shows the symmetry elements that are present and the left-hand one shows the *equivalent positions* that are generated by the presence of these symmetry elements.

6.1.1.1 *222*

This orthorhombic point group has three mutually perpendicular twofold axes. That axis perpendicular to the plane of the paper is represented by the symbol in the centre of the circle (a, ii). The axes lying horizontally and vertically in the plane of the paper are represented by the two respective pairs of symbols lying on the circumference of the circle.

The presence of three mutually perpendicular twofold axes gives sets of four *equivalent positions* (a, i). An equivalent position is really the same as the 'identical orientation' used in defining symmetry elements. Thus the presence of a single twofold axis means that an object possessing such symmetry has two identical orientations (separated by rotation of 180°). In other words, a twofold axis has associated with it two equivalent positions. The dots and circles in the left-hand diagrams represent equivalent positions that are not in the plane of the paper; let us say that dots are above the plane and open circles are an equal distance below the plane.

The sequence of steps that is used in deriving the equivalent positions of point group 222 are set out in Fig. 6.2. Commencing with a single position in (i) (the thin vertical line that bisects the circle is merely a construction line), the effect of adding a twofold axis perpendicular to the plane of the paper is to generate a second position (ii). Both positions in (ii) must be at the same height, above the

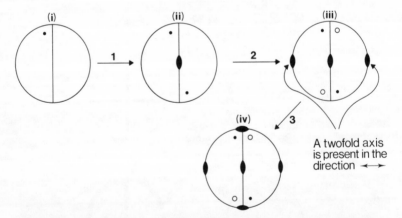

Fig. 6.2 Equivalent positions in the point group 222. In step 1, a twofold axis perpendicular to the plane of the paper is added. In 2, a second twofold axis, in the plane of the paper and lying horizontal, is added. In 3, a third twofold axis, lying vertically in the plane of the paper has been automatically created by step 2

plane of the paper, since the twofold axis is perpendicular to this plane. On adding a second twofold axis, in the plane of the paper and, say, in the horizontal direction, each of the positions in (ii) generates another to give the four positions shown in (iii). Since this second rotation axis is in the plane of the paper, the two new positions generated by this axis must be below the plane of the paper and so are represented as open circles. Comparison of (iii) and (iv) shows that the addition of a third twofold axis, in the plane of the paper and lying vertical, does not lead to any extra equivalent positions in addition to the four already present. In fact, this third axis is also present in (iii) and is generated automatically by the other two axes. The point group 222 could therefore be represented in the shortest possible notation, as 22 because the third twofold axis is not independent. The longer notation is normally used in order to show consistency with the essential symmetry requirements (Table 5.2) for orthorhombic unit cells.

6.1.1.2 *mm2*

This orthorhombic point group contains two mirror planes at right angles to each other with a twofold axis passing along the line of intersection of the mirror planes. In (b, ii) of Fig. 6.1, the twofold axis is perpendicular to the paper and the mirror planes are indicated in projection as the thick lines lying horizontally and vertically. This point group also has four equivalent positions and all are at the same height relative to the plane of the paper (b, i). Labelling the starting position as 1, the effect of adding the twofold axis is to generate position 3. The vertical mirror plane then generates position 2 from 1 and position 4 from 3. The horizontal mirror plane relates positions 1 and 4 and also 2 and 3; it does not create any new positions. As in the previous example, the third symmetry element is not independent but is generated by the combined operation of the other two elements. The choice of order of the symmetry elements is immaterial; any two out of the three, in combination, will generate the third element.

6.1.1.3 *mmm*

This orthorhombic point group contains, as essential symmetry elements, three mirror planes mutually perpendicular to each other and, as a consequence of the mirror planes, three mutually perpendicular twofold axes are generated. (Note that the reverse process does not occur. The three twofold axes in 222 do not lead to the generation of mirror planes.) These symmetry elements are shown in (c, ii) of Fig. 6.1. The symbols are the same as in the two previous examples, with the addition of the thick circle which indicates the presence of a mirror plane in the plane of the paper.

Eight equivalent positions occur in the point group *mmm*, four at the same height above the plane of the paper and four at the same height and below the plane (c, i). To see how these positions may be generated, the action of any two mirror planes generates four positions in a plane, as shown in (b, i) for the point group *mm2*. The action of the third mirror plane (the one lying in the plane of the paper) gives the eightfold set of positions in (c, i).

Fig. 6.3 Triclinic point group, $\bar{1}$, with a centre of symmetry

Only three orthorhombic point groups are possible. If other combinations of symmetry elements are tried, many will turn out to be equivalent to one of the three allowed point groups. For instance, the combination $22m$ can readily be shown to yield the same set of positions and symmetries as *mmm*.

Of the three orthorhombic point groups, only *mmm* possesses a centre of symmetry ($\bar{1}$). The appearance of a centre of symmetry is shown in Fig. 6.3 for the triclinic point group $\bar{1}$. Inversion through the centre of the circle converts position 1 to 2, and vice versa. Each position in *mmm* has a centrosymmetrically related partner, but this is not the case for 222 and *mm2*. The relevance of centrosymmetry is discussed in Section 6.2.

6.1.1.4 *32*

Let us consider one example of trigonal symmetry which is characterized by a single threefold axis (Fig. 6.4). As the threefold axis is the unique axis it is arranged to be perpendicular to the plane of the paper (Fig. 6.4 (ii)). There are also three twofold axes lying in the plane of the paper and at 60° to each other. In fact, only one of these is independent and so only one appears in the symbol 32. To find the equivalent positions in this point group, start with position 1 and consider the effect of the threefold axis (rotation by 120°). Positions 3 and 5 result. Then consider the effect of one of the twofold axes, say XX′ in (ii). This generates three new positions: 1 → 4, 3 → 2 and 5 → 6, and two more twofold axes YY′ and ZZ′ are also automatically generated, e.g. axis YY′ relates positions 1 and 6, 2 and 5, 3 and 4.

Of the thirty two crystallographic point groups, twenty seven are non-cubic and we have looked at five of these. The remaining twenty two can be treated along similar lines and should cause no problem. The main difficulty likely to be encountered concerns the orientation of the different symmetry elements in a point group. Some guidelines are as follows. In monoclinic, hexagonal, trigonal and tetragonal point groups, the unique axis is arranged to be perpendicular to

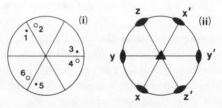

Fig. 6.4 The trigonal point group 32

the plane of the paper (stereogram). A slashed line, as in 4/*mmm*, indicates that, in this case the fourfold axis has a mirror plane perpendicular to it; it would be clearer if the symbol were written as (4/*m*)*mm*. In tetragonal, trigonal and hexagonal point groups, the twofold axes, as in $\bar{4}2m$, are always in the plane perpendicular to the unique axis (in this case perpendicular to $\bar{4}$).

The five cubic point groups are rather more complicated to work with as they are difficult to represent by simple two-dimensional projections. This is because there are so many symmetry elements present and many are not perpendicular to each other, e.g. threefold and fourfold axes are at 45° to each other. Whereas non-cubic point groups may be drawn with their symmetry axes either in the plane or perpendicular to the plane of the stereograms, this is not generally possible for cubic point groups and oblique projections are needed to represent the threefold axes (see Appendix 5). No further discussion of cubic point groups will be given.

6.1.2 Examples of point symmetry of molecules: general and special positions

The relationships between point symmetry and structure are best seen by some examples taken from small molecules. Consider the methylene dichloride molecule, CH_2Cl_2 (Fig. 6.5). This possesses a single twofold axis which bisects the H—C—H and Cl—C—Cl bond angles (a) and two mirror planes (b and c). The twofold axis is parallel to the line of intersection of the mirror planes. These symmetry elements may be represented as in (d), in which the twofold axis is perpendicular to the plane of the paper and the mirror planes appear in projection as horizontal and vertical lines. From inspection of Fig. 6.1, it is seen that CH_2Cl_2 belongs to the point group *mm2*. However, the number of equivalent positions in *mm2* (b, i) is four, and this does not appear to tally with the realities of the CH_2Cl_2 molecule. If we take a hydrogen atom as one equivalent position, there are only two hydrogens present and therefore only two possible equivalent

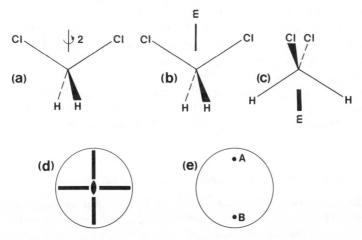

Fig. 6.5 The point group *mm2* of the methylene dichloride molecule

194

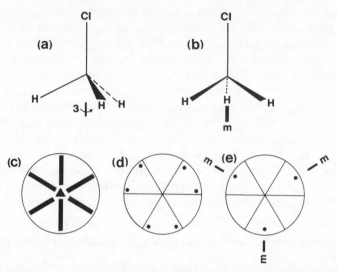

Fig. 6.6 The point group $3m$ of the methyl chloride molecule

positions in the molecule. The anomaly is resolved by letting the equivalent positions in Fig. 6.1 (b, i) lie on the vertical mirror plane instead of to either side of the mirror. This yields the arrangement shown in Fig. 6.5(e), which has only two equivalent positions. Thus, positions 1 and 2 in Fig. 6.1 (b, i) become the single position A in Fig. 6.5(e). We can now distinguish between the *general equivalent positions* of a point group and the *special equivalent positions*; the latter arise when the general positions lie on a symmetry element such as a mirror plane or rotation axis. Thus A and B in Fig. 6.5(e) are special positions.

As a further example, consider the point symmetry of the methyl chloride molecule, CH_3Cl (Fig. 6.6). The molecule possesses one threefold axis along the direction of the C—Cl bond (a). It does not have any twofold axes but has three mirror planes oriented at 60° to each other; one is shown in (b). The threefold axis coincides with the line of intersection of the mirror planes. The symmetry elements are shown as a stereogram in (c) and by comparison of this with Appendix 5 we see that the point group is $3m$. The six general equivalent positions in $3m$ are given in (d). We again have the problem that there are more equivalent positions than possible atoms, and this is overcome by allowing the general positions to lie on the mirror planes (e); the number of positions is thereby reduced to three.

6.1.3 Centrosymmetric and non-centrosymmetric point groups

Of the thirty two point groups, twenty one do not possess a centre of symmetry. The absence of a centre of symmetry is an essential but not sufficient requirement for the presence in crystals of optical activity, pyroelectricity and piezoelectricity (Chapter 15). Optical activity is confined to fifteen of the twenty one non-centrosymmetric point groups and piezoelectricity to twenty of these. This is of use in, for example, the search for new materials with piezoelectric activity; it is a waste of time trying to detect piezoelectricity in crystals whose point group is not among the twenty active groups! Crystallographers also make some use of the piezoelectric effect in structure determinations. It is a considerable help in solving an unknown structure to know the space group at the outset. If a test for piezoelectricity is carried out with positive results, this limits the choice of space group to the non-centrosymmetric ones. The absence of piezoelectricity does not necessarily mean, however, that the point group and space group are centrosymmetric.

6.2 Space groups

The combination of the thirty-two possible point groups and the fourteen Bravais lattices (which in turn are combinations of the seven crystal systems, or unit cell shapes, and the different possible lattice types) gives rise to 230 possible space groups. All crystalline materials have a structure which belongs to one of these space groups. This does not, of course, mean that only 230 different crystal structures are possible. For the same reason, the human body (from its external appearance) is not the only object to belong to point group $\bar{2}$—teapots also do.

Space groups are formed by adding elements of translation to the point groups. The space symmetry elements, screw axes and glide planes are derived from their respective point symmetry elements, rotation axes and mirror planes by adding a translation step in between each operation of rotation or reflection (see Section 5.3.4). A complete tabulation of all the possible screw axes and glide planes and their symbols is not given here. Instead, symbols are explained as they arise. Also, there is space to discuss only a few of the simpler space groups. The interested reader is recommended to acquire his own copy of *International Tables for X-ray Crystallography*, Vol. 1; once the basic rules have been learned, by working through the examples given here, there should be no difficulty in understanding and using any space group.

The written symbol of a space group is a list of between two and four characters. The first character is always a capital letter which corresponds to the lattice type—P, I, A, etc. The remaining characters correspond to some of the symmetry elements that are present. If the crystal system has a unique or principal axis (e.g. the fourfold axis in tetragonal crystals), the symbol for this axis appears immediately after the lattice symbol. For the remaining characters, there are no universal rules but, instead, different rules for different crystal systems. As these rules are not essential to an understanding of space groups and are not usually of interest to the non-specialist, they are not repeated here.

Space groups are usually drawn as parallelograms with the plane of the parallelogram corresponding to the xy plane of the unit cell. By convention (Fig. 6.7), the origin is taken as the top left-hand corner, with y horizontally, x vertically (downwards) and the positive z direction pointing up out of the plane of the paper. For each space group, two parallelograms are used, the left-hand one

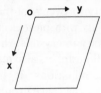

Fig. 6.7 Convention used to label axes in a space group

to give the equivalent positions and the right-hand one the symmetry elements that are present. Let us see some examples. Each one will introduce at least one new feature.

6.2.1 Triclinic P$\bar{1}$

This space group is primitive and centrosymmetric; it is shown in Fig. 6.8. The right-hand diagram shows the symmetry elements: there are centres of symmetry at the origin (t), midway along the a and b edges and in the middle of the C face (i.e. the face bounded by a and b). Additional centres of symmetry, not shown, occur in the middle of the other faces, halfway along the c edge and at the body centre of the unit cell.

The left-hand diagram gives the equivalent positions in the space group P$\bar{1}$. To derive them, it is necessary to choose a starting position and operate on this position with the various symmetry elements that are present. The conventional starting position is at 1, close to the origin and with small positive values of x, y and z (the latter indicated by the $+$ sign). This position must be present in all other unit cells (definition of the unit cell) and three of these are shown as 1$'$, 1$''$ and 1$'''$.

Consider now the effect of the centre of symmetry, t, at the origin of the unit cell. This acts upon position 1 to create position 2. The minus sign at 2 indicates a negative z height and the comma shows that position 2 is enantiomorphous relative to position 1. The effect of any reflection or inversion operation is to

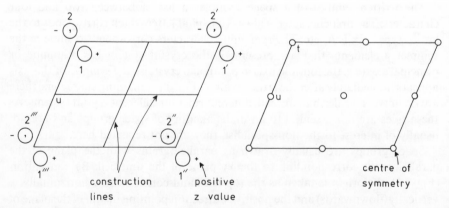

Fig. 6.8 Space group P$\bar{1}$. Coordinates of equivalent positions: xyz, $\bar{x}\bar{y}\bar{z}$

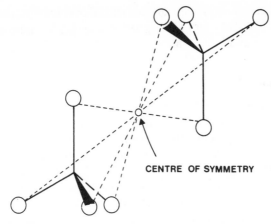

Fig. 6.9 Two tetrahedra related by a centre of symmetry.

convert a left-handed object into a right-handed one. This is shown in Fig. 6.9 for two tetrahedra which are positioned so as to be related to each other by inversion through a centre of symmetry. Thus, although individual tetrahedra do not possess a centre of symmetry, groupings of tetrahedra may possess one. Positions 2′, 2″ and 2‴ in Fig. 6.8 are automatically generated from position 2 because they are equivalent positions in neighbouring cells.

The next step is to write down the coordinates of the equivalent positions. This is done in the form x, y, z where x, y and z are the fractional distances, relative to the unit cell edge dimensions, from the origin of the cell. Let position 1 have fractional coordinates x, y and z. Position 2 is then $-x$, $-y$, $-z$. Position 2″ is at $1 - x$, $1 - y$, $-z$, etc. If a position lies outside the unit cell under consideration, an equivalent position within the unit cell can be found, usually by adding or subtracting 1 from one or more of the fractional coordinates. Position 2″ is outside the cell because it has a negative z value; the equivalent position inside the cell is given by a displacement of one unit cell in the z direction and has coordinates $1 - x, 1 - y, 1 - z$. In shorthand, these coordinates are written as $\bar{x}, \bar{y}, \bar{z}$. The unit cell in the space group $P\bar{1}$ has two equivalent positions, x, y, z (position 1) and $\bar{x}, \bar{y}, \bar{z}$ (height c above 2″).

Although only one centre of symmetry is necessary to generate the equivalent positions in $P\bar{1}$ other centres of symmetry are automatically created. For example, the centre of symmetry at u arises because pairs of positions such as 1 and 2‴, 2 and 1‴, etc., are centrosymmetrically related through u. This may be seen from the diagram or may be proven by comparing coordinates of the three positions. Positions 2‴ and 1 are equidistant from u and lie on a straight line that passes through u.

The positions x, y, z and $\bar{x}, \bar{y}, \bar{z}$ are called *general positions* and apply to any value of x, y and z between 0 and 1. In certain circumstances, x, y, z and $\bar{x}, \bar{y}, \bar{z}$ coincide, e.g. if $x = y = z = \frac{1}{2}$. In this case, there is only one position, called a *special position*. The special positions in $P\bar{1}$ arise when the general position lies on a centre of

198

symmetry. The coordinates of the onefold special positions are, therefore, $(0, 0, 0)$, $(\frac{1}{2}, 0, 0), (0, \frac{1}{2}, 0), (0, 0, \frac{1}{2}), (\frac{1}{2}, \frac{1}{2}, 0), (\frac{1}{2}, 0, \frac{1}{2}), (0, \frac{1}{2}, \frac{1}{2})$ and $(\frac{1}{2}, \frac{1}{2}, \frac{1}{2})$, and correspond to the corner, edge, face and body centres of the unit cell.

6.2.2 Monoclinic C2

The convention adopted here for labelling the unique twofold axis in monoclinic space groups is that which is in most common use, namely, with b as the unique axis. It is a pity that this is different to the use of c as the unique axis in tetragonal, trigonal and hexagonal cells but this usage for monoclinic cells is now so well established that it is unlikely to be altered. With b as the unique axis the unit cell projects onto the xy plane as a rectangle (because $\gamma = 90°$), as shown in Fig. 6.10. Since $\beta \neq 90°$, the z axis is not perpendicular to the plane of the paper but is inclined to the vertical.

The C-centring in space group C2 means that the Bravais lattice has a lattice point at the origin (with coordinates $0, 0, 0$) and a lattice point in the middle of the side bounded by a and b, at $\frac{1}{2}, \frac{1}{2}, 0$. For any position x, y, z in this space group, there will, therefore, be an equivalent position at $x + \frac{1}{2}$, $y + \frac{1}{2}$, z (i.e. (x, y, z) $+ (\frac{1}{2}, \frac{1}{2}, 0)$). This C-centring has no representation in the right-hand diagram of Fig. 6.10 but can be seen in the left-hand diagram. The main symmetry element that is present is a twofold rotation axis (d), parallel to b and coincident with b (i.e. passing through $x = 0$, $z = 0$). A twofold rotation axis in the plane of the paper is indicated by an arrow. Other symmetry elements are generated by a combination of the twofold rotation axis and the C-centring, namely, a twofold rotation axis (e) parallel to b, cutting the x axis at $\frac{1}{2}$ and the z axis at 0, and two twofold screw axes (f and g), again parallel to b and cutting the x axis at

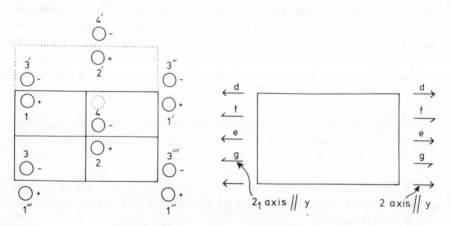

Fig. 6.10 Monoclinic space group C2
Coordinates of equivalent positions
$000: xyz, \bar{x}y\bar{z}$
$\frac{1}{2}\frac{1}{2}0: x + \frac{1}{2} y + \frac{1}{2} z, \frac{1}{2} - x \frac{1}{2} + y \bar{z}$

$\frac{1}{4}$ and $\frac{3}{4}$ and z at 0. Screw axes in the plane of the paper are represented as half arrows.

Space group C2 has four equivalent positions. To see how they arise, let us take position 1 as a general position; 1', 1″ and 1‴ are the equivalent positions in neighbouring unit cells. The effect of C-centring is to create position 2 which is displaced by $(\frac{1}{2}, \frac{1}{2}, 0)$ from position 1. The effect of the twofold rotation axis (d) is to rotate position 1 about the b edge by 180° and create position 3′. As 1 has a positive z coordinate, 3′ must have a corresponding negative z value. Similarly, positions 2 and 4′ are related by this same twofold rotation axis; 4 and 4′ are identical positions in adjacent cells. Alternatively, 4 may be regarded as generated by the action of the C-centring on 3′.

The new twofold rotation axis (e) that is created relates, for example, positions 1 and 3, 2 and 4, 1‴ and 3′, etc. The twofold screw axis (f) relates, for example, 1 and 4: in a combined operation, 1 is translated halfway along y, retaining its x and z values, to the position shown as the dotted circle and then rotated 180° about the axis parallel to y and at $x = \frac{1}{4}$, $z = 0$, to arrive at 4. Repetition of the process takes position 4 to 1′ which effectively takes us back to the starting position. Similarly, related sets of positions are 3′, 2 and 3″; 3, 2′ and 3‴; 1‴, 4′ and 1″, etc. By similar reasoning, screw axis (g) relates positions 3, 2 and 3‴, and so on.

The coordinates of positions 1 to 4 are as follows: x, y, z; $x + \frac{1}{2}, y + \frac{1}{2}, z$; $\bar{x}, y, -z$; $\frac{1}{2} - x, \frac{1}{2} + y, -z$. As positions 3 and 4 lie below the plane of the paper, they are outside the chosen unit cell. The positions equivalent to 3 and 4 that are inside the cell are displaced by one cell along z and have coordinates $\bar{x}, y, \bar{z}$ and $\frac{1}{2} - x, \frac{1}{2} + y, \bar{z}$. These four positions can be grouped into two sets: x, y, z; $\bar{x}, y, \bar{z}$ and $x + \frac{1}{2}, y + \frac{1}{2}, z$; $\frac{1}{2} - x, \frac{1}{2} + y, \bar{z}$, such that the second set is related to the first by the lattice centring (i.e. by adding $\frac{1}{2}, \frac{1}{2}, 0$ to the coordinates). It is common practice (e.g. in *International Tables of X-ray crystallography*) to list only those positions that belong to the first set and at the same time specify that the other positions are created by the lattice centring. This leads to considerable shortening and simplification in labelling the equivalent positions of the more complex and higher symmetry space groups.

The general positions in space group C2 are fourfold, i.e. there are four of them per cell, but if they lie on the twofold rotation axes their number is reduced to two and they become special positions. Thus, if $x = z = 0$, the two positions have coordinates $0, y, 0$ and $\frac{1}{2}, y + \frac{1}{2}, 0$. A second set of special positions arises when $x = 0$, $z = \frac{1}{2}$ (the reader may like to check that there is a twofold axis parallel to b and at $x = 0$, $z = \frac{1}{2}$ not indicated in Fig. 6.10: it is at height $c/2$ above axis d).

We have seen earlier how the presence of lattice centring or elements of space symmetry lead to systematically absent reflections from the X-ray patterns. For space group C2, the C-centring imposes the condition that only those reflections that satisfy the rule, for hkl: $h + k = 2n$, may be observed. The 2_1 screw axes parallel to b impose the condition for reflection that for $0k0$: $k = 2n$. However, this is also a consequence of the C-centring condition, for the special case that $h = l = 0$ and so does not lead to any extra absences.

6.2.3 Monoclinic C2/m

This space group, shown in Fig. 6.11, is also C-centred and has, as its principal symmetry elements, a mirror plane perpendicular (/) to a twofold axis. The twofold axis is parallel to b, by convention, and therefore the mirror plane is the xz plane. Two mirror planes are present in the cell; they intersect b at 0 and $\frac{1}{2}$ and are shown as thick vertical lines in Fig. 6.11. As in the space group C2, there are two twofold rotation axes, parallel to b and intersecting a at 0 and $\frac{1}{2}$, and two 2_1 screw axes parallel to b and intersecting a at $\frac{1}{4}$ and $\frac{3}{4}$. All of these 2 and 2_1 axes are at c height equal to zero. An additional set of axes, not shown, occurs at $c = \frac{1}{2}$. Also present, as will be discussed later, are centres of symmetry and glide planes.

The space group C2/m contains eight general equivalent positions, all of which may be generated from position 1 by the combined action of the C-centring, twofold axis and mirror plane. Thus, the C-centring creates an equivalent position, 2, after translation by $\frac{1}{2}, \frac{1}{2}, 0$. Action of the twofold axis passing through the origin generates 6′ from 1. Position 3 is similarly related to 2 by the action of the twofold axis passing through $a = \frac{1}{2}$, $c = 0$. Alternatively, 3 may be generated from 6′ by the C-centring condition. The mirror plane at $b = 0$ generates positions 8″ from 1 and 7‴ from 6′. Note that 8″ and 1 are at the same positive c value and that 8″ contains a comma to indicate its enantiomorphous relation to 1. Positions 4 and 5 are related to 3 and 2 by the mirror plane that cuts b at $\frac{1}{2}$; alternatively, 4 and 5 are generated from 7‴ and 8″ by the centring.

The coordinates of the eight equivalent positions within the cell, together with their number, if shown, are x, y, z (1); $x + \frac{1}{2}, y + \frac{1}{2}, z$ (2); $\frac{1}{2} - x, \frac{1}{2} + y, \bar{z}$; $\frac{1}{2} - x, \frac{1}{2} - y, \bar{z}; \frac{1}{2} + x, \frac{1}{2} - y, z$ (5); $\bar{x}, y, \bar{z}; \bar{x}, \bar{y}, \bar{z}; x, \bar{y}, z$ (8). These eight positions may be grouped into two sets of four positions that are related by the C-centring. The coordinates of both sets are given in Fig. 6.11. Several sets of special positions are possible in this space group, e.g. if $y = 0$, a fourfold set occurs which contains $x, 0,$

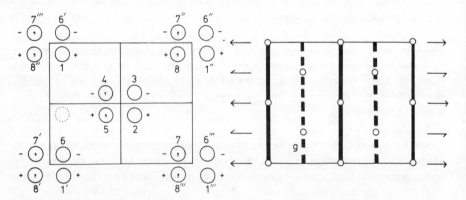

Fig. 6.11 Monoclinic space group C2/m. Coordinates of equivalent positions

$000 : xyz,\ x\bar{y}z,\ \bar{x}y\bar{z},\ \bar{x}\bar{y}\bar{z}$

$\frac{1}{2}\frac{1}{2}0 : \frac{1}{2} + x\ \frac{1}{2} + y\ z, \frac{1}{2} + x\ \frac{1}{2} - y\ z, \frac{1}{2} - x\ \frac{1}{2} + y\ \bar{z}, \frac{1}{2} - x\ \frac{1}{2} - y\ \bar{z}$

z; $\bar{x}, 0, \bar{z}$; $x + \frac{1}{2}, \frac{1}{2}, z$; $\frac{1}{2} - x, \frac{1}{2}, \bar{z}$. If both x and $y = 0$ and $z = \frac{1}{2}$, a twofold set arises which contains $0, 0, \frac{1}{2}$ and $\frac{1}{2}, \frac{1}{2}, \frac{1}{2}$.

The combination of a mirror plane perpendicular to a twofold axis, together with the C-centring, leads to the generation of several other symmetry elements. These include 2_1 screw axes parallel to b, centres of symmetry and glide planes. For example, the centre of symmetry created at the origin relates positions 1 and 7''', 6' and 8''. The thick dashed line g in the right-hand diagram indicates a glide plane for which the translation component is $a/2$ and reflection is across a plane lying perpendicular to b. Such a glide plane is called 'an a glide perpendicular to b'. Thus, position 1 is translated by $a/2$ to the position shown as the dotted circle; reflection across the plane, g, which cuts b at $\frac{1}{4}$ leads to position 5. Repetition of the process converts 5 into 1', which is equivalent to the starting position, 1. Similarly, positions 8, 2 and 8''' are related by the glide plane which cuts b at $\frac{3}{4}$.

The presence of glide planes in a crystal may sometimes by detected by the absence of a set of X-ray reflections. For an a glide perpendicular to b, the condition limiting the $h0l$ reflections is that $h = 2n$ (i.e. only even h values may be observed). In the space group C2/m, this condition is part of the more general condition for C-centring, namely, that for hkl, $h + k = 2n$. Independent evidence for the existence of the glide plane is therefore not immediately available from the X-ray patterns.

6.2.4 Orthorhombic P222₁

This primitive orthorhombic space group has twofold rotation axes parallel to x and y and a twofold screw axis parallel to z. The feature of this space group, shown in Fig. 6.12, which makes the generation of the equivalent positions a little difficult to visualize, is that the twofold rotation axes parallel to y occur at a c height of $\frac{1}{4}$. Consider first the effect of the axis parallel to y at $a = 0$ and $c = \frac{1}{4}$. The starting position 1 has a small positive z coordinate of $+ z$. The twofold axis is at $z = \frac{1}{4}$; therefore, position 1 is at $(\frac{1}{4} - z)$ *below* the twofold axis. The new position, 2',

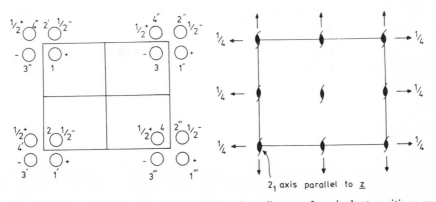

Fig. 6.12 Orthorhombic space group P222₁. Coordinates of equivalent positions xyz, $\bar{x}y\frac{1}{2} - z$, $x\bar{y}\bar{z}$, $\bar{x}\bar{y}\frac{1}{2} + z$

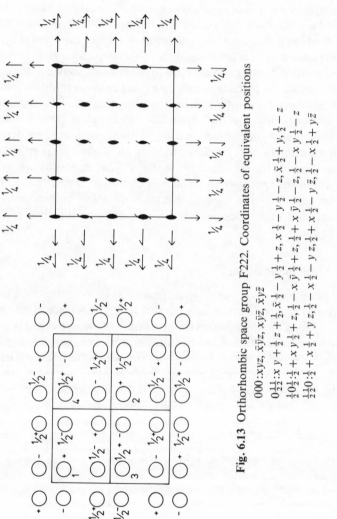

Fig. 6.13 Orthorhombic space group F222. Coordinates of equivalent positions

$000; x y z, \bar{x}\bar{y}z, x\bar{y}\bar{z}, \bar{x}y\bar{z}$

$0\frac{1}{2}\frac{1}{2}: x y + \frac{1}{2} z + \frac{1}{2}, \bar{x}\frac{1}{2} - y\frac{1}{2} + z, x\frac{1}{2} - y\frac{1}{2} - z, \bar{x}\frac{1}{2} + y, \frac{1}{2} - z$

$\frac{1}{2}0\frac{1}{2}:\frac{1}{2} + x y\frac{1}{2} + z, \frac{1}{2} - x \bar{y}\frac{1}{2} + z, \frac{1}{2} + x \bar{y}\frac{1}{2} - z, \frac{1}{2} - x y\frac{1}{2} - z$

$\frac{1}{2}\frac{1}{2}0:\frac{1}{2} + x\frac{1}{2} + y z, \frac{1}{2} - x\frac{1}{2} - y z, \frac{1}{2} + x\frac{1}{2} - y \bar{z}, \frac{1}{2} - x\frac{1}{2} + y\bar{z}$

formed by rotation about this axis is, therefore, at $(\frac{1}{4} - z)$ *above* the twofold axis, i.e. it has z coordinate $\frac{1}{4} + (\frac{1}{4} - z) = \frac{1}{2} - z$. This is shortened to $\frac{1}{2} -$ in Fig. 6.12. Consider now the twofold axis parallel to x and at $b = c = 0$ (i.e. passing through the origin). This axis generates positions $3''$ from 1 and $4'''$ (its equivalent in the cell below) from $2'$. With these two axes we have generated all four equivalent positions in this space group. The third axes, the 2_1 axis parallel to z, is automatically generated by the combined action of the other two axes and is not independent of them. This 2_1 axis relates, for example, positions 1 and $4'''$ (i.e. translation of position 1 by $c/2$ followed by 180° rotation about c gives $4'''$). Positions $2'$ and $3''$ are similarly related. The coordinates of the equivalent positions are given in Fig. 6.12. The only symmetry element which causes systematically absent reflections in P222$_1$ is the 2_1 axis parallel to z, i.e. only $00l$ reflections for which $l = 2n$ may be observed. Several sets of twofold special positions arise, e.g. if $y = z = 0$ (i.e. $x,0,0; \bar{x},0,\frac{1}{2}$).

6.2.5 Orthorhombic F222

The new feature of this space group is that it has a face centred lattice which, as can be seen from Fig. 6.13, leads to a considerable increase in the number of symmetry elements and equivalent positions. The basic symmetry elements are three intersecting twofold axes, parallel to x, y and z and passing through the origin. Many other twofold axes occur automatically, e.g. one intersecting the cell at $a = \frac{1}{4}, c = \frac{1}{4}$ and parallel to b and another at $a = \frac{1}{4}, b = \frac{1}{4}$ and parallel to c. Also many 2_1 axes are created, e.g. one at $a = 0, b = \frac{1}{4}$ parallel to c and another at $b = \frac{1}{4}$, $c = 0$ and parallel to a.

There are sixteen general equivalent positions which fall into four groups related by the face centring condition. The four sets are related as $(0, 0, 0); (\frac{1}{2}, \frac{1}{2}, 0);$ $(\frac{1}{2}, 0, \frac{1}{2})$ and $(0, \frac{1}{2}, \frac{1}{2})$. Thus, position 1, (x, y, z), is related to positions 2 to 4: $(x + \frac{1}{2}, y + \frac{1}{2}, z); (x + \frac{1}{2}, y, z + \frac{1}{2})$ and $(x, y + \frac{1}{2}, z + \frac{1}{2})$. Generation of the remaining equivalent positions by the action of the twofold axes should be a straight forward matter. Coordinates of the equivalent positions are given in Fig. 6.13.

6.2.6 Tetragonal I4$_1$

The principal axis in the space group I4$_1$ (Fig. 6.14) is a 4_1 screw axis parallel to z. There are four such axes which intersect the unit cell at $x = \frac{1}{4}, y = \frac{1}{4}; x = \frac{3}{4},$ $y = \frac{1}{4}; x = \frac{1}{4}, y = \frac{3}{4}$ and $x = \frac{3}{4}, y = \frac{3}{4}$. The operation of a 4_1 screw axis involves translation of $\frac{1}{4}$ combined with rotation by 90° about the axis. The positions 1 to 4 are related to each other by the 4_1 axis at s, and it can be seen that these positions lie on a spiral about s. The symbols of screw axes s and t are reversed because their direction of rotation is different; s involves a clockwise rotation and t an anticlockwise rotation (e.g. the sequence of positions $7', 2, 5, 4'$). The body centring relates positions 1–4 to 5–8 and the cell contains eight general equivalent positions. These are listed in Fig. 6.14. Several twofold axes parallel to z are also generated. Two conditions are imposed on the reflections which are possible for this space group. For the body centring, only the reflec-

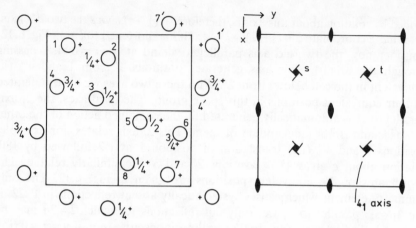

Fig. 6.14 Tetragonal space group $I4_1$. Coordinates of equivalent positions

$$000:xyz, \bar{x}\bar{y}z, \bar{y}\tfrac{1}{2}+x\tfrac{1}{4}+z, y\tfrac{1}{2}-x\tfrac{1}{4}+z$$

$$\tfrac{1}{2}\tfrac{1}{2}\tfrac{1}{2}:x+\tfrac{1}{2}y+\tfrac{1}{2}z+\tfrac{1}{2}, \tfrac{1}{2}-x\tfrac{1}{2}-y\tfrac{1}{2}+z, \tfrac{1}{2}-y x\tfrac{3}{4}+z, \tfrac{1}{2}+y \bar{x}\tfrac{3}{4}+z$$

tions $hkl : h + k + l = 2n$ may be observed. The 4_1 screw axis places on the $00l$ reflections the condition that $l = 4n$.

6.3 Space groups and crystal structures

The purpose of this section is to show how drawings or models of crystal structures may be made if one knows the space group and essential atomic coordinates of the structure. As examples, two simple structures are considered in some detail. This is then followed by a more systematic approach to crystal chemistry in Chapter 7.

6.3.1 The perovskite structure, $SrTiO_3$

The basic information that we need to know is the following:

Unit cell: cubic, $a = 3.905\,\text{Å}$

Space group: $Pm3m$ (number 221 in *International Tables for X-ray crystallography*, Vol. 1)

Atomic coordinates: Ti in 1(a) at $0, 0, 0$

Sr in 1(b) at $\tfrac{1}{2}, \tfrac{1}{2}, \tfrac{1}{2}$

O in 3(d) at $0, 0, \tfrac{1}{2}$

This is, in fact, a very simple example since although the space group $Pm3m$ is complicated, as are all cubic space groups, all the atoms in perovskite lie on special positions. There are forty-eight general equivalent positions in the space group $Pm3m$ but a large number of special positions arise when atoms lie on symmetry elements. Titanium occupies a onefold special position at the origin of the unit cell; the symbol 1(a) indicates that there is only one position in this set and (a) is a label, the Wyckoff lablel, for this (set of) position(s). Strontium also occupies a onefold special position, 1(b), at the body centre of the cell. Oxygen occupies a threefold special position 3(d); the coordinates of one of these positions are given—$0, 0, \frac{1}{2}$—and the only remaining information that is needed from the space group is the coordinates of the other two oxygen positions. From *International Tables*, these are $0, \frac{1}{2}, 0$ and $\frac{1}{2}, 0, 0$.

From this information, the unit cell and atomic positions may be drawn, first as a projection down one of the cubic cell axes (Fig. 6.15a) and then as an oblique projection to show the atomic positions more clearly (Fig. 6.15b). The coordination environment of each atom may be seen in (b) and interatomic distances calculated by simple geometry. The octahedral coordination by oxygen of one of

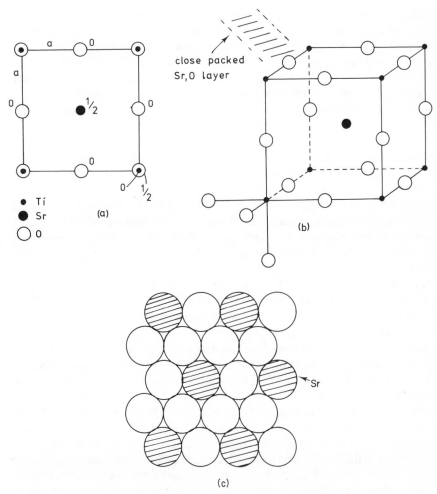

Fig. 6.15 The perovskite structure of $SrTiO_3$

the corner titaniums is shown. The Ti—O bond length $= a/2 = 1.953\,\text{Å}$. The strontium in the cube centre is equidistant from all twelve oxygens at the centres of the edges of the unit cell. The strontium–oxygen distance is equal to half the diagonal of any cell face, i.e. $a/\sqrt{2}$ or $2.76\,\text{Å}$ (from the geometry of triangles, the length of a cell face diagonal is equal to $\sqrt{a^2 + a^2}$).

Each oxygen has two titaniums as its nearest cationic neighbours, at a distance of $1.953\,\text{Å}$, and four strontium atoms, coplanar with the oxygen at a distance of $2.76\,\text{Å}$. However, eight other oxygens are at the same distance, $2.76\,\text{Å}$, as the four strontiums. It is debatable whether the oxygen coordination number is best regarded as two (linear) or as six (a grossly squashed octahedron with two short and four long distances) or as fourteen (six cations and eight oxygens). No firm recommendation is made!

Having arrived at the unit cell of $SrTiO_3$, the atomic coordinates, coordination numbers and bond distances, we now wish to view the structure on a rather larger scale. There are several questions which may be asked. Does the structure have a close packed anion arrangement? The close packing approach provides a good way of classifying many crystal structures, Chapter 7. Can the structure be regarded as some kind of framework with atoms in the interstices? Many silicate structures may be thought of in this way. Can smaller polymeric units in the structure be identified? Orthosilicates have, for example, discrete SiO_4^{4-} tetrahedra. Some answers to these questions are as follows.

Perovskite does not contain close packed oxide ions as such but the oxygens and strontiums, considered together, do form a cubic close packed array with the layers parallel to the (111) planes (Fig. 6.15b and c). To see this, compare the perovskite structure with that of NaCl(Fig. 5.9). The latter contains Cl^- ions at the edge centre and body centre positions of the cell and is cubic close packed. By comparison, perovskite contains O^{2-} ions at the edge centres and Sr^{2+} at the body centre. The structure of the mixed Sr, O close packed layers in perovskite is such that one quarter of the atoms are strontium, arranged in a regular fashion (Fig. 6.15c).

It is quite common for fairly large cations, such as Sr^{2+} ($r = 1.13\,\text{Å}$) to play apparently different roles in different structures, i.e. as twelve coordinate packing ions in perovskite or as octahedrally coordinated cations within a close packed oxide array, as in SrO (rock salt structure).

The formal relation between rock salt and perovskite also includes the Na^+ and Ti^{4+} cations as both are in octahedral sites. Whereas in NaCl all octahedral sites are occupied (corners and face centres), in perovskite only one quarter (the corner sites) are occupied.

Perovskite may also be regarded as a framework structure constructed from corner-sharing (TiO_6) octahedra and with Sr^{2+} ions placed in twelve-coordinate interstices. The octahedral coordination of one titanium is shown in Fig. 6.15b; each oxygen of this octahedron is shared with one other octahedron, such that the Ti–O–Ti arrangement is linear. In this way, octahedra are linked at their corners to form sheets, Appendix 2, Fig. A2.4c, and neighbouring sheets are linked similarly to form a three-dimensional framework.

A further discussion of crystal structures, including close packing and linked polyhedra, is given in Chapter 7, while perovskites and their use as ferroelectric and dielectric materials are discussed in Chapter 15.

6.3.2 The rutile structure, TiO_2

We need to know the following information:

Unit cell: tetragonal, $a = 4.594$, $c = 2.958 \text{Å}$

Space group: $P4_2/mnm$ (number 136)

Atomic coordinates: Ti in 2(a) at $(0, 0, 0), (\frac{1}{2}, \frac{1}{2}, \frac{1}{2})$

O in 4(f) at $(x, x, 0), (\bar{x}, \bar{x}, 0),$

$(\frac{1}{2} + x, \frac{1}{2} - x, \frac{1}{2}), (\frac{1}{2} - x, \frac{1}{2} + x, \frac{1}{2})$

As in the perovskite structure, only special positions are used to accommodate atoms and the general positions (sixteenfold) are unoccupied. The titanium positions are fixed at the corner and body centre but the oxygen has a variable parameter, x, whose value must be determined experimentally. Crystal structure determination and refinement gives $x = 0.30$ for TiO_2. The unit cell of rutile is shown projected onto the xy plane in Fig. 6.16(a).

We next need to determine the coordination environment of the atoms. The body centre titanium at $(\frac{1}{2}, \frac{1}{2}, \frac{1}{2})$ is coordinated octahedrally to six oxygens. Four of these—two at 0 and two at 1 directly above the two at 0—are coplanar with titanium. Two oxygens at $z = \frac{1}{2}$ are collinear with titanium and form the axes of the octahedron. The corner titaniums are also octahedrally coordinated but the orientation of their octahedra is different (Fig. 6.16b). The oxygens are coordinated trigonally to three titaniums: e.g. oxygen at 0 in (a) is coordinated to titanium atoms at the corner, at the body centre and at the body centre of the cell below.

Since the oxygen atoms form the corners of TiO_6 octahedra this means that each corner oxygen is shared between three octahedra. The octahedra are linked by sharing edges and corners to form a three-dimensional framework. Consider the TiO_6 octahedron in the centre of the cell in (b); a similar octahedron in identical orientation occurs in the cells above and below such that octahedra in adjacent cells share edges to form infinite chains parallel to c (see Appendix 2, Fig. A2.4a). For example, titanium atoms at $z = +\frac{1}{2}$ and $-\frac{1}{2}$ in adjacent cells are both coordinated to two oxygens at $z = 0$. Chains of octahedra are similarly formed by the octahedra centred at the corners of the unit cell. The two types of chains, which differ in orientation about z by 90° and which are $c/2$ out of step with each other, are linked by their corners to form a three-dimensional framework (Fig. 6.16c).

The rutile structure is also commonly described as a distorted hexagonal close packed oxide array with half the octahedral sites occupied by titanium. A 3 × 3 block of unit cells is shown in Fig. 6.16(d) with only the oxygen positions marked.

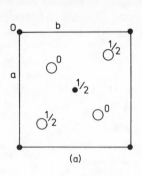 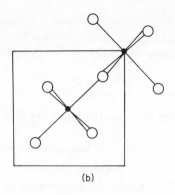

(a)　　　　　　　　　(b)

- Ti at 000; 0.5, 0.5, 0.5

○ 0 at 0.3, 0.3, 0; 0.7, 0.7, 0; 0.8, 0.2, 0.5; 0.2, 0.8, 0.5

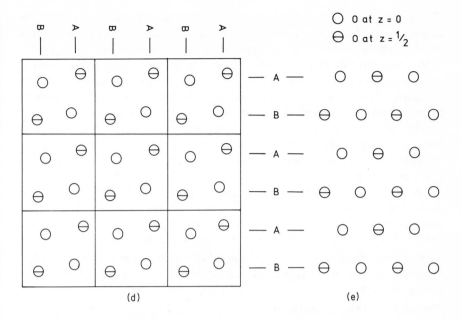

(d)

(e)

○ 0 at z = 0
⊖ 0 at z = ½

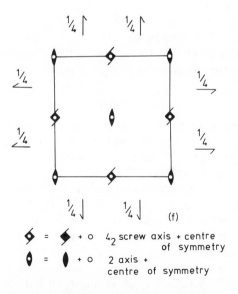

(f)

◆ = ◆ + ○ 4_2 screw axis + centre of symmetry

◖ = ◖ + ○   2 axis + centre of symmetry

Fig. 6.16 The rutile structure, TiO_2

Corrugated close packed layers occur, both horizontally and vertically. This contrasts with the undistorted hexagonal close packed arrangement (Fig. 6.16e), in which the layers occur in one orientation only (horizontally).

Very recently, an alternative way of describing the packing arrangement of oxide ions in rutile has been proposed. The oxide ion arrangement is a slightly distorted version of a new type of packing called *primitive tetragonal packing* (p.t.p.), which is characterized by fourfold symmetry and a sphere coordination number of *eleven*. This contrasts with hexagonal and cubic close packing which have a packing sphere coordination number of twelve and body centred tetragonal packing which has a coordination number of ten. More details are given in Chapter 7.

The bond lengths in TiO_2 may be calculated either graphically if (a) is drawn to scale or by geometry; e.g. for the Ti—O bond length between titanium at $(\frac{1}{2}, \frac{1}{2}, \frac{1}{2})$ and oxygen at $(0.3, 0.3, 0)$ the difference in a and b parameters of titanium and oxygen is $(\frac{1}{2} - 0.3)a = 0.92\,\text{Å}$. From a right-angled triangle calculation, the titanium–oxygen distance in projection down c (Fig. 6.16a) is $\sqrt{0.92^2 + 0.92^2}$. However, titanium and oxygen have a difference in c height of $(\frac{1}{2} - 0)c = 1.48\,\text{Å}$ and the Ti—O bond length is therefore equal to $\sqrt{0.92^2 + 0.92^2 + 1.48^2} = 1.97\,\text{Å}$. The axial Ti—O bond length between, for example, $Ti(\frac{1}{2}, \frac{1}{2}, \frac{1}{2})$ and $O(0.8, 0.2, 0.5)$ is easier to calculate because both atoms are at the same c height. It is equal to $\sqrt{2(0.3 \times 4.594)^2} = 1.95\,\text{Å}$.

Some of the symmetry elements in the space group $P4_2/mnm$ are shown in (Fig. 6.16f); most of them should be readily apparent on inspection of (a). Thus, the 4_2 axes are located halfway along the cell edges although no atoms lie on these 4_2 axes. The oxygen atoms are arranged on spirals around the 4_2 axes such that translation by $c/2$ and rotation by $90°$ converts one oxygen position to another. Centres of symmetry are present, for example, at the cell corners and also 2 and 2_1 axes and (not shown) mirror planes and glide planes.

Questions

6.1 What point groups result on adding a centre of symmetry to point groups (a) 1, (b) 2, (c) 3, (d) $\bar{4}$, (e) 4, (f) 222, (g) $mm2$, (h) $4mm$, (i) 6, (j) $\bar{6}$, (k) $6m2$?

6.2 What point groups result from the combination of two mirror planes at (a) $90°$ to each other, (b) $60°$, (c) $45°$, (d) $30°$?

6.3 What point groups result from the combination of two intersecting twofold axes at (a) $90°$ to each other, (b) $60°$, (c) $45°$, (d) $30°$?

6.4 An atom in an orthorhombic unit cell has fractional coordinates 0.1, 0.15 and 0.2. Give the coordinates of a second atom in the unit cell that is related to the first by each of the following, separately: (a) body centring, (b) a centre of symmetry at the origin, (c) a 2 axis parallel to z and passing through the origin, (d) a 2_1 axis parallel to z and passing through the origin, (e) A-centring.

6.5 Li_2PdO_2 has an orthorhombic unit cell, $a = 3.74$, $b = 2.98$, $c = 9.35\,\text{Å}$, $Z = 2$, space group $Immm$. Atomic coordinates are: Pd:2(a) 000; Li:4(i) $00z:z = 0.265$; O:4 (j) $0\frac{1}{2}z:z = 0.143$. Draw projections of the unit cell, determine coordination numbers and bond lengths and describe the structure. (*J. Solid State Chem.*, **6**, 329, 1973.)

6.6 Repeat the above question but for the structures of (a) ilmenite (Table 16.8), (b) garnet (Table 16.7) and (c) spinel (Table 16.5).

References

F. Donald Bloss (1971). *Crystallography and Crystal Chemistry*, Holt, Rinehart, Winston.

Lesley S. Dent Glasser (1977). *Crystallography and Its Applications*, Van Nostrand Reinhold.

N. F. M. Henry and K. Lonsdale (Eds) (1952). *International Tables for X-ray Crystallography*, Vol. 1, Kynoch Press.

M. F. C. Ladd and R. A. Palmer (1978). *Structure Determination by X-ray Crystallography*, Plenum Press.

Helen, D. Megaw (1973). *Crystal Structures, A Working Approach*, Saunders.

A. R. West and P. G. Bruce (1982). Tetragonal Packed Crystal Structures, *Acta Cryst.*, **B38**, 1891–6.

R. W. G. Wyckoff (1971). *Crystal Structures*, Vols 1 to 6, Wiley.

Chapter 7
Descriptive Crystal Chemistry

Crystal chemistry is concerned with the structures of crystals. It is to be distinguished from crystallography which is mainly concerned with the experimental methods used to solve crystal structures. Crystal chemistry includes topics such as: the description and classification of crystal structures, the factors that govern which type of structure is observed for a particular composition, and vice versa, the conditions under which a particular type of crystal structure is observed and the relation between crystal structure and physical or chemical properties.

This chapter and the next aim to cover the pertinent points of crystal chemistry. It is not intended to give a comprehensive survey of crystal structures as excellent reference works are available for this (e.g. Wells, 1984; Wyckoff, 1971; *Structure Reports*) and, anyway, space is strictly limited. Instead, a few of the more important structure types are discussed in some detail in this chapter and emphasis is given to the different ways in which a particular structure may be described and classified. In the next chapter, the factors which influence crystal structures are considered.

7.1 Description of crystal structures

Crystal structures may be described in various ways. The most common way, and one which gives all the necessary information, is to refer the structure to the unit cell (Section 5.3.1). In this approach, the structure is given by the size and shape of the cell and the positions of the atoms inside the cell. However, a knowledge of the unit cell and atomic coordinates alone is often insufficient to give a revealing picture of the structure in three dimensions. This latter is obtained only by considering a larger part of the structure, comprising perhaps several unit cells, and by considering the arrangement of atoms relative to each other, their coordination numbers, interatomic distances, types of bonding, etc. From this more general view of structures, it is possible to find alternative ways of visualizing them and also to compare and contrast different types of structure.

Two of the most useful ways of describing structures are the *close packing* approach and the *space filling polyhedron* approach. Neither of these can be applied to all types of structure and both have their advantages and limitations. They do, however, provide a much greater insight into crystal chemistry than is obtained using unit cells alone.

7.1.1 Close packed structures—cubic close packing (c.c.p.) and hexagonal close packing (h.c.p.)

Many metallic, ionic, covalent and molecular crystal structures can be described using the concept of close packing. The guiding factor is that structures are arranged so as to have the maximum density, although this needs some qualification when ionic structures are being discussed. The principles involved can be understood by considering the most efficient way of packing together in three dimensions, equal-sized spheres.

The most efficient way of packing spheres in *two* dimensions is as shown in Fig. 7.1. Each sphere, e.g. A, is surrounded by, and is in contact with, six others. By regular repetition, infinite sheets called *close packed layers* are formed. The coordination number of six is the maximum possible for a planar arrangement of contacting, equal-sized spheres. Lower coordination numbers are, of course, possible, as shown in the inset in Fig. 7.1, but in such cases the layers are no longer close packed. Note also that within a close packed layer, three *close packed directions* occur. In Fig. 7.1, rows of spheres in contact occur in the directions XX′, YY′ and ZZ′ and sphere A belongs to each of these rows.

The most efficient way of packing spheres in *three* dimensions is to stack close packed layers on top of each other to give *close packed structures*. There are two simple ways in which this may be done, resulting in *hexagonal close packed* and *cubic close packed* structures. These are derived as follows.

The most efficient way for two close packed layers A and B to be in contact is

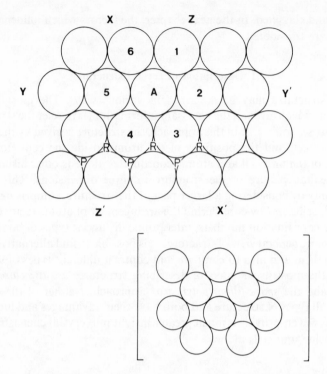

Fig. 7.1 A close packed layer of equal-sized spheres. Inset shows a non-close packed layer

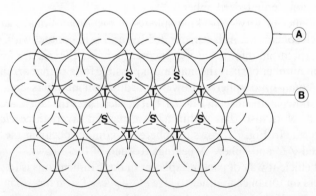

Fig. 7.2 Two close packed layers arranged in A and B positions. The B layer occupies the P positions shown in Fig. 7.1

for each sphere of one layer to rest in a hollow between three spheres in the other layer, e.g. at P or R in Fig. 7.1. Two layers in such a position relative to each other are shown in Fig. 7.2. Atoms in the second layer may occupy either P or R positions, but not both together, nor a mixture of the two. Any B (dashed) sphere is therefore seated between three A (solid) spheres, and vice versa.

Addition of a third close packed layer to the two shown in Fig. 7.2 can be done in two possible ways and herein lies the distinction between hexagonal and cubic close packing. In Fig. 7.2, suppose that the A layer lies underneath the B layer and we wish to place a third layer on top of B. There is a choice of positions for this third layer as there was for the second layer: the spheres can occupy either of the new sets of positions S or T but not both together nor a mixture of the two. If the third layer is placed at S, then it is directly over the A layer. As subsequent layers are added, the following sequence arises:

$$...ABABAB...$$

This is known as *hexagonal close packing*. If, however, the third layer is placed at T, then all three layers are staggered relative to each other and it is not until the fourth layer is positioned (at A) that the sequence is repeated. If the position of the third layer is called C, this gives (Fig. 7.3):

$$...ABCABCABC...$$

This sequence is known as *cubic close packing*.

Hexagonal close packing (h.c.p.) and cubic close packing (c.c.p.) are the two simplest stacking sequences and are by far the most important in structural chemistry. Other more complex sequences with larger repeat units, e.g. ABCACB or ABAC, do occur in a few materials; some of these larger repeat units are related to the phenomenon of polytypism (see later).

In a close packed structure each sphere is in contact with *twelve* others and this is the maximum coordination number possible for contacting and equal-sized spheres. (A common non-close packed structure is the body centred cube, e.g. in α-Fe, with a coordination number of eight; see Fig. 5.24a.) Six of these neighbours

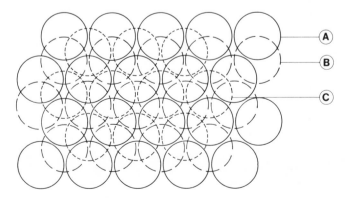

Fig. 7.3 Three close packed layers in c.c.p. sequence

216

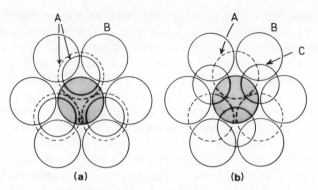

Fig. 7.4 Coordination number 12 of shaded sphere in (a) h.c.p.
and (b) c.c.p. structures

are coplanar with the central sphere (Fig. 7.1) and from Figs 7.2 and 7.3 the
remaining six are in two groups of three spheres, one in the plane above and one
in the plane below (Fig. 7.4); h.c.p. and c.c.p. differ only in the relative orientations
of these two groups of three neighbours.

The unit cells of cubic close packed and hexagonal close packed structures are
given in Figs 7.5 and 7.6. The unit cell of a c.c.p. arrangement is the familiar face
centred cubic (f.c.c.) unit cell (Fig. 7.5a) with spheres at corner and face centre
positions. The relation between c.c.p. and f.c.c. is not immediately obvious since
the faces of the f.c.c. unit cell do not correspond to close packed (c.p.) layers: thus,
each face centre sphere has only four equidistant neighbours at the corners of the
unit cell, as in the inset in Fig. 7.1. The c.p. layers are, instead, parallel to the {111}
planes (Section 5.3.6) of the f.c.c. unit cell. This is shown in Fig. 7.5(b) by removing
sphere 1 from its corner position to reveal part of a close packed layer underneath
(compare (b) with Fig. 7.1): the orientations of (a) and (b) are the same but the
spheres in (b) are shown larger and almost contacting. A similar arrangement to
that shown in (b) would be seen on removing any corner sphere and, therefore, in

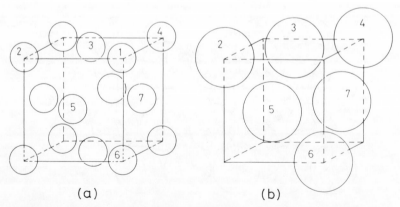

Fig. 7.5 Face centred cubic unit cell of a c.c.p. arrangement of spheres

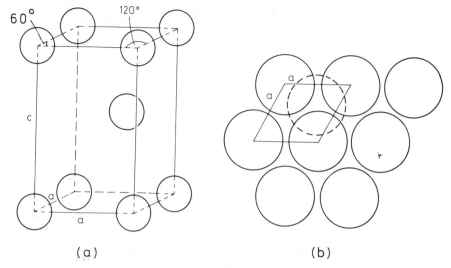

Fig. 7.6 Hexagonal unit cell of an h.c.p. arrangement of spheres

a c.c.p. structure, c.p. layers occur in four orientations perpendicular to the body diagonals of the cube (the cube has eight corners but each layer orientation is then counted twice, giving four different orientations).

The hexagonal unit cell of an h.c.p. arrangement of spheres (Fig. 7.6) is simpler in that the *basal plane* of the cell coincides with a c.p. layer of spheres (b). The unit cell contains only two spheres, one at the origin (and hence at all corners) and one inside the cell at position $\frac{1}{3}, \frac{2}{3}, \frac{1}{2}$ (dashed circle in b). Close packed layers occur in only one orientation in a h.c.p. structure.

In c.p. structures, 74.05 per cent of the total volume is occupied by spheres. This is the maximum density possible in structures constructed of spheres of only one size. This density value may be calculated by considering the volume and contents of unit cells of c.p. structures. In a c.c.p. array of spheres, the unit cell is f.c.c. and contains four spheres, one at a corner and three at face centre positions (Fig. 7.5) (this is equivalent to the statement that a f.c.c. unit cell contains four lattice points).

Close packed directions (XX', YY', ZZ' in Fig. 7.1), in which spheres are in contact, occur parallel to the face diagonals of the unit cell, e.g. spheres 2, 5 and 6 in Fig. 7.5(b) form part of a c.p. row and are parallel to directions of the type $\langle 110 \rangle$. If the diameter of a sphere is $2r$, the length of the diagonals is $4r$. The length of the cell edge is then $2\sqrt{2}r$ and the cell volume is $16\sqrt{2}r^3$. The volume of each sphere is $1.33\,\pi r^3$ and so the ratio of the total sphere volume to the unit cell volume is given by:

$$\frac{4 \times 1.33\pi r^3}{16\sqrt{2}\,r^3} = 0.7405$$

Similar results are obtained for hexagonal close packing by considering the contents and volume of the appropriate hexagonal unit cell (Fig. 7.6).

In non-close packed structures, densities lower than 0.7405 are obtained, e.g. the density of body centred cubic (b.c.c.) is 0.6802 (to calculate this it is necessary to know that the c.p. directions in b.c.c. are $\langle 111 \rangle$, i.e. parallel to the body diagonals of the cube).

7.1.2 Materials that can be described as close packed structures

7.1.2.1 *Metals*

Most metals crystallize in one of the three arrangements, c.c.p., h.c.p. and b.c.c., the first two of which are c.p. structures. The distribution of structure type among the metals is irregular (Table 7.1 and Appendix 9) and no clear-cut trends are observed. It is still not well understood why particular metals prefer one structure type to another. Calculations reveal that the lattice energies of h.c.p. and c.c.p. metal structures are comparable and, therefore, the structure observed in a particular case probably depends on fine details of the bonding requirements or band structures of the metal.

Some metals are *polymorphic* and exhibit more than one structure type, e.g. iron can be either c.c.p. or b.c.c. depending on temperature, cobalt can be either c.c.p. or h.c.p. but other forms also exist which possess more complex stacking sequences of c.p. layers. *Polytypism* is the name given to a special kind of polymorphism in that the structural differences between *polytypes* are confined to one dimension only. In c.p. metal structures, the arrangement of atoms within one c.p. layer is the same as that in any other c.p. layer and structural differences are confined to the way in which layers are stacked relative to each other. There are two simple stacking sequences, h.c.p., AB, and c.c.p., ABC, but an infinite number of more complex sequences, and some of these larger repeat units are observed in cobalt metal polytypes. In some polytypic materials, very long sequences have been observed which contain several hundred layers in the repeat unit. It is an

Table 7.1 *Structures and unit cell dimensions* (Å) *of some common metals*

Cubic close packed, *a*		Hexagonal close packed, *a, c*			Body centred cubic, *a*	
Cu	3.6147	Be	2.2856,	3.5832	Fe	2.8664
Ag	4.0857	Mg	3.2094,	5.2105	Cr	2.8846
Au	4.0783	Zn	2.6649,	4.9468	Mo	3.1469
Al	4.0495	Cd	2.9788,	5.6167	W	3.1650
Ni	3.5240	Ti	2.506,	4.6788	Ta	3.3026
Pd	3.8907	Zr	3.312,	5.1477	Ba	5.019
Pt	3.9239	Ru	2.7058,	4.2816		
Pb	4.9502	Os	2.7353,	4.3191		
		Re	2.760,	4.458		

intriguing question as to how such regular and long range structures form; for instance, after a repeat unit involving several hundred layers and a distance of, say, 500 Å, what kind of forces dictate that this sequence shall be repeated in the next 500 Å? As yet, this has not been explained satisfactorily although it has been suggested that screw dislocations may provide a mechanism of spiral crystal growth which could conceivably lead to these very large repeat units.

Stacking disorder is mentioned in Chapter 9 and occurs quite commonly; in this, pairs of layers in an otherwise regular stacking sequence are reversed, as in:

$$... ABCABCABCAC B ABCABC ...$$

This is an example of a two-dimensional or planar crystal defect. In some highly disordered materials, random stacking sequences occur.

7.1.2.2 *Alloys*

Alloys are intermetallic phases or solid solutions and, as is the case for pure metals, many can be regarded as c.p. structures. As an example, copper and gold both crystallize in c.c.p. structures and at high temperatures a complete range of *solid solutions* between copper and gold forms. In these the copper and gold atoms are statistically distributed over the lattice points of the f.c.c. unit cell and therefore the c.c.p. layers contain a random mixture of copper and gold atoms. On annealing the alloy compositions AuCu and $AuCu_3$ at lower temperatures, the gold and copper atoms order themselves; c.c.p. layers still occur but the arrangement of gold and copper atoms within the layers is no longer statistical. Order–disorder phenomena are discussed further in Chapter 12.

7.1.2.3 *Ionic structures*

The structures of materials such as NaCl, Al_2O_3, Na_2O, ZnO, etc., in which the anion is somewhat larger than the cation, can be regarded as built of c.p. layers of anions with the cations placed in interstitial sites. A considerable variety of structure types are possible in which the variables are the stacking sequence of the anions, either h.c.p. or c.c.p., and the number and type of interstitial sites that are occupied by cations. Before these structures are considered, some general comments (discussed in more detail in the next chapter) should be made.

In c.p. metallic structures, all the atoms are of one type and it may be assumed that nearest neighbour atoms are in contact. With ionic structures, ions of opposite charge are present and the resulting structure is a balance of attractive and repulsive electrostatic forces. Further, since there are two (at least) types of ion present, their relative sizes affect the structure that is adopted. Although, for many purposes, it is convenient to regard ionic structures such as NaCl as containing c.p. anion layers with cations in interstitial sites, in fact the cations are often too large for the prescribed interstitial sites. The structure can accommodate them only by expanding the anion array. Consequently, the arrangement of anions is the same as in a c.p. array of spheres, but the anions may not be in

220

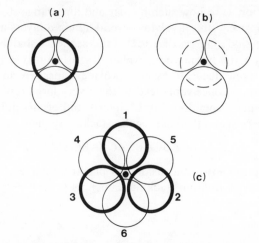

Fig. 7.7 Interstitial sites in a c.p. structure. Heavy circles are above and the dashed circles below the plane of the paper: (a) T_+ site, (b) T_- site, (c) O site

contact. O'Keeffe has suggested the term *eutactic* for structures such as these in which the arrangement of ions is the same as that in a c.p. array but in which the ions are not necessarily contacting. In the subsequent discussion of ionic structures use of the terms h.c.p. and c.c.p. for the anion arrays does not necessarily imply that the anions are in contact but rather that the structures are eutactic.

Two types of interstitial sites, tetrahedral and octahedral, are present in c.p. structures (Fig. 7.7). For the tetrahedral sites, three anions that form the base of the tetrahedron belong to one c.p. layer with the apex of the tetrahedron in the layer above (a) or below (b). This gives two types of tetrahedral site, T_+ and T_-, in which the apex is up and down, respectively. Because the centre of gravity of a tetrahedron is nearer to the base than to the apex (Appendix 1) cations in tetrahedral sites are not located midway between adjacent anion layers but are nearer to one layer than the other. Octahedral sites, on the other hand, are coordinated to three anions in each layer (Fig. 7.7c) and are placed midway between the anion layers. A more common way of regarding octahedral coordination is as four coplanar atoms with one atom at each apex above and below the plane. In (c), atoms 1, 2, 4 and 6 are coplanar; 3 and 5 form apices of the octahedron. Alternatively, atoms 2, 3, 4, 5 and 1, 3, 5, 6 are coplanar.

The distribution of interstitial sites between any two adjacent layers of c.p. anions is shown in Fig. 7.8. Counting up the numbers of each, for every anion there is one octahedral site and two tetrahedral sites, one T_+ and one T_-. It is rare that all the interstitial sites in a c.p. structure are occupied; often one set is fully or partly occupied and the remaining two sets are empty. A selection of c.p. ionic structures, classified according to the anion layer stacking sequence and the occupancy of the interstitial sites, is given in Table 7.2. Individual structures are

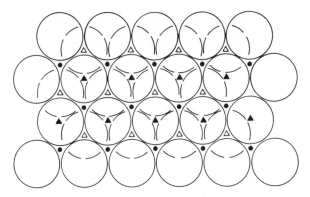

● octahedral sites
▲ T+ tetrahedral sites
△ T− tetrahedral sites

Fig. 7.8 Distribution of interstitial sites between two c.p.
layers. Dashed circles are below the plane of the paper

described in more detail later; here we note how a wide range of diverse structures
can be grouped into one large family and this helps to bring out the similarities
and differences between them. For example:

(a) The rock salt and nickel arsenide structures both have octahedrally
coordinated cations and differ only in the anion stacking sequence. A similar
relation exists between the structures of olivine and spinel.

Table 7.2 *Some close packed structures*

Anion arrangement	Interstitial sites			Examples
	T_+	T_-	Oct	
c.c.p.	—	—	1	NaCl, rock salt
	1	—	—	ZnS blende or sphalerite
	$\frac{1}{8}$	$\frac{1}{8}$	$\frac{1}{2}$	$MgAl_2O_4$, spinel
	—	—	$\frac{1}{2}$	$CdCl_2$
	1	—	—	$CuFeS_2$
	—	—	$\frac{1}{3}$	$CrCl_3$
	1	1	—	K_2O antifluorite
h.c.p.	—	—	1	NiAs
	1	—	—	ZnS, wurtzite
	—	—	$\frac{1}{2}$	CdI_2
	—	—	$\frac{1}{2}$	TiO_2^*, rutile
	—	—	$\frac{2}{3}$	Al_2O_3
	$\frac{1}{8}$	$\frac{1}{8}$	$\frac{1}{2}$	Mg_2SiO_4, olivine
	1	—	—	β-Li_3PO_4
	$\frac{1}{2}$	$\frac{1}{2}$	—	γ-$Li_3PO_4^*$
c.c.p. 'CaO₃' layers	—	—	$\frac{1}{4}$	$CaTiO_3$ perovskite

*The h.c.p. oxide layers in rutile and γ-Li_3PO_4 are not planar but are buckled. The oxide
ion arrangement in these may alternatively be described as tetragonal packed (t.p.).

β – Li₂BeSiO₄ , 'Pn'

β – LiGaO₂ , Pbn2₁

β_II – Li₂Zn SiO₄ Pmn2₁

β_II – Li₃PO₄ , Pmn2₁

ZnO , P6₃mc

Fig. 7.9 Ordered tetrahedral structures related to the wurtzite structure of ZnO. One oxide layer is shown together with one layer of cations in T₊ tetrahedral sites. Note the different cation ordering sequences. (From West, 1975)

(b) Rutile, TiO_2 and CdI_2 both have hexagonal close packed anions (although the layers are buckled in rutile) with half the octahedral sites occupied by cations but they differ in the manner of occupancy of these octahedral sites. In rutile, half the octahedral sites between any pair of c.p. anion layers are occupied by Ti^{4+}, in ordered fashion; in CdI_2, layers of fully occupied octahedral sites alternate with layers in which all sites are empty and this gives CdI_2 an obvious layered structure which is reflected in its physical properties.

(c) Both β and γ polymorphs of Li_3PO_4 have h.c.p. oxide ions (although the layers are buckled, especially in γ) with half the tetrahedral sites occupied by cations; in β-Li_3PO_4, all the T_+ sites are occupied with T_- sites empty, but in γ-Li_3PO_4 half of both the T_+ and T_- sites are occupied, in ordered fashion.

(d) A large number of more complex c.p. structures are possible. For example, in the series; ZnO, $LiGaO_2$, β-Li_3PO_4, β-Li_2ZnSiO_4, all the compounds have structures related to that of ZnO. The T_+ sites are fully occupied in each but different structures arise as the cations order themselves in different ways over the T_+ sites (Fig. 7.9).

(e) In a few structures, it is useful to regard the *cation* as forming the c.p. layers with the anions occupying the interstitial sites. The most common example is the fluorite structure, CaF_2, which may be regarded as cubic close packed Ca^{2+} ions with all the T_+ and T_- tetrahedral sites occupied by F^- ions. The antifluorite structure of, for example, K_2O is the exact inverse of fluorite and is included in Table 7.2.

(f) The close packing concept may be extended to include structures in which a mixture of anions and large cations form the packing layers and smaller cations occupy the interstitial sites. In perovskite, $CaTiO_3$, c.c.p. layers of composition 'CaO_3' occur and one quarter of the octahedral sites between these layers are occupied by titanium. Some of the octahedral sites must contain Ca^{2+} ions at the corners, but Ti^{4+} occupies those octahedral sites in which all six corners are O^{2-} ions, Fig. 6.15.

(g) Some structures may be regarded as anion-deficient c.p. structures in which the anions form an essentially c.p. array but in which some are missing. The ReO_3 structure may be regarded as having c.c.p. oxide layers with one quarter of the O^{2-} ion sites empty. It is analogous to the perovskite structure just described in which the Ti^{4+} ions are replaced by Re^{6+} and the Ca^{2+} ions are removed and their sites left vacant. The structure of β-alumina, nominally of formula $NaAl_{11}O_{17}$, contains c.p. oxide layers in which every fifth layer has approximately three quarters of the O^{2-} ions missing.

7.1.2.4 Covalent network structures

Materials such as diamond and silicon carbide, which have very strong, directional, covalent bonds, can also be described as c.p. structures or eutactic structures; many have the same structures as ionic compounds (Section 7.1.2.3). Thus, one polymorph of SiC has the wurtzite structure and it is immaterial

224

whether silicon or carbon is regarded as the packing atom since the net result—a three-dimensional framework of corner sharing tetrahedra—is the same. Diamond can be regarded as a sphalerite structure in which half the carbon atoms form a c.c.p. array and the other half occupy T_+ interstitial sites, but again the two types of atom are equivalent. Classification of diamond as a eutactic structure is useful since in diamond all the atoms are of the same size and it is unrealistic to distinguish between packing atoms and interstitial atoms.

Many structures have mixed ionic–covalent bonding, e.g. ZnS and $CrCl_3$ in Table 7.2, and one advantage of describing their structures in terms of close packing is that this can be done, if necessary, without reference to the type of bonding that is present.

7.1.2.5 *Molecular structures*

Since c.p. structures provide an efficient means of packing atoms together, many molecular compounds crystallize as c.p. structures even though the bonding forces between adjacent molecules are weak van der Waals forces. If the molecules are roughly spherical in shape or become spherical because they can rotate or occupy different orientations at random, then simple h.c.p. or c.c.p. structures result, e.g. in crystalline H_2, CH_4 and HCl. Non-spherical molecules,

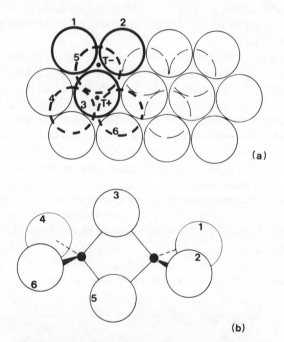

(a)

(b)

Fig. 7.10 Hexagonal close packing arrangement of bromine atoms in crystalline Al_2Br_6 molecules; aluminium atoms occupy T_+ and T_- sites. Dashed circles are below the plane of the paper

especially if they are built of tetrahedra and octahedra, can also fit into a c.p. arrangement. For example, Al_2Br_6 is a dimeric molecule which can be regarded as two $AlBr_4$ tetrahedra sharing a common edge (Fig. 7.10b). In crystalline Al_2Br_6, the bromine atoms form an h.c.p. arrangement and aluminium atoms occupy $\frac{1}{6}$ of the available tetrahedral sites. One molecule, with the bromine atoms in heavy outline is shown in Fig. 7.10(a); aluminium atoms occupy one pair of adjacent T_+ and T_- sites. Bromine atoms 3 and 5 are common to both tetrahedra and are the bridging atoms in (b). Adjacent Al_2Br_6 molecules are arranged so that each bromine in the h.c.p. array belongs to only one molecule. $SnBr_4$ is a tetrahedral molecule and also crystallizes with an h.c.p. bromine array but now only one eighth of the tetrahedral sites are occupied.

7.1.3 Other packing arrangements: tetragonal packing

The two c.p. arrangements, h.c.p. and c.c.p., are the most efficient ways of packing spheres in three dimensions. They are characterized by a coordination number of 12 and a density of 74.05 per cent. Only slightly less dense is the *primitive tetragonal packed* (p.t.p.) arrangement with a coordination number of 11 and a packing density of 71.87 per cent, and the *body centred tetragonal* (b.c.t.) arrangement with a coordination number of 10 and a packing density of 69.81 per cent. Although the p.t.p. arrangement has been discovered only recently, it is, in fact, commonly occurring and forms the basis of the anion array in structures such as rutile, TiO_2, ramsellite, MnO_2, and γ tetrahedral structures such as γ-Li_3PO_4 and γ-$LiAlO_2$.

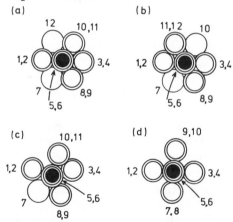

Fig. 7.11 Coordination number of shaded atom of (a) 12 in h.c.p., (b) 12 in c.c.p., (c) 11 in primitive tetragonal packing (p.t.p.) and (d) 10 in body centred tetragonal (b.c.t.). It is found that (c) occurs in various structures including rutile, TiO_2, ramsdellite, MnO_2, and γ tetrahedral oxides, γ-$LiAlO_2$, γ-Li_3PO_4. (From West and Bruce, 1982)

The coordination number and environment in p.t.p. and b.c.t. are shown in Fig. 7.11, together with those for h.c.p. and c.c.p. The p.t.p. arrangement (Fig. 7.11c) may be regarded as derived from h.c.p. in that sphere 12 in Fig. 7.11(a) is no longer part of the coordination sphere of the shaded atom in Fig. 7.11(c) and atoms 1 to 6, which together with the shaded atom are coplanar in Fig. 7.11(a), are slightly non-coplanar in Fig. 7.11(c). A more extended version of Fig. 7.11(c) can be seen in Fig. 6.16(d) which shows the oxide ion arrangement in rutile.

The interstitial sites in a p.t.p. array fall into two categories: (a) tetrahedral and octahedral sites that are undistorted, as in h.c.p.; (b) tetrahedral and octahedral sites that are distorted. The undistorted sites can be occupied by a variety of cations, e.g. octahedral sites are occupied by $Ti(TiO_2)$, $Mn(MnO_2)$ and tetrahedral sites by Li, $Al(LiAlO_2)$, Li, $P(Li_3PO_4)$. The distorted sites, on the other hand, are usually empty.

The b.c.t. arrangement (Fig. 7.11d) appears not to occur in known crystal structures. This may be because it does not have undistorted interstitial sites suitable for occupation by cations.

7.1.4 Structures built of space filling polyhedra

This approach to describing crystal structures emphasizes the coordination number of the cations and the way that structures may be regarded as built of polyhedra linked together by sharing corners, edges or faces. For example, in NaCl, each Na^+ ion has six Cl^- ions as its nearest neighbours, arranged octahedrally; this is represented as an octahedron with Cl^- ions at the corners and Na^+ at the centre. A three-dimensional overview of the structure is obtained by looking at the way in which neighbouring octahedra are linked to each other. In NaCl, each octahedron edge is shared between two octahedra, resulting in an infinite framework of edge-sharing octahedra. Although the polyhedra may link up to form a three-dimensional framework, not all the available space is usually filled; e.g. in NaCl, empty tetrahedral cavities remain within the framework of octahedra. These cavities are shown in Appendix 2 for a single layer of octahedra. Also shown in the Appendix are pictures of various polyhedral linkages and hints on making one's own models from card paper are given.

Although it is legitimate to estimate the efficiency of packing of spheres in c.p. structures, it is incorrect to do this for space filling polyhedra since the anions, usually the largest ions in the structure, are represented by points at the corners of the polyhedra. In spite of this obvious misrepresentation, the space filling polyhedron approach has the advantage that it shows the topology or connectivity of a framework structure and indicates clearly the location of empty interstitial sites. Some examples of structures that can be viewed as built of space filling polyhedra are given in Table 7.3. A variety of polyhedra occur in inorganic structures, although tetrahedra and octahedra appear to be the most common.

A complete scheme for classifying polyhedral structures has been proposed by Wells and others in which the initial problem is a geometrical one—what types of network built of linked polyhedra are possible? The variables to be considered

Table 7.3 *Some structures built of space filling polyhedra*

Octahedra only

---

12 edges shared	NaCl
6 corners shared	ReO_3
3 edges shared	$CrCl_3$, BiI_3
2 edges and 6 corners shared	TiO_2
4 corners shared	$KAlF_4$

Tetrahedra only

---

4 corners shared (between 4 tetrahedra)	ZnS
4 corners shared (between 2 tetrahedra)	SiO_2
1 corner shared (between 2 tetrahedra)	$Si_2O_7^{6-}$
2 corners shared (between 2 tetrahedra)	$(SiO_3)_n^{2n-}$, chains or rings

---

are the following. Polyhedra may share some or all of their corners, edges and faces with adjacent polyhedra, which may or may not be of the same type. The corners and edges may be common to more than two polyhedra (obviously, only two polyhedra can share a common face); e.g. in SiO_2, each corner (oxygen) is shared between two (SiO_4) tetrahedra but in ZnS each corner is shared between four tetrahedra; in spinel, $MgAl_2O_4$, each corner is shared between three octahedra and one tetrahedron. An enormous number of structures are feasible, at least theoretically, and it is an interesting exercise to categorize real structures on this basis.

This classification scheme for non-metallic materials has not, as yet, received a wide acceptance, perhaps because by its very nature it is all-embracing and indicates similarities between different structures whereas physically or chemically perhaps none exists. The topological problem of what kinds of structure may be generated by arranging polyhedra in various ways takes no account of the bonding forces between atoms or ions. Such information must come from elsewhere. Also the description of structures in terms of polyhedra does not necessarily imply that such entities exist in the structure as separate species. Thus, in NaCl, the bonding is mainly ionic and physically distinct $NaCl_6$ octahedra do not occur. Similarly, SiC has a covalent network structure and separate SiC_4 tetrahedral entities do not exist. Polyhedra *do* have a separate existence in structures of (a) molecular materials, e.g. Al_2Br_6 contains pairs of edge-sharing tetrahedra, and (b) compounds that contain complex ions, e.g. the silicate structures are built up of SiO_4 tetrahedra which link up to form complex anions ranging in size from isolated monomeric units to infinite chains and sheets and, finally, three-dimensional frameworks.

In considering the types of polyhedral linkage that are likely to occur in crystal structures, Pauling's third rule for the structures of complex ionic crystals (next chapter) provides a very useful guideline. This rule states that the presence of shared polyhedron edges and, especially, shared faces decreases the stability of a structure. The effect is particularly large for cations of high valence and small coordination number, i.e. for small polyhedra, especially tetrahedra, that contain highly charged cations. When polyhedra share edges and, particularly, faces, the cation–cation distances (i.e. the distances between the centres of the polyhedra) decrease and the cations repel each other electrostatically. In Fig. 7.12 are shown pairs of (a) corner-sharing and (b) edge-sharing octahedra. The cation–cation distance is clearly less in the latter and for face-sharing octahedra, not shown, the cations are even closer. Comparing edge-shared tetrahedra and edge-shared octahedra, the cation–cation distance M–M (relative to the cation–anion distance M–X) is less between the edge-shared tetrahedra because the MXM bridging angle is (c) 71° for the tetrahedra and (b) 90° for the octahedra. A similar effect occurs for face-sharing tetrahedra and octahedra. In Table 7.4, the M–M distances for the various polyhedral linkages are given as a function of the distance M–X. The M–M distance is greatest for corner sharing of either tetrahedra or octahedra and is least for face-sharing tetrahedra. For the sharing of corners and edges, the M–M distances given are the maximum possible distances. Reduced distances occur if the polyhedra are rotated about the shared corner or edge (e.g. so that the M–X–M angle between corner-shared polyhedra is less than 180°).

From Table 7.4, the M–M distance in face-shared tetrahedra is considerably less than the M–X bond distance. This represents an unstable situation, therefore,

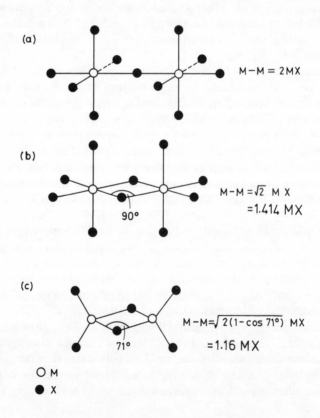

(a) $M-M = 2MX$

(b) $M-M = \sqrt{2}\ M\ X$
        $= 1.414\ MX$
        90°

(c) $M-M = \sqrt{2(1-\cos 71°)}\ MX$
        $= 1.16\ MX$
        71°

○ M
● X

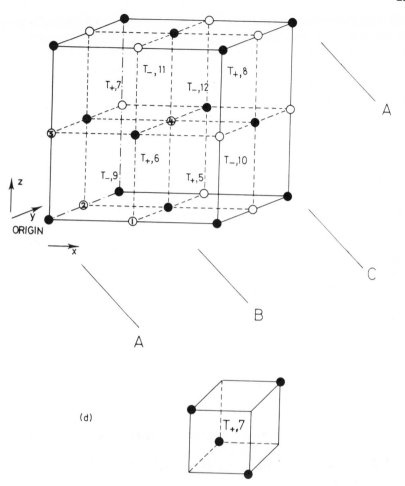

Fig. 7.12 Cation–cation separation in octahedra which share (a) corners and (b) edges and in tetrahedra which (c) share edges. (d) Available cation sites in an f.c.c. anion array

Table 7.4 *The distance M–M between centres of regular* MX_4 *or* MX_6 *groups sharing X atom(s).* (Data taken from Wells, 1975)

		Sharing a common	
	corner*	edge*	face
Two tetrahedra	2.00 MX(tet.)	1.16 MX(tet.)	0.67 MX(tet.)
Two octahedra	2.00 MX(oct.)	1.41 MX(oct.)	1.16 MX(oct.)

*Maximum possible value.

because of the large cation–cation repulsions which would arise, and does not normally occur. A good example is provided by the absence of any crystal whose structure is the h.c.p. equivalent of the fluorite (or antifluorite) structure. In, for example, Na_2O which has the antifluorite structure, the oxide ions are in a c.c.p. array and the NaO_4 tetrahedra share edges. However, in a structure with an h.c.p. anion array and all the tetrahedral cation sites occupied, the MX_4 tetrahedra would share faces. Edge sharing of tetrahedra in which the M–M separation is only 16 per cent larger than the M–X bond distance appears to be energetically acceptable in principle, as shown by the common occurrence of compounds with a fluorite type of structure, but face sharing of tetrahedra is unacceptable.

For tetrahedra containing cations of high charge, edge sharing appears to be unacceptable and only corner sharing occurs. Thus, in silicate structures, which are built of SiO_4 tetrahedra, edge sharing of SiO_4 tetrahedra never occurs*. In an ideally ionic structure the charge on silicon would be $4+$, but the actual charges are considerably less due to partial covalency of the Si—O bonds.

An additional factor to be taken into account when comparing linkages of tetrahedra and octahedra (Table 7.4) is that, for a given M and X, the M–X distance varies, depending on the nature of the polyhedra. In a c.p. structure, the tetrahedral sites are smaller than the octahedral sites, i.e.

$$\frac{\text{M–X in } MX_4(\text{tet.})}{\text{M–X in } MX_6(\text{oct.})} = \frac{\sqrt{3}}{2} = 0.866$$

This may be shown by, for example, calculating and comparing the M–X distances present in the f.c.c. cells of NaCl and CaF_2 structures (see Table 7.8).

7.2 Some important structure types

7.2.1 Rock salt (NaCl), zinc blende or sphalerite (ZnS) and antifluorite (Na_2O)

These structures are considered together because they have in common a cubic close packed (i.e. a face centred cubic) arrangement of anions and differ only in the positions of the cations. In Fig. 7.12(d) are shown the anions (•) in a f.c.c. unit cell with octahedral O, tetrahedral T_+ and tetrahedral T_- sites for the cations. Octahedral sites occur at the edge centre 1, 2, 3 and body centre 4 positions. In order to see the T_+ and T_- sites clearly it is convenient to divide the unit cell into eight smaller cubes by bisecting each cell edge (dashed lines). These smaller cubes contain anions at only four of the eight corners and in the middle of each small cube is a tetrahedral site (inset, Fig. 7.12d). Note that this is a useful way to represent tetrahedra and permits ready calculation of bond distances, bond angles, etc., in tetrahedra. A tetrahedral site occurs in the middle of each smaller cube and parallel to any of the cell axes, x, y and z, T_+ and T_- sites alternate.

*By contrast, in SiS_2, edge sharing of SiS_4 tetrahedra occurs. The Si—S bond is longer than and more covalent than the Si—O bond and both effects act to reduce the Si—Si repulsion; consequently edge sharing becomes acceptable.

The c.p. anion layers are oriented parallel to the $\{111\}$ planes of the unit cell (Fig. 7.5) and four such layers, ABCA, are indicated in Fig. 7.12(d). The coordinates of the anions in the unit cell (Chapter 6) are

$$000, \quad \tfrac{1}{2}\tfrac{1}{2}0, \quad \tfrac{1}{2}0\tfrac{1}{2}, \quad 0\tfrac{1}{2}\tfrac{1}{2}$$

Only one corner atom is included in this list, at 000, since the other seven corner atoms, at 100, 110, etc., are equivalent and can be regarded as the corner atoms of adjacent unit cells. Likewise for each pair of opposite faces, e.g. $\tfrac{1}{2}\tfrac{1}{2}0$ and $\tfrac{1}{2}\tfrac{1}{2}1$, only one of these is specified since the other is equivalent and is generated by translation of one unit cell dimension in the z direction.

The various cation positions have the following coordinates:

Octahedral:	$1 - \tfrac{1}{2}00,$	$2 - 0\tfrac{1}{2}0,$	$3 - 00\tfrac{1}{2},$	$4 - \tfrac{1}{2}\tfrac{1}{2}\tfrac{1}{2}$
Tetrahedral T_+:	$5 - \tfrac{3}{4}\tfrac{1}{4}\tfrac{1}{4},$	$6 - \tfrac{1}{4}\tfrac{3}{4}\tfrac{1}{4},$	$7 - \tfrac{1}{4}\tfrac{1}{4}\tfrac{3}{4},$	$8 - \tfrac{3}{4}\tfrac{3}{4}\tfrac{3}{4}$
Tetrahedral T_-:	$9 - \tfrac{1}{4}\tfrac{1}{4}\tfrac{1}{4},$	$10 - \tfrac{3}{4}\tfrac{3}{4}\tfrac{1}{4},$	$11 - \tfrac{1}{4}\tfrac{3}{4}\tfrac{3}{4},$	$12 - \tfrac{3}{4}\tfrac{1}{4}\tfrac{3}{4}$

Note that there are four of each type of cation site, O, T_+, T_-, in the unit cell together with four anions. By having different sites occupied, different structures are generated, as follows:

Rock salt: O sites occupied by cations; T_+, T_- empty

Zinc blende: T_+ (or T_-) sites occupied; O, T_- (or T_+) empty

Antifluorite: T_+, T_- occupied; O empty.

Unit cells of these three structures are shown in Fig. 7.13(a), (b) and (c). In Fig. 7.13(d), (e) and (f), the structures are shown in the same orientation but now the coordination numbers of the ions are emphasized. In rock salt (d), both anions and cations are octahedrally coordinated, whereas in zinc blende (e) both are tetrahedrally coordinated. In antifluorite (f), cations are tetrahedrally coordinated and anions are eight-coordinated (not shown).

A general rule regarding coordination numbers is that in any structure of formula $A_x X_y$, the coordination numbers of A and X must be in the ratio of $y:x$. In both rock salt and zinc blende, $x = y$ and therefore, in each, anions and cations have the same coordination number. In antifluorite, of formula A_2X, the coordination numbers of cation and anion must be in the ratio of 1:2. Since the cations occupy tetrahedral sites, the anion coordination number must therefore be eight.

The tetrahedral cation coordination in antifluorite is shown in 7.13(f). In order to see the anion coordination number of eight, it is convenient to redefine the origin of the unit cell so that the origin coincides with a cation rather than an anion. This may be done by displacing the unit cell along one of the body diagonals and by one quarter of the length of the diagonal. The cation at X in (c), with coordinates $\tfrac{1}{4}\tfrac{1}{4}\tfrac{1}{4}$, may be chosen as the new origin of the unit cell. The coordinates of the atoms in the new cell are given by subtracting $\tfrac{1}{4}\tfrac{1}{4}\tfrac{1}{4}$ from their coordinates in the old cell, i.e.

232

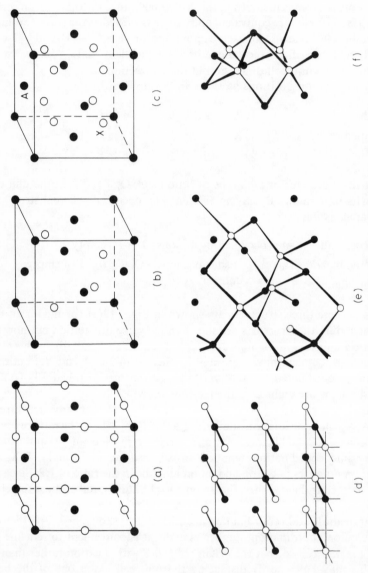

Fig. 7.13 Unit cell of (a) NaCl, (b) ZnS, sphalerite, and (c) Na$_2$O. Open circles are cations; closed circles anions. Coordination numbers of ions in (d) NaCl, (e) ZnS and (f) Na$_2$O

	Old cell	New cell
Anions	000	$-\frac{1}{4}-\frac{1}{4}-\frac{1}{4} \to \frac{3}{4}\frac{3}{4}\frac{3}{4}$
	$\frac{1}{2}\frac{1}{2}0$	$\frac{1}{4}\ \ \frac{1}{4}-\frac{1}{4} \to \frac{1}{4}\frac{1}{4}\frac{3}{4}$
	$\frac{1}{2}0\frac{1}{2}$	$\frac{1}{4}-\frac{1}{4}\ \ \frac{1}{4} \to \frac{1}{4}\frac{3}{4}\frac{1}{4}$
	$0\frac{1}{2}\frac{1}{2}$	$-\frac{1}{4}\ \ \frac{1}{4}\ \ \frac{1}{4} \to \frac{3}{4}\frac{1}{4}\frac{1}{4}$
Cations	$\frac{1}{4}\frac{1}{4}\frac{1}{4}$	$0\ \ 0\ \ 0$
	$\frac{1}{4}\frac{1}{4}\frac{3}{4}$	$0\ \ 0\ \ \frac{1}{2}$
	$\frac{1}{4}\frac{3}{4}\frac{1}{4}$	$0\ \ \frac{1}{2}\ \ 0$
	$\frac{3}{4}\frac{1}{4}\frac{1}{4}$	$\frac{1}{2}\ \ 0\ \ 0$
	$\frac{1}{4}\frac{3}{4}\frac{3}{4}$	$0\ \ \frac{1}{2}\ \ \frac{1}{2}$
	$\frac{3}{4}\frac{1}{4}\frac{3}{4}$	$\frac{1}{2}\ \ 0\ \ \frac{1}{2}$
	$\frac{3}{4}\frac{3}{4}\frac{1}{4}$	$\frac{1}{2}\ \ \frac{1}{2}\ \ 0$
	$\frac{3}{4}\frac{3}{4}\frac{3}{4}$	$\frac{1}{2}\ \ \frac{1}{2}\ \ \frac{1}{2}$

In cases where negative coordinates occur as a result of this subtraction, e.g. $-\frac{1}{4}-\frac{1}{4}-\frac{1}{4}$, the position lies outside the new unit cell and it is necessary to find an equivalent position within the unit cell. In this particular case, one is added to each coordinate, giving $\frac{3}{4}\frac{3}{4}\frac{3}{4}$. Addition of one to, say, the x coordinate is equivalent to moving to a similar position in the next unit cell in the x direction. This is illustrated further in Appendix 4. The new unit cell of the antifluorite structure with its origin at cation X, is shown in Fig. 7.14(a). It contains cations at corners, edge centres, face centres and body centre. In order to see the anion coordination more clearly, the unit cell may be imagined as divided into eight smaller cubes (as in Fig. 7.12). Each of these smaller cubes has cations at all eight corners and hence at the centre of each is an eight-coordinate site. Anions occupy four of these eight smaller cubes such that parallel to the cell axes the eight

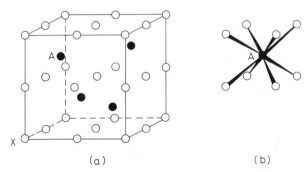

(a) (b)

Fig. 7.14 Alternative view of the fluorite and antifluorite structures

234

coordinate sites are alternately occupied and empty. The eight fold coordination for one anion, A, is shown in Fig. 7.14(b).

In the antifluorite structure, the effect of changing the origin from an anion position to a cation position is to show the structure in a completely different light. This does not happen with the rock salt and zinc blende structures, however. In these the cation and anion positions are interchangeable and it is immaterial whether the origin coincides with an anion or a cation.

So far the NaCl, ZnS and Na_2O structures have been described in two ways: (a) as c.p. structures and (b) in terms of their unit cells. A third way is to regard them as built of space filling polyhedra. In Fig. 7.13(d), (e) and (f) and Fig. 7.14(b), the coordination environment of each ion is shown. Each ion and its nearest neighbours may be represented by the appropriate polyhedron, e.g. a tetrahedron represents a zinc ion and its four sulphide neighbours (or vice versa) in zinc blende. It is then necessary to consider how neighbouring polyhedra are linked— whether by sharing corners, edges or faces.

In the rock salt structure, $NaCl_6$ octahedra share common edges. Each octahedron has twelve edges and each edge is shared between two octahedra. In Fig. 7.15, the unit cell of NaCl is outlined (dotted) and is in the same orientation as in Fig. 7.13(a). Octahedra centred on sodium ions in edge centre positions 1, 2 and 3 are outlined. Octahedra 1 and 2 share a common edge, indicated by the thick dashed line, and octahedra 2 and 3 share a common edge, shown by the thick solid line. Since each octahedron is linked by its edges to twelve other octahedra it is difficult to represent this satisfactorily in a drawing.

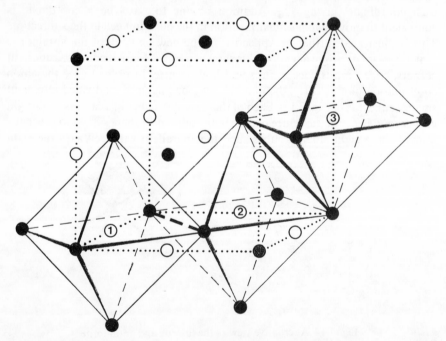

Fig. 7.15 Unit cell of the rock salt structure showing edge sharing of octahedra

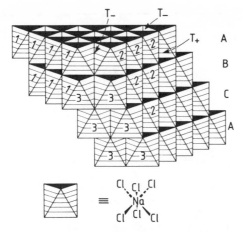

Fig. 7.16 The rock salt structure depicted as
an array of edge-sharing octahedra

A schematic drawing showing the rock salt structure as an array of octahedra is shown in Fig. 7.16. Also shown are tetrahedral interstices (arrowed) which are normally empty in NaCl. In this structure, each octahedron face is parallel to layers of c.p. anions. This is emphasized in the drawing in that coplanar octahedron faces are numbered or shaded. Parts of four different sets of faces are shown, corresponding to the four c.p. orientations in a.c.c.p. array.

A large number of AB compounds possess the rock salt structure. A selection is given in Table 7.5 together with values of the a dimension of the cubic unit cell. Most halides and hydrides of the alkali metals and Ag^+ have this structure, as do also a large number of the chalcogenides (oxides, sulphides, etc.) of divalent metals such as alkaline earths and divalent transition metals. Many of these are ionic but others, e.g. TiO, have a metallic character.

Table 7.5 *Some compounds with the NaCl structure*

	$a(\text{Å})$		$a(\text{Å})$		$a(\text{Å})$		$a(\text{Å})$
MgO	4.213	MgS	5.200	LiF	4.0270	KF	5.347
CaO	4.8105	CaS	5.6948	LiCl	5.1396	KCl	6.2931
SrO	5.160	SrS	6.020	LiBr	5.5013	KBr	6.5966
BaO	5.539	BaS	6.386	LiI	6.00	KI	7.0655
TiO	4.177	αMnS	5.224	LiH	4.083	RbF	5.6516
MnO	4.445	MgSe	5.462	NaF	4.64	RbCl	6.5810
FeO	4.307	CaSe	5.924	NaCl	5.6402	RbBr	6.889
CoO	4.260	SrSe	6.246	NaBr	5.9772	RbI	7.342
NiO	4.1769	BaSe	6.600	NaI	6.473	AgF	4.92
CdO	4.6953	CaTe	6.356	NaH	4.890	AgCl	5.549
SnAs	5.7248	SrTe	6.660	ScN	4.44	AgBr	5.7745
TiC	4.3285	BaTe	7.00	TiN	4.240	CsF	6.014
UC	4.955	LaN	5.30	UN	4.890		

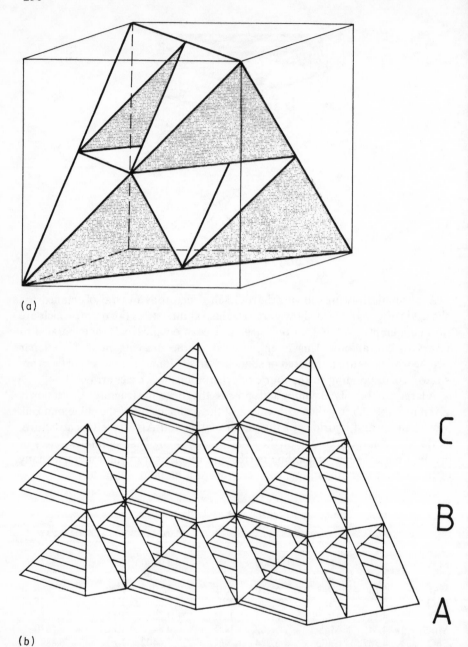

C

B

A

(b)

Fig. 7.17 The sphalerite (zinc blende) structure showing (a) the unit cell contents and (b) a more extended network of corner-sharing tetrahedra

Table 7.6 Some compounds with the zinc blende (sphalerite) structure

	$a(\text{Å})$		$a(\text{Å})$		$a(\text{Å})$		$a(\text{Å})$		$a(\text{Å})$
CuF	4.255	BeS	4.8624	β-CdS	5.818	BN	3.616	GaP	5.448
CuCl	5.416	BeSe	5.07	CdSe	6.077	BP	4.538	GaAs	5.6534
γ-CuBr	5.6905	BeTe	5.54	CdTe	6.481	BAs	4.777	GaSb	6.095
γ-CuI	6.051	β-ZnS	5.4060	HgS	5.8517	AlP	5.451	InP	5.869
γ-AgI	6.495	ZnSe	5.667	HgSe	6.085	AlAs	5.662	InAs	6.058
β-MnS,red	5.600	ZnTe	6.1026	HgTe	6.453	AlSb	6.1347	InSb	6.4782
β-MnSe	5.88	β-SiC	4.358						

238

The zinc blende structure contains ZnS_4 tetrahedra or SZn_4 tetrahedra, which are linked at their corners (Fig. 7.13e). Each corner is common to four such tetrahedra; this must be so since anion and cation have the same coordination number in ZnS. The unit cell of zinc blende (Fig. 7.13b and e) is shown again in Fig. 7.17(a), but in terms of corner-sharing ZnS_4 tetrahedra. The faces of the tetrahedra are parallel to the c.p. anion layers, i.e. the {111} planes, and this is emphasized in a more extensive model of the structure (Fig. 7.17b), in which the model is oriented so that one set of tetrahedron faces is approximately horizontal.

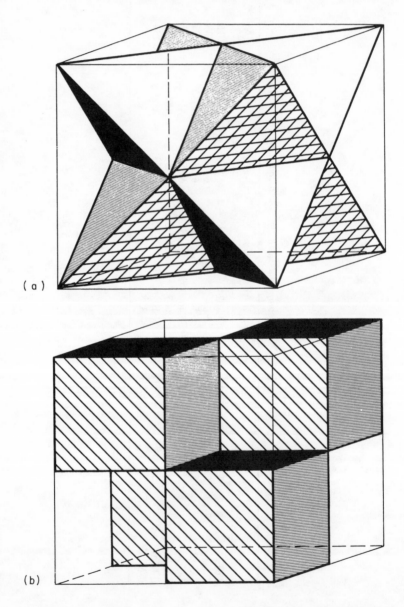

(a)

(b)

Fig. 7.18 The antifluorite structure showing the unit cell in terms of (a) NaO_4 tetrahedra or (b) ONa_8 cubes. A more extended array of cubes is shown in (c); This model resides on a roundabout in Mexico City

Conventionally, the ZnS structure is regarded as built of c.p. layers of sulphide anions with the smaller zinc cations in tetrahedral sites. Since the same structure is generated by interchanging the Zn^{2+} and S^{2-} ions, the structure could also be described as a c.p. array of Zn^{2+} ions with S^{2-} ions occupying one set of tetrahedral sites. A third, equivalent description is as an array of c.c.p. ZnS_4 (or SZn_4) tetrahedra.

Some compounds which exhibit the zinc blende structure are listed in Table 7.6 together with their cubic lattice parameter, a. The bonding in these compounds is less ionic than in corresponding AB compounds which have the rock salt structure. Thus, oxides do not usually have the zinc blende structure (ZnO, not included in Table 7.6, is an exception; it is dimorphic with zinc blende and wurtzite structure polymorphs). Chalcogenides (S, Se, Te) of the alkaline earth metals have the rock salt structure whereas the more covalent corresponding compounds of Be, Zn, Cd and Hg have the zinc blende structure. Most metal(I) halides are like rock salt, with the exception of the more covalent copper(I) halides and γ-AgI.

The antifluorite structure contains tetrahedrally coordinated cations and eight-coordinate anions (Figs 7.13f and 7.14). This leads to two ways of describing the structure, either as a three-dimensional network of tetrahedra or as a three-dimensional network of cubes. The results of these two methods are shown in Fig. 7.18. The unit cell contains either (a) eight NaO_4 tetrahedra or (b) four ONa_8 cubes. In (a), tetrahedra share edges and each edge is common to two tetrahedra. In (b), cubes share corners, but each corner is common to only four cubes (up to a maximum number of eight cubes may share a common corner). Also, the cubes share edges, but each edge is common to only two cubes (the maximum possible is four). A more extended network of corner-sharing cubes is shown in (c). This must surely rate as one of the world's largest models of the antifluorite structure!

The two ways of describing the arrangement of polyhedra in the antifluorite structure coincide with the two classes of compound which possess this structure. So far, we have considered the *antifluorite* structure which is shown by a large number of oxides and other chalcogenides of the alkali metals (Table 7.7), i.e. compounds of the general formula $A_2^+ X^{2-}$. A second group of compounds, which includes fluorides of large, divalent cations and oxides of large tetravalent cations, has the inverse, *fluorite* structure, i.e. $M^{2+} F_2$ and $M^{4+} O_2$. In the fluorite structure, the cations are eight coordinate and the anions are four coordinate, which is the inverse of the antifluorite structure. The arrangement of cubes in Fig. 7.18(b) and (c) also show the MF_8 and MO_8 coordination in the fluorite structure.

A common way of describing the fluorite structure is as a primitive cubic array of anions in which the eight-coordinate sites at the cube body centres are alternately empty and occupied by a cation: this description is consistent with (b) and (c). It should be stressed, however, that the true lattice type of the fluorite structure is face centred cubic and not primitive cubic, since the primitive cubes represent only a small part (one eighth) of the f.c.c. unit cell. Description of the fluorite structure as a primitive cubic array of anions with alternate cube body centres occupied by cations shows a similarity to the CsCl structure (next section). This also has a primitive cubic array of anions, but, instead, cations occupy all the cube body centre sites.

It is very often desirable to be able to make calculations of bond and other interatomic distances in crystal structures. This is usually a straighforward exercise, especially for crystals which have orthogonal unit cells (i.e. $\alpha = \beta =$

Table 7.7 *Some compounds with fluorite and antifluorite structure*

	Fluorite structure				Antifluorite structure		
	a(Å)		a(Å)		a(Å)		a(Å)
CaF_2	5.4626	PbO_2	5.349	Li_2O	4.6114	K_2O	6.449
SrF_2	5.800	CeO_2	5.4110	Li_2S	5.710	K_2S	7.406
$SrCl_2$	6.9767	PrO_2	5.392	Li_2Se	6.002	K_2Se	7.692
BaF_2	6.2001	ThO_2	5.600	Li_2Te	6.517	K_2Te	8.168
$BaCl_2$	7.311	PaO_2		Na_2O	5.55	Rb_2O	6.74
CdF_2	5.3895	UO_2	5.372	Na_2S	6.539	Rb_2S	7.65
HgF_2	5.5373	NpO_2	5.4334	Na_2Se	6.823		
EuF_2	5.836	PuO_2	5.386	Na_2Te	7.329		
β-PbF_2	5.940	AmO_2	5.376				
		CmO_2	5.3598				

$\gamma = 90°$), and involves simple trigonometric calculations. For example, in the rock salt structure, the anion–cation distance is $a/2$ and the anion–anion distance is $a/\sqrt{2}$. As an aid to rapid calculation of interatomic distances, formulae are given in Table 7.8 for the more important structure types. These may be used in conjunction with the tables of unit cell dimensions (Table 7.5, etc.) to make calculations on specific compounds.

Table 7.8 *Calculation of interatomic distances in some simple structures*

Structure type	Distance	Number of such distances	Magnitude of separation in terms of unit cell dimensions
Rock salt (cubic)	Na–Cl	6	$a/2 = 0.5a$
	Cl–Cl	12	$a/\sqrt{2} = 0.707a$
	Na–Na	12	$a/\sqrt{2} = 0.707a$
Zinc blende (cubic)	Zn–S	4	$a\dfrac{\sqrt{3}}{4} = 0.433a$
	Zn–Zn	12	$a/\sqrt{2} = 0.707a$
	S–S	12	$a/\sqrt{2} = 0.707a$
Fluorite (cubic)	Ca–F	4 or 8	$a\dfrac{\sqrt{3}}{4} = 0.433a$
	Ca–Ca	12	$a/\sqrt{2} = 0.707a$
	F–F	6	$a/2 = 0.5a$
Wurtzite* (hexagonal)	Zn–S	4	$a\sqrt{\dfrac{3}{8}} = 0.612a = \dfrac{3c}{8} = 0.375c$
	Zn–Zn	12	$a = 0.612c$
	S–S	12	$a = 0.612c$
Nickel arsenide* (hexagonal)	Ni–As	6	$a/\sqrt{2} = 0.707a = 0.433c$
	As–As	12	$a = 0.612c$
	Ni–Ni	2	$c/2 = 0.5c = 0.816a$
	Ni–Ni	6	$a = 0.612c$
Cesium chloride (cubic)	Cs–Cl	8	$a\dfrac{\sqrt{3}}{2} = 0.866a$
	Cs–Cs	6	a
	Cl–Cl	6	a
Cadmium iodide (hexagonal)	Cd–I	6	$a/\sqrt{2} = 0.707a = 0.433c$
	I–I	12	$a = 0.612c$
	Cd–Cd	6	$a = 0.612c$

*These formulae do not necessarily apply when c/a is different from the ideal value of 1.633.

Consideration of the anion and cation arrangements in the three structure types described above shows that the concept of c.p. anions with cations in interstitial sites begins to break down in the fluorite structure. Thus, while the antifluorite structure of, for example, Na_2O may be regarded as containing cubic close packed O^{2-} ions with Na^+ ions in tetrahedral sites, in the fluorite structure of, for example, CaF_2, it is necessary to regard the Ca^{2+} ions as forming the c.c.p. array with the F^- ions occupying tetrahedral interstitial sites. In CaF_2, although the Ca^{2+} ions have the same arrangement as in a c.c.p. arrangement of spheres, the Ca^{2+} ions are well separated from each other; from Tables 7.7 and 7.8, Ca–Ca = 3.86 Å, which is much larger than the diameter of a Ca^{2+} ion (depending on which table of ionic radii is consulted, the diameter of a Ca^{2+} ion is in the range ~ 2.2 to 2.6 Å). Therefore, CaF_2 is a good example of a eutactic structure.

The fluorine–fluorine distance in CaF_2 is 2.73 Å, which indicates that the fluorines are approximately contacting ($r_{F^-} = 1.2$ to 1.4 Å). Although the primitive cubic array of F^- ions in CaF_2 is not a c.p. arrangement, the anions are approximately in contact and this is perhaps a more realistic way of describing the structure than as containing a c.c.p. array of Ca^{2+} ions.

7.2.2 Wurtzite (ZnS) and nickel arsenide (NiAs)

These structures have in common a hexagonal close packed arrangement of anions and differ only in the positions of the cations, as follows:

Wurtzite: T_+(or T_-) sites occupied; T_-(or T_+), O empty

Nickel arsenide: O sites occupied; T_+, T_- empty.

These structures are the hexagonal close packed analogues of the cubic close packed, sphalerite and rock salt structures. Note that there is no hexagonal equivalent of the fluorite and antifluorite structures.

Both wurtzite and nickel arsenide have hexagonal symmetry and unit cells. A unit cell containing hexagonal close packed anions is shown in Fig. 7.19(a). It is less easy to visualize and draw on paper than a cubic cell because of the γ angle of $120°$. The unit cell contains two anions, one at the origin and one inside the cell. Their coordinates are:

$$0, 0, 0; \quad \tfrac{1}{3}, \tfrac{2}{3}, \tfrac{1}{2}$$

In Fig. 7.19(b) is shown a projection down c of the same structure. Close packed layers occur in the basal plane, i.e. at $c = 0$ (open circles) and at $c = \tfrac{1}{2}$ (shaded circles). The layer arrangement at $c = 0$ is repeated at $c = 1$ and therefore the stacking sequence is hexagonal...ABABA.... Atoms 1 to 4 in the basal plane outline the base of the unit cell (dashed).

The contents of one unit cell are shown in Fig. 7.19(c). Dashed circles represent atoms at the top four corners of the unit cell, i.e. at $c = 1$.

In metals which have hexagonal close packed structures, metal atoms are in contact, e.g. 1 and 2, 1 and 4, 1 and 5. In eutactic close packed ionic structures, however, the anions may be pushed apart by the cations in the interstitial sites

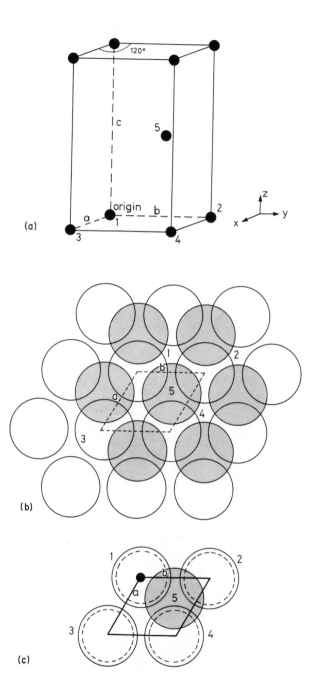

(a)

(b)

(c)

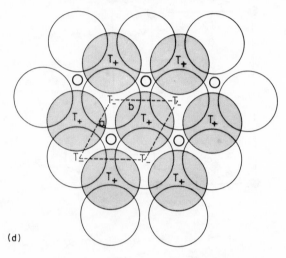

(d)

T_-	:	$0, 0, \tfrac{3}{8}$	;	$\tfrac{1}{3}, \tfrac{2}{3}, \tfrac{7}{8}$	
T_+	:	$\tfrac{1}{3}, \tfrac{2}{3}, \tfrac{1}{8}$	;	$0, 0, \tfrac{5}{8}$	
0	:	$\tfrac{2}{3}, \tfrac{1}{3}, \tfrac{1}{4}$	;	$\tfrac{2}{3}, \tfrac{1}{3}, \tfrac{3}{4}$	
ANION	:	$0, 0, 0$		$\tfrac{1}{3}, \tfrac{2}{3}, \tfrac{1}{2}$	

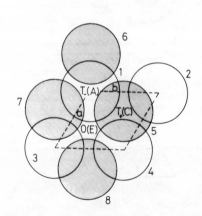

(e)

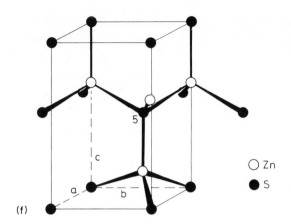

(f)

○ Zn
● S

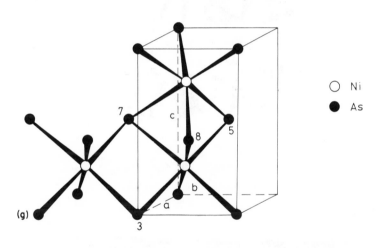

(g)

○ Ni
● As

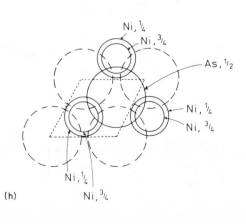

(h)

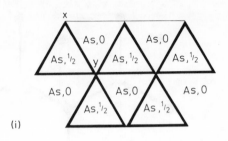

(i)

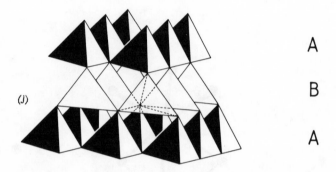

(J)

A

B

A

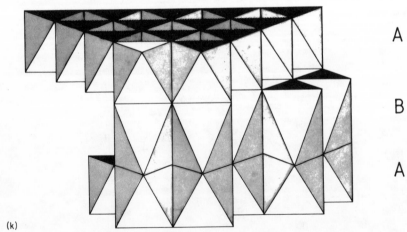

A

B

A

(k)

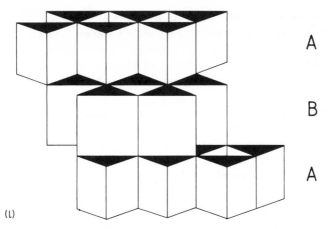

A

B

A

(l)

Fig. 7.19 The wurtzite and nickel arsenide structures: (a), (b), (c) the hexagonal unit cell of an h.c.p. anion array; (d), (e) interstitial sites in an h.c.p. array; (f), (g) structures of wurtzite and NiAs; (h), (i) trigonal prismatic coordination of arsenic in NiAs; (j), (k), (l), models of the ZnS and NiAs structures showing the arrangement and linkages of the polyhedra

and anion–anion contacts do not usually occur. Assuming for the moment that the anions are in contact, then the hexagonal unit cell has a definite shape given by the ratio $c/a = 1.633$. This is because a is equal to the shortest distance X–X, i.e. the diameter of an anion, and c is equal to twice the vertical height of a tetrahedron comprising four anions. The ratio c/a may then be calculated by geometry (Appendix 1).

The interstitial sites available for cations in a hexagonal close packed anion array are shown in Fig. 7.19(d). Since the cell contains two anions, there must be two of each type of interstitial site, T_+, T_- and O. Thus, in Fig. 7.19(e), a T_- site, A, occurs along the c edge of the cell at height $\frac{3}{8}$ above anion 1 at the origin. This T_- site is coordinated to three anions, shaded (5 to 7), at $c = \frac{1}{2}$ and to one anion (1) at the corner, $c = 0$. The tetrahedron so formed therefore points downwards. The position of the interstitial site inside this tetrahedron is at the centre of gravity of

Table 7.9 *Some compounds with the wurtzite structure.* (Data taken from Wyckoff, 1971, Vol. 1)

	$a(\text{Å})$	$c(\text{Å})$	u	c/a		$a(\text{Å})$	$c(\text{Å})$	u	c/a
ZnO	3.2495	5.2069	0.345	1.602	AgI	4.580	7.494		1.636
ZnS	3.811	6.234		1.636	AlN	3.111	4.978	0.385	1.600
ZnSe	3.98	6.53		1.641	GaN	3.180	5.166		1.625
ZnTe	4.27	6.99		1.637	InN	3.533	5.693		1.611
BeO	2.698	4.380	0.378	1.623	TaN	3.05	4.94		1.620
CdS	4.1348	6.7490		1.632	NH_4F	4.39	7.02	0.365	1.600
CdSe	4.30	7.02		1.633	SiC	3.076	5.048		1.641
MnS	3.976	6.432		1.618					
MnSe	4.12	6.72		1.631					

the tetrahedron, i.e. at one quarter of the vertical distance from base to apex (Appendix 1). Since the apex and base are at $c = 0$ and $c = \frac{1}{2}$, this T_- site is at $c = \frac{3}{8}$. In practice the occupant of this T_- site in the wurtzite structure may not be at exactly $0.375c$. For those structures which have been studied accurately (Table 7.9), values range from 0.345 to 0.385; the letter u is used to represent the fractional c value.

The three anions (5 to 7) at $c = \frac{1}{2}$ that form the base of this T_- site, A, also form the base of a T_+ site, B (not shown), centred at 0, 0, $\frac{5}{8}$. The apex of the latter tetrahedron is the anion at the top corner with coordinates 0, 0, 1. Another T_+ site, C, is shown in Fig. 7.19(e) at $\frac{1}{3}, \frac{2}{3}, \frac{1}{8}$. It is coordinated to anions 1, 2 and 4 in the basal plane at three corners of the unit cell and to anion 5 at $\frac{1}{3}, \frac{2}{3}, \frac{1}{2}$. The triangular base of this site, at $c = 0$, is shared with a T_- site underneath (not shown) at $\frac{1}{3}, \frac{2}{3}, -\frac{1}{8}$. The equivalent T_- site that lies inside the unit cell, D (not shown) is at $\frac{1}{3}, \frac{2}{3}, \frac{7}{8}$.

Octahedral site E, in Fig. 7.19(e) is coordinated to anions 1, 3 and 4 at $c = 0$ and to anions 5, 7 and 8 at $c = \frac{1}{2}$. The centre of gravity of the octahedron lies midway between these two groups of anions and has coordinates $\frac{2}{3}, \frac{1}{3}, \frac{1}{4}$. The second octahedral site, F (not shown) in the cell lies above E at $c = \frac{3}{4}$ and has coordinates $\frac{2}{3}, \frac{1}{3}, \frac{3}{4}$. The three anions 5, 7 and 8 are therefore common to the two octahedra.

The coordination environments of the cations in wurtzite and NiAs are emphasized in Fig. 7.19(f) and (g). Zinc is shown in T_+ sites and forms ZnS_4 tetrahedra, (f), which link up at their corners to form a three-dimensional network, as in (j). A similar structure results on considering the tetrahedra formed by four zinc atoms around a sulphur and the manner in which these SZn_4 tetrahedra are linked. The tetrahedral environment of the sulphur ion (5) is shown in (f). The SZn_4 tetrahedron which it forms points downwards, in contrast to the ZnS_4 tetrahedra which all point upwards; on turning the SZn_4 tetrahedra upside down, however, the same structure results.

Comparing larger scale models of the zinc blende (Fig. 7.17b) and wurtzite (Fig. 7.19j) structures, they are clearly very similar and can both be regarded as networks of tetrahedra. In zinc blende, layers of tetrahedra form an ABC stacking sequence and the orientation of the tetrahedra within each layer is identical. in wurtzite, however, the layers form an AB sequence and alternate layers are rotated by 180° about c relative to each other.

The $NiAs_6$ octahedra in NiAs are shown in Fig. 7.19(g). They share one pair of opposite faces (e.g. the face formed by arsenic ions 5, 7 and 8) to form chains of face-sharing octahedra that run parallel to c. In the ab plane, however, the octahedra share only common edges: arsenic ions 3 and 7 are shared between two octahedra such that chains of edge-sharing octahedra form parallel to b. Similarly, chains of edge-sharing octahedra form parallel to a (not shown). A more extended view of the octahedra and their linkages is shown in Fig. 7.19(k).

The NiAs structure is unusual in that the anions and cations have the same coordination number but do not have the same coordination environment. In the other AB structures—rock salt, sphalerite, wurtzite and CsCl—anions and cations are interchangeable and have the same coordination number and environment. Since the cation : anion ratio is 1:1 in NiAs and the nickel coordination is octahedral, the arsenic ions must also be six coordinate. However, the six nickel neighbours are arranged as in a trigonal prism and not octahedrally. This is shown for arsenic at $c = \frac{1}{2}$ in Fig. 7.19(h), which is coordinated to three nickel ions at $c = \frac{1}{4}$ and to three at $c = \frac{3}{4}$. The two sets of nickel ions are superposed in projection down c and give the trigonal prismatic coordination for arsenic. (Note that in a similar projection for octahedral coordination, the two sets of three outer atoms appear to be staggered relative to each other, as in Fig. 7.19(e) for octahedral site E.)

The NiAs structure may also be regarded as built of $AsNi_6$ trigonal prisms, therefore, which link up by sharing edges to form a three-dimensional array. In Fig. 7.19(i) each triangle represents a prism in projection down c. The prism edges that run parallel to c, i.e. those formed by nickel ions at $c = \frac{1}{4}$ and $\frac{3}{4}$ in Fig. 7.19(h), are shared between three prisms. Thus the vertical edge at y (in projection it is a point) is shared between three prisms and therefore three arsenic ions at $c = \frac{1}{2}$. Prism edges that lie in the ab plane are shared between only two prisms, however. In Fig. 7.19(i), the edge xy is shared between arsenic ions at $c = \frac{1}{2}$ and $c = 0$.

The structure may be regarded as built up of layers of prisms; two layers are shown in Fig. 7.19(i) centred at $c = \frac{1}{2}$ and $c = 0$ and are rotated by 180° about c relative to each other. The next layer of prisms, at $c = 1$, is in the same orientation as the layer at $c = 0$ and hence, an ...ABABA... hexagonal stacking sequence arises. This is shown further in Fig. 7.19(l).

A selection of compounds which have wurtzite and NiAs structures is given in Tables 7.9 and 10 with values of their hexagonal cell parameters a and c. The wurtzite structure is formed mainly by chalcogenides of some divalent metals and may be regarded as a fairly ionic structure (Chapter 8). The NiAs structure is a more metallic structure and is adopted by a variety of intermetallic compounds and by some transition metal chalcogenides (S, Se, Te). The value of the ratio c/a

Table 7.10 *Some compounds with the NiAs structure.* (Data taken from Wyckoff, 1971, Vol. 1)

	$a(\text{Å})$	$c(\text{Å})$	c/a		$a(\text{Å})$	$c(\text{Å})$	c/a
NiS	3.4392	5.3484	1.555	CoS	3.367	5.160	1.533
NiAs	3.602	5.009	1.391	CoSe	3.6294	5.3006	1.460
NiSb	3.94	5.14	1.305	CoTe	3.886	5.360	1.379
NiSe	3.6613	5.3562	1.463	CoSb	3.866	5.188	1.342
NiSn	4.048	5.123	1.266	CrSe	3.684	6.019	1.634
NiTe	3.957	5.354	1.353	CrTe	3.981	6.211	1.560
FeS	3.438	5.880	1.710	CrSb	4.108	5.440	1.324
FeSe	3.637	5.958	1.638	MnTe	4.1429	6.7031	1.618
FeTe	3.800	5.651	1.487	MnAs	3.710	5.691	1.534
FeSb	4.06	5.13	1.264	MnSb	4.120	5.784	1.404
δ'-NbN*	2.968	5.549	1.870	MnBi	4.30	6.12	1.423
PtB*	3.358	4.058	1.208	PtSb	4.130	5.472	1.325
PtSn	4.103	5.428	1.323	PtBi	4.315	5.490	1.272

* Anti-NiAs structure.

is approximately constant in the family of wurtzite structures but varies considerably between the different compounds with the NiAs structure. This is associated with the presence of metallic bonding which arises from metal–metal interactions in the c direction, as follows:

Let us consider the environment of nickel and arsenic ions.
Each arsenic is surrounded by (Table 7.8):
  6 nickel ions (in a trigonal prism) at distance $0.707a$
  12 arsenic ions (h.c.p. arrangement) at distance a
Each nickel is surrounded by:
  6 arsenic ions (octahedrally) at distance $0.707a$
  2 nickel ions (linearly, parallel to c) at distance $0.816a$ (i.e. $c/2$)
  6 nickel ions (hexagonally, in ab plane) at distance a

The main effect of changing the value of the c/a ratio is to alter the nickel–nickel distance parallel to c. Thus, in FeTe, $c/a = 1.49$, and hence the iron–iron distance parallel to c is reduced to $0.745a$ (i.e. $c/2 = \frac{1}{2}(1.49a)$), thereby bringing these iron ions into closer contact and increasing the metallic bonding in the c direction. Quantitative calculations of the effect of changing the c/a ratio are more difficult to make since it is not readily possible to distinguish between, for example, an increase in a and a decrease in c, either of which could cause the same effect on the c/a ratio.

The non-occurrence of AX_2 compounds whose structures are the hexagonal equivalent of the cubic fluorite and antifluorite structures has already been mentioned. This may be understood by considering the various interatomic distances that would be present in such a structure. A hexagonal fluorite-like structure, AX_2, would have hexagonal packed A cations with X anions fully occupying T_+ and T_- sites. From Fig. 7.19(d), X ions would occupy, for example, sites at $0, 0, \frac{3}{8}(T_-)$ and $0, 0, \frac{5}{8}(T_+)$, thereby giving an X–X distance of $c/4 = 0.25c$. This compares with the A–X distance (as in wurtzite, Table 7.8) of $0.375c$. Since the shortest interatomic separations in ionic structures are almost always cation–anion contacts, it is most unlikely that a structure would exist in which anion–anion distances were much shorter than anion–cation distances (and vice versa for the equivalent antifluorite structure).

7.2.3 Caesium chloride, CsCl

The unit cell of CsCl is shown in Fig. 7.20. It is primitive cubic, containing Cl^- ions at the corners and a Cs^+ ion at the body centre, or vice versa (note that it is *not* body centred cubic since there are different ions at corner and body centre positions). The coordination numbers of both Cs^+ and Cl^- are eight with interatomic distances of $0.866a$ (Table 7.8). The CsCl structure cannot be regarded as close packed. In a c.p. structure, e.g. NaCl, each Cl^- has twelve other Cl^- ions as next nearest neighbours, which is a characteristic feature of c.p. (or eutectic c.p.) structures. In CsCl, however, each Cl has only six Cl^- ions as next nearest neighbours (arranged octahedrally). Some compounds which exhibit the

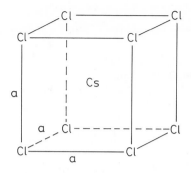

Cl : 0 , 0 , 0

Cs : ½ , ½ , ½

Fig. 7.20 The primitive cubic unit cell of CsCl

Table 7.11 *Some compounds with the CsCl structure*

	$a(\text{Å})$		$a(\text{Å})$
CsCl	4.123	CuZn	2.945
CsBr	4.286	CuPd	2.988
CsI	4.5667	AuMg	3.259
CsCN	4.25	AuZn	3.19
NH_4Cl	3.8756	AgZn	3.156
NH_4Br	4.0594	LiAg	3.168
TlCl	3.8340	AlNi	2.881
TlBr	3.97	LiHg	3.287
TlI	4.198	MgSr	3.900

CsCl structure are given in Table 7.11. They fall into two groups, halides of large monovalent elements and a variety of intermetallic compounds.

7.2.4 Other AX structures

There are five main AX structure types: rock salt, CsCl, NiAs, sphalerite and wurtzite, each of which is found in a large number of compounds. There is also a considerable number of less common AX structures. Some may be regarded as distorted variants of one of the main structure types, e.g.:

(a) FeO at low temperatures, < 90 K, has a rock salt structure which has undergone a slight rhombohedral distortion (the α angle is increased from 90 to 90.07° by a slight compression along one threefold axis). This rhombohedral distortion is associated with magnetic ordering in FeO at low temperatures, Chapter 16.

(b) TlF has a rock salt related structure in which the f.c.c. cell is distorted into a face centred orthorhombic cell by changing the lengths of all three cell axes by different amounts.

(c) NH_4CN has a distorted CsCl structure (as in NH_4Cl) in which the CN^- ions do not assume spherical symmetry but are oriented parallel to [110] directions (i.e. face diagonals). This distorts the symmetry to tetragonal and effectively increases a relative to c.

Other AX compounds have completely different structures, e.g.:

(a) Compounds of the d^8 ions, Pd and Pt (in PdO, PtS, etc.) often have a square planar coordination for the cation with tetragonal or orthorhombic symmetry for the structure as a whole; d^9 ions show this effect as well, e.g. Cu in CuO.

(b) Compounds of heavy p-block ions in their lower oxidation states (Pb^{2+}, Bi^{3+}, etc.) often have distorted polyhedra in which the cation exhibits the *inert pair effect*. Thus PbO and SnO have structures in which the M^{2+} ion has four O^{2-} neighbours to one side giving a square pyramidal arrangement; the O^{2-} coordination is a more regular tetrahedron of M^{2+} ions. InBi has a similar structure in which Bi^{3+} is the ion with the inert pair effect and the irregular coordination.

7.2.5 Rutile (TiO_2), CdI_2, $CdCl_2$ and Cs_2O structures

The title structures together with the fluorite structure (Section 7.2.2) constitute the main AX_2 structure types. Rutile, TiO_2, has been described in some detail in Chapter 6; it may be regarded as having a distorted hexagonal close packed oxide array or, alternately, as a tetragonal packed oxide array. In either description, one half of the octahedral sites are occupied by Ti^{4+}. The TiO_6 octahedra thus formed link up by sharing one pair of opposite edges with neighbouring octahedra to form infinite chains parallel to c. These chains link at their corners with neighbouring chains to form a three-dimensional network of octahedra.

The octahedral sites in an ideal h.c.p. anion array are shown in Fig. 7.21(a). While all these sites are occupied in NiAs (Fig. 7.19h), only half are occupied in rutile and in such a manner that alternate rows of octahedral sites are full and empty. It should be emphasized again that this is an idealized situation since, in rutile, the oxide layers (parallel to the plane of the paper in Fig. 7.21a) are not planar but are buckled. The orientation of the tetragonal unit cell in rutile is shown in Fig. 7.21(a).

Parallel to the tetragonal c axis, the TiO_6 octahedra share edges. This is shown in Fig. 7.21(b) for two octahedra with oxygens 1 and 2 forming the common edge and in Appendix 2 for several octahedra similarly linked to form a chain.

Two main groups of compounds exhibit the rutile structure (Table 7.12): oxides of some tetravalent metal ions and fluorides of small divalent metal ions.

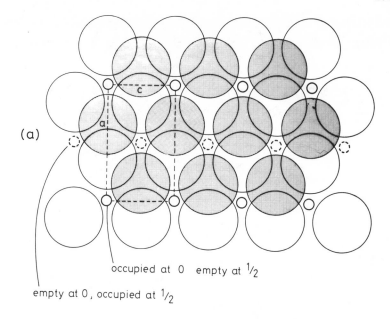

(a)

occupied at 0   empty at $^1/_2$

empty at 0, occupied at $^1/_2$

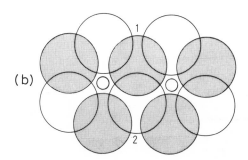

(b)

Fig. 7.21 The rutile structure: (a) in idealized form, with planar oxide layers; (b) edge-sharing octahedra

In both cases, these M^{4+} and M^{2+} ions are too small to form the fluorite structure with O^{2-} and F^-, respectively. The rutile structure may be regarded as an essentially ionic structure.

The CdI_2 structure is nominally very similar to rutile because it also may be described as a hexagonal close packed anion array in which half the octahedral sites are occupied by M^{2+} ions. The manner of occupancy of the octahedral sites is quite different, however, since in CdI_2, entire layers of octahedral sites are occupied and these alternate with layers of empty sites (Fig. 7.22). CdI_2 is therefore a layered material in both its crystal structure and properties, in contrast to rutile which has a more rigid, three-dimensional character.

Two iodide layers in a hexagonal close packed array are shown in Fig. 7.22(a) with the octahedral sites in between occupied by Cd^{2+}. To either side of the

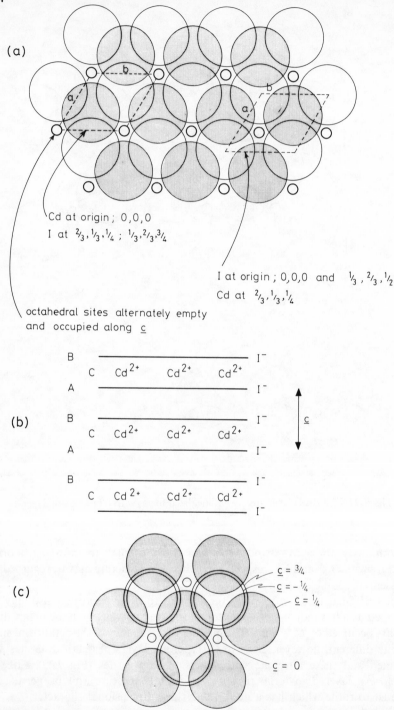

(a)

b

a

a

b

Cd at origin ; 0,0,0
I at $^2/_3$, $^1/_3$, $^1/_4$; $^1/_3$, $^2/_3$, $^3/_4$

I at origin ; 0,0,0 and $^1/_3$, $^2/_3$, $^1/_2$
Cd at $^2/_3$, $^1/_3$, $^1/_4$

octahedral sites alternately empty
and occupied along c

B ———————————————— I^-
C Cd^{2+} Cd^{2+} Cd^{2+}
A ———————————————— I^-

B ———————————————— I^-
C Cd^{2+} Cd^{2+} Cd^{2+}
A ———————————————— I^-

B ———————————————— I^-
C Cd^{2+} Cd^{2+} Cd^{2+}
———————————————— I^-

c

(b)

(c)

$c = ^3/_4$
$c = -^1/_4$
$c = ^1/_4$

$c = 0$

Fig. 7.22 The CdI_2 structure: (a) the unit cell, with two possible choices of origin, (b) the layer stacking sequence, (c) the coordination environment of I

Table 7.12 *Some compounds with the rutile structure.* (Data taken from Wyckoff, 1971, Vol. 1)

	$a(Å)$	$c(Å)$	x		$a(Å)$	$c(Å)$	x
TiO_2	4.5937	2.9581	0.305	CoF_2	4.6951	3.1796	0.306
CrO_2	4.41	2.91		FeF_2	4.6966	3.3091	0.300
GeO_2	4.395	2.859	0.307	MgF_2	4.623	3.052	0.303
IrO_2	4.49	3.14		MnF_2	4.8734	3.3099	0.305
$\beta\text{-}MnO_2$	4.396	2.871	0.302	NiF_2	4.6506	3.0836	0.302
MoO_2	4.86	2.79		PdF_2	4.931	3.367	
NbO_2	4.77	2.96		ZnF_2	4.7034	3.1335	0.303
OsO_2	4.51	3.19					
PbO_2	4.946	3.379					
RuO_2	4.51	3.11					
SnO_2	4.7373	3.1864	0.307				
TaO_2	4.709	3.065					
WO_2	4.86	2.77					

iodide layers, the octahedral sites are empty. Compare this with NiAs (Fig. 7.19d and h) which has the same anion arrangement but with all the octahedral sites occupied. The layer stacking sequence along c in CdI_2 is shown schematically in Fig. 7.22(b) and emphasizes the layered nature of the CdI_2 structure. I^- layers form an ...ABABA... stacking sequence. Cd^{2+} ions occupy octahedral sites which may be regarded as the C positions relative to the AB positions for I^-. The CdI_2 structure may be regarded as a sandwich structure in which Cd^{2+} ions are sandwiched between layers of I^- ions and adjacent sandwiches are held together by weak van der Waals bonds between the layers of I^- ions. In this sense, CdI_2 has certain similarities to molecular structures. For example, solid CCl_4 has strong C—Cl bonds within the molecule but only weak Cl—Cl bonds between adjacent molecules. Because the intermolecular forces are weak, CCl_4 is volatile with a low melting and boiling point. In the same way, CdI_2 may be regarded as an infinite sandwich 'molecule' in which there are strong Cd—I bonds within the molecule but weak van der Waals bonds between adjacent molecules.

The coordination of the I^- ion in CdI_2 is shown in Fig. 7.22(c). An I^- ion at $c = \frac{1}{4}$ (shaded) has three close Cd^{2+} neighbours to one side at $c = 0$. The next nearest neighbours are the twelve I^- ions that form the h.c.p. array.

The layered nature of CdI_2 is emphasized further by constructing a model from polyhedra: CdI_6 octahedra link up at their edges to form infinite sheets (Appendix 2), but there are no direct polyhedral linkages between adjacent sheets. A self-supporting, three-dimensional model of octahedra cannot be made for CdI_2, therefore, unlike, for example, rutile.

Some compounds which have the CdI_2 structure are listed in Table 7.13. This structure occurs mainly in transition metal iodides and also with some bromides, chlorides and hydroxides.

The cadmium chloride structure is closely related to that of CdI_2 and differs only in the nature of the anion packing: Cl^- ions are cubic close packed in $CdCl_2$

Table 7.13 *Some compounds with the* CdI$_2$ *structure.* (Data taken from Wyckoff, 1971, Vol. 1)

	a(Å)	c(Å)		a(Å)	c(Å)
CdI$_2$	4.24	6.84	VBr$_2$	3.768	6.180
CaI$_2$	4.48	6.96	TiBr$_2$	3.629	6.492
CoI$_2$	3.96	6.65	MnBr$_2$	3.82	6.19
FeI$_2$	4.04	6.75	FeBr$_2$	3.74	6.17
MgI$_2$	4.14	6.88	CoBr$_2$	3.68	6.12
MnI$_2$	4.16	6.82	TiCl$_2$	3.561	5.875
PbI$_2$	4.555	6.977	VCl$_2$	3.601	5.835
ThI$_2$	4.13	7.02	Mg(OH)$_2$	3.147	4.769
TiI$_2$	4.110	6.820	Ca(OH)$_2$	3.584	4.896
TmI$_2$	4.520	6.967	Fe(OH)$_2$	3.258	4.605
VI$_2$	4.000	6.670	Co(OH)$_2$	3.173	4.640
YbI$_2$	4.503	6.972	Ni(OH)$_2$	3.117	4.595
ZnI$_2$(I)	4.25	6.54	Cd(OH)$_2$	3.48	4.67

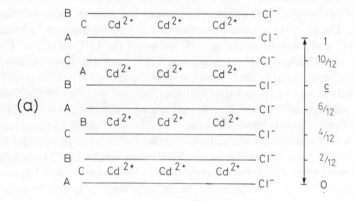

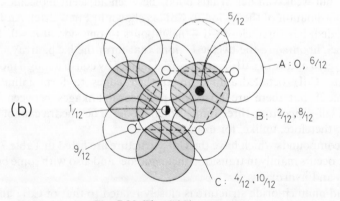

Fig. 7.23 The CdCl$_2$ structure

whereas I^- ions are hexagonal close packed in CdI_2. $CdCl_2$ and CdI_2 form a pair of closely related structures, therefore, in the same way that wurtzite and zinc blende or rock salt and nickel arsenide differ only in the anion stacking sequence.

The $CdCl_2$ structure may be represented by a hexagonal unit cell, although a smaller rhombohedral cell can also be chosen. The base of the hexagonal cell is of similar size and shape to that in CdI_2 but the c axis of $CdCl_2$ is three times as long as c in CdI_2. This is because in $CdCl_2$, the Cd^{2+} positions, and the $CdCl_6$ octahedra, are staggered along c and give rise to a three-layer repeat for Cd^{2+} ions (CBA) and a six-layer repeat for Cl^- ions (ABCABC) (Fig. 7.23). In contrast, in CdI_2, the Cd^{2+} positions and the CdI_6 octahedra are stacked on top of each other and the c repeat contains only two I^- layers (AB) and one Cd^{2+} layer (C).

The unit cell of $CdCl_2$ in projection down c is shown in Fig. 7.23(b). Chloride layers occur at $c = 0$(A), $\frac{2}{12}$(B) and $\frac{4}{12}$(C) and this sequence is repeated at $c = \frac{6}{12}$, $\frac{8}{12}$ and $\frac{10}{12}$. Between those Cl^- layers at 0 and $\frac{2}{12}$ are Cd^{2+} ions in the octahedral sites at $\frac{1}{12}$. However, the octahedral sites between Cl^- layers at $\frac{2}{12}$ and $\frac{4}{12}$ are empty (these sites, at $c = \frac{3}{12}$, are directly below the Cd^{2+} ions at $\frac{9}{12}$).

The $CdCl_2$ structure is a layered structure, similar to CdI_2, and many of the comments made about structure and bonding in CdI_2 apply equally well to $CdCl_2$. Some compounds which have the $CdCl_2$ structure are given in Table 7.14. It also occurs with a variety of transition metal halides.

The structure of Cs_2O is unusual since it may be regarded as an anti-$CdCl_2$ structure (as in fluorite and antifluorite structures). Cs^+ ions form cubic close packed layers and oxide ions occupy the octahedral sites between alternate pairs of caesium layers. This raises some interesting questions because caesium is the most electropositive element and caesium salts are usually regarded as highly ionic. However, the structure of Cs_2O clearly shows that Cs^+ ions are not surrounded by anions, as expected for an ionic structure, but have only three oxide neighbours, all located at one side. The structure is held together, in three dimensions, by bonding between caesium ions in adjacent layers.

It may be that the structure of Cs_2O does not reflect any peculiar type of bonding but rather that it is the only structural arrangement which is feasible for a compound of this formula and for ions of this size. Thus, from the formula, the

Table 7.14 *Some compounds with the $CdCl_2$ structure.* (Data taken from Wyckoff, 1971, Vol. 1)

	$a(\text{Å})$	$c(\text{Å})$		$a(\text{Å})$	$c(\text{Å})$
$CdCl_2$	3.854	17.457	$NiCl_2$	3.543	17.335
$CdBr_2$	3.95	18.67	$NiBr_2$	3.708	18.300
$CoCl_2$	3.544	17.430	NiI_2	3.892	19.634
$FeCl_2$	3.579	17.536	$ZnBr_2$	3.92	18.73
$MgCl_2$	3.596	17.589	ZnI_2	4.25	21.5
$MnCl_2$	3.686	17.470	Cs_2O^*	4.256	18.99

*Cs_2O has an anti-$CdCl_2$ structure.

coordination numbers of Cs^+ and O^{2-} must be in the ratio of 1:2; since Cs^+ is considerably larger than O^{2-}, the maximum possible coordination number of oxygen by caesium may be six, which then leads to a coordination number of three for Cs^+.

A related question arises with the structures of the other alkali oxides, in particular K_2O and Rb_2O. These have the antifluorite structure with coordination numbers of four and eight for M^+ and O^{2-}, respectively. These compounds are unusual since Rb^+ is normally far too large a cation to enter into tetrahedral coordination with oxygen. However, if there is no feasible alternative structure, then perhaps Rb^+ ions have no choice but to enter the tetrahedral sites. With Cs_2O, tetrahedral coordination of Cs^+ by O^{2-} is probably impossible and hence it adopts the anti-$CdCl_2$ structure rather than the antifluorite structure. Thermodynamic data support these observations since neither Cs_2O nor Rb_2O are very stable: they oxidize readily to give peroxides, M_2O_2, and superoxides, MO_2, which contain much larger anions.

7.3 Silicate structures—some tips to understanding them

Silicates, especially many minerals, often have very complicated formulae. The purpose of this section is not to give a review of the crystal structures of silicates but simply to show that a considerable amount of structural information may be obtained from their chemical formulae. Using certain guidelines one can appreciate, without the necessity of remembering a large number of complex formulae, whether a particular silicate is a three-dimensional framework structure, whether it is sheet-like or chain-like, etc.

It is common practice to regard many silicate structures as composed of cations and silicate anions. Various types of silicate anion are possible, ranging from the extremes of isolated SiO_4^{4-} tetrahedra in orthosilicates such as olivine (Mg_2SiO_4), to infinite three-dimensional frameworks, as in silica itself(SiO_2). The structures of the various silicate anions are based on certain principles:

(1) Almost all silicate structures are built of SiO_4 tetrahedra.
(2) The SiO_4 tetrahedra may link up by sharing corners to form larger polymeric units.
(3) No more than two SiO_4 tetrahedra may share a common corner (i.e. oxygen)
(4) SiO_4 tetrahedra never share edges or faces with each other.

Exceptions to (1) are structures in which silicon is octahedrally coordinated to oxygen as in, for instance, one of the polymorphs of SiP_2O_7. The number of these exceptions is very small, however, and we can regard SiO_4 tetrahedra as the normal building block in silicate structures. Guidelines (3) and (4) are concerned respectively with maintaining local electroneutrality and with ensuring that highly charged cations, such as Si^{4+}, are not too close together. They are discussed in more detail in Chapter 8.

Let us see now how the structures of silicate anions are related, in a simple way to their formulae. The important factor in relating formula to structure type is the silicon to oxygen ratio. This ratio is variable since two types of oxygen may be

Table 7.15 *Relation between chemical formula and silicate anion structure.*

Si:O ratio‡	Number of oxygens per Si		Type of silicate anion	Examples
	bridging	non-bridging		
1:4	0	4	isolated SiO_4^{4-}	Mg_2SiO_4 olivine, Li_4SiO_4
1:3.5	1	3	dimer $Si_2O_7^{6-}$	$Ca_3Si_2O_7$ rankinite, $Sc_2Si_2O_7$ thortveite
1:3	2	2	chains $(SiO_3)_n^{2n-}$	Na_2SiO_3, $MgSiO_3$ pyroxene
			rings, eg $Si_3O_9^{6-}$	$CaSiO_3^*$, $BaTiSi_3O_9$ benitoite
			$Si_6O_{18}^{12-}$	$Be_3Al_2Si_6O_{18}$ beryl
1:2.5	3	1	infinite sheets $(Si_2O_5)_n^{2n-}$	$Na_2Si_2O_5$
1:2	4	0	3D framework	$SiO_2^†$

* $CaSiO_3$ is dimorphic. One polymorph has $Si_3O_9^{6-}$ rings. The other polymorph has infinite $(SiO_3)_n^{2n-}$ chains.

† The three main polymorphs of silica, quartz, tridymite and cristobalite each have a different kind of 3D framework structure.

‡ In some structures, as in sphene, $CaTiSiO_5$ and Ca_3SiO_5, the Si:O ratio is less than 1:4; these contain SiO_4^{4-} tetrahedra together with extra oxygens entirely unconnected to any silicon.

distinguished in the silicate anions: *bridging oxygens* and *non-bridging oxygens*. Bridging oxygens are those that link up, or are common to, two tetrahedra. Effectively, they may be regarded as belonging half to one Si and half to another Si. In evaluating the net Si:O ratio, bridging oxygens count as $\frac{1}{2}$. Non-bridging oxygens are those that are linked to only one silicon or silicate tetrahedron. They may also be called terminal oxygens. In order to maintain charge balance, non-bridging oxygens must also be linked to other cations in the crystal structure. In evaluating the overall Si:O ratio, non-bridging oxygens count as 1.

The overall Si:O ratio in a silicate crystal structure depends on the relative number of bridging and non-bridging oxygens. Some examples are given in Table 7.15. In these, the alkali and alkaline earth cations do not form part of the silicate anion whereas in certain other cases, e.g. in aluminosilicates, cations such as Al^{3+} may substitute for silicon in the silicate anion. The examples given in the Table are all straightforward and one may deduce the type of silicate anion directly from the chemical formula.

Many other more complex examples could be given. In these, although the detailed structure cannot be deduced from the formula, one can at least get an approximate idea of the type of silicate anion. For example, in the phase $Na_2Si_3O_7$, the Si:O ratio is 1:2.33. This corresponds to a structure in which, on average, two thirds of an oxygen per SiO_4 tetrahedron are non-bridging. Clearly, therefore, some of the SiO_4 tetrahedra in this structure must be composed entirely of bridging oxygens whereas others contain one non-bridging oxygen. The structure of the silicate anion would therefore be expected to be something between an infinite sheet and a 3D framework. In fact, the structure contains an infinite, double-sheet silicate anion in which two thirds of the silicate tetrahedra have one non-bridging oxygen.

The relation between formula and anion structure becomes more complex in cases where ions such as Al^{3+} may substitute for Si^{4+} in the silicate anion. Examples are as follows.

The plagioclase feldspars are a family of aluminosilicates typified by albite, $NaAlSi_3O_8$ and anorthite, $CaAl_2Si_2O_8$. In both of these Al is partly replacing Si in the silicate anion. It is therefore appropriate to consider the overall ratio $(Si + Al):O$. In both cases this ratio is $1:2$ and therefore, a 3D framework structure is expected. Framework structures also occur in orthoclase, $KAlSi_3O_8$, kalsilite, $KAlSiO_4$, eucryptite, $LiAlSiO_4$ and spodumene, $LiAlSi_2O_6$.

Substitution of Al for Si occurs in many sheet structures such as micas and the clay minerals. The mineral talc has the formula $Mg_3(OH)_2Si_4O_{10}$ and, as expected for an $Si:O$ ratio of $1:2.5$, the structure contains infinite silicate sheets. In the mica phlogopite, one quarter of the Si atoms in talc are effectively replaced by Al and extra K^+ ions are added to preserve electroneutrality. Hence, phlogopite has the formula $KMg_3(OH)_2(Si_3Al)O_{10}$. In talc and phlogopite, Mg^{2+} ions occupy octahedral sites between silicate sheets; K^+ ions occupy 12-coordinate sites.

Further complications arise in other aluminosilicates in which Al^{3+} ions may occupy octahedral sites as well as tetrahedral ones. In such cases, one has to have information on the coordination number of aluminium since this cannot be deduced from the formula. One example is the mica muscovite, $KAl_2(OH)_2(Si_3Al)O_{10}$. This is structurally similar to phlogopite, with infinite sheets of constitution $(Si_3Al)O_{10}$ and a $(Si + Al):O$ ratio of $1:2.5$. However, the other two Al^{3+} ions replace the three Mg^{2+} ions of phlogopite and occupy octahedral sites. By convention, only ions that replace Si in tetrahedral sites are included as part of the complex anion. Hence octahedral Al^{3+} ions are formally regarded as cations in much the same way as alkali and alkaline earth cations, Table 7.15.

Questions

7.1 Sodium oxide, Na_2O, has the antifluorite structure (Table 7.7). Calculate (a) the sodium–oxygen bond length, (b) the oxygen–oxygen bond length, (c) the density of Na_2O.

7.2 Starting with a cubic close packed array of anions, what structure types are generated by (a) filling all the tetrahedral sites with cations, (b) filling one half of the tetrahedral sites, e.g. the T_+ sites, with cations, (c) filling all the octahedral sites with cations, (d) filling alternate layers of octahedral sites with cations.

7.3 Repeat the above question but with a hexagonal close packed array of anions. Comment on the absence of any known structure in category (a).

7.4 $SrTiO_3$ has the perovskite structure, $a = 3.905$ Å. Calculate (a) the Sr—O bond length, (b) the Ti—O bond length, (c) the density of $SrTiO_3$. What is the lattice type?

7.5 Metallic gold and platinum both have face centred cubic unit cells with dimensions 4.08 and 3.91 Å, respectively. Calculate the metallic radii of the gold and platinum atoms.

7.6 Identify the following cubic structure types from the information on unit cells and atomic coordinates:

(i) MX: M $\frac{1}{2}00$, $0\frac{1}{2}0$, $00\frac{1}{2}$, $\frac{1}{2}\frac{1}{2}\frac{1}{2}$

   X 000, $\frac{1}{2}\frac{1}{2}0$, $\frac{1}{2}0\frac{1}{2}$, $0\frac{1}{2}\frac{1}{2}$

(ii) MX: M 000, $\frac{1}{2}\frac{1}{2}0$, $\frac{1}{2}0\frac{1}{2}$, $0\frac{1}{2}\frac{1}{2}$

   X $\frac{111}{444}$, $\frac{313}{444}$, $\frac{331}{444}$, $\frac{133}{444}$

(iii) MX: M 000

   X $\frac{111}{222}$

(iv) MX_2: M 000, $\frac{1}{2}\frac{1}{2}0$, $\frac{1}{2}0\frac{1}{2}$, $0\frac{1}{2}\frac{1}{2}$

   X $\frac{111}{444}$, $\frac{311}{444}$, $\frac{131}{444}$, $\frac{113}{444}$,

   $\frac{331}{444}$, $\frac{313}{444}$, $\frac{133}{444}$, $\frac{333}{444}$

7.7 Starting from the rock salt structure, what structure types are generated by the following operations:

   (i) removal of all atoms or ions of one type;
   (ii) removal of half the atoms or ions of one type in such a way that alternate layers only are present;
   (iii) replacement of all the cations in the octahedral sites by an equal number of cations in one set of tetrahedral sites.

7.8 Explain why the NiAs structure is commonly found with metallic compounds but not with ionic ones.

7.9 Compare the packing density of the NaCl and CsCl structures for which, in both cases, anion–anion and anion–cation direct contacts occur.

7.10 What kind of complex anion do you expect in the following: (a) Ca_2SiO_4; (b) $NaAlSiO_4$(Al tetrahedral); (c) $BaTiSi_3O_9$; (d) $Ca_2MgSi_2O_7$ melilite; (e) $CaMgSi_2O_6$ diopside; (f) $Ca_2Mg_5Si_8O_{22}(OH, F)_2$ amphibole, tremolite (the OH, F ions are not bonded to Si); (g) $CaAl_2(OH)_2(Si_2Al_2)O_{10}$ mica, margarite (two Al tetrahedral, two Al octahedral); (h) $Al_2(OH)_4Si_2O_5$ kaolinite (Al octahedral, OH not bonded to Si).

References

D. M. Adams (1974). *Inorganic Solids, An Introduction to Concepts in Solid-State Structural Chemistry*, Wiley.

L. Bragg, G. F. Claringbull and W. H. Taylor (1965). *Crystal Structure of Minerals*, Cornell University Press, Ithaca, N.Y.

G. O. Brunner (1971). An unconventional view of the 'closest sphere packings', *Acta Cryst.*, A**27**, 388.

G. M. Clark (1972). *The Structures of Non-Molecular Solids. A Coordinated Polyhedron Approach*, Applied Science Publishers.

R. C. Evans (1964). *An Introduction to Crystal Chemistry*, 2nd ed., Cambridge University Press.

S. M. Ho and B. E. Douglas (1969). *J. Chem. Ed.*, **46**, 207.

H. Krebs (1968). *Inorganic Crystal Chemistry*, McGraw-Hill.

I. Naray-Szabo (1969). *Inorganic Crystal Chemistry*, Akademiai Szabo.

E. Parthé (1964). *Crystal Chemistry of Tetrahedral Structures*, Gordon and Breach.

Structure Reports, International Union of Crystallography.

Ajit R. Verma and P. Krishna (1966). *Polymorphism and Polytypism in Crystals*, Wiley, New York.

R. Ward (1959). Mixed metal oxides, *Prog. Inorg. Chem.*, **1**, 465.

A. F. Wells (1984). 5th Ed. *Structural Inorganic Chemistry*, Oxford.

A. R. West (1975). *Z. Krist*, **141**, 422–436.

A. R. West and P. G. Bruce (1982). *Acta Cryst.*, **B38**, 1891–1896.

R. W. G. Wyckoff (1971). *Crystal Structures*, Vols 1 to 6, Wiley.

Chapter 8

Some Factors which Influence Crystal Structures

The previous chapter has dealt with the description and classification of crystal structures without paying too much attention to the reasons why a particular compound prefers one structure type to another. Crystal structures are influenced by a considerable number of factors—atom size, bonding type, electron configuration, etc.—and while each factor is understood fairly well in isolation, it is more difficult to assess the effect of all the factors in combination. Thus, it is a difficult, if not impossible, task to predict the structure of a new or unknown compound unless it falls into an obvious category such as a new spinel or perovskite phase. In this chapter, some of the factors that influence crystal structures are considered and an attempt made to review current ideas in crystal chemistry.

8.1 Preliminary survey

The structure adopted by a particular crystalline compound depends, to a first approximation, on three main factors: the general formula of the compound and the valencies of the elements present, the nature of the bonding between the atoms and the relative size of the atoms or ions.

8.1.1 General formulae, valencies and coordination numbers

Use of the term 'general formula' here refers to the relative number of atoms of each type that are present, without specifying what the atoms are, i.e. for a compound $A_x B_y$, the general formula gives the values of x and y without identifying A and B. For such a compound $A_x B_y$, the coordination numbers of A and B are related directly to the general formula. A general rule is that: *the coordination numbers of A and B are in the ratio $y:x$, provided that direct A–A or B–B contacts do not occur.* This applies to most ionic, polar and covalent polymeric materials but not to catenated compounds, such as polymers which have C—C bonds. Thus, for a compound AB_2, the coordination numbers of A(by B) and B(by A) are in the ratio of 2:1, as in $SiO_2(4:2)$, $TiO_2(6:3)$ and CaF_2 (8:4). Proof of this rule is not given here but after a little thought it should be obvious that it applies, at least to simple formulae, AB, AB_2, etc. The rule does not predict absolute coordination numbers for a given formula but it does place restrictions on the combination of coordination numbers that are possible in a structure.

The rule may be extended to more complex structures. In a compound $A_x B_y C_z$, in which A and B are cations coordinated only to anions, C, the *average* cation coordination number (CN) is related to the anion coordination number by

$$\frac{\text{Average cation CN}}{\text{Anion CN}} = \frac{z}{x+y} \tag{8.1}$$

in which the average cation CN is given by

$$\frac{x(\text{CN of A}) + y(\text{CN of B})}{x+y} \tag{8.2}$$

Substitution of (8.2) into (8.1) gives

$$x(\text{CN of A}) + y(\text{CN of B}) = z(\text{CN of C}) \qquad (8.3)$$

The application of (8.3) can be seen in the following examples:

(a) Perovskite, $CaTiO_3$, contains octahedral Ti^{4+} and twelve coordinate Ca^{2+} ions. From (8.3), therefore, the anion CN is calculated to be six. The actual structure of perovskite is in agreement with this since oxygen is coordinated octahedrally to two Ti^{4+} and four Ca^{2+} ions.

(b) Spinel, $MgAl_2O_4$, contains tetrahedral Mg^{2+} and octahedral Al^{3+} ions. From (8.3), the oxygen CN is calculated to be four. This is correct since, in spinel, oxygen is tetrahedrally coordinated to three Al^{3+} and one Mg^{2+} ions.

This relationship between general formulae and coordination number is of little predictive value alone since it cannot be used in the absence of structural information. However, it does allow a certain degree of rationalization of formulae and coordination numbers and is useful for checking the anion CN in complex structures. It breaks down when bonding occurs between atoms of the same type, e.g. in $CdCl_2$ in which chlorine–chlorine contacts occur.

The above comments apply to the *relative* coordination numbers in a compound and take no account of the valency of the atoms. In molecular materials, the *absolute* coordination numbers are obviously controlled by valency since electron pair covalent bonds hold the molecules together. Unless multiple or partial bonds occur, the number of bonds to a particular atom in a molecule is equal to the coordination number and hence to the valency of that atom.

In non-molecular materials, however, the valency of an atom or ion does not have a direct bearing on coordination numbers and structure, apart from its obvious importance in controlling the general formula of the compound. Thus, the compounds in the series, LiF, MgO, ScN, TiC, all have the same general formula, AB, and the same crystal structure, that of rock salt. Coordination numbers are 6:6, therefore, but the valencies of the atoms increase from one in LiF to four in TiC. The type of bonding present certainly varies across the series, from ionic in LiF to essentially covalent in TiC, but the structure, given by the relative arrangement of atoms, is independent of atom valence.

8.1.2 Bonding

The nature of the bonding between atoms affects considerably the coordination numbers of the atoms and hence has a major influence on the crystal structure that is adopted. Broadly speaking, ionic bonding leads to structures with high symmetry and in which the coordination numbers are as high as possible. In this way, the net electrostatic attractive force which holds crystals together (and hence the lattice energy) is maximized. Covalent bonding, on the other hand, gives highly directional bonds in which one or all of the atoms present has a definite preference for a certain coordination environment, irrespective of the other atoms that are present. The coordination numbers in covalently bonded structures are usually small and may be less than those in corresponding ionic structures which contain atoms of similar size to those in the covalent structure.

The type of bonding that occurs in a compound correlates fairly well with the position of the component atoms in the periodic table and, especially, with their electronegativity. Alkali and alkaline earth elements usually form essentially ionic structures (beryllium is sometimes an exception), especially in combination with smaller, more electronegative anions such as O^{2-} and F^-. Covalent structures occur especially with (a) small atoms of high valency which, in the cationic state, would be highly polarizing, e.g. B^{3+}, Si^{4+}, P^{5+}, S^{6+}, etc., and to a lesser extent with (b) large atoms which in the anionic state are highly polarizable, e.g. I^-, S^{2-}.

Most non-molecular compounds have bonding which is a mixture of ionic and covalent and, as discussed later, it is becoming possible to make quantitative assessments of the *ionicity* of a particular bond, i.e. the percentage of ionic character in the bond. An additional factor in some transition metal compounds is the occurrence of metallic bonding.

Some clear-cut examples of the influence of bonding type on crystal structure are as follows:

(a) *SrO, BaO, HgO*. SrO and BaO both have the rock salt structure with octahedrally coordinated M^{2+} ions. Based on size considerations alone, and if it were ionic, HgO would also be expected to have the same structure. However, mercury is only two coordinate in HgO and the structure may be regarded as covalent. Linear O—Hg—O segments occur in the structure and may be rationalized on the basis of *sp* hybridization of mercury. The ground state of atomic mercury is

$$\text{(Xe core)} \ 4f^{14} \quad 5d^{10} \quad 6s^2$$

The first excited state, corresponding to mercury (II), is

$$\text{(Xe core)} \quad 4f^{14} \quad 5d^{10} \quad 6s^1 \quad 6p^1$$

Hybridization of the 6s and one 6p orbital gives rise to two, linear *sp* hybrid orbitals, each of which forms a normal, electron pair, covalent bond by overlap with an orbital on oxygen. Hence mercury has a CN of two in HgO.

(b) *AlF₃, AlCl₃, AlBr₃ and AlI₃*. These compounds show a smooth transition from ionic to covalent bonding as the electronegativity difference between the two elements decreases. Thus, AlF_3 is a high melting, essentially ionic solid with a distorted octahedral coordination of the Al^{3+} ions; its structure is related to that of ReO_3. $AlCl_3$ has a layered, polymeric structure in the solid state similar to the structure of $CrCl_3$, which is related to the $CdCl_2$ and CdI_2 structures. The bonds may be regarded as part ionic/part covelent. $AlBr_3$ and AlI_3 have molecular structures with dimeric units of formula Al_2X_6. Their structure and shape is shown in Fig. 7.10 and the bonding between aluminium and bromine or iodine is essentially covalent.

Halides of other elements, e.g. Be, Mg, Ga, In, also show variations in bond type and structure, depending on the halide. The trends are always the same; with the fluorides there is the largest difference in electronegativity between the two elements and these structures are the most ionic. With the other halides increasing covalent character occurs in the series, chloride–bromide–iodide.

8.1.3 Size

The relative size of the atoms in a compound has a major influence on the structure adopted, especially for more ionic structures. A guiding principle in ionic structures is that the coordination number of a particular ion is as large as possible, provided that it can be in contact with all of its neighbouring ions of opposite charge. The limiting situation occurs when a cation is too small to fit snugly into a particular hole in the anion array and a hypothetical structure in which a cation can rattle inside its hole is regarded as unstable. The limiting size of the interstitial hole in various anion arrays, e.g. f.c.c. and b.c.c., can be estimated, in theory, using the radius ratio rules (Section 8.2.3), but it must be stated that in practice there are many exceptions to the rules. The general relation between size and coordination number is clear, however; as the value of the ratio (cation radius/anion radius) increases, so the coordination number of the cation increases. A good example of this, and to which the radius ratio rules may apparently be applied successfully, is oxides, MO_2. With increasing size of M, the structure and coordination number changes in the sequence:

$$CO_2 \quad \text{(CN of C} = 2)$$
$$SiO_2 \quad \text{(CN of Si} = 4)$$
$$TiO_2 \quad \text{(CN of Ti} = 6)$$
$$PbO_2 \quad \text{(CN of Pb} = 8)$$

Other examples of this trend are:

$BeF_2(SiO_2 \, \text{str.}, CN = 4)$; $\quad MgF_2(TiO_2 \, \text{str.}, CN = 6)$; $\quad CaF_2(CN = 8)$
$Rb_2O(CN = 4)$; $\quad Cs_2O(CN = 6)$
$BeO(ZnS \, \text{str.}, CN = 4)$; $\quad MgO(NaCl \, \text{str.}, CN = 6)$

In molecular materials, however, size considerations are less important. This is partly because the coordination numbers in molecular materials are controlled by valency and partly because the covalent radii of elements do not show the same spread of values as do the ionic radii. Usually, the radii of a particular element are in the following sequence:

Cation radius < covalent radius < anion radius

although few elements can exist in all three states. Thus magnesium may be cationic or covalent in its compounds but never anionic, whereas fluorine may be covalent or anionic but never cationic. Examples of elements which may exist in all three states are hydrogen and iodine.

Because elements do not differ too much in their covalent radii (see Table 8.10 later), it is usually possible to satisfy the valence and bonding requirements of atoms in molecules without interference from size problems. For example, consider CI_4 which is a covalently bonded, tetrahedral molecule. The covalent radius of carbon is large enough that four covalent C—I bonds can form. If the structure were ionic, however, the C^{4+} ion (which does not exist, chemically) would be too small to be tetrahedrally coordinated by the large I^- ions.

8.2 Ionic structures

Purely ionic bonding in crystalline compounds is an idealized or extreme form of bonding which is rarely attained in practice. Even in structures that are regarded as essentially ionic, e.g. NaCl and CaO, there is usually a certain amount of covalent bonding between cation and anion which acts to reduce the charge on each. The degree of covalent bonding increases with increasing valence of the ions to the extent that ions with a *net* charge greater than $+1$ or -1 appear unlikely to exist. Thus, while NaCl may reasonably be represented as Na^+Cl^-, TiC (which also has the NaCl structure) certainly does not contain Ti^{4+} and C^{4-} ions and the main bonding type in TiC must be non-ionic. This brings us to a dilemma. Do we continue to use the ionic model in the knowledge that for many structures, e.g. Al_2O_3, $CdCl_2$, a large degree of covalent bonding must be present? If not, we must find an alternative model for the bonding. In this chapter, ionic bonding is given considerable prominence because of its apparent wide applicability and its usefulness as a *starting point* for describing structures which in reality have a considerable amount of covalent bonding. In Section 8.3 and 8.4, two methods of assessing the degree of covalent character in 'ionic structures' are discussed.

8.2.1 Ions and ionic radii

It is difficult to imagine discussing crystal chemistry without having information available on the sizes of ions in crystals. However, crystal chemistry is currently undergoing a minor revolution in that the long-established tables of ionic radii of Pauling (1928), Goldschmidt and others are now thought to be seriously in error; at the same time, our concepts of ions and ionic structures are also undergoing revision. In recent compilations of ionic radii, e.g. of Shannon and Prewitt (1969, 1970), cations are shown as being larger and anions smaller than previously thought. For example, Pauling radii of Na^+ and F^- are 0.98 and 1.36 Å, respectively, whereas Shannon and Prewitt give values of 1.14 to 1.30 Å, depending on the coordination number, for Na^+ and 1.19 Å for F^-.

These changes have arisen largely because, with modern, high quality X-ray diffraction work it is possible to obtain fairly accurate maps of the distribution of electron density throughout ionic crystals. Thus, one can effectively 'see' ions and tell something about their size, shape and nature. In Fig. 8.1 is shown an electron density 'contour map' (see Chapter 5, Section 5.5.6) for LiF for a section passing through the structure parallel to one unit cell face. The map therefore passes

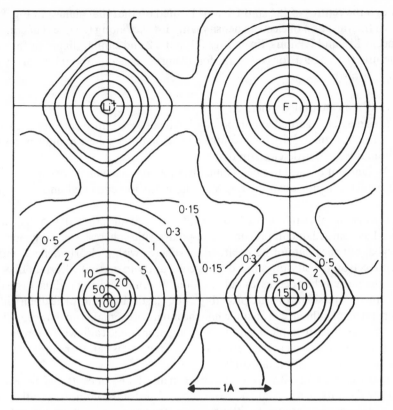

Fig. 8.1 Electron density contour map of LiF: a section through part of the unit cell face. The electron density (electrons Å^{-3}) is constant along each of the contour lines. (From Krug, Witte and Welfel, 1955)

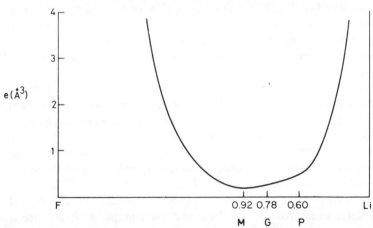

Fig. 8.2 Variation of electron density along the line connecting adjacent Li and F nuclei in LiF. (From Krebs, 1968). P = Pauling radius of Li^+, G = Goldschmidt radius, M = minimum in electron density

through the centres of Li^+ and F^- ions located on the (100) planes. In Fig. 8.2 is shown the variation of electron density with distance along the line that connects adjacent Li^+ and F^- ions. From Figs 8.1 and 8.2 and similar diagrams for other structures (Fig. 5.39 a, for NaCl), the following conclusions about ions in crystals may be drawn:

(a) Ions are essentially spherical.
(b) Ions may be regarded as composed of two parts: a central core in which most of the electron density is concentrated and an outer sphere of influence which contains very little electron density.
(c) Assignment of radii to ions is difficult; even for ions which are supposedly in contact, it is not obvious (Fig. 8.2) where one ion ends and another begins.

Conclusion (b) is in contrast to the oft-stated assumption that 'ions can be treated as charged, incompressible, non-polarizable spheres'. Certainly, ions are charged, but they cannot be regarded as spheres with a clearly defined radius. Their electron density does not decrease abruptly to zero at a certain distance from the nucleus but decreases only gradually with increasing radius. Instead of being incompressible, ions are probably quite elastic, by virtue of flexibility in the outer sphere of influence of an ion while the inner core remains unchanged. This is necessary in order to explain variations of apparent ionic radii with coordination number and environment (see later). Within limits, ions can therefore expand or contract as the situation demands.

From Figs 8.1 and 8.2, most of the electron density is concentrated close to the nuclei of the ions; in a crystal, therefore, most of the total volume is essentially free space and contains relatively little electron density.

The difficulties involved in determining ionic radii arise because, between adjacent anions and cations, the electron density passes through a broad minimum. For LiF (Fig. 8.2), the radii for Li^+ given by Pauling and Goldschmidt are marked together with the value which corresponds to the minimum in the electron density along the line connecting Li^+ and F^-. Although the values of these radii vary from 0.60 to 0.92 Å, all lie within the broad electron density minimum of Fig. 8.2.

The many methods that have been used in the past to estimate ionic radii will not be discussed here. In spite of the difficulties involved in determining absolute radii, it is necessary to have a set of radii for reference. Fortunately, most sets of radii are *additive* and *self-consistent*; provided one does not mix radii from different tabulations it is possible to use any set of radii to evaluate interionic distances in crystals with reasonable confidence. Shannon and Prewitt give two sets of radii: one is based on $r_0 = 1.40$ Å and is similar to Pauling, Goldschmidt, etc.; the other is based on $r_{F^-} = 1.19$ Å (and $r_{O^{2-}} = 1.26$ Å) and is related to the values determined from X-ray electron density maps. Both sets are comprehensive for cations in their different coordination environments but only pertain to oxides and fluorides. We choose here to use the Shannon and Prewitt set based on $r_{F^-} = 1.19$ Å and $r_{O^{2-}} = 1.26$ Å. Cation radii are shown in graphical form in

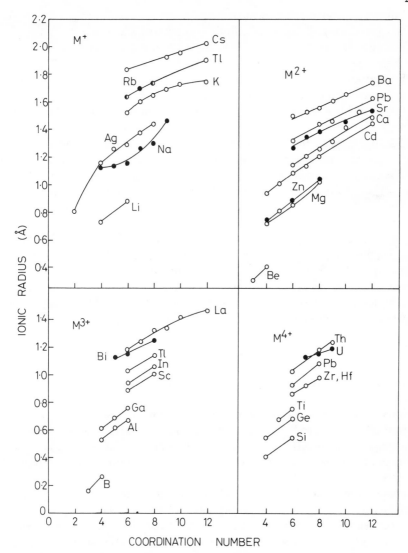

Fig. 8.3 Ionic radii as a function of coordination number for cations M^+ to M^{4+}. (From Shannon and Prewitt, 1969, 1970.) Data based on $r_{F^-} = 1.19\,\text{Å}$ (and $r_{O^{2-}} = 1.26\,\text{Å}$)

Fig. 8.3 as a function of cation coordination number. Radii are given for ions M^+ to M^{4+}; it should be stressed that the more highly charged ions are unlikely to exist as such in the purely ionic state but probably have their positive charged reduced by polarization of the anion and consequent partial covalent bonding between cation and anion.

The following trends in ionic radii, with position in the periodic table, formal charge and coordination number, occur:

(a) For the s- and p-block elements, radii increase with atomic number for any vertical group, e.g. octahedrally coordinated alkali ions.

(b) For any isoelectronic series of cations, the radius decreases with increasing charge, e.g. Na^+, Mg^{2+}, Al^{3+} and Si^{4+}.

(c) For any element which can have variable oxidation states, the cation radius decreases with increasing oxidation state, e.g. V^{2+}, V^{3+}, V^{4+}, V^{5+}.

(d) For an element which can have various coordination numbers, the cationic radius increases with increasing coordination number.

(e) The 'lanthanide contraction' occurs as follows: across the lanthanide series, ions with the same charge but increasing atomic number show a reduction in size due to the ineffective shielding of the nuclear charge by the d and, especially, f electrons, e.g. octahedral radii, La^{3+} (1.20 Å)... Eu^{3+} (1.09 Å)... Lu^{3+} (0.99 Å). Similar effects occur across some series of transition metal ions.

(f) The radius of a particular transition metal ion is smaller than that of the corresponding main group ion for the reasons given in (e), e.g. Rb^+ (1.63 Å) and Ag^+ (1.29 Å) or Ca^{2+} (1.14 Å) and Zn^{2+} (0.89 Å).

(g) Certain pairs of elements positioned diagonally to one another in the periodic table have similar ionic size (and chemistry), e.g. Li^+ (0.88 Å) and Mg^{2+} (0.86 Å). This is due to a combination of effects (a) and (b).

8.2.2 Ionic structures—general principles

Consider the following as a guide to ionic structures:

(a) Ions are regarded as charged, elastic .and polarizable spheres.

(b) Ionic structures are held together by electrostatic forces and, therefore, are arranged so that cations are surrounded by anions, and vice versa.

(c) In order to maximize the net electrostatic attraction between ions in a structure (i.e. the lattice energy), coordination numbers are as high as possible, provided that the central ion maintains contact (via its sphere of influence) with all its neighbouring ions of opposite charge.

(d) Next nearest neighbour interactions are of the anion–anion and cation–cation type and are repulsive. Like ions arrange themselves to be as far apart as possible, therefore, and this leads to structures of high symmetry with a maximized volume.

(e) Local electroneutrality prevails, i.e. the valence of an ion is equal to the sum of the electrostatic bond strengths between it and adjacent ions of opposite charge.

Point (a) has been considered in the previous section; ions are obviously charged, are elastic because their size varies with coordination number and are polarizable when departures from purely ionic bonding occurs. For example, the electron density map for LiF (Fig. 8.1) shows a small distortion from spherical shape in the outer part of the sphere of influence of the Li^+ ion, and this may be attributed to the occurrence of a small amount of covalent bonding between Li^+ and F^-.

Points (b) and (d) infer that the forces which hold ionic crystals together and the net energy of the interaction between the ions are the same as would be obtained by regarding the crystal as a three-dimensional array of point charges and considering the net coulombic energy of the array. From Coulomb's Law, the force F between two ions of charge Z_+e and Z_-e, separated by distance r, is given by

$$F = \frac{(Z_+e)(Z_-e)}{r^2} \tag{8.4}$$

A similar equation applies to each pair of ions in the crystal and evaluation of the resulting force between all the ions leads to the lattice energy of the crystal (Section 8.2.5).

Point (c) includes the proviso that nearest neighbour ions should be 'in contact'. Given the nature of electron density distributions in ionic crystals (Figs 8.1 and 8.2), it is hard to quantify what is meant by 'in contact'. It is nevertheless an important factor since, although the apparent size of ions varies with coordination number, most ions, smaller ones especially, appear to have a maximum coordination number; for Be^{2+} this is four and for Li^+ it is six. Ions are flexible, therefore, but expand or contract only within fairly narrow limits.

The idea of *maximizing* the volume of ionic crystals, point (d) is rather unexpected (Brunner; O'Keeffe) since one is accustomed to regarding ionic structures and derivative close packed structures, expecially, as having *minimum* volume. There is no conflict, however. The prime bonding force in ionic crystals is the nearest neighbour cation–anion *attractive* force and this force is maximized at a small cation–anion separation (when ions become too close, additional repulsive forces come into play, Section 8.2.5, thereby reducing the net attractive force). Superposed on this is the effect of next nearest neighbour *repulsive* forces between like ions. With the constraints that (a) cation–anion distances be as short as possible and (b) coordination numbers be as large as possible, like ions arrange themselves to be as far apart as possible in order to reduce their mutual repulsion. This leads to regular and highly symmetrical arrays of like ions. It has been shown (O'Keeffe) that such regular arrays of like ions tend to have maximized volumes and that, by distorting the structures, a reduction in volume is possible, at least in principle.

An excellent example of a structure whose volume is maximized is rutile (see Chapters 6 and 7). The buckling of the oxide layers (Fig. 6.16) causes the coordination number of oxygen to be reduced from 12 (as in h.c.p.) to 11 (as in p.t.p.). The coordination of titanium by oxygen, and vice versa, is unaffected by this distortion but the overall volume of the structure increases by 2 to 3 per cent.

Point (e) is Pauling's electrostatic valence rule, the second of a set of rules formulated by Pauling for ionic crystals. Basically, the rule means that the charge on a particular ion, e.g. an anion, must be balanced by an equal and opposite charge on the surrounding cations. However, since these cations are also shared with other anions, it is necessary to estimate the amount of positive charge that is effectively associated with each cation–anion bond. For a cation M^{m+} surrounded by n anions, X^{x-}, the *electrostatic bond strength* (e.b.s.) of the cation–anion bonds is defined as

$$\text{e.b.s.} = \frac{m}{n} \qquad (8.5)$$

For each anion, the sum of the electrostatic bond strengths of the surrounding cations must balance the negative charge on the anion, i.e.

$$\sum \frac{m}{n} = x \qquad (8.6)$$

For example:

(a) Spinel, $MgAl_2O_4$, contains octahedral Al^{3+} and tetrahedral Mg^{2+} ions; each oxygen is surrounded tetrahedrally by three Al^{3+} ions and one Mg^{2+} ion. We can check that this must be so, as follows:

For Mg^{2+}:   e.b.s. $= \frac{2}{4} = \frac{1}{2}$

For Al^{3+}:   e.b.s. $= \frac{3}{6} = \frac{1}{2}$

Therefore,

$$\sum \text{e.b.s.} (3Al^{3+} + 1Mg^{2+}) = 2$$

(b) We can show that three SiO_4 tetrahedra cannot share a common corner in silicate structures:

For Si^{4+}:   e.b.s. $= \frac{4}{4} = 1$

Therefore, for an oxygen that bridges two SiO_4 tetrahedra, $\sum$ e.b.s. $= 2$, which is acceptable. However, three tetrahedra sharing a common oxygen would give $\sum$ e.b.s. $= 3$ for that oxygen, which is quite unacceptable.

This rule of Pauling's provides an important guide to the kinds of polyhedral linkages that are and are not possible in crystal structures. In Table 8.1 is given a list of some common cations with their formal charge, coordination number and electrostatic bond strength. In Table 8.2 are listed some allowed and unallowed combinations of polyhedra about a common oxide ion, together with some examples of the allowed combinations. Many other combinations are possible and the reader may like to deduce some, bearing in mind that there are also topological restrictions on the number of possible polyhedral combinations; thus the maximum number of octahedra that can share a common corner is six (as in rock salt), etc.

Pauling's third rule is concerned with the topology of polyhedra and has been considered in the previous chapter. Pauling's first rule states: 'A coordinated polyhedron of anions is formed about each cation, the cation–anion distance

Table 8.1 *Electrostatic bond strengths of some cations*

Cation with formal charge	Coordination number	Electrostatic bond strength
Li^+	4, 6	$\frac{1}{4}, \frac{1}{6}$
Na^+	6, 8	$\frac{1}{6}, \frac{1}{8}$
Be^{2+}	3, 4	$\frac{2}{3}, \frac{1}{2}$
Mg^{2+}	4, 6	$\frac{1}{2}, \frac{1}{3}$
Ca^{2+}	8	$\frac{1}{4}$
Zn^{2+}	4	$\frac{1}{2}$
Al^{3+}	4, 6	$\frac{3}{4}, \frac{1}{2}$
Cr^{3+}	6	$\frac{1}{2}$
Si^{4+}	4	1
Ge^{4+}	4, 6	$1, \frac{2}{3}$
Ti^{4+}	6	$\frac{2}{3}$
Th^{4+}	8	$\frac{1}{2}$

Table 8.2 *Allowed and unallowed combinations of corner-sharing oxide polyhedra*

Allowed	Example	Unallowed
$2SiO_4$ tet.	Silica	$> 2SiO_4$ tet.
$1MgO_4$ tet. + $3AlO_6$ oct.	Spinel	$3AlO_4$ tet.
$1SiO_4$ tet. + $3MgO_6$ oct.	Olivine	$1SiO_4$ tet. + $2AlO_4$ tet.
$8LiO_4$ tet.	Li_2O	$4TiO_6$ oct.
$2TiO_6$ oct. + $4CaO_{12}$ dod.	Perovskite	
$3TiO_6$ oct.	Rutile	

being determined by the radius sum and the coordination number of the cation by the radius ratio.' The idea that cation–anion distances are determined by the radius sum is implicit to every tabulation of ionic radii since a major objective of such tabulations is to be able to predict, correctly, interatomic distances. Let us now consider coordination numbers and *the radius ratio rules*.

8.2.3 The radius ratio rules

In ideally ionic crystal structures, the coordination numbers of the ions are determined largely by electrostatic considerations. Cations surround themselves with as many anions as possible, and vice versa. This maximizes the electrostatic attractions between neighbouring ions of opposite charge and hence maximizes the lattice energy of the crystal (see later). This requirement led to the formulation of the *radius ratio rules* for *ionic structures* in which the ions and the structure adopted for a particular compound depend on the *relative* sizes of the ions. There are two guidelines to be followed. First, a cation must be in contact with its anionic neighbours and so this places a *lower* limit on the size of cation which may occupy a particular site. A situation in which a cation may 'rattle' inside its anion polyhedron is assumed to be unstable. Second, neighbouring anions may or may not be in contact. Using these guidelines, one may calculate the range of cation sizes that can occupy the various interstitial sites in an anion array, as follows.

For a face centred cubic array (e.g. NaCl structure), in which the anions are in contact, octahedral cation sites have a minimum radius r_m, given from Fig. 8.4 by

$$(2r_x)^2 + (2r_x)^2 = [2(r_m + r_x)]^2$$

and therefore,

$$\frac{r_m}{r_x} = \sqrt{2} - 1 = 0.414$$

Atoms 1, 2, 3 and 4, 5, 6 belong to adjacent close packed layers and an octahedral site lies midway between. Atoms 1, 2 and 3 are in contact, as also are 4, 5 and 6. Between the layers, atoms 2 and 3 are in contact with 4 and 5. For smaller cations (i.e. $r_m/r_x < 0.414$) the cation could not be in contact with all six anionic nearest neighbours and, therefore, according to the theory, a structure with lower cation coordination number would result. (Note that, in practice, structures *do* occur in

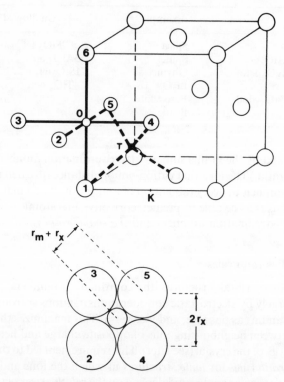

Fig. 8.4 Octahedral and tetrahedral cation sites in a face centred cubic (cubic close packed) anion array

which a cation is obviously too small for its particular site, e.g. the sodium ions in the solid electrolyte, β-alumina, occupy very large sites.)

For radius ratios > 0.414 the cation would push the anions apart and this happens increasingly up to a radius ratio of 0.732. At and above this value, the cation is sufficiently large to have eight anionic neighbours, all of which are in contact with the cation. The CsCl structure (Fig. 7.20) has eight-coordinate ions. Here,

$$[2(r_m + r_x)]^2 = (\text{cube body diagonal})^2$$
$$= 3(2r_x)^2$$

Therefore,

$$\frac{r_m}{r_x} = \sqrt{3} - 1 = 0.732$$

For tetrahedral coordination (Fig. 8.4), distance $5 - K$ is the body diagonal of a small cube and is equal to $2(r_m + r_x)$. Therefore,

$$(2r_x)^2 + (\sqrt{2}r_x)^2 = [2(r_m + r_x)]^2$$

and

$$\frac{r_m}{r_x} = \tfrac{1}{2}(\sqrt{6} - 2) = 0.225$$

The minimum radius ratios for various coordination numbers are given in Table 8.3. With the exception of CN $= 8$, all are applicable to close packed structures. Note that CN $= 5$ is absent from the table; in close packed structures, it is not possible to have a coordination number of five in which all M—X bonds are of the same length.

The radius ratio rules have had a limited amount of success in predicting trends in coordination number and structure type and at best can only be used as a general guideline. Radius ratios depend very much on which table of ionic radii is consulted and there appears to be no clear advantage in using either one of the more traditional sets or the modern set of values based on X-ray diffraction results. For example, for RbI, $r + /r - \, = 0.69$ or 0.80, according to the tables based on $r_{O^{2-}} = 1.40$ and 1.26 Å, respectively. Thus one value would predict six-coordination (rock salt), as observed, but the other predicts eight-coordination (CsCl). On the other hand, LiI has $r + /r - \, = 0.28$ and 0.46, according to the

Table 8.3 *Minimum radius ratios for different cation coordination numbers*

Coordination	Minimum $r_m : r_x$
Linear, 2	—
Trigonal, 3	0.155
Tetrahedral, 4	0.225
Octahedral, 6	0.414
Cubic, 8	0.732

Table 8.4 *Structures and radius ratios of oxides, MO_2*

Oxide	Calculated radius ratio*		Observed structure type	
CO_2	~ 0.1	(CN = 2)	Molecular	(CN = 2)
SiO_2	0.32	(CN = 4)	Silica	(CN = 4)
GeO_2	$\begin{cases} 0.43 \\ 0.54 \end{cases}$	(CN = 4) (CN = 6)	Silica Rutile	(CN = 4) (CN = 6)
TiO_2	0.59	(CN = 6)	Rutile	(CN = 6)
SnO_2	0.66	(CN = 6)	Rutile	(CN = 6)
PbO_2	0.73	(CN = 6)	Rutile	(CN = 6)
HfO_2	$\begin{cases} 0.68 \\ 0.77 \end{cases}$	(CN = 6) (CN = 8)	Fluorite	(CN = 8)
CeO_2	$\begin{cases} 0.75 \\ 0.88 \end{cases}$	(CN = 6) (CN = 8)	Fluorite	(CN = 8)
ThO_2	0.95	(CN = 8)	Fluorite	(CN = 8)

* Since cation radii vary with coordination number, as shown in Fig. 8.3, radius ratios may be calculated for different coordination numbers. The coordination numbers used here are shown in parentheses. Calculations are based on $r_{O^{2-}} = 1.26\,\text{Å}$.

tables based on $r_{O^{2-}} = 1.40$ and $1.26\,\text{Å}$, respectively. One value predicts tetrahedral coordination and the other octahedral (as observed). For the larger cations, especially caesium, $r + /r - > 1$, and it is perhaps more realistic to consider instead the inverse ratio, $r - /r +$, for CsF.

A more convincing example of the relevance of radius ratio rules is provided by oxides and fluorides of general formula, MX_2. Possible structure types, with their cation coordination numbers, are silica (4), rutile (6) and fluorite (8). A selection of oxides in each group is given in Table 8.4, together with the radius ratios calculated from Fig. 8.3 (based on $r_{O^{2-}} = 1.26\,\text{Å}$). Changes in coordination number are expected to occur at radius ratios of 0.225, 0.414 and 0.732. Bearing in mind that the calculated radius ratio values depend on the particular table of radii that is consulted, the agreement between theory and practice is reasonable. For example, GeO_2 is polymorphic and can have both silica and rutile structures; the radius ratio calculated for tetrahedral coordination of germanium is borderline between the values predicted for CN = 4 and 6.

8.2.4 Borderline radius ratios and distorted structures

The structural transition from CN = 4 to 6 which occurs with increasing cation size is often clear-cut. A good example is provided by GeO_2 which has a border line radius ratio and which also exhibits polymorphism. Both polymorphs have highly symmetric structures; one has a silica-like structure with CN = 4 and the other has a rutile structure with CN = 6. Polymorphs with CN = 5 do not occur with GeO_2.

In other cases of borderline radius ratios, however, distorted polyhedra and/or coordination numbers of 5 are observed. Thus V^{5+} (radius ratio $= 0.39$ for $CN = 4$ or radius ratio $= 0.54$ for $CN = 6$) has an environment in one polymorph of V_2O_5 which is a gross distortion of octahedral; five V—O bonds are of reasonable length, in the range 1.5 to 2.0 Å, but the sixth bond is much longer, 2.8 Å, and the coordination is better regarded as distorted square pyramidal. It appears that V^{5+} is rather small to happily occupy an octahedral site and that, instead, a structure occurs which is transitional between tetrahedral and octahedral. Similar types of distortions occur between $CN = 6$ and 8. Thus, ZrO_2 has a borderline radius ratio (0.68 for $CN = 6$; 0.78 for $CN = 8$) and although it may have the fluorite structure at very high temperatures ($> 2000\,^\circ C$), with a CN of 8 for zirconium, in its normal form at room temperature as the mineral baddleyite, zirconium has a CN of 7.

Less severe distortions occur in cases where a cation is only slightly too small for its anion environment. The regular anion coordination is maintained but the cation may rattle or undergo small displacements within its polyhedron. In, for example, $PbTiO_3$ (radius ratio for Ti $= 0.59$ for $CN = 6$), titanium may undergo displacement by ~ 0.2 Å off the centre of its octahedral site towards one of the corner oxygens. The direction of displacement may be reversed under the action of an applied electric field and this gives rise to the important property of ferroelectricity (Chapter 15).

The concept of 'maximum contact distance' has been proposed by Dunitz and Orgel (1960). If the metal–anion distance increases above this distance then the cation is free to rattle. If the metal–anion distance decreases the metal ion is subjected to compression. However, the maximum contact distance does not correspond to the sum of ionic radii, as they are usually defined, and this is a difficult concept to quantify.

8.2.5 Lattice energy of ionic crystals

Ionic crystals may be regarded as regular three-dimensional arrangements of point charges. The forces that hold the crystals together are entirely electrostatic in origin and may be calculated by summing all the electrostatic repulsions and attractions in the crystal. The *lattice energy*, U, of a crystal is defined as the net potential energy of the arrangement of charges that forms the structure. It is equivalent to the energy required to sublime the crystal and convert it into a collection of gaseous ions, e.g.

$$NaCl(s) \rightarrow Na^+(g) + Cl^-(g), \qquad \Delta H = U$$

The value of U depends on the crystal structure that is adopted, the charge on the ions and the internuclear separation between the anion and cation.

Two principal kinds of force are involved in determining the crystal structure of ionic materials:

(a) The electrostatic forces of attraction and repulsion between ions. Two ions M^{z+} and X^{z-} separated by a distance, r, experience an attractive force, F, given by Coulomb's Law:

$$F = \frac{Z_+ Z_- e^2}{r^2} \qquad (8.4)$$

and their coulombic potential energy, V, is given by

$$V = \int_\infty^r F \, dr = -\frac{Z_+ Z_- e^2}{r} \qquad (8.7)$$

(b) Short range repulsive forces which are important when atoms or ions are so close together that their electron clouds begin to overlap. Born suggested that this repulsive energy has the form:

$$V = \frac{B}{r^n} \qquad (8.8)$$

where B is a constant and the Born exponent, n, has a value in the range 5 to 12. Because n is large, V falls rapidly to zero with increasing r.

The lattice energy, U, of a crystal may be calculated by combining the net electrostatic attraction and the Born repulsion energies and finding the internuclear separation, r_e, which gives the maximum U value. The procedure is as follows:

Consider the NaCl structure (Fig. 7.13a). Between each pair of ions in a crystal there is an electrostatic interaction given by equation (8.4). We wish to sum all such interactions which occur in the crystal and calculate the net attractive energy. Let us first consider one particular ion, e.g. Na$^+$ in the body centre of the unit cell (Fig. 7.13a), and calculate the interaction between it and its neighbours. Its nearest neighbours are six Cl$^-$ ions in the face centre positions and at a distance r ($2r$ is the value of the unit cell edge). The attractive potential energy (ignoring for the moment Cl$^-$–Cl$^-$ repulsions) is given by

$$V = -6\frac{e^2 Z_+ Z_-}{r} \qquad (8.9)$$

The next nearest neighbours to the Na$^+$ ion are twelve Na$^+$ ions arranged at the edge centre positions of the unit cell, i.e. at a distance $\sqrt{2}r$; this gives a repulsive potential energy term

$$V = +12\frac{e^2 Z_+ Z_-}{\sqrt{2}r} \qquad (8.10)$$

The third nearest neighbours are eight Cl$^-$ ions at the cube corners and distance $\sqrt{3}r$; these are attracted to the central Na$^+$ ion according to

$$V = -8\frac{e^2 Z_+ Z_-}{\sqrt{3}r} \qquad (8.11)$$

Table 8.5 *Madelung constants for some simple structures*

Structure type	A
Rock salt	1.748
CsCl	1.763
Wurtzite	1.641
Sphalerite	1.638
Fluorite	5.039
Rutile	4.816

The net attractive energy between our Na^+ ion and all other ions in the crystal is given by an infinite series:

$$V = -\frac{e^2 Z_+ Z_-}{r}\left(6 - \frac{12}{\sqrt{2}} + \frac{8}{\sqrt{3}} - \frac{6}{\sqrt{4}} + \cdots\right) \tag{8.12}$$

This summation is repeated for each ion in the crystal, i.e. for $2N$ ions per mole of NaCl. Since each ion pair interaction is thereby counted twice it is necessary to divide the final value by 2, giving

$$V = -\frac{e^2 Z_+ Z_-}{r}NA \tag{8.13}$$

where the *Madelung constant*, A, is the numerical value of the series summation in parentheses in equation (8.12). The Madelung constant depends only on the geometrical arrangement of point charges. It has the same value, 1.748, for all compounds with the rock salt structure. Values of A for some other simple structure types are given in Table 8.5.

If equation (8.13) represented the only factor involved in the lattice energy, the structure would collapse in on itself since $V \propto 1/r$ (Fig. 8.5). This catastrophe is avoided by the mutual repulsion between ions, of whatever charge, when they become too close, and is given by equation (8.8). The dependence of this repulsive force on r is also shown schematically in Fig. 8.5. The total energy of the crystal, the lattice energy U, is given by summing equations (8.8) and (8.13) and differentiating with respect to r to find the maximum U value and equilibrium interatomic distance, r_e; i.e.

$$U = -\frac{e^2 Z_+ Z_- NA}{r} + \frac{BN}{r^n} \tag{8.14}$$

Therefore,

$$\frac{dU}{dr} = \frac{e^2 Z_+ Z_- NA}{r^2} - \frac{nBN}{r^{n+1}} \tag{8.15}$$

When

$$\frac{dU}{dr} = 0$$

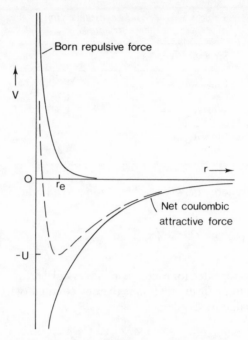

Fig. 8.5 Lattice energy (dashed line) of ionic crystals as a function of internuclear separation

then

$$B = \frac{e^2 Z_+ Z_- A r^{n-1}}{n} \tag{8.16}$$

and therefore

$$U = -\frac{e^2 Z_+ Z_- N A}{r_e}\left(1 - \frac{1}{n}\right) \tag{8.17}$$

The dashed line in Fig. 8.5 shows schematically the variation of U with r and gives the minimum U value when $r = r_e$.

For most practical purposes, equation (8.17) is entirely satisfactory, but in more refined treatments certain modifications are made:

(a) The Born repulsive energy term is better represented by an exponential function:

$$V = B \exp\left(\frac{-r}{\rho}\right) \tag{8.18}$$

where ρ is a constant, typically 0.35. When r is small ($r \ll r_e$), equations (8.8) and (8.18) give very different values for V, but for realistic interatomic distances, i.e. $r \simeq r_e$, the two values are similar. Use of equation (8.18) in the expression for U gives the *Born–Mayer equation*:

$$U = -\frac{e^2 Z_+ Z_- AN}{r_e}\left(1 - \frac{\rho}{r_e}\right) \tag{8.19}$$

(b) The zero point energy of the crystal should be included in the calculation of U. This is equal to $2.25\, hv_{O_{max}}$, where $v_{O_{max}}$ is the frequency of the highest occupied vibrational mode in the crystal. Its inclusion leads to a small reduction in U.

(c) Van der Waals attractive forces exist between ions due to induced dipole–induced dipole interactions between them. These are of the form NC/r^6 and lead to an increase in U.

A more complete equation for U, after correcting for these factors, is

$$U = -\frac{Ae^2 Z_+ Z_- N}{r} + BNe^{-r/\rho} - CNr^{-6} + 2.25Nhv_{O_{max}} \tag{8.20}$$

Typical values for these four terms, in kilojoules per mole, are (from Greenwood):

Substance	$NAe^2 Z_+ Z_- r^{-1}$	$NBe^{-r/\rho}$	NCr^{-6}	$2.25Nhv_{O_{max}}$	U
NaCl	$-\ 859.4$	98.6	$-\ 12.1$	7.1	$-\ 765.8$
MgO	$-\ 4631$	698	$-\ 6.3$	18.4	$-\ 3921$

from which it can be seen that the Born repulsive term contributes 10 to 15 per cent to the value of U whereas the zero point vibrational and van der Waals terms contribute about 1 per cent each and, being of opposite sign, tend to cancel each other out. For most purposes, therefore, we can use the simplified equation (8.17); let us consider each of the terms in equation (8.17) and evaluate their significance.

The magnitude of U depends on six parameters A, N, e, Z, n and r_e, four of which are constants for a particular ionic structure type. This leaves just two variables, the charge on the ions, Z_+, Z_-, and the internuclear separation, r_e. Of the two, the charge is by far the most important since the value of the product $(Z_+ Z_-)$ is capable of much larger variation than is r_e. For instance, a material with divalent ions should have a lattice energy that is four times as large as an isostructural crystal with the same r_e but containing monovalent ions (compare alkaline earth oxides and alkali halides, in the above table). For a series of isostructural phases with the same Z values but increasing r_e, a decrease in U is expected (e.g. alkali fluorides, alkaline earth oxides with NaCl structure). A selection of lattice energies for materials with the rock salt structure and showing these two trends is given in Table 8.6.

Since the lattice energy of a crystal is equivalent to its heat of dissociation, a correlation exists between U and the melting point of the crystal (a better

Table 8.6 *Some lattice energies* $(kJ\,mol^{-1})$. (Data from Ladd and Lee, 1963)

MgO	3938	LiF	1024	NaF	911
CaO	3566	LiCl	861	KF	815
SrO	3369	LiBr	803	RbF	777
BaO	3202	LiI	744	CsF	748

correlation may be sought between U and the sublimation energy, but such data are rarely available). The effect of $(Z_+ Z_-)$ on the melting point is shown by the refractoriness of the alkaline earth oxides (m.p. of CaO = 2572 °C) compared with the alkali halides (m.p. of NaCl = 800 °C). The effect of r_e on melting points may be seen in series such as:

$$\text{MgO}(2800\,°C), \qquad \text{CaO}(2572\,°C) \qquad \text{and} \qquad \text{BaO}(1923\,°C)$$

8.2.6 Kapustinskii's equation

Kapustinskii (1956) noted an empirical increase in the value of the Madelung constant, A, as the coordination number of the ions in the structure increased, e.g. in the series ZnS, NaCl, CsCl (Table 8.5). Since, for a particular anion and cation, r_e also increases with coordination number (e.g. Fig. 8.3), Kapustinskii proposed a general equation for U in which variations in A and r_e are auto-compensated. He suggested using the rock salt value for A and octahedral ionic radii (Goldschmidt) in calculating r_e; substituting $r_e = r_c + r_a$, $\rho = 0.345$, $A = 1.745$ and values for N and e into equation (8.19) gives the Kapustinskii equation:

$$U = \frac{1200.5\,V Z_+ Z_-}{r_c + r_a}\left(1 - \frac{0.345}{r_c + r_a}\right) \qquad kJ\,mol^{-1} \tag{8.21}$$

where V is the number of ions per formula unit (two in NaCl, three in PbF_2, etc.). This formula may be used to calculate the lattice energy of any known or hypothetical ionic compound and in spite of the assumptions involved, the answers obtained are surprisingly accurate.

The Kapustinskii equation has been used to successfully predict the stable existence of several previously unknown compounds. In cases where U is known from Born–Haber cycle calculations (see later), it has been used to derive values for ionic radii. This has been particularly useful for complex anions, e.g. SO_4^{2-}, PO_4^{3-}, whose effective size in crystals is difficult to measure by other means.

Table 8.7 *Thermochemical radii* (Å) *of complex anions.* (Data from Kapustinskii, 1956)

BF_4^-	2.28	CrO_4^{2-}	2.40	IO_4^-	2.49
SO_4^{2-}	2.30	MnO_4^-	2.40	MoO_4^{2-}	2.54
ClO_4^-	2.36	BeF_4^-	2.45	SbO_4^{3-}	2.60
PO_4^{3-}	2.38	AsO_4^{3-}	2.48	BiO_4^{3-}	2.68
OH^-	1.40	O_2^{2-}	1.80	CO_3^{2-}	1.85
NO_2^-	1.55	CN^-	1.82	NO_3^-	1.89

Radii determined in this way are known as *thermochemical radii* and some values are given in Table 8.7. It should be noted that radii for non-spherical ions such as CN^- represent gross simplifications and are only really applicable to other lattice energy calculations.

8.2.7 The Born–Haber cycle and thermochemical calculations

The lattice energy of a crystal is equivalent to its heat of formation from one mole of its ionic constituents in the gas phase:

$$Na^+(g) + Cl^-(g) \rightarrow NaCl(s), \qquad \Delta H = U$$

It cannot be measured experimentally. However, the heat of formation of a crystal, ΔH_f, can be measured relative to the reagents in their standard states:

$$Na(s) + \tfrac{1}{2}Cl_2(g) \rightarrow NaCl(s), \qquad \Delta H = \Delta H_f$$

ΔH_f may be related to U by constructing a thermochemical cycle known as a Born–Haber cycle, in which ΔH_f is given by the summation of energy terms in a hypothetical reaction pathway. For NaCl, the individual steps in the pathway, commencing with the elements in their standard states, are:

Sublimation of solid Na	$\Delta H = S$
Ionization of gaseous Na atoms	$\Delta H = IP$
Dissociation of Cl_2 molecules	$\Delta H = \tfrac{1}{2}D$
Formation of the Cl^- ion	$\Delta H = EA$
Coalescence of gaseous ions to	$\Delta H = U$
give crystalline NaCl	

Addition of these five terms is equivalent to forming crystalline NaCl from solid Na and Cl_2 molecules, as shown:

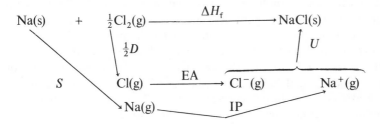

From Hess' Law,

$$\Delta H_f = S + \tfrac{1}{2}D + IP + EA + U \tag{8.22}$$

Applications. The Born–Haber cycle and equation (8.22) have various uses:

(a) Six enthalpy terms are present in equation (8.22). If all six can be determined independently for a particular compound, then the cycle gives a check on the internal consistency of the data. The values for NaCl are as follows:

S	109 kJ mol^{-1}
IP	$493.7 \text{ kJ mol}^{-1}$
$\frac{1}{2}D$	121 kJ mol^{-1}
EA	-356 kJ mol^{-1}
U	$-764.4 \text{ kJ mol}^{-1}$
ΔH_f	$-410.9 \text{ kJ mol}^{-1}$

Summation of the first five terms gives a calculated H_f of $-396.7 \text{ kJ mol}^{-1}$, which compares reasonably well with the measured ΔH_f value of $-410.9 \text{ kJ mol}^{-1}$.

(b) If only five of the energy terms are known, then the sixth may be evaluated using equation (8.22). An early application ($\sim$ 1918) was in the calculation of electron affinities, for which data were not then available.

(c) The possible stability of an unknown compound may be estimated. It is necessary to assume a structure for the compound in order to calculate U and while there are obviously errors involved, e.g. in choosing r_e, these are usually unimportant compared with the effect of some of the other energy terms involved in equation (8.22). Having estimated U, ΔH_f may then be calculated. If ΔH_f is large and positive then this explains why the compound is unknown—it is unstable relative to its elements. If ΔH_f(calc.) is negative, however, it may be worth while to try and prepare the compound under certain conditions. Examples are given in Section 8.2.8.

(d) Differences between values of lattice energies obtained by the Born–Haber cycle using thermochemical data and theoretical values calculated from an ionic model of the crystal structure may be used as evidence for non-ionic bonding effects. Data for the silver halides (Table 8.8) and for thallium and copper halides (not given) show that the differences between the two lattice energies are least for the fluorides and greatest for the iodides. This is attributed to the presence of a strong covalent contribution to the bonding in the iodides which leads to an increase in the values of the thermochemical lattice energy. A correlation also exists between the insolubility of the Ag salts, especially AgI, in water and the presence of partial covalent bonding. Data for the corresponding alkali halides show that the differences between thermochemical and calculated lattice energies are small and indicate that the ionic bonding model may be applied satisfactorily to them.

While covalent bonding is present in, for example, AgCl and AgBr, as evidenced by the lattice energies, it is not strong enough to change the crystal structure from that of rock salt to one of lower coordination number. AgI does have a different structure, however; it is polymorphic and can exist in at least three structure types, all of which have low coordination numbers, usually four. Changes in structure type and coordination number due to increased covalent bonding are described in Section 8.1.2.

Table 8.8 *Lattice energies* $(kJ\,mol^{-1})$ *of some Group I halides.*
(Data from Waddington, 1959)

	U_{calc}	$U_{Born-Haber}$	ΔU
AgF	920	953	33
AgCl	832	903	71
AgBr	815	895	80
AgI	777	882	105

(e) Certain transition metals have crystal field stabilization energies due to their d electron configuration (Section 8.6.1.1) and this gives an increased lattice energy in their compounds. For example, the difference between experimental and calculated lattice energies in CoF_2 is $83\,kJ\,mol^{-1}$, which is in fair agreement with the CFSE value calculated for the high spin state of Co in CoF_2 of $104\,kJ\,mol^{-1}$. Ions which do not exhibit CFSE effects are those with configurations d^0 (e.g. Ca^{2+}), d^5 high spin (e.g. Mn^{2+}) and d^{10} (e.g. Zn^{2+}). A further discussion is given in Section 8.6.1.1.

(f) The Born–Haber cycle has many other uses, e.g. in solution chemistry to determine energies of complexation and hydration of ions. These usually require a knowledge of the lattice energy of the appropriate solids, but since these applications do not provide any new information about solids, they are not discussed further.

8.2.8 Stabilities of real and hypothetical compounds

8.2.8.1 *Inert gas compounds*

One may ask, is it worth while trying to synthesize, for example, ArCl? Apart from ΔH_f, the only unknown in equation (8.22) is U. Suppose that hypothetical ArCl had the rock salt structure and the radius of the Ar^+ ion is between that of Na^+ and K^+. An estimated lattice energy for ArCl is, then, $-745\,kJ\,mol^{-1}$ ($NaCl = -764.4$; $KCl = -701.4$). Substitution in equation (8.22) gives, in kilojoules per mole:

S	$\frac{1}{2}D$	IP	EA	U	ΔH_f(calc.)
0	121	1524	-356	-745	$+544$

from which it can be seen that ArCl has a large positive heat of formation and would be thermodynamically unstable, by a large amount, relative to the elements.

Such a calculation also tells us *why* ArCl is unstable and cannot be synthesized. Comparing the calculations for ArCl and NaCl, it is clear that the instability of ArCl is due to the very high ionization potential of argon (stability is strictly governed by free energies of formation, but ΔS is small and hence $\Delta G \simeq \Delta H$). The heats of formation calculated for several other hypothetical compounds are given in Table 8.9.

Table 8.9 *Enthalpies of formation* $(kJ\,mol^{-1})$ *of some hypothetical* (*) *and real compounds*

HeF*	+ 1066	NeCl*	+ 1028	CsCl$_2^*$	+ 213	CuI$_2$	− 21
ArF*	+ 418	NaCl	− 411	CsF$_2^*$	− 125	CuBr$_2$	− 142
XeF*	+ 163	MgCl*	− 125	AgI$_2^*$	+ 280	CuCl$_2$	− 217
MgCl$_2$	− 639	AlCl*	− 188	AgCl$_2$	+ 96	CuF$_2$	− 890
NaCl$_2^*$	+ 2144	AlCl$_3$	− 694	AgF$_2$	− 205		

There is now a large number of inert gas compounds known, following on from the preparation of XePtF$_6$ by Bartlett in 1962. Consideration of lattice energies and enthalpies of formation led Bartlett to try and synthesize XePtF$_6$ by direct reaction of Xe and PtF$_6$ gases. He had previously prepared O$_2$PtF$_6$ as an ionic salt, $(O_2)^+(PtF_6)^-$, by reacting (by accident) O$_2$ with PtF$_6$. From a knowledge of the similarity in the first ionization potentials of molecular oxygen $(1176\,kJ\,mol^{-1})$ and xenon $(1169\,kJ\,mol^{-1})$, he reasoned, correctly, that the corresponding xenon compound should be stable.

8.2.8.2. *Lower and higher valence compounds*

Consider alkaline earth compounds. In these, the metal is always divalent. Since a great deal of extra energy is required to doubly ionize the metal atoms, it is reasonable to ask why monovalent compounds, such as MgCl, do not form. Data in Table 8.9 show that MgCl is indeed stable relative to the elements (ΔH(calc.) = $-125\,kJ\,mol^{-1}$) but that MgCl$_2$ is much more stable ($\Delta H = -639\,kJ\,mol^{-1}$). This is shown by the following sequence:

$$2Mg + Cl_2 \xrightarrow[\;\;-639\;\;]{\overset{-250}{} 2MgCl \xrightarrow{-389}} Mg + MgCl_2$$

In any attempt to synthesize MgCl, attention should be directed towards keeping the reaction temperature low and/or isolating the MgCl product, in order to prevent it from reacting further or disproportionating. Similar trends are observed for other hypothetical compounds such as ZnCl, Zn$_2$O, AlCl and AlCl$_2$.

From a consideration of the factors that affect the stability of compounds, the following conclusions may be drawn about compounds with metals in unusual oxidation states:

(a) The formation and stability of compounds with lower than normal valence states appears to be favourable when (i) the second, and higher, ionization potentials of the metal are particularly high and (ii) the lattice energy of the corresponding compounds with the metal in its normal oxidation state is reduced.

(b) Conversely, in order to prepare compounds in which the metal has a higher than normal oxidation state and in which it is probably necessary to break into a closed electron shell, it is desirable to have (i) low values for the second (or higher) ionization potential of the metal atoms and (ii) large lattice energies of the resulting higher valence compounds.

As examples of these trends, calculations for the alkaline earth monohalides show that while all are unstable relative to the dihalides, the enthalpy of disproportionation is least in each case for the iodide ($U_{(MI_2)} < U_{(MBr_2)}$, etc.—effect a, ii). Higher valence halogen compounds of the Group I elements are most likely to occur with caesium and the copper subgroup elements (effect b, i), in combination with fluorine (effect b, ii). Thus, from Table 8.9, all caesium dihalides, apart from CsF_2, have positive ΔH_f values and would be unstable. CsF_2 is stable in principle, with its negative ΔH_f value, but has not been prepared because its disproportionation to CsF has a large negative ΔH:

$$Cs + F_2 \xrightarrow[\qquad\qquad -530 \qquad\qquad]{\quad -125 \quad} CsF_2 \xrightarrow{\quad -405 \quad} CsF + \tfrac{1}{2}F_2$$

For the silver dihalides, ΔH_f becomes less positive and finally negative across the series AgI_2 to AgF_2. This again shows the effect of r_e on U and hence on ΔH_f (effect b, ii). Unlike CsF_2, AgF_2 is a stable compound since AgF and AgF_2 have similar enthalpies of formation and therefore the disproportionation enthalpy of AgF_2 (to $AgF + \tfrac{1}{2}F_2$) is ~ zero.

The copper halides are particularly interesting. The divalent state of copper, in which the d^{10} shell of copper is broken into, is the most common state and again the dihalides show decreasing stability across the series CuF_2 to CuI_2 (Table 8.9): CuI_2 appears not to exist and its calculated ΔH_f is barely negative. On the other hand, in the monovalent state the situation is reversed and all the halides apart from CuF are known. CuF is calculated to be stable relative to the elements but not relative to CuF_2:

$$2Cu + F_2 \xrightarrow[\qquad\qquad -127 \qquad\qquad]{\quad -36 \quad} 2CuF \xrightarrow{\quad -91 \quad} CuF_2 + Cu$$

The above examples show that several factors affect the formulae and stability of compounds: ionization potentials, lattice energies (via internuclear distances and the charges on ions) and the relative stability of elements in different oxidation states. Often a delicate balance between opposing factors controls the stability or instability of a compound and, as with the copper halides, detailed calculations are needed to assess the factors involved.

8.2.9 Polarization and partial covalent bonding

Covalent bonding, partial or complete, occurs when the outer electronic charge density on an anion is polarized towards and by a neighbouring cation. The net effect is that an electron pair which would be associated entirely with the anion in a purely ionic structure is displaced to occur between the anion and cation such that the electron pair is common to both. In cases of partial covalent bonding, some of the electron density is common to both atoms but the rest is still associated with the more electronegative atom.

The occurrence of partial covalent bonding in an otherwise ionic structure may sometimes be detected by abnormally large values of the lattice energy (Section 8.2.7). In other cases, it is clear from the nature of the structure and the coordination numbers of the atoms that the bonding cannot be purely ionic (Section 8.1.2). Until comparatively recently, it has been difficult to quantify the degree of partial covalence in a particular structure, although one has intuitively felt that most so-called, ionic structures must have a considerable degree of covalent character. Two new approaches which have had considerable success are the coordinated polymeric model of Sanderson and the ionicity plots of Mooser and Pearson.

8.3 Coordinated polymeric structures—Sanderson's model

A fresh approach to the theory and description of non-molecular crystal structures has been developed by Sanderson (1967, 1976). Basically, he regards all bonds in non-molecular crystal structures as being polar covalent. The atoms contain partial charges whose values may be calculated readily using a new scale of electronegativities developed by Sanderson. From the partial charges, the relative contributions of ionic and covalent bonding to the total bond energy may be estimated. Since all non-molecular crystals contain only partially charged atoms, the ionic model represents an extreme form of bonding that is not attained in practice. Thus in, for example, KCl, the charges on the atoms are calculated to be ± 0.76 instead of ± 1.0 for a purely ionic structure.

The essential features of this new approach to non-molecular solids, which is part of a more general and widely applicable theory of bonding, are as follows. The features of an atom which control its physical and chemical properties are (a) its electronic configuration and (b) the effective nuclear charge felt by the valence electrons. It is recognition of the importance of the latter effect that provides the starting point for Sanderson's theory.

8.3.1 Effective nuclear charge

The *effective nuclear charge* of an atom is the positive charge that would be felt by a foreign electron on arriving at the periphery of the atom. Atoms are, of course, electrically neutral overall but, nevertheless, the valence electrons of an atom are not very effective in shielding the outside world from the positive charge on the nucleus of the atom. Consequently, an incoming electron (e.g. an electron that belongs to a neighbouring atom and is coming to investigate the possibility of bond formation) feels a positive, attractive charge. Were this not the case and the surface of an atom was completely shielded from the nuclear charge, then atoms would have zero electron affinity and no bonds, ionic or covalent, would ever form.

The effective nuclear charge is greatest in elements which have a single vacancy in their valence shell, i.e. in the halogens. In the inert gases, the outermost electron shell is full and no foreign electrons can enter it. Consequently, incoming electrons would have to occupy vacant orbitals which are essentially 'outside' the atom and the effective nuclear charge experienced in such orbitals would be greatly reduced. Calculations of 'screening constants' were made by Slater who found that outermost electrons are much less efficient at screening nuclear charge than are electrons in inner shells. The screening constants of outermost electrons are calculated to be approximately one third; this means that for each unit increase in atomic number and positive nuclear charge across the series, e.g. sodium to chlorine, the additional positive charge is screened by only one third. Therefore, the effective nuclear charge increases in steps of two thirds and the valence electrons experience an increasingly strong attraction to the nucleus on going from sodium to chlorine. Similar effects occur throughout the periodic table: the effective nuclear charge is small for the alkali metals and increases to a maximum in the halogens.

Many atomic properties may be correlated with effective nuclear charge:

(a) Ionization potentials gradually increase from left to right across the periodic table.
(b) Electron affinites become increasingly negative in the same direction.
(c) Atomic radii progressively decrease from left to right.
(d) Electronegativities increase from left to right.

Let us consider two of these properties, atomic radii and electronegativities, in more detail.

8.3.2 Atomic radii

Atomic radii vary considerably for a particular atom depending on bond type and coordination number and many tabulations of radii are available; indeed, the subject of ionic radii is still a controversial subject. Fortunately, however, the *non-polar covalent radii* of atoms can be measured accurately and represent a point of reference with which to compare other radii. Thus the atomic, non-polar covalent radius of carbon, given by half the C—C single bond length is constant at 0.77 Å in materials as diverse as diamond and gaseous, paraffin hydrocarbons. Non-polar radii of atoms are listed in Table 8.10; from these radii, it is possible, using Sanderson's method, to estimate the effect that partial charges on atoms have on their radii. The general trends are that with increasing amounts of partial positive charge, the radii become smaller (i.e. as electrons are removed from the valence shell, but with the nuclear charge unchanged, so the remaining valence shell electrons feel a stronger attraction to the nucleus and the atom contracts). Conversely, with increasing negative charge on the atom, the radius becomes larger. Sanderson developed a simple, empirical formula to quantify the variation of radius, r, with partial charge:

$$r = r_c - B\delta \qquad (8.23)$$

where r_c is the non-polar covalent radius, B is a constant for a particular atom and δ is its partial charge. B values are also given in Table 8.10. The partial

Table 8.10 *Some electronegativity and size parameters of atoms.* (*After Sanderson,* 1976)

Element	S	$r_c(\text{Å})$	$B(\text{solid})$	ΔS_c	$r_i(\text{Å})$
H	3.55	0.32		3.92	
Li	0.74	1.34	0.812	1.77	0.53
Be	1.99	0.91	0.330	2.93	0.58
B	2.93	0.82		2.56	
C	3.79	0.77		4.05	
N	4.49	0.74		4.41	
O	5.21	0.70	4.401	4.75	1.10
F	5.75	0.68	0.925	4.99	1.61
Na	0.70	1.54	0.763	1.74	0.78
Mg	1.56	1.38	0.349	2.60	1.03
Al	2.22	1.26		3.10	
Si	2.84	1.17		3.51	
P	3.43	1.10		3.85	
S	4.12	1.04	0.657	4.22	1.70
Cl	4.93	0.99	1.191	4.62	2.18
K	0.42	1.96	0.956	1.35	1.00
Ca	1.22	1.74	0.550	2.30	1.19
Zn	2.98			3.58	
Ga	3.28			3.77	
Ge	3.59	1.22		3.94	
As	3.90	1.19		4.11	
Se	4.21	1.16	0.665	4.27	1.83
Br	4.53	1.14	1.242	4.43	2.38
Rb	0.36	2.16	1.039	1.25	1.12
Sr	1.06	1.91	0.429	2.14	1.48
Ag	2.59	1.50	0.208		1.29
Cd	2.84	1.46	0.132	3.35	1.33
Sn	3.09	1.40		3.16, 3.66	
Sb	3.34	1.38		3.80	
Te	3.59	1.35	0.692	3.94	2.04
I	3.84	1.33	1.384	4.08	2.71
Cs	0.28	2.35	0.963	1.10	1.39
Ba	0.78	1.98	0.348	1.93	1.63
Hg	2.93			3.59	
Tl	3.02	1.48		2.85	
Pb	3.08	1.47		3.21, 3.69	
Bi	3.16	1.46		3.74	

charges on atoms cannot be measured directly but may be estimated from Sanderson's electronegativity scale, as follows.

8.3.3 Electronegativity and partially charged atoms

The electronegativity of an atom is a measure of the net attractive force experienced by an outermost electron interacting with the nucleus (Gordy, Allred and Rochow). The concept of electronegativity was originated by Pauling as a parameter that would correlate with the observed polarity of bonds between unlike atoms. Atoms of high electronegativity attract electrons (in a covalent bond) more than do atoms of low electronegativity and, hence, they acquire a partial negative charge. The magnitude of the partial charge depends on the initial difference in electronegativity between the two atoms. The difficulty in working with electronegativities in the past has been that electronegativity has not had a precise definition and, therefore, it has not been possible to use an absolute scale of electronegativities and make accurate calculations. Pauling observed a correlation between the strengths of polar bonds and the degree of polarity in the bonds. He proposed that bond strengths are a combination of (a) a homopolar bond energy and (b) an 'extra ionic energy' due to bond polarity and, hence, electronegativity difference. He then used this correlation between polarity and extra ionic energy to establish his scale of electronegativities. These ideas have since been extended and modified by Sanderson to permit quantitative calculations of bond energies to be made for a wide variety of compounds. However, Sanderson used a different method to derive a scale of electronegativities. Since electronegativity is a measure of the attractive force between the effective nuclear charge and an outermost electron, it is related to the compactness of an atom. He used the relation:

$$S = \frac{D}{D_a} \tag{8.24}$$

to evaluate the electronegativity, S, in which D is the electron density of the atom (given by the ratio of atomic number: atomic volume) and D_a is the electron density that would be expected for the atom by linear interpolation of the D values for the inert gas elements. The electronegativity values so obtained are listed in Table 8.10, with some minor modifications made by Sanderson.

An important contribution to our understanding of bond formation is *the principle of electronegativity equalization* proposed by Sanderson. It may be stated: 'When two or more atoms initially different in electronegativity combine chemically, they become adjusted to the same intermediate electronegativity within the compound.' It is found in practice that the value of the intermediate electronegativity is given by the geometric mean of the individual electronegativities of the component atoms, e.g. for NaF:

$$S_b = \sqrt{S_{Na} S_F} = 2.006 \tag{8.25}$$

This means that in a bond between unlike atoms, the bonding electrons are preferentially and partially transferred from the less electronegative to the more electronegative atom. A partial negative charge on the more electronegative atom results whose magnitude is defined, by Sanderson, as follows: '*Partial charge* is defined as the ratio of the change in electronegativity undergone by an atom on bond formation to the change it would have undergone on becoming completely ionic with charge + or − 1.'

In order to make calculations of partial charges, a point of reference was necessary and for this it was assumed that the bonds in NaF are 75 per cent ionic;

this assumption appears to have been well justified by subsequent developments. Further, it was necessary to assume that electronegativity changes linearly with charge. It can then be shown that the change in electronegativity, ΔS_c, of any atom on acquiring a unit positive or unit negative charge is given by

$$\Delta S_c = 2.08\sqrt{S} \qquad (8.26)$$

Partial charge, δ, may therefore be defined as

$$\delta = \frac{\Delta S}{\Delta S_c} \qquad (8.27)$$

where $\Delta S = S - S_b$. Values of ΔS_c are also given in Table 8.10.

Returning now to the radii of partially charged atoms, the problem in assigning individual radii to atoms or ions has been the question of how to divide an experimentally observed internuclear distance into its component radii. Many methods have been tried and various tabulations of radii are available. All of them are additive in that they correctly predict bond distances. The method adopted by Pauling (and Sanderson) has been to divide the experimental internuclear distance in an isoelectronic 'ionic' crystal, such as NaF, according to the inverse ratio of the effective nuclear charges on the two 'ions'. The effective nuclear charges can be calculated from the screening constants. From a knowledge of the partial charges on the atoms and assuming that radii change systematically with partial charge (according to the relation $r = r_c - B\delta$), the ionic radii of Na^+ and F^- may be calculated; these then serve as a point of reference for calculating radii of other ions in materials that are not isoelectronic. In Table 8.10 are listed the radii of singly charged ions in the solid state calculated by the above method; also given are B and r_c data to enable the radii of partially charged atoms to be calculated and electronegativity data S, ΔS_c to enable the partial charges, δ, to be calculated.

Let us consider the use of these data for one example, BaI_2. From Table 8.10, $S_{Ba} = 0.78$ and $S_I = 3.84$; the intermediate electronegativity, S_b, is given by $S_b = \sqrt[3]{S_{Ba}S_I^2} = 2.26$. Therefore, for barium, $\Delta S = 2.26 - 0.78 = 1.48$ and for iodine, $\Delta S = 3.84 - 2.26 = 1.58$. The values of ΔS_c are 1.93 (Ba) and 4.08(I). Hence $\delta_{Ba} = 1.48/1.93 = 0.78$ and $\delta_I = 1.58/4.08 = -0.39$, i.e. BaI_2 is ~ 39 per cent ionic, 61 per cent covalent (using the δ_I value). The radii of the partially charged atoms may now be calculated. For barium, $r_c = 1.98$, $B = 0.348$ and δ is calculated to be 0.78; hence $r_{Ba} = r_c - B\delta = 1.71\,\text{Å}$. For iodine, $r_c = 1.33$, $B = 1.384$ and $\delta = -0.39$; hence $r_I = 1.87\,\text{Å}$. Therefore, the barium–iodine distance is calculated to be $1.87 + 1.71 = 3.58\,\text{Å}$, which compares very well with the experimental value of $3.59\,\text{Å}$.

Using the above methods, Sanderson has evaluated the partial charges and atomic radii in a large number of solid compounds. For example, data are given in Table 8.11 for a number of mono- and divalent chlorides; the charge on the chloride atom varies from -0.21 in $CdCl_2$ to -0.81 in CsCl and at the same time the calculated radius of the chlorine atom varies from 1.24 to $1.95\,\text{Å}$. These

Table 8.11 *Partial charge and radius of the chlorine atom in some solid chlorides*

Compound	$-\delta_{Cl}$	$r_{Cl}(\text{Å})$
$CdCl_2$	0.21	1.24
$BeCl_2$	0.28	1.26
$CuCl$	0.29	1.34
$AgCl$	0.30	1.35
$MgCl_2$	0.34	1.39
$CaCl_2$	0.40	1.47
$SrCl_2$	0.43	1.50
$BaCl_2$	0.49	1.57
$LiCl$	0.65	1.76
$NaCl$	0.67	1.79
KCl	0.76	1.90
$RbCl$	0.78	1.92
$CsCl$	0.81	1.95

compare with a non-polar covalent radius of 0.99 Å and an ionic radius of 2.18 Å (Table 8.10). While these radii and partial charges may not be quantitatively correct, because some of the assumptions involved in their calculation are rather empirical, they nevertheless appear to be realistic. Most of the compounds in Table 8.11 are normally regarded as ionic, and if the data are in any way correct, they clearly show that it is unrealistic and misleading to assign a radius to the chloride ion which is constant for all solid chlorides.

A similar but more extensive list of the partial charge on oxygen in a variety of oxides in given in Table 8.12, where values cover almost the entire range between 0 and -1. Although oxides are traditionally regarded as containing the oxide

Table 8.12 *Partial charge on oxygen in some solid oxides*

Compound	$-\delta_O$	Compound	$-\delta_O$	Compound	$-\delta_O$	Compound	$-\delta_O$
Cu_2O	0.41	HgO	0.27	Ga_2O_3	0.19	CO_2	0.11
Ag_2O	0.41	ZnO	0.29	Tl_2O_3	0.21	GeO_2	0.13
Li_2O	0.80	CdO	0.32	In_2O_3	0.23	SnO_2	0.17
Na_2O	0.81	CuO	0.32	B_2O_3	0.24	PbO_2	0.18
K_2O	0.89	BeO	0.36	Al_2O_3	0.31	SiO_2	0.23
Rb_2O	0.92	PbO	0.36	Fe_2O_3	0.33	MnO_2	0.29
Cs_2O	0.94	SnO	0.37	Cr_2O_3	0.37	TiO_2	0.39
		FeO	0.40	Sc_2O_3	0.47	ZrO_2	0.44
		CoO	0.40	Y_2O_3	0.52	HfO_2	0.45
		NiO	0.40	La_2O_3	0.56		
		MnO	0.41				
		MgO	0.50				
		CaO	0.56				
		SrO	0.60				
		BaO	0.68				

ion, O^{2-}, the calculations show that the actual charge carried by an oxygen never exceeds -1 and is usually much less than -1.

These ideas and calculations on partial charges which have been developed by Sanderson enable many correlations to be made between partial charge and chemical properties. To give one example, the acidic, amphoteric and basic properties of oxides correlate nicely with the partial charge on the oxide ion and the changeover in behaviour occurs with a partial charge of ~ -0.30. We are concerned here, however, with the structures of solids. Since the ionic model appears to be inappropriate or only partially correct for most solids, Sanderson proposed the *coordinated polymeric model* for solids. This is essentially a blend of the two extreme forms of bonding: ionic and covalent.

8.3.4 Coordinated polymeric structures

From the principle of electronegativity equalization, electrons in a hypothetical covalent bond are partially transferred to the more electronegative atom. This removal of electrons from the electropositive atom leads to an increase in its effective nuclear charge, decrease in its size and hence an increase in its effective electronegativity. Likewise, as the electronegative atom acquires electrons, so its ability to attract still more electrons diminishes and its electronegativity decreases. In this way the electronegativities of the two atoms adjust themselves until they are equal. This principle of electronegativity equalization may be applied equally to diatomic gas molecules, in which only one bond is involved, or to three-dimensional solid structures in which each atom is surrounded by and bonded to several others. This argument illustrates how covalent bonds that are initially non-polar may become polar due to the electronegativity equalization. An alternative approach is to start with purely ionic bonding and consider how the bonds may acquire some covalent character. In an ionic structure, M^+X^-, the cations are surrounded by anions (usually 4, 6 or 8). However, the cations have empty valence shells and are potential electron pair acceptors; likewise, anions with their filled valence shells are potential electron pair donors. The cations and anions therefore interact in the same way as do Lewis acids and bases: the anions, with their lone pairs of electrons, coordinate to the surrounding cations. The strength of this interaction, and hence the degree of covalent bonding which results, is again related to the electronegativities of the two atoms. Thus electronegative cations such as Al^{3+} are much stronger electron pair acceptors (and Lewis acids) than are electropositive cations such as K^+. The coordinated polymeric model of structures proposed by Sanderson is based on this idea of acid–base interactions between ions. It therefore forms a bridge between the ionic and covalent extremes of bond type.

8.3.5 Bond energy calculations

In most solids, the bonds are a blend of covalent and ionic. For simplicity and convenience in calculations, it is possible to regard the bonds as being wholly ionic for part of the time and wholly covalent for the remainder of the time. The relative proportions of the two components are then directly related to the partial charges on the atoms. In Pauling's view of polar bonds, the ionic contribution to the bond energy, due to the electronegativity difference between the atoms, is regarded as *adding to* the energy of a 100 per cent covalent bond. However, Sanderson argues that the ionic contribution *replaces* part of the covalent contribution; since ionic bonds always have higher energy than covalent bonds this automatically leads to an increasing bond energy with increasing bond polarity.

The ionic contribution to bond energy may be calculated using the Born–Mayer equation (8.19) of Section 8.2.5; the lattice energy for a purely ionic crystal is calculated in this way and multiplied by the fractional ionic character of the bonds. Covalent bond energies are estimated as follows.

For a homopolar, covalent bond between like atoms, Sanderson proposed the relation:

$$E = CrS \qquad (8.28)$$

between the covalent bond energy, E, the electronegativity, S, and non-polar covalent radius, r, of the atoms; C is an empirical constant. The value of C is the same across, for example, the series, Li...F, for which a plot of E against (rS) is shown in Fig. 8.6. The linearity of the plot was shown by the experimental data for Li, Be, B and C; these values were obtained from the dissociation energies of gas phase molecules such as Li_2. By extrapolation, single bond energies of N, O and F were then estimated. It is known that the experimentally determined single bond energies of these atoms, N, O and F, are anomalous, e.g. the F_2 bond

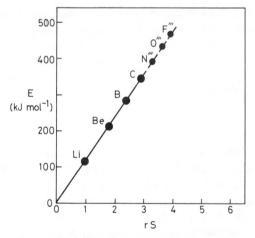

Fig. 8.6 Plot of homopolar bond energy, E, against rS, the product of non-polar covalent radius, r, and Sanderson electronegativity, S. E values for N, O and F are obtained by extrapolation

dissociation energy is much less than expected by comparison with the values for the other halogens. The experimental bond dissociation energies for O, N and F are reduced by what Sanderson calls the 'lone pair bond weakening effect' (Section 8.3.6).

From Fig. 8.6 it is possible to estimate bond energies for these elements which would be free from the lone pair bond weakening effect. Thus the value for fluorine, F''', extrapolated from Fig. 8.6 is 466 kJ mol^{-1} whereas the experimental value obtained from the dissociation of the F_2 molecule is only 165 kJ mol^{-1}. Using a similar procedure, single covalent bond energies have been estimated for most of the main group elements in cases where such data were not already available.

The next step is to calculate the covalent bond energy, E_c of a heteronuclear bond. Sanderson used Pauling's equation, but added a correction factor for bond length: for a bond between atoms A and B, in a molecule,

$$E_c = \frac{R_c}{R_0} \sqrt{E_{AA} E_{BB}} \tag{8.29}$$

i.e. the bond energy is the geometric mean of the two homonuclear bond energies, E_{AA} and E_{BB}, multiplied by the ratio R_c/R_0 in which R_c is the sum of the covalent radii (tabulated) and R_0 is the observed bond length.

For crystalline solids, a correction factor is needed for the total number of electron pairs, n, that are involved in the bonding. Thus, in gaseous NaCl molecules, there is only one electron pair and one bond to be considered, whereas in crystalline NaCl each atom is coordinated to six others and, on the coordinated polymeric model of bonding, there are four electron pairs to be distributed over the six bonds around any particular atom. For most solid oxides and halides on which calculations have been made, use of $n = 4$ gives good agreement between experimental and calculated bond energies. In some cases, especially with the smaller atoms, and for reasons which are not understood, it appears necessary to use $n = 3$ and occasionally $n = 6$.

The total bond energy, E, of a crystalline solid is given by the sum of the ionic and covalent contributions, i.e.

$$E = t_c E_c n + \frac{t_i U}{f} \tag{8.30}$$

where t_c and t_i are the fractional covalent and ionic characters, respectively, in the bonds. The term U/f is a modified lattice energy; part of the difficulty in making bond energy calculations for ionic solids is that the energy required to completely dissociate the solid into the gas phase corresponds to the atomization energy of the solid. This is not the same as the lattice energy of the solid, which refers to the energy of the crystal relative to *ions* in the gas phase. However, by having a correction factor, f, in equation (8.30) which is constant at $f = 1$ for halides and at $f = 0.63$ for oxides, Sanderson obtained good agreement between experimental and calculated bond energies using equation (8.30). Some examples are given in Table 8.13.

Table 8.13 *Bond energies*

Compound	n	$-\delta_x^*$	E_{calc} (kJ mol^{-1})	E_{exp} (kJ mol^{-1})
Li_2O	6	0.80	1172	1168
Na_2O	4	0.80	851	881
K_2O	4	0.84	777	791
Rb_2O	4	0.86	762	743
Cs_2O	4	0.90	721	723
BeO	4	0.42	1174	1175
MgO	4	0.50	1020	1040
CaO	4	0.57	1038	1061
SrO	4	0.60	1007	1002
BaO	4	0.67	958	982
LiF	3	0.74	869	852
NaF	3	0.75	766	760
KF	4	0.84	736	735
RbF	4	0.86	711	710
CsF	4	0.90	688	688
BeF_2	3	0.34	1462	1502
MgF_2	4	0.41	1499	1471
CaF_2	4	0.47	1546	1549
SrF_2	4	0.50	1522	1534
BaF_2	4	0.57	1532	1533

*Partial charge on anion.

8.3.6 Bond energies and structure

In order to illustrate how bond energy may influence the structure adopted by a particular compound some clear-cut examples are needed. Consider the long-standing question mark over the enormous difference in the structure of CO_2 and SiO_2. One is a gas, the other a high melting solid. And yet carbon and silicon are adjacent elements in Group IV of the periodic table and are expected to show many similarities in their physical and chemical properties. Clearly, CO_2 must be more stable as a molecule that contains two carbon–oxygen double bonds than as some other structure that contains four carbon–oxygen single bonds.

Although double bonds are stronger than single bonds they are not normally regarded as being at least twice as strong; however, clearly, they must be so in the case of CO_2. Why should this be? Sanderson's calculations indicate the answer to this and other questions, only a brief summary of which can be given here.

As mentioned in Section 8.3.5, the elements N, O and F exhibit the 'lone pair bond weakening effect' such that their single bond energies are considerably less than expected. The explanation that has previously been given is that the presence of lone pairs on adjacent atoms causes a kind of steric hindrance and hence repulsion between the atoms which leads to a weakening of the bond. Sanderson argues that this explanation must be incorrect—e.g. the N—H bonds in ammonia are also weakened but the hydrogen atom has no lone pairs—and that, instead, the lone pairs must act towards screening off the effective nuclear charge of the atoms from the valence electrons. This leads to a reduction in bond energy by effectively reducing the electronegativity of the atoms (equation 8.28). Using correlations such as shown in Fig. 8.6, unweakened single bond energies were estimated. For oxygen this value, E''', 434 kJ mol^{-1}, is about three times the value of the oxygen–oxygen single bond energy determined experimentally from the dissociation of H_2O_2, 142 kJ mol^{-1}. Further, the dissociation energy of oxygen gas molecules, requiring cleavage of the double bond is 498 kJ mol^{-1}.

Recognition of the lone pair bond weakening effect in N, O and F leads to explanations of many phenomena:

(a) Oxygen is a diatomic gas whereas sulphur is a polymeric solid: one $O{=}O$ double bond is stronger than two O—O single bonds. The lone pair bond weakening effect is much reduced in sulphur (and P, Cl, etc.), for which two single bonds are stronger than one double bond.

(b) Similarly, N_2 is a diatomic gas whereas P is a polymeric solid.

(c) The bond dissociation energy of fluorine is much less than that of chlorine: here there is no possibility of forming double bonds and direct comparison between the halogen single bond energies is possible. The reduced F—F bond energy accounts in large part for the high reactivity of fluorine.

(d) Carbon dioxide is a molecular gas whereas silica is a polymeric solid. Using data on the electronegativities and double bond energies of carbon and oxygen, the total dissociation energy of CO_2 was calculated by Sanderson, using the sequence of steps outlined above, to be 1608 kJ mol^{-1}. This is in close agreement with the experimental atomization energy of CO_2. On the other hand, for a hypothetical polymeric structure containing four C—O single bonds per carbon, the calculated dissociation energy is less, 1413 kJ mol^{-1}. Hence CO_2 prefers to exist as a triatomic molecule. (The dissociation energy of O_2 is 498 kJ mol^{-1}; this value is also regarded as being reduced, by the presence of one lone pair on each oxygen. The unweakened double bond energy is estimated as 574 kJ mol^{-1} and hence the value per oxygen atom, E'', is half, i.e. 287 kJ mol^{-1}.)

For SiO_2, the lone pair bond weakening effect is less important for two reasons: (i) the bond weakening effect influences only the covalent contribution and as silicon is more electropositive than carbon the Si—O bond is more ionic than the C—O bond; (ii) silicon has vacant d orbitals which allow partial double bonding to occur by donation of the lone pair of electrons on oxygen. Hence the single bond energy of oxygen is less weakened when bonded to silicon than when bonded to carbon. For calculations on SiO_2, Sanderson found it necessary to use the E'' value of 287 kJ mol^{-1} in order to obtain a dissociation energy for polymeric SiO_2 of 1859 kJ mol^{-1}, which was in good agreement with the experimental value. By contrast, calculations on the 'SiO_2 molecule' containing $Si{=}O$ double bonds gave a much smaller dissociation energy, 1264 kJ mol^{-1}. Hence SiO_2 shows a clear preference for a polymeric structure with Si—O single bonds.

8.3.7 Some final comments on Sanderson's approach

Sanderson has extended his semi-empirical calculations on electronegativities and partial charges of atoms to the calculation of bond energies in polar materials (Section 8.3.6). This is a considerable advance on what has been possible using other methods. However, it has not been possible to make these calculations on a purely *ab initio* basis and the values of certain parameters have been chosen so as to give good agreement between experiment and calculation; one example is the n value used in covalent bond energy calculations on solids. Sanderson acknowledges that his calculations suffer from certain drawbacks and that, to a certain extent, some of his arguments are circular. However, these disadvantages appear to be more than outweighed by the increased understanding of bonding in solid materials which has resulted from these calculations and in particular by the quantization of concepts such as partial charge on atoms, bond ionicities and electronegativity equalization.

8.4 Mooser–Pearson plots and ionicities

While the radius ratio rules appear to be rather unsatisfactory for predicting and explaining the structure adopted by, for instance, particular AB compounds, an alternative approach by Mooser and Pearson (1959) has had considerable success. This approach focuses on the directionality or covalent character of bonds. The two factors which are regarded as influencing covalent bond character in crystals are (a) the average principle quantum number, $\bar{n}$, of the atoms involved and (b) their difference in electronegativity, Δx. In constructing

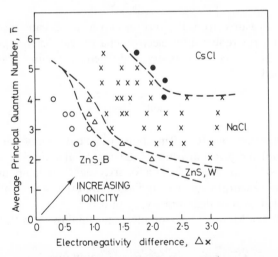

Fig. 8.7 Mooser–Pearson plot for AB compounds containing A group cations. (From Mooser and Pearson, 1959.) Arrow indicates direction of increasing bond ionicity

Mooser–Pearson plots, these two parameters are plotted against each other, as shown for AB compounds in Fig. 8.7. The most striking feature of the plot is the almost perfect separation of compounds into four groups corresponding to the structure: zinc blende (ZnS, B), wurtzite (ZnS, W), rock salt and CsCl. Similarly, successful diagrams have been presented for other formula types—AB_2, AB_3, etc. Of the four simple AB structure types, one's intuitive feeling is that zinc blende (or sphalerite) is the most covalent and either rock salt or CsCl is the most ionic. Mooser–Pearson plots present this intuitive feeling in diagrammatic form.

The term *ionicity* has been used to indicate the degree of ionic character in bonds; in Fig. 8.7, ionicity increases from the bottom left to the top right of the diagram, as shown by the arrow. Thus it appears that ionicity is not governed by electronegativity alone but also depends on the principal valence shell of the atoms and hence on atomic size. There is a general trend for highly directional covalent bonds to be associated with lighter elements, i.e. at the bottom of Fig. 8.7, and with small values of Δx, i.e. at the left-hand side of Fig. 8.7.

The fairly sharp crossover between the different 'structure fields' in Fig. 8.7 suggests that for each structure type there are critical ionicities which place a limit on the compounds that can have that particular structure. Theoretical support for Mooser–Pearson plots has come from the work of Phillips (1970) and van Vechten and Phillips (1970) who measured optical absorption spectra of some AB compounds and, from these, calculated electronegativities and ionicities. The spectral data gave values of band gaps, E_g (Chapter 14). For isoelectronic series of compounds, e.g. ZnSe, GaAs, Ge, the band gaps have contributions from (a) a homopolar band gap, E_h, as in pure germanium and (b) a charge transfer, C, between A and B, termed the 'ionic energy'. These are related by

$$E_g^2 = E_h^2 + C^2 \tag{8.31}$$

E_g and E_h are measured from the spectra and hence C can be calculated. C is related to the energy required for electron charge transfer in a polar bond and hence is a measure of electronegativity, as defined by Pauling. A scale of ionicity has been devised:

$$\text{Ionicity}, f_i = \frac{C^2}{E_g^2} \tag{8.32}$$

Values of f_i range from zero ($C = 0$ in a homopolar covalent bond) to one ($C = E_g$ in an ionic bond) and give a measure of the fractional ionic character of a bond. Phillips analysed the spectroscopic data for sixty-eight AB compounds with either octahedral or tetrahedral structures and found that the compounds fall into two groups separated by a critical ionicity, f_i, of 0.785.

The link between Mooser–Pearson plots and Phillips–Van Vechten ionicities is that

$$\Delta x \, (\text{Mooser–Pearson}) \simeq C \, (\text{Phillips})$$

and

$$\bar{n} \, (\text{Mooser–Pearson}) \simeq E_h \, (\text{Phillips})$$

The explanation of the latter is that as the principal quantum number of an element increases, the outer orbitals become larger and more diffuse and the energy differences between outer orbitals (s, p, d and /or f) decrease; the band gap, E_h, decreases until metallic behaviour occurs at $E_h = 0$. Use of $\bar{n}$ gives the average behaviour of anion and cation. The Phillips–Van Vechten analysis has so far been restricted to AB compounds but this has provided a theoretical justification for the more widespread use and application of the readily constructed Mooser–Pearson plots. Further developments in this area of crystal chemistry are anticipated; an enthusiastic and more detailed account is given by Adams (1974).

8.5 Bond valence and bond length

The structures of most *molecular* materials—organic and inorganic—may be satisfactorily described using valence bond theory in which single, double, triple and occasionally partial bonds occur between atoms. Difficulties rapidly arise, however, when valence bond theory is applied to crystalline, *non-molecular* inorganic materials, even though the bonding in them may be predominantly covalent. This is because there are usually insufficient bonding electrons available for each bond to be treated as an electron pair single bond. Instead, most bonds must be regarded as partial bonds.

An empirical but nevertheless useful approach to describing such bonds has been developed by Pauling, Brown, Shannon, Donnay and others and involves the evaluation of *bond orders* or *bond valences* in a structure. Bond valences are defined in a similar way to electrostatic bond strengths in Pauling's electrostatic valence rule for ionic structures (Section 8.2.2). As such, bond valences represent an extension of Pauling's rule to structures that are not necessarily ionic. Bond valences are defined empirically, using information on atom valences and experimental bond lengths; no reference is made, at least not initially, to the nature of the bonding, whether it be covalent or ionic or some blend of the two.

Pauling's electrostatic valence rule requires that the sum of the electrostatic bond strengths between an anion and its neighbouring cations should be equal in magnitude to the formal charge on the anion (equations 8.5 and 8.6). This rule may be modified to include structures which are not necessarily ionic by replacing (a) the electrostatic bond strength by the bond valence and (b) the formal charge on the anion by the valence of that atom (valence being defined as the number of electrons that take part in bonding). This leads to the *valence sum rule* which relates the valence, V_i, of atom i to the bond valence, b_{ij}, between atom i and neighbouring atom j; i.e.

$$V_i = \sum_j b_{ij} \qquad (8.33)$$

Thus, the valence of an atom must be equal to the sum of the bond valences for all the bonds that it forms. For cases in which b_{ij} is an integer, this rule becomes the familiar rule that is used for evaluating the number of bonds around an atom in molecular structures, i.e. the valence of an atom is equal to the number of bonds

304

that if forms (counting double bonds as two bonds, etc.). In non-molecular, inorganic structures, however, integral bond valences are the exception rather than the rule.

The electrostatic bond strength in Pauling's rule is given simply by the ratio of cation charge: cation coordination number. Thus for structures in which the cation coordination is irregular or the cation–anion bonds are not all of the same length, only an average electrostatic bond strength is obtained. One advantage of the bond valence approach is that each bond is treated as an individual and hence irregularities or distortions in coordination environments can be taken into account.

For a given pair of elements, an inverse correlation between bond valence and bond length exists, as shown in Fig. 8.8 for bonds between oxygen and atoms of the second row in their group valence states, i.e. Na^I, Mg^{II}, Al^{III}, Si^{IV}, P^V and S^{VI}. While each atom and oxidation (or valence) state has its own bond valence–bond length curve, it is a considerable simplification that 'universal curves' such as Fig. 8.8 may be used for isoelectronic series of ions. Various analytical expressions have been used to fit curves such as Fig. 8.8, including

$$b_{ij} = \left(\frac{R_0}{R} \right)^N \tag{8.34}$$

where R is the bond length and R_0, N are constants (R_0 is the value of the bond length for unit bond valence); for the elements represented by Fig. 8.8, $R_0 = 1.622$ and $N = 4.290$.

From Fig. 8.8, bond length clearly increases with decreasing bond valence and since, for a given atom, bond valence inevitably decreases with increasing coordination number, a correlation also exists between increasing coordination

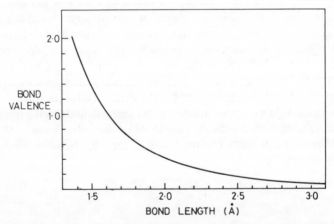

Fig. 8.8 Bond valence–bond length universal correlation curve for bonds between oxygen and second row atoms: Na, Mg, Al, Si, P and S. (From Brown, 1978)

number and increasing bond length. This correlation has already been presented, in a slightly different form, in Fig. 8.3, in which the ionic radii of cations increase with increasing coordination number. Since the data of Fig. 8.3 assume a constant radius of the fluoride or oxide ion, the ordinate in Fig. 8.3 could be changed from the ionic radius to the metal–oxygen bond length.

Curves such as Fig. 8.8 are important for several reasons, the most important of which is that they lead to an increased rationalization and understanding of crystal structures. They have several applications that are specifically associated with the determination of crystal structures; for example:

(a) As a check on the correctness of a proposed structure, the valence sum rule should be obeyed by all atoms of the structure to within a few per cent.
(b) To locate hydrogen atoms. Hydrogen atoms are often 'invisible' in X-ray structure determinations because of the very low scattering power of hydrogen. It may be possible to locate them by evaluating the bond valence sums around each atom and noting which atoms (e.g. oxygen atoms in hydrates) show a large discrepancy between the atom valence and the bond valence sum. The hydrogen is then likely to be bonded to such atoms.
(c) To distinguish between Al^{3+} and Si^{4+} positions in aluminosilicate structures. By X-ray diffraction, Al^{3+} and Si^{4+} cannot be distinguished because of their very similar scattering power, but in sites of similar coordination number they give different bond valences, e.g. in regular MO_4 tetrahedra, Si—O bonds have a bond valence of one but Al—O bonds have a bond valence of 0.75. Site occupancies may therefore be determined using the valence sum rule and/or by comparing the M—O bond lengths with values expected for Si—O and Al—O.

8.6 Non-bonding electron effects

In this section, the influence on structure of two types of non-bonding electrons is considered: the d electrons in transition metal compounds and the s^2 pair of electrons in compounds of the heavy p-block elements in low oxidation states. These two types of electrons do not take part in bonding as such, but nevertheless exert a considerable influence on the coordination number and environment of the metal atom.

8.6.1 d Electron effects

In transition metal compounds, the majority of the d electrons on the metal atom do not usually take part in bond formation but do influence the coordination environment of the metal atom and are responsible for properties such as magnetism. For present purposes, basic crystal field theory (CFT) is adequate to describe qualitatively the effects that occur. It is assumed that the reader is acquainted with CFT and only a summary will be given here.

8.6.1.1 *Crystal field splitting of energy levels*

In an octahedral environment, the five d orbitals on a transition metal atom are no longer degenerate but split into two groups, the t_{2g} group of lower energy and the e_g group of higher energy (Fig. 8.9a). If possible, electrons occupy orbitals singly, according to Hund's rule of maximum multiplicity. For d^4 to d^7 atoms or ions, two possible configurations occur, giving low spin (LS) and high spin (HS) states; these are shown for a d^7 ion in Fig. 8.10. In these, the increased energy, Δ, required to place an electron in an e_g orbital, and hence maximize the multiplicity, has to be balanced against the repulsive energy or pairing energy, P, which arises when two electrons occupy the same t_{2g} orbital. The magnitude of Δ depends on the ligand or anion to which the metal his bonded: for weak field anions, Δ is small and the HS configuration occurs, and vice versa for strong field ligands. Δ also depends on the metal and, in particular, to which row it belongs: generally

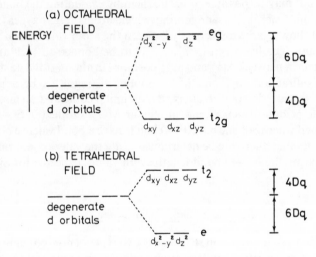

Fig. 8.9 Splitting of d energy levels in (a) an octahedral and (b) a tetrahedral field

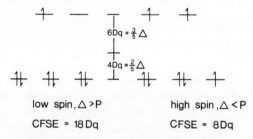

Fig. 8.10 Low spin and high spin states for a d^7 transition metal ion in an octahedral coordination environment

Table 8.14 *d Electron configuration in octahedrally coordinated metal atoms*

Number of electrons	Low spin, $\Delta > P$. t_{2g}	e_g	High spin, $\Delta < P$ t_{2g}	e_g	Gain in orbital energy for low spin	Example
1	↑		↑			V^{4+}
2	↑ ↑		↑ ↑			Ti^{2+}, V^{3+}
3	↑ ↑ ↑		↑ ↑ ↑			V^{2+}, Cr^{3+}
4	↑↓ ↑ ↑		↑ ↑ ↑	↑	Δ	Cr^{2+}, Mn^{3+}
5	↑↓ ↑↓ ↑		↑ ↑ ↑	↑ ↑	2Δ	Mn^{2+}, Fe^{3+}
6	↑↓ ↑↓ ↑↓		↑↓ ↑ ↑	↑ ↑	2Δ	Fe^{2+}, Co^{3+}
7	↑↓ ↑↓ ↑↓	↑	↑↓ ↑↓ ↑	↑ ↑	Δ	Co^{2+}
8	↑↓ ↑↓ ↑↓	↑ ↑	↑↓ ↑↓ ↑↓	↑ ↑		Ni^{2+}
9	↑↓ ↑↓ ↑↓	↑↓ ↑	↑↓ ↑↓ ↑↓	↑↓ ↑		Cu^{2+}
10	↑↓ ↑↓ ↑↓	↑↓ ↑↓	↑↓ ↑↓ ↑↓	↑↓ ↑↓		Zn^{2+}

$\Delta(5d) > \Delta(4d) > \Delta(3d)$. Consequently HS behaviour is rarely observed in the $4d$ and $5d$ series. Δ values may be determined experimentally from electronic spectra. The possible spin configurations for the different numbers of d electrons are given in Table 8.14.

The radii of transition metal ions depend on their d electron configuration, as shown in Fig. 8.11(a) for the octahedrally coordinated divalent ions. With increasing atomic number, several trends occur. First, there is a gradual, overall decrease in radius as the d shell is filled, as shown by the dashed line that passes through Ca^{2+}, Mn^{2+}(HS) and Zn^{2+}. For these three ions, the distribution of d electron density around the M^{2+} ion is spherically symmetrical because the d orbitals are either empty (Ca), singly occupied (Mn) or doubly occupied (Zn). The gradual decrease in radius with increasing atomic number is associated with poor shielding of the nuclear charge by the d electrons; hence a greater effective nuclear charge is experienced by the outer, bonding electrons which results in a steady contraction in radius with increasing atomic number. Similar effects occur across any horizontal row of the periodic table as the valence shell is filled, but are particularly well documented for the transition metal series.

For the other ions d^1 to d^4 and d^6 to d^9, the d electron distribution is not spherical. The shielding of the nuclear charge by these electrons is reduced even further and the radii are smaller than expected. Thus, the Ti^{2+} ion has the configuration $(t_{2g})^2$, which means that two of the three t_{2g} orbitals are singly occupied. In octahedrally coordinated Ti^{2+}, these electrons (which are non-bonding) occupy regions of space that are directed away from the (Ti^{2+}–anion) axes. Comparing Ti^{2+} with Ca^{2+}, for instance, Ti^{2+} has an extra nuclear charge of $+2$ but the two extra electrons in the t_{2g} orbitals do not shield the bonding electrons from this extra charge. Hence, Ti—O bonds in, for example, TiO, are shorter than Ca—O bonds in CaO due to the stronger attraction between Ti^{2+} and the bonding electrons. This trend continues in V^{2+}, Cr^{2+}(LS), Mn^{2+}(LS) and Fe^{2+}(LS), all of which contain only t_{2g} electrons (Table 8.14). Beyond

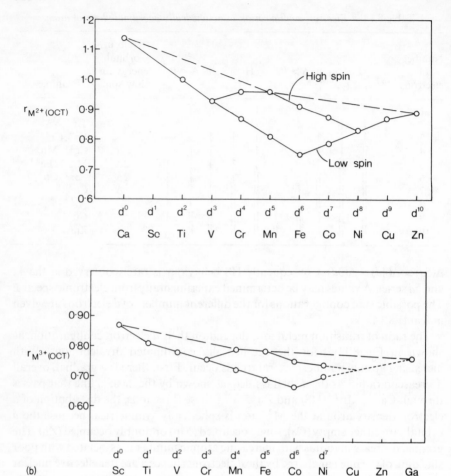

Fig. 8.11 (a) Radii of octahedrally coordinated divalent transition metal ions. (b) Radii of octahedrally coordinated trivalent transition metal ions. Data from Shannon and Prewitt, relative to $r_{F^-} = 1.19\,\text{Å}$

Fe^{2+}(LS), the electrons begin to occupy e_g orbitals and these electrons do shield the nuclear charge much more effectively. The radii then begin to increase again in the series Fe^{2+}(LS), Co^{2+}(LS), Ni^{2+}, Cu^{2+} and Zn^{2+}.

For the high spin ions a different trend is observed. On passing from V^{2+} to Cr^{2+}(HS) and Mn^{2+}(HS), electrons enter the e_g orbitals, thereby shielding the nuclear charge and giving rise to an increased radius. However, on passing from Mn^{2+}(HS) to Fe^{2+}(HS), Co^{2+}(HS) and Ni^{2+}, the additional electrons occupy t_{2g} orbitals and the radii decrease once again.

Trivalent transition metal ions show a similar trend but of reduced magnitude (Fig. 8.11b). However, the various effects that occur are now effectively transferred to higher atomic number and hence to the right by one atom; thus, the ion with the smallest radius is Co^{3+}(LS) instead of Fe^{2+}(LS).

So far, only octahedral coordination of transition metal ions has been considered. Tetrahedral coordination is also quite common but a different energy level diagram applies to the d electrons. A tetrahedral field also splits the d orbitals into two groups, but in the opposite manner to an octahedral field. Thus, three orbitals have higher energy d_{xy}, d_{xz} and d_{yz} whereas the other two, $d_{x^2-y^2}$ and d_{z^2} have lower energy (Fig. 8.9b).

It was mentioned in Section 8.2.7 that crystal field splitting of d orbitals in transition metal ions may result in *crystal field stabilization energies* (CFSE) and increased lattice energies of ionic compounds. For example, CoF_2 has the rutile structure with octahedrally coordinated $Co^{2+}(d^7)$ ions in which Co^{2+} adopts the high spin configuration (Fig. 8.10). The energy difference, Δ, between t_{2g} and e_g orbitals is set equal to 10 Dq and the t_{2g} orbitals are stabilized by an amount 4 Dq whereas the e_g orbitals are destabilized by an amount 6 Dq. Relative to the situation of five degenerate orbitals without crystal field splitting (Fig. 8.9), the crystal field stabilization energy of Co^{2+} in HS and LS states may be calculated. For the LS state, the CFSE is 6×4 Dq $- 1 \times 6$ Dq $= 18$ Dq. For the HS state, the CFSE is 5×4 Dq $- 2 \times 6$ Dq $= 8$ Dq.

The occurrence of CFSE leads to an increased lattice energy. The value of the CFSE calculated for $Co^{2+}(HS)$ in rutile is 104 kJ mol^{-1}. This compares fairly well with the value of 83 kJ mol^{-1} given by the difference between the lattice energy determined from a Born–Haber cycle (2959 kJ mol^{-1}) and that calculated using the Born–Mayer equation (2876 kJ mol^{-1}).

The lattice energies of the divalent fluorides of the first row transition elements, determined from a Born–Haber cycle, are shown in Fig. 8.12. Similar trends are

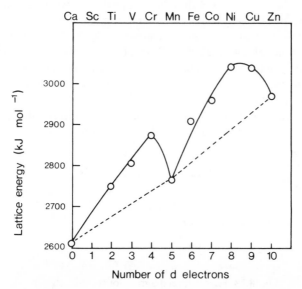

Fig. 8.12 Lattice energies of divalent transition metal fluorides determined from Born–Haber cycle calculations. (From Waddington, 1959)

observed for other halides. Ions that do not exhibit CFSE are d^0(Ca), d^5(HS)(Mn) and d^{10}(Zn), and their lattice energies fall on the lower, dashed curve. Most ions do show some degree of CFSE, however, and their lattice energies fall on the upper, solid curve. For the fluorides (Fig. 8.12), the agreement between the calculated CFSE and the difference in lattice energy ΔU [i.e. U(Born–Haber) $- U$(Born–Mayer)] is reasonable and indicates that the bonding may be treated as ionic. For the other halides, however, $\Delta U \gg$ CFSE and indicates that other effects, perhaps covalent bonding, must be present.

8.6.1.2 Jahn–Teller distortions

In many transition metal compounds, the metal coordination is distorted octahedral and the distortions are such that the two axial bonds are either shorter than or longer than the other four bonds. The Jahn–Teller effect is responsible for these distortions in d^9, d^7(LS) and d^4(HS) ions. Consider the d^9 ion Cu^{2+} whose configuration is $(t_{2g})^6(e_g)^3$. One of the e_g orbitals contains two electrons and the other contains one. The singly occupied orbital can be either d_{z^2} or $d_{x^2-y^2}$ and in a free ion situation both would have the same energy. However, since the metal coordination is octahedral the e_g levels, with one doubly and one singly occupied orbitals, are no longer degenerate. The e_g orbitals are high energy orbitals (relative to t_{2g}) since they point directly towards the surrounding ligands and the doubly occupied orbital will experience stronger repulsions and hence have somewhat higher energy than the singly occupied orbital. This has the effect of lengthening the metal–ligand bonds in the directions of the doubly occupied orbital, e.g. if the d_{z^2} orbital is doubly occupied, the two metal–ligand bonds along the z axis will be longer than the other four metal–ligand bonds. The energy level diagram for this latter situation is shown in Fig. 8.13(a). Lengthening of the metal–ligand bond along the z axis leads to a *lowering* of energy of the d_{z^2} orbital. The distorted structure is stabilized by an amount $\frac{1}{2}\delta_2$ relative to the regular octahedral arrangement and, hence, the distorted structure becomes the observed, ground state.

High spin d^4 and low spin d^7 ions also have odd numbers of e_g electrons and show Jahn–Teller distortions. It is not clear which type of distortion is preferred

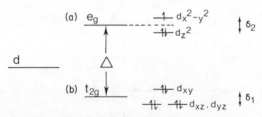

Fig. 8.13 Energy level diagram for the *d*-levels in a d^9 ion experiencing a Jahn–Teller distortion. The two bonds parallel to z are longer than the other four

(i.e. two short and four long bonds, or vice versa) and the actual shapes in a particular structure must be determined experimentally. The degeneracy of the t_{2g} levels may also be removed by the Jahn–Teller effect, but the magnitude of the splitting, δ_1 in Fig. 8.13(b), is small and the effect is relatively unimportant.

The normal coordination environment of the Cu^{2+} ion is distorted octahedral with four short and two long bonds. The distortion varies from compound to compound. In, for example, CuF_2 (distorted rutile structure), the distortion is fairly small (four fluorine atoms at 1.93 Å, two at 2.27 Å), but is larger in $CuCl_2$ (four chlorine atoms at 2.30 Å, two at 2.95 Å) and extreme in tenorite, CuO, which is effectively square planar (four oxygen atoms at 1.95 Å, two at 2.87 Å).

The importance of Jahn–Teller distortions in Cu^{2+} and Cr^{2+} (d^4 ion) compounds is seen by comparing the structures of the oxides and fluorides of the first-row divalent, transition metal ions. For the oxide series, $MO(M^{2+}$ is Ti, V, Cr, Mn, Fe, Co, Ni and Cu), all have the rock salt structure with regular octahedral coordination apart from (a) CuO which contains grossly distorted (CuO_6) octahedra and possibly (b) CrO, whose structure is not known. For the fluoride series, MF_2, all have the regular rutile structure apart from CrF_2 and CuF_2 which have distorted rutile structures.

Other examples of distorted octahedral coordination due to the Jahn–Teller effect are found in compounds of Mn^{3+}(HS) and Ni^{3+}(LS).

8.6.1.3 Square planar coordination

The d^8 ions—Ni^{2+}, Pd^{2+}, Pt^{2+}—commonly have square planar or rectangular planar coordination in their compounds. In order to understand this, consider the d energy level diagram for such ions in (a) octahedral and (b) distorted octahedral fields:

(a) The normal configuration of a d^8 ion in an octahedral field is $(t_{2g})^6(e_g)^2$. The two e_g electrons singly occupy the d_{z^2} and $d_{x^2-y^2}$ orbitals, which are degenerate, and the resulting compounds, with unpaired electrons, are paramagnetic.

(b) Consider, now, the effect of distorting the octahedron and lengthening the two metal–ligand bonds along the z axis. The e_g orbitals lose their degeneracy, and the d_{z^2} orbital becomes stabilized by an amount $\frac{1}{2}\delta_2$ (Fig. 8.13). For small elongations along the z axis, the pairing energy required to doubly occupy the d_{z^2} orbital is larger than the energy difference between d_{z^2} and $d_{x^2-y^2}$, i.e. $P > \delta_2$. There is no gain in stability if the d_{z^2} orbital is doubly occupied, therefore, and no reason why small distortions from octahedral coordination should be stable. With increasing elongation along the z axis, however, a stage is reached where $P < \delta_2$ and in which the doubly occupied d_{z^2} orbital becomes stabilized and is the preferred ground state for a d^8 ion. The distortion from octahedral coordination is now sufficiently large that the coordination is regarded as square planar; in many cases, e.g. PdO, there are, in fact, no axial ligands along z and, hence, the transformation from octahedral to square planar coordination is complete. Because they have no unpaired electrons, square planar compounds are diamagnetic.

Square planar coordination is more common with $4d$ and $5d$ transition elements than with $3d$ elements because the $4d$ and, especially, $5d$ orbitals are more diffuse and extend to greater radial distances from the nucleus. Consequently, the magnitude of the crystal (or ligand) field splitting (Δ, δ) caused by a particular ligand, e.g. O^{2-}, increases in the series $3d < 4d < 5d$. Thus, NiO has the rock salt structure with regular octahedral coordination of $Ni^{2+}(3d$ ion) whereas PdO and PtO both have square planar coordination for the metal atoms ($4d$ and $5d$). The only known compound of palladium with octahedral Pd^{2+} is PdF_2 (rutile structure) and no octahedral Pt^{2+} or Au^{3+} compounds are known.

8.6.1.4 *Tetrahedral coordination*

As stated earlier, a tetrahedral field causes splitting of the d energy levels, but in the opposite sense to an octahedral field (Fig. 8.9b). Further, the magnitude of the splitting, Δ, is generally less in a tetrahedral field since none of the d orbitals point directly towards the four ligands. Rather, the d_{xy}, d_{xz} and d_{yz} orbitals are somewhat closer to the ligands than are the other two orbitals, Fig. 8.14 (only two orbital directions, d_{yz} and $d_{x^2-y^2}$ are shown). Jahn–Teller distortions again occur, especially when the upper t_2 orbitals contain 1, 2, 4 or 5 electrons (i.e. d^3(HS), d^4(HS), d^5 and d^9). Details will not be given since various types of distortion are possible (e.g. tetragonal or trigonal distortions) and these have not been as well studied as have the octahedral distortions. A common type of distortion is a flattening or elongation of the tetrahedron in the direction of one of the twofold axes of the tetrahedron (e.g. along z in Fig. 8.14). An example is the flattened $CuCl_4$ tetrahedron in Cs_2CuCl_4.

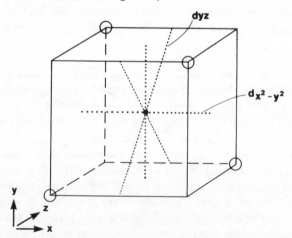

Fig. 8.14 The orientation of d orbitals in a tetrahedral field

8.6.1.5 *Tetrahedral versus octahedral coordination*

Most transition metal ions prefer octahedral coordination or a distorted variant of octahedral and an important factor is their large CFSE in octahedral sites. This can be estimated as follows. In octahedral coordination, each t_{2g} electron experiences a stabilization of $(4/10)\Delta^{oct}$ and each e_g electron a destabilization of $(6/10)\Delta^{oct}$. Thus Cr^{3+}, $d^3(t_{2g}^3)$ has a CFSE of $1.2\Delta^{oct}$ whereas $Cu^{2+} d^9(t_{2g})^6(e_g)^3$, has a CFSE of $0.6\Delta^{oct}$. In tetrahedral coordination, each e electron has a stabilization of $(6/10)\Delta^{tet}$ and each t_2 electron has a destabilization of $(4/10)\Delta^{tet}$ (Fig. 8.9). Thus, Cr^{3+} would have a CFSE of $0.8\Delta^{tet}$ and Cu^{2+} would have a CFSE of $0.4\Delta^{tet}$. With the guideline that

$$\Delta^{tet} \simeq 0.4\Delta^{oct}$$

the values of Δ^{oct} and Δ^{tet} for ions may be used to predict site preferences. More accurate values may be obtained spectroscopically and are given in Table 8.15 for some oxides of transition metal ions. It can be seen that high spin d^5 ions, as well as d^0 and d^{10} ions, have no particular preference for octahedral or tetrahedral sites insofar as crystal field effects are concerned. Ions such as Cr^{3+}, Ni^{2+} and Mn^{3+} show the strongest preference for octahedral coordination: thus tetrahedral coordination is rare for Ni^{2+}.

The coordination preferences of ions are shown by the type of spinel structure that they adopt. Spinels have the formula AB_2O_4 and may be:

(a) normal—A tetrahedral, B octahedral;
(b) inverse—A octahedral, B tetrahedral and octahedral;
(c) some intermediate between normal and inverse.

The parameter, γ, is the fraction of A ions on octahedral sites. For normal spinels, $\gamma = 0$; for inverse, $\gamma = 1$; and for a random arrangement of A and B ions, $\gamma = 0.67$. Lattice energy calculations show that, in the absence of CFSE effects, spinels of the type 2, 3 (i.e. $A = M^{2+}$, $B = M^{3+}$, e.g. $MgAl_2O_4$) tend to be normal whereas

Table 8.15 *Crystal field stabilization energies $(kJ\,mol^{-1})$ estimated for transition metal oxides.* (Data from Dunitz and Orgel, 1960)

Ion		Octahedral stabilization	Tetrahedral stabilization	Excess octahedral stabilization
Ti^{3+}	d^1	87.4	58.5	28.9
V^{3+}	d^2	160.1	106.6	53.5
Cr^{3+}	d^3	224.5	66.9	157.6
Mn^{3+}	d^4	135.4	40.1	95.3
Fe^{3+}	d^5	0	0	0
Mn^{2+}	d^5	0	0	0
Fe^{2+}	d^6	49.7	33.0	16.7
Co^{2+}	d^7	92.8	61.9	30.9
Ni^{2+}	d^8	122.1	35.9	86.2
Cu^{2+}	d^9	90.3	26.8	63.5

Table 8.16 *The γ parameters of some spinels.* (Data from Greenwood, 1968 and Dunitz and Orgel, 1960)

M^{3+} \ M^{2+}	Mg^{2+}	Mn^{2+}	Fe^{2+}	Co^{2+}	Ni^{2+}	Cu^{2+}	Zn^{2+}
Al^{3+}	0	0.3	0	0	0.75	0.4	0
Cr^{3+}	0	0	0	0	0	0	0
Fe^{3+}	0.9	0.2	1	1	1	1	0
Mn^{3+}	0	0	0.67	0	1	0	0
Co^{3+}	—	—	—	0	—	—	0

spinels of type 4, 2 (i.e. $A = M^{4+}$, $B = M^{2+}$, e.g. $TiMg_2O_4$) tend to be inverse. However, these preferences may be changed by the intervention of CFSE effects, as shown by the γ parameters of some 2, 3 spinels in Table 8.16. Examples are:

(a) All chromate spinels contain octahedral Cr^{3+} and are normal. This is consistent with the very large CFSE of Cr^{3+} and ions such as Ni^{2+} are forced into tetrahedral sites in $NiCr_2O_4$.

(b) Most 2, 3 Mg^{2+} spinels are normal apart from $MgFe_2O_4$ which is essentially inverse. This reflects the lack of any CFSE for Fe^{3+}.

(c) $Co_3O_4 (\equiv CoO \cdot Co_2O_3)$ is normal because low spin Co^{3+} gains more CFSE by going into the octahedral site than Co^{2+} loses by occupying the tetrahedral site. Mn_3O_4 is also normal. Magnetite, Fe_3O_4, however, is inverse because whereas Fe^{3+} has no CFSE in either tetrahedral or octahedral coordination, Fe^{2+} has a preference for octahedral sites.

Spinels usually have cubic symmetry but some show tetragonal distortions in which one of the cell edges is of a different length to the other two. The Jahn–Teller effect gives rise to such distortions in Cu^{2+} containing spinels, $CuFe_2O_4$ (the tetragonal unit cell parameter ratio, $c/a = 1.06$) and $CuCr_2O_4(c/a = 0.91)$. $CuFe_2O_4$ is an inverse spinel with octahedral Cu^{2+} ions and the Jahn–Teller effect distorts the CuO_6 octahedra so that the two Cu—O bonds along z are longer than the four Cu—O bonds in the xy plane. On the other hand, $CuCr_2O_4$ is normal and the CuO_4 tetrahedra are flattened in the z direction, again due to the Jahn–Teller effect, thereby causing a shortened c axis.

8.6.2 Inert pair effect

The heavy, post-transition elements, especially Tl, Sn, Pb, Sb and Bi exhibit, in some compounds, a valence that is two less than the group valence (e.g. the divalent state in Group IV elements, Sn and Pb). This is the so-called 'inert pair effect' and manifests itself structurally by a distortion of the metal ion coordination environment. Thus, Pb^{2+} has the configuration: (Xe core) $4f^{14}5d^{10}6s^2$, and the $6s^2$ pair is 'stereochemically active' in that these electrons are not in a spherically symmetrical orbital but stick out to one side of the Pb^{2+} ion (perhaps in some kind of s–p hybridized orbital).

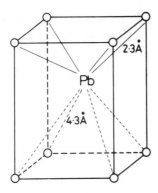

Fig. 8.15 The structure of red PbO showing the presence of the inert pair effect by the variation in Pb—O bond distances

Various kinds of distorted coordination polyhedra occur. Sometimes, the lone pair comes between the metal ion and some of its immediate anionic neighbours and causes a variation of bond length about the metal ion, e.g. red PbO has a structure that is a tetragonal distortion of the CsCl structure (Fig. 8.15). Four oxygens are situated at 2.3 Å from the Pb^{2+} ion, which is a reasonable Pb—O bond length, but the other four oxygens are at 4.3 Å. Although the lone pair is not directly visible, its presence is apparent from the distorted nature of the cubic coordination of Pb^{2+}.

A related distortion occurs in SnS which has a distorted rock salt structure. In this case, the SnS_6 octahedra are distorted along a [111] direction such that three sulphur atoms on one side of the tin are at ~ 2.64 Å but the other three are repelled by the lone pair to a distance of ~ 3.31 Å.

Another common kind of distortion occurs when the lone pair simply takes the plane of an anion and its associated pair of bonding electrons. Five-coordinated structures result in, for example, TlI, in which one corner (anion) of the octahedron 'TlI_6' is missing.

Questions

8.1 The oxides MnO, FeO, CoO, NiO all have the cubic rock salt structure with octahedral coordination of the cations. The structure of CuO is different and contains grossly distorted CuO_6 octahedra. Explain.

8.2 While the d electrons in many transition metal compounds may not be involved directly in bond formation, they nevertheless exert a considerable influence on structure. Explain.

8.3 The mineral grossular, $Ca_3Al_2Si_3O_{12}$ has the garnet structure, Chapter 16, with 8-coordinate Ca, octahedral Al and tetrahedral Si. Determine the probable coordination number and environment of oxygen and show that the structure obeys Pauling's electrostatic valency rule.

8.4 The most stable oxide of lithium is Li_2O but for rubidium and caesium, the peroxides M_2O_2 and superoxides MO_2 are more stable than the simple oxides M_2O. Explain.

8.5 BeF_2 has the same structure as SiO_2, MgF_2 is the same as rutile and CaF_2 has the fluorite structure. Does this seem reasonable?

8.6 Account for the observation that whereas CuF_2 and CuI are stable compounds, CuF and CuI_2 are not stable.

8.7 In Table 7.5 are given unit cell constants for some oxides MO with the rocksalt structure. Assuming that (i) $r_{o^{2-}} = 1.26\,\text{Å}$; (ii) $r_{o^{2-}} = 1.40\,\text{Å}$, calculate for each (a) two values of the cation radius, $r_{M^{2+}}$, (b) two values for the radius ratio, $r_{M^{2+}}/r_{o^{2-}}$. Assess the usefulness of the radius ratio rules in predicting octahedral coordination for M^{2+} for these oxides. Repeat the calculations for two oxides with the wurtzite structure, Table 7.9.

8.8 Calculate lattice energy values for the alkaline earth oxides using Kapustinskii's equation and the data given in Table 7.5. Compare your results with those given in Table 8.6. Estimate the enthalpies of formation of these oxides.

8.9 Using Sanderson's methods, estimate the partial ionic character of the sodium halides. Hence calculate the radii of the atoms involved and the unit cell a value (all have the rock salt structure). Compare your answers with the data given in Table 7.5.

References

D. M. Adams (1974). *Inorganic Solids*, Wiley.

L. H. Ahrens (1952). The use of ionisation potentials. I, Ionic radii of the elements, *Geochim. Cosmochim. Acta*, **2**, 155–169.

I. D. Brown (1978). Bond valences—a simple structural model for inorganic chemistry, *Chem. Soc. Revs.*, **7** (3), 359.

I. D. Brown and R. D. Shannon (1973). Empirical bond strength–bond length curves for oxides, *Acta Cryst.*, A**29**, 266.

G. O. Brunner (1971). An unconventional view of the 'closest sphere packings', *Acta Cryst.*, A**27**, 388.

J. D. Dunitz and L. E. Orgel (1960). Stereochemistry of ionic solids, *Adv. Inorg. Radiochem.*, **2**, 1–60.

F. G. Fumi and M. P. Tosi (1964). Ionic sizes and Born repulsive parameters in the NaCl type alkali halides, *J. Phys. Chem. Solids*, **25**, 31–43.

N. N. Greenwood (1968). *Ionic Crystals, Lattice Defects and Nonstoichiometry*, Butterworths.

A. F. Kapustinskii, Lattice energy of ionic crystals, *Quart. Rev.*, **10**, 283–294.

H. Krebs (1968). *Fundamentals of Inorganic Crystal Chemistry*, McGraw-Hill.

J. Krug, H. Witte and E. Wölfel (1955). *Zeit. Phys. Chem., Frankfurt*, **4**, 36.

M. F. C. Ladd and W. H. Lee (1963, 1965). Lattice energies and related topics, *Progr. Solid State Chem.*, **1**, 37–82; **2**, 378–413.

G. J. Moody and J. D. R. Thomas, Lattice energy and chemical prediction. Use of the Kapustinskii equations and the Born–Haber cycle, *J. Chem. Educ.*, **42** (4), 204.

E. Mooser and W. B. Pearson (1959). On the crystal chemistry of normal valence compounds, *Acta Cryst.*, **12**, 1015–1022.

L. Pauling (1928). The sizes of ions and their influence on the properties of salt-like compounds, *Z. Krist.*, **67**, 377–404.

J. C. Phillips (1970). Ionicity of the chemical bond in crystals, *Rev. Modern Phys.*, **42**, 317–356.

R. T. Sanderson (1967). The nature of 'ionic' solids, *J. Chem. Ed.*, **44** (9), 516.

R. T. Sanderson (1976). *Chemical Bonds and Bond Energy*, Academic Press.

R. D. Shannon and C. T. Prewitt (1969, 1970). Effective ionic radii in oxides and fluorides, *Acta Cryst.*, **B25**, 725–945; **B26**, 1046.

Shih-Ming Ho and Bodie E. Douglas (1969). A system of notation and classification for typical close packed structures, *J. Chem. Ed.*, **46**, 207–216.

K. H. Stern and E. S. Amis (1959). Ionic size, *Chem. Revs.*, **59**, 1.

J. A. van Vechten and J. C. Phillips (1970). New set of tetrahedral covalent radii, *Phys. Rev.*, **B2**, 2160–2167.

T. C. Waddington (1959). Lattice energies and their significance in inorganic chemistry, *Adv. Inorg. Chem. Radiochem.*, **1**, 157–221.

T. C. Waddington (1966). Ionic radii and the method of the undetermined parameter, **62**, 1482–1492.

A. F. Wells (1975). *Structural Inorganic Chemistry*, 4th ed., Oxford.

Chapter 9

Crystal Defects and Non-Stoichiometry

9.1 Perfect and imperfect crystals

A perfect crystal may be defined as one in which all the atoms are at rest on their correct lattice positions in the crystal structure. Such a perfect crystal can be obtained, hypothetically, only at absolute zero. At all real temperatures, crystals are imperfect. Apart from the fact that atoms are vibrating, which may be regarded as a form of defect, a number of atoms are inevitably misplaced in a real crystal. It is such imperfections or defects in an otherwise perfectly regular atomic array that concern us here. In some crystals, the number of defects present may be very small, $\ll 1$ per cent, as in, for example, high purity diamond or quartz crystals. In other crystals, very high defect concentrations, > 1 per cent, may be present. In the latter, highly defective crystals, the question arises as to whether or not the defects themselves should be regarded as forming a fundamental part of the crystal structure rather than as some imperfection in an otherwise ideal structure.

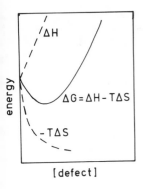

[defect]

Fig. 9.1 Energy changes on introducing defects into a perfect crystal

Crystals are imperfect because the presence of defects up to a certain concentration leads to a reduction of free energy (Fig. 9.1). Let us consider the effect on the free energy of a perfect crystal of creating a single defect, say a vacant cation site. This requires a certain amount of energy, ΔH. Creation of this single defect causes a considerable increase in entropy, ΔS, of the crystal, because of the large number of positions which this defect can occupy. Thus, if the crystal contains 1 mole of cations, there are $\sim 10^{23}$ possible positions for the vacancy. The entropy gained by having this choice of positions is called *configurational entropy* and is given by the Boltzmann formula, $S = k \ln W$, where the probability, W, is proportional to 10^{23}; other, smaller, entropy changes are also present due to the disturbance of the crystal structure in the neighbourhood of the defect. As a result of this considerable increase in entropy the enthalpy required to form the defect initially is more than offset by the gain in entropy of the crystal. Consequently, the free energy, given by $\Delta G = \Delta H - T\Delta S$, decreases (Fig. 9.1).

If we go now to the other extreme where, say, 10 per cent of the cation sites are vacant, the change in entropy on introducing yet more defects is small because the crystal is already very disordered in terms of occupied and vacant cation sites. The energy required to create more defects may be larger than any subsequent gain in entropy and hence such a high defect concentration would not be stable. In between these two extremes lie most real materials. A minimum in free energy occurs at a certain defect concentration which represents the number of defects present in the crystal under conditions of thermodynamic equilibrium (Fig. 9.1). At this defect concentration, the increase in entropy which arises on creating further defects is no longer sufficient to offset the energy of creation of these extra defects.

Although this is a simplified explanation, it serves to illustrate why crystals are imperfect. It also follows that the equilibrium number of defects present in a crystal increases with increasing temperature; assuming that ΔH and ΔS are independent of temperature, the $-T\Delta S$ term is larger and so the free energy minimum is displaced to higher defect concentrations, with increasing temperature.

For a given crystal or material, curves such as shown in Fig. 9.1 can be drawn for every possible type of crystal defect; the main difference between them will be

Table 9.1 *Predominant point defects in various crystals*

Crystal	Crystal structure	Predominant intrinsic defect
Alkali halides (not Cs)	Rock salt, NaCl	Schottky
Alkaline earth oxides	Rock salt	Schottky
AgCl, AgBr	Rock salt	Cation Frenkel
Cs halides, TlCl	CsCl	Schottky
BeO	Wurtzite, ZnS	Schottky
Alkaline earth fluorides, CeO_2, ThO_2	Fluorite, CaF_2	Anion Frenkel

in the position of the free energy minimum. The type of defect that predominates in a particular material is clearly the defect which is easiest to form, i.e. the defect with the smallest ΔH and for which the free energy minimum is associated with the highest defect concentration. Thus, for NaCl it is easiest to form vacancies (the Schottky defect) and so these are the predominant point defects, whereas in AgCl the reverse is true and interstitial (Frenkel) defects predominate. In Table 9.1 the defects which predominate in a variety of inorganic solids are summarized.

9.2 Types of defect

Various schemes have been proposed for the classification of defects, each of which have their uses and none of which is entirely satisfactory. Defects can be broadly divided into two groups: *stoichiometric defects* in which the crystal composition is unchanged on introducing the defects and *non-stoichiometric defects* which are a consequence of a change in crystal composition. Alternatively, the size and shape of the defect can be used as a basis for classification: *point defects* involve only one atom or site, e.g. vacancies or interstitials, although the atoms immediately surrounding the defect are also somewhat perturbed; *line defects*, i.e. *dislocations*, are effectively point defects in two dimensions but in the third dimension the defect is very extensive or infinite; in *plane defects*, whole layers in a crystal structure can be defective. Sometimes the name *extended defects* is used to include all those which do not come in the category of point defects.

It is convenient here to begin with point defects. These are the 'classical' defects; they were proposed in the 1930s following the work of Schottky, Frenkel, Wagner and others, and it was several decades before direct experimental evidence for their existence was forthcoming.

9.3 Point defects

9.3.1 Schottky defect

The Schottky defect is a stoichiometric defect in ionic crystals. It is a pair of vacant sites, an anion vacancy and a cation vacancy. To compensate for the

```
Cℓ  Na  Cℓ  Na Cℓ  Na  Cℓ  Na  Cℓ
Na  Cℓ  Na  Cℓ Na  Cℓ  Na  Cℓ  Na
Cℓ  Na  Cℓ  Na Cℓ  Na  Cℓ  Na  Cℓ
Na  Cℓ  ☐   Cℓ Na  Cℓ  Na  Cℓ  Na
Cℓ  Na  Cℓ  Na Cℓ  Na  ☐   Na  Cℓ
Na  Cℓ  Na  Cℓ Na  Cℓ  Na  Cℓ  Na
Cℓ  Na  Cℓ  Na Cℓ  Na  Cℓ  Na  Cℓ
Na  Cℓ  Na  Cℓ Na  Cℓ  Na  Cℓ  Na
```

Fig. 9.2 Two-dimensional representation of a Schottky defect with cation and anion vacancies

vacancies, there should be two extra atoms at the surface of the crystal for each Schottky defect. The Schottky defect is the principal point defect in the alkali halides and is shown forNaCl in Fig. 9.2. There are equal numbers of anion and cation vacancies in order to preserve local electroneutrality as much as possible, both inside the crystal and at the surface of the crystal.

The vacancies may be distributed at random in the crystal or may occur in pairs. An anion vacancy in NaCl has a net positive charge of $+1$ because the vacancy is surrounded by six Na^+ ions, each with partially unsatisfied positive charge. Alternatively, the anion vacancy has charge $+1$ because, on placing an anion of charge -1 in the vacancy, local electroneutrality is restored. Similarly, the cation vacancy has a net charge of -1. Since vacancies are charged, vacancies of opposite sign attract each other and may 'associate' to form pairs. In order to dissociate the pairs, the enthalpy of association, $1.30\,eV$ for NaCl ($\sim 120\,kJ\,mol^{-1}$) must be provided.

The number of Schottky defects in a crystal of NaCl is either very small or very large, depending on one's point of view. In NaCl at room temperature, typically only one in 10^{15} of the possible anion and cation sites is vacant, a number which is insignificant in terms of the average crystal structure of NaCl as determined by X-ray diffraction. On the other hand, a grain of salt weighing 1 milligram (and containing approximately 10^{19} atoms) contains $\sim 10^4$ Schottky defects, hardly an insignificant number! Point defects such as Schottky defects are responsible for the optical and electrical properties of NaCl (Chapters 13 and 14).

9.3.2 Frenkel defect

This is also a stoichiometric defect and involves an atom displaced off its lattice site into an interstitial site that is normally empty. Silver chloride (which also has the NaCl crystal structure) has predominantly this defect, with silver as the interstitial atom (Fig. 9.3). The nature of the interstitial site is shown in Fig. 9.3(b) and Fig. 7.12(d). The interstitial Ag^+ is surrounded tetrahedrally by four Cl^- ions but also, and at the same distance, by four Ag^+ ions. The interstitial Ag^+ ion is in

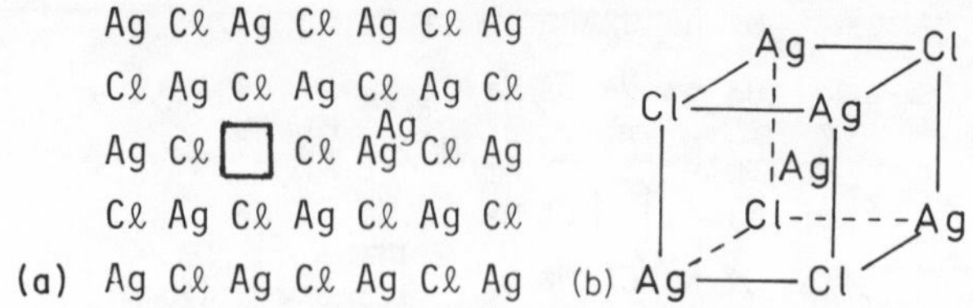

Ag Cℓ Ag Cℓ Ag Cℓ Ag
Cℓ Ag Cℓ Ag Cℓ Ag Cℓ
Ag Cℓ □ Cℓ Ag̗Cℓ Ag
Cℓ Ag Cℓ Ag Cℓ Ag Cℓ
(a) Ag Cℓ Ag Cℓ Ag Cℓ Ag   (b)

Fig. 9.3 (a) Two-dimensional representation of a Frenkel defect in AgCl, (b) interstitial site showing tetrahedral coordination by both silver and chlorine

an eight-coordinate site, therefore, with four Ag^+ and four Cl^- nearest neighbours. There is probably some covalent interaction between the interstitial Ag^+ ion and its four Cl^- neighbours which acts to stabilize the defect and give Frenkel defects, in preference to Schottky defects, in AgCl. On the other hand, Na^+, with its 'harder', more cationic character, would not find much comfort in a site which was tetrahedrally surrounded by four other Na^+ ions, Frenkel defects therefore do not occur to any significant extent in NaCl.

Calcium fluoride, CaF_2, also has predominantly Frenkel defects but in this case it is the anion, F^-, which occupies the interstitial site. These interstitial sites (empty cubes) can be seen in Fig. 7.18. Other materials with fluorite and antifluorite structures have similar defects, e.g. ZrO_2 (O^{2-} interstitial) and Na_2O (Na^+ interstitial).

As with Schottky defects, the vacancy and interstitial in a Frenkel defect are oppositely charged and may attract each other to form a pair. These pairs are electrically neutral overall, in both Schottky and Frenkel disorder, but they are dipolar. Pairs can therefore attract each other to form larger aggregates or clusters. Clusters similar to these may act as nuclei for the precipitation of phases of different composition in non-stoichiometric crystals.

9.3.3 Thermodynamics of Schottky and Frenkel defect formation

Both Schottky and Frenkel defects are *intrinsic defects*, i.e. they are present in the pure material and a certain minimum number of these defects must be present from thermodynamic considerations. It is possible and, indeed, very common for crystals to have more defects present than corresponds to the thermodynamic equilibrium concentration. This arises because crystals are usually prepared at high temperatures and intrinsically, more defects are present the higher the temperature because of the increasing importance of the $T\Delta S$ term in the free energy (Fig. 9.1). On cooling the crystals to room temperature, a small number of the defects may be eliminated, by various mechanisms, but in general and unless the cooling rate is extremely slow, the defects present at high temperature are preserved on cooling and are then present in excess of the equilibrium concentration.

An excess of defects may also be generated, deliberately, by bombarding a crystal with high energy radiation. Atoms may be knocked out of their normal lattice sites and the reverse reaction, involving elimination or recombination of defects, usually takes place only slowly.

Two approaches are used to study point defect equilibria. Statistical thermodynamics can be applied to the problem; the complete partition function for a rational model of the defective crystal is constructed, the free energy is expressed in terms of the partition function and then the free energy is minimized in order to obtain the equilibrium condition. This method can also be applied to non-stoichiometric equilibria. Alternatively, the Law of Mass Action can be applied to Schottky and Frenkel equilibria and the concentration of defects expressed as an exponential function of temperature. Because of the simplicity and ease of application of the Law of Mass Action to stoichiometric crystals, this latter method is used here. The notation used for representing crystal defects is similar to that proposed by Kröger (1974).

The Schottky equilibria in a crystal of, for example, NaCl can be represented by the following equation;

$$Na^+ + Cl^- + V_{Na}^s + V_{Cl}^s \rightleftharpoons V_{Na} + V_{Cl} + Na^{+,s} + Cl^{-,s} \qquad (9.1)$$

where V_{Na}, V_{Cl}, V_{Na}^s, V_{Cl}^s, Na^+, Cl^-, $Na^{+,s}$ and $Cl^{-,s}$ represent cation and anion vacancies, vacant cation and anion surface sites, normally occupied cation and anion sites and occupied cation and anion surface sites, respectively. The equilibrium constant for formation of Schottky defects is given by

$$K = \frac{[V_{Na}][V_{Cl}][Na^{+,s}][Cl^{-,s}]}{[Na^+][Cl^-][V_{Na}^s][V_{Cl}^s]} \qquad (9.2)$$

where square brackets represent concentrations of the species involved. The number of surface sites is always constant in a crystal of constant total surface area. Hence the number of Na^+ and Cl^- ions that occupy surface sites is always constant. On formation of Schottky defects, Na^+ and Cl^- ions move out of the crystal to occupy surface sites but, at the same time an equal number of fresh surface sites are created. (In fact, the total surface area of a crystal must increase slightly as Schottky defects are created but this effect can be ignored.)

Therefore, $[Na^{+,s}] = [V_{Na}^s]$ and $[Cl^{-,s}] = [V_{Cl}^s]$. Equation (9.2) then reduces to

$$K = \frac{[V_{Na}][V_{Cl}]}{[Na^+][Cl^-]} \qquad (9.3)$$

Let N be the total number of sites of each kind. Let N_V be the number of vacancies of each kind and hence the number of Schottky defects. The number of occupied sites of each kind is, therefore, $N - N_V$. Substituting into equation (9.3) gives

$$K = \frac{(N_V)^2}{(N - N_V)^2} \qquad (9.4)$$

For small concentrations of defects:

$$N \simeq N - N_V$$

and

$$N_V \simeq N\sqrt{K} \tag{9.5}$$

The equilibrium constant, K, can be expressed as an exponential function of temperature:

$$K \propto \exp(-\Delta G/RT) \tag{9.6}$$

$$\propto \exp(-\Delta H/RT)\exp(\Delta S/R) \tag{9.7}$$

$$= \text{constant} \times \exp(-\Delta H/RT) \tag{9.8}$$

Therefore,

$$N_V = N \times \text{constant} \times \exp(-\Delta H/2RT) \tag{9.9}$$

where ΔG, ΔH and ΔS are the free energy, enthalpy and entropy of formation of one mole of defects.

A similar expression can be derived for the concentration of vacancies in a monoatomic crystal of, for example, a metal. The difference is that, because only one type of vacancy is present, equations (9.2) to (9.4) simplify to give

$$N_V = NK$$

The factor of 2 therefore drops out of the exponential term in equation (9.9).

The Frenkel equilibria in a crystal of, for example, AgCl can be represented by

$$Ag^+ + V_i \rightleftharpoons Ag_i^+ + V_{Ag} \tag{9.10}$$

where V_i and Ag_i^+ represent empty and occupied interstitial sites. The equilibrium constant for this is given by

$$K = \frac{[Ag_i^+][V_{Ag}]}{[Ag^+][V_i]} \tag{9.11}$$

Let N be the number of lattice sites that would be occupied in a perfect crystal. Let N_i be the number of occupied interstitial sites, i.e.

$$[V_{Ag}] = [Ag_i^+] = N_i$$

and

$$[Ag^+] = N - N_i$$

For most regular crystal structures,

$$[V_i] = \alpha N$$

i.e. the number of available interstitial sites is simply related to the number of occupied lattice sites. For AgCl, $\alpha = 2$ because there are two tetrahedral interstitial sites for every octahedral site that is occupied by Ag^+ (in the cubic close packed rock salt-like structure of AgCl, there are twice as many tetrahedral sites as octahedral sites, see for example Fig. 7.13) Substituting into equation (9.11):

$$K = \frac{N_i^2}{(N - N_i)(\alpha N)} \simeq \frac{N_i^2}{\alpha N^2} \tag{9.12}$$

Using the Arrhenius expression for the temperature dependence of the number of Frenkel defects:

$$[V_{Ag}] = [Ag_i^+] = N_i = N\sqrt{\alpha}\exp(-\Delta G/2RT) \tag{9.13}$$

$$= \text{constant} \times N\exp(-\Delta H/2RT) \tag{9.14}$$

For formation of both Schottky and Frenkel defects, a factor of 2 appears in the exponential part of the expression for the number of defects (equations 9.9 and 9.14). This is because, for both of these, there are two defective sites per defect, i.e. two vacancies per Schottky defect and one vacancy, one interstitial site per Frenkel defect. In each case, therefore, the overall ΔH for defect formation can be regarded as the sum of two component enthalpies.

Experimental values for the number of Frenkel defects in AgCl are given in Fig. 9.4. The method of presentation is the one that is normally used for the Arrhenius (or Boltzmann) equation, i.e. taking logs in equation (9.14):

$$\text{Log}_{10}(N_i/N) = \log_{10}(\text{constant}) - (\Delta H/2RT)\log_{10}e$$

A plot of $\log_{10}(N_i/N)$ against $1/T$ should be a straight line of slope $-(\Delta H\log_{10}e)/2R$. The experimental data for AgCl fit the Arrhenius equation reasonably well although there is a slight but definite upward departure from linearity at high temperatures. From Fig. 9.4, and ignoring for the moment the slight curvature at high temperatures, the numbers of vacancies and interstitials increase rapidly with temperature: on extrapolating the data in Fig. 9.4 to just below the melting point of AgCl, 456 °C the concentration of Frenkel defects in equilibrium at ~ 450 °C is estimated as 0.6 per cent, i.e. approximately 1 in 200 of the silver ions move off their octahedral lattice sites to

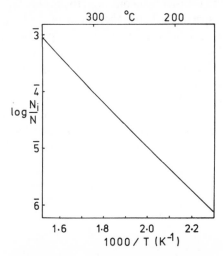

Fig. 9.4 Atomic fraction of Frenkel defects in AgCl as a function of temperature. (Data from Abbink and Martin, 1966)

occupy interstitial tetrahedral sites. This defect concentration is 1 to 2 orders of magnitude higher than observed for most ionic crystals, with either Frenkel or Schottky disorder, just below their melting points. The enthalpy of formation of Frenkel defects in AgCl is $\sim 1.35\,eV$ ($130\,kJ\,mol^{-1}$) and of Schottky defects in NaCl $\simeq 2.3\,eV$ ($\sim 220\,kJ\,mol^{-1}$); these values are fairly typical for ionic crystals.

The departure from linearity at high temperatures in Fig. 9.4 is attributed to the existence of a long-range type of Debye–Hückel attractive force between defects of opposite charge, e.g. between vacancies and interstitials in a Frenkel defect. This attraction reduces somewhat the energy of formation of the defects and so the number of defects increases, especially at high temperatures.

9.3.4 Colour centres

A considerable area of solid state physics is concerned with the study of colour centres in alkali halide crystals. The best known example is the F-centre (from the German *Farbenzentre*), shown in Fig. 9.5; it is an electron trapped on an anion vacancy. F-centres can be prepared by, for instance, heating an alkali metal halide in vapour of an alkali metal. Sodium chloride heated in sodium vapour becomes slightly non-stoichiometric due to the uptake of sodium to give $Na_{1+\delta}Cl : \delta \ll 1$, which has a greenish-yellow colour. The process must involve the absorption of sodium atoms, which subsequently ionize on the surface of the crystals. The resulting Na^{+} ions stay at the surface but the ionized electrons can diffuse into the crystal where they encounter and occupy vacant anion sites. To preserve charge balance throughout the crystals, an equal number of Cl^{-} ions must find their way out to the surface. The trapped electron provides a classic example of an 'electron in a box'. A series of energy levels are available for the electron within this box and the energy required to transfer from one level to another falls in the visible part of the electromagnetic spectrum; hence the colour of the F-centre. The magnitude of the energy levels and the colour observed depend on the host crystal and not on the source of the electron. Thus, NaCl heated in potassium vapour has the same yellowish colour as NaCl heated in sodium vapour, whereas KCl heated in potassium vapour is violet.

Another means of producing F-centres in NaCl is by irradiation. Using one of the normal methods of recording an X-ray diffraction pattern, given in Chapter 5, powdered NaCl has a greenish-yellow colour after bombardment with X-rays for half an hour or so. The cause of the colour is again trapped electrons, but in this

```
Cℓ  Na  Cℓ  Na  Cℓ

Na  Cℓ  Na  Cℓ  Na

Cℓ  Na  e   Na  Cℓ

Na  Cℓ  Na  Cℓ  Na

Cℓ  Na  Cℓ  Na  Cℓ
```

Fig. 9.5 The F-centre, an electron trapped on an anion vacancy

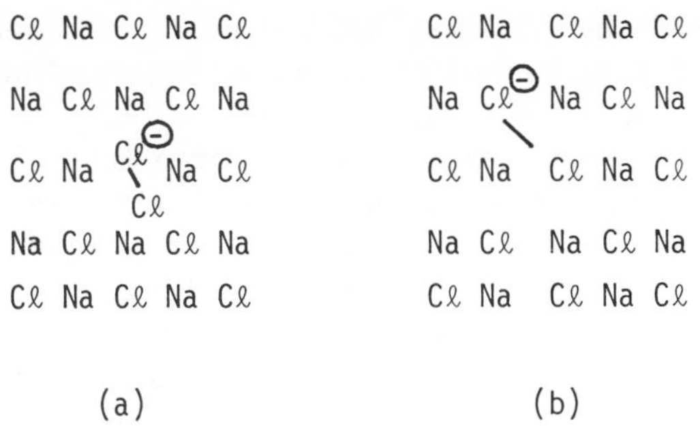

```
Cℓ  Na  Cℓ  Na  Cℓ              Cℓ  Na   Cℓ  Na  Cℓ

Na  Cℓ  Na  Cℓ  Na              Na  Cℓ⊖  Na  Cℓ  Na
                                        \
Cℓ  Na  Cℓ⊖ Na  Cℓ              Cℓ  Na   Cℓ  Na  Cℓ
        Cℓ
Na  Cℓ  Na  Cℓ  Na              Na  Cℓ   Na  Cℓ  Na

Cℓ  Na  Cℓ  Na  Cℓ              Cℓ  Na   Cℓ  Na  Cℓ


       (a)                            (b)
```

Fig. 9.6 (a) H-centre and (b) V-centre in NaCl

case they cannot arise from a non-stoichiometric excess of sodium. They probably arise from ionization of some chloride anions within the structure.

The F-centre is a single trapped electron which has an unpaired spin and, therefore, an electron paramagnetic moment. The most powerful method for studying colour centres is e.s.r. spectroscopy (Chapter 3), which detects unpaired electrons. The structure of the F-centre and the delocalized nature of the electron inside the octahedral hole was shown by e.s.r. spectroscopy. A hyperfine interaction was observed between the magnetic moment due to the electron spin and the magnetic moments of the sodium ions surrounding the trapped electron.

Many other colour centres have been characterized in alkali halide crystals; two of these, the H-centre and V-centre, are shown in Fig. 9.6. Both contain the chloride molecule ion, Cl_2^-, but this occupies one site in the H-centre and two sites in the V-centre; in both cases the axis of the Cl_2^- ion is parallel to the [101] direction. The V-centre occurs on irradiation of NaCl with X-rays. The mechanism of formation presumably involves ionization of a Cl^- ion to give a neutral chlorine atom which then covalently bonds with a neighbouring Cl^- ion.

One method by which defects can be eliminated from crystals is by annihilating each other. For example, if F- and H-centres come together they may cancel each other out, leaving a region of perfect crystal. Other defect centres which have been identified in the alkali halides include:

(a) the F′-centre, which is two electrons trapped on an anion vacancy;
(b) the F_A-centre, which is an F-centre, one of whose six cationic neighbours is a foreign monovalent cation, e.g. K^+ in NaCl;
(c) the M-centre, which is a pair of nearest neighbour F-centres;
(d) the R-centre which is three nearest neighbour F-centres located on a (111) plane;
(e) ionized or charged cluster centres also exist, such as M^+, R^+ and R^-.

9.3.5 Vacancies and interstitials in non-stoichiometric crystals

Some of the colour centres described in the previous section are non-stoichiometric defects. Non-stoichiometric crystals can also be prepared by doping pure crystals with *aliovalent impurities*, i.e. impurity atoms which have a different valency to those in the host crystal. For instance, NaCl may be doped with $CaCl_2$ to give non-stoichiometric crystals of formula, $Na_{1-2x}Ca_x V_{Na_x} Cl$, where V_{Na} represents a cation vacancy. In these crystals, the cubic close packed chloride arrangement is retained but the Na^+, Ca^{2+} ions and the cation vacancies are distributed over the octahedral cation sites. The overall effect of doping NaCl with Ca^{2+} ions is to increase the number of cation vacancies. Those vacancies that are controlled by the impurity level are termed *extrinsic* defects, in contrast to the thermally created *intrinsic* defects such as Schottky pairs.

For crystals with dilute ($\ll 1$ per cent) defect concentrations, the law of mass action can be applied. From equation (9.3), the equilibrium constant K for Schottky defects is proportional to the product of anion and cation vacancy concentrations, i.e.

$$K \propto [V_{Na}][V_{Cl}]$$

It is assumed that small additions of impurities such as Ca^{2+} do not affect the value of K. As the cation vacancy concentration increases with increasing $[Ca^{2+}]$, the anion vacancy concentration must decrease correspondingly.

The practice of doping crystals with aliovalent impurities, coupled with the study of mass transport or electrical conductivity, has proved to be a powerful method for studying point defect equilibria. Mass transport or electrical conductivity in NaCl occurs by migration of vacancies. In reality, an ion adjacent to a vacancy moves into that vacancy, thereby leaving its own site vacant, but this process can effectively be regarded as vacancy migration. Measurements are made of the dependence of conductivity on temperature and defect concentration; from a suitable analysis, various thermodynamic parameters, such as the enthalpies of creation and migration of defects, may be determined. Further details are given in Chapter 13.

The subject of doping crystals with aliovalent impurities is treated more generally in Chapter 10. Non-stoichiometric crystals can usually be regarded either as pure crystals to which a dopant has been added or as *solid solutions* based on the structure of the pure crystals. These two approaches are essentially equivalent.

9.4 Defect clusters or aggregates

Research into crystal defects is an active and rapidly advancing area of solid state science. As defects are being studied in more detail, using high resolution electron microscopy and other techniques, coupled with computer assisted modelling of defect structures, it is becoming clear that the apparently simple point defects such as vacancies and interstitials are often, in fact, complex. Instead of single atom defects, larger defect clusters form. Take the example of an interstitial metal atom in a face centred cubic metal. If the assumption is made that creation of defects, such as interstitial atoms, does not perturb the host metal structure then there are two possible sites for the interstitial atoms, tetrahedral and octahedral. Recent research shows, however, that

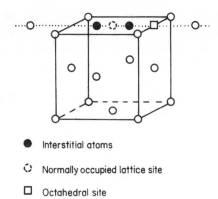

● Interstitial atoms

◌ Normally occupied lattice site

□ Octahedral site

Fig. 9.7 Split interstitial defect in a face centred cubic metal

interstitial atoms *do* perturb the host structure, especially in the immediate vicinity of the interstitial atom. An example is shown in Fig. 9.7 for platinum metal containing an interstitial platinum atom. Instead of the interstitial platinum atom occupying the octahedral site, shown □, it is displaced about 1 Å off the centre of this site and in the direction of one of the face centre atoms. The platinum atom on this face centre position also suffers a corresponding displacement in the same [100] direction. Thus, the defect involves *two* atoms, both of which are on distorted interstitial sites. This defect is known as a *split interstitial* or *dumbell-shaped interstitial*.

A similar split interstitial defect is present in body centred cubic metals such as α-Fe. It involves displacement of a normal lattice atom in the [110] direction, together with introduction of the extra iron atom (Fig. 9.8). The 'ideal' site for the interstitial would be in the centre of a cube face, but instead it is displaced off the centre of this site in the direction of one of the corners. (Note that by changing the origin of the cell to the cube body centre it would appear as if the defect involved was displacing a body centred atom in the direction of one of the cube edges. The two descriptions are equivalent.)

The exact structure of interstitial defects in alkali halides is not clear. Although Schottky defects predominate, interstitials are also present but in much smaller

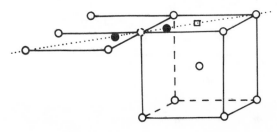

Fig. 9.8 Split interstitial in a body centred cubic metal, e.g. α-Fe. Symbols as in Fig. 9.7

330

quantities. Calculations indicate that, in some materials, occupation of undistorted interstitial sites is favoured whereas in others a split interstitial is preferred. However, these results await experimental confirmation.

The presence of vacancies in metals and ionic crystals appears to cause a relaxation of the structure in the immediate environment of the vacancy. In metals, the atoms surrounding the vacancy appear to relax inward by a few per cent, i.e. the vacancy becomes smaller, whereas in ionic crystals the reverse occurs and as a result of an imbalance in electrostatic forces the atoms relax outwards.

Vacancies in ionic crystals are charged and therefore vacancies of opposite charge can attract each other to form clusters. The smallest cluster would be either an anion vacancy/cation vacancy pair or an aliovalent impurity (e.g. Cd^{2+})/cation vacancy pair. These pairs are dipolar, although overall electrically neutral, and so can attract other pairs to form larger clusters.

One of the most studied and best understood defect systems is wüstite, $Fe_{1-x}O : 0 \lesssim x \lesssim 0.1$. Stoichiometric FeO has the rock salt structure with Fe^{2+} ions on octahedral sites. Density measurements (Chapter 10) have shown that the crystal structure of non-stoichiometric $Fe_{1-x}O$ contains a deficiency of iron rather than an excess of oxygen, relative to stoichiometric FeO. Using ideas of point defects, one would anticipate that non-stoichiometric $Fe_{1-x}O$ would have a structure represented by the formula $Fe^{2+}_{1-3x}Fe^{3+}_{2x}V_xO$ in which Fe^{2+}, Fe^{3+} ions and cation vacancies were distributed at random over the octahedral sites in the cubic close packed oxide ion array. The defect structure must be different to this, however, since neutron and X-ray diffraction studies have shown that Fe^{3+} ions are in *tetrahedral* sites. In spite of much work, there is still controversy over the

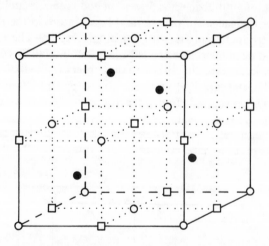

○ c.c.p. oxide ions

□ Vacant octahedral sites

● Fe^{3+} in tetrahedral sites

Fig. 9.9 Koch cluster postulated in wüstite,
$Fe_{1-x}O$

actual structure of non-stoichiometric $Fe_{1-x}O$. It appears that defect clusters must be present. The structure of one possibility, the so-called Koch cluster, is shown in Fig. 9.9. The oxygen ions are cubic close packed throughout the crystals and the cluster involves all the cation sites in a cube the size of the normal f.c.c. rock salt unit cell. The twelve edge centre and one body centre octahedral sites are all empty; four of the eight possible tetrahedral sites contain Fe^{3+} ions (dividing the cube into eight smaller cubes, these tetrahedral sites occupy the body centres of the small cubes). This cluster has a net negative charge of fourteen, since there are thirteen vacant M^{2+} sites (26 −) and only four interstitial Fe^{3+} ions (12 +). Extra Fe^{3+} ions are distributed over the normal octahedral sites around the clusters to preserve electroneutrality. It has been suggested that these Koch clusters are present in wüstites of different x values. Their number increases with x and hence the average separation between clusters decreases. Evidence from diffuse neutron scattering indicates that ordering of the clusters into a regularly repeating pattern may occur and this results in a superstructure for wüstite.

Another well-studied defect system is oxygen-rich uranium dioxide. X-ray diffraction is virtually useless for studying this material because information about the oxygen positions is lost in the heavy scattering from uranium. Instead, neutron diffraction has been used. The formula of this non-stoichiometric material is UO_{2+x}:$0 < x \lesssim 0.25$. Stoichiometric UO_2 has the fluorite structure in which interstitial positions are present at the centres of alternate cubes containing oxygen ions at the corners. In the defect cluster proposed for UO_{2+x}, an interstitial oxygen is displaced off a cube centre interstitial position in the [110] direction, i.e. towards one of the cube edges. At the same time, the two nearest corner oxygens are displaced into adjacent empty cubes along [111] directions (Fig. 9.10). Thus, in place of a single interstitial atom, the cluster appears to contain three interstitial oxygens and two vacancies.

At this point, it is worth while to emphasize why it is difficult to determine the exact structures of defects in crystals. Diffraction methods (X-ray, neutron, electron), as they are normally used in crystallographic work, yield *average* structures for crystals. For crystals that are pure and relatively free from defects, the averaged structure is usually a close representation of the true structure. For non-stoichiometric and defective crystals, however, the averaged structure may give a very poor and even an incorrect representation of the actual structure in the region of the defect. In order to determine defect structures, one essentially needs techniques that are sensitive to local structure. While study of local structure is a principal application of spectroscopic techniques (Chapter 3), they have been of limited use for studying defects other than point defects. For instance, the sites occupied by impurity atoms can often be determined if the impurity is, in some way, spectroscopically active. However, the spectra are not usually informative about structure extending beyond the immediate coordination environment of a particular atom. What is needed, therefore, are techniques that give information on local structure, but extending over a distance of, say, 10 to 20Å. Such techniques are very scarce, although there is hope that improved EXAFS methods may be suitable.

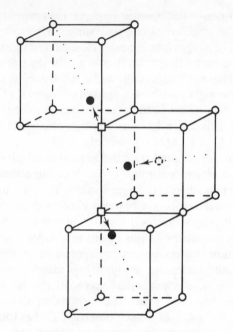

○   Oxygen

◌   Ideal interstitial site for oxygen

●   Interstitial oxygen

□   Vacant lattice site

Fig. 9.10 Interstitial defect cluster in UO_{2+x}. Uranium positions (not shown) are in the centre of every other cube

9.5 Interchanged atoms

In certain crystalline materials, it is common to find that some pairs of atoms or ions have swapped places. This may occur in alloys that contain two or more different elements, each arranged on a specific set of sites. It also occurs in certain ionic structures that contain two or more types of cation, again, each being located on a specific set of sites. If the number of interchanged atoms is large, and especially if it increases significantly as a function of temperature, this takes us into the realm of *order–disorder phenomena* (Chapter 12). The limiting situation is reached when sufficient pairs have swapped places that they no longer show any preference for particular sites. The structure is then *disordered* as regards the particular sets of atoms or ions that are involved.

Alloys, by their very nature, involve a distribution of atoms of two or more different metals over one (sometimes more) set of crystallographically equivalent sites. Such alloys are therefore examples of substitutional solid solutions (Chapter 10). Alloys can either be disordered, with the atoms distributed at random over the available sites, or can be ordered with the different atoms

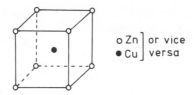

o Zn] or vice
● Cu] versa

Fig. 9.11 Ordered, primitive cubic
unit cell of β'-brass, CuZn

occupying distinct sets of sites. Ordering is generally accompanied by the formation of a supercell which is detected by X-ray diffraction from the presence of extra reflections. The ordered superstructure which is present in β'-brass, CuZn, below $\sim 450\,°C$ is shown in Fig. 9.11. Copper atoms occupy the body centre positions in a cube with zinc atoms at the corners, as in the CsCl structure; the lattice type is therefore primitive. In the disordered alloy of the same composition, β-brass, the copper and zinc atoms are distributed at random over the corner and body centre positions and therefore the lattice is body centred, as in the structure of α-Fe. The main point of interest to this chapter is that it is possible, within the ordered, β'-brass structure, to mix up *some* of the copper and zinc atoms while still retaining the long range order and superstructure. This disordering can be regarded as the introduction of defects into an otherwise perfect structure.

Good examples of interchanged atoms in non-metallic structures are provided by materials with the spinel structure. Spinels have the general formula AB_2O_4 in which A and B can be a wide variety of cations. Spinel itself is $MgAl_2O_4$. It contains a cubic close packed array of oxide ions with Mg^{2+} in tetrahedral sites and Al^{3+} is octahedral sites; as such, it is known as a *normal* spinel for which the general formula is written: $A^{tet}B_2^{oct}O_4$. Spinels can also be *inverse* in which the A ions and half the B ions swap positions to give $B^{tet}[AB]^{oct}O_4$. An example of an inverse spinel is Mg_2TiO_4, i.e. $Mg^{tet}[MgTi]^{oct}O_4$. *Intermediate* spinels can have any cation arrangement between the extremes of normal and inverse spinels. The factors that govern the manner of site occupancy in spinels have received a great deal of attention, especially because many of the cations are transition metal ions. Cation size, oxidation state, spin state and the degree of ligand field stabilization are some of the factors involved. More details are given in Chapters 8 and 16.

9.6 Extended defects

9.6.1 The crystallographic shear structures

For a long time it was known that certain transition metal oxides could be prepared with an apparent wide range of non-stoichiometry, e.g. WO_{3-x}, MoO_{3-x}, TiO_{2-x}. Following the work of Magneli, it was recognized that in certain of these systems, instead of continuous solid solution formation, a series of closely related phases with very similar formulae and structures existed. In oxygen-deficient rutile, a homologous series of phases was prepared of formula

334

Ti_nO_{2n-1} with $n = 4, \ldots, 10$. Thus, $Ti_8O_{15}(TiO_{1.875})$ and $Ti_9O_{17}(TiO_{1.889})$ are each homogeneous, physically separate phases. The structural principal underlying these phases was worked out by Magneli, Wadsley and others and the term *crystallographic shear* was coined for the type of 'defect' involved. In the oxygen-deficient rutiles, regions of normal rutile structure occur. These are separated from each other by crystallographic shear planes (CS planes) which are thin lamellae of rather different structure and composition. All of the oxygen deficiency is concentrated within these CS planes. With increased reduction, the variation in stoichiometry is accommodated by increasing the number of CS planes and decreasing the thickness of the blocks of rutile structure between adjacent CS planes.

For the purpose of understanding CS structures, it is useful to consider schematically how they might form on reduction of TiO_2. The first step in this schematic reduction involves the formation of vacant oxygen sites together with the reduction of Ti^{4+} ions to either Ti^{3+} or Ti^{2+}; the vacant oxygen sites are not distributed at random but are located on certain planes within the crystal. To avoid having whole layers of vacant sites, condensation of the structure occurs, eliminating the vacancies and forming CS planes. As a result of condensation, octahedra within the CS planes share some faces whereas in unreduced regions of rutile the corresponding linkages are by edge sharing only. Similarly, the CS planes that occur in reduced WO_3 contain edge-sharing octahedra whereas only corner sharing is observed in stoichiometric WO_3.

The structures of the CS planes in reduced rutiles are rather difficult to visualize from a drawing; instead the simpler example of reduced tungsten and molybdenum trioxides is used (Fig. 9.12). Stoichiometric MoO_3 is built of corner-sharing MoO_6 octahedra that link up to form a three-dimensional array (Fig. 2.13) as in the ReO_3 structure. In reduced MoO_3; e.g. Mo_8O_{23}, groups of octahedra link up to form edge-sharing units; such units, comprised of four octahedra, are seen in Fig. 9.12. These repeat at regular intervals in a direction that runs obliquely through the structure.

In the different members of a homologous series of phases, such as the reduced rutiles, the integrity and composition of each phase is given by the fixed spacing between adjacent CS planes. With increased reduction, i.e. as the value of n in the general formulae decreases, the distance between adjacent CS planes decreases.

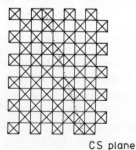

CS plane

Fig. 9.12 CS plane in the oxides M_nO_{3n-1}: $n = 8$, M = Mo. Each crossed square represents, in projection, a chain of MoO_6 octahedra linked by sharing corners. The parent, unreduced part of the structure is similar to that of cubic ReO_3

The separation of the CS planes changes in a stepwise manner in a homologous series. Each phase in the series can be regarded as a 'line phase' of essentially fixed composition and adjacent phases are separated by a narrow two-phase region. At high temperatures disorder can be introduced and each phase may begin to show a range of compositions and in some cases, such as in reduced ceria, CeO_{2-x}, the distinction between phases may genuinely disappear and a complete range of solid solutions may form above a certain critical temperature. There are various possible means by which the latter may be accomplished, one of which is by randomization of the positions of CS planes relative to the rest of the structure. Randomly spaced CS planes are known as *Wadsley defects*.

The reduced rutiles have received a lot of experimental attention and it is now known that two series of phases form, both with the general formula Ti_nO_{2n-1}. The two series differ in that the orientations of the CS planes are separated by an angular rotation of $11.53°$. One series has n values between 3 and 10 and the second series has n values between 16 and 36. Compositions which correspond to n values between 10 and 14 are particularly interesting because, in these, the orientation of the shear plane gradually changes with composition. Each composition in this range has its own value for the angular orientation of the CS planes; the latter are well ordered, as in the two homologous series of discrete phases. Since each composition in the range $n = 10$ to 14 has a distinct structure, this raises interesting questions as to exactly what is meant by the term 'a phase'. Further discussion on this rather philosophical point is deferred to Section 9.7.

Reduced vanadium dioxides, $V_nO_{2n-1}:4 \leq n \leq 8$, and mixed oxides of chromium and titanium, $Cr_2Ti_{n-2}O_{2n-1}:6 \leq n \leq 11$, also have structures containing CS planes. They appear to be isostructural with the corresponding reduced rutiles.

In the examples considered so far, the crystals contain one set of parallel and regularly spaced CS planes. The regions of unreduced structure between adjacent CS planes are limited to thin slabs or sheets. In reduced Nb_2O_5 and mixed oxides of Nb, Ti and Nb, W, the CS planes occur in *two* orthogonal sets (i.e. at 90° to each other) and the regions of perfect (unreduced) structure are reduced in size from infinite sheets to infinite columns or blocks. These 'block' or 'double shear' structures are characterized by the length, width and manner of connectivity of the blocks of unreduced ReO_3 structure. As well as having phases which are built of blocks of only one size, the complexity can be much increased by having blocks of two or three different sizes arranged in an ordered fashion. Examples of phases built on these principles are $Nb_{25}O_{62}$, $Nb_{47}O_{116}$, $W_4Nb_{26}O_{77}$ and $Nb_{65}O_{161}F_3$. The formulae of these phases can also be written in general terms, as members of homologous series, but the formulae are rather clumsy, involving several variables.

In principle, it should be possible to have structures that contain *three* sets of mutually orthogonal CS planes, in which cases the regions of unreduced material would have diminished in size to small blocks of finite length. As yet, there appear to be no known members of this group.

Structures that contain CS planes are normally studied by X-ray diffraction and high resolution electron microscopy. Single crystal X-ray diffraction is, of course, by far the most powerful method for solving crystal structures and in this sense electron microscopy usually plays a minor role; its use is limited to the determination of unit cells and space groups for very small crystals and to studying defects such as stacking faults and dislocations. However, with the technique of direct lattice imaging, electron microscopy has found great application for structural studies of CS phases. In favourable cases, an image of the projected structure at about 3Å resolution can be obtained. This usually takes the form of fringes or lines which correspond to the more strongly diffracting heavy metal atoms. The separation of the fringes can be measured and, in a 'perfect' crystal, should be absolutely regular. Whenever a CS plane is imaged, irregularity in the fringe spacing occurs because, as a result of the condensation to form a shear plane, the metal atom separation across the CS plane is reduced. A pair of fringes that is more closely spaced than normal therefore indicates a CS plane. By counting the number of normal fringes in each block, the n value of the phase in the homologous series can usually be determined. If the crystal structure of one of the members of the series has been worked out in detail using X-ray methods, it is a relatively simple matter to deduce the structures of the remainder from the electron microscope results.

For studying defects in CS structures, the electron microscope is indispensable. Thus Wadsley defects (random CS planes) can be recognized immediately; heterogeneities within supposedly single crystals can be detected, e.g. if a crystal is zoned and has a slight variation in composition with position or if the crystal is composed of intergrowths of two or more phases.

9.6.2 Stacking faults

Stacking faults occur commonly in materials that have layered structures, especially those which also exhibit polytypism. Stacking faults are examples of two-dimensional or plane defects, as also are CS planes. A metal which exhibits both polytypism and stacking faults is cobalt. It can be prepared in two main forms (polytypes) in which the arrangement of the metal atoms is either cubic close packing (...ABCABC...) or hexagonal close packing (...ABABAB...). In these two polytypes, the structures are the same in two dimensions, i.e. within the layers, and differ only in the third dimension, i.e. in the sequence of layers. *Stacking disorder* occurs when the normal stacking sequence is interrupted at irregular intervals by the presence of 'wrong' layers, e.g. schematically ...ABABABA*BCA*BABA.... The letters in italics correspond to layers that either are completely wrong (*C*) or do not have their normal neighbouring layers (*A* and *B*) on either side. Graphite is another element which exhibits polytypism (usually h.c.p. but sometimes c.c.p. of carbon atoms) and stacking disorder (mixed h.c.p. and c.c.p.). A further discussion of polytypism is given in Chapter 12.

9.6.3 Subgrain boundaries and antiphase domains (boundaries)

One type of imperfection in so-called single crystals is the presence of a domain or mosaic texture. Within the domains, which are typically ~ 10000Å in size, the

Fig. 9.13 Domain texture in a single crystal

structure is relatively perfect, but at the interface between domains there is a structural mismatch (Fig. 9.13). This mismatch may be very small and involve a difference in angular orientation between the domains that is several orders less than 1°. The interfaces between grains are called *subgrain boundaries* and can be treated in terms of dislocation theory (Section 9.8).

Subgrain boundaries involve a difference in the relative angular orientation between two parts of essentially the same crystal. Another type of boundary, called an *antiphase boundary*, involves a relative lateral displacement of two parts of the same crystal. This is shown schematically for a two-dimensional crystal, AB, in Fig. 9.14. Across the antiphase boundary, like atoms face each other and the ...ABAB... sequence (in the horizontal rows) is reversed. The two domains are said to be in antiphase relationship to each other. The term arises because, if the A and B atoms are regarded as the positive and negative parts of a wave, then a phase change of π occurs at the boundary.

The occurrence of antiphase domains in metals has been known for about thirty years. By dark field imaging in the electron microscope, the boundaries can be seen as fringes. If the boundaries are regularly spaced, unusual diffraction effects occur: a superlattice is associated with the ordering of domains but because, at the boundaries, an expansion of the structure occurs due to repulsions between like atoms satellite spots appear on either side of the supercell positions in the reciprocal lattice. This has been observed in alloys such as CuAu and in some silicate minerals such as plagioclase feldspars.

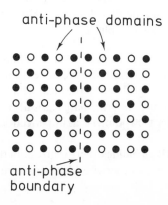

Fig. 9.14 Antiphase domains and boundaries in an ordered crystal AB: A open circles. B closed circles

9.7 Non-stoichiometry and defects—general comments

The recent advances in understanding the structures of non-stoichiometric systems such as those involving the crystallographic shear structures have necessitated a revision of some of the well-established concepts of solid state chemistry. For the past fifty years and until very recently, defects in crystals and non-stoichiometry have been treated entirely from the point of view of the classical point defects, vacancies and interstitials of Schottky and Frenkel. However, it is now becoming apparent that these isolated point defects are the exception rather than the rule and that, instead, larger defect clusters form. One problem is how to treat these new defects using thermodynamics and statistical mechanics.

Along with the discovery that more and more systems do not have simple point defects, there is a growing awareness that solid solutions which were thought to be random have in fact a heterogeneous structure on a very fine scale. Experimental evidence on the nature of the heterogeneities is difficult to obtain and may be of an indirect nature. Some examples are as follows:

(a) In high temperature solid solution systems that have immiscibility domes at lower temperatures (Chapter 11), the high temperature solid solutions may show evidence of premonitory unmixing phenomena in which clustering of atoms of the end-member phases occurs.

(b) In compositions which undergo phase changes on cooling, such as disorder–order transformations, nuclei of the low temperature form may be present in the high temperature solid solutions. In solid solutions that are stable at high temperatures but which become supersaturated and precipitate out a second phase on cooling, tiny nuclei of the precipitate may be present within the high temperature solid solutions. An example taken from doped alkali halides is the clustering of defects in $Na_{1-2x}Cd_xCl$ solid solutions prior to precipitation of $CdCl_2$ (formation of Suzuki phases).

(c) Wadsley defects (random CS planes) provide excellent examples of micro-heterogeneity, as they have different structure and composition from the remainder of the solid solution in which they occur. An interesting question arises as to whether or not Wadsley defects can be regarded as a separate phase. If they can, the structure in which they occur would be a two-phase mixture of the end-member (e.g. TiO_2) and the Wadsley defect. However, this causes difficulties in defining exactly what is a phase.

The structural status of some crystal defects has become a debatable issue following work on highly non-stoichiometric systems. Many cases exist in which the defects should clearly be regarded as an integral part of the ideal structure and not as some perturbation of the ideal structure. Intermediate cases also exist in which the defects can be regarded from either viewpoint, i.e. as imperfections or as essential structural components.

In very dilute defect systems such as the alkali halides, there is no doubt that the defects can be regarded as imperfections in an otherwise ideal structure. Even though the presence of defects is based on sound thermodynamic principles, nevertheless from a structural point of view the vacancies and interstitials are imperfections and are not essential to the ideal structure.

What happens, however, when defects begin to interact significantly and to order themselves, as do crystallographic shear planes in the various homologous series of reduced oxides? In these, the integrity of each phase is given solely by the manner of ordering of the CS planes. The question is, then, how could we possibly regard the CS planes as defects when they are essential to the ideal structure of the phase?

To understand how the conflict can arise, we need to consider the nature of non-stoichiometric systems. In the 'classical' systems such as NaCl, the predominant defects are vacancies and we can introduce these either by increasing the temperature or by doping with an aliovalent impurity. The crystal structure of non-stoichiometric, doped NaCl can readily be referred back to the stoichiometric structure without the need to introduce any new structural units. Thus, if Cd^{2+} is the aliovalent cation, non-stoichiometry is accommodated by varying the number of Cd^{2+}, Na^+ ions and cation vacancies which are distributed over the octahedral cation sites. However, even in dilute defect systems such as this, there is evidence of clustering between, for example, Cd^{2+} ions and cation vacancies (the clusters may be dissociated with an appropriate amount of energy and both associated and unassociated Cd^{2+} ions commonly occur). This now brings us to the crux of the problem. If defects can interact with each other in solid solutions the possibility exists that clusters of defects may form and either grow or arrange themselves in ordered fashion such that a new phase results.

It would seem, therefore, that we cannot automatically assume that, say, a vacancy or CS plane is a defect without first stating what is the standard of reference. A CS plane is a defect only if it is not part of the ordered arrangement of CS planes which constitutes the identity of a certain phase. Otherwise, it merits the status of being regarded as a fundamental part of a structure and must not be relegated to the level of defect or imperfection. Vacancies are usually regarded as defects but examples are known where this cannot be so. In Pr_2O_3, a superstructure forms which is derived from the PrO_2 structure by ordering of the vacant oxygen sites. PrO_2 has the fluorite structure and in Pr_2O_3 every fourth site in the fluorite structure is vacant, $Pr_2O_3 V_0$. The vacancies are therefore essential to the formation of the superstructure. (Further studies may show that a structure with simple vacancies is not an adequate description for Pr_2O_3, but this would not affect the conclusion drawn here.)

The definition of a phase, in the phase rule sense (Chapter 11), normally presents no difficulties in crystalline systems, but with the elucidation of the CS structures the situation is no longer clear-cut. With the ordered CS structures, e.g. Ti_nO_{2n-1}, $n = 4, \ldots, 10$, each n value corresponds to a definite arrangement of CS planes and there is no doubt that each one constitutes a separate phase. Thus, compositions of intermediate n values, e.g. 5.5, at equilibrium, can be regarded as physical mixtures of the two appropriate CS phases ($n = 5$ and 6). However, in the composition region $n = 10, \ldots, 14$, the situation is rather different because *every* composition has its own unique structure. Each structure is unique because the angular orientation of the CS planes changes continuously with composition. Thus, each composition has its own value for the angular orientation of the ordered CS planes. Consequently, a range of compositions exists which is not a solid solution (because the structures are not based on a single average structure). But what is it? To say that every possible composition (and therefore the number of possibilities is infinite) constitutes a separate phase is rather hard to swallow, but what alternative is there? Although the theory and application of the phase rule is unlikely to disintegrate into chaos as a result of these discoveries, the question is, nevertheless, very relevant.

9.8 Dislocations, mechanical properties and reactivity of solids

Dislocations are an extremely important class of crystal defect. They are responsible for the relative weakness of pure metals and in certain cases (after work-hardening) for just the opposite effect of extra hardness. The mechanism of crystal growth from either solution or vapour appears to involve dislocations. Reactions of solids often occur at specific active surface sites where dislocations emerge from the crystal.

Dislocations are stoichiometric line defects. Their existence was postulated long before direct experimental evidence of their occurrence was obtained. There were several types of observation which indicated that defects other than point defects must be present in crystals:

(a) Metals are generally much softer than expected. Calculations of the shearing stress of metals gave values of $\sim 10^6$ p.s.i. whereas experimental values for many metals are as low as $\sim 10^2$ p.s.i. This indicated that there must be some kind of weak link in their structures which allows metals to cleave so easily.

(b) Many well-formed crystals were seen under the microscope or even with the naked eye to have spirals on their surfaces which clearly provided a mechanism for crystal growth. Such spirals could not occur in perfect crystals, however.

(c) The malleable and ductile properties of metals were difficult to explain without invoking dislocations. Thus, ribbons of magnesium metal can be stretched out to several times their original length, almost like chewing gum, without rupture.

(d) The process of work-hardening of metals was difficult to explain without invoking dislocations.

Dislocations can be one of two extreme types, edge or screw dislocations, or can have any degree of intermediate character. An *edge dislocation* is shown schematically in Fig. 9.15 and constitutes an extra half-plane of atoms, i.e. a plane of atoms that goes only part of the way through a crystal structure. Planes of atoms within the crystal structure are shown in projection as lines. These lines are parallel except in the vicinity of the region where the extra half-plane terminates.

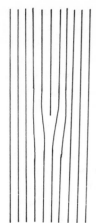

Fig. 9.15 Edge dislocation in projection

The centre of the distorted region is a line that passes right through the crystal, perpendicular to the paper, and approximates to the end of the extra half-plane; this is the *line* of the dislocation. Outside this stressed region, the crystal is essentially normal; the top half in the drawing must be slightly wider than the bottom half in order to accommodate the extra half-plane.

In order to understand the effect of dislocations on the mechanical properties of crystals, consider the effect of applying a shearing stress to a crystal that contains an edge dislocation, Fig. 9.16. The top half of the crystal is being pushed

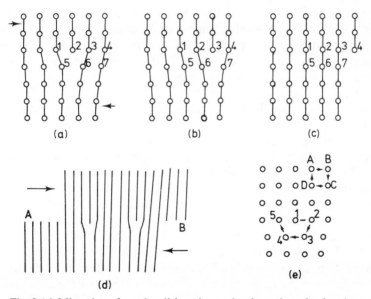

Fig. 9.16 Migration of an edge dislocation under the action of a shearing stress

to the right and the bottom half to the left. Comparing (a) and (b) in Fig. 9.16, the extra half-plane which terminates at 2 in (a) can effectively move simply by breaking the bond 3–6 and forming a new linkage 2–6. Thus, with a minimum of effort, the half-plane has moved one unit of distance in the direction of the applied stress. If this process continues, the extra half-plane eventually arrives at the surface of the crystal as in Fig. 9.16(c). All that is now needed is an easy means of generating half-planes and by a process of repetition the crystal will eventually completely shear. In Fig. 9.16(d), it is assumed that half-planes are generated at the left-hand end of the crystal (mechanisms of generation are discussed later). With each one that is generated, another one, equal and opposite in orientation and sign, is left behind. At the left-hand side, five half-planes in the bottom part of the crystal (negative dislocations with the symbol T) have accumulated. Balancing these in the top part of the crystal, three half-planes have arrived at the right-hand end and two more (positive dislocations with the symbol $\perp$) are still inside the crystal and in the process of moving out.

The easy motion of dislocations has an analogy in one's experience of trying to rearrange a carpet, preferably a large one. The method of lifting up one end and tugging needs a great deal more effort than making a ruck in one end of the carpet and gliding this through to the other end.

The process of movement of dislocations is called *slip*, the pile-up of half-planes at opposite ends gives ledges or *slip steps*. The line AB in (d) represents the projection of the plane over which the dislocation moves and is called the *slip plane*.

Dislocations are characterized by a vector, the *Burgers vector*, **b**. To find the magnitude and direction of **b**, it is necessary to make an imaginary atom-to-atom circuit around the dislocation (Fig. 9.16e). In normal regions of the crystal a circuit such as ABCDA, involving one unit of translation in each direction, is a closed loop and the starting point and finishing point are the same, A. However, the circuit 12345 which passes round the dislocation is not a closed circuit because 1 and 5 do not coincide. The magnitude of the Burgers vector is given by the distance 1–5 and its direction by the direction 1–5 (or 5–1). For an edge dislocation, **b** is perpendicular to the line of the dislocation and parallel to the direction of motion of the line of the dislocation under the action of an applied stress. It is also parallel to the direction of shear.

The *screw dislocation* is a little more difficult to visualize and is shown in Fig. 9.17. In (b), the line SS′ represents the line of the screw dislocation. In front of this line the crystal has undergone slip but behind the line it has not. The effect of continued application of a shearing stress, arrowed in (b), is such that the slip step gradually extends across the whole face (side faces) of the crystal at the same time as the line SS′ moves towards the back face (Fig. 9.17c). To find the Burgers vector of a screw dislocation, consider the circuit 12345 (Fig. 9.17a) which passes around the dislocation. The magnitude and direction of the distance 1–5 defines **b**. For a screw dislocation, the Burgers vector is parallel to the line of the dislocation (SS′) and perpendicular to the direction of motion of this line. This contrasts with the edge dislocation and it can be seen that edge and screw dislocations are effectively

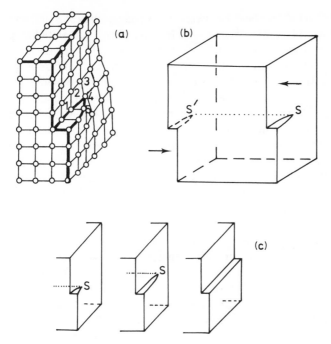

Fig. 9.17 Screw dislocation

at 90° to each other. For both edge and screw dislocations, **b** is parallel to the direction of shear or slip. The origin of the term screw is easy to see by considering the atoms 54321 in Fig. 9.17(a). They lie on a spiral which passes right through the crystal and emerges with opposite hand or sign at S'. As in the case of an edge dislocation, it is necessary to break only a few bonds in order for a screw dislocation to move. Thus in (a), the bond between atoms 2 and 5 has just broken and 2 has joined up with 1; the bond between 3 and 4 will be the next to break. Note that it is useful to consider bonds as breaking and forming although in practice it is not nearly as clear-cut as this; in metals and ionic crystals the bonds are certainly not covalent.

The process of generation of dislocations is complicated but seems always to involve the formation of *dislocation loops*. Consider the crystal shown in Fig. 9.18. On looking at the right-hand face it is obvious that a screw dislocation energes near to point S, where the slip step terminates. However, this dislocation does not

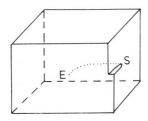

Fig. 9.18 A quarter dislocation loop

reach the left-hand face because there is no corresponding slip step on it. Looking now at the front face, in isolation, it is clear that there is an extra half plane and a positive edge dislocation in the top part of the crystal which emerges, say, at E. However, once again this dislocation does not extend to the appropriate point on the opposite face, i.e. the back face. If a dislocation enters a crystal it must reappear somewhere because, by their nature, dislocations cannot just terminate inside a crystal. What happens in this case is that the two dislocations change direction inside the crystal and meet up to form a *quarter dislocation loop*. Thus, the same dislocation is pure edge at one end, pure screw at the other and in between has a whole range of intermediate properties. It is difficult without the aid of a three-dimensional model to picture the structural distortions which must occur around the part of the dislocation which has a mixed edge and screw character.

The origin of dislocation loops is by no means well understood. One source, discussed later, is by means of clustering of vacancies on to a plane, followed by an inward collapse of the structure. This generates loops which are entirely inside the crystal. For purpose of illustration, an alternative source is shown in Fig. 9.19. To start a loop, all that is needed is a nick in one edge of the crystal; the displacement of a few atoms is then sufficient to create a small quarter loop (Fig. 9.19a). Once created, the loops can expand very easily (Fig. 9.19b). Usually loops do not expand symmetrically because edge dislocations move easier and more rapidly than corresponding screw dislocations. The result is shown in Fig. 9.19(c); the edge component has sped across to the opposite, left-hand face and the slip process for the front few layers of the crystal is complete. The dislocation which remains is mainly or entirely screw in character and this continues to move slowly to the back of the crystal, thus terminating the slip process for this dislocation.

As a consequence of the very easy generation of dislocation loops, the mechanical strength of materials, especially metals, is greatly reduced. This is obviously very serious for metals which are used in construction. The impact resistance of metals to high stresses is easy to measure because the experiments can be carried out rapidly. However, much more difficult to assess is the resistance over a long period of time to much smaller stresses. These small stresses

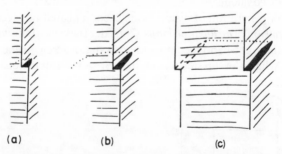

(a)   (b)   (c)

Fig. 9.19 Generation and motion of a dislocation loop

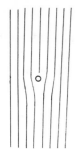

Fig. 9.20 Locking of an edge dislocation at an impurity atom

may be sufficient to generate and move dislocations only very slowly but the results are cumulative and, generally, non-reversible, so that one day catastrophic failure of the metal may occur, perhaps for no apparent reason.

As well as dislocations being a great weakener of materials they can also have the reverse effect and greatly increase the strength or hardness. One process that occurs is the 'locking' of dislocations at certain impurity atoms (Fig. 9.20), e.g. interstitial carbon atoms dissolved in iron. The dislocation moves freely until it encounters the impurity atom and then becomes trapped by the impurity atom(s).

A very important strengthening process is that which occurs in the work-hardening of metals. On hammering a metal, an enormous number of dislocations are generated which in a polycrystalline material are in a large number of orientations. These start to move through the crystal but sooner or later, depending on the metal and its crystal structure, movement stops. Grain boundaries provide an effective means of stopping dislocations inside the grains. This is because, as a dislocation passes out of a grain, the surface of the grain becomes deformed and the resulting stress imposed upon neighbouring grains may act to prevent the dislocation from ever reaching the surface. Because the area around a dislocation is stressed, two dislocations may repel each other if they get too close. Thus, once the leading dislocations become trapped at grain boundaries or against dislocations arriving from other directions, the succeeding dislocations pile up. A log-jam of dislocations then rapidly forms and the individual dislocations can neither move forwards nor backwards, so that a considerable increase in the strength of the metal results. This process is called *strain-hardening*.

Strain-hardened metals can be rendered malleable and ductile once again by high temperature annealing. At high temperatures, atoms have considerable thermal energy which enables them to move. Dislocations may therefore be able to reorganize themselves or annihilate each other. If a positive and a negative edge dislocation meet on the same slip plane they cancel each other out, leaving behind a strain-free area in the crystal. The process can be very rapid; e.g. platinum crucibles used in laboratory experiments may be softened in a few minutes by placing them at, say, 1200 °C.

9.8.1 Observation of dislocations

Optical and electron microscopy are by far the most important methods for the study of dislocations. Using the technique of lattice imaging, dislocations may be seen directly in electron micrographs of thin crystals. An older and more established procedure is to etch the surface of the material under study and view the resulting etch patterns by microscopy.

Much of the early work on dislocations (in the 1950s) was done by metallurgists using crystals of LiF and the technique of etching followed by examination with optical microscopy. Metals are difficult to study by this technique because it is not usually possible to prepare metals that are free from dislocations. LiF is an excellent material to study, however, because it can be prepared relatively free from dislocations, in single crystal form and, on the application of moderate stresses, dislocations can be introduced in a controlled manner. The dislocations are observed, indirectly, by etching the LiF crystals in H_2O_2 solution, in which LiF is normally only very sparingly soluble. At the sites of emergent dislocations on the crystal surface, the crystal is in an abnormally stressed condition, the atoms or ions are not located in deep free energy minima and this region of the crystal dissolves rapidly in H_2O_2. This results in the formation of deep, pointed etch pits, in the form of inverted pyramids. The method is rather destructive because, on an atomic scale, enormous holes are created just so that one tiny dislocation can be observed. However, it clearly demonstrates the effect of dislocations on the reactivity, in this case solubility, of crystals. If the pits are etched to a size of 10 μm at their base (i.e. at the surface of the crystal) the dislocations must not be closer than 10 μm (i.e. 10^5 Å) in order for the etch pits to be seen in isolation. During slip, dislocations may be much closer than this, in which case a continuous linear etch pit may form. The pits occur in lines or sets of lines because of the repeated generation of dislocations at certain surface sites.

With this combined approach of etching followed by microscopic examination, the number and distribution of dislocations in a crystal can be assessed. This also gives an indirect, yet sensitive method of studying the movement of dislocations. Etch pits have pointed bottoms only as long as there is an emergent dislocation at the bottom of the pit. If the dislocation moves away and etching is continued, the pit continues to grow, but only laterally; flat-bottomed pits result. In favourable cases, it is therefore possible to study the migration of dislocations under the action of an applied stress by periodically releasing the stress, re-etching the crystal and observing the dislocation network from the pattern of etch pits; new dislocation sites can be distinguished from old ones by the shape of the pits. One interesting result is that, in crystals that have been studied in this way, edge dislocations appear to move much more rapidly than do screw dislocations.

An overall picture of the presence of dislocations may be obtained by viewing the crystal between crossed polars in the optical microscope. LiF is cubic and therefore dark in crossed polars. The presence of dislocations caused by a compressive stress distorts the structure from cubic symmetry in the vicinity of the dislocations and this shows up as *stress birefringence*. The dislocations in LiF are concentrated on the planes {110} and {100} and this gives a criss-cross or tartan appearance to the stress birefringence pattern.

9.8.2 Dislocations and crystal structure

For a given crystal there is usually a preferred plane or set of planes on which dislocations can occur and also preferred directions for dislocation motion. The energy, E, that is required to move a dislocation by one unit of translation is proportional to the square of the magnitude of the Burgers vector, $\mathbf{b}$, i.e.

$$E \propto |\mathbf{b}|^2$$

Thus, the dislocations that generally are most important in a particular material are those that have the smallest $\mathbf{b}$ value.

In metals, the direction of motion of dislocations is usually parallel to one of the directions of close packing in the metal crystal structure. In Fig. 9.21(a), two rows of spheres are shown and each represents a layer of atoms in projection; the rows are close packed because the individual atoms of each row are touching neighbouring atoms in the same row. Under the action of a shearing stress and with the aid of a dislocation (not shown) the top row shears over the bottom row by one unit. After shear, Fig. 9.21(b), atom $1'$ occupies the position that was previously occupied by atom $2'$, and so on. The distance $1'–2'$ corresponds to the unit of translation for slip and for dislocation motion. It is equivalent to the magnitude of the Burgers vector, i.e. $|\mathbf{b}'| = d$, where d is the diameter of the spheres. Therefore $E_\mathbf{b} \propto d^2$.

The corresponding example of dislocation motion in a non-close packed direction is shown in Fig. 9.21 (c) and (d). The distance $1'–2'$ is now longer than in

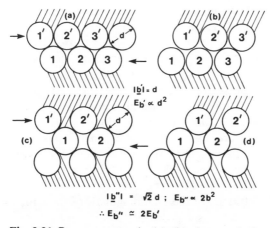

Fig. 9.21 Burgers vector in (a), (b) close packed and (c), (d) non-close packed directions in a crystal

348

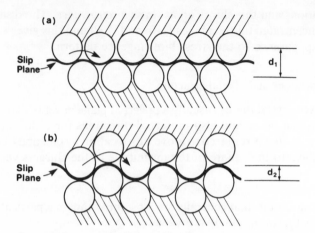

Fig. 9.22 Slip occurs more easily if the slip plane is a close packed plane as in (a)

(a) and (b); by a simple geometrical construction, it can be shown that $\mathbf{b}'' = d\sqrt{2}$. Therefore, $E_b'' \propto 2d^2$. Consequently, motion of dislocations in a close packed direction requires only half the energy needed for motion in the particular non-close packed direction shown in (c).

It has been shown that the preferred direction of motion of dislocations is parallel to close packed directions. It can also be shown that the preferred plane of shear or slip is usually a close packed plane. This is because the distance of separation between two close packed planes, d_1, is greater than the corresponding distance between two non-close packed planes, d_2 (Fig. 9.22). During shear, the atoms to either side of the slip plane move, relatively, in opposite directions and the distortion of the structure for intermediate or saddle positions is much less in (a) than in (b). Alternatively, the energy barrier to be surmounted by an atom in moving from one position to an equivalent one (represented by the curved arrows) is much greater for example (b).

Metals which are face centred cubic—Cu, Ag, Au, Pt, Pb, Ni, Al, etc.—are generally more malleable and ductile than either hexagonal close packed—Ti, Zr, Be—or body centred cubic metals—W, V, Mo, Cr, Fe—although there are notable exceptions such as Mg (h.c.p.) and Nb (b.c.c.) which are malleable and ductile. Many factors influence malleability and ductility. In part, malleability and ductility depend on the numbers of close packed planes and directions possessed by a structure. Face centred cubic metals have four different sets of close packed planes; these can be visualized by removing, in turn, each of the four corners in the top part of the cubic unit cell, Fig. 7.5. Each close packed layer possesses three close packed directions, $x - x'$, $y - y'$, $z - z'$, Fig. 7.1. These close packed directitons are the face diagonals of the cube and have indices [110]; there are a total of six close packed directions.

A hexagonal close packed structure contains only one set of close packed layers, which is parallel to the basal plane of the unit cell shown in Fig. 7.6. There are also only three close packed directions; these are in the plane of the close packed layers. It is rather difficult to show in a drawing that there is only one set of c.p. layers in hexagonal close packing and the reader is asked either to take this on trust or to make a three-dimensional model for himself to verify it.

A body centred cubic unit cell does not contain any close packed layers. The atomic coordination number in b.c.c. is eight whereas close packed structures have a coordination number of twelve. The b.c.c. structure does contain four close packed directions, however; these correspond to the four body diagonals of the cube, i.e. $\langle 111 \rangle$.

The behaviour of a metal under stress depends very much on the direction of the applied stress relative to the direction and orientation of the slip directions and slip planes. The effect of tensile stress on a single crystal rod of magnesium metal, whose crystallographic orientation is such that the basal plane of the hexagonal unit cell is at 45° to the rod axis, is shown in Fig. 9.23. For stresses below ~ 100 p.s.i., the crystal undergoes *elastic deformation*, i.e. no permanent elongation occurs. Above the *yield point*, ~ 100 p.s.i., *plastic flow* begins to occur and the crystal suffers an irreversible elongation. Magnesium metal is quite remarkable in that it can be stretched out to an elongation of up to several times its original length. How this can occur is shown schematically in Fig. 9.23(c). Slip occurs on a massive scale and the resulting slip steps are so large that they can be observed directly with an optical microscope. If the passage of a dislocation gives rise to a slip step that is, say, 2 Å high (or wide), then in order that such features are visible in the microscope the slip steps must have dimensions of at least 2 μm. At

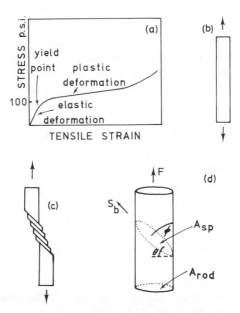

Fig. 9.23 (a) Tensile stress–strain curve for single crystal Mg. (b), (c) Tensile stress resulting in slip and elongation. (d) Definition of the resolved shear stress

least 10000 dislocations must therefore have passed on the slip plane for every slip step that can be seen.

If the basal planes of magnesium are orientated at other than 45° to the rod axis, it is found experimentally that a larger applied stress is needed in order for plastic flow to occur. This is because the important factor is the magnitude of the applied shear stress when resolved parallel to the basal plane. This is shown as follows (Fig. 9.23 d).

For an applied force, F, the stress S_A, on the slip plane is given by

$$S_A = F/A_{sp} = F(\cos \theta/A_{rod}) \qquad (9.15)$$

where A_{rod} is the cross-sectional area of the rod, A_{sp} is the area of the slip plane and θ is the angle between A_{rod} and A_{sp}. The stress S_A on the slip plane has a component S_B parallel to the direction of slip in the slip plane, given by

$$S_B = S_A \cos \phi = (F/A_{rod}) \cos \theta \cos \phi \qquad (9.16)$$

where ϕ is the angle between the slip direction and the stress axis. S_B is the resolved shear stress and the value which is needed to cause plastic flow is the *critical resolved shear stress*. From the interrelation that exists between θ and ϕ, the maximum value of S_B occurs when $\theta = \phi = 45°$, i.e.

$$S_B = \tfrac{1}{2}(F/A_{rod}) = \tfrac{1}{2}S \qquad \text{for } \theta = \phi = 45° \qquad (9.17)$$

where S is the stress applied to the crystal. It follows from equation (9.16) that when the slip plane is either perpendicular or parallel to the direction of applied stress, the resolved shear stress is zero and therefore slip cannot occur.

The ability of face centred cubic metals to undergo severe plastic deformation in contrast to hexagonal and especially body centred cubic metals can be qualitatively explained. Face centred cubic metals have four close packed planes and six close packed directions which are suitable for slip. For any given applied stress, at least one and usually more of these slip planes and directions is suitably oriented, relative to the direction of applied stress, such that slip can occur. In contrast, hexagonal metals slip easily only when in a certain range of orientations relative to the applied stress. The difference between these groups of metals shows up markedly in the mechanical properties of polycrystalline pieces of metal. Polycrystalline hexagonal metals almost certainly contain grains whose slip planes are either parallel or perpendicular to the direction of applied stress. The amount of plastic deformation that is possible in these is therefore more limited. On the other hand, face centred cubic metals are malleable and ductile whether in the form of single crystals or polycrystalline pieces.

The critical resolved shear stresses for face centred cubic metals are usually small, e.g. in pounds per square inch Cu 92, Ag 54, Au 132 and Al 148. Values for the hexagonal metals fall into two groups. One group is similar to values for the face centred cubic metals, e.g. Zn 26, Cd 82, Mg 63. The other group has much higher values, e.g. Be 5700, Ti 16000 (slip in titanium occurs more easily on a non-basal plane, but one which has a close packed direction; it has a critical resolved shear stress of 7100 p.s.i.). The reason for the much higher values of Be, Ti (and Zr) probably has something to do with the fact that the unit cell dimensions of these metals show that the structures are somewhat compressed in the c direction (perpendicular to the basal plane). An ideal hexagonal unit cell has a $c:a$ ratio of 1.632; values of $c:a$ for Zn, Cd and Mg are 1.856, 1.886 and 1.624, respectively, and are similar to or greater than the ideal value. Values for Be, Ti and Zr are 1.586, 1.588 and 1.590, respectively, and are all consistently *less* than the ideal c/a ratio. Thus, it would seem that this decrease in distance between adjacent basal planes makes it much more difficult for basal slip to occur.

The critical resolved shear stress for body centred metals is also high, e.g. α-Fe $\simeq 4000$ p.s.i., because, although they possess four close packed directions, they do not have any close packed planes.

9.8.3 Dislocations, vacancies and stacking faults

It is commonly assumed that although point defects and dislocations are both examples of crystal imperfections they do not really belong under the same heading. Vacancies, interstitials and the crystallographic shear structures can be regarded as chemical defects; they are an intrinsic part of the crystal structure and are important in chemical reactions and mass transport. On the other hand, dislocations are very much a physical defect, being related more to the mechanical properties of materials. With advances in our understanding of both types of defect it is clear that in fact they have much in common. For example, vacancies can contribute to the motion of dislocations by the process of *climb* (Fig. 9.24). An atom at the end of an extra half-plane, which constitutes a dislocation, can move if there is a vacancy adjacent to it. The net effect of this is that the dislocation effectively gets smaller or begins to climb out of the crystal.

An intimate relation exists between dislocations and stacking faults (discussed in section 9.6.2). Two layers of a face centred cubic metal are shown in projection in Fig. 9.25. The upper layer (white circles) has a missing zigzag row of atoms. This corresponds to a negative edge dislocation in the lower layer, immediately beneath the row of missing atoms. Thus, the presence of an extra half-plane in the bottom part of the crystal, and which terminates as an edge dislocation in the lower layer, forces apart the atoms in the immediately succeeding (upper) layer such that a row of atoms is apparently missing. This is shown in a different orientation in Fig. 9.25(c). The bottom part of the crystal in (a) terminates in (c) at

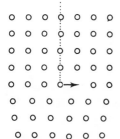

Fig. 9.24 Climb of an edge dislocation by vacancy migration

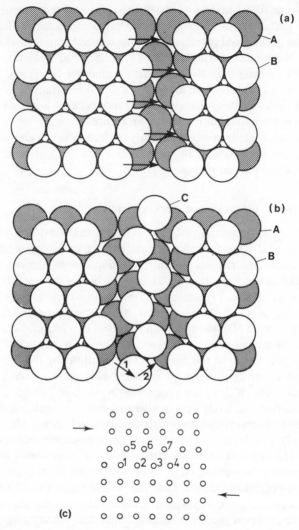

Fig. 9.25 (a) An edge dislocation and (b) a partial dislocation in a face centred cubic metal; (c) the edge dislocation in projection

the layer which in projection contains the atoms 1, 2, 3 and 4; the upper layer with the missing row is equivalent to the layer that contains atoms 5, 6 and 7. Motion of the negative dislocation in (c) effectively involves motion of the vacant row immediately above the dislocation such that atom 6 and the row of atoms beneath it undergo a sideways displacement. This is equivalent to movement of one row of atoms in the direction indicated by the arrows in (a). The stacking sequence of layers is unchanged by passage of the dislocation, as atoms in the upper layer occupy B positions before and after slip. The Burgers vector is given by the direction of the arrows and has magnitude approximately equal to one atomic diameter.

The atomic motions shown in Fig. 9.25(a) are relatively difficult to effect because atoms in the upper layer virtually have to climb over the top of atoms in the layer below in order to follow the direction indicated by the arrows. A much easier alternative route is shown in Fig. 9.25(b) in which the migration is divided into two smaller steps (arrows 1 and 2). For each of these, the B atoms traverse a low pass between two A atoms; after the first pass (arrow 1), the atoms find themselves in a C position; Fig. 9.25(b) shows a row of atoms in C positions. This process is repeated (arrow 2) and after the second pass the atoms enter the adjacent set of B positions. In this manner a dislocation is divided into two *partial dislocations*. The formation of partial dislocations can be regarded from two viewpoints. In terms of the energy barriers to be surmounted, two small passes are easier than one big hump, even though the combined distance is longer. In terms of the Burgers vector for the two alternatives, the longer, divided pathway is favoured, provided the distance is not too much longer. If the direct jump is across a distance b, the energy E_1 is given by $E \propto |b|^2$. For each partial jump, the distance involved is $b/\sqrt{3}$, because the direction is at 30° to the horizontal. Thus for the two partial jumps the combined energy E_2 is proportional to $(b/\sqrt{3})^2 + (b/\sqrt{3})^2$, i.e.

$$E_2 \propto \tfrac{2}{3}|b|^2$$

Therefore,

$$E_2 < E_1$$

Separation of two partial dislocations may occur as in Fig. 9.26, due to their mutual repulsion. In between the two partial dislocations the stacking sequence is 'wrong' because all the white atoms occupy C positions; we now have the makings of a stacking fault. If the two partial dislocations can be encouraged to

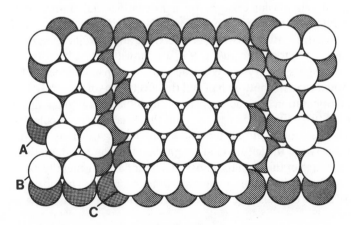

Fig. 9.26 Separation of two partial dislocations to give a stacking fault

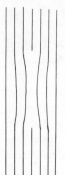

Fig. 9.27 Collapse of the structure around a cluster of vacancies thereby generating a dislocation loop

go to the edges of the crystal the whole layer will be faulty. Whether or not this can happen in practice is unknown but, nevertheless, a clear relationship exists between dislocations, partial dislocations and stacking faults.

One suggested way to generate dislocations is by clustering vacancies onto a certain plane in a structure followed by collapse of the structure, thereby generating a dislocation loop. This mechanism would be important only at high temperatures where vacancies both form readily and are relatively mobile. If in a pure metal vacancies cluster onto a certain plane in the structure such that every atomic site within, say, a certain radius is empty, then a disc-shaped hole would be present in the structure. The two sides of the disc would then cave in, as shown in cross-section in Fig. 9.27, and it can be seen that the two opposite ends have the appearance of edge dislocations of opposite sign. This therefore gives a means of generating dislocations that has nothing to do with mechanical stresses.

9.8.4 Dislocations and grain boundaries

An elegant relationship exists between dislocations and grain boundaries such that the interface between two grains in a polycrystalline material may be regarded as a dislocation network, provided the angular difference in orientation of the grains is not too large. In Fig. 9.28 are shown six positive edge dislocations at different heights in a crystal. With each one, the top half of the crystal becomes slightly wider than the bottom half and the effect of having several dislocations in a similar orientation and position is to introduce an angular misorientation between the two halves of the crystal. Thus the surface perpendicular to the paper on which these dislocations terminate can be regarded as an interface or boundary between the left-hand and right-hand grains, although the continuity in the structures of the crystals across this boundary is generally excellent. This type of interface is known as a *low angle grain boundary*. The mosaic or domain texture of single crystals (Fig. 9.13) can be described in terms of low angle grain boundaries.

The angular misorientation of two grains can be calculated readily. If across a

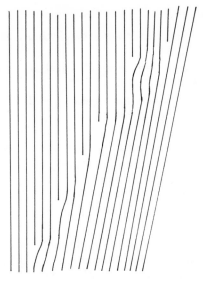

Fig. 9.28 Array of edge dislocations which constitutes a low angle grain boundary

piece of crystal which is say 1 μm (i.e. 10000 Å) wide there are five positive edge dislocations forming a low angle grain boundary, then one end of the crystal will be five atomic planes, say 10 Å, wider than the other end. The angular misorientation of the grains, $\theta = \tan^{-1} (10/10000) = 0.057°$. Low angle grain boundaries can sometimes be revealed by appropriate etching of the surface, although for the pits to be seen in isolation in the optical microscope they must be at least 1 to 2 μm in size and separated by the same distance. The dislocation model for grain boundaries is useful for angular misorientations of up to several degrees. For misorientations > 10 to 20°, however, the model is probably no longer meaningful because the edge dislocations would have to be spaced too closely and would lose their separate identity. The domain texture of 'single crystals' is well known from X-ray diffraction experiments, where it can be shown that the reflections are broader than expected for a single crystal.

The angular mismatch of grains or domains can thus be determined by two independent methods—by X-ray spot broadening and by counting etch pits; the results which have been obtained are in good agreement. Although surface energies of crystals are beyond the scope of this book, it is worth while noting that there is good agreement between experimental and calculated values of grain boundary energies. Read and Shockley used the dislocation model for grain boundaries and calculated the strain energies associated with the dislocations. The values obtained agreed fairly well with the experimental data for a polycrystalline FeSi alloy that had angles of mismatch between adjacent grains of up to ~ 30°.

356

Questions

9.1 Explain why crystalline solids are generally more defective as a result of increasing temperature.

9.2 Calculate the enthalpy of formation of Frenkel defects in AgCl using the data given in Fig. 9.4.

9.3 What kind of defects would you expect to predominate in crystals of the following: (a) NaCl doped with $MnCl_2$; (b) ZrO_2 doped with Y_2O_3 (Chapters 10, 13); (c) CaF_2 doped with YF_3; (d) Si doped with As (Chapter 14); (e) a piece of aluminium that has been hammered into a thin sheet; (f) WO_3 after heating in a reducing atmosphere?

9.4 Explain why copper is a much softer metal than tungsten.

9.5 What effect would you expect the ordering of Cu, Zn atoms in β' brass, Fig. 9.11, to have on the X-ray powder pattern; i.e. how do you expect the X-ray powder patterns of β and β' brass to differ?

9.6 The Law of Mass Action can be used to analyse defect equilibria in systems of low defect concentrations. What difficulties are likely to be encountered if this method is applied to systems with larger defect concentrations?

9.7 What kind of dislocations are characterized by the following: (a) the Burger's vector is parallel to the direction of shear and perpendicular to the line of the dislocation; (b) the Burger's vector is perpendicular to the direction of shear and parallel to the line of the dislocation.

9.8 In which directions would you expect slip to occur most readily in (a) Zn, (b) Cu, (c) α-Fe, (d) NaCl?

9.9 Assuming that the enthalpy of creation of Schottky defects in NaCl is 2.3 eV and that the ratio of vacancies to occupied sites at 750 °C is 10^{-5}, estimate the equilibrium concentration of Schottky defects in NaCl at (a) 300 °C, (b) 25 °C.

References

H. C. Abbink and D. S. Martin (1966). *J. Phys. Chem. Solids.*, **27**, 205.

J. S. Anderson (1969). *Bull. Soc. Chim. France.*, **7**, 2203.

J. S. Anderson (1973). *J. Chem. Soc. (Dalton)*, **1973**, 1107.

L. W. Barr and A. B. Lidiard (1970). Defects in ionic crystals, in *Physical Chemistry* (Ed. W. Jost), Vol. 10, Academic Press, N. Y.

R. J. Brook (1974). Defect structure of ceramic materials, chap. 3 in *Electrical Conductivity in Ceramics and Glass* (Ed. N. M. Tallan), Part A, Marcel Dekker, N. Y.

C. R. A. Catlow and A. N. Cormack (1982). Computer modelling of Solids. *Chem. Brit.*, **18**, 627–35.

LeRoy Eyring and M. O'Keeffe (Ed.) (1970). *The Chemistry of Extended Defects in Non-Metallic Compounds*, North Holland.

Morris E. Fine (1973). Introduction to chemical and structural defects in crystalline solids, in *Treatise on Solid State Chemistry* (Ed. N. B. Hannay), Vol. I, Plenum Press.

N. N. Greenwood (1968). *Ionic Crystals, Lattice Defects and Nonstoichiometry*, Butterworths, London.

R. Hoppe (1970). The coordination number—an inorganic chameleon, *Angew Chemie, Intern. Ed.*, **9**, 25.

D. Hull (1965). *Introduction to Dislocations*, Pergamon.

A. Kelly (1966). *Strong Solids*, Clarendon Press, Oxford.

P. Kofstad (1972). *Non-Stoichiometry, Electrical Conductivity and Diffusion in Binary Metal Oxides*, Wiley, N. Y.

F. A. Kröger (1974). *The Chemistry of Imperfect Crystals*, North Holland.

George G. Libowitz (1973). Defect equilibria in solids, in *Treatise on Solid State Chemistry* (Ed. N. B. Hannay), Vol I, Plenum Press.

E. Mandelcorn (Ed.) (1964). *Non-stoichiometric Compounds*, Academic Press.

W. J. Moore (1967). *Seven Solid States*, W. A. J. Benjamin Inc.

E. Mooser and W. B. Pearson (1959). *Acta Cryst.*, **12**, 1015.

F. R. N. Nabarro (1967). *Theory of Crystal Dislocations*, Clarendon Press, Oxford.

J. Nolting (1970). Disorder in solids, *Angew Chemie, Intern. Ed.*, **9**, 989.

A. Rabenau (1970). *Problems of Non-stoichiometry*, North Holland.

W. T. Read Jr. (1953). *Dislocations in Crystals*, McGraw-Hill, N.Y.

R. E. Reed-Hill (1964). *Physical Metallurgy Principles*, Van Nostrand Reinhold.

A. L. G. Rees (1954). *The Defect Solid State*, Methuen.

C. A. Wert and R. M. Thomson (1970). *Physics of Solids*, 2nd ed., McGraw-Hill.

Chapter 10

Solid Solutions

Solid solutions are very common in crystalline materials. A solid solution is basically a crystalline phase that can have variable composition. Often, certain properties of materials, e.g. conductivity, ferromagnetism, are modified by changing the composition in such a way that a solid solution forms and great use may be made of this in designing new materials that have specific properties.

Simple solid solution series are one of two types: in *substitutional* solid solutions, the atom or ion that is being introduced directly replaces an atom or ion of the same charge in the parent structure; in *interstitial* solid solutions, the introduced species occupies a site that is normally empty in the crystal structure and no ions or atoms are left out. Starting with these two basic types, a considerable variety of more complex solid solution mechanisms may be derived, by having both substitution and interstitial formation occurring together and/or by introducing ions of different charge to those in the host structure.

10.1 Substitutional solid solutions

An example of a substitutional solid solution is the series of oxides formed on reacting together Al_2O_3 and Cr_2O_3 at high temperatures. Both of these end-member phases have the corundum crystal structure (approximately hexagonal close packed oxide ions with Al^{3+}, Cr^{3+} ions occupying two-thirds of the available octahedral sites) and the solid solution may be formulated as $(Al_{2-x}Cr_x)O_3: 0 \leq x \leq 2$. At intermediate values of x, Al^{3+} and Cr^{3+} ions are distributed at random over those octahedral sites that are normally occupied in Al_2O_3. Thus, while any particular site must contain either a Cr^{3+} or an Al^{3+} ion, the probability that it is one or the other is related to the composition x. When the structure is considered as a whole and the occupancy of all the sites is averaged out, it is useful to think of each site as being occupied by an 'average cation' whose properties, atomic number, size, etc., are intermediate between those of Al^{3+} and Cr^{3+}.

If a range of simple substitutional solid solutions is to form, there are certain minimum requirements that must be met. The ions that are replacing each other must have the same charge (otherwise vacancies or interstitials would also be created) and be fairly similar in size. From a review of the experimental results on alloy formation, it has been suggested that a difference of 15 per cent in the radii of the metal atoms that are replacing each other is the most that can be tolerated if a substantial range of solid solutions is to form. For solid solutions in non-metallic systems, the limiting difference in size that is acceptable appears to be rather larger than 15 per cent, although it is difficult to quantify this. To a large extent this is because it is difficult to quantify the sizes of the ions themselves (Chapter 8). Let us use the Pauling crystal radii (in angstroms) for the alkali cations as an example, i.e. Li^+ 0.60, Na^+ 0.95, K^+ 1.33, Rb^+ 1.48, Cs^+ 1.69. The radii of K^+, Rb^+ and Rb^+, Cs^+ pairs are both within 15 per cent of each other and it is common to get solid solutions between, say, corresponding Rb^+ and Cs^+ salts. However, Na^+ and K^+ salts also commonly form solid solutions with each other (e.g. KCl and NaCl at high temperatures) and the K^+ ion is ~ 40 per cent larger than the Na^+ ion. Sometimes, Li^+ and Na^+ replace each other over limited ranges of compositions and Na^+ is ~ 60 per cent larger than Li^+. The difference in size of Li^+ and K^+ appears to be too large for any significant ranges of solid solution to form, however. If instead, one uses the Shannon and Prewitt radii based on $r_{O^{2-}} = 1.26\,\text{Å}$ (Fig. 8.3), similar effects are seen regarding the differences in size of the alkali cations.

In systems that exhibit complete ranges of solid solution, it is essential that the two end-member phases be isostructural. The reverse is not necessarily true, however, and just because two phases are isostructural it does not follow that they form solid solutions with each other; e.g. LiF and CaO both have the rock salt structure but they are not miscible with each other (i.e. they do not react together to form solid solutions) in the crystalline state.

While complete ranges of solid solution form in favourable cases, as for example, with $Al_2O_3–Cr_2O_3$ at high temperatures, it is far more common to have only partial or limited ranges of solid solution. In such cases, the restriction that the end-member phases be isostructural no longer holds. For example, the mineral forsterite, Mg_2SiO_4 (an olivine), is partially soluble in willemite, Zn_2SiO_4 and vice versa, as shown by their phase diagram (Fig. 10.1). The crystal structures of olivine and willemite are quite different; olivine contains

360

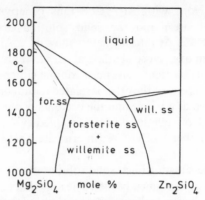

Fig. 10.1 Phase diagram for the system, forsterite (Mg_2SiO_4)–willemite (Zn_2SiO_4). (Data from Segnit and Holland, 1965)

approximately hexagonal close packed oxide layers but close packed oxide layers are not present in willemite. Both contain SiO_4 tetrahedra but magnesium is coordinated octahedrally in olivine and zinc is coordinated tetrahedrally in willemite. Both magnesium and zinc are flexible ions in their coordination requirements, however, and are happy in either tetrahedral or octahedral coordination. Thus, in the forsterite solid solutions, $(Mg_{2-x}Zn_x)SiO_4$, zinc replaces magnesium in octahedral sites whereas in the willemite solid solutions, $(Zn_{2-x}Mg_x)SiO_4$, magnesium replaced zinc in the tetrahedral sites. Mg^{2+} is a slightly larger cation than Zn^{2+} and this is reflected in the observation that, in oxide structures, magnesium shows a slight preference for octahedral coordination whereas zinc appears to prefer tetrahedral coordination.

Aluminium is also capable of being four or six coordinate to oxygen and this is shown in the system $LiAlO_2$–$LiCrO_2$, in which $LiCrO_2$ forms an extensive range of solid solutions, $Li(Cr_{2-x}Al_x)O_2 : 0 \leq x \lesssim 0.6$; in these, Cr^{3+} and Al^{3+} are in octahedral sites. $LiAlO_2$ contains tetrahedrally coordinate Al^{3+}, however, and the dislike of Cr^{3+} for tetrahedral coordination is shown by the complete absence of solid solutions of $LiCrO_2$ in $LiAlO_2$.

In systems where the two ions that are replacing each other are of considerably different size, it is usually found that a larger ion may be partially replaced by a smaller one, but it is much more difficult to do the reverse and replace a small ion by a larger one. For example, in the alkali metasilicates, rather more than half the Na^+ ions in Na_2SiO_3 may be replaced by Li^+ at high temperatures ($\sim 800\,°C$) to give solid solutions $(Na_{2-x}Li_x)SiO_3$, but only ~ 10 per cent of Li^+ in Li_2SiO_3 may be replaced by Na^+.

Many types of atom or ion may replace each other to form substitutional solid solutions. Silicates and germanates are often isostructural and form solid solutions with each other by $Si^{4+} \rightleftharpoons Ge^{4+}$ replacement. The lanthanide elements, because of their similarity in size, are notoriously good at forming solid solutions with each other in, say, their oxides. Indeed, one cause of the great difficulty experienced by the early chemists in trying to separate the lanthanides was this very easy solid solution formation. Anions may also replace each other in substitutional solid solutions, e.g. AgCl–AgBr solid solutions, but these are not nearly as common as the solid solutions formed by cation substitution, probably because there are not many pairs of anions that have similar size and coordination/bonding requirements. Many alloys are nothing more than substitutional solid solutions, e.g. in brass, copper and zinc atoms replace each other over a wide range of compositions.

10.2 Interstitial solid solutions

Many metals form interstitial solid solutions in which small atoms, e.g. hydrogen, carbon, boron, nitrogen, etc., can enter into empty interstitial sites within the host structure of the metal. Palladium metal is well known for its ability to 'occlude' enormous volumes of hydrogen gas and the product hydride is an interstitial solid solution of formula PdH_x: $0 \leq x \lesssim 0.7$, in which hydrogen atoms occupy interstitial sites within the face centred cubic palladium metal structure. There is still uncertainty as to whether hydrogen is in octahedral or tetrahedral holes and it appears that the sites occupied may depend on the composition x.

Possibly the technologically most important interstitial solid solution is that of carbon in the octahedral sites of face centred cubic γ-Fe. This solid solution is the starting point for the manufacture of steel and is discussed in Chapter 11 in more detail.

10.3 More complex solid solution mechanisms

Consider now what happens when cation substitution occurs but the two cations are of different charge. There are four possibilities, summarized in Fig. 10.2. A similar scheme is possible for anion substitution but is not considered further because anion substitution occurs rather infrequently in solid solutions.

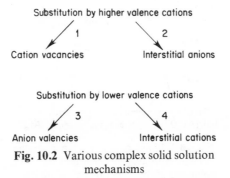

Fig. 10.2 Various complex solid solution mechanisms

10.3.1 Creating cation vacancies

If the replaceable cation of the host structure has a lower charge than that of the cation which is replacing it, additional changes are needed in order to preserve electroneutrality. One way is to create cation vacancies. Thus, NaCl is able to dissolve a small amount of, for example, $CaCl_2$ and the mechanism of solid solution formation involves the replacement of two Na^+ ions by one Ca^{2+} ion; one Na^+ site therefore remains vacant. The formula may be written as $Na_{1-2x}Ca_xV_xCl:0 \le x \lesssim 0.15$ at 600 °C, in which V represents a vacant cation site (see Chapter 9).

Spinel, $MgAl_2O_4$, forms an extensive range of solid solution with Al_2O_3 at high temperatures (Fig. 10.3). In these, Mg^{2+} ions on tetrahedral sites are replaced by Al^{3+} ions, in the ratio of 3:2 and the solid solution formula may be written $Mg_{1-3x}Al_{2+2x}O_4$; x vacant cation sites, presumably tetrahedral ones, must therefore be created.

Many transition metal compounds are non-stoichiometric and may exist over a range of compositions because the transition metal ion is present in more than one oxidation state. This then gives rise to a range of solid solutions, e.g. wüstite, $(Fe^{2+}_{1-3x}, Fe^{3+}_{2x})O$, has the overall stoichiometry $Fe_{1-x}O:0 < x \lesssim 0.1$. The actual structure of wüstite is in fact rather more complicated than having Fe^{2+}, Fe^{3+} and cation vacancies distributed over the octahedral cation sites of the rock salt structure and, instead defect complexes form (see Chapter 9).

10.3.2 Creating interstitial anions

The other mechanism by which a cation of higher charge may substitute for one of lower charge is to create, at the same time, interstitial anions. Calcium

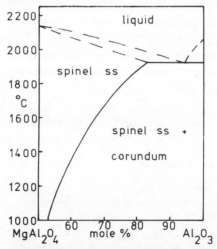

Fig. 10.3 Phase diagram for part of the system $MgAl_2O_4$–Al_2O_3 showing solid solutions of the spinel phase. (Data from Roy, Roy and Osborn, 1953)

fluoride can dissolve small amounts of yttrium fluoride; the total number of cations remains constant and, therefore, fluoride interstitials are created to give the solid solution formula $(Ca_{1-x}Y_x)F_{2+x}$. These F^- interstitial ions occupy large sites in the fluorite structure which are surrounded by eight other F^- ions at the corners of a cube (Figs 7.14 and 7.18).

Uranium dioxide has the fluorite structure. A similar solid solution to yttrium-doped CaF_2 forms on oxidation of UO_2. The product is UO_{2+x} and contains x interstitial O^{2-} ions together with xU^{6+} cations to preserve charge balance, i.e. the solid solution formula may be written as $(U_{1-x}^{4+}U_x^{6+})O_{2+x}$ (see Chapter 9).

10.3.3 Creating anion vacancies

If the replaceable cation of the host structure has a higher charge than that of the replacing cation, charge balance may be maintained by creating either anion vacancies or interstitial cations. Anion vacancies occur in cubic, lime-stabilized zirconia, $(Zr_{1-x}Ca_x)O_{2-x}$: $0.1 \lesssim x \lesssim 0.2$. Cubic zirconia has the fluorite structure and in the solid solutions with lime the total number of cations stays constant; replacement of Zr^{4+} by Ca^{2+} therefore requires the creation of oxide vacancies. These materials are important as refractories (Chapter 20) and as oxide ion conducting solid electrolytes (Chapter 13).

10.3.4 Creating interstitial cations

The alternative mechanism to Section 10.3.3 is to create interstitial cations at the same time as a cation of lower charge substitutes for one of higher charge. 'Stuffed silica' phases are aluminosilicates in which the structure of silica (quartz,

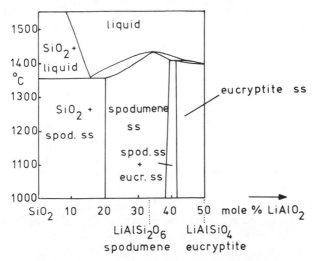

Fig. 10.4 Phase diagram for part of the system SiO_2–$LiAlO_2$. (Data from Hatch, 1943)

tridymite or cristobalite) may be modified by partial replacement of Si^{4+} by Al^{3+} and, at the same time, alkali metal cations enter normally empty interstitial holes in the silica framework.

Stuffed quartz structures have the formula, $Li_x(Si_{1-x}Al_x)O_2$ for $0 < x \lesssim 0.5$. The phase diagram of the join, SiO_2–$LiAlO_2$ (Fig. 10.4), shows that wide ranges of stuffed quartz solid solutions form but that special compositions exist at $x = 0.5$ (i.e. $LiAlSiO_4$, eucryptite) and $x = 0.33$ (i.e. $LiAlSi_2O_6$, spodumene). β-spodumene has the unusual property of a very small, perhaps even slightly negative, coefficient of thermal expansion; ceramics containing β-spodumene as a major constituent are therefore dimensionally stable and resistant to thermal shock. As such, they find many high temperature applications (Chapter 18). The interstitial holes in the quartz structure are too small to accommodate cations larger than Li^+. Tridymite and cristobalite have lower densities than quartz with larger interstices in their structures. Stuffed tridymite and cristobalite solid solutions, similar to the stuffed quartz solid solutions, form but in these the interstitial or stuffing cations are Na^+ and K^+.

A variety of other complex solid solution mechanisms occur, two of which are now given.

10.3.5 Double substitution

In such processes two subtitutions take place simultaneously. For example, in synthetic olivines, Mg^{2+} may be replaced by Fe^{2+} at the same time as Si^{4+} is replaced by Ge^{4+} to give solid solutions $(Mg_{2-x}Fe_x)(Si_{1-y}Ge_y)O_4$. Silver bromide and sodium chloride form a complete range of solid solutions in which both anions and cations replace each other: $(Ag_{1-x}Na_x)(Br_{1-y}Cl_y): 0 < x, y < 1$. The substituting ions may be of different charge, providing that overall electroneutrality prevails, e.g. in the plagioclase feldspars a complete range of solid solutions forms between anorthite, $CaAl_2Si_2O_8$, and albite, $NaAlSi_3O_8$. Their formulae may be written $(Ca_{1-x}Na_x)(Al_{2-x}Si_{2+x})O_8: 0 < x < 1$, and the two substitutions $Na \rightleftharpoons Ca$ and $Si \rightleftharpoons Al$ must occur simultaneously and to the same extent; the plagioclase phase diagram is given later in Fig. 11.12.

Double substitutional processes occur in *sialons*, which are solid solutions in the system Si—Al—O—N and based on the Si_3N_4 parent structure. β-Silicon nitride is built of SiN_4 tetrahedra linked at their corners to form a 3-D network. Each nitrogen is in planar coordination and forms the corner of three SiN_4 tetrahedra. In the sialon solid solutions, Si^{4+} is partly replaced by Al^{3+} and N^{3-} is partly replaced by O^{2-}. In this way charge balance is retained. The structural units in the solid solutions are $(Si, Al)(O, N)_4$ tetrahedra and the solid solution mechanism may be written as $(Si_{3-x}Al_x)(N_{4-x}O_x)$. Silicon nitride is potentially, a very useful high temperature ceramic (Chapter 20). The recent discovery of sialon and its derivatives by Jack and coworkers at Newcastle has opened up a new field of crystal chemistry and increased the possible applications of nitrogen-based ceramics.

10.4 General comments on the requirements for solid solution formation

The factors that govern whether or not solid solutions, especially the more complex ones, form are understood only qualitatively. For a given system, it is not usually possible to predict whether solid solutions will form or, if they do form, what is their compositional extent. Instead, this has to be determined experimentally. If we are restricted to solid solutions that exist under equilibrium conditions (Fig. 9.1) and are represented on the appropriate phase diagram, then solid solutions form only if they have lower free energy than any other phase or assemblage of phases with the same overall composition. Under non-equilibrium conditions, however, and by using 'chemie douce' or other preparative techniques (Chapter 2), it is often possible to prepare solid solutions that are much more extensive than those existing, if at all, under equilibrium conditions. A simple example is provided by the β-aluminas, $Na_2O \sim 8\,Al_2O_3$. Part or all of the Na^+ ions may be ion exchanged for a variety of other monovalent ions, including Li^+, K^+, Ag^+ and Cs^+, even though most of these ion exchanged materials are not thermodynamically stable (Chapter 13).

The limitations on the relative sizes of ions that either substitute directly for each other or enter interstitial sites have been referred to above. In the more complex solid solution mechanisms, the charges of the ions that take part are often different to each other. Clearly charge balance must be preserved overall, but within this limitation and provided the size requirements are met, there is often much scope for introducing ions of different charge. A rather extreme example of this is Li_2TiO_3. This has, at high temperatures, a rock salt structure in which Li^+, Ti^{4+} ions are disordered over the octahedral sites in a cubic close packed oxide ion array. It can form two series of solid solutions with either excess Li_2O or excess TiO_2 and of respective formulae

$$\text{(a)} \quad Li_{2+4x}Ti_{1-x}O_3: \quad 0 < x \lesssim 0.08$$
$$\text{(b)} \quad Li_{2-4x}Ti_{1+x}O_3: \quad 0 < x \lesssim 0.19.$$

Both of these involve the interchange of mono- and tetravalent ions with the creation of either interstitial Li^+ ions (a) or Li^+ vacancies (b) to maintain electroneutrality. The large difference in charge of Li^+ and Ti^{4+} obviously does not prevent solid solution formation. Part of the reason why solid solutions form appears to be that both Li^+ and Ti^{4+} are able to occupy similar-sized octahedral sites with metal–oxygen distances in the range 1.9 to 2.2 Å.

There are many other examples in which ions of similar size but very different charge can replace each other in solid solution formation. The ilmenite-like phase $LiNbO_3$ forms a limited range of solid solution by $5Li^+ \rightleftharpoons Nb^{5+}$ substitution on octahedral sites (Fig. 3.10a). The Zr^{4+} ion can be replaced by Ca^{2+}, Y^{3+} ions in the eight coordinate sites of the cubic stabilized zirconia structure, e.g. $Zr_{1-x}Ca_xO_{2+x}$. Na^+ is also of a similar size to Zr^{4+} and these ions replace each other in the solid solution series, $Na_{5-4x}Zr_{1+x}P_3O_{12}$: $0 < x < 0.15$.

10.5 Experimental methods for studying solid solutions

10.5.1 X-ray powder diffraction

There are two main ways in which powder diffraction may be used to study solid solutions. One is as a simple fingerprint method in which qualitative phase analysis is carried out. The objective is to determine the crystalline phases that are present in a sample without necessarily measuring the patterns very accurately. The second way is to measure the powder pattern accurately in order to obtain information about the composition of the solid solution. Usually, the unit cell undergoes a small contraction or expansion as the composition varies across a solid solution series and once a calibration graph of d-spacing or cell volume against composition has been drawn, the compositions of solid solutions may be obtained from an accurate measurement of their unit cell parameters or the d-spacings of certain lines in the powder X-ray pattern.

The use of the qualitative fingerprint method may be shown by reference to the phase diagram of $MgAl_2O_4$–Al_2O_3 (Fig. 10.3). The spinel solid solutions are much more extensive at 1800 °C than at 1000 °C, as shown by the *solvus*, which is the curve limiting the maximum compositional extent of the solid solutions. Let us do some imaginary experiments on a sample of composition 65 mol% Al_2O_3, 35 mol% MgO. According to the phase diagram, such a composition, under equilibrium conditions, would give a single phase, homogeneous spinel solid solution of the same composition (i.e. 65:35) above about 1550 °C. Below 1550 °C, two phases would be present, essentially stoichiometric Al_2O_3 and a spinel solid solution containing less than 65% Al_2O_3; e.g. at 1200 °C, the spinel solid solution composition is given by the position of the solvus at 1200 °C and is about 55% Al_2O_3. The relative amounts of the two phases, spinel solid solution and corundum (i.e. the phase composition), are given by a lever rule calculation (Chapter 11) and it can be seen that with decreasing temperature the proportion of alumina present increases gradually. Powder X-ray diffraction may be used to detect the presence or absence of alumina in samples that have been subjected to various heat treatment schedules i.e. alumina should be present in samples heated at e.g. 1500 °C but not in samples heated at 1600 °C. The method may therefore be used to determine phase diagrams, i.e. to determine whether or not solid solutions form and, if so, their compositional extent, perhaps as a function of temperature. This is practicable only if the solid solutions that exist at the temperature of the experiment are preserved to room temperature on rapid quenching. In many cases, however, precipitation from supersaturated solid solution occurs during cooling.

Information about the composition of solid solutions may be obtained if the d-spacings of the powder lines can be measured accurately. Methods for doing this are discussed in Chapter 5. Usually, the unit cell expands if a small ion is being replaced by a larger one, and vice versa. From Bragg's Law and the d-spacing formulae, an increase in the unit cell parameters leads to an increase in the d-spacings of the powder lines; the whole pattern shifts to lower values of 2θ,

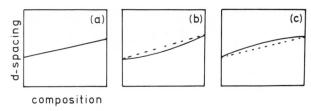

Fig. 10.5 (a) Vegard's Law behaviour, (b) negative departures and (c) positive departures (From Castellanos and West, 1980)

therefore, although all the lines do not usually move by the same amount. In non-cubic crystals, the expansion or contraction of the unit cell with changing composition may not be the same for all three axes and sometimes one axis may expand while the others contract (or vice versa).

According to *Vegard's Law*, unit cell parameters should change linearly with composition. In practice, Vegard's Law is often obeyed only approximately and accurate measurements reveal departures from linearity.

Vegard's Law is not really a law but rather is a generalization that applies to solid solutions formed by random substitution or distribution of ions. It assumes implicitly that the changes in unit cell parameters with composition are governed purely by the relative sizes of the atoms or ions that are 'active' in the solid solution mechanism, e.g. the ions that replace each other in a simple substitutional mechanism.

Departures from Vegard's Law behaviour have been observed in many solid solution series, especially in metals. In metals, there appears to be no systematic trend or correlation between the direction of the departure (i.e. positive or negative) from Vegard's Law and structural features of the solid solutions. In non-metallic solid solutions, a correlation has been observed between positive departures from Vegard's Law (Fig. 10.5c) and the occurrence of immiscibility domes inside the temperature–composition diagram of the solid solutions. For example, the phase diagram of Al_2O_3–Cr_2O_3 solid solutions (Fig. 10.6) shows a complete range of solid solutions between the solidus temperatures, $\sim 2100\,°C$ and about 950 °C, but below 950 °C an immiscibility dome is present, inside which two crystalline phases are present. Thus, a mixture of 50:50 composition should, under equilibrium conditions, be a single phase solid solution above 950 °C, but contain a mixture of alumina-rich and chrome-rich solid solutions at lower temperatures. The actual decomposition of a homogeneous solid solution into two phases at, for example, 800 to 900 °C takes place only very slowly, but may be speeded up by using hydrothermal or high pressure treatment. It is easy, therefore, to preserve a complete range of Al_2O_3–Cr_2O_3 solid solutions, prepared at, say, 1300 °C to room temperature. From their X-ray powder patterns, the variation with composition in the values of the hexagonal unit cell parameters, a and c, may be determined. Results, given in Fig. 10.7, show a positive departure from linearity and Vegard's Law. The explanation of this is that the Cr^{3+} and Al^{3+} ions are not arranged at random but cluster together to

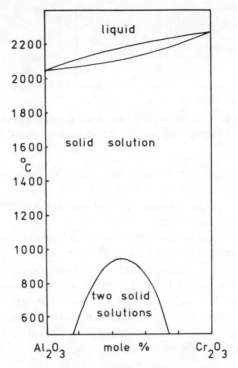

Fig. 10.6 Phase diagram for Al_2O_3–Cr_2O_3. (Data from Bunting, 1931, and Roy and Barks, 1972)

form small alumina-rich and chrome-rich domains, even though on a macro-scopic scale the solid solutions appear to be homogeneous. This dislike of Al^{3+} and Cr^{3+} for each other's company in the corundum structure and the segregation of 'like with like' in the solid solutions causes a small increase of the unit cell parameters compared to the values expected for a random arrangement of non-interacting Cr^{3+} and Al^{3+} ions.

Various other solid solution systems show both positive departures from Vegard's Law and immiscibility domes. Sometimes, a positive departure from Vegard's Law has been used as a basis for predicting the occurrence of a previously unsuspected immiscibility dome.

Negative departures from Vegard's Law in non-metallic systems (Fig. 10.5b) may be evidence for a net attractive interaction between unlike ions (i.e. in an A–B system, the A–B interactions may be stronger than the average of A–A and B–B interactions). In cases where the A–B interactions are quite strong, cation ordering may occur to give a periodic superstructure which may be detected by X-ray diffraction (e.g. ordering of copper and zinc atoms in β'-brass, CuZn). Superstructures usually occur at special compositions, such as at 1:1 ratios. At other compositions or in cases where the A–B interactions are less strong, cation ordering may occur over only short distances (i.e. a few atomic diameters), in

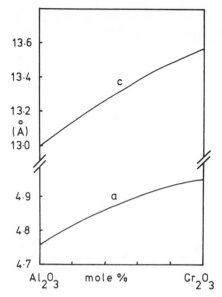

Fig. 10.7 Unit cell parameters against composition for Al_2O_3–Cr_2O_3 solid solutions. (Data from Graham, 1960)

which case the solid solution is apparently still disordered and homogeneous. Short range ordering such as this is difficult to detect experimentally and, therefore, only a very tentative correlation between cation ordering and negative departures from Vegard's Law can be made at this stage.

So far, we have considered examples of smooth departures from Vegard's Law. More abrupt changes or discontinuities may occur at certain compositions if either a change in symmetry of the solid solutions or a change in the solid solution mechanism occurs. An example of the latter is provided by a (partial) range of solid solutions between Li_4SiO_4 and Zn_2SiO_4; variations of a and b of the orthorhombic unit cell with composition at $700\,°C$ are shown in Fig. 10.8. A change in slope occurs at the 1:1 composition, γ-Li_2ZnSiO_4. The c dimension is virtually unchanged at $5.10\,Å$ across the entire series and is omitted from the diagram. The interpretation of Fig. 10.8 is that a different solid solution mechanism is operative on either side of the Li_2ZnSiO_4 composition. For zinc-rich compositions, the solid solution mechanism is thought to involve cation replacement plus vacancy creation, with a formula $(Li_{2-2x}Zn_x)SiO_4 : 0 < x \lesssim 0.5$. For lithia-rich compositions, a combination of cation replacement and formation of interstitial Li^+ ions occurs, with a formula $(Li_{2+2x}Zn_{1-x})SiO_4 : 0 < x \lesssim 0.5$. The structure of γ-Li_2ZnSiO_4 is rather complicated; it is distantly related to the wurtzite structure in that the oxide ion arrangement is intermediate between hexagonal close packing and tetragonal packing (Chapter 7). The cations are distributed over various sets of tetrahedral sites.

370

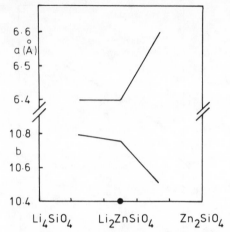

Fig. 10.8 Unit cell parameters against composition at 700 °C for solid solutions based on Li_2ZnSiO_4. (Data from West and Glasser, 1970)

10.5.2 Density measurements

The mechanism of solid solution formation may sometimes be inferred by a combination of density and unit cell volume measurements for a range of compositions. Broadly speaking, an interstitial mechanism leads to an increase in density because extra atoms or ions are added to the unit cell, whereas a mechanism involving vacancy creation may lead to a decrease in density.

As an example, consider the stabilized zirconia solid solutions formed between ZrO_2 and CaO over the compositional range ~ 10 to $25\%CaO$. Two simple mechanisms could be postulated for the solid solutions: (a) the total number of oxide ions remains constant and, therefore, interstitial Ca^{2+} ions are created according to the formula $(Zr_{1-x}Ca_{2x})O_2$; (b) the total number of cations remains constant and, therefore, O^{2-} vacancies are created according to the formula $(Zr_{1-x}Ca_x)O_{2-x}$. In mechanism (a), two calcium ions replace one zirconium and the formula unit decreases in mass by 11 g as x varies, hypothetically, from 0 to 1. In (b), one zirconium and one oxygen is replaced by one calcium with a decrease in mass of the formula unit by 67 g as x varies from 0 to 1. Assuming that the unit cell volume does not change with composition (this is not strictly true), mechanism (b) would lead to a larger decrease in density with increasing x than would mechanism (a).

Experimental results (Fig. 10.9) confirm that mechanism (b) is operative, at least for samples heated at 1600 °C. It is possible, in theory at least, to propose alternative and more complex mechanisms than (a) and (b); e.g. the total number of zirconium ions remains constant, in which case both interstitial Ca^{2+} and O^{2-} ions are needed. Usually, however, simple mechanisms operate and there is no need to invoke more far-fetched possibilities.

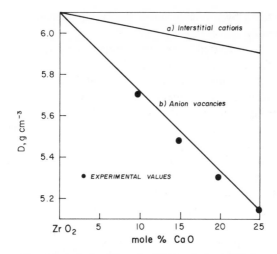

Fig. 10.9 Density data for cubic CaO-stabilized zirconia solid solutions for samples quenched from 1600 °C. Data from Diness and Roy, *Solid State Commun.*, **3**, 123 (1965)

Density data for the CaF_2-YF_3 solid solutions described in section 10.3.2 are given in Fig. 10.10. These data clearly show that a model based on interstitial fluoride ions fits the data rather than a model based on cation vacancies.

Density measurements do not, of course, give any atomistic details of the vacancies or interstitials involved, but only a bulk mechanism. Other techniques, such as diffuse neutron scattering are needed to probe the defect structure, and as more systems are studied in detail it is becoming increasingly apparent that simple point defects such as vacancies and interstitials do not occur. Instead,

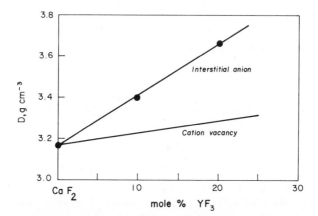

Fig. 10.10 Density data for solid solutions of YF_3 in CaF_2. From Kingery, Bowen and Uhlmann, *Introduction to Ceramics*, Wiley, New York, 1976

defect clusters form by a relaxation of the crystal structure in the immediate vicinity of the point defect (Chapter 9).

Densities can be measured by several simple techniques. The volume of a few grams of material may be measured by displacement of liquid from a specific gravity bottle whose volume is known accurately. From the difference in weight of the bottle filled with displacement liquid, e.g. CCl_4, and the bottle containing the solid topped up with liquid, the volume of the solid may be calculated if the density of the displacement liquid is known. In the float–sink method, a few crystals of the material are suspended in liquids of a range of densities until a liquid is found in which the crystals neither sink nor float. The density of the crystals then equals that of the liquid. A variant of this method is to use a density gradient column which is a column of liquid of gradually increasing density. Crystals are dropped in at the top and sink until their density equals that of the liquid. The crystal density is then obtained from a calibration curve of height in the column against density. In all of the above methods, it is important that air bubbles are not trapped on the surface of the crystals, otherwise anomalously low density values may be obtained.

A good method for measuring the density of larger samples (10 to 100 g) is by gas displacement pycnometry. In this, the sample is compressed in a gas-filled chamber by a piston until a certain pressure, e.g. 2 atmospheres, is reached. The volume of the sample is obtained from the difference in position of the piston when the sample is in the chamber compared to when the chamber is empty of solid and contains only gas compressed to the same pressure.

10.5.3 Changes in other properties—thermal activity and DTA

Many materials undergo abrupt changes in structure or property on heating and, if the material forms a solid solution, the temperature of the change usually varies with composition. The changes, which may be, for example, ferroelectric-paraelectric transitions at the Curie temperature or straightforward polymorphic transitions, such as quartz → tridymite, can usually be studied readily by DTA since most phase transitions have an appreciable enthalpy of transition. This provides a very sensitive way of studying solid solutions because often a transition temperature varies over tens or hundreds of degrees as the composition changes, e.g. addition of carbon to iron causes the temperature of the $\alpha \rightleftharpoons \gamma$ transition to drop rapidly from 910 to 723 °C with addition of only 0.02 wt % carbon. Further examples are given in Chapter 11.

Questions

10.1 For the solid solutions of YF_3 in CaF_2, calculate the density as a function of composition for (a) a cation vacancy model, and (b) an interstitial F^- model. Unit cell data for CaF_2 are given in Chapter 7. Assume that the unit cell volume is independent of solid solution composition. Compare your results with those given in Fig. 10.10.

10.2 Give probable formulae for the following:
    (a) partial solid solution of $MnCl_2$ in KCl;
    (b) partial solid solution of Y_2O_3 in ZrO_2;
    (c) partial solid solution of Li in TiS_2;
    (d) partial solid solution of Al_2O_3 in $MgAl_2O_4$.

References

E. N. Bunting (1931). *J. Res. Natl. Bur. Stds.*, **6** (6), 948.

M. Castellanos and A. R. West (1980). *J. Chem. Soc. Faraday*, **76**, 2159–2169.

B. E. Fender (1972). Theories of non-stoichiometry, in *Solid State Chemistry* (Ed. L. E. J. Roberts), Butterworths.

J. Graham (1960). *J. Phys. Chem. Solids*, **17**(1/2), 18–25.

N. N. Greenwood (1968). *Ionic Crystals, Lattice Defects and Non-stoichiometry*, Butterworths, London.

R. A. Hatch (1943). *Amer. Mineral*, **28**, 471–496.

E. Mandelcorn (Ed.) (1964). *Non-stoichiometric Compounds*, Academic Press.

D. M. Roy and R. E. Barks (1972). *Nature Phys. Sci.*, **235**, 118–119.

D. M. Roy, R. Roy and E. F. Osborn (1953). *J. Amer. Ceram. Soc.*, **36**(5), 149.

E. R. Segnit and A. E. Holland (1965). *J. Amer. Ceram. Soc.*, **48**(8), 412.

O. Toft Sorensen (Ed.) (1981). *Nonstoichiometric Oxides*, Academic Press.

A. R. West and F. P. Glasser (1970). *J. Materials Sci.*, **5**, 676–688.

Chapter 11

Interpretation of Phase Diagrams

The subject of phase equilibria and phase diagrams is one of the cornerstones of solid state chemistry. Phase diagrams are plots of temperature (occasionally pressure) against composition. They summarize in graphical form the ranges of temperature and composition over which certain phases or mixtures of phases exist under conditions of thermodynamic equilibrium. Thus the effect of temperature on solids and the reactions that may or may not occur between solids may often be deduced from the appropriate phase diagram. This chapter is concerned primarily with the interpretation of phase diagrams. Thermodynamic background is largely excluded and only a brief outline is given of the methods that are used to determine phase diagrams. Instead, the approach used is to emphasize the interpretation and practical applications of phase diagrams.

The fundamental rule on which phase diagrams are based is the *phase rule*, derived by W. J. Gibbs. The phase rule applies strictly only to conditions of thermodynamic equilibrium but in practice it also has great value in some non-equilibrium situations. The phase rule itself will not be derived but is presented and used as an empirical statement. One big difference from the treatment given here to that found in many physical chemistry textbooks is that here the solid state is given primary importance whereas in most textbooks attention is focused on the gaseous and liquid states with rather cursory treatment of the solid state. Discussion of the vapour state has been largely omitted from this chapter. This leads to much simplification, and even to a simplified form of the phase rule itself. However, it does mean that certain topics such as the behaviour of systems containing variable valence metal atoms and decomposition reactions have been omitted.

11.1 Definitions

The phase rule is given by the equation

$$P + F = C + 2$$

where P is the number of phases present in equilibrium, C is the number of components needed to describe the system and F is the number of degrees of freedom or independent variables taken from temperature, pressure and composition of the phases present. Each of these terms is now explained more fully.

The number of *phases* is the number of physically distinct and mechanically separable (in principle) portions of a system, each phase being itself homogeneous. The distinction between different *crystalline* phases is usually clear. For example, the differences between chalk, $CaCO_3$, and sand, SiO_2, are obvious. The distinction between crystalline phases made from the same components but of different composition is also usually clear. Thus, the magnesium silicate minerals enstatite, $MgSiO_3$, and forsterite, Mg_2SiO_4, are different phases. They have very different composition, structures and properties. With solids it is also possible to get different crystalline phases having the *same* chemical composition. This is known as *polymorphism*. For example, two polymorphs of Ca_2SiO_4 can be prepared at room temperature, the stable γ form and the metastable β form, but these have quite distinct physical and chemical properties and crystal structures (see Chapter 19).

One complicating factor in classifying solid phases is the occurrence of solid solutions (Chapter 10). For example, $\alpha - Al_2O_3$ and Cr_2O_3 have the same crystal structure (corundum) and form a continuous range of solid solutions at high temperature. Any mixture of Al_2O_3 and Cr_2O_3 can react at high temperature to form a single, homogeneous phase whose composition may be altered without changing the integrity or homogeneity of the single phase.

In recent years, with advances in our knowledge of defects in crystals, especially extended defects such as the crystallographic shear structures (Chapter 9), it has become difficult in certain cases to decide exactly what constitutes a separate phase. This is because a minute change in composition can lead to a different arrangement of defects in a structure. In the oxygen-deficient tungsten oxides, WO_{3-x}, what was previously thought to be a range of homogeneous solid solutions is now known to be a large number of phases that are very close in composition and similar, but distinct, in structure. Some of these phases have formulae belonging to the homologous series W_nO_{3n-1}. Thus $W_{20}O_{59}$ and $W_{19}O_{56}$ are physically distinct phases (Chapter 9). Although these crystallographic shear structures are an interesting new area of solid state chemistry, their occurrence is so far restricted mainly to a small number of transition metal compounds; with by far the majority of solid compounds there is no problem in deciding exactly what constitutes a phase.

In the *liquid* state the number of possible, separate homogeneous phases that can exist is much more limited than in the solid state. This is because single phase liquid solutions form much more readily and over wider compositional ranges than do single phase solid solutions. Take the Na_2O–SiO_2 system for instance. In the liquid state at high temperatures, Na_2O and SiO_2 are completely miscible to give a single, liquid, sodium silicate phase. In the crystalline state, however, the number of phases is quite large. Crystalline sodium silicate phases form at five different compositions and at least one of these shows polymorphism.

In the *gaseous* state, the maximum number of possible phases appears always to be 1; there are no known cases of immiscibility between two gases, if the effects of gravity are ignored.

The number of *components* of a system is rather more difficult to visualize. It is the number of constituents of the system that can undergo independent variation in the different phases; alternatively, it is the *minimum* number of constituents needed in order to describe completely the compositions of the phases present in the particular system. This is best understood with the aid of examples:

(a) All of the crystalline calcium silicates can be considered to be built from CaO and SiO_2 in varying proportions. CaO–SiO_2 is therefore a two-component system even though there are three elements present, Ca, Si and O. Compositions between CaO and SiO_2 can be regarded as forming a binary (i.e. two-component) join in the ternary system Ca–Si–O (Fig. 11.1).

(b) The system MgO is a unary (one-component) system, at least up to the melting point 2700 °C, because the composition of MgO is always fixed.

(c) The composition 'FeO' is part of the two-component system, iron–oxygen because wüstite is actually a non-stoichiometric, iron-deficient phase, $Fe_{1-x}O$, caused by having some Fe^{3+} present (Chapter 9). The bulk composition 'FeO' in fact contains a mixture of two phases at equilibrium: $Fe_{1-x}O$ and Fe metal.

The number of *degrees of freedom* of a system is the number of independently variable factors taken from temperature, pressure and composition of phases, i.e. it is the number of these variables that must be specified in order that the system be completely defined. Again, let us see some examples:

(a) A system that consists of boiling water, i.e. water and steam in equilibrium, does not have a composition variable since both water and steam contain molecules of the same fixed formula, H_2O. To define the system it is necessary

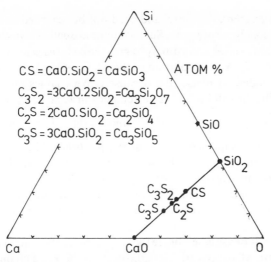

Fig. 11.1 Binary join $CaO-SiO_2$ in the ternary system Ca–Si–O. Note the method used for the labelling of phases, $C = CaO$, etc. This type of abbreviation is widely used in oxide chemistry

to specify only the steam pressure because then the temperature of boiling is automatically fixed (or vice versa). Application of the phase rule to this system gives:

$$P + F = C + 2; \quad C = 1 \text{(i.e. } H_2O), P = 2 \text{ (vapour and liquid) and so } F = 1$$
(either temperature or pressure but not both).

At sea level, water boils at $100\,°C$ but at the high altitude of Mexico City atmospheric pressure is only $580\,mm$ Hg and water in equilibrium with steam at this pressure boils at $92\,°C$. The system water–steam is therefore *univariant* because only one degree of freedom, either P or T, is needed to describe completely the system at equilibrium. It should be emphasized that the relative amounts of water and steam are not given by the phase rule. As long as there is sufficient steam present to maintain the equilibrium pressure, the volume of vapour is unimportant.

(b) A solid solution in the system $Al_2O_3-Cr_2O_3$ has one composition variable because the $Al_2O_3 : Cr_2O_3$ ratio can be varied and the same homogeneous phase obtained. Temperature can also be varied within the single phase solid solution field. In the temperature versus composition phase diagram of the $Al_2O_3-Cr_2O_3$ system the solid solutions occupy an area. Two degrees of freedom are therefore needed in order to characterize a certain solid solution composition and temperature (Fig. 10.6).

In refractory systems with very high melting temperatures, such as $Al_2O_3-Cr_2O_3$, the vapour pressure of the solid phases and even that of the liquid phase is negligible in comparison with atmospheric pressure. The vapour phase is effectively non-existent, therefore, and need not be regarded as a possible phase for work at atmospheric pressure. Such systems are called *condensed systems* and the phase rule is modified accordingly to give the *condensed phase rule*, $P + F = C + 1$ (instead of $+ 2$). A solid solution in the Al_2O_3–Cr_2O_3 system is *bivariant* in terms of the condensed phase rule because two variables must be specified in order to describe the solid solution, i.e. temperature and composition. Therefore,

$$P + F = C + 1; \qquad C = 2 \,(Al_2O_3 \text{ and } Cr_2O_3), \; P = 1 \text{ and so } F = 2.$$

It is important to define what is meant by *equilibrium*. Thermodynamically, the conditions for equilibrium can be defined precisely—at equilibrium, no useful energy may pass into or out of the system. Experimentally, it may be difficult to assess whether or not a system is in a state of equilibrium. There are, however, several criteria that may help one to decide in a particular situation:

(a) Is the phase or phase assemblage that is present in the system unchanged with time, keeping all other variables constant? If there is a change with time, we can definitely say that equilibrium has not been reached, but the converse is not necessarily true. Thus, lime and sand may coexist together indefinitely without reacting but we know from the CaO–SiO_2 phase diagram that they should react to give, at equilibrium, various calcium silicates, depending on the lime: sand ratio. Probably a more thought-provoking example is the thermodynamic instability of the human body in the presence of oxygen; fortunately for us, oxidation of the particular carbon compounds involved is kinetically very slow at room temperature!

Another example of kinetic stability but thermodynamic instability is the occurrence of diamond at room temperature and pressure. From the condensed phase rule, a one-component system, such as carbon, can contain either two phases in equilibrium at a single fixed temperature ($P = 2$, and so $F = 0$), or one phase coexisting over a range of temperatures ($P = 1$, $F = 1$). At ambient temperature and pressure, the equilibrium behaviour of carbon belongs to the second category with graphite as the equilibrium phase. At high pressures, the situation reverses and diamond is stable relative to graphite (Fig. 2.19). However, once diamond has been prepared at high pressures, the pressure can be released without danger of conversion to graphite occurring at measurable rates (and provided the temperature is $\lesssim 900\,°C$).

There are countless other examples of phases or materials that coexist together only because reaction between them is kinetically slow. The phase rule tells us absolutely nothing about the kinetics of reaction, only the general direction in which reactions may proceed. Bearing in mind that the phase rule pertains only to equilibrium, there are no known examples of systems in which the phase rule is violated.

(b) For systems that react at observable rates, a useful guideline is to observe the approach to equilibrium from two different directions and see whether the products obtained are identical. For example, at a certain temperature, is the same phase assemblage produced by either raising or lowering the temperature of the reaction mixture to this value?

An interesting example of this is provided by the crystalline phase Ca_3SiO_5, which is unstable below $\sim 1250\,°C$. Ca_3SiO_5 may be prepared by reacting lime and silica at $1300\,°C$ but not by reaction at $1100\,°C$. Is this because the kinetics of reaction are too slow at $1100\,°C$? This question may be answered by taking Ca_3SiO_5 prepared at, for example, $1300\,°C$ and subsequently heating it at $1100\,°C$. It is then found to decompose slowly to give a mixture of CaO and Ca_2SiO_4. This result proves that Ca_3SiO_5 has a lower limit of stability somewhere between 1100 and $1300\,°C$. Consequently, CaO and Ca_2SiO_4 can never react to form Ca_3SiO_5 at $1100\,°C$.

(c) A related test is to use different starting materials and see whether the products of reaction are the same. For example, crystalline sodium silicates, e.g. $Na_2Si_2O_5$, may be prepared in several ways. Na_2CO_3 and SiO_2 in a $1:2$ molar ratio can be heated at subsolidus temperatures (700 to $750\,°C$) for several days to give $Na_2Si_2O_5$. Alternatively, aqueous solutions of a sodium salt and silica (usually as ethyl orthosilicate) may be mixed and dried to give a hydrated gelatinous mixture and fired at $\sim 700\,°C$. A third pathway is to melt the starting materials, cool them rapidly to form a glass and crystallize the glass by heating for several hours at $\sim 700\,°C$. Since all these methods yield $Na_2Si_2O_5$, it is likely to be a thermodynamically stable phase.

It is important to use different preparative methods where possible, as in the above example, because the processes of reacting together crystalline solids or of crystallizing glasses often occur in several stages (but different stages in the two methods) and the final approach to equilibrium may be slow.

The equilibrium state is always that which has the lowest free energy. It may be thought of as lying at the bottom of a free energy well (Fig. 11.2). The problem in determining whether equilibrium has been reached is that other free energy minima may exist but not be as deep as the equilibrium well. There may be a considerable energy barrier involved in moving from a *metastable* to the *stable* state and under many conditions this barrier may be prohibitively high. Such an example is the metastability of diamond relative to graphite at room tempera-

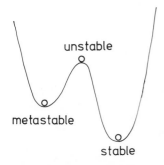

unstable

metastable

stable

Fig. 11.2 Schematic diagram showing stable, unstable and metastable conditions

380

ture. The energy barrier or activation energy for the diamond → graphite reaction is so high that, once formed, diamond is kinetically stable although thermodynamically metastable.

The thermodynamic meaning of the term *unstable* is shown in Fig. 11.2. If a ball is perched on a hill-top, the slightest movement is sufficient to cause it to start rolling down one side or the other. In the same way, there is no activation energy involved in changing from a thermodynamically unstable to either a stable or a metastable state. Examples of unstable equilibrium are difficult to find (because of their instability!) but one can point to their would-be occurrence. Inside a region of liquid immiscibility (see later and Chapter 18) exists an area bounded by a dome called a *spinodal*. Within the spinodal, a homogeneous liquid would be unstable and would spontaneously separate into two liquids by the process known as *spinodal decomposition*.

11.2 One-component systems

The independent variables in a one-component system are limited to temperature and pressure because the composition is fixed. From the phase rule, $P + F = C + 2 = 3$. The system is *bivariant* ($F = 2$) if one phase is present, *univariant* ($F = 1$) if two are present and *invariant* ($F = 0$) if three are present. Schematic phase relations are given in Fig. 11.3 for a one-component system in which the axes are the independent variables, pressure and temperature. Possible phases are two crystalline modifications or polymorphs (sometimes called allotropes), X and Y, liquid and vapour. Each of these phases occupies an area or *field* on the diagram when $F = 2$ (both pressure and temperature are needed to describe a point in one of these fields). Each single phase region is separated from the neighbouring single phase regions by univariant curves ($P = 2$ and so $F = 1$). Thus if one variable, say pressure, is fixed, then the other, temperature, is automatically fixed. The univariant curves on the diagram represent the following equilibria:

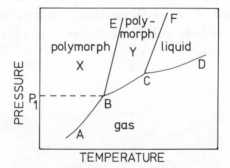

Fig. 11.3 Schematic pressure versus temperature phase diagram of a one-component system

(a) BE—transition curve for polymorphs X and Y; it gives the change of transition temperature with pressure.
(b) FC—change of melting point of polymorph Y with pressure.
(c) AB, BC—sublimation curves for X and Y, respectively.
(d) CD—vapour pressure curve for the liquid.

From Fig. 11.3, crystals of polymorph X can never melt directly under equilibrium conditions because the fields of X and liquid never meet on the diagram. On heating, crystals of X can either sublime at a pressure below p_1 or transform to polymorph Y at pressures above p_1. They cannot melt directly. Also present in Fig. 11.3 are two invariant points B and C for which $P = 3$ and $F = 0$. The three phases that coexist at point B are: polymorph X, polymorph Y and vapour. Points B and C are also called *triple points*.

11.2.1 The system H_2O

This important system, shown in Fig. 11.4, gives examples of solid–solid and solid–liquid transitions. Ice I is the polymorph that is stable at atmospheric pressure; several high pressure polymorphs are also known—ice II to ice VI. At first sight, there is little similarity between the diagram for a schematic one-component system (Fig. 11.3) and that for water (Fig. 11.4), but this is mainly because of the location of the univariant curve XY that separates the fields of ice I and water. It is well known that ice I has the unusual property of being less dense than liquid water at $0\,^{\circ}C$. The effect of pressure on the ice I–water transition temperature can be understood from Le Chatelier's principle which states: 'When a constraint is applied to a system in equilibrium the system adjusts itself so as to nullify the effects of this constraint.' The melting of ice I is accompanied by a decrease in volume; increased pressure makes melting easier and so melting temperatures decrease with increased pressure, in the direction YX. The water system (Fig. 11.4) is also more complex than Fig. 11.3 since additional invariant points exist which correspond to three solid phases in equilibrium (e.g. point Z). The rest of the diagram should be self-explanatory; thus the curves Y X A B C give the variation of melting point with pressure for some of the different ice

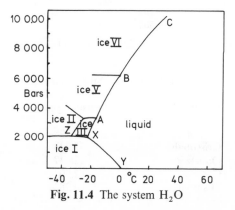

Fig. 11.4 The system H_2O

polymorphs. Liquid–vapour equilibria are omitted from Fig. 11.4 because, with the pressure scale used, these equilibria lie very close to the temperature axis and in the high temperature corner.

11.2.2 The system SiO$_2$

Silica is the main component of many ceramic materials as well as being the most common oxide, apart from H$_2$O, in the earth's crust. The polymorphism of SiO$_2$ is complex with major, first-order phase changes such as quartz–tridymite and minor changes such as α(low)–β(high) quartz. The polymorphism at atmospheric pressure can be summarized by the following sequence of reactions on heating:

$$\alpha\text{-Quartz} \xrightarrow{573^\circ C} \beta\text{-quartz} \xrightarrow{870^\circ C} \beta\text{-tridymite} \xrightarrow{1470^\circ C} \beta\text{-cristobalite} \xrightarrow{1710^\circ C} \text{liquid}$$

With increasing pressure, two main changes are observed (Fig. 11.5); first, the contraction of the field of tridymite and its eventual disappearance, at ~ 900 atm; second, the disappearance of the field of cristobalite at ~ 1600 atm. Above 1600 atm, quartz is the only stable crystalline polymorph and exists up to much higher pressures. The disappearance of tridymite and cristobalite with increasing pressure can be correlated with the lesser density of these phases relative to that of quartz (Table 11.1); the effect of pressure generally is to produce polymorphs that have a higher density and therefore smaller volume. Above 20000 to 40000 atm (depending on temperature), quartz transforms to another polymorph, coesite,

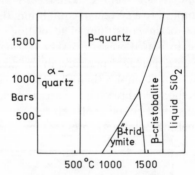

Fig. 11.5 The system SiO$_2$

Table 11.1 *Densities of* SiO$_2$ *polymorphs*

Polymorph	Density (g cm^{-3})
Tridymite	2.298
Cristobalite	2.334
Quartz	2.647
Coesite	2.90
Stishovite	4.28

and above 90000 to 120000 atm yet another polymorph, stishovite, is the stable polymorph of SiO_2.

It should be noted that there are many metastable polymorphs of SiO_2 which are absent from Fig. 11.5; e.g. it is very easy to undercool cristobalite and to observe a reversible α(low)–β(high) transformation at $\sim 270\,°C$. However, at these temperatures, cristobalite is metastable relative to quartz and so this transformation is omitted from Fig. 11.5.

11.2.3 Condensed one-component systems

For most systems and applications of interest in solid state chemistry, the condensed phase rule is applicable, pressure is not a variable and the vapour phase is not important. The phase diagram for a condensed, one-component system then reduces to a line since temperature is the only degree of freedom. It is not normal practice to represent such a line phase diagram in graphical form, unless it forms part of, say, a binary system. For instance, the condensed phase diagram at 1 atmosphere pressure for SiO_2 would simply be a line showing the polymorphic changes that occur with changing temperature. In such cases it is easier to represent the changes as a 'flow diagram', as indicated above for SiO_2.

11.3 Two-component condensed systems

Two-component or binary systems have three independent variables: pressure, temperature and composition. In most systems of interest in the general sphere of solid state chemistry, the vapour pressure remains low for large variations in temperature and so, for work at atmospheric pressure, the vapour phase and the pressure variable need not be considered. In almost all of what follows, the condensed phase rule $P + F = C + 1$ is used. In binary systems under these conditions an invariant point occurs when three phases coexist in equilibrium: a univariant curve for two phases and a bivariant condition for one phase. Conventionally, temperature is the vertical scale and composition the horizontal one in binary phase diagrams.

11.3.1 A simple eutectic system

The simplest possible type of two-component condensed system is the simple eutectic system shown in Fig. 11.6. In the solid state there are no intermediate compounds or solid solutions but only a mixture of the end-member crystalline phases, A and B. In the liquid state, at high temperatures, a complete range of single phase, liquid solutions occurs. At intermediate temperatures, regions of partial melting appear on the diagram. These regions contain a mixture of a crystalline phase and a liquid of different composition to the crystalline phase.

The phase diagram shows several regions or areas which contain either one or two phases. These areas are separated from each other by solid curves or lines. The area 'liquid' at high temperatures is single phase and bivariant. Every point

384

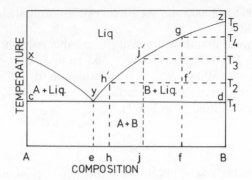

Fig. 11.6 Simple eutectic binary system

within this area represents a different state for the liquid, i.e. a different temperature and composition. In this region, $P = 1$ and $F = 2$. The other three areas shown in Fig. 11.6 contain two phases—A + B, A + liquid, B + liquid— and are univariant, since $P = 2$ and $F = 1$ in these areas. Consider the region B + liquid and let the mixture have overall composition f. Let temperature be the degree of freedom and fix this at T_2. The compositions of the two coexisting phases, B and liquid, are then automatically fixed, although one would have to determine experimentally what their compositions were.

In order to determine the compositions of these two phases from the phase diagram, construction lines (dashed) are drawn. First, an *isotherm* is drawn at the temperature of interest, T_2. This is the horizontal dashed line that terminates on the *liquidus* curve at point h'. This point, h', represents the liquid that is present in the mixture of B + liquid at temperature T_2. The composition of this liquid is given by drawing a vertical line or *isopleth* which intersects the composition (horizontal) axis at h, at which point the composition of the liquid may be simply read off the composition scale. The other phase that is present in the mixture is B, whose composition is fixed, as pure B, in this example.

An important distinction to be made here is between different meanings of the word 'composition'. It has at least three meanings:

(a) The composition of a particular phase. In the above example, the liquid phase has composition h.
(b) The relative amounts of the different phases present in a mixture. This may be referred to as the *phase composition*. In the above example, B and liquid are present in the ratio $\sim 3:2$ (see later for an explanation of the lever rule used to determine phase compositions).
(c) The overall composition of a mixture, in terms of the components and irrespective of the phases present. This may be termed the *component composition*. In the above example, the component composition of mixture f is $\sim 22\%$ A, 78% B.

Since there is no universally adopted convention over the use of the word 'composition', one can only say, be careful!

In the sense to which the phase rule is applied, composition may be regarded as a degree of freedom only when it refers to the actual compositions of the phases involved, category (a). The relative amounts of the different phases in a mixture, category (b), does not constitute a degree of freedom. Thus, along the isotherm, $h'f'T_2$, the relative amount of the phases B and liquid varies but the compositions of the two individual phases do not vary. The component composition, category (c), is included in the phase rule, not as a degree of freedom but as the number of components.

The liquidus curve, xyz, gives the highest temperature at which crystals can exist as a function of overall composition. Liquids whose compositions lie between A and e cross the liquidus curve between points x and y on cooling and enter the two-phase region: A + liquid. For these compositions A is the *primary phase* because it is the first phase to crystallize on cooling. The line cyd is the *solidus* and gives the lowest temperature at which liquids can exist in equilibrium over this composition range.

Point y is an invariant point at which three phases coexist: A, B and liquid. It is a *eutectic* and its temperature is the lowest temperature at which a composition (in this case e) can be completely liquid. Alternatively, it is the lowest temperature at which liquid may be present for any composition in this system. In simple eutectic systems such as this one, the solidus and eutectic temperatures are the same.

The melting or crystallization behaviour of different mixtures of A and B on heating or cooling can be understood by reference to Fig. 11.6. Consider composition f. It contains a mixture of A and B below temperature T_1 and is completely liquid above T_4. Between T_1 and T_4, varying amounts of crystalline B and liquid coexist. Thus at T_2 the liquid composition is at h'. The relative amounts of liquid h' and crystals B, in equilibrium, are given by the *lever rule*. These relative amounts are inversely proportional to their distances on the composition axis from the bulk composition f. The fraction of liquid present at T_2 is given by the ratio fB/Bh and the fraction of B by fh/Bh. The lever rule is the same as the principle of moments but instead of having weights balanced on a beam at different distances from the fulcrum we have different amounts of two phases giving an overall bulk or component composition. The lever rule may be derived by application of the principle of moments, as follows: For composition f at temperature T_2, the phases present in equilibrium are B and liquid, h'. The relative amounts of B and liquid are given by:

(Fractional amount of liquid, h') × (distance hf) = (fractional amount of B) × (distance Bf)

i.e.

$$\frac{\text{Fractional amount of liquid, } h'}{\text{Fractional amount of B}} = \frac{\text{fractional amount of liquid, } h'}{1 - \text{fractional amount of liquid, } h'}$$

$$= \frac{Bf}{hf}$$

i.e.

$$\text{Fractional amount of liquid, } h' = \frac{Bf}{Bf + fh}$$

$$= \frac{Bf}{Bh}$$

The lever rule may be used to determine how the relative amounts of phases in a mixture change, if at all, with temperature. Thus at temperature T_2, the amount of liquid present in composition f is given by Bf/Bh, i.e. ~ 0.38. At a higher temperature T_3, the amount of liquid is given by Bf/Bj, i.e. ~ 0.5. At a lower temperature, just above the solidus T_1, the amount of liquid is given by Bf/Be, i.e. ~ 0.33. Clearly, therefore, the effect of raising the temperature above T_1 is to cause an increase in the degree of melting from ~ 0.33 at T_1 to ~ 0.5 at T_3. The limit is reached at T_4 where the fraction of liquid is 1 and melting is complete. As the degree of melting increases with increasing temperature, so the composition of the liquid phase must change accordingly: since crystals of B disappear into the liquid phase on melting, the liquid must become richer in B. Thus the first liquid that appears on heating, at temperature T_1, has composition e, i.e. $\sim 33\%$ B, 67% A. As melting continues, the liquid follows the liquidus curve $yh'j'g$ until, when melting is complete at T_4, the liquid has composition f, i.e. $\sim 78\%$ B, 22% A. On cooling the liquid of composition f, the reverse process should be observed under equilibrium conditions. At T_4, crystals of B begin to form and with falling temperature the liquid composition moves from g to y as more crystals of B precipitate.

The *eutectic reaction* which occurs on cooling through temperature T_1 gives a good example of the use of the lever rule. Just above T_1, the fraction of B present is given by fe/Be and is roughly 0.67. Just below T_1 the fraction of B is fA/BA and is roughly 0.78. Thus, the residual liquid, of composition e, has crystallized to a mixture of A and B, i.e. the quantity of B present has increased even further and crystals of A are formed for the first time.

A solid mixture of A and B of overall composition e undergoes complete melting at temperature T_1 and conversely, on cooling, a homogeneous liquid of this composition completely crystallizes to a mixture of A and B at T_1.

The reactions described above are those that should occur under equilibrium conditions. This usually means that slow rates of heating and, especially, cooling are necessary. Rapid cooling rates often lead to different results, especially in systems which have more complicated phase diagrams. However, the equilibrium diagram can often be very useful in rationalizing these non-equilibrium results (see later).

The liquidus curve xyz may be regarded in various ways. As well as giving the maximum temperature at which crystals can exist, it is also a *saturation solubility curve*. Thus, curve yz could be regarded as giving the solubility limit with temperature for crystals of B dissolved in liquid. Above yz a homogeneous solution occurs but below this curve undissolved crystals of B are present. On cooling, precipitation of crystals B would therefore occur below the temperatures of curve yz; otherwise a metastable undercooled, supersaturated solution would be present.

Another interpretation of the liquidus is that it shows the effect of soluble impurities on the melting points of pure compounds. If a specimen of B is held at temperature T_4 and a small amount of A is added, then some liquid of composition g will form. As the amount of added A increases, so the amount of liquid g increases until, when sufficient *flux*, A, has been added to bring the overall composition to g, the solid phase disappears and the sample will be all liquid. Therefore a small amount of soluble impurity A has lowered the melting point of B from T_5 to T_4. A familiar practical example of this is the addition of salt to icy roads. In the binary system H_2O–$NaCl$, addition of $NaCl$ lowers the melting point of ice below $0\,°C$; the system contains a low temperature eutectic at $\simeq -21\,°C$.

11.3.2 Binary systems with compounds

Three types of binary system with a compound AB are shown in Fig. 11.7. A stoichiometric binary compound such as AB is represented on the phase diagram by a vertical line. This shows the range of temperatures over which compound AB is stable. Compound AB melts congruently in (a) because it changes directly from solid AB to liquid of the same composition. Figure 11.7(a) may be conveniently divided into two parts, given by the composition ranges A–AB and AB–B; each part may be treated as a simple eutectic system in exactly the same manner as Fig. 11.6. Although the horizontal lines at T_1 and T_2, corresponding to the two eutectic temperatures, meet the vertical line representing crystalline AB, no changes would be observed at T_1 and T_2 on heating pure AB. This is because these horizontal lines should peter out as composition AB is approached; composition AB, by itself, is a one-component system (but not in Fig. 11.7b and c) and only when another component, A or B, is added are changes observed at T_1 or T_2.

In Fig. 11.7(b), compound AB melts *incongruently* at T_2 to give a mixture of crystals A and liquid of composition x. The relative amounts of liquid and

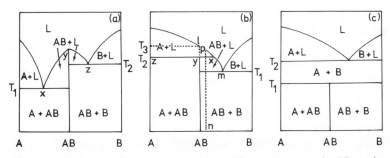

Fig. 11.7 Three types of diagram with a binary compound. AB melts congruently in (a), incongruently in (b) and has an upper limit of stability in (c). L = liquid

crystals A just above T_2 are given by the lever rule; fraction liquid = yz/xz. On further heating, crystals of A gradually dissolve as the liquid becomes richer in A and moves along the liquidus curve in the direction xl. At T_3, the liquid composition has reached l and the last A crystals should disappear. Point x is an invariant point at which three phases coexist: A, AB and liquid. It is a *peritectic* point because the composition of the liquid cannot be represented by positive quantities of the two coexisting solid phases, i.e. composition x does not lie between A and AB, as is the case for a eutectic point (Fig. 11.7a). A characteristic feature of a peritectic is that it is not a minimum point on the liquidus, as is a eutectic. Phase AB has a primary phase field. It is the first phase to crystallize on cooling liquids in the composition range $x-m$. However, the composition of AB is separated from its primary phase field. This is different to the case for congruently melting AB in Fig. 11.7(a), where the composition AB lies *within* the range xyz over which AB is the primary phase.

The behaviour of liquid of composition n on cooling is worth describing. At point p, crystals of A start to precipitate; more A crystals form as the temperature drops and the liquid composition moves from p to x. At T_2, the *peritectic reaction* liquid (x) + A → liquid (x) + AB occurs. Thus, the crystalline phase changes from A to AB and the amount of liquid present must diminish. From the lever rule, just above T_2, the mixture is $\sim$ 90 per cent liquid and just below T_2 only $\sim$ 50 per cent liquid. Therefore, *all* of phase A has reacted with *some* of the liquid to give AB. On further cooling from T_2 to T_1, more AB crystallizes as the liquid composition moves from x to m; finally, at T_1, the residual liquid of composition m crystallizes to a mixture of AB and B. Just above T_1, the mixture is $\sim$ 30% liquid and 70% AB and just below T_1, $\sim$ 10% B and 90% AB.

The behaviour on cooling of any liquid of composition between A and AB is similar but with one important difference. At T_2, the peritectic reaction for these compositions involves *some* of A reacting with *all* the liquid to give AB. Below T_2, a mixture of A and AB coexists and no further changes occur on cooling.

In systems that contain incongruently melting compounds, such as Fig. 11.7(b), it is very easy to get non-equilibrium products on cooling. This is because the peritectic reaction that should occur between A and liquid is slow, especially if the crystals of A are much more dense than the liquid and have settled to the bottom of the sample. What commonly happens in practice is that the crystals of A which have formed are effectively lost to the system and there is not time for much peritectic reaction to occur at T_2. The liquid of composition x then effectively begins crystallizing again below T_2, but this time crystals of AB form; at the eutectic temperature T_1, the residual liquid m crystallizes to a mixture of AB and B, as usual. Thus, it is quite common to obtain a mixture of *three* crystalline phases on cooling: A, AB and B, at least one of which should be absent under equilibrium conditions.

Another common type of non-equilibrium assemblage occurs when the intermediate, incongruently melting compound AB fails completely to crystallize on cooling. If this happened on cooling liquids in Fig. 11.7(b), the crystalline products would be A and B with no AB. Hence the peritectic reaction A + liquid → AB has been suppressed entirely.

In some systems, the liquids may fail to crystallize at all on cooling rapidly. The resulting products are *glasses*. These are effectively supercooled liquids whose viscosity has increased to the extent that the material is rigid and has the mechanical properties of a solid rather than a liquid (Chapter 18).

The first two examples above show how phase diagrams can be applied to non-equilibrium situations. If the equilibrium phase diagram is known the occurrence of non-equilibrium effects may be understood and, in many cases, predicted. Glass formation cannot be predicted from the phase diagram although, in some cases, it is found that glass-forming compositions are in the vicinity of low melting eutectics.

Sometimes, compounds decompose before their melting point is reached, as shown for AB in Fig. 11.7(c). Compound AB has an *upper limit of stability* and at temperature T_1 disproportionates into a mixture of crystalline A and B; at higher temperatures the system is simple eutectic in character.

There are also many examples of systems which contain compounds with a *lower limit of stability*, i.e. below a certain temperature, compound AB decomposes into a mixture of A and B. The behaviour of AB at higher temperatures can then be any of the three types described above.

A phase diagram which contains most of the binary features discussed above and which is one of the most important diagrams in silicate technology is that of

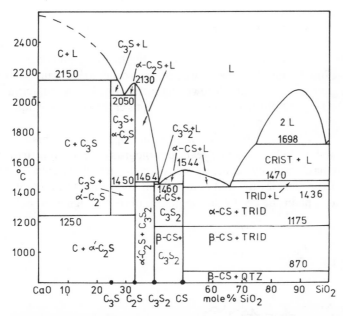

Fig. 11.8 Phase diagram for the binary system CaO–SiO_2. Data from B. Philips and A. Muan, J. Am. Ceram. Soc., **42** 414 (1959) $C = CaO$, $C_3S = Ca_3SiO_5$, $C_2S = Ca_2SiO_4$, $C_3S_2 = Ca_3Si_2O_7$, $CS = CaSiO_3$, $CRIST = $ cristobalite, $TRID = $ tridymite, $QTZ = $ quartz, $L = $ liquid

390

the system $CaO-SiO_2$, shown in Fig. 11.8. Two congruently melting compounds are present, $C_2S(Ca_2SiO_4)$, $CS(CaSiO_3)$, and two compounds that melt incongruently, $C_3S(Ca_3SiO_5)$ and C_3S_2 $(Ca_3Si_2O_7)$. In addition, C_3S has a lower limit of stability and decomposes to CaO and C_2S below 1250 °C. At the silica-rich end of the phase diagram, a range of liquid immiscibility exists between 1698 and ~ 2100 °C. Liquid immiscibility is discussed in more detail next.

11.3.3 Binary systems with immiscible liquids

In some systems, two liquid phases can coexist over a range of compositions and temperatures. This usually gives rise to an immiscibility dome, as shown by the area abc in Fig. 11.9. The dome has effectively made space for itself on the phase diagram by interrupting the liquidus curve of primary phase A. To see this, imagine the effect of shrinking the immiscibility dome; points a and c move closer together until, when they meet, the immiscibility dome has disappeared completely and the diagram has the appearance of a simple eutectic system (Fig. 11.6).

Point a is an invariant point because three phases are in equilibrium at this point: liquid a, liquid c and crystals A($P = 3$, $C = 2$, and so $F = 0$). It is called a *monotectic*.

With increasing temperature, for compositions between a and c, the two liquids become progressively more soluble in each other until, above temperature T_4, complete liquid miscibility occurs for all compositions. Point b is known as the *upper consolute temperature* of the immiscibility dome.

Liquid immiscibility is a feature of many silicate and borate systems. In binary silicate systems, such as $CaO-SiO_2$ (Fig. 11.8), the two-liquid region exists only above ~ 1700 °C, but sometimes the addition of other oxides causes a stable two-liquid region to exist at much lower temperatures. In the quaternary system $K_2O-Al_2O_3-CaO-SiO_2$, certain compositions enter a region of liquid immiscibility at temperatures as low as 1100 °C.

Silicate liquid immiscibility has important implications in glass technology, ceramics and geology. For many years, petrologists have toyed with the idea that some magmas, naturally occurring silicate melts, may have split up because at one time their temperature and composition lay within an immiscibility dome. This could explain the occurrence in nature of pairs of igneous rock types, one

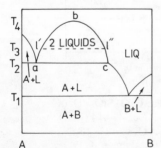

A   B   **Fig. 11.9** Binary system with liquid immiscibility dome

rich in silica and the other much poorer in silica, with essentially no rocks of intermediate composition present. Irrefutable evidence that liquid immiscibility was the cause of such occurrences has not been found in earth rocks, but was found in lunar samples obtained in the Apollo II mission to the moon. It therefore seems quite likely that liquid immiscibility also played a role in the early history of the earth.

Liquid immiscibility is common in systems of importance as glasses and glass-ceramics. This is because both glass-forming compositions and the compositions over which immiscibility domes exist are usually located at the SiO_2 — rich end of phase diagrams. During the cooling of a liquid to form a glass (Chapter 18), many compositions enter a dome of liquid immiscibility and a process of liquid unmixing or *phase separation* occurs. Often, the immiscibility dome is not present on the equilibrium phase diagram, however, and is an entirely metastable immiscibility dome (Fig. 11.10b). In other cases, the top of the dome is an

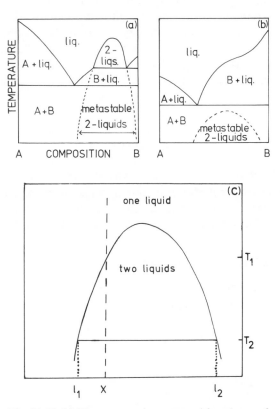

Fig. 11.10 (a) Binary eutectic system with a dome of liquid immiscibility; metastable extension of the stable immiscibility dome is shown by dashed curves. (b) An S-shaped liquidus curve which indicates the presence of a metastable immiscibility dome at lower temperatures. (c) A liquid immiscibility dome

392

equilibrium feature and is present on the phase diagram, as in Figs. 11.9 and 11.10(a).

The region of two liquids in Fig. 11.10(a) is terminated at its lower temperature end by the precipitation of crystals of B. What would happen if precipitation of B were a kinetically slow process? The two-liquid dome would then extend, metastably, to lower temperatures and in so doing it would probably expand to include a greater range of compositions. This extension is shown by the dashed lines. We therefore have *two* phase diagrams which are appropriate to different sets of experimental conditions. The equilibrium phase diagram (solid lines) is observed at slow rates of cooling. At faster cooling rates, the only feature to affect a liquid is the dome of liquid immiscibility, part of which is stable and part metastable. This is redrawn in Fig. 11.10(c). Composition x is a single phase liquid above the temperature T_1, but below T_1 a mixture of two liquids occurs. At T_2, these liquids have compositions l_1 and l_2.

Although liquids may avoid crystallizing during cooling and give glasses instead, they spontaneously separate into two liquids once they get an appreciable distance inside an immiscibility dome. This unmixing cannot be suppressed, even with rapid cooling. The unmixing is called *spinodal decomposition* and is discussed in Chapter 18.

In other silicate systems, such as the alkali silicates, an immiscibility dome occurs that is entirely metastable and does not extend above the liquidus (Fig. 11.10b). A pointer to the occurrence of such a dome is usually given by an irregular or S-shaped liquidus curve, which indicates that in the stable melt at high temperature there is probably clustering or premonitory unmixing phenomena. In some systems such as $BaO-SiO_2$, the deviation from ideality is so great that the liquidus is almost horizontal over a considerable range of compositions. Unmixing or phase separation is common in glass-forming compositions and sometimes this is put to commercial use, as in the preparation of Vycor glass. It is also believed to be important in the nucleation of glass-ceramics (Chapter 18).

11.3.4 Binary systems with solid solutions

The simplest form of solid solution system is one that shows complete miscibility in both solid and liquid states (Fig. 11.11). The melting point of A is

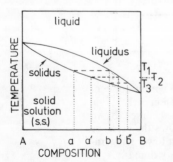

Fig. 11.11 Binary system with a complete range of solid solutions

depressed by the addition of B and that of B is increased by the addition of A. The liquidus and solidus are both smooth curves which meet only at the end-member compositions A and B. At low temperatures, a single phase solid solution exists and is bivariant ($C = 2$, $P = 1$, and so $F = 2$). At high temperatures, a single phase liquid solution exists and is similarly bivariant. At intermediate temperatures, a two-phase region of solid solution + liquid exists. Within this two-phase region, the compositions of the two phases in equilibrium are found by drawing isotherms or *tie-lines* at the temperature of interest, e.g. T_1. The intersection of the tie-line and the solidus gives the composition of the solid solution, a, and the intersection of the tie-line and the liquidus gives the liquid composition, b.

On cooling liquids in a system such as this, the crystallization pathways are complicated. A liquid of bulk composition b begins to crystallize a solid solution of composition a at temperature T_1. At a lower temperature, T_2, and at equilibrium, the amount of solid solution present increases but also its composition changes to a'. The fraction of solid solution a' is given by the lever rule and is equal to $bb'/a'b'$, i.e. the equilibrium mixture is approximately one-third solid solution and two-thirds liquid at T_2. Crystallization is therefore a complex process because with decreasing temperature, the composition of the solid solution has to change continuously in order to maintain equilibrium. With falling temperature, both crystals and liquid become progressively richer in B but the quantities of the two phases change in accord with the lever rule; the overall composition must obviously always be b. Finally, at temperature T_3, the solid solution composition reaches the bulk composition b and the last remaining liquid, of composition b'', disappears.

In systems with phase diagrams such as this, metastable or non-equilibrium products are often produced by a process of *fractional crystallization*. This occurs unless cooling takes place very slowly such that equilibrium is reached at each temperature. The crystals that form first on cooling liquid b have composition a. If these crystals do not have time to re-equilibrate with the liquid on further cooling they are effectively lost from the system. Each new crystal that precipitates will be a little bit richer in B and the result is that crystals form which have composition ranging from a to somewhere between b and B. In practice, the crystals that precipitate during cooling are often 'cored'. The central part that formed first may have composition a and on moving out radially from the centre the crystal becomes increasingly rich in B.

Coring occurs often in many rocks and metals. The plagioclase feldspars, which are solid solutions of anorthite, $CaAl_2Si_2O_8$, and albite, $NaAlSi_3O_8$, have the simple phase diagram shown in Fig. 11.12. Igneous rocks contain plagioclase feldspars and form by slow cooling of liquids. Such feldspar liquids and crystals are notoriously slow to equilibrate and although the cooling of melts in nature may have been very slow, it is common nevertheless to find rocks in which fractional crystallization has occurred. In these, the plagioclase crystals have calcium-rich centres and sodium-rich outer regions.

Coring may occur in metals during the manufacture of bars and ingots. The molten metal is poured into moulds (or 'sand cast' in moulds of sand) and allowed

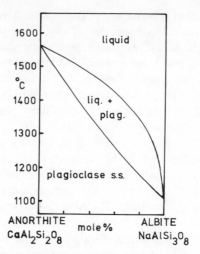

Fig. 11.12 The plagioclase feldspar
system, anorthite–albite

to cool. If the metal composition is part of a solid solution then coring may occur. Coring is usually deleterious to the properties of the metal and has to be eliminated. This can be done by subsequently heating the bars to just below the solidus temperature at which homogenization of the metal, with the elimination of coring, occurs rapidly.

It has been noted that the equilibrium subsolidus assemblage is obtained if cooling rates are sufficiently slow, but that with somewhat faster cooling fractional crystallization may occur. This is a property of the phase diagram and can occur with any kind of material, whether it be rocks, metals, or synthetic inorganic or organic materials. At still faster cooling rates, other types of pathway may be followed. With both equilibrium and fractional crystallization, no undercooling or supersaturation of the liquid occurs and crystallization begins as soon as the temperature reaches that of the liquidus. With faster cooling rates, however, undercooling of the liquid is usually possible and this may lead to various products. It may be possible to go directly from a homogeneous liquid to homogeneous single phase crystals of the same composition in one step; e.g. if a liquid of composition 50 per cent albite, 50 per cent anorthite (Fig. 11.12) is cooled quickly to just below the solidus, 1200 to 1250 °C, and held at this temperature, homogeneous plagioclase of the same composition may crystallize directly. Alternatively, if the liquid is cooled rapidly to room temperature, there may not be time for any crystallization to occur and a glass forms. Glass formation is common in inorganic materials such as silicates. Recently there has been much scientific and technological interest in glassy semiconductors and metals. These are materials that have unusual electrical and mechanical properties (see Chapters 14 and 18).

The simplest type of solid solution phase diagram is that shown in Fig. 11.11. Other relatively simple types of phase diagram are possible that show complete

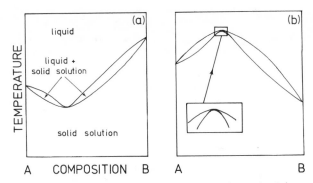

Fig. 11.13 Binary solid solutions with (a) thermal minima
and (b) thermal maxima in the liquidus and solidus curves

solubility in both the solid and liquid states but have either a thermal minimum
or a thermal maximum in the liquidus and solidus curves (Fig. 11.13). These
thermal maxima and minima are called *indifferent points* because they are not
true invariant points. For an invariant point, three phases are needed in
equilibrium ($F = 0$, $P = C + 1 = 3$), but this condition can never exist in solid
solution systems such as these because there are never more than two phases
present, i.e. solid solution and liquid solution. The liquidus and solidus are
therefore continuous through the thermal maximum or minimum and do not
show a discontinuity such as is observed for peritectics and eutectics.

The melting points of congruently melting compounds in binary systems can
similarly be regarded as indifferent points because only two phases are present in
equilibrium. However, an alternative in these cases is to regard each congruently
melting compound as a separate one-component system, in which case the
melting point does become an invariant point ($P = 2$, and so $F = 0$). With thermal
maxima and minima in solid solution series it is not usual to regard the
composition of the solid solution which has the maximum or minimum as a
special composition and which can be treated as a one-component system in its
own right. This would be done only if this composition had a simple ratio of the
components, e.g. 1:1, 1:2, etc., or there was other evidence that this composition
was special, e.g. if ordering of the solid solutions occurred, giving evidence of a
superstructure in the X-ray diffraction patterns.

Complete solid solubility, such as shown in Figs 11.11 to 11.13, occurs only
when the cations or anions that are replacing each other are similar in size, e.g.
Al^{3+} and Cr^{3+}. It is far more common to have phase diagrams in which the
crystalline phases have only partial solubility in each other. The simplest possible
case, shown in Fig. 11.14, is a straightforward extension of the simple eutectic
system, shown in inset (a). Crystals of B dissolve in crystals of A to form a solid
solution whose maximum extent depends on temperature and is given by the
curves *xmp*. Just below the melting point of A, point *p*, A cannot form a solid
solution (s.s.) with B but melts instead (and enters the region A s.s. + liquid). With
falling temperature the range of A solid solution gradually grows; the extent of

396

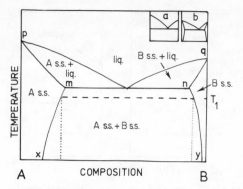

Fig. 11.14 Simple binary eutectic system with partial solid solution formation

the solid solution is a maximum at the solidus temperature, point m. This is usually the case in systems containing a solid solution. At lower temperatures the range of solid solutions diminishes along the curve mx. Crystalline B is also able to dissolve A and its maximum limit of solid solution is given by the curve ynq. The range of B solid solutions is less extensive then that of A solid solutions and, again, the maximum extent occurs at the solidus, point n.

In the two-phase region (A s.s. + B s.s.), the composition of the A s.s. is given by the intersection of the tie-line at the temperature of interest, say T_1, and the curve $m-x$ which limits the extent of A s.s. The composition of B s.s. is similarly fixed by the intersection of the tie-line and curve $n-y$.

In many phase diagrams, the solid phases are shown as line phases, i.e. as being stoichiometric and without a range of homogeneity or solid solution formation

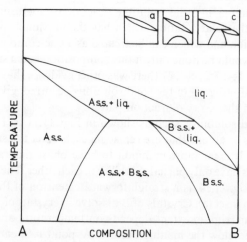

Fig. 11.15 Binary system with partial solid solution formation

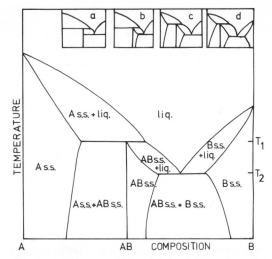

Fig. 11.16 Binary system with incongruently melting compound and partial solid solution formation

(inset a). In practice, however, the phases may have a slightly variable composition, as in inset (b), but it may be difficult to detect or measure this.

Examples of real systems that are similar to Fig. 11.14 are forsterite (Mg_2SiO_4)–willemite (Zn_2SiO_4), shown in Fig. 10.1, and spinel ($MgAl_2O_4$)–corundum (Al_2O_3), shown in Fig. 10.3.

Another type of simple binary system with partial solid solubility is shown in Fig. 11.15. This rather strange looking diagram can be derived from a simple system showing complete solubility (inset a). First, suppose that an immiscibility dome exists within the solid solutions which has an upper consolute temperature as shown in inset (b). Above the upper consolute temperature, a single phase solid solution exists, but below it a mixture of two phases exists.

Second, let the dome expand to higher temperatures until it intersects the melting curves. The result is shown in inset (c) and on an expanded scale as Fig. 11.15. The phase diagram for silica (SiO_2)–eucryptite ($LiAlSiO_4$), shown in Fig. 10.4, for compositions near the eucryptite end is similar to that in Fig. 11.15. The Al_2O_3–Cr_2O_3 diagram (Fig. 10.6) is similar to that shown in inset (b).

A more complex phase diagram containing an incongruently melting phase that forms a range of limited solid solutions is shown in Fig. 11.16. The progressive introduction of solid solutions is shown in insets (a), (b), (c) and (d), again working on the principle that the maximum extent of solid solution occurs at the solidus (temperatures T_1 and T_2).

11.3.5 Binary systems with solid–solid phase transitions

The representation of solid–solid phase transitions on phase diagrams depends on the nature of the phase transition. Transitions that thermodynami-

398

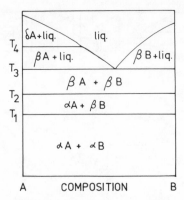

Fig. 11.17 Simple binary system
with polymorphic transitions

cally are first order (Chapter 12) and involve a change in some property such as
volume or enthalpy, or which, crystallographically, are reconstructive and
involve the breaking and forming of many primary bonds, can be treated in much
the same way as melting phenomena. In a one-component condensed system, e.g.
pure A or pure B, two solid polymorphs may coexist in equilibrium at only one
fixed point. In binary systems that do not contain solid solutions, phase
transitions in either of the end-member phases are represented by horizontal (i.e.
isothermal) lines, there being one line for each phase transition. In Fig. 11.17, the
low temperature polymorphs of A and B are labelled αA and αB, respectively.
Transition temperatures are αB$\rightleftharpoons\beta$B at T_1, αA$\rightleftharpoons\beta$A at T_2 and βA$\rightleftharpoons\delta$A at T_4. In
the absence of solid solutions, these transition temperatures are the same for the
pure phase as for the phase mixed with other phase(s). Several examples of solid–
solid phase transitions occur in the CaO–SiO$_2$ phase diagram (Fig. 11.8); e.g.
CaSiO$_3$(CS) undergoes a transformation between the low temperature β
polymorph and the high temperature α polymorph at 1175 °C. This transformation is observed in CaSiO$_3$ alone or when mixed with either SiO$_2$ or
Ca$_3$Si$_2$O$_7$(C$_3$S$_2$).

In systems that exhibit complete solid solubility, as well as phase transitions,
three types of phase diagram are possible (Fig. 11.18). For the end-member
phases, A and B, the transitions occur at a fixed temperature, as indeed they must
according to the phase rule ($C = 1$, $P = 2$, and so $F = 0$). However, for the
intermediate compositions, two phases can coexist over a range of temperatures
or bulk compositions because there is now one degree of freedom ($C = 2$, $P = 2$,
and so $F = 1$). Thus, two-phase regions containing two solid phases, e.g. ($\alpha + \beta$),
are generally observed. The treatment of the $\alpha\rightleftharpoons\beta$ change in Fig. 11.18(a) is
exactly the same as for the melting of β solid solutions, as discussed in detail for
Fig. 11.11. The melting relations in Fig. 11.18(a) are the same as in Fig. 11.11 but
in fact could be any of the three types given in Figs 11.11 and 11.13. In
Fig. 11.18(a), both end-members A and B and the entire range of solid solutions
show both α and β polymorphs.

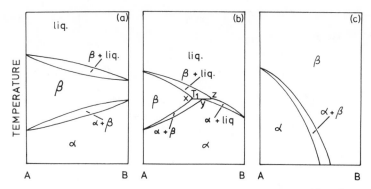

Fig. 11.18 Binary solid solution systems with polymorphic phase transformations

In Fig. 11.18(b), the $\alpha \rightleftharpoons \beta$ phase transition curves intersect the solidus curve because the $\alpha \rightleftharpoons \beta$ transition in B and in solid solutions rich in B now occurs, hypothetically, above the melting point of B and the B-rich solid solutions. The nature of the intersection of the three one-phase fields (α, β and liquid) and the three two-phase regions at T_1 is typical of solid solution systems. Always, two one-phase areas such as α and β must be separated from each other by a two-phase area ($\alpha + \beta$), although in practice the width of the two-phase regions may be difficult to detect experimentally. Three phases can coexist at only one temperature, T_1, and on the horizontal line xyz. There is an apparent contradiction with the phase rule here because, although three coexisting phases constitute an invariant condition, they coexist over the *range* of compositions xyz. However, there is no contradiction because the compositions of the individual phases are fixed (as x, y and z for β, α and liquid, respectively). The only variable is the relative amounts of these three phases; these relative amounts are not a degree of freedom in the phase rule.

A similar situation exists in simple eutectic systems (Fig. 11.6). The eutectic is point y but the invariant condition extends along the line cyd.

In Fig. 11.18(c), the temperature of the $\alpha \rightleftharpoons \beta$ transition decreases increasingly rapidly as the solid solutions become more rich in B. For pure B, the α polymorph does not exist at any real temperature. The melting behaviour in (c) is not shown. It could be any of the types shown in Figs 11.11 and 11.13.

A typical binary system that has both phase transitions and partial solid solubility is shown in Fig. 11.19. The $\alpha \rightleftharpoons \beta$ transition occurs at one fixed temperature in pure A and pure B but, in the solid solutions, two-phase regions of ($\alpha A + \beta A$) and ($\alpha B + \beta B$) solid solution exist.

We have already seen that there are analogies between melting behaviour and phase transition behaviour in solid solution systems (Fig. 11.18a). A further analogy exists between the eutectic, E, and the *eutectoid*, B, in Fig. 11.19. The line A–B–C represents an invariant condition over which three phases coexist, βA s.s (composition A), βB s.s. (composition B) and αB s.s. (composition C). The *eutectic reaction* at E on cooling is 1 liquid → 2 solids (βA s.s. + βB s.s.). The *eutectoid*

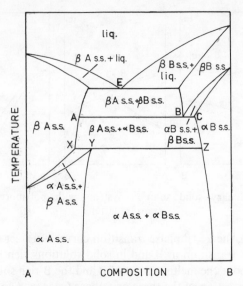

Fig. 11.19 Binary eutectic system with polymorphic phase transformations and partial solid solution formation

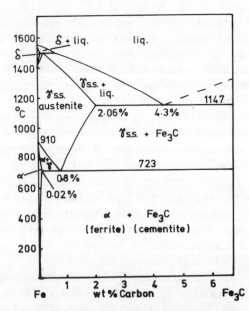

Fig. 11.20 The Fe–C phase diagram. The diagram is not an equilibrium diagram because cementite, Fe$_3$C, is metastable thermodynamically although kinetically stable

reaction at B on cooling is 1 solid (βB s.s.) → 2 solids (βA s.s. + αB s.s.). Thus, both the eutectic and eutectoid reactions are disproportionation reactions.

Point Y is a *peritectoid*. The line X–Y–Z represents an invariant condition over which three phases coexist, αA s.s. (composition Y), βA s.s. (composition X) and αB s.s. (composition Z). On heating, the reaction 1 solid (αA s.s.) → 2 solids (βA s.s. + αB s.s.) occurs. This is analogous to the melting of an incongruent compound at a peritectic temperature (Fig. 11.7b), which involves the reaction 1 solid → 1 solid + 1 liquid.

A eutectoid reaction in the system Fe–C is of great importance to steel making. The iron-rich part of this phase diagram is shown in Fig. 11.20. The diagram is, in fact, a metastable phase diagram because the iron carbide phase, cementite, Fe_3C, is not an equilibrium phase and should, thermodynamically, decompose to a mixture of iron and graphite. However, kinetically, decomposition of cementite is slow and is not observed under normal conditions of steel making. The changes that occur on thermal cycling of Fe–C alloys can therefore be studied with the aid of Fig. 11.20.

Iron exists in three polymorphic forms: body centred cubic α, stable below 910 °C; face centred cubic γ, stable between 910 and 1400 °C; and body centred cubic (again!) δ, stable between 1400 °C and the melting point 1534 °C. γ-Iron can dissolve appreciable amounts of carbon, up to 2.06 wt%, in solid solution formation, whereas the α and δ forms dissolve very much less carbon, up to a maximum of 0.02 and 0.1 wt%, respectively.

There is a simple explanation for the very different solubilities of carbon in the γ and α polymorphs of iron. Although the face centred cubic γ structure is more densely packed than the body centred cubic α structure, the interstitial holes (suitable for occupation by carbon) are larger although much less numerous in γ-Fe. Unit cells of the two forms are shown in Fig. 11.21 together with the octahedral sites that are available for occupation by carbon. These sites are at the cube face centres in α-Fe and at the body centre in γ-Fe. The iron–carbon distances and, hence, the sizes of the interstitial sites are considerably larger in γ-Fe than in α-Fe. In α-Fe, these sites are distorted. The cubic unit cell edge in α-Fe is 2.866 Å. Two iron–carbon distances would be 1.433 Å. The other four iron–carbon distances would be 2.03 Å. In γ-Fe, the octahedral sites are undistorted and the iron–carbon distance would be half the cell edge, $a = 3.591$ Å. Therefore,

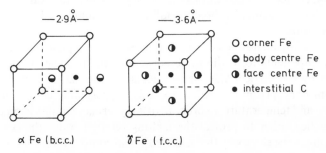

Fig. 11.21 Interstitial sites for carbon in α-Fe and γ-Fe

Fe–C = 1.796 Å. (Values are for γ-Fe at 22 °C; values at, for example, 900 °C would be a few per cent greater.)

Most carbon steels contain less than 1 wt % carbon, viz. 0.2 to 0.3 per cent for use as structural members. On cooling from the melt, and in the temperature range 800 to 1400 °C, these steels form a solid solution of carbon in γ-Fe called austenite (Fig. 11.20). However, the austenite solid solutions are unstable at lower temperatures (< 723 °C) because, when the structure changes from that of γ-Fe to α-Fe, exsolution or precipitation of the carbide phase, Fe_3C, occurs. This decomposition starts at the boundaries of the austenite grains; the ferrite (α-Fe) and cementite (Fe_3C) crystals grow side by side to give a lamellar texture known as pearlite.

If the steel is cooled quickly, however, there is not time for decomposition to ferrite and cementite to occur; instead martensite forms. Martensite has a deformed austenite structure in which the carbon atoms are retained in solid solution. It is possible to release these carbon atoms, as cementite, by tempering, i.e. reheating, to give a fine scale pearlite texture.

The hardness of steel depends very much on the cooling conditions and/or tempering treatment. Martensitic steels are hard largely, it seems, because of the stressed state of the martensite crystals which prevents easy motion of dislocations. In steels with pearlite texture, the hardness depends on the size, amount and distribution of the very hard cementite grains; a finer texture with a large number of closely spaced grains gives harder steel. If the steel is cooled slowly or held just below 723 °C, the decomposition is slow, giving a coarse pearlite texture. A finer texture is obtained by using a faster cooling rate to produce martensite which is then subsequently tempered at a low temperature, e.g. 200 °C.

11.4 Three-component condensed systems

Three-component or ternary systems have four independent variables: pressure, temperature and the composition of two of the components. If the composition or concentration of two of the components in a phase is fixed, the third is automatically fixed by difference. In this chapter, for simplicity, we are not considering volatile systems so we can use the condensed phase rule, i.e. $P + F = C + 1 = 4$. Under these conditions the ternary system is, for example, invariant when four phases—generally three crystalline phases and liquid—coexist at one fixed temperature.

Ternary systems are nowadays usually represented by equilateral triangles and the three components form the three corners of the triangle. Temperature is represented by the vertical axis, perpendicular to the plane of the triangle. A three-dimensional prism is needed to fully display the effects of varying composition and temperature. In order to display ternary equilibria on paper it is normal practice either to project the melting relations onto the composition triangle, in much the same way that a geographical contour map is a projection of

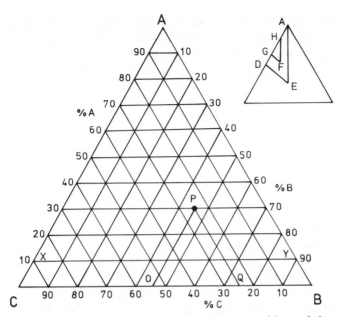

Fig. 11.22 Triangular grid for representing compositions of three component systems

features of the earth's surface, or to construct isothermal sections. Examples of both types of diagram are discussed.

Ternary compositions within equilateral triangles are given by reference to a grid (Fig. 11.22). Each of the three binary edges is divided into a hundred divisions (only ten are shown for clarity). Point A corresponds to 100% A; the edge BC corresponds to 0% A and the line XY to 10% A. Thus, the A content of any ternary composition is given by drawing a line through that composition and parallel to the BC edge. The A content is then read off from the intersection of this line with either the AB or AC edges. The contents of B and C are given similarly by drawing lines parallel to the AC and AB edges, respectively. Point P in Fig. 11.22 has a composition of 30% A, 45% B and 25% C. Compositions can be either in atom per cent, in mole per cent or in weight per cent (but obviously not a mixture). In practice, it is usually preferrable to use mole per cent for inorganic systems because the formulae of binary and ternary phases are then clearly and simply related to the composition.

An alternative method of determining composition is, by making appropriate construction lines, to read off the concentration of all three components from one of the edges. For composition P (Fig. 11.22), lines PO and PQ are drawn, parallel to AC and AB, respectively. The length CO represents the percentage of B (45), the length QB represents the percentage C (25) and the difference, length OQ, the percentage A (30). Thus, the middle section (OQ) represents the concentration of the phase or component that is not located on the line.

404

This method is very useful since it can be applied to determining phase compositions in triangles that are not equilateral. In most ternary systems which have binary and ternary phases present, as well as liquid, the triangles that represent the equilibria involved are not equilateral. An example is shown as an inset to Fig. 11.22 for point F, which is a mixture of three phases of composition A, D and E. Lines GF and HF are drawn parallel to DE and AE, respectively. The percentages of A, D and E present in F are calculated from the relative lengths of DG, AH and GH, respectively.

11.4.1 Simple eutectic system without binary of ternary compounds

A simple eutectic system is shown in Fig. 11.23 in which the liquidus relations are projected onto the composition plane. Each of the three binary edges, AB, AC and BC, is simple eutectic in character, as shown for the AB edge. The binary edges have eutectic temperatures T_a, T_b and T_c, respectively. For each phase, the liquidus surface of the primary phase field in the ternary system is a curved surface that forms part of a dome. For a congruently melting phase such as B, the apex of the dome corresponds with the composition and melting point of B. Isotherms $T_1,...,T_5$ on the surface of the liquidus describe the shape of the dome. The primary phase field of B is bounded by B, T_a, T_d and T_c.

Neighbouring primary phase fields, e.g. of A and B, intersect in a sloping boundary line or valley, in this case T_a-T_d. When three neighbouring primary phase fields meet, as at T_d, the point of intersection is a ternary invariant point. As T_d is the lowest temperature at which liquid can exist in this triangle it is a ternary eutectic (at T_d, $P = 4$: A, B, C and liquid; and so $F = 0$). The line T_a-T_d, as well as

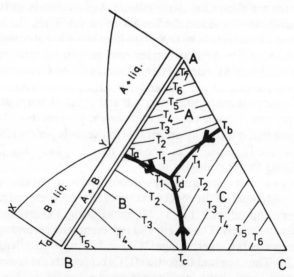

Fig. 11.23 Simple ternary eutectic system showing uni-variant curves and liquidus isotherms

the lines T_b-T_d and T_c-T_d, is a *univariant curve* ($P = 3$: A, B and liquid; and so $F = 1$). One degree of freedom, either temperature or composition of the liquid, is needed to define the phases that are in equilibrium on this univariant curve. Arrows on the univariant curves indicate directions of falling temperature. It should be clear that an invariant point in a binary system, such as the eutectic at y, becomes a univariant curve in the ternary system when a third component is added.

The equilibrium assemblage present in ternary compositions at any temperature between the solidus and liquidus can best be understood from a knowledge of the crystallization pathways followed on cooling. Consider the changes that are expected to occur in composition a (Fig. 11.24) as it is cooled from the liquid state. Composition a lies within the primary phase field of B. Hence B is the first crystalline phase to appear on cooling, once the temperature has fallen inside the liquidus dome. As B progressively crystallizes, the liquid becomes deficient in B and it should be apparent that composition B and the locus of the changing liquid composition must lie on a straight line that passes through the bulk composition a. Thus, with falling temperature, the liquid composition moves away from B, on an extension of the line Ba and towards b. For temperatures in this range, such that the liquid composition is between a and b, the relative amounts of B and liquid are given by a simple lever rule calculation. For instance, when the liquid has almost reached b, the fraction of liquid present is given approximately by aB$/b$B.

Once the temperature has fallen sufficiently that the liquid composition has reached b, crystals of A begin to form. With a further drop in temperature the liquid composition is constrained to move down the univariant boundary curve bcd and the equilibrium assemblage over this temperature range is (A + B + liquid). The relative quantities of the three phases can be found from a lever rule construction similar to that shown in the inset in Fig. 11.22. For example, with liquid of composition c, the bulk composition a lies in the triangle with A, B and c as corners; lines passing through a and parallel to the Bc and Ac edges may

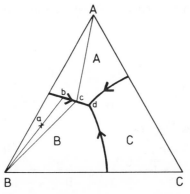

Fig. 11.24 Crystallization pathway in a simple ternary eutectic system

be drawn (not shown) and the phase composition determined from the intercepts of these lines on the AB edge.

In addition to the precipitation of A crystals that occurs as the liquid composition moves from b to d, precipitation of B crystals also continues. This can be checked by making lever rule calculations for different liquid compositions between b and d. Alternatively, it can be seen that if crystals of A precipitate from liquid, say of composition c, then crystals of B must also precipitate in order to maintain the liquid composition on the line cd.

Point d is a ternary eutectic. Its temperature, T_d, is the lowest at which liquid can exist in equilibrium in this system. When this temperature is reached on cooling and the liquid composition has arrived at d, the residual liquid must crystallize to give a mixture of A, B and C. Crystals of C are therefore forming for the first time. The final amounts of A, B and C in equilibrium are determined by lever rule calculations in the triangle A, B, C.

The behaviour of other compositions on cooling can be treated similarly and a sequence of reactions is generally observed. Differences occur only for compositions that happen to coincide with either univariant curves, e.g. c, or invariant points, e.g. d. Thus liquid of composition d must change completely from 100 per cent liquid to a solid mixture of A, B and C at one temperature. The behaviour of solid mixtures on heating is simply the reverse of the crystallization pathway followed on cooling and is not described further.

11.4.2 Ternary systems containing binary compounds

Ternary systems that contain one congruently melting binary compound, but no ternary compounds or solid solutions, can be one of the two types shown in Fig. 11.25. In (a), the triangle ABC can be divided into two smaller triangles, A–B–BC and A–BC–C. Both of these are simple eutectic in character, similar to Figs 11.23 and 11.24. Each triangle can be treated separately and no new principles are involved. The only feature that is worthy of comment is the join A–BC.

The join A–BC is a true *binary join* because any composition on this join,

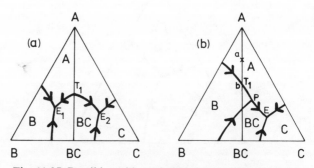

Fig. 11.25 Possible melting relations in ternary systems that contain a congruently melting binary phase

whether solid, liquid or a mixture of solid and liquid, can be represented completely by positive quantities of A and BC alone. The join is simple eutectic in character with a eutectic temperature, T_1. However, whereas T_1 is the *lowest* temperature at which liquid can exist on the join A–BC, T_1 is in fact a *thermal maximum* on the univariant boundary curve which separates the primary phase fields of A and BC in the ternary A–B–C system, i.e. $T_1 > E_1, E_2$. Drawing an analogy with geographical contour maps, T_1 represents the height of a pass between the two mountains A and BC.

In Fig. 11.25(b), compound BC again melts congruently but the ternary melting behaviour is different for certain compositions. The univariant curves that separate the primary phase fields of (i) A and B and (ii) B and BC now cross the join A–BC, whereas this does not happen in Fig. 11.25(a). As a result, the invariant point located at the intersection of these two univariant curves lies *outside* the triangle A–B–BC. Every triangle, such as A–B–BC has an associated invariant point. The three crystalline phases that coexist at this point are the three phases at the corners of the associated triangle. If an invariant point lies outside its own triangle, automatically it is a ternary peritectic point. If it lies inside its own triangle, it must be a ternary eutectic point. The direction of falling temperature along the univariant curve that separates the fields of A and BC must be in the direction arrowed since the peritectic temperature P must be higher than the eutectic temperature E.

The join A–BC is not a binary join, as it is in the previous example, because over a range of compositions, B is the primary phase and the composition B does not lie on the join A–BC. The join A–BC is shown in Fig. 11.26. It is much more complicated than a simple eutectic system, such as the join A–BC in Fig. 11.25(a). Consider the crystallization of the liquid of composition a on cooling (Figs. 11.25b and 11.26). This composition is within the primary phase field of A and so A is the first crystalline phase to appear. As the temperature drops, A continues to precipitate until temperature T_1 is reached, by which time the liquid composition has arrived at point b. At b, crystals of B begin to form and the liquid

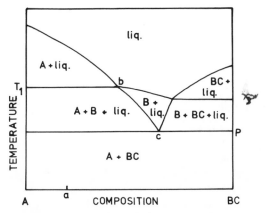

Fig. 11.26 The join A–BC in Fig. 11.25(b)

composition departs from the line A–BC because it is constrained to follow the univariant curve b–P that separates the fields of A and B. In Fig. 11.26 the composition has entered the region, A + B + liquid, although neither B nor the liquid composition (represented by the line bc) lie on the join A–BC. As the temperature drops further, A and B continue to crystallize until the peritectic temperature P is reached. At P, all of B and all of the liquid react (this may be kinetically slow) to give more crystals of A and, for the first time, crystals of BC. Below P no further changes occur as the equilibrium subsolidus assemblage A + BC is now present. Thus, the final subsolidus assemblage has returned to the join A–BC although for most of the crystallization process the compositions of some of the phases present do not lie on the join A–BC.

For all compositions that lie within the triangle A–B–BC (Fig. 11.25b), the processes of crystallization that occur on cooling from the liquid state terminate at temperature P: i.e. P is the solidus temperature for the triangle A–B–BC. For compositions that lie within the triangle A–BC–C, the solidus temperature is the eutectic temperature E.

Two other types of ternary system that contain only one binary compound are shown in Fig. 11.27. The compound BC melts incongruently in Fig. 11.27(a) and the BC edge has a phase diagram that is similar to that shown in Fig. 11.7(b). The ternary diagram A–B–C (Fig. 11.27a) contains one peritectic and one eutectic and is similar to that described above (Fig. 11.25b).

In Fig. 11.27(b), the binary compound BC has an upper temperature limit of stability. The phase diagram of the B–C edge is similar to that shown in Fig. 11.7(c). Above temperature T_1, BC decomposes to give a solid mixture of B and C; only at the considerably higher solidus temperature, T_2, does melting begin to occur. The binary eutectic point, T_2, on the BC edge becomes a univariant curve in the ternary system. This curve separates the primary phase fields of B and C and its temperature falls with increasing A content. When the temperature falls to below T_1, B and C are no longer stable together but react to give BC. This is true in both binary B–C and ternary A–B–C compositions. As a consequence, a ternary primary phase field for phase BC appears at temperatures

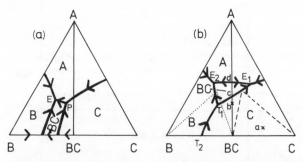

Fig. 11.27 Ternary systems containing (a) an incongruently melting binary phase and (b) a binary phase with an upper limit of stability

below T_1. Three ternary invariant points are shown in Fig. 11.27(b); two of these are eutectics, E_1 and E_2, and belong to the triangles A–BC–C and A–B–BC, respectively. The third invariant point at temperature T_1 corresponds to the decomposition of compound BC and at this point the primary phase fields of B, C and BC meet. This invariant point is neither a eutectic nor a peritectic; it does not belong to either of the two triangles but to the straight line B–BC–C. Sometimes, such a point is called a *distribution point*.

Some of the crystallization pathways for compositions within the triangles in Fig. 11.27(b) are quite complicated and involve several stages. For example, on cooling of a liquid of composition *a*, the crystallization pathway under equilibrium conditions can be summarized in the following scheme of reactions:

$$\text{Liquid } a \rightarrow \text{C} + \text{liquid} \rightarrow \text{C} + \text{B} + \text{liquid} \xrightarrow{T_1} \text{C} + \text{BC} + \text{liquid} \xrightarrow{E_1} \text{C} + \text{BC} + \text{A}$$

As always, the crystallization process ends when the invariant point is reached that belongs to the final, desired subsolidus assemblage, which in this case is A + BC + C. One feature of this sequence that is worth discussing is the reaction that occurs at temperature T_1. Given that once a liquid composition has arrived at a univariant curve it continues moving down that curve, or another similar univariant curve, with falling temperature (there are exceptions to this, see below), then a liquid arriving at the invariant point and temperature T_1 has apparently a choice of univariant curves to follow. It would appear to be able to go either in the direction of eutectic E_1 or in the direction of eutectic E_2. However, if the liquid composition followed the univariant curve leading to eutectic E_2 an impossible situation would arise. Clearly the bulk composition *a* must be represented by positive amounts of the three phases that are in equilibrium at a particular temperature. Suppose the temperature was just below T_1, but above those of E_1 and E_2. If the liquid had decided to follow the curve in the direction of E_2, then the three phases present in equilibrium would be B, BC and liquid. The compositions of these three phases, B, BC and liquid, form a triangle (dotted) and, quite clearly, *a* lies outside this triangle. This therefore is an impossible situation. The alternative is for the liquid to follow the curve T_1–E_1, in which case the three phases present are BC, C and liquid. Bulk composition *a* does lie within this triangle (dashed).

Although liquid compositions follow a univariant curve on cooling, if at all possible, sometimes this is not possible. Consider liquid of bulk composition *b* in Fig. 11.27(b). On cooling, primary phase crystals of C appear first. On further cooling, a three-phase equilibrium arises when the liquid composition meets the curve T_1–E_1: the phases in equilibrium are BC, C and liquid. This situation continues on further cooling until the liquid composition reaches point *c*. At this stage, composition *b* lies on the straight line that connects the compositions of crystals BC and liquid *c*. Therefore, the amount of crystals C that are present in equilibrium must have decreased to zero. Further movement of the liquid composition towards E_1 *is now prohibited* because *b* would lie outside the triangle BC–C–liquid. Instead the liquid composition must depart from the univariant curve T_1E_1 and enter the primary phase field of BC.

This effect may be explained with the aid of the phase rule. A liquid is constrained to follow a univariant curve when the reaction mixture is in a condition of univariant equilibrium, i.e. it has only one degree of freedom. This occurs when the liquid composition lies somewhere between T_1 and c. When, however, the liquid composition reaches c, one of the phases, crystalline C, disappears; consequently an extra degree of freedom is created, in accordance with the phase rule. The liquid is therefore no longer constrained to follow the univariant curve.

As the liquid leaves the univariant curve at point c, it enters the primary phase field of BC, i.e. the assemblage present is BC + liquid. As the temperature drops, BC continues to precipitate until the liquid reaches point d. Here, the residual liquid crystallizes to give a mixture of A and BC, which is the subsolidus assemblage for composition b. The equilibrium pathway followed by composition b on cooling may therefore be summarized:

$$Liquid \rightarrow C + liquid \rightarrow C + BC + liquid \rightarrow BC + liquid \rightarrow BC + A$$

The stage has now been reached where the basic principles of ternary melting behaviour have been set out and illustrated with simple examples. If several binary compounds or one or more ternary compounds are present, the diagrams may be very complicated, but, even so, no new principles are necessary in order to understand the diagrams. It does not seem worth while to try and discuss more complicated systems here; instead it is recommended that the keen reader choose a system and try to work out crystallization pathways for various liquid compositions. In this way, one can begin to find out about a system in much the same manner that a hill-walker can become acquainted with the hills and valleys of a new region by studying an ordinance survey map.

11.4.3 Subsolidus equilibria

The behaviour of systems at subsolidus temperatures is very important, especially in more complicated systems. A system that contains several binary or ternary phases is, at subsolidus temperatures, divided up into a number of smaller triangles. On paper, there is usually more than one possible set of triangles and one of the experimental problems in studying ternary equilibria is to determine the equilibrium arrangement of triangles. The problem is illustrated in Fig. 11.28. Two binary compounds AC and BC are present. The subsolidus triangle AC–BC–C must exist as there are no alternative triangles in this part of the diagram. For the rest of the diagram, however, there is a choice between having either A and BC as *compatible phases* or B and AC. If we assume that A and BC are compatible, a line is drawn between them that indicates their stability in the presence of each other. Two three-phase triangles result, A–B–BC and A–BC–AC. This means that B and AC would be incompatible phases; mixtures of B and AC should therefore react to give a mixture of A and BC, although the reaction rate may be slow.

The compatibility or incompatibility of phase mixtures or assemblages is of

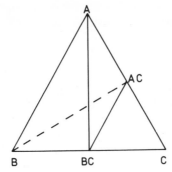

Fig. 11.28 Compatibility triangles in a ternary system at subsolidus temperatures

great practical importance in, for example, cement and refractories. In the manufacture of Portland cement clinker, the desired calcium silicates are Ca_2SiO_4 and Ca_3SiO_5; undesired ones are $Ca_3Si_2O_7$ and $CaSiO_3$. Additives must clearly be avoided which would react with either Ca_3SiO_5 or Ca_2SiO_4 to produce one of the other calcium silicates. Alternatively, using Fig. 11.28 as an example, if AC is the phase with the desirable properties, it is important to exclude B from the system because B and AC are incompatible. Similar considerations affect the lifetime of refractory bricks, i.e. molten slag and the refractory brick should be compatible.

11.4.4 Ternary systems containing binary solid solutions

Ternary phase equilibria may be quite complicated if solid solutions are present. A full treatment of such systems is not attempted here; instead only some of the more important points in simple systems are highlighted. Subsolidus equilibria for some simple systems are shown in Fig. 11.29. In (a), a complete range of solid solutions between B and C is shown by the hatching of the B–C join. These solid solutions coexist with A but there are never more than two

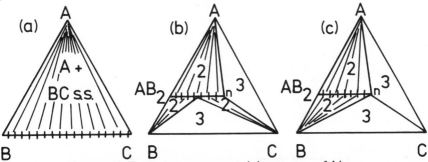

Fig. 11.29 Ternary systems containing ranges of binary solid solution. Numbers refer to the numbers of phases in equilibrium in each region

phases present in subsolidus equilibrium for any composition. Tie-lines radiate from A towards the BC edge; the whole ternary system can therefore be divided into an infinite number of closely spaced tie-lines, each of which represents a different two-phase assemblage.

Partial solid solution between a binary phase AB_2 and a non-existent compound AC_2 is shown in Fig. 11.29(b) and (c); the formula of this solid solution may be written $AB_{2-x}C_x$. These solid solutions must coexist with A, forming a two-phase region, as there is no other possible arrangement for compositions in this part of the triangle. Similarly, the three-phase triangle (A + C + solid solutions of composition n) must exist. For the remaining compositions which involve the coexistence of B, C and AB_2 solid solutions, however, there is more than one possible compatibility arrangement; two of these possibilities are shown in Fig. 11.29(b) and (c). The AB_2 solid solutions may coexist almost entirely with B (Fig. 11.29c) or almost entirely with C (not shown) or partly with B and partly with C (Fig. 11.29b). The correct, equilibrium, arrangement must be determined experimentally.

The presence of solid solutions in a phase diagram complicates melting behaviour. For the subsolidus diagram shown in Fig. 11.29(a), melting relations are typically as shown in Fig. 11.30. The edges AB and AC are simple eutectic in character; the BC edge (Fig. 11.30b) is similar to Fig. 11.11 and has a smoothly changing solidus and liquidus without thermal maxima, minima or invariant points. In the ternary system (Fig. 11.30a), there are no invariant points and only one univariant curve. This separates the primary phase fields of A and BC solid solutions; on this curve, temperatures fall in the direction XY. Liquid on the univariant curve XY coexists with both A and BC solid solutions; each liquid composition coexists with a particular BC solid solution composition. Tie-lines

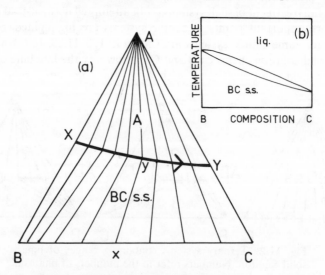

Fig. 11.30 Solid–liquid compatibility relations in ternary systems with binary solid solution

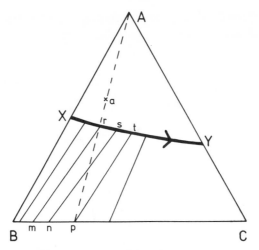

Fig. 11.31 Crystallization pathway in ternary system with binary solid solutions

that connect A and liquid on the curve XY radiate from A; however, tie-lines between BC solid solutions and the same liquids XY do not radiate from either B or C but move smoothly across the diagram. For example, liquid of composition y coexists with A and BC solid solution of composition x.

Consider the cooling of liquid of composition a (Fig. 11.31). The equilibrium subsolidus assemblage is A and solid solution p. On cooling, A starts to precipitate from the liquid and the liquid composition moves away from a on the extension of the line Aa. When the liquid composition reaches r on the univariant curve, BC solid solution crystals start to form; these have composition m. With further cooling, the liquid composition follows the curve XY in the direction rst. As it does so, the composition of the BC solid solution must change continuously at equilibrium and the amount of crystalline material, both A and BC solid solutions, increases. For liquid s the composition of the equilibrium solid solution is n. With further cooling, crystallization continues until the liquid reaches t and the solid solution composition reaches p. This is now the final equilibrium subsolidus condition and so the last trace of liquid disappears at t.

Unless cooling rates are very slow, a non-equilibrium assemblage is usually produced, by a process of fractional crystallization. Thus, if the first solid solution crystals to form, of composition m, do not re-equilibrate with the liquid on lowering the temperature a little further, then they may effectively be lost from the system. If this happens with all of the solid solution crystals that precipitate, a wide range of cored solid solution compositions results, extending from m, through p to some composition more rich in C. Hence, the crystals that form may be inhomogeneous, having B-rich centres and C-rich surfaces. Examples of such zoning or coring in metals and minerals were mentioned in the section on binary systems.

Crystallization of liquids within the primary phase field of BC solid solutions is

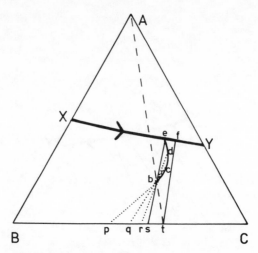

Fig. 11.32 Crystallization pathway in ternary system with binary solid solutions

rather more complex (Fig. 11.32). Consider liquid b; the first BC solid solution crystals to form have composition p. As the liquid moves down the liquidus surface, precipitation of more C-rich solid solution occurs and also the previously precipitated solid solution, p, re-equilibrates with the liquid to become more C-rich. With falling temperature, the liquid composition follows a curved pathway $bcde$ as the solid solution composition moves along $pqrs$. Thus, the liquid and solid solution compositions rotate about the bulk composition, b. Each liquid composition that lies on the tie-line se follows a unique pathway on cooling, but all of these pathways meet at the common point e on the univariant curve XY. At point e, crystals of A start to form and the liquid composition is constrained to follow the curve XY until point f is reached, at which the last of the liquid disappears. The subsolidus assemblage is A + solid solution t.

An apparently simple ternary diagram such as shown in Figs 11.30 to 11.32 requires a great deal of experimental work for its complete determination. Each composition within the primary phase field of BC solid solutions has its own particular crystallization pathway (e.g. $bcde$ for composition b; see Fig. 11.32) and these pathways have to be determined experimentally. Few systems have therefore been completely evaluated.

Questions

11.1 What is (i) the mole per cent and (ii) the weight per cent of (a) Al_2O_3 in mullite, $Al_6Si_2O_{13}$ (b) Na_2O in devitrite, $Na_2Ca_3Si_6O_{16}$ (c) Y_2O_3 in yttrium iron garnet, $Y_3Fe_5O_{12}$?

11.2 Sketch the phase diagram for the system Al_2O_3–SiO_2 using the following information. Al_2O_3 and SiO_2 melt at 2060 and 1720 °C. One congruently melting compound, $Al_6Si_2O_{13}$, forms between Al_2O_3 and SiO_2 with a melting point of 1850 °C. Eutectics occur at ~ 5 mol% Al_2O_3, 1595 °C and ~ 67 mol% Al_2O_3, 1840 °C. Compare your diagram with that shown in Fig. 20.3(a).

11.3 Explain with the aid of examples the difference between the concepts of phases and components. Under what conditions can a component be a phase?

11.4 Phase diagrams are drawn with the compositions represented usually as either weight per cent or mole per cent. Derive expressions for a binary AB system for converting from one to the other.

11.5 Sketch a phase diagram for a system A–B that has the following features. Three binary compounds are present A_2B, AB and AB_2. Both A_2B and AB_2 melt congruently. AB melts incongruently to give A_2B and liquid. AB also has a lower limit of stability.

11.6 For the $MgAl_2O_4$–Al_2O_3 phase diagram (Fig. 10.3) describe the reactions that would be expected to occur, under equilibrium conditions, on cooling a liquid of composition 40 mol% MgO, 60% Al_2O_3. Using rapid cooling rates, how might the product(s) differ?

11.7 The system Mg_2SiO_4–Zn_2SiO_4 is a simple eutectic system in which the two end-member phases form limited ranges of solid solution. Sketch a probable phase diagram for this system. How would you determine experimentally: (i) the compositions of the solid solution limits; (ii) the mechanism of solid solution formation in each case; (iii) the eutectic temperature.

11.8 Pure iron undergoes the $\alpha \rightleftharpoons \gamma$ transformation at 910 °C. The effect of added carbon is to reduce the transformation temperature from 910 to 723 °C. Sketch the general appearance of the Fe-rich end of the Fe–C phase diagram using this information.

11.9 Construct a triangular grid for representing three-component phase diagrams. Let the three components be Na_2O, CaO and SiO_2. Mark on your triangle, using a mole% scale, the compositions of the following phases: Na_2SiO_3, $Na_2Si_2O_5$, $CaSiO_3$, $Ca_3Si_2O_7$, Ca_2SiO_4, Ca_3SiO_5, Na_2CaSiO_4, $Na_2Ca_3Si_6O_{16}$, $Na_2Ca_2Si_3O_9$, $Na_4CaSi_3O_9$, $Na_2CaSi_5O_{12}$.

11.10 The ternary system A–B–C contains no binary compounds and only one ternary compound, X. (i) Sketch the layout of the subsolidus compatibility triangles. (ii) Assuming that X melts congruently, sketch the melting relations. Identify three ternary eutectics, three thermal maxima and six univariant curves.

References

A. M. Alper (Ed.) (1971). *High Temperature Oxides*, Vols 1 to 4, Academic Press, New York.

A. M. Alper (1976). *Phase Diagrams*, Vols 1 to 5, Academic Press.

W. G. Ernst (1976). *Petrologic Phase Equilibria*, W. H. Freeman & Co., San Francisco.

A. Findlay (1951). *The Phase Rule and Its Applications*, 9th Ed Dover N.Y.

P. Gordon (1968). *Principles of Phase Diagrams in Materials Systems*, McGraw-Hill, New York.

Phase Diagrams for Ceramists 1964 ed., 1969 Suppl., 1975 Suppl., 1981 Suppl., American Ceramic Society, Columbus, Ohio. The standard work of reference for phase diagrams of non-metallic inorganic materials.

A. Reisman (1970). *Phase Equilibria*, Academic Press, New York.

J. E. Ricci (1966). *The Phase Rule and Heterogeneous Equilibrium*, Dover, New York.

F. Tamas and I. Pal (1970). *Phase Equilibria Spatial Diagrams*, Butterworths 1970.

Chapter 12

Phase Transitions

Phase transitions are important in most areas of solid state science. They are interesting academically, e.g. a considerable slice of current research in solid state physics concerns soft mode theory, which is one aspect of phase transitions, and they are important technologically, e.g. in the synthesis of diamond from graphite, the processes for strengthening of steel and the properties of ferroelectricity and ferromagnetism. This chapter discusses structural, thermodynamic and kinetic aspects of phase transitions and their classification. A few of the more important phase transitions are described; others are mentioned elsewhere in this book.

12.1 What is a phase transition?

If a crystalline material is capable of existing in two or more polymorphic forms (e.g. diamond and graphite), the process of transformation from one polymorph to another is a phase transition. The terms *transition* and *transformation* are both used to describe this and are interchangeable. In the narrowes sense, phase transitions are restricted to changes in structure only, without any changes in composition, i.e. to changes in elements or single phase materials. A much wider definition that is sometimes used includes the possibility of compositional changes, in which case more than one phase may be present before and/or after the transition. However, one then has to try and draw a dividing line between polymorphic transitions, on the one hand, and chemical reactions, on the other. The easiest solution is probably to try and avoid giving a precise definition of phase transitions!

Phase transitions are affected by both thermodynamic and kinetic factors. Thermodynamics gives the behaviour that should be observed under equilibrium conditions and, for a particular material or system, this information is represented by the phase diagram. Phase transitions occur as a response to a change in conditions, usually temperature or pressure but sometimes composition. The rates at which transitions occur i.e. kinetics, are governed by various factors. Transitions that proceed by a nucleation and growth mechanism are often slow because the rate controlling step, which is usually nucleation, is difficult. In martensitic and displacive phase transitions, nucleation is easy, occurs spontaneously and the rates of transition are usually fast.

12.2 Buerger's classification: reconstructive and displacive transitions

We can begin with the classification scheme of Buerger (1961) which initially divides phase transitions into two groups: reconstructive and displacive transitions. *Reconstructive* transitions involve a major reorganization of the crystal structure, in which many bonds have to be broken and new bonds formed. The transition, graphite $\rightleftharpoons$ diamond, is reconstructive and involves a complete change in crystal structure, from the hexagonal sheets of three-coordinated carbon atoms in graphite to the infinite framework of four-coordinated carbon atoms in diamond, and vice versa. The quartz $\rightleftharpoons$ cristobalite transition in SiO_2 is also reconstructive because although there is no difference in coordination between the two polymorphs—both structures are built of SiO_4 tetrahedra linked at their corners to form a three-dimensional framework—the polymorphs have different types of framework structure and many Si—O bonds must break and reform in order that the transition may take place. Because many bonds must break, reconstructive transitions usually have high activation energies and, therefore, take place only slowly.

Often, reconstructive transitions may be prevented from occurring, in which case the untransformed phase is kinetically stable although thermodynamically it is metastable. A classic example is the occurrence of diamond at normal temperatures and pressures. At 25 °C and 1 atmosphere, graphite is the stable polymorph of carbon, but for kinetic reasons the transition diamond $\rightarrow$ graphite does not occur at detectable rates under ambient conditions.

Since there is often no structural relationship between two polymorphs separated by a reconstructive phase transition, there may also be no relation between the symmetry and space groups of the two polymorphs.

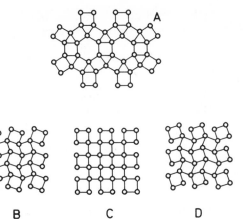

Fig. 12.1 Transformation from structure A to any other structure requires the breaking of first coordination bonds. Transformations among B, C and D are distortional only. (After Buerger, 1961)

Displacive phase transitions involve the distortion of bonds rather than their breaking and the structural changes that occur are usually small. For this reason, displacive transitions take place readily, with zero or small activation energies, and cannot usually be prevented from occurring. As well as a structural similarity, a symmetry relationship exists between the two polymorphs such that the symmetry of the low temperature polymorph is lower than, and belongs to a subgroup of, that of the high temperature polymorph. Examples are provided by the three main polymorphs of silica: quartz, tridymite and cristobalite, all of which undergo displacive, low–high transitions. These transitions involve small distortions or rotations of the SiO_4 tetrahedra, without breaking any primary Si—O bonds.

The distinction between reconstructive and displacive phase transitions is shown schematically in Fig. 12.1. In order to convert structure A into any of the other structures, B, C and D, bond breaking is necessary and the transition is reconstructive. On the other hand, interconversions between structures B, C and D do not involve bond breaking but only small rotational movements. These transitions are therefore displacive.

A more detailed and specific classification scheme, also due to Buerger, is given in Table 12.1. First coordination refers to bonds between nearest neighbour atoms (e.g. Si and O in SiO_4 tetrahedra), i.e. to the first coordination sphere of a particular atom. Second coordination refers to interactions between next nearest neighbour atoms (it is probably not true to regard these interactions as bonds), e.g. between adjacent silicon atoms in a chain of corner-sharing SiO_4 tetrahedra.

Transformations involving first coordination may occur by two mechanisms: (a) by completely disrupting the crystal structure of the original polymorph, as in graphite⇌diamond, or (b) by a much more subtle and easier method involving

Table 12.1 *Classification of phase transitions*

Type of transition	Examples
1. Transitions involving first coordination	
(a) Reconstructive (Slow)	Diamond $\rightleftharpoons$ graphite
(b) Dilational (Rapid)	Rock salt $\rightleftharpoons$ CsCl
2. Transitions involving second coordination	
(a) Reconstructive (Slow)	Quartz $\rightleftharpoons$ cristobalite
(b) Displacive (Rapid)	Low $\rightleftharpoons$ high quartz
3. Transitions involving disorder	
(a) Substitutional (Slow)	Low $\rightleftharpoons$ high $LiFeO_2$
(b) Orientational Rotational } (Rapid)	Ferroelectric $\rightleftharpoons$ Paraelectric $NH_4H_2PO_4$
4. Transitions of bond type (Slow)	Grey $\rightleftharpoons$ white Sn

dilation. An example of the latter is the rock salt $\rightleftharpoons$ CsCl transition which occurs in several alkali and ammonium halides at high temperature and/or high pressure. Although the unit cell of rock salt is face centred cubic, $Z = 4$, a rhombohedral cell ($a = b = c, \alpha = \beta = \gamma = 60°$) that has one quarter the volume of the cubic cell, with $Z = 1$, can be defined (Fig. 5.16). The rhombohedral cell is primitive in that it has Na^+ ions at the corners and Cl^- at the body centre (or vice versa). If the rhombohedral cell is now compressed along its threefold axis, the angle α increases above 60° until, when $\alpha = 90°$, the structure has changed to that of CsCl (Fig. 12.2). The crystal structures of rock salt and CsCl are therefore interconvertible by this mechanism of dilation. A change in primary coordination number between 6 and 8 occurs. The bonds from chlorine to cations 2 to 7

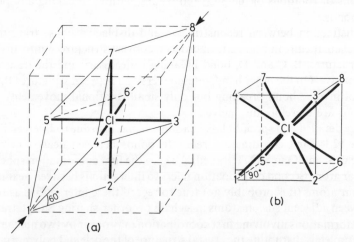

Fig. 12.2 Displacive phase transition between (a) rock salt and (b) CsCl structure types. Chlorine is octahedrally coordinate in (a) and eight-coordinate in (b). The cations are numbered 1 to 8

are unchanged but cations 1 and 8 move in and out of the first coordination environment of chlorine. The degree of bond breaking is much less than for reconstructive transitions; also there is no intermediate state of high energy and hence transition rates are rapid. For example, CsCl transforms to the rock salt structure above 479 °C but on cooling the high temperature, rock salt-like polymorph, it spontaneously reverts back to the CsCl structure. A similar dilational mechanism may be postulated for the f.c.c. → b.c.c. transition that occurs in some metals, e.g. $\gamma \rightarrow \delta$ Mn at 1134 °C, $\alpha \rightarrow \gamma$ Fe at 910 °C and $\gamma \rightarrow \delta$ Fe at 1400 °C.

Transformations involving second coordination are reconstructive only if the mechanism involves the breaking and forming of bonds of first coordination. Thus quartz, tridymite and cristobalite all have three-dimensional network structures built of corner-sharing SiO_4 tetrahedra and the structures differ in the patterns of linkage of the tetrahedra, i.e. they differ in second coordination only. In order to transform from one polymorph to another, however, it is necessary to break and reform primary Si—O bonds.

Order–disorder transitions that involve atoms or ions exchanging places (i.e. substitutional effects) are usually sluggish. The structures of ordered and disordered polymorphs of $LiFeO_2$, stable below and above ~ 700 °C, respectively, are shown in Fig. 12.3. The disordered polymorph may be readily preserved to room temperature where it is kinetically stable. It has a rock salt structure with Li^+ and Fe^{3+} ions distributed at random over the octahedral sites of the face centred cubic unit cell (Fig. 12.3b). On heating disordered $LiFeO_2$ at, for example, 600 °C, the oxide ion arrangement remains unchanged but the cations order themselves over the octahedral sites (Fig. 12.3a), resulting in a larger unit cell of lower symmetry (tetragonal). The ordering reaction involves cation migration and takes place only slowly. As in all reconstructive tran-

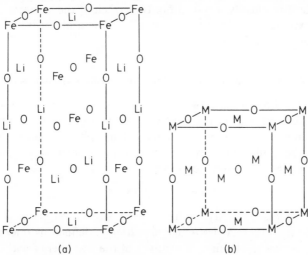

(a) (b)

Fig. 12.3 (a) Ordered cation arrangement in tetragonal $LiFeO_2$; $a = 4.057$, $c = 8.759$ Å. (b) Disordered cation arrangement in cubic $LiFeO_2$ with the rock salt structure; $a = $ ca. 4 Å; $M = Li^+$ or Fe^{3+}, at random

sitions, however, the transition rates are very temperature dependent. The ordering transition in $LiFeO_2$ is slow since 600 °C is, relatively, a low temperature. On the other hand, Li_2TiO_3 exhibits a similar order–disorder transition based on the rock salt structure, but since the equilibrium transition temperature is 1213 °C the ordering reaction in Li_2TiO_3 proceeds very rapidly on cooling below 1213 °C. Thus, the disordered polymorph of $LiFeO_2$ may be readily preserved to room temperature, but this is difficult to achieve with Li_2TiO_3.

A good example of an orientational order–disorder transition is the ferroelectric–paraelectric transition in $NH_4H_2PO_4$. Displacement of hydrogen atoms within —O—H—O— hydrogen bonds leads to an apparent change in the orientation of $PO_2(OH)_2$ tetrahedra (see Fig. 15.17). In the low temperature ferroelectric phase, these tetrahedra have a similar orientation, giving a net alignment of dipoles, but in the paraelectric phase they are randomized. Since the transition is accomplished by small displacements of hydrogen atoms, it takes place easily and rapidly.

The fourth category in Buerger's scheme involves transitions of bond type:

(a) grey⇌white tin, which involves a change from semiconducting to metallic character;
(b) diamond⇌graphite, which is insulting to semiconducting.

However, in addition to the change in bond type, major structural changes occur—in both tin and carbon, the primary coordination number changes at the transition—and hence these transitions could also be included in category 1(a).

In summary, the dividing lines between the different categories of phase transition, as classified by Buerger, are not rigid and in some cases it is difficult to decide how best to classify a transition. Nevertheless, the schemes of Buerger have been extremely useful in providing a structural basis on which to understand phase transitions.

12.3 Thermodynamic classification of phase transitions

Ehrenfest classified phase transitions into *first order* and *second order* by considering the behaviour of thermodynamic quantities such as entropy, heat capacity, volume, etc., on passing from one polymorph to the other through the transition. At the equilibrium temperature (or pressure) of a phase transition, the Gibbs free energies of the two polymorphs are equal: i.e.

$$\Delta G = \Delta H - T\Delta S = 0 \qquad (12.1)$$

Therefore, no discontinuity in free energy occurs on passing from one polymorph to the other. A *first-order transition* is defined as one in which a discontinuity occurs in the first derivatives of the free energy with respect to temperature and pressure. These derivatives correspond to entropy and volume, respectively, i.e.

$$\frac{\mathrm{d}G}{\mathrm{d}T} = -S \tag{12.2}$$

and

$$\frac{\mathrm{d}G}{\mathrm{d}P} = V \tag{12.3}$$

from

$$H = U + PV \tag{12.4}$$

Usually, first-order transitions are easy to detect. A discontinuity in volume corresponds to a change in crystal structure such that the density and the unit cell volume per formula unit are different in the two polymorphs. The change in volume may be followed by dilatometry (also called thermomechanical analysis) or sometimes by visual observation, e.g. the increase in volume associated with the tetragonal to monoclinic transition in ZrO_2 causes bodies containing tetragonal zirconia to shatter. Associated with a change in volume there is usually a change in enthalpy, ΔH (equation 12.4), which can be detected by DTA, Chapter 4; exothermic or endothermic peaks are observed as the transition proceeds. Direct measurement of entropy changes are less easily made but can also be inferred by the occurrence of DTA peaks; at the transition temperature, $\Delta G = 0$ and, therefore,

$$\Delta S = \frac{\Delta H}{T} \tag{12.5}$$

or by X-ray diffraction studies in the case of order–disorder transitions. Some examples of first-order phase transitions and their thermodynamic characteristics are given in Table 12.2.

Second-order transitions are characterized by discontinuities in the second derivatives of the free energy, i.e. in the heat capacity, C_p, thermal expansion, α, and compressibility, β:

$$\frac{\partial^2 G}{\partial P_T^2} = \frac{\partial V}{\partial P_T} = -V\beta \tag{12.6}$$

$$\frac{\partial^2 G}{\partial P \partial T} = \frac{\partial V}{\partial T_P} = V\alpha \tag{12.7}$$

$$\frac{\partial^2 G}{\partial T^2} = -\frac{\partial S}{\partial T_P} = \frac{-C_p}{T} \tag{12.8}$$

Higher order transitions can be defined, in principle, by differentiating further. Detection of second-order transitions is not so easy as for first-order ones since the changes involved are usually much smaller. The best method is probably to measure heat capacities, by calorimetry. Heat capacities usually increase as the

Table 12.2 *Characteristics of some first-order phase transitions.* (Taken in part from a compilation by Rao and Rao, 1966)

Compound	Transition	$T_c(°C)$	$\Delta V(cm^3)$	$\Delta H(kJ\,mol^{-1})$
Quartz, SiO_2	Low $\rightleftharpoons$ high	573	1.33	0.360
CsCl	CsCl structure $\rightleftharpoons$ rock salt structure	479	10.3	2.424
AgI	Wurtzite structure $\rightleftharpoons$ b.c. cubic structure	145	−2.2	6.145
NH_4Cl	CsCl structure $\rightleftharpoons$ rock salt structure	196	7.1	4.473
NH_4Br	CsCl structure $\rightleftharpoons$ rock salt structure	179	9.5	3.678
Li_2SO_4	Monoclinic $\rightleftharpoons$ cubic	590	3.81	28.842
$RbNO_3$	Trigonal structure $\rightleftharpoons$ CsCl structure	166	6.0	3.971
	CsCl structure $\rightleftharpoons$ hexagonal structure	228	3.12	2.717
	Hexagonal structure $\rightleftharpoons$ NaCl structure	278	3.13	1.463

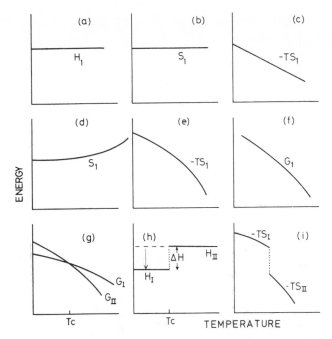

Fig. 12.4 (a) to (f) Thermodynamic properties of phases and (g) to (i) variations in these properties at first-order phase transitions

transition temperature, T_c, is approached and show a discontinuity at T_c (equation 12.8).

The temperature dependences of various thermodynamic functions in a polymorphic material are shown in Figs 12.4 and 12.5. In 12.4(a), the enthalpy H_1 of polymorph 1 is shown as being temperature independent, although in practice this is not strictly true. In Fig. 12.4(b), a temperature independent entropy S_1 is assumed which, again, is not strictly true in practice. This leads to a linear decrease in $-TS_1$ with increasing temperature (Fig. 12.4c). In Fig. 12.4(d), the entropy is shown as being markedly temperature dependent, especially at higher temperatures. This behaviour is characteristic of structures which experience some disorder with rising temperature. Consequently, $-TS_1$ decreases increasingly rapidly with rising temperature (Fig. 12.4e). For most materials, (a) and (e) are fairly good representations of their behaviour. These then give rise to a free energy that decreases increasingly rapidly with rising temperature (Fig. 12.4f).

In materials that are polymorphic, each polymorph has its own $G-T$ curve, e.g. G_I and G_{II} (Fig. 12.4g); by definition, the polymorph that is stable under a particular set of conditions is the one of lower free energy, i.e. polymorph I at temperatures below T_c and polymorph II above T_c. The two curves cross over at the equilibrium transition temperature, T_c, at which $G_I = G_{II}$ (equation 12.1). The entropies of polymorphs I and II are given by the slopes of the $G-T$ curves, i.e. $dG/dT = -S$. Hence, it follows that polymorph II has a higher entropy than

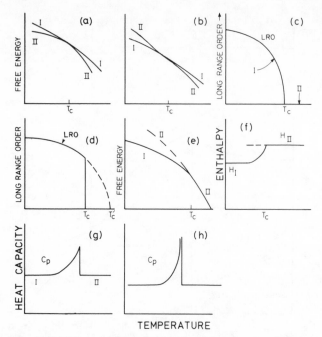

Fig. 12.5 Thermodynamic properties of phases involved in second-order phase transitions

polymorph I, i.e. $-TS_{II} < -TS_I$. From equation (12.5) polymorph II must also have a higher heat content than polymorph I, i.e. $H_I < H_{II}$. Broadly speaking, then, the polymorph which is stable at lower temperatures is the one with both the lower heat content and the lower entropy. Figures 12.4(h) and (i) show that discontinuities in H and S occur at T_c since although $\Delta G = 0$, $\Delta H = T\Delta S \neq 0$; these are the characteristics of a first-order phase transition. In order to get transformation from a low to a high temperature polymorph the latent heat of transformation, ΔH, must be provided. This explains a general principle of DTA that, on heating, transformation from one stable polymorph to another is endothermic and the reverse transition, on cooling, is exothermic (Chapter 4).

At a first-order transition, the $G–T$ curves for the two polymorphs (Fig. 12.4g) intersect. This cannot be so for a second-order phase transition and the graphical and thermodynamic representation of second-order phase transitions presents some difficulties. Since, for a second-order transition, there is no discontinuity in entropy at T_c, the slope of the $G–T$ curves $(\mathrm{d}G/\mathrm{d}T = -S)$ for polymorphs I and II must be the same at T_c. Attempts to represent this graphically are shown in Fig. 12.5(a) and (b). Situation (a) is satisfactory from the point of view that the tangents to curves I and II at T_c are parallel and coincide (hence $\Delta S = S_{II} - S_I = 0$), but it is unsatisfactory because polymorph II always has the lower free energy and hence no reversible transition between I and II could ever be observed! This problem is partly overcome in Fig. 12.5(b), but at the cost of distorting the $G–T$ curve of polymorph II and/or that of polymorph I so that they are parallel in the region of the transition temperatures. Since the $G-T$ curves of polymorphs I and II are independent of each other, it seems too much of a coincidence that one curve should distort over exactly the temperature range that corresponds to a phase transition.

A way out of this impasse in trying to represent second-order transitions comes from a consideration of the changes in thermodynamic properties for a practical example. Many order–disorder transitions are second-order transitions. Consider an alloy A_xB_y that is ordered at low temperatures but whose crystal structure is disordered (i.e. A and B atoms are placed at random) at high temperatures, $> T_c$. At low temperatures, below T_c, the alloy exhibits long range order (LRO) in which, for instance, the A atoms prefer a certain set of sites throughout the crystal. At absolute zero, the alloy must be perfectly ordered and hence LRO is complete. As the temperature rises, atoms begin to disorder and change places; hence the LRO decreases. This process continues increasingly rapidly as temperature increases until, at T_c, the LRO has disappeared altogether (Fig. 12.5c). Above T_c, only short range order (SRO) exists. In SRO, A atoms may prefer to be surrounded by B atoms, and vice versa, but this gives rise to regions that are ordered only on a very small scale. Disorder may be equated to entropy and it can be seen that, although an enormous increase in entropy occurs between 0 K and T_c, there is no discontinuity in entropy at T_c. Such a transition may be regarded as a second-order transition.

Consider now the form of the $G-T$ curves for the ordered, I, and disordered, II, polymorphs in Fig. 12.5(c). These are shown schematically in Fig. 12.5(e). The dashed extension for curve II represents the disordered polymorph that has been supercooled to temperatures below T_c. However, it is impossible to do the reverse and superheat polymorph I above T_c because the process of continuous transition from I to II that begins at absolute zero has terminated at T_c. Temperature T_c represents the upper temperature limit of existence of I and, in this sense, it is a critical point, as, for example, the water–steam critical point. A thermodynamic theory due to Tizza treats phase transitions as critical points. The crucial difference between Fig. 12.5(e) and (b) is that, whereas it is possible to supercool II in both cases, it is impossible to superheat I in Fig. 12.5(e).

Thus far, the distinction between first order and second order transitions is clear cut and is based on thermodynamic principles. In practice, however, many transitions do not belong simply to one category or the other but may have hybrid character. This is illustrated graphically in Fig. 12.5(f). At some temperature well below T_c, there is a clear difference between the enthalpy of polymorph I, H_I and that of undercooled polymorph II, H_{II} (dashed line). On heating, the enthalpy of polymorph I shows a gradual, anomalous increase until, at temperature T_c, $H_I = H_{II}$. How are we to regard this transition? It is clearly not first order since at T_c, $H_I = H_{II}$. It may be possible to regard the change at T_c as a second order transition, especially if there is a discontinuity in C_p, as shown in Fig. 12.5(g). This is not very satisfactory, however since the classification of this transition as second order, based on its behaviour at T_c, ignores completely the large, anomalous increase in H_I below T_c.

A further way of classifying transitions which is both interesting and useful is that proposed by Ubbelohde. In this scheme, transitions are grouped into *continuous* and *discontinuous*. A continuous transition is one such as shown in Fig. 12.5(f): there is no discontinuity in enthalpy at T_c, but perhaps more importantly (and not shown), the crystal structure changes smoothly and continuously from that of polymorph I to that of polymorph II. A discontinuous transition is one such as quartz–cristobalite or diamond–graphite. In these, the structures of the two polymorphs are quite clearly different and it is not possible for the structure to change smoothly from one polymorph to the other.

Returning now to first and second order transitions, these represent extreme, ideal cases. Both are concerned only with the changes in thermodynamic properties at T_c. In practice, first order transitions usually show some 'premonitory' phenomena, such as an increase in disorder, as T_c is approached. However, such effects may be conveniently ignored, especially if there is a large change in enthalpy at T_c. In second order transitions, especially of the type order–disorder, almost all of the changes in structure and thermodynamic properties are associated with the 'premonitory' changes below T_c and surely cannot be ignored. For these, T_c merely represents the temperature at and above which the structural changes are complete.

In conclusion, then, most transitions, of whatever kind, show premonitory phenomena, which appear as an increase in heat capacity or a baseline drift on DTA, as T_c is approached (Fig. 12.5g). These premonitory phenomena may terminate with a discontinuity in e.g. enthalpy or LRO at T_c, in which case the transition is first order and discontinuous. Alternatively, if the transition did not take place with a discontinuity at T_c, then behaviour similar to that shown in Fig. 12.5(c) would be observed, with a continuous or second order transition at an extrapolated temperature T_c' (dashed line, Fig. 12.5d). We can see therefore, that premonitory phenomena, such as increasing disorder or defect concentrations, as T_c is approached, provide the link between first-order and second-order transitions. Premonitory phenomena also mark the onset of a continuous transition that may or may not be interrupted by a discontinuity.

A favourable philosophical point among physicists concerns the changeover from first-order to second-order behaviour: as a transition becomes increasingly second order in character, so the enthalpy of the transition at T_c decreases, but at what stage does this enthalpy become zero? Even though a transition has an extremely small ΔH, it must retain some first order character.

Many transitions that show finite or infinite discontinuities in C_p at T_c (Fig. 12.5g and h) are called *lambda transitions*, because the shape of the C_p curve resembles the Greek letter lambda. These are second-order transitions, if the change in C_p is finite (Fig. 12.5g), but are first order if ΔC_p is infinite (Fig. 12.5h). An example of a lambda transition is the low–high transition in quartz at 573 °C (Fig. 12.6). Changes in ΔH and ΔV occur at 573 °C, as given in Table 12.2, and therefore the transition has some first-order character. The rapid increase in C_p between ~ 500 and 573 °C shows the premonitory disordering effects that occur prior to the transition at T_c.

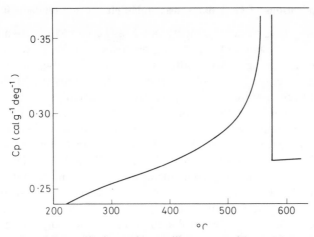

Fig. 12.6 Specific heat of crystalline quartz. (From Moser, 1936)

Most attention is given in this chapter to transitions that occur as a consequence of changing the temperature, but pressure-induced phase transitions also occur; some examples are given in Table 12.3. All of these transitions are accompanied by a decrease in volume and provide examples of the relevance of Le Chatelier's principle. In pressure-induced phase transitions, the effect of increased pressure is to cause a change in crystal structure such that the high pressure polymorph has a higher density and hence smaller volume than the low pressure polymorph. Schematic free energy–pressure diagrams may be constructed similar to the G–T relations given in Figs 12.4 and 12.5. For example, the C_p–T diagram in Fig. 12.5(g) would be replaced by a compressibility–P diagram.

The pressure dependence of phase transitions as a function of temperature is given by the *Clausius–Clapeyron equation*, which is a quantitative statement of Le Chatelier's principle:

$$\frac{dP}{dT} = \frac{\Delta H}{T\Delta V} \tag{12.9}$$

Table 12.3 *Some pressure-induced phase transitions.* (Data taken from Rao and Rao, 1966)

Compound	Transition	P_c(k bars)	ΔV(cm)^3	ΔH(kJ mol^{-1})
KCl	Rock salt to CsCl structure	19.6	− 4.11	8.03
KBr	Rock salt to CsCl structure	18.0	− 4.17	7.65
RbCl	Rock salt to CsCl structure	5.7	− 6.95	3.39
ZnO	Wurtzite to rock salt	88.6	− 2.55	19.23
SiO$_2$	Quartz to coesite	18.8	− 2.0	2.93
	Coesite to stishovite	93.1	− 6.6	57.27
CdTiO$_3$	Ilmenite to perovskite	40.4	− 2.9	15.88
FeCr$_2$O$_4$	Spinel to Cr$_3$S$_4$ structure	36.0	− 6.5	− 30.51

Transitions are at room temperature. P_c is the critical transformation pressure.

430

12.4 Applications of G–T diagrams; stable phases and metastable phases

Plots of free energy against temperature, Figs 12.4 and 12.5, are a useful and simple way of representing the polymorphism and stability or metastability of phases. The free energy axis is always schematic because, only rarely, are sufficient free energy data available with which to construct a quantitative diagram. However, transition temperatures are usually known accurately, together with the order of stability of the different polymorphs, and hence the *G–T* diagrams are qualitatively correct. Some examples are given in Fig. 12.7. In (a) is shown the $\beta \rightleftharpoons \alpha$ transition in AgI which occurs at 145 °C. Below 145 °C, β-AgI has lower free energy and is stable; above 145 °C, α-AgI is the equilibrium polymorph. Dashed lines represent metastable extensions of stable states; (i) represents superheated β and (ii) supercooled α. Neither of these *metastable* states is long-lived since the transition takes place rapidly in both directions. The arrows indicate that if it were possible to prepare either polymorph outside its range of stability, it would revert to the stable form, with a consequent decrease in free energy. This transition is important in the field of solid electrolytes. Much time has been spent studying doped AgI materials with the hope that either (a) the $\alpha \rightleftharpoons \beta$ transition temperature can be lowered so that the highly conducting α phase is stable to lower temperatures, especially to room temperature, or (b) the

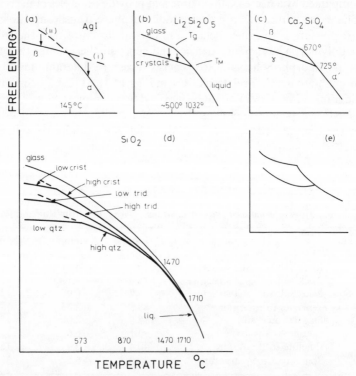

Fig. 12.7 Free energy–temperature diagrams showing polymorphism

kinetics of the $\alpha \to \beta$ transition can be modified so that the α form could be quenched to room temperature without transforming to β during cooling. Success has been achieved with several systems, the best known of which is $RbAg_4I_5$. Both effects (a) and (b) are observed since the $\alpha \rightleftharpoons \beta$ transition temperature is reduced from 145 °C in AgI to 27 °C in $RbAg_4I_5$ and the transition from α to β occurs only slowly below 27 °C and needs the presence of free iodine to act as a catalyst.

$G-T$ diagrams can represent liquid as well as crystalline substances, as shown for $Li_2Si_2O_5$ (Fig. 12.7b). Lithium disilicate, $Li_2Si_2O_5$, melts at 1032 °C and the rather viscous liquid that forms can be readily undercooled without crystallizing. With decreasing temperature, the viscosity of this supercooled liquid increases until the glass transformation temperature, T_g, is reached, below which the liquid freezes to a rigid, amorphous glass. (Some people claim that the glass transformation is an example of a second-order phase transition.) At room temperature, then, $Li_2Si_2O_5$ can exist in two forms, as crystals or as glass. Since the glass has higher free energy it is metastable and will crystallize, provided the conditions are kinetically favourable. This metastability of glass, and its subsequent crystallization on heating at high temperatures ($\sim$ 450 to 700 °C), forms the basis of the manufacture of glass-ceramics (pyrosil, pyroceram, slagceram, etc.). Glass-ceramics withstand high temperatures, unlike most glasses which soften or crystallize, and are resistant to thermal shock (see Chapter 18). Figure 12.7(b) is somewhat simplified in that crystalline $Li_2Si_2O_5$ undergoes minor polymorphic phase transitions at $\sim$970°C, but these have been omitted for clarity.

Examples of phase transitions that must be avoided if possible are the $\alpha' \to \gamma$ and $\beta \to \gamma$ transitions in Ca_2SiO_4, which is present as a major constituent of cement. Under equilibrium conditions, the α' polymorph of Ca_2SiO_4 should transform to γ below 725 °C (Fig. 12.7c). However, with rapid cooling and/or the addition of suitable additives, the $\alpha' \to \gamma$ transition does not occur and, instead, undercooled α' transforms to β below 670 °C. β-Ca_2SiO_4 is entirely metastable since at all temperatures over which it can exist it has a higher free energy than γ-Ca_2SiO_4. β-Ca_2SiO_4 is one of the major components of Portland cement clinker and it sets hard on reaction with water. On the other hand γ-Ca_2SiO_4 has very little cementitious value and, hence, in the manufacture of Portland cement, transformation to give the γ polymorph must be avoided.

A material with a complex $G-T$ diagram is silica, SiO_2. The equilibrium forms, as a function of temperature, are:

$$\text{Low quartz} \overset{573\,°C}{\rightleftharpoons} \text{high quartz} \overset{870\,°C}{\rightleftharpoons} \text{high tridymite} \overset{1470\,°C}{\rightleftharpoons}$$

$$\text{high cristobalite} \overset{1710\,°C}{\rightleftharpoons} \text{liquid}$$

The three main crystalline polymorphs, quartz, tridymite and cristobalite, all undergo high $\to$ low transitions on cooling, but both low tridymite and low cristobalite are entirely metastable relative to low quartz (Fig. 12.7d). Once formed, however, both tridymite and cristobalite are kinetically stable outside their ranges of equilibrium existence (i.e. below 870 and 1470 °C, respectively) and their subsequent conversion to quartz on lowering the temperature proceeds only slowly because these transitions are reconstructive. Silica glass can be prepared by supercooling silica liquid. The glass transformation temperature, T_g, is high, $\sim 1200\,°C$. Silica also forms high pressure polymorphs, coesite and stishovite, but these do not appear on Fig. 12.7(d).

With complex $G-T$ diagrams, such as Fig. 12.7(d), it is often difficult to show the transition points clearly. In order to make the drawings clearer, some authors use diagrams such as Fig. 12.7(e) in which each $G-T$ curve is shown as being concave upwards, rather than concave downwards as in Fig. 12.7 (d). While the drawings are certainly clearer, they are thermodynamically incorrect and their use is not to be encouraged. For instance, the $G-T$ curve for each polymorph must become horizontal as absolute zero is approached since $dG/T = -S$ and $S \to 0$ at 0 K. Figure 12.7(e) shows the opposite of this.

All the examples considered in Fig. 12.7 are for one-component systems in which all polymorphs have the same composition. Similar representations can be used for multicomponent systems although it becomes more difficult to show the effect of three variables—G, T, composition—on a single diagram. One application in multicomponent systems is in the phenomenon of spinodal decomposition, which is an important effect associated with liquid immiscibility (Chapter 18).

DTA may be used in favourable circumstances to distinguish between phases that are stable and metastable. The change from one stable polymorph to another on heating should appear on DTA as an endothermic event. The change from a quenched, metastable polymorph to a stable polymorph on heating should appear as an exothermic event. An example would be the crystallization of $Li_2Si_2O_5$ glass on heating; crystallization would occur at ~ 600 to $800\,°C$ and give an exotherm on DTA (Fig. 4.7b).

12.5 Ubbelohde's classification: continuous and discontinuous transitions

As mentioned in the previous section separation of phase transitions into first order and second order according to thermodynamic principles is fine in theory but does not work so well in practice because many transitions have an intermediate character. An alternative scheme, due to Ubbelohde (1957), classifies transitions into two groups, *continuous* and *discontinuous*. Broadly speaking, discontinuous and continuous transitions correspond to thermodynamic first-and second-order transitions, respectively. Ubbelohde's scheme has been particularly useful for transitions that experimentally involve only minor structural changes. In many cases, these transitions proceed by the formation of 'hybrid crystals' in which domains of the product phase grow inside the parent crystal. At the interface between parent and product crystals, one or both phases will be in a stressed condition since it is unlikely that the molar volume of the two phases is identical. The free energies of the two phases are therefore modified by a strain energy term ζ and this leads to a modified phase rule in which the strain energy contributes an extra degree of freedom:

$$P + F = C + 2 + \sum \pi$$

where $\sum \pi$ refers to the number of additional degrees of freedom introduced (as well as strain energy, surface energy may also be important, there by contributing a further degree of freedom). Some transitions have been studied directly by high temperature single crystal X-ray diffraction and the presence of hybrid crystals observed over a range of temperatures. Using the modified phase rule, these observations may be rationalized and it is not necessary to invoke any violations of the phase rule. It seems highly probable that there is a close relation between the occurrence of hybrid crystals and the phenomenon of martensitic transformations (Section 12.8.2).

12.6 Representation of phase transitions on phase diagrams

Phase diagrams are treated systematically in Chapter 11. Some further points regarding the representation of phase transitions on phase diagrams are worth making here:

(a) In one-component systems, e.g. C, SiO_2, which are subjected to changes in temperature and pressure, first-order phase transitions represent univariant conditions. From the phase rule, for a one-component system, $P + F = C + 2 = 3$. At the transition point, two phases are in equilibrium, $P = 2$ and so $F = 1$. The transition temperature therefore changes if the pressure is varied, and vice versa, e.g. Figs 11.3, 11.4 and 11.5. The condensed phase rule, $P + F = C + 1$, is used in condensed systems where the vapour phase is unimportant and for transitions that take place at fixed (often ambient) pressure. In such cases, first-order transitions occur at fixed temperatures, i.e. at invariant points.

(b) In two-component solid solution systems that exhibit phase transitions, addition of the extra component, composition, generates, from the phase rule, an extra degree of freedom. Whereas two phases coexist at a fixed point in a condensed, one-component system, two phases may coexist over a range of temperatures in a binary solid solution (Fig. 11.18a). These two phase regions, such as $\alpha + \beta$, must be present in theory, although in practice they are sometimes narrow and difficult to detect.

(c) Second-order phase transitions, strictly speaking, cannot be represented on equilibrium phase diagrams. In a second-order or continuous transition, the critical temperature represents the condition under which the low–high transition is complete. At no stage do two phases coexist in equilibrium. Since $P = 1$ throughout, it is impossible to represent a second-order phase transition on a conventional phase diagram because the latter demands that, at a transition point, $P = 2$. It is necessary, of course, to be able to represent such phase transitions and this may be done by using a single curve to represent the variation of transition temperature with, for example, composition.

12.7 Kinetics of phase transitions

Thermodynamics tells us the temperature (and pressure) at which a transition occurs under equilibrium conditions but gives no information about the rates at which transitions occur. The latter is the subject of kinetics. The rates at which transitions occur vary enormously. At one extreme are transitions that take place very rapidly in both forward and back directions and without any hysteresis (i.e. the transition temperature is the same on heating and cooling). At the other extreme are transitions that occur only on geological timescales; e.g. obsidian, a glassy mineral, should transform to one of the crystalline forms of SiO_2 but clearly, from the occurrence of obsidian as a mineral, the transition rates are very slow. Most transitions are intermediate between these extremes and occur with some hysteresis, i.e. the high temperature polymorph may be undercooled to varying degrees before transforming to the low temperature polymorph.

Transition rates vary enormously and are controlled by several factors. In Fig. 12.8 the temperature dependence of the rate of transition between a low temperature polymorph, I, and a high temperature polymorph, II, is shown schematically. Rates are slow in either direction at temperatures that are close to the equilibrium transition temperature, T_c (region A). Since $\Delta G \sim 0$ close to T_c, there is little driving force for the transition to occur in either direction. At temperatures further removed from T_c, reaction rates increase (regions B, C). A maximum in the rate of the $II \rightarrow I$ transition may occur at a certain degree of undercooling, at T_M, and if it is possible to undercool the high form II to below T_M then the rate of the $II \rightarrow I$ transition again decreases (region D). If the maximum

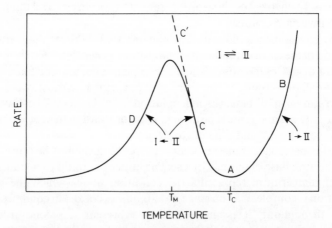

Fig. 12.8 Temperature dependence of the transition rates for a typical first-order transition between a low temperature polymorph, I, and a high temperature polymorph, II

cannot be detected experimentally, then the $II \rightarrow I$ transition rate continues to increase (region C'), although, in theory, a maximum may still occur at some lower temperature. There is no corresponding maximum in the $I \rightarrow II$ transition above T_c; instead the rate increases increasingly rapidly (region B).

The general form of Fig. 12.8 and the occurrence of the maximum at T_M may be understood by considering the combination of (a) the effect of temperature on reaction rates, as given by the Arrhenius equation and (b) the relative free energies of the polymorphs I and II as a function of temperature. These two effects are as follows:

(a) The Arrhenius equation, as applied to kinetics, is of the form:

$$\text{Rate} = A \exp\left(\frac{-E}{RT}\right) \tag{12.10}$$

where E is the activation energy of the transition. This equation predicts a rapid increase in rate with increasing temperature, as observed in regions D and B of Fig. 12.8. The usual method of analysing results in terms of the Arrhenius equation is, of course, to take logs:

$$\log_{10}\text{rate} = \log_{10}A - \frac{E}{RT}\log_{10}e \tag{12.11}$$

and plot $\log_{10}$ rate against $1/T$. If the data fit the Arrhenius equation, a straight line is obtained of slope $(-E/R)\log_{10}e$ and intercept (extrapolated) equal to $\log_{10}A$.

(b) The magnitude of the difference in free energy ΔG_{I-II} between the two polymorphs I and II gives a measure of the driving force for the transition to occur (Fig. 12.9). At T_c, $G_I = G_{II}$ and there is no driving force for the transition to occur in either direction (Fig. 12.9a). At any other temperature, $G_I \neq G_{II}$ and the transition takes place preferentially in one direction (Fig. 12.9b). For idealized transitions in which H and S of the two polymorphs are independent of temperature (Fig. 12.4a and b), the magni-

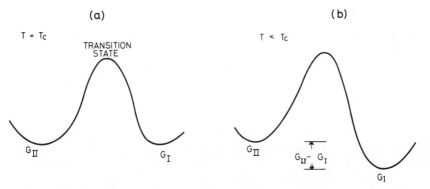

Fig. 12.9 Difference in free energy between polymorphs I and II

tude of ΔG_{I-II} at temperature T is a simple function of the difference in temperature $(T_c - T)$:

$$\Delta G_{I-II} = \Delta H_{I-II} - T\Delta S_{I-II} \tag{12.12}$$

$$= (T_c - T)\Delta S_{I-II} \tag{12.13}$$

or

$$\Delta G_{I-II} = \frac{T_c - T}{T}\Delta H_{I-II} \tag{12.14}$$

Both factors (a) and (b) outlined above are important in controlling the rates at which phase transitions occur. Qualitatively, factor (b) is dominant at temperatures close to T_c where there is little thermodynamic driving force for the transition to occur. At temperatures further away from T_c, however, factor (a) becomes more important, especially in regions D and B of Fig. 12.8, where the transition rate increases increasingly rapidly with rising temperature.

Difficulties arise when attempts are made to quantify the effects of factor (a) and, especially, factor (b). The traditional approach is to relate the thermodynamic driving force, given by ΔG_{I-II}, factor (b), to the problem of forming nuclei of the product phase. Most first-order or reconstructive transitions take place by a mechanism of *nucleation and growth*, in which the slow step is the initial nucleation of the product phase. Although the theory of nucleation is well developed, it is difficult, if not impossible, to apply it quantitatively since the magnitudes of certain parameters, such as the surface energies of nuclei, are not known. This theory is now briefly described.

12.7.1 Critical size of nuclei

Let us begin with a phase which is in its field of thermodynamic stability. Transformation or reaction to give another polymorph cannot occur to any appreciable extent, because this would lead to an overall increase in free energy (e.g. transformation in the direction I → II in Fig. 12.9b).

Suppose now that the conditions, T or P, of the original polymorph are changed so that it moves outside its field of equilibrium existence. Transformation to a polymorph of lower free energy must therefore occur according to thermodynamics, but if the transformation mechanism is one of nucleation and growth then, for kinetic reasons, the rate of transformation may be very slow. In such a mechanism, small nuclei of the product phase form initially, either at the surface and/or throughout the bulk of the original polymorph, and these nuclei subsequently grow. The difficulty with nucleation is that nuclei have surfaces and their surface energy makes a positive contribution to the free energy of the system. The net change in free energy, ΔG_n, on nucleation is therefore a combination of a decrease in free energy due to formation of the product phase in the nuclei and an increase in free energy due to the surface energy of the nuclei. A strain energy term may also be present if the starting phase and product have different volumes; here we assume $\Delta V = 0$ and the strain

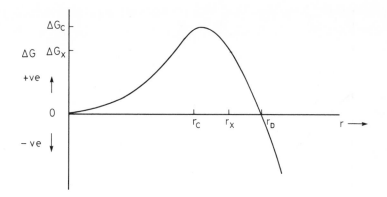

Fig. 12.10 Change in free energy of nuclei as a function of raduis

energy is zero. Let us assume that ΔG_v represents the free energy per unit volume of the nucleus relative to the parent phase and ΔG_a represents the surface free energy per unit area of the nucleus. For a spherical nucleus of radius r,

$$\Delta G_n = 4\pi r^2 \Delta G_a - \tfrac{4}{3}\pi r^3 \Delta G_v \qquad (12.15)$$

For small values of r, ΔG_n is positive since $4\pi r^2 \Delta G_a > \tfrac{4}{3}\pi r^3 \Delta G_v$ (Fig. 12.10), but with increasing r, ΔG_n passes through a maximum ΔG_c at $r = r_c$, and decreases to zero at $r = r_D$. For $r > r_D$, ΔG_n is negative. This then gives rise to the notion that a nucleus should have a certain minimum size in order for it to be stable. At first sight, it might be thought that r_D represents the critical radius of the nucleus, above which the nucleus is stable because ΔG_n is negative. Kinetically, however, the critical radius corresponds to r_c since for $r > r_c$, ΔG_n decreases with increasing r. In order to understand this, consider a nucleus of intermediate radius, r_X and free energy $+ \Delta G_X$. While such a nucleus is unstable thermodynamically, it is stable kinetically since, if it were to start dissolving, r would decrease and the free energy would begin to rise towards the value ΔG_c. Once nuclei of size $r > r_c$ form, therefore, they are kinetically stable and continue to grow. In fact, nuclei of radius r_X, where $r_c < r_X < r_D$, are *metastable* since they require an activation energy, given by $\Delta G_c - \Delta G_X$, for their dissolution, whereas nuclei of size $r < r_c$ are *unstable* since they have no activation barrier to their dissolution; nuclei of size $r > r_D$ are *stable*.

The value of r_c (Fig. 12.10) occurs when $\mathrm{d}\Delta G/\mathrm{d}r = 0$; i.e.

$$\frac{\mathrm{d}\Delta G}{\mathrm{d}r} = 8\pi r \Delta G_a - 4\pi r^2 \Delta G_v$$

and

$$r_c = \frac{2\Delta G_a}{\Delta G_v} \qquad (12.16)$$

The critical excess free energy, ΔG_c, is given by substituting r_c into equation

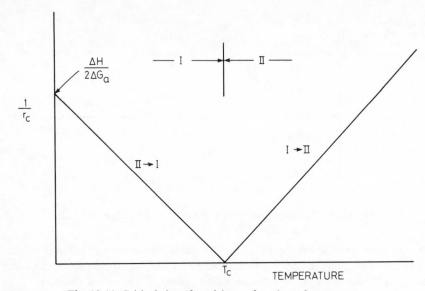

Fig. 12.11 Critical size of nuclei as a function of temperature

(12.15), i.e.

$$\Delta G_c = \frac{16}{3} \frac{\pi \Delta G_a^3}{\Delta G_v^2} \tag{12.17}$$

We can now see why nucleation is difficult at temperatures close to T_c. Since $\Delta G_v \to 0$ as $T \to T_c$ then, from equations (12.16) and (12.17), both r_c and ΔG_c become increasingly large as T_c is approached. The variation of r_c with temperature is given by substituting equation (12.14) into (12.16), i.e.

$$r_c = \frac{2\Delta G_a T_c}{(T_c - T)\Delta H} \tag{12.18}$$

A plot of r_c^{-1} against T (Fig. 12.11) should be linear, intersecting the T axis at T_c as $r \to \infty$. The two lines on Fig. 12.11 correspond to the two transformation directions, $I \rightleftharpoons II$.

12.7.2 Rate equations

12.7.2.1 *Nucleation rate*

Nuclei of the product phase form as a consequence of the thermal motion of atoms; the rate of nucleation is given by

$$R = A \exp\left[\frac{-(\Delta G_c + \Delta G_a)}{kT}\right] \tag{12.19}$$

The overall activation energy, $\Delta G_c + \Delta G_a$, is a combination of the critical free

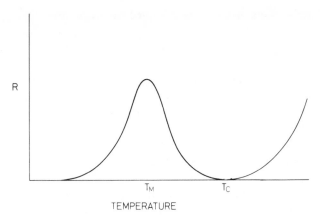

TEMPERATURE

Fig. 12.12 Effect of temperature on nucleation rate, R

energy, ΔG_c, which must be surmounted in order for the nuclei to be stable and the activation energy, ΔG_a, for the individual atomic jumps involved in nucleation. Since ΔG_c is temperature dependent, equations (12.17) and (12.14), the nucleation rate, R, varies with temperature as shown in Fig. 12.12. For transformation in the direction $II \rightarrow I$, R passes through a maximum at a certain temperature T_M, below T_c, and tends to zero both at 0 K and at T_c. For transformation in the direction $I \rightarrow II$, R increases rapidly with temperature above T_c.

There is considerable evidence that equation (12.19) and Fig. 12.12 represent, at least qualitatively, the kinetics of nucleation in many transitions. However, it is difficult to measure nucleation rates experimentally and test the theories. Part of the difficulty is that nucleation is very dependent on the presence of impurities, dislocations, surfaces, etc. It is possible to distinguish two types of nucleation. *Homogeneous nucleation* occurs when all parts of the parent phase are identical. This is a random process and depends only on thermal fluctuations in atomic positions. Usually, however, it is much easier for *heterogeneous nucleation* to occur in which nucleation takes place preferentially at defect centres and sites of higher local free energy.

12.7.2.2 *Overall transformation rate—Avrami equation*

Experimentally, it is much easier to measure the overall rate of transformation than to try and isolate the nucleation and growth stages, especially if the substance under study is a powder. Many data are analysed using the Avrami equation (also called the Avrami–Johnson–Mehl–Erofeev equation):

$$\alpha = 1 - \exp(-kt)^n \qquad (12.20)$$

in which α is the volume fraction of the product phase, k is the rate constant and n is a constant whose value depends on the nature of the nucleation and growth

440

process. Values of n and k may be obtained by taking logs twice:

$$\log \cdot \log \left(\frac{1}{1-\alpha} \right) = n \log k + n \log t + \log \cdot \log e \qquad (12.21)$$

and plotting $\log \cdot \log [1/(1-\alpha)]$ against $\log t$.

Use of the Avrami equation provides a convenient way of treating experimental data; no assumptions are needed initially and in favourable cases, e.g. if an integral n value is obtained, then insight into the transformation mechanism may be obtained. Thus, for polymorphic transitions, $n = 3$ may indicate a mechanism in which nucleation occurs only at the start of the transformation whereas $n = 4$ indicates that nucleation continues to occur in untransformed material. In situations where it is not possible to evaluate the significance of the n value that is obtained, the Avrami analysis does nevertheless give a value of the rate constant parameter k. If measurements on the transformation rate are made over a range of temperatures, an activation energy may then be obtained from an Arrhenius plot of $\log k$ against T^{-1}. An example is shown in Fig. 12.13 for the polymorphic transition $\beta \rightleftharpoons \gamma$ in Li_2ZnSiO_4. The kinetics were studied over a large temperature range, 480 to 940 °C, and in both transformation directions. At temperatures well below T_c, 883 °C, the data fall on a straight line and give an activation energy of 185 kJ mol^{-1}. Between 780 and 950 °C, however, the transformation rates are reduced, especially in the region of T_c. This result serves to emphasize the importance of thermodynamic factors, i.e. the relative free energies of the two polymorphs, to the kinetics of phase transitions.

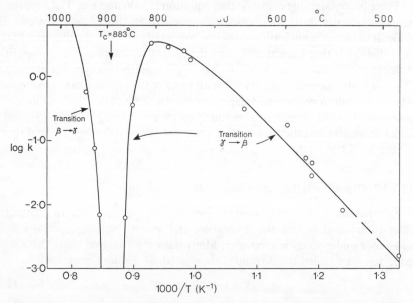

Fig. 12.13 Arrhenius plot for the rate of transition $\beta \rightleftharpoons \gamma$ in Li_2ZnSiO_4. (Data from Villafuerte-Castrejon and West, 1981)

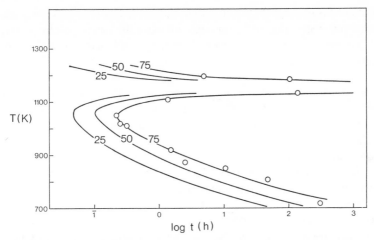

Fig. 12.14 Time–temperature–transformation (TTT) diagram for the transition $\beta \rightleftharpoons \gamma$ in Li_2ZnSiO_4

12.7.2.3 *Time–temperature–transformation (TTT) diagrams*

In experimental studies on phase transitions, one is often interested in a general knowledge of reaction rates as a function of temperature without being particularly concerned with details of the mechanism or the correct rate equation to use. It is then convenient to plot kinetic data as TTT diagrams, in which the time taken to achieve a certain degree of conversion, say 25 per cent, is ascertained over a range of temperatures; a graph of temperature against log time is constructed. As an example, data for the $\beta \rightleftharpoons \gamma$ transition in Li_2ZnSiO_4 (Fig. 12.13) are replotted as a TTT diagram (Fig. 12.14) for 25, 50 and 75 per cent conversion. The graphs show that transition rates for the $\gamma \to \beta$ direction pass through a maximum at ~ 1060 K (i.e. $\sim 790\,^\circ$C) and that at ~ 1150 K, transition rates are extremely slow in either direction. A rapid increase in the rate of $\beta \to \gamma$ transition occurs above ~ 1180 K.

TTT curves similar to Fig. 12.14 have been observed in many metals and alloys. They are probably characteristic of many inorganic phase transitions as well.

12.7.3 Factors that influence the kinetics of phase transitions

In most books on kinetics, attention is focused on reactions in the gaseous and liquid states with barely a mention of solids, apart from the use of solids to catalyse liquid and/or gaseous reactions. The reason for this is not that reactions involving solids are unimportant—they obviously are extremely important—but that solid state reactions are highly complex; rigorous, quantitative interpretation of the results of such reactions is difficult, if not impossible. For example, the concept of reaction order, which is a central, indispensable feature of the study of gaseous reactions, is meaningless in most solid reactions.

Although this chapter is concerned with phase transitions and not with solid state reactions in general, the factors that influence the kinetics of the two types of process are very similar. Phase transitions are usually simpler than solid state reactions for two reasons:

(a) There is usually no change in composition of the phases during a phase transition; in solid state reactions, however, compositional changes must occur and the diffusion of ions through solids is an important factor in rate studies.
(b) Liquid and gas phases are usually not important in phase transitions, unless they have catalytic effects on the transition rates. In reactions between solids, however, liquid and gas phases are often important, either because gases are absorbed or evolved during the reaction or, more generally, because they provide a medium for transferring matter from one solid particle to another. A further discussion of the reactions between solids is given in Chapter 2.

Comparison of the kinetics and mechanism of phase transitions, on the one hand, and reactions involving liquid or gas phases, on the other, shows that there are major differences between them. The most striking difference is one of scale. Liquid and gas reactions involve only a small number of atoms in each self-contained reaction, e.g. reaction of a hydrogen atom and a chlorine atom to form a molecule of hydrogen chloride:

$$H + \cdot Cl \rightarrow HCl$$

In most phase transitions, however, a much larger number of atoms are involved in forming a stable nucleus of the product phase. For example, in the transformation of anatase, TiO_2, to rutile, nuclei of rutile form both at the surface and throughout the bulk of the anatase crystals and these nuclei grow with increasing time. Let us suppose that under a certain set of conditions, the smallest nucleus of rutile capable of independent existence has a diameter of 50 Å. Such a nucleus contains about 5000 atoms!

Another difference between gas and solid reactions is that the latter are strongly influenced by surface effects: nucleation of the product phase often occurs at sites on the surface of the original crystals. The transformation kinetics depend on the total surface area of the original crystals, therefore, and on whether, for example, a single crystal or a powdered sample is used. Surface effects also control whether or not a nucleus of the product phase is stable, as discussed above. The surface energy of a nucleus gives a positive contribution to the free energy and if this more than cancels the decrease in free energy in the bulk of the nuclei, then such nuclei are unstable and redissolve. In contrast, surfaces are unimportant in gas and liquid reactions unless the reactions occur with the aid of solid catalysts.

Many other factors influence the kinetics of phase transitions, some of which are summarized in Table 12.4. The surface area of the sample has already been mentioned as well as the effect of the temperature of study relative to the

Table 12.4 *Factors that influence kinetics of phase transitions*

---

1. Nature of sample—e.g. single crystal or powder—and surface area
2. Temperature of kinetic study, relative to equilibrium transition temperature
3. Activation energy and strength of bonds that are broken
4. Pre-exponential factor, A
5. Change in volume, ΔV
6. Pressure, P
7. Transition mechanism

---

equilibrium transition temperature, T_c (e.g. Figs 12.8 and 12.13). The effect of activation energy, E, and the prefactor, A, can be seen from Fig. 12.13. A high value of A ($\log A = \log k$ when $T^{-1} \to 0$) ensures rapid rates at high temperatures. If, at the same time, E is small then these rapid rates extend to lower temperatures. The value of E gives a measure of the strengths of the bonds that are broken in order to form the transition state. Hence, major reconstructive transitions have large E values and are slow whereas minor displacive transitions have much smaller E values and are rapid.

Transition rates also depend on the difference in volume, ΔV, between the two polymorphs. From absolute reaction rate theory,

$$\log(\text{rate}) = \text{constant} - \frac{\Delta V^* P}{RT} \tag{12.22}$$

where ΔV^* is the difference in volume between the initial polymorph and the transition state.

It can be seen that the rate decreases with increasing ΔV. Many transitions are accomplished under high pressure and equation (12.22) shows that the reaction rate also decreases with increasing pressure. This has consequences in, for example, the synthesis of diamond from graphite. Diamond is thermodynamically stable only at high pressures. The effect of increasing pressure is to increase its stability, but at the same time its rate of formation in reduced.

The mechanism of transition is a major factor that effects the transition rates, e.g. compare dilational and reconstructive transition mechanisms in Table 12.1. Nucleation is the rate limiting step in many transitions and depends on many factors, including the nature of the solid (i.e. single crystal or powder), crystal defects (i.e. vacancies, dislocations, impurities, etc.) and atmosphere. The difference in crystal structure between the two polymorphs also greatly affects the ease of nucleation. If the structural differences are small and involve changes in second coordination only, then nucleation is very easy. At the other extreme, if there is no structural similarity between the two phases, nucleation is difficult. An intermediate class of transitions are *topotactic transitions* in which a definite orientation relation exists between the two phases but in which, nevertheless, considerable structural reorganization is necessary. Examples include transitions

444

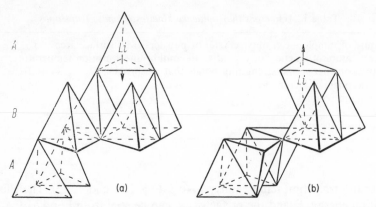

Fig. 12.15 Topotactic mechanism for the transformation $\beta \rightleftharpoons \gamma$ in Li$_2$ZnSiO$_4$. (From West, 1975)

in which the anion arrangement is unchanged throughout the transition but cation reorganization occurs. In such cases, nucleation is moderately difficult. A topotatic mechanism has been proposed for the transition $\beta \rightleftharpoons \gamma$ in Li$_2$ZnSiO$_4$ (Fig. 12.15). β-Li$_2$ZnSiO$_4$ has a wurtzite derivative structure with the cations ordered over one set of tetrahedral sites in a hexagonal close packed oxide array. On transformation to γ-Li$_2$ZnSiO$_4$ the oxygen layers are unchanged apart from a slight buckling, but half the cations move from filled to empty tetrahedral sites. This is represented by inversion of some of the MO$_4$ tetrahedra in Fig. 12.15.

An additional factor that is well appreciated in gas phase reactions and is likely to be important also in solid state reactions and transitions is the *principle of microscopic reversibility*. This states that, for a reaction that may be represented simply by

$$A \underset{k_{-1}}{\overset{k_1}{\rightleftharpoons}} B$$

both forward and back reactions occur, with rate constants k_1 and k_{-1}. In the gas phase reactions, this has two consequences. First, the overall rate constant is given by the difference between k_1 and k_{-1}. Second, reaction cannot go to completion in either direction but an equilibrium situation must be reached in which the rates of forward and back reaction are equal. If the idea of microscopic reversibility is applicable also to solid state reactions, and it seems entirely reasonable that it should be then an additional constraint must be present. Thus the first of the above consequences could still apply and the net rate be given by the difference between the rates of forward and back reaction. However, the second consequence could not apply. Under conditions of thermodynamic equilibrium, solid state reactions and transitions must, in general, proceed to completion in one direction or another. If not, this would incur a violation of the phase rule. To show this, consider the example of the $\beta \rightleftharpoons \gamma$ transition in Li$_2$ZnSiO$_4$, mentioned above. Since Li$_2$ZnSiO$_4$ is a congruently melting, thermodynamically stable phase, it may be treated as a one-component system

$(C = 1)$ and therefore in the absence of the vapour phase and at ambient pressure, the system is subject to the condensed phase rule, $P + F = C + 1 = 2$. Hence the coexistence of two phases $(P = 2)$ can occur *only* at a fixed point, $T_c(F = 0)$. At all other temperatures, one or other of the polymorphs β, γ is stable. Such considerations have therefore led to the suggestion that in solid state reactions and transitions microscopic reversibility may be involved, but subject to the constraint that the phase rule is obeyed and that, in general, reactions proceed to completion.

12.8 Crystal chemistry and phase transitions

12.8.1 Structural changes with increasing temperature and pressure

From an understanding of the thermodynamic changes that occur at phase transitions (Fig. 12.4), it is possible to understand, and to a certain degree predict, the changes in crystal structure that occur at phase transitions. Thermodynamic considerations tell us that: *an increase in volume and entropy accompany first-order transitions from low temperature to high temperature polymorphs*. There are several structural consequences of this:

(a) High temperature phases have more open structures and often the atoms or ions have a lower coordination number.
(b) High temperature structures are more disordered.
(c) High temperature structures often have higher symmetry.

A similar set of guidelines may be given for pressure-induced transitons. From Le Chatelier's principle, *a decrease in volume accompanies first-order transitions from low pressure to high pressure polymorphs*. When the pressure term is not negligible, a PV term must be included in the Gibbs free energy and at the transition we have:

$$\Delta G = \Delta U + P\Delta V - T\Delta S = 0$$

Therefore,

$$\Delta U + P\Delta V = T\Delta S$$

In order to balance a decrease in volume, the internal energy, U, must increase and/or the entropy, S, must decrease. The structural consequences of this are, therefore:

(a) High pressure phases have more dense structures and often the atoms or ions have increased coordination number.
(b) High pressure phases have more ordered structures.

On comparing the structural effects of P and T at a transition, we can say that, to a first approximation, the structural consequences of increasing temperature are similar to those of decreasing pressure. Quantitative predictions are difficult to make, especially with more complex structures, but these guidelines are, nevertheless, very useful.

As examples, for simple AB compounds, the following types of change may be expected:

CsCl structure, CN $= 8 \xrightarrow[\xleftarrow{\text{incr.} P\text{(c)}}]{\text{incr.} T\text{(a)}}$ NaCl structure, CN $= 6 \xrightarrow[\xleftarrow{\text{incr.} P\text{(d)}}]{\text{incr.} T\text{(b)}}$ ZnS structure, CN $= 4$

Examples, taken from Rao and Rao (1978), are:

(a) CsCl (479 °C), NH_4Br (179 °C)
(b) MnS
(c) KCl (19.6 kbar), RbCl (5.7 kbar)
(d) ZnO (88.6 kbars), CdS (17.4 kbar).

For AB_2 compounds the following sequences are possible:

Distorted rutile structure (cation CN ~ 5, monoclinic)

$\downarrow$ incr. T(a)

rutile structure (cation CN $= 6$, tetragonal) $\underset{(d)}{\overset{\text{incr.} P \;\text{(c)}}{\rightleftharpoons}}$ quartz structure (cation

$\downarrow$ incr. T(b)   incr. T CN $= 4$)

fluorite structure (cation CN $= 8$)

Examples are:

(a) VO_2, NbO_2
(b) SiO_2 (120 kbar)
(c) GeO_2 (1049 °C)

These latter examples are not quite as straight forward as for the AB compounds because coordination numbers may either increase, or decrease (c) with temperature, but the other guidelines still apply.

12.8.2 Martensitic transformations

Martensitic transformations are a special kind of transformation which occur in a variety of metallic and non-metallic systems. Martensite was the name originally given to the hard material obtained during the quenching of steels; it forms by transformation of austenite, the face centred cubic solid solutions of carbon in γ-Fe (Fig. 11.20). Austenite is unstable below 723 °C and should, under equilibrium conditions, decompose to a mixture of α-Fe and cementite, Fe_3C. On quenching austenite, this eutectoid decomposition reaction is suppressed and, instead, the undercooled, cubic austenite transforms to a metastable tetragonal phase martensite.

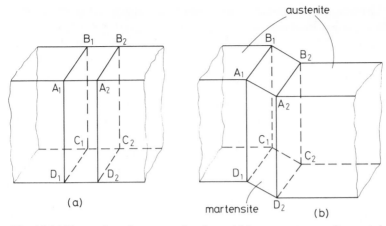

Fig. 12.16 Formation of a martensite plate within a parent austenite crystal

The austenite–martensite transformation is depicted in Fig. 12.16. A crystal of austenite, γ-Fe, is shown in Fig. 12.16 (a). Part of the crystal, between the cross-sections $A_1B_1C_1D_1$ and $A_2B_2C_2D_2$, changes shape by a shearing process on transforming to the martensite structure (Fig. 12.16b). Martensitic transformations have the following characteristics, some or all of which are usually observed:

(a) Transformation occurs by a shearing mechanism to give plates of product crystal within the parent crystal. At the parent–product interfaces, $A_1B_1C_1D_1$ and $A_2B_2C_2D_2$, which are known as *habit planes*, the structures match well and there is a definite orientation relationship between the crystal structures of the two phases. The sizes of the martensite plates are often large enough to be seen with an optical microscope.

(b) The parent and product phases have the same composition and their crystal structures are closely related. Small atomic displacements, often less than one bond length, are necessary to accomplish the transformation and hence the transformations do not involve diffusion.

(c) Because there is no activation energy for diffusion involved, the transformation rates are very high. The parent–product interfaces are said to be *glissile* since they can move without thermal activation. Transformation rates are often independent of temperature, in which case the transformation is said to be *athermal*, but may be affected by applied stresses and strains.

(d) Fig. 12.16(b) shows a partly formed martensite crystal. Unlike other phase transitions, martensitic transformations do not proceed to completion at a constant temperature but take place over a wide range of temperatures. On cooling, martensitic transformations begin to occur at a temperature, M_S, and the extent of transformation usually depends on the degree of cooling below M_S. At a certain lower temperature, M_f, the transition is complete. At temperatures between M_S and M_f, the degree of transformation can be increased by applying shearing stresses to the crystal.

448

(e) The reverse transformation, e.g. martensite–austenite, can be accomplished on reheating martensite, but it occurs at temperatures well above M_S. The transformation has a large hysteresis (i.e. difference in temperature dependence in the two directions), therefore, which is often as large as several hundred degrees.

Martensitic transformations have been studied most in alloy systems but are probably widespread in inorganic systems as well. The dilational rock salt–CsCl transformation in alkali and ammonium halides (Fig. 12.2) may proceed by a martensitic mechanism. Probably the most studied martensitic transformation in non-metallic systems is the monoclinic–tetragonal transformation in zirconia, ZrO_2. Both polymorphs have distorted fluorite structures and the transformation takes place by a diffusionless, shear mechanism over a range of temperatures around $\sim 1000\,°C$, as shown in Fig. 12.17. On heating the monoclinic phase which is stable at low temperatures, transformation to the tetragonal phase begins above $\sim 1000\,°C$ but is not complete until $\gtrsim 1120\,°C$. The transformation exhibits a hysteresis of about $200\,°C$ and the reverse transformation on cooling begins at only $\lesssim 930\,°C$. Since the high temperature tetragonal phase cannot normally be quenched to room temperature, the transformation characteristics, as shown in Fig. 12.17, have to be determined directly at high temperatures. Various methods may be used, including high temperature X-ray diffraction, dilatometry, resistivity measurements and DTA. The transformation is classified as athermal because it takes place over a range of temperatures and the percentage transformation within that range does not change with time as long as the temperature remains constant. On changing the temperature, the new 'equilibrium' state is reached extremely rapidly; growth of the product phase by movement of the coherent interface between monoclinic and tetragonal domains occurs at velocities approaching the speed of sound.

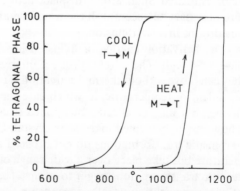

Fig. 12.17 Monoclinic (M)–tetragonal (T) martensitic transformation in zirconia determined by high temperature powder X-ray diffraction (After Wolten, 1963)

12.8.3 Order–disorder transitions

Order–disorder transitions have been mentioned in Section 12.2 and, as examples, the structures of cation ordered and disordered polymorphs of $LiFeO_2$ are given in Fig. 12.3. Order–disorder transitions may be thermodynamically first order and discontinuous if the long range order in the crystals changes abruptly at the transition. Alternatively, they may be second order and continuous if the disordering takes place over a large temperature range and without any discontinuity at T_c. Most transitions have mixed character, however, and show premonitory disordering as T_c is approached from below (or, equally, do not give a completely ordered phase as the temperature is reduced below T_c), together with discontinuities in ΔH, ΔS and long range order at T_c.

The magnitude of the changes in entropy correlates well with the structural changes that occur and can often be used to determine the nature of the disorder. The change in entropy at a transition is made up of contributions from configurational, rotational, vibrational and electronic effects but usually configurational entropy is the major factor. Its value may be calculated if the structures of ordered and disordered polymorphs are known, e.g. AgI transforms from wurtzite-like (β) to a body centred cubic (α) structure at 145 °C and the entropy increases by $\sim 14.5\,J\,mol^{-1}\,K^{-1}$ (Section 13.2.2.1) at the transition. In β-AgI, the hexagonal unit cell contains two Ag^+ ions located on specific tetrahedral sites but in α-AgI, the two Ag^+ ions are distributed at random over twelve tetrahedral positons. The change in entropy at the transition is given by

$$\Delta S = k\ln\left(\frac{n_2}{n_1}\right)^N = R\ln\left(\frac{n_2}{n_1}\right) \tag{12.23}$$

where k is Boltzman's constant, R the gas constant, N Avogadro's number and n_2, n_1 the number of configurations in the two polymorphs. In α-AgI, an Ag^+ ion can be placed in any one of twelve positions but in β AgI, only two positions are available. Therefore, $n_2 = 12$, $n_1 = 2$ and $\Delta S = R\ln 6 = 14.7\,J\,mol^{-1}K^{-1}$, in good agreement with the experimental value.

Entropy measurements have been particularly useful in evaluating the type of disorder in orientational order–disorder transitions. For example, crystalline KCN contains the cigar-shaped CN^- ion. KCN undergoes two transitions with measured entropies approximately as shown:

$$\text{III} \xrightarrow[\Delta S = R\ln 2]{83\,K} \text{II} \xrightarrow[\Delta S = R\ln 4]{168\,K} \text{I}_{\text{cubic}}$$

For III → II, $n_2/n_1 = 2$ and the CN^- ion can adopt two possible orientations in polymorph II. For II → I, $n_2/n_1 = 4$ and, therefore, CN^- can adopt any of eight orientations. Polymorph I has a CsCl derivative structure and it appears that the CN^- at the cube body centre can orient along any of eight $\langle 111 \rangle$ body diagonal directions.

NH_4Cl undergoes a phase transition that involves reorientation of the

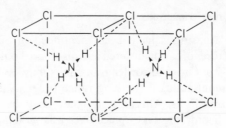

Fig. 12.18 Two possible orientations for NH_4^+ ions in phase II of NH_4Cl (After Rao and Rao, 1978)

NH_4^+ ions; it was thought originally that free rotation of the NH_4^+ ion occurred in the high temperature phase but entropy measurements indicate that this is unlikely. The transition is

$$\text{III} \xrightarrow[\Delta S = R\ln 2]{243\,\text{K}} \text{II}$$
$$\text{CsCl derivative} \qquad\qquad\qquad \text{CsCl derivative}$$

In phase III, NH_4^+ ions at the body centre position of the CsCl structure adopt the same orientation in all unit cells, e.g. one of the two orientations shown in Fig. 12.18. In phase II, two different orientations of NH_4^+ occur, as in Fig. 12.18, and these are arranged at random throughout the structure. The measured entropy change corresponds to $\Delta S = R \ln 2$, indicating two orientations of NH_4^+ in phase II. If free rotation of NH_4^+ occurred in phase II, a much larger entropy change would be required. Further polymorphic changes in NH_4Cl occur at higher temperatures.

Questions

12.1 How would you classify the following phase transitions: (a) quartz → cristobalite, SiO_2; (b) rutile → quartz, GeO_2; (c) tetragonal → monoclinic ZrO_2; (d) diamond → graphite; (e) ferroelectric → paraelectric $BaTiO_3$ (Chapter 15).

12.2 Using the data given in Table 12.2, calculate the entropies of the transitions: (a) low → high quartz; (b) $\beta \to \alpha$ AgI; (c) monoclinic → cubic Li_2SO_4. Comment on the relative magnitudes of your results.

12.3 What kind of structural changes, if any, might you expect the following to undergo as a consequence of (a) increasing temperature, (b) increasing pressure: (i) SiO_2; (ii) ZnO; (iii) SnO_2 (rutile structure); (iv) NH_4I?

12.4 The heat of fusion of tin metal is $61\,\text{J gm}^{-1}$. What is the change in free energy when 1 mole of tin melts at the equilibrium melting temperature? What is the corresponding entropy change?

References

M. J. Buerger (1961, 1972). Polymorphism and phase transformations, *Fortschr. Miner.*, **39** (1), 9–24; *Soviet Physics—Crystallography*, **16** (6) 959–968.

S. Flandrois (1974). Kinetics of phase changes in the solid state, *J. Chim. Physique*, **71** (6), 979–991.

H. Henisch, R. Roy and L. E. Cross (Eds) (1973). *Phase Transitions*, Pergamon Press, New York.

H. Lipson (1950). Order–disorder changes in alloys, in *Progress in Metal Physics* (Ed. B. Chalmers), Vol. II, Butterworths.

H. Moser (1936). *Phys. Z.*, **37**, 737.

C. N. R. Rao and K. J. Rao (1978). *Phase Transitions in Solids*, McGraw-Hill.

K. J. Rao and C. N. R. Rao (1966). *J. Materials Sci.*, **1**, 238.

R. Smoluchowski (Ed.) (1951). *Phase Transformations in Solids*, Wiley, New York.

E. C. Subbarao, H. S. Maiti and K. K. Srivastava (1974). Martensitic transformation in zirconia, *Phys. Stat. Sol.*, **A21**, 9.

A. R. Ubbelohde (1957). Thermal transformations in solids, *Quart. Rev. (London)*, **11**, 246.

M. E. Villafuerte-Castrejon and A. R. West (1981). *J. Chem. Soc. Faraday 1*, **77**, 2297.

A. R. West (1975). *Z. Krist.*, **141**, 422.

G. M. Wolten (1963). *J. Amer. Ceram. Soc.*, **46**, 418.

Chapter 13

Ionic Conductivity and Solid Electrolytes

Electrical conduction occurs by the long range migration of either electrons or ions. Usually conduction by one or other type of charge carrier predominates, but in some inorganic materials both ionic and electronic conduction are appreciable in the same material.

In solid materials, one is usually interested in the *specific conductivity*, σ, which is the conductivity of a crystal of pellet that has a cell constant of unity, i.e. unit cross-sectional area and unit length. Commonly used units of specific conductivity are $ohm^{-1}cm^{-1}$, $ohm^{-1}m^{-1}$, and Sm^{-1} where 1 siemen, $S \equiv 1\,ohm^{-1}$.

452

Table 13.1 *Typical values of electrical conductivity*

	Material	$\sigma(\text{ohm}^{-1}\,\text{cm}^{-1})$
Ionic conduction	Ionic crystals	$< 10^{-18}\text{--}10^{-4}$
	Solid electrolytes	$10^{-3}\,\text{--}10^{1}$
	Strong (liquid) electrolytes	$10^{-3}\,\text{--}10^{1}$
Electronic conduction	Metals	$10^{1}\,\text{--}10^{5}$
	Semiconductors	$10^{-5}\,\text{--}10^{2}$
	Insulators	$< 10^{-12}$

For any material and charge carrier, the specific conductivity is given by

$$\sigma = \sum_i n_i e_i \mu_i \tag{13.1}$$

where n_i is the number of charge carriers of species i, e_i is their charge and μ_i their mobility. For electrons and monovalent ions, e is the charge on an electron, 1.6×10^{-19} C. Typical conductivities for a range of materials are given in Table 13.1. Conductivities are usually temperature dependent and for all materials, except metals, the conductivity increases with increasing temperature. In metals, σ is highest at low temperatures and in some metals the phenomenon of superconductivity occurs close to absolute zero.

Migration of ions does not occur to any appreciable extent in most ionic and covalent solids such as oxides and halides. Rather, the atoms tend to be essentially fixed on their lattice sites and can only move via crystal defects. Only at high temperatures, where the defect concentrations become quite large and the atoms have a lot of thermal energy, does the conductivity become appreciable, e.g. the conductivity of NaCl at $\sim 800\,°C$, just below its melting point, is $\sim 10^{-3}\,\text{ohm}^{-1}\text{cm}^{-1}$, whereas at room temperature, pure NaCl is an insulator.

In contrast, there exists a small group of solids called, variously, *solid electrolytes, fast ion conductors* and *superionic conductors*, in which one of the sets of ions can move quite easily. Such materials often have rather special crystal structures in that there are open tunnels or layers through which the mobile ions may move. The conductivity values, e.g. $10^{-3}\,\text{ohm}^{-1}\,\text{cm}^{-1}$ for Na^+ ion migration in β-alumina at $25\,°C$, are comparable to those observed for strong liquid electrolytes. There is currently great interest in studying the properties of solid electrolytes, developing new ones and extending their range of applications in solid state electrochemical devices.

This chapter is concerned only with ionic conduction; electronic conductivity is treated in Chapter 14. The behaviour of simple materials such as NaCl and AgCl is first described in some detail. Although this is normally regarded as a branch of solid state physics, an understanding of certain aspects of the subject is fundamental to solid state chemistry, in areas such as solid electrolytes, crystal defects and reactivity of solids (see also Chapters 2 and 9). The remainder of the chapter is devoted to solid electrolytes and their applications and experimental methods for measuring conductivity.

13.1 Typical ionic crystals

13.1.1 Alkali halides

In crystals of the alkali halides, e.g. NaCl, cations are usually more mobile than anions. A section through the NaCl structure is shown in Fig. 13.1 in which a Na^+ ion is moving into an adjacent vacant cation site and thereby leaving its own site vacant. The Na^+ ion that has moved can travel no further because (a) there are no other vacant sites in the vicinity for it to move into and (b) interstitial migration of Na^+ in NaCl appears not to occur to any appreciable extent (see Section 13.1.2). The cation vacancy may continue to move, however, because it is always surrounded by twelve Na^+ ions, one of which can jump in such a way as to occupy and effectively change places with the vacancy. It is convenient, therefore, to regard cation vacancies as the main current carriers in NaCl. Anion vacancies are present, too, in NaCl but they appear to be rather less mobile than the cation vacancies.

The magnitude of the ionic conductivity of NaCl depends on the number of cation vacancies present, which in turn depends very much on the purity and thermal history of the crystals. Vacancies are normally created by one of two methods. On heating a crystal, the number of vacancies present in thermodynamic equilibrium increases exponentially (see Fig. 9.1 and equation 9.9), and this is the number that is *intrinsic* to the pure crystals. If, on the other hand, aliovalent impurities are introduced, vacancies may be created so as to preserve charge balance, e.g. addition of a small amount of $MnCl_2$ yields, at equilibrium, a solid solution of formula, $Na_{1-2x}Mn_x V_{Na_x}Cl$ where, for each Mn^{2+} ion, there is an associated cation vacancy, V_{Na}. Such vacancies are *extrinsic* because they would not be present in pure NaCl. At low temperatures (e.g. 25 °C), the number of thermally generated intrinsic vacancies is very small and, unless the crystal is very pure, is much less than the concentration of extrinsic vacancies. The

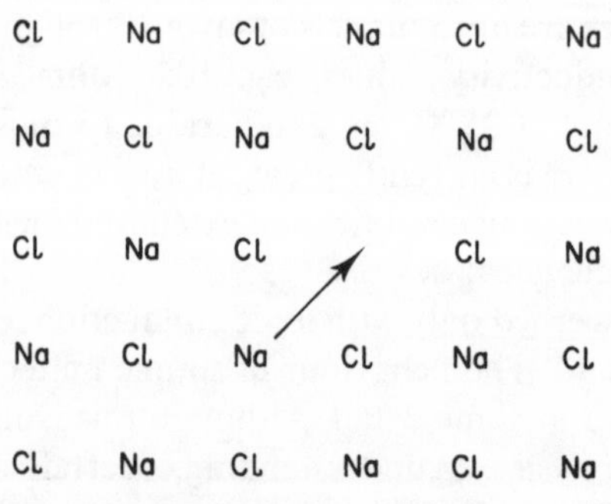

Fig. 13.1 Migration of cation vacancies (i.e. Na^+ ions) in NaCl

changeover from extrinsic to intrinsic behaviour occurs at some higher temperature that is determined by the impurity concentration.

The temperature dependence of ionic conductivity is usually given by the Arrhenius equation

$$\sigma = A\exp\left(\frac{-E}{RT}\right) \tag{13.2}$$

where E is the activation energy, R the gas constant and T the absolute temperature. The pre-exponential factor, A, contains several constants, including the vibrational frequency of the potentially mobile ions. Graphs of $\log_e \sigma$ against T^{-1} should give straight lines of slope $-E/R$. In some treatments, a reciprocal temperature term is included in the prefactor, A, in which cases, it is normal to plot $\log \sigma T$ against T^{-1}. This may make a small difference to the value of the slope, $-E/R$. A schematic Arrhenius plot for NaCl is shown in Fig. 13.2. In the low temperature extrinsic region, the number of vacancies is dominated by the impurity level and is constant for a given impurity concentration. A set of parallel lines is shown, each of which corresponds to the conductivity of a crystal with a different amount of dopant. In the extrinsic region, the temperature dependence of σ depends only on the cation mobility, μ (equation 13.1), whose temperature dependence is also given by an Arrhenius expression:

$$\mu = \mu_0 \exp\left(\frac{-E_m}{RT}\right) \tag{13.3}$$

where E_m is the activation energy for migration of cation vacancies.

In order to understand the origin of the activation energy for migration, consider the possible paths that may be taken by a Na^+ ion in jumping from its

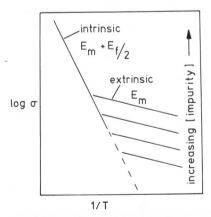

Fig. 13.2 Schematic ionic conductivity of doped NaCl crystals. Parallel lines in the extrinsic region correspond to different dopant concentrations

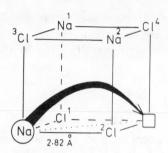

Fig. 13.3 Pathway for Na^+ migration in NaCl

lattice site into an adjacent vacancy. A small part of the NaCl structure is shown in Fig. 13.3. It is a simple cube with Na^+, Cl^- ions at alternate corners and corresponds to one-eighth of the unit cell of NaCl. One corner Na^+ site is shown empty and a Na^+ ion from one of the other three corners moves to occupy it. The direct jump (dotted line) across the cube face is not possible because Cl^- ions 1 and 2 are very close, if not in actual contact, and an Na^+ ion would not be able to squeeze between them. Instead, Na^+ must take an indirect route (curved arrow) that passes through the middle of the cube. At the cube centre is an interstitial site that is equidistant from the eight corners. Four of the corners are occupied by Cl^- ions which are arranged tetrahedrally about the cube centre. Before the moving Na^+ ion arrives at this central interstitial site, it has to pass through a triangular window formed by the Cl^- ions 1, 2 and 3. Let us now calculate the size of this window in order to appreciate how difficult it is for Na^+ to squeeze through.

For NaCl, the unit cell parameter, a, is equal to 5.64 Å. The Na—Cl bond length in Fig. 13.3 is $a/2 = 2.82$ Å. This is equal to $(r_{Na^+} + r_{Cl^-})$, assuming the anions and cations to be in contact. Tabulated ionic radii (which vary somewhat, depending on the table that is consulted) of Na^+ and Cl^- are ~ 0.95 and ~ 1.85 Å; the value of the Na—Cl bond lengths calculated from the sum of these radii is ~ 2.80 Å, close to the experimentally measured value.

In close packed structures, such as NaCl, the anions are either in contact with each other or are in close proximity. The Cl^- ions 1, 2 and 3 form part of a close packed layer and the distance Cl(1)–Cl(3) is given by $[(a/2)^2 + (a/2)^2]^{1/2} = 3.99$ Å. This is about 0.3 Å larger than given by $2r_{Cl^-}$ and so adjacent Cl^- ions in NaCl are not quite in contact.

The radius r' of the triangular window formed by Cl^- ions 1, 2 and 3 may be calculated from Fig. 13.4 as follows:

$$\cos 30° = \frac{x/2}{y} = \frac{1.995}{(r_{Cl^-} + r')}$$

Therefore,

$$(r_{Cl^-} + r') = \frac{1.995}{\cos 30°} = 2.30 \text{ Å}$$

If

$$r_{Cl^-} = 1.85 \text{ Å}, \qquad \text{then} \qquad r' = 0.45 \text{ Å}$$

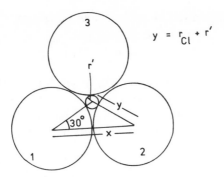

Fig. 13.4 Triangular interstice of radius r' through which a moving Na^+ ion must pass in NaCl. Circles 1 to 3 are Cl^- ions of radii $x/2$

The radius r'' of the interstitial site at the cube centre may be calculated similarly. The body diagonal of the cube (Fig. 13.3), effectively given by $(2r_{Cl^-} + 2r'')$ is equal to

$$2(r_{Cl^-} + r'') = [(a/2)^2 + (a/2)^2 + (a/2)^2]^{1/2}$$
$$= 4.88\,\text{Å}$$

Therefore,

$$r'' = 0.59\,\text{Å}$$

We can see, therefore, that the jumping of a Na^+ ion in NaCl is a difficult and complicated process. First, the Na^+ ion has to squeeze through a narrow triangular gap of radius $0.45\,\text{Å}$, whereupon it finds itself in a small tetrahedral interstitial site of radius $0.59\,\text{Å}$. The residence time here is quite short since this site has a hostile environment with two Na^+ ions, 1 and 2, at a distance of $2.44\,\text{Å}$ as well as four Cl^- ions at the same distance. The Na^+ ion leaves by squeezing through another gap of radius $0.45\,\text{Å}$ (formed by Cl^- ions 1, 2 and 4) to occupy the vacant octahedral site on the other side. These calculations are inevitably somewhat idealized since relaxation or distortion of the structure must occur in the vicinity of the defects, thereby modifying the distances involved. They do nevertheless show that the migration of Na^+ ions is difficult and is associated with a considerable activation energy barrier.

In the extrinsic regions of Fig. 13.2, then, the conductivity depends on both the vacancy concentration and the mobility. It is given by combining equations (13.1) and (13.3) to give

$$\sigma = ne\mu_0 \exp\left(\frac{-E_m}{RT}\right) \tag{13.4}$$

In the intrinsic conductivity region at higher temperatures, the concentration of thermally induced vacancies is greater than the vacancy concentration associated with the dopant. Hence, the number of vacancies, n, is temperature

458

dependent and is also given by an Arrhenius equation:

$$n = N \times \text{constant} \times \exp\left(\frac{-E_f}{2RT}\right) \qquad (13.5)$$

This equation is the same as equation (9.9), in which $E_f/2$ is the activation energy for formation of one mole of cation vacancies, i.e. half the energy required to form one mole of Schottky defects. The vacancy mobility is again expressed by equation (13.3) and so the overall conductivity in the intrinsic region is given by

$$\sigma = N \times \text{constant} \times e\mu_0 \exp\left(\frac{-E_m}{RT}\right)\exp\left(\frac{-E_f}{2RT}\right)$$

i.e.

$$\sigma = A \exp -\frac{E_m + E_f/2}{RT} \qquad (13.6)$$

In Fig. 13.2 are shown schematic Arrhenius plots for NaCl crystals with varying degrees of purity. The set of parallel lines in the extrinsic region corresponds to different impurity levels of, for example, Mn^{2+}, whereas the single line in the intrinsic region shows that the conductivity is unaffected by the impurity content. This latter effect is reasonable if the actual impurity concentration is very small (< 1 per cent Mn^{2+}) since the presence of Mn^{2+} ions at this level does not affect significantly the activation energy for migration of the cation vacancies. The gradient of the intrinsic line is greater than that of the extrinsic lines and if both intrinsic and extrinsic gradients can be measured then E_m and E_f may be determined separately.

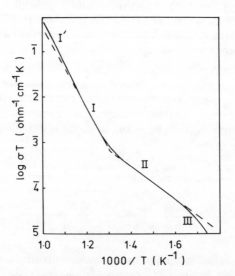

Fig. 13.5 Ionic conductivity of 'pure' NaCl as a function of temperature. (From Kirk and Pratt, 1967)

Good quality experimental data for NaCl single crystals (Fig. 13.5) indicate that the simple behaviour shown in Fig. 13.2 is somewhat idealized and that several complicating factors are present. Stages I and II in Fig. 13.5 correspond to the regions of intrinsic and extrinsic conductivity, respectively, of Fig. 13.2. The dashed lines in Fig. 13.5 represent extrapolations of regions I and II and show where, experimentally, deviations occur. Stage 1' occurs close to the melting point (802 °C) and has been variously attributed to two possible causes. One explanation is that the upward curvature in conductivity in I' arises because anion vacancies are now becoming increasingly mobile and make a significant contribution to σ. The other is that long range Debye–Hückel interactions between cation and anion vacancies become appreciable as the vacancy concentration increases at high temperatures. An attractive force results (similar to the Debye–Hückel interaction between ions in solutions) which has the effect of partially compensating for the energy of formation of the vacancies. Thus, vacancy formation is easier, the vacancy concentration increases and σ increases. It is not known which explanation is correct for NaCl, although the Debye–Hückel effect is probably the cause of a similar deviation in σ at high temperatures in AgCl single crystals (see the discussion of Fig. 9.4).

Below ~ 390 °C (in this particular NaCl crystal), σ deviates downwards from the ideal extrinsic line, stage III. This is attributed to the formation of defect complexes, e.g. cation vacancy/anion vacancy pairs or cation vacancy/aliovalent cation impurity pairs. These complexes are formed by short range attractions between defects on neighbouring or next nearest neighbour sites and are quite distinct from the Debye–Hückel interactions mentioned above which are long range interactions and are associated with preserving electroneutrality. In stage III, cation vacancies that are present in defect complexes must first acquire the additional energy needed to dissociate themselves from the defect complexes before they can move. As a result the activation energy of stage III is greater than E_m (stage II).

Although the conductivity of NaCl crystals has been measured many times and in many different laboratories, there is not even good agreement on the value of E_m, the activation energy for Na$^+$ migration. Values range from 0.65 to 0.85 eV (~ 60 to 80 kJ mol^{-1}). Data for the various activation energies associated with ionic conduction in NaCl are given in Table 13.2. The probable reason for the variation in E_m values is that other defects, especially dislocations, are inevitably present in crystals and have a considerable influence on cation migration. The regions of the crystal structure that are close to a dislocation are in a stressed and

Table 13.2 *Conductivity in NaCl crystals*

Process	Activation energy (eV)
Migration of Na$^+$, E_m	0.65–0.85
Migration of Cl$^-$	0.90–1.10
Formation of Schottky pair	2.18–2.38
Dissociation of vacancy pair	~ 1.3
Dissociation of cation vacancy $-$ Mn^{2+} pair	0.27–0.50

distorted condition and migration of cations along 'dislocation pipes' may take place more easily than through regions of perfect crystal. If this is so, conductivity and the magnitude of E_m may depend on the number and distribution of dislocations and hence on the thermal history of the crystals. It is, however, difficult to make quantitative measurements of the effect of dislocations on σ because their number cannot be controlled and measured accurately.

The effect of impurities on σ is more amenable to study since their number may be controlled and measured and they have a dramatic effect on σ in the extrinsic region. The concentration of Schottky defects in NaCl in the intrinsic region and at thermodynamic equilibrium is given by the Law of Mass Action (equation 9.3):

$$K = \frac{[V_{Na}][V_{Cl}]}{[Na^+][Cl^-]}$$
(13.7)

where V_{Na} and V_{Cl} refer to vacant cation and anion sites, respectively. It is assumed that the equilibrium constant, K, is unaffected by the presence of small amounts of aliovalent impurity and that the denominator, the number of occupied sites, is essentially constant and equal to 1. Therefore,

$$[V_{Na}][V_{Cl}] = \text{constant} = x_0^2$$
(13.8)

where, in the intrinsic region, $x_0 = [V_{Na}] = [V_{Cl}]$. If the number of cation vacancies in the extrinsic region is increased, e.g. by adding aliovalent cation impurity, then the number of anion vacancies must decrease in accord with equation (13.8). Let x_a and x_c be the concentration of anion and cation vacancies under extrinsic conditions (and so $x_a \neq x_c$) and c be the concentration of divalent impurity cations. Then

$$x_c = x_a + c$$
(13.9)

This condition arises because both anion vacancies and divalent cation impurities carry a net positive charge of $+1$, whereas the cation vacancy has an effective charge of -1. Overall, charge balance must be retained. On combining equations (13.8) and (13.9) and solving the resulting quadratic, the following positive values of x_c and x_a are obtained:

$$x_c = \frac{c}{2}\left[1 + \left(1 + \frac{4x_0^2}{c^2}\right)^{1/2}\right]$$
(13.10)

$$x_a = \frac{c}{2}\left[\left(1 + \frac{4x_0^2}{c^2}\right)^{1/2} - 1\right]$$
(13.11)

If $x_0 \ll c$, then $x_c \to c$ and $x_a \to 0$; this result applies to the extrinsic region and is the same as that deduced above from equation (13.8). If $x_0 \gg c$, then $x_c = x_0 = x_a$; this applies to the intrinsic region of conductivity.

Not all impurities lead to an increase in σ in the extrinsic region. Impurities of the same valence, e.g. K^+ or Br^- in NaCl, normally have no effect on σ unless they are present in large concentrations. Impurities that lead to a reduction in concentration of the mobile species cause a reduction in σ. Thus, if divalent anions could be dissolved in NaCl crystals, the concentration of the less mobile anion vacancies would increase at the expense of the more mobile cation vacancies. This effect is not observed to any large extent in NaCl because divalent anion impurities are not soluble, but it is an important effect in AgCl, as described below.

13.1.2 Silver chloride

The predominant defects in AgCl are Frenkel defects, i.e. interstitial Ag^+ ions associated with Ag^+ ion vacancies (Chapter 9). Experiments have shown that the interstitial Ag^+ ions are more mobile than the Ag^+ vacancies. Two possible mechanisms for migration of interstitial Ag^+ are shown schematically in Fig. 13.6(a). In the direct *interstitial* mechanism (1), the interstitial Ag^+ ion jumps to an adjacent empty interstitial site. In the indirect or *interstitialcy* mechanism, a knock-on process (2) occurs. The interstitial Ag^+ ion causes one of its four Ag^+ neighbours to move off its normal site into an adjacent interstitial site and itself occupies the vacant lattice site thereby created. It is possible to distinguish between the interstitialcy and direct interstitial mechanisms if accurate data for

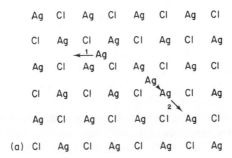

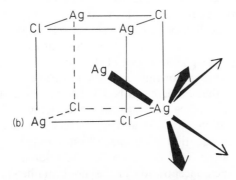

Fig. 13.6 (a) Migration of interstitial Ag^+ ions by (1) direct interstitial jump and (2) indirect interstitialcy mechanism. (b) Pathway for migration of Ag^+ in AgCl by an interstitialcy mechanism

both diffusion and conductivity are available. In diffusion measurements, the crystal is doped with radioactive Ag^{+*} ions and the migration of these radioactive or tracer Ag^+ ions is followed. In conductivity measurements, all the Ag^+ ions, not only the radioactive ones, contribute to the net conductivity. A relationship exists between the coefficient for self-diffusion, D, and the conductivity, σ. This is given by the Nernst–Einstein equation:

$$D = \frac{kT}{f n (Ze)^2} \sigma \qquad (13.12)$$

where (Ze) is the charge on the mobile ions and n is their concentration; f is a correlation factor, the *Haven ratio*, whose value depends on the mechanism of ion migration. The value of the Haven ratio is different for the two mechanisms shown in Fig. 13.6(a). In mechanism 2, the net distance of charge displacement is greater than either of the jump distances of the individual Ag^+ ions. In mechanism I, however, the jump distance of the Ag^+ ion equals the distance of the overall charge migration. Since diffusion measurements determine the movement of the ions themselves and conductivity measure the overall charge displacement, the Haven ratio is different for the two mechanisms. It has been found experimentally that the interstitialcy mechanism (2) is the operative mechanism in AgCl.

The detailed structure of an interstitial ion and its immediate surroundings are not known but it could be similar to the dumb-bell or split interstitial which has been observed in metals (Figs 9.7 and 9.8), i.e. with two Ag^+ ions arranged symmetrically in interstitial positions about a vacant octahedral site. The formation and dissociation of split interstitials is consistent with the observed interstitialcy mechanism of migration.

The difference between the vacancy migration mechanism that occurs in NaCl and the interstitialcy mechanism in AgCl may be seen by comparing Figs 13.3 and 13.6(b). Both NaCl and AgCl have the same rock salt crystal structure. In the vacancy mechanism, a Na^+ ion moves from one corner of the cube to another, via an interstitial site at the cube centre which is occupied only transiently (Fig. 13.3). In the interstitialcy mechanism, a Ag^+ ion effectively moves from the interstitial site in the centre of one cube to the site in the centre of an adjacent cube by knocking on a Ag^+ ion placed at one of the corner sites (Fig. 13.6b).

The effect of aliovalent cation impurities on the extrinsic conductivity region of AgCl is different to that observed in NaCl. The presence of, say, Cd^{2+} again increases the number of cation vacancies, but because the product of the concentrations of cation vacancies and Ag^+ interstitials is constant (for dilute impurity levels, see equation 9.11) the interstitial Ag^+ concentration must decrease with increasing Cd^{2+} concentration. Addition of Cd^{2+} therefore leads to a reduction in the concentration of the more mobile species; the resulting Arrhenius conductivity plot is shown schematically in Fig. 13.7. An extrinsic region at lower temperatures is again observed but it is displaced *downwards* to lower σ values. The degree of downward displacement increases with increasing Cd^{2+} concentration until a minimum conductivity is reached at which the conductivity due to the more numerous but less mobile cation

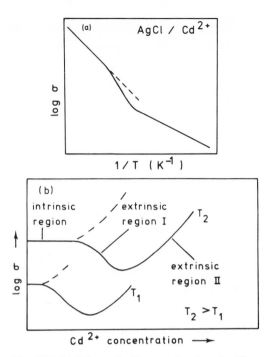

Fig. 13.7 (a) Schematic diagram showing the effect of Cd^{2+} on conductivity of AgCl crystals. (b) Effect of Cd^{2+} impurity on isothermal conductivity of AgCl. Dashed lines represent the effect of adding a divalent anionic impurity

vacancies equals that due to the less numerous but more mobile interstitial Ag^+ ions. At still higher defect concentrations, cation vacancy migration predominates and σ increases. This is shown in Fig. 13.8 as a plot of conductivity versus defect concentration for two temperatures. In the intrinsic region, the conductivity is independent of Cd^{2+} concentration. In extrinsic region I, interstitial conduction predominates and σ decreases as $[Ag_i^+]$ decreases with increasing $[Cd^{2+}]$. In extrinsic region II, vacancy conduction predominates and σ increases as $[V_{Ag}]$ increases with increasing $[Cd^{2+}]$.

The relevant equations for the dependence of σ on impurity concentration, c, are obtained as follows. In order to preserve charge balance:

$$x_c = c + x_i \tag{13.13}$$

where x_i and x_c are the concentrations of interstitial Ag^+ ions and Ag^+ ion vacancies, respectively. From eq. 9.11, i.e. $x_c \cdot x_i = x_0^2$ and eq. 13.13, the conductivity in the the extrinsic region is given by

$$\sigma = e(x_c \mu_c + x_i \mu_i)$$

$$= e\mu_c \frac{c}{2}\left[1 + \left(1 + \frac{4x_0^2}{c^2}\right)^{1/2}\right] + e\mu_1 \frac{c}{2}\left[\left(1 + \frac{4x_0^2}{c^2}\right)^{1/2} - 1\right] \tag{13.14}$$

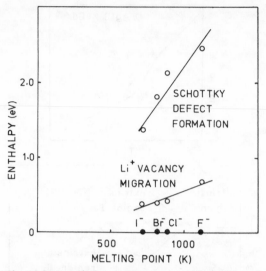

Fig. 13.8 Correlation between enthalpies of Schottky defect formation, cation vacancy migration and melting temperature for the lithium halides. (Data from Barr and Lidiard, 1970)

Accurate measurements of σ for AgCl show that Fig. 13.7 is somewhat idealized and that deviations do occur. At high temperatures, long range Debye–Hückel interactions become important and give an upward departure from the intrinsic slope, whereas at low temperatures 'complexes' form between cation vacancies and aliovalent cation impurities to give a downward deviation in the extrinsic region. Defect energies, in electronvolts, for AgCl are as follows:

Formation of Frenkel defect	1.24
Migration of cation vacancy	0.27–0.34
Migration of interstitial Ag^+	0.05 – 0.16

13.1.3 Alkaline earth fluorides

The most important defect in this group is probably the anion Frenkel defect in which an interstitial F^- ion occupies the centre of a cube that has eight F^- ions at the corners (Fig. 7.18). Conductivity measurements have shown that the anion vacancy is more mobile than the interstitial F^- ion. This contrasts with AgCl in which the interstitial Ag^+ is more mobile than the cation vacancy. In some materials that have the fluorite structure, e.g. PbF_2, the conductivity at high temperatures becomes quite large (Section 13.2.3).

13.1.4 Simple stoichiometric oxides

The conductivity of oxides such as MgO is difficult to study because (a) they have high melting points ($\sim 2500\,°C$) and it is difficult to grow pure crystals that are free from contamination at these temperatures and (b) the various energies for formation and migration of defects are several times larger than those for NaCl.

Consequently, the conductivities are very low, even at high temperatures, and are often dominated by impurity conduction.

In both ionic halides and metals, an approximate linear correlation has been found to exist between the melting temperature and the energy of formation of Schottky defects (alkali halides) or vacancies (metals), as shown for the lithium halides in Fig. 13.8. This correlation is reasonable because the melting temperature depends on the lattice energy of the crystals and is related to the energy required to break up the crystal structure. The energy of Schottky defect or vacancy creation can be regarded as the energy required to break up a very small part of the crystal and remove an atom or an ion pair from the structure. Similar bonds must therefore be broken. The similarity could be taken one stage further since some of the theories of melting involve the creation of a significant proportion of crystal defects as the melting temperature is approached. Correlations have also been noted between the melting temperature and energy of migration of, say, cation vacancies, Fig. 13.8. Again this seems a reasonable correlation. On the assumption that a similar correlation exists for oxides, the energies of defect creation and migration can be estimated. In this way, the energy of formation of Schottky defects in MgO has been estimated as 6.5 eV and the energy of migration of the cation vacancy as 2.5 eV. Note that these values are very high which accounts for the very low conductivity of MgO. Experimental measurements of the self-diffusion of Mg^{2+} in MgO gave an activation energy of 3.4 eV. These results probably applied, therefore, to an extrinsic region. There is also evidence that dislocations are important in oxides and that oxygen vacancies appear to migrate preferentially down dislocation 'pipes'.

13.2 Solid electrolytes (or fast ion conductors, superionic conductors)

Most crystalline materials, like NaCl or MgO, have low ionic conductivities because, although the atoms or ions undergo thermal vibrations, they cannot usually escape from their lattice sites. The small group of solid electrolytes are an exception. In solid electrolytes, one component of the structure, cationic or anionic, is not confined to specific lattice sites but is essentially free to move throughout the structure. Solid electrolytes are, therefore, intermediate in structure and property between, on the one hand, normal crystalline solids with regular three-dimensional structures and immobile atoms or ions and, on the other, liquid electrolytes which do not have regular structures but do have mobile ions. Often solid electrolytes are stable only at high temperatures. At lower temperatures they may undergo a phase transition to give a polymorph with a low ionic conductivity and a more usual type of crystal structure (Fig. 13.9). For example, Li_2SO_4 and AgI are both poor conductors at 25 °C but at temperatures of 572 and 146 °C, respectively, their crystal structures change to give polymorphs, α-Li_2SO_4 and α-AgI, that have mobile Li^+ and Ag^+ ions ($\sigma \sim 1$ ohm^{-1} cm^{-1}). On heating, the conductivity therefore increases dramatically at the phase transition.

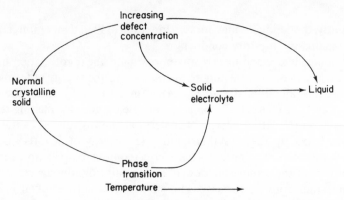

Fig. 13.9 Solid electrolytes as intermediate between normal crystalline solids and liquids

Other solid electrolytes form as a consequence of a gradual increase in defect concentration with increasing temperature. For example, in ZrO_2, the concentration of anion vacancies above $\sim 600\,^{\circ}C$ is sufficiently large that zirconia is a good, high temperature oxide ion conductor. The distinction between normal ionic solids and solid electrolytes is often not well defined, especially for materials such as ZrO_2 which undergo a gradual change in behaviour with increasing temperature.

It is now becoming apparent from both theoretical studies and experimental results on a wide variety of materials that ionic conductivities of between 0.1 and $10\,ohm^{-1}\,cm^{-1}$ are the maximum that are likely to be obtained for any solid material. These values are obtained when a large proportion of the ions are moving at any one time. According to some authors, the names 'superionic conductor' and 'fast ion conductor' should be reserved for materials that belong to this latter category of optimized conductivity. Although these names are in common usage, they are misnomers because the mobile ions do not have any 'super' properties nor are they exceptionally mobile. Rather, the high conductivities are associated with a large concentration of mobile species and a relatively small activation energy for ion migration.

The classification of solid electrolytes as intermediate between normal ionic solids and ionic liquids (Fig. 13.9) is supported by data on the relative entropies of polymorphic transitions and of melting. For normal, monovalent ionic materials such as NaCl, disordering of both cations and anions occurs on melting; a typical entropy or fusion is $24\,J\,mol^{-1}\,K^{-1}$. For AgI, the $\beta \rightarrow \alpha$ transition at $146\,^{\circ}C$ may be regarded as quasi-melting of the silver ions. It has an entropy of transition of $14.5\,J\,mol^{-1}\,K^{-1}$. At the melting point of AgI, only the iodide atoms are left to become disordered. This fits in with a much reduced value for the entropy of fusion, $11.3\,J\,mol^{-1}\,K^{-1}$. The combined entropies of transition and fusion in AgI are close to the fusion entropy of NaCl. Similar effects are observed in fluorides of some divalent metals, e.g. PbF_2. It has an entropy of fusion of only $16.4\,J\,mol^{-1}\,K^{-1}$ whereas that for MgF_2, a typical ionic solid with low conductivity, is $\sim 35\,J\,mol^{-1}\,K^{-1}$! This is because the fluoride ions in PbF_2 are

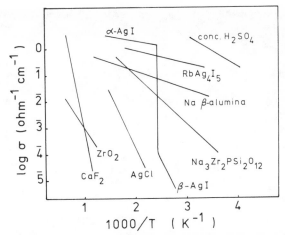

Fig. 13.10 Ionic conductivity of some solid electrolytes with concentrated H_2SO_4 for comparison. In searching for new materials, one aims for the top right-hand corner of the diagram (high σ, low T)

disordered above $\sim 500\,^{\circ}C$ and the entropy of fusion corresponds to disordering of the lead ions alone.

Conductivity values for a variety of solid electrolytes are given in Fig. 13.10 in the form of Arrhenius diagrams. A selection of the more important solid electrolytes are discussed next. Applications are deferred to Section 13.5.

13.2.1 β-alumina

13.2.1.1 *Structure*

β-Alumina is the name for a family of compounds of general formula $M_2O.nX_2O_3$ where n can have various values in the range 5 to 11, M is a monovalent cation, such as alkali$^+$, Cu^+, Ag^+, Ga^+, In^+, Tl^+, NH_4^+, H_3O^+, and X is a trivalent cation, Al^{3+}, Ga^{3+} or Fe^{3+}. The most important member of this family is sodium β-alumina (M = Na$^+$, X = Al^{3+}), which has been known for many years as a byproduct of the glass-making industry. It forms in the refractory lining of furnaces by reaction of soda from the melt with alumina in the refractory bricks. Its name is a misnomer because, although it was originally thought to be a polymorph of Al_2O_3, it is now known that additional oxides such as Na_2O are essential in order to stabilize its crystal structure.

Interest in β-alumina as a solid electrolyte began with the pioneering work of the Ford Motor Co. who, in 1966, found that the Na$^+$ ions are very mobile at room temperature and above. Ford also found that other cations could be ion exchanged for Na$^+$ and that these cations, too, are mobile. Since that time, interest in solid electrolytes has mushroomed. In our energy-conscious society much of the impetus for further research comes from the possibility of developing new, high density energy storage systems.

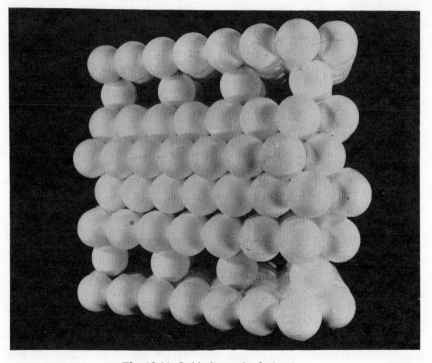

Fig. 13.11 Oxide layers in β-alumina

The high conductivity of the monovalent ions in β-alumina is a consequence of its unusual crystal structure, shown in Fig. 13.11. It is built of close packed layers of oxide ions, stacked in three dimensions, but every fifth layer has three-quarters of its oxygens missing. The Na^+ ions reside in these oxygen-deficient layers and are able to move very easily because (a) there are more sites available than there are Na^+ ions to occupy them and (b) the radius of Na^+ is less than that of the O^{2-} ion. β-alumina exists in two structural modifications, named β and β″, which differ in the stacking sequence of the layers (Fig. 13.12). The β″ form occurs with more soda-rich crystals, $n \simeq 5 - 7$, whereas β occurs for $n \simeq 8 - 11$. Both the β and β″ structures are closely related to that of spinel, $MgAl_2O_4$; Al^{3+} ions occupy a selection of both tetrahedral and octahedral interstices between pairs of adjacent close packed oxide layers. Both the β- and β″-alumina structures may be regarded as built of 'spinel blocks' that are four oxide layers thick and in which the oxide layers are in cubic stacking sequence ABCA. Adjacent spinel blocks are separated by the oxygen-deficient layers or 'conduction planes' in which the Na^+ ions reside. The unit cells are hexagonal with $a = 5.60$ Å and $c = 22.5$ Å (β), 33.8 Å (β″). In the c direction, perpendicular to the oxide layers, there are two spinel blocks in the unit cell of β and three blocks in the unit cell of β″. The structure of the 'spinel blocks' must be considered as defective in comparison with the ideal spinel structure. Thus, spinel contains both Mg^{2+} and Al^{3+} ions, in the ratio of 1:2, but the spinel blocks of β, β″ alumina contain only Al^{3+}, apart from small

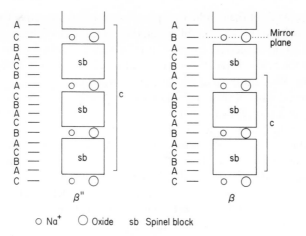

○ Na⁺   ○ Oxide   sb Spinel block

Fig. 13.12 Oxide packing arrangements in β- and β''-
alumina

amounts of dopant ions such as Li^+, Mg^{2+} that are often added. In order to
maintain charge balance, Al^{3+} vacancies must therefore also be present in the
spinel blocks. The overall stacking sequence of oxide ions including both spinel
blocks and conduction planes is cubic, ABC, in β'' but is a more complex, ten-
layer sequence in β (Fig. 13.12).

The details of the atomic structure in the region of the conduction plane have
been the subject of much crystallographic work but are still not well understood.
In Fig. 13.13 is shown one layer of close packed O^{2-} ions, A, B, that forms one
wall of the conduction plane, and the 'column' or 'spacer' O^{2-} ions, C, of the
conduction plane in the layer immediately above. It should be apparent that in
the conduction plane only one-quarter of the O^{2-} sites are occupied, i.e. for every
O^{2-} ion, C, there are three empty sites, m. In the β modification, a mirror plane of
symmetry passes through and parallel to each conduction plane (Fig. 13.12);
hence the layers of close packed O^{2-} ions, A, B, on either side of the conduction
plane are superposed in the projection in Fig. 13.13. In the β'' modification, the
conduction plane does not coincide with a mirror plane and the oxide layers that

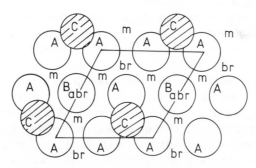

Fig. 13.13 Conduction plane in β-alumina

form the two walls of the conduction plane are staggered relative to each other.

For the Na^+ ions in the β modification there is a choice of three possible types of site: (a) the mid oxygen positions, m, (b) the Beevers–Ross sites, br, which were favoured in the original structure determination of Beevers and Ross, and (c) the anti-Beevers–Ross sites, abr. From the crystallographic results, it appears that Na^+ ions spend most of their time in br and m sites, but in order to undergo any long range migration they must pass through abr sites. Both br and m sites are large sites, e.g. Na^+ in a br site is coordinated to three oxygens in the oxide plane, A, below, three in the plane above and three, C, within the conduction plane. Na—O bond distances are large, ~ 2.8 Å, compared with normal values of ~ 2.4 Å. The abr site is much smaller than br and m sites because two oxygens, B, are quite close, one directly above and one directly below, giving two short Na—O distances of 2.3 Å. Most other monovalent cations also prefer to occupy the br and m sites in β-alumina with the exception of Ag^+ and Tl^+, which show a considerable preference for the abr sites. This is probably because Ag^+ and Tl^+ prefer covalent bonding and sites of low coordination number, such as the abr site.

Both β-and β''-alumina exist over a range of compositions, i.e. they are solid solution phases. The mechanism of solid solution appears to be different for each, however. It is often suggested that the ideal stoichiometry for β-alumina is $Na_2O \cdot 11Al_2O_3$, i.e. $NaAl_{11}O_{17}$, although, in practice, β-aluminas usually contain considerably more soda than this. In the original structure determination of β-alumina by Beevers and Ross, the structure of $NaAl_{11}O_{17}$ was deduced to contain one Na^+ ion, in each conduction plane of the unit cell, located on the br site. More recent studies on β-alumina, with a typical stoichiometry $Na_2O \cdot 8Al_2O_3$, show that the Na^+ ions, $1\frac{1}{3}$ per unit cell per conduction plane are distributed over br and m sites. In order to maintain charge balance, extra O^{2-} ions are also present, in some of the m positions in the conduction plane. Compared with the structure of $NaAl_{11}O_{17}$, therefore, the soda-rich β-alumina contains interstitial Na^+ and O^{2-} ions.

In β''-alumina, with a typical stoichiometry of $Na_2O \cdot 6.6Al_2O_3$, there are $\sim 1\frac{2}{3}$ Na^+ ions for each conduction plane of the unit cell. The extra $\frac{2}{3} Na^+$ ions are compensated not by interstitial O^{2-} ions in the conduction planes, however, but by Al^{3+} vacancies in the spinel block.

The two solid solution mechanisms described above for β, β''-alumina are often, in practice, still more complicated since other additives, especially Li_2O and/or MgO, may be present. Both of these appear to enter the spinel blocks by substitution of Li^+, Mg^{2+} for Al^{3+}.

The two mechanisms outlined above for accommodating the extra Na^+ ions in β, β'' alumina both give rise to an improvement in local electroneutrality. In the 'ideal' structure of $NaAl_{11}O_{17}$, there is too much positive charge in the spinel blocks and too little in the conduction planes; the proposed structure does not satisfy Paulings' rule of local electroneutrality. In the soda-rich β-alumina, the introduction of extra O^{2-} ions into the conduction planes leads to the creation of new tetrahedral sites, two per extra O^{2-} ion. It appears that Al^{3+} ions from the spinel block are able to transfer to these new sites, thereby lessening the departures from local electroneutrality. In β''-alumina, the extra Na^+ ions in the conduction plane are compensated by either Al^{3+} vacancies in the spinel block or by partial replacement of Al^{3+} by Li^+, Mg^{2+} in the spinel block. Either way, departures from local electroneutrality are also reduced.

13.2.1.2 Conductivity and conduction mechanisms

β-alumina is a two-dimensional conductor. Alkali ions are able to move freely within the conduction planes but cannot penetrate the dense, spinel blocks. The conductivity of different single crystal β-aluminas in directions parallel to the conduction planes is shown in Fig. 13.14. Much of the basic scientific work has been carried out on β rather than β'' because, although the conductivity of β'' is greater than that of β by a factor of 2 to 3, large, good quality single crystals have until very recently, been available only for the β form. The conductivity is highest and the activation energy lowest for Na^+ and Ag^+ β-alumina. With increasing cation size (K^+, Tl^+), conduction becomes more difficult since the larger cations cannot move as readily within the conduction planes. Ag^+ and Na^+ appear to have the optimum size because Li^+ β-alumina (not shown) also has a higher activation energy and lower conductivity than Na^+ or Ag^+ β-alumina. In this

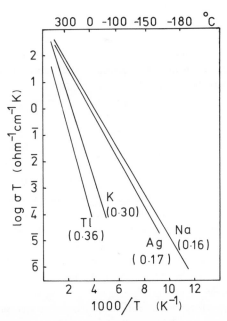

Fig. 13.14 Conduction of some single crystal β-aluminas. Activation energies, in electronvolts, are in parentheses. (From Whittingham and Huggins, 1972, p. 139)

case, Li^+ ions appear to occupy sites in the walls of the conduction plane; Li^+ is a small, highly polarizing cation and, unlike Na^+, is not happy in large sites of high coordination number.

The conductivity data for the β-aluminas fit the Arrhenius equation very well over large ranges of temperature (up to 1000 °C) and conductivity (up to seven orders of magnitude). Contrast this with the NaCl data (Fig. 13.5) which show changes of slope several times over a range of ~ 400 °C. This simple behaviour seen in β-alumina is characteristic of solid electrolytes in general and also of many complex oxides, silicates, etc., which are not particularly good ionic conductors. It is a remarkable fact that many materials with simple crystal structures, such as NaCl, exhibit complex conductivity behaviour whereas materials with complex structures and stoichiometries, such as β-alumina, exhibit simple conductivity behaviour. This simple behaviour is observed independently of the magnitude of σ because β-alumina is equally well behaved at 300 °C where $\sigma \sim 10^{-1}$ ohm^{-1} cm^{-1} or at -180 °C where $\sigma \sim 10^{-8}$ ohm^{-1} cm^{-1}.

Another important feature of the conductivity of solid electrolytes is that, unlike in, for example NaCl, the σ values are usually reproducible between different samples and laboratories and appear to be largely insensitive to the presence of small amounts of impurities. The conductivity of β-alumina has been measured in many laboratories and in the large majority the data agree very well with an activation energy of 0.16 ± 0.01 eV.

It is not feasible to apply defect equilibria considerations and the Law of Mass Action (Section 13.1) to, for example, β-alumina for several reasons. First, the equations that govern the concentration of Frenkel and Schottky defects apply only to very small defect concentrations (< 0.1 per cent of the lattice sites defective) whereas the evidence is that a considerable number, if not all, of the Na^+ ions in β-alumina are mobile. Second, the absence of doping effects in β-alumina indicates that the separation of the conductivity activation energy into components due to formation and migration of defects cannot be carried out readily and, indeed, it may be a mistake to try and effect such a separation. The number of mobile ions is so large that they must be regarded as part of the normal structure rather than as defects, and, as such, their number is insensitive to the presence of small amounts of impurity. Thus, whereas the addition of, say, 0.1 per cent $MnCl_2$ to NaCl may lead to an increase in concentration of cation vacancies by several orders of magnitude and, therefore, have a dramatic effect on σ, the corresponding effect of impurities on the conductivity of β-alumina is negligible. Small amounts of impurity may influence the performance of β-alumina ceramics used in the Na/S battery, but that is a different story.

The conduction pathways in β-alumina are shown in Figs 13.13 and 13.15. The br, abr and m sites form a hexagonal network and for long range migration to occur an Na^+ ion must pass through sites in the sequence–br–m–abr–m–br–m–. The activation energy for conduction, which is 0.16 eV, represents the overall value for migration of a Na^+ ion from, say, one br site to the next.

Measurements of the Haven ratio, consideration of the pathways for conduction in β-alumina and theoretical calculations have all indicated that a

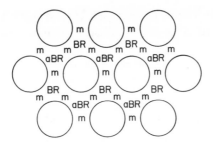

Fig. 13.15 Conduction pathway in β-alumina

knock-on or interstitialcy mechanism is operative. Consider first the idealized situation with β-alumina crystals of composition $NaAl_{11}O_{17}$. In this, the Na^+ ions occupy the *br* sites; *abr* and *m* sites are unoccupied. Next allow one Na^+ ion to leave its *br* site. It must pass through an adjacent *abr* site and, on the other side, it encounters an Na^+ ion on its *br* site. There is not room for the moving Na^+ to squeeze past the Na^+ on the *br* site. Neither is there room for it to stop short in the nearest *m* site because it would be too close to the Na^+ ion on the *br* site. The moving Na^+ ion has two choices therefore. It can return to its own *br* site, in which case no net conduction occurs. Alternatively, it can knock the 'blocking' Na^+ ion out of its *br* site. If the latter occurs, the ejected Na^+ ion may simply move into one of the two adjacent *m* sites, in which case, the incoming Na^+ ion remains in the third *m* site; the net result is that a split interstitial is created. Alternatively, the ejected Na^+ ion may escape completely from its *br* and adjacent *m* sites, in which case a chain reaction begins. While the exact mechanistic details are a matter of speculation it is clear that Na^+ ions cannot move independently of each other in β-alumina but that, instead, a cooperative process must occur.

An interesting effect that has been observed in β-aluminas containing a mixture of two alkali cations is the so-called *mixed alkali effect*. In this, the mobility of both alkali ions, e.g. Na^+ and K^+, is less than that in either pure end-member. Consequently, the conductivity passes through a minimum and the activation energy for conduction passes through a maximum value for intermediate compositions. The mixed alkali effect is a well known but poorly understood phenomenon in glasses (Chapter 18) and has not been previously encountered in crystalline materials. The occurrence of this rather unusual effect and the absence of any satisfactory explanation for it shows how difficult it is to understand the nature of the forces that control ion mobility, even in a crystalline material whose structure is apparently well understood.

Much of the work on β-alumina has been directed towards maximizing the conductivity of polycrystalline β-alumina ceramics. Since the conductivity of β'' is several times larger than that of β, compositions are desired that both stabilize the β'' phase (it is not very stable in the $Na_2O-Al_2O_3$ system without additives) and maximize its content relative to that of the β phase. It is found that this is best achieved by doping with small amounts of Li_2O and MgO. These appear to enter into solid solution with β''-alumina and stabilize its crystal structure.

13.2.2 AgI and Ag⁺ ion solid electrolytes

13.2.2.1 *AgI*

It has been known for many years that AgI undergoes a phase transition at 146 °C and that the high temperature form, α-AgI, has an exceptionally high conductivity, ~ 1 ohm^{-1} cm^{-1}, which is about four orders of magnitude larger than the room temperature value (Fig. 13.10). The activation energy for conduction in α-AgI is only 0.05 eV and the structure of α-AgI is so suited for easy motion of Ag⁺ that the ionic conductivity actually *decreases* slightly on melting at 555 °C.

β-AgI, stable below 146 °C, has the wurtzite structure with hexagonal close packed I⁻ ions and Ag⁺ in tetrahedral sites. Another low temperature polymorph is γ-AgI; its structure is that of sphalerite. The highly conducting, high temperature α polymorph is body centred cubic. In it, iodide ions lie at the corner and body centre positions and Ag⁺ ions appear to be distributed statistically over a total of thirty-six sites of tetrahedral and trigonal coordination. The tetrahedral sites are linked together by sharing faces and the trigonal sites lie at the centres of the faces of the AgI_4 tetrahedra. The iodide ions are essentially fixed and Ag⁺ ions can readily move from one site to the next in a liquid-like manner. The disordered Ag⁺ arrangements and the easy motion of Ag⁺ between sites must be related to the nature of the bonding between silver and iodine. Silver is a polarizing cation since its outer $4d$ electrons are relatively ineffective in shielding nuclear charge. Iodide is a large and polarizable anion and so covalent bonds readily form between Ag⁺ and I⁻ that are characterized by structures with low coordination numbers. During conduction, silver can readily move from one tetrahedral site to the next via an intermediate, three-coordinate site; covalent bonding at the intermediate site helps to stabilize it and reduce the activation energy for conduction. It is interesting that AgCl and AgBr have reasonable conductivities at high temperatures—but nothing like the values exhibited by AgI. It is expected that the bonding in these is less covalent than in AgI. However, they have a different crystal structure to that of α-AgI (they have the rock salt structure) and this is also undoubtedly an important factor.

13.2.2.2 *RbAg₄I₅*

In an attempt to stabilize the highly conducting α-AgI phase at lower temperatures, various anionic and cationic substitutions have been tried. The most successful so far has been the partial replacement of silver by rubidium in $RbAg_4I_5$. This material has the highest ionic conductivity at room temperature of any known crystalline substance, 0.25 ohm^{-1} cm^{-1}. Its activation energy for conduction is 0.07 eV (Fig. 13.16). The amount of electronic conductivity in $RbAg_4I_5$ is negligibly small, $\sim 10^{-9}$ ohm^{-1} cm^{-1} at 25 °C.

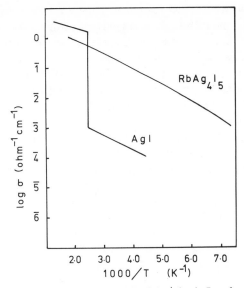

Fig. 13.16 Conductivity of Ag$^+$ in AgI and RbAg$_4$I$_5$

The phase diagram for the system RbI–AgI is shown in Fig. 13.17. Two binary compounds are present, Rb$_2$AgI$_3$ and RbAg$_4$I$_5$. The latter melts incongruently to AgI and liquid at $\sim 230\,°C$ and has a lower limit of stability at 27 °C. Decomposition of RbAg$_4$I$_5$ to AgI and Rb$_2$AgI$_3$ below 27 °C takes place only slowly although it is hastened by the presence of moisture or iodine vapour. With care, RbAg$_4$I$_5$ may be readily cooled to below room temperature without

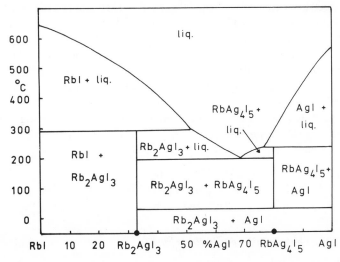

Fig. 13.17 Phase diagram for AgI–RbI. (From Takahashi, 1973, p. 84)

decomposing and its conductivity can then be measured over a wide temperature range. In order to prepare $RbAg_4I_5$, a 1:4 molar mixture of RbI and AgI is melted in vacuum at $\sim 500\,°C$ and then quenched to room temperature. During cooling, the liquid crystallizes, to give a fine grained and macroscopically homogeneous solid. On subsequent annealing at $\sim 165\,°C$ for $\sim 10\,hr$ this reacts further to give $RbAg_4I_5$. It is also possible to react mixtures of RbI and AgI directly at ~ 100 to $200\,°C$ without first melting them, but unless the reagents are finely powdered, well mixed and pressed into a compact, the reaction proceeds much more slowly.

The crystal structure of $RbAg_4I_5$ is rather different to that of α-AgI, but it also contains a random arrangement of silver atoms distributed over a network of face-sharing tetrahedral sites. Again there are many more available sites than there are silver atoms to fill them. Rubidium atoms are immobilized in sites that have a distorted octahedral environment of I^- ions.

Conductivity data for $RbAg_4I_5$ (Fig. 13.16) fall on a smooth curve that shows a small departure from linear Arrhenius behaviour.

13.2.2.3 *Other AgI derivatives*

A disordered, α-AgI-like structure can be stabilized at low temperatures by a variety of cations, notably large alkalis, NH_4^+, substituted NH_4^+ and certain organic cations. Some examples, all of which have conductivities at $25\,°C$ in the range 0.02 to 0.20 $ohm^{-1}\,cm^{-1}$ are $[(CH_3)_4N]_2Ag_{13}I_{15}$, $PyAg_5I_6$ where Py is the pyridinium ion $(C_5H_5NH)^+$, and $(NH_4)Ag_4I_5$.

A range of anions may be partially substituted for iodine and some of the phases formed have high conductivities, e.g. Ag_3SI, $Ag_7I_4PO_4$ and $Ag_6I_4WO_4$. The latter mixed iodide/oxysalt phases have considered thermal stability and are unaffected by iodine vapour or moisture. They are suitable materials for use as the solid electrolyte in electrochemical cells. An interesting new development is the preparation of glassy solid electrolytes in which molten mixtures of AgI with salts such as Ag_2SeO_4, Ag_3AsO_4 and $Ag_2Cr_2O_7$ can be preserved to room temperature, as glasses, by rapid quenching. The Ag^+ ion conductivities are again very high, e.g. 0.01 $ohm^{-1}\,cm^{-1}$, for the composition $Ag_7I_4AsO_4$. Obviously, it is not possible to know the structure of the iodide and oxysalt framework in non-crystalline materials such as these, but it seems reasonable to assume that silver can again move through a network of face-sharing tetrahedra.

It is also possible to partially replace both Ag^+ and I^- at the same time and prepare materials that have a high conductivity at room temperature; e.g.

(a) Replacement of Ag^+ by Hg^{2+} and I^- by Se^{2-}: the composition $Ag_{1.8}Hg_{0.45}Se_{0.70}I_{1.30}$ has a conductivity of 0.1 $ohm^{-1}\,cm^{-1}$ at $25\,°C$.
(b) Replacement of Ag^+ by Rb^+ and I^- by CN^- in $RbAg_4I_4CN$.

13.2.2.4 *Silver chalcogenide derivative phases*

All of the silver iodide derivative phases described above have high silver ion mobilities but very small electronic conductivities. They may therefore be used as solid electrolytes without any danger of short circuits through the electrolyte due to electronic conductivity. On the other hand, a very desirable if not essential attribute of solid *electrode* materials is that they should have high ionic *and* electronic conductivities. Silver chalcogenides, such as α-Ag_2S which is stable at high temperatures, exhibit mixed ionic/electronic conductivity and, as in the case of α-AgI, it is possible to stabilize the high temperature, disordered phase at lower temperatures by making various substitutions. For example, in the system Ag_2Se–Ag_3PO_4, α-Ag_2Se is able to dissolve 5 to 10 per cent Ag_3PO_4 in solid solution formation and the α phase is then stable down to room temperature, where it has an ionic conductivity of $0.13\,ohm^{-1}\,cm^{-1}$ and an electronic conductivity of between 10^4 and $10^5\,ohm^{-1}\,cm^{-1}$.

13.2.3 Halide ion conductors

Several halides that have the fluorite (CaF_2) structure may be classified as solid electrolytes at high temperatures because they have high halide ion conductivity. One of the best examples is PbF_2 in which $\sigma \simeq 5\,ohm^{-1}\,cm^{-1}$ above $\sim 500\,°C$. At room temperature, PbF_2 has a very low ionic conductivity and as such is a typical ionic solid. With rising temperature its conductivity increases smoothly and rapidly until at $\sim 500\,°C$ a limiting value of $\sim 5\,ohm^{-1}\,cm^{-1}$ is reached (Fig. 13.18). Above this temperature σ increases only slowly and there is little, if any, change in σ on melting at $822\,°C$. It is interesting that some materials such as PbF_2 arrive in the highly conducting condition gradually on increasing the temperature, whereas other materials such as AgI do so by undergoing an abrupt change in crystal structure (Fig. 13.16). Other materials that behave like PbF_2 are $SrCl_2$, which has a very high σ between $\sim 700\,°C$ and the melting point $873\,°C$,

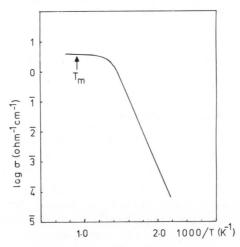

Fig. 13.18 Conductivity of PbF_2. (From Derrington and O'Keeffe, 1973)

and CaF_2, which arrives in the highly conducting condition just as the melting point, $1418\,^\circ C$, is approached.

The fluorite structure may be described in various ways, one of which is as primitive cubes of F^- ions with Ca^{2+} ions at the body centres of alternate cubes (Fig. 7.18). The sites available for interstitial F^- are then at the centres of the set of empty alternate cubes; these sites are normally coordinated by six calciums in an octahedral arrangement and eight fluorines at the cube corners. In creating an interstitial F^- ion, one of the corner F^- ions must move off its corner site and into the body of the cube. Defect complexes probably form (Chapter 9), but the details of the sites occupied are not known, especially in the more disordered structures at high temperatures.

13.2.4 Oxide ion conductors

The high temperature, cubic polymorph of zirconia has the fluorite crystal structure shown in Fig. 7.18 and may be stabilized to room temperature by formation of solid solutions with CaO, Y_2O_3, etc. Such 'stabilized zirconias' are good O^{2-} ion conductors at high temperatures, mainly because the mechanism of solid solution formation involves the creation of vacant O^{2-} sites in order to preserve electroneutrality, e.g. lime-stabilized zirconia has the formula $(Ca_xZr_{1-x})O_{2-x}$:$0.1 \lesssim x \lesssim 0.2$. With each Ca^{2+} ion that is introduced one O^{2-} ion vacancy is created (Chapter 10). The fluorite crystal structure appears to be particularly suitable for high ionic conductivity since the F^- ion conductors such as PbF_2 (Section 13.2.3) have this structure also. Typical conductivities in stabilized zirconias (e.g. 85 mol % ZrO_2, 15% CaO) are $5 \times 10^{-2}\,ohm^{-1}\,cm^{-1}$ at $1000\,^\circ C$ with an activation energy for conduction of $\sim 1.3\,eV$. At lower temperatures, of course, stabilized zirconias have conductivities that are many order of magnitude less than those of the Na^+ and Ag^+ ion solid electrolytes. Their usefulness stems from the fact that they are refractory materials, can be used up to very high temperatures (e.g. $1500\,^\circ C$) and have good oxide ion conductivity, which is an unusual property. Thoria (ThO_2) and hafnia (HfO_2) may be doped and are good O^{2-} ion conductors at high temperatures, similar to ZrO_2.

13.2.5 Search for new solid electrolytes

Since the mid 1960s, a very active field of research has been the investigation of known and new materials for solid electrolyte behaviour. Many of these searches have been carried out on a 'try it and see' basis because the crystal structures of the materials were not known. In others, it has been possible to select for study materials whose crystal structures contain open channels, layers, etc., in the hope that these channels may provide pathways for easy ionic transport. Although many theories have been proposed to account for solid electrolyte behaviour, it is still very difficult to make *a priori* reliable predictions about the value of σ in a particular material whose crystal structure is known. The best we can do at present is to have a set of guidelines which show us the likely structural characteristics that are a prerequisite for high ionic conductivity. These are:

(a) A large number of the ions of one species should be mobile (i.e. a large value of n in the equation $\sigma = ne\mu$).
(b) There should be a large number of empty sites available for the mobile ions to jump into. This is essentially a corollary of (a) because ions can be mobile only if there are empty sites available for them to occupy.
(c) The empty and occupied sites should have similar potential energies with a low activation energy barrier for jumping between neighbouring sites. It is no use having a large number of available interstitial sites if the moving ion cannot get into them.
(d) The structure should have a framework, preferably three dimensional, permeated by open channels through which mobile ions may migrate.
(e) The anion framework should be highly polarizable.

β-Alumina meets the first four of these conditions, as does stabilized zirconia. The good Ag^+ ion conductors meet all five conditions. Materials which are not good solid electrolytes may satisfy some of the conditions but not all. For example, many silicates have framework structures but the cations are trapped in relatively deep potential wells. Poorly conducting β- and γ-AgI meet condition (e) but, more important, do not meet condition (c).

In the search for new solid electrolytes, several framework structures have been found to give high cation mobility. One of the most interesting and potentially useful is $Na_3Zr_2PSi_2O_{12}$ discovered by Hong, Kafalas and Goodenough, which has a framework built of corner-sharing ZrO_6 octahedra and $(P, Si)O_4$ tetrahedra and a three-dimensional network of tunnels in which Na^+ ions reside. This material has been named NASICON (from sodium superionic conductor). The Na^+ ion conductivity is comparable to that in β-alumina.

The zeolites with their large cavities would appear to be obvious candidates for solid electrolyte behaviour but they appear to have only moderately high conductivities. The cations are usually present as hydrated complexes, which allows for ready ion exchange but not necessarily for high cation mobility. It appears that in, for example, dehydrated zeolites the channels are too large and the cations tend to stick at sites in the channel walls. A similar effect occurs in $Li^+\beta$-alumina. The Li^+ conductivity is considerably less than in Na^+ β-alumina and the Li^+ ions occupy sites in the walls of the conduction plane; at high pressures the conductivity of Li^+ β-alumina increases because the size of the channels in the conduction plane is reduced.

There is still a great need for new materials that have high conductivity of Li^+ ions. This is because cells that contain lithium anodes have a higher e.m.f. than, for example, corresponding cells containing sodium anodes. Li_2SO_4 undergoes a phase transition at $572\,^\circ C$ and has $\sigma \simeq 1\,ohm^{-1}\,cm^{-1}$ above this temperature; below this temperature σ is much smaller and Li_2SO_4 is of no interest as a low temperature solid electrolyte. Various substituted lithium sulphates have been studied but so far it has not been possible to reduce the temperature of the phase transition below about 400 °C.

The structures of Li_4SiO_4 and Li_4GeO_4 have moderately good Li^+ conductivity, $\sim 10^{-4}\,ohm^{-1}\,cm^{-1}$ at 300 to 400 °C, and are versitile host structures for doping. They are built of isolated SiO_4 and GeO_4 tetrahedra and the Li^+ ions are distributed through a network of face-sharing polyhedral sites. Substitutions such as

$$Li^+ + Si^{4+} \rightleftharpoons P^{5+} \qquad \text{in } Li_4SiO_4$$
$$Si^{4+} \rightleftharpoons Al^{3+} + Li^+ \qquad \text{in } Li_4SiO_4$$

and

$$2Li^+ \rightleftharpoons Zn^{2+} \qquad \text{in } Li_4GeO_4$$

have all led to an improvement in conductivity of several orders of magnitude (Fig. 13.19), especially at lower temperatures (~ 25 to 300 °C). The best conductor at medium temperatures that has been found so far is $Li_{14}ZnGe_4O_{16}$ for which $\sigma = 10^{-1}\,ohm^{-1}\,cm^{-1}$ at 300 °C. It has been named LISICON. The highest room temperature conductivity occurs in $Li_{3.5}V_{0.5}Ge_{0.5}O_4$ for which $\sigma \simeq 5 \times 10^{-5}\,ohm^{-1}\,cm^{-1}$.

An unusual effect has been found in mixtures of LiI and Al_2O_3. Although there is no evidence that LiI and Al_2O_3 react chemically, the conductivity of equimolar mixtures of LiI and Al_2O_3 is several orders of magnitude higher than that in pure LiI (pure Al_2O_3 is an insulator). The conductivity at 25 °C is $\sim 10^{-5}\,ohm^{-1}\,cm^{-1}$. The origin of this effect is not understood but is probably due to surface

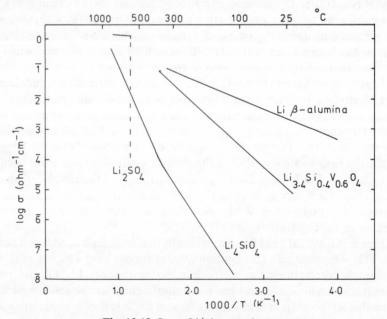

Fig. 13.19 Some Li^+ ion conductors

conductivity at the interface between LiI and Al_2O_3 grains. It may also involve the presence of moisture: LiI is an extremely hygroscopic material.

Various materials have been examined for protonic conductivity. Two of the best that have been found are H_3O^+ β-alumina and hydrogen uranyl phosphate, $HUO_2PO_4 \cdot 4H_2O$, which has $\sigma \sim 4 \times 10^{-2}$ ohm^{-1} cm^{-1} at 25 °C.

13.3 Conductivity measurements

13.3.1 D.C. methods

Measurement of accurate and meaningful conductivity values, especially in polycrystalline materials, is often quite difficult. Ideally, the d.c. conductivity should be measured, in order to be sure that the values pertain to long range ion migration and not to dielectric losses such as would be associated with limited or localized rattling of ions within cages. The difficulty in making d.c. measurements is in finding an electrode material that is compatible with the solid electrolyte and that does not give polarization effects at the electrode–solid electrolyte interface. For example, if gold metal electrodes are attached to a β-alumina crystal (Fig. 13.20) and a small voltage, e.g. 100 mV, is applied across the crystal, Na$^+$ ions migrate preferentially towards the cathode, but pile up, without being discharged, at the gold/β-alumina interface. A Na$^+$ ion-deficient layer forms at the β-alumina/gold anode interface. The cell therefore behaves as a capacitor; a large instantaneous current I_0 is observed when the cell is switched on, whose magnitude is related to the applied voltage and the resistance of the β-alumina crystals, but the current then falls exponentially with time (Fig. 13.20b).

Interfacial problems in d.c. measurements may be overcome by using *reversible electrodes*, i.e. electrodes that allow conduction by both electrons and the mobile ions of the solid electrolyte. Using reversible electrodes, there is no polarization problem because, say, Na$^+$ ions can move across the interface from electrode to solid electrolyte, and vice versa. Molten sodium is a suitable reversible electrode

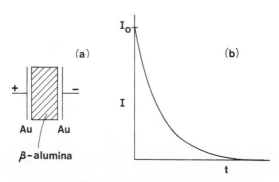

Fig. 13.20 Polarization at blocking electrodes in d.c. measurements

482

for β-alumina because the reaction

$$Na^+ + e \rightleftharpoons Na$$

occurs at the sodium/β-alumina interface.

There is much interest in *solid solution electrodes*, first suggested by Steele, which are solid reversible electrodes in which the electrode composition may vary by addition or removal of ions. Tungsten bronzes, e.g. Na_xWO_3, have good Na^+ mobility because the parent WO_3 structure has a framework built of corner-sharing WO_6 octahedra (Fig. 2.13); a three-dimensional network of interconnected channels permeates the structure along which Na^+ (or other alkali cations) may migrate. The oxidation state of tungsten may vary between $+V$ and $+VI$ and the chemical reactions involved in the operation of the tungsten bronze solid solution electrode may be written as

$$xNa^+ + WO_3 + xe^- \rightleftharpoons Na_xW_x^VW_{1-x}^{VI}O_3$$

Layered, transition metal dichalcogenide structures, such as TiS_2, are promising solid solution electrode materials because 'intercalation' of alkali metal ions may occur. Alkali ions enter the structure by pushing the TiS_2 layers apart to give solid solutions such as Li_xTiS_2 (Fig. 2.9), in which $0 \lesssim x \lesssim 1$. The reaction is reversible and the TiS_2 is recovered as Li^+ ions diffuse out.

13.3.2 A.C. methods

The alternative to d.c. conductivity measurements is to use a.c. methods and make measurements over a wide range of frequencies; d.c. conductivity values can usually be extracted from the a.c. data and in favourable cases it is possible to obtain information about electrode capacitance, grain boundary resistances and capacitances and the amount of electronic conductivity present.

A.C. measurements are often made with a Wheatstone bridge type of apparatus in which the resistance, R, and capacitance, C, of the sample are balanced against variable resistors and capacitors (Fig. 13.21). In an *admittance*

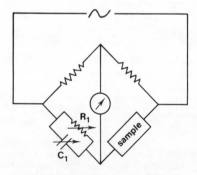

Fig. 13.21 Measurement of R and C
with a Wheatstone bridge

bridge, the variable R and C components are in parallel with each other, as shown; in an *impedance* bridge, R and C are in series. The central problem with a.c. measurements arises over the interpretation of the data. This is because the sample and electrode arrangement is electrically a 'black box' whose *equivalent circuit* (i.e. its representation by some combination of R and C elements) is often unknown. This means that, for any particular fixed frequency, the values of R and C that are obtained at the balance point of the bridge do not necessarily correspond to real R and C values of the sample or cell. For this reason, it is necessary to make measurements over a wide frequency range and find the range in which the measured R corresponds to the true bulk resistance of the crystals.

Many a.c. measurements are made with blocking gold electrodes. In these measurements no discharge or reaction occurs at the electrode–electrolyte interface, and the gold–solid electrolyte interface may be represented as a double-layer capacitance, C_{dl}, with a typical magnitude of 1×10^{-6}F (i.e. 1 μF) cm^{-2}. This double-layer capacitance is effectively in series with the sample resistance. In polycrystalline materials the overall sample resistance may be a combination of the intragranular resistance, or bulk crystal resistance, R_b, and the intergranular or grain boundary resistance, R_{gb} (Fig. 13.22). Grain boundary resistances have an associated capacitance, C_{gb}, in parallel with R_{gb} whose magnitude depends inversely on the thickness of the grain boundary layer. For a parallel plate capacitor the capacitance, C, is given by $C = \varepsilon' e_0 A d^{-1}$, where A is the area of the plates, d their separation, e_0 the permittivity of free space, 8.85×10^{-14}F cm^{-1}, and ε' the dielectric constant or permittivity of the material between the plates. A typical value of C_{gb} is 10^{-9}F (or 1 nF). It is more difficult to give a typical value for R_{gb}; usually the resistivity (i.e. resistance per unit cell constant) of the grain boundary is larger than that of the bulk crystal, but since the grain boundary layer may be several orders of magnitude thinner than the crystal dimensions, the actual grain boundary resistance, R_{gb}, may not necessarily be larger than R_b. Also, R_b and R_{gb} are usually very temperature dependent whereas capacitances change little with changing temperature. The bulk crystal resistance, R_b is in parallel with an associated bulk capacitance, C_b (Fig. 13.22); C_b corresponds to

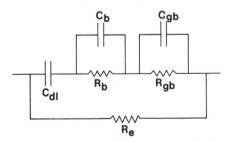

Fig. 13.22 Equivalent circuit for a polycrystalline solid electrolyte: R_{gb}, C_{gb}—grain boundaries; R_b, C_b—bulk crystals; R_e—electronic resistance; C_{dl}—electrode double-layer capacitance

the bulk or geometric capacitance of the sample or cell and is related to the dielectric constant, ε', of the solid electrolyte, i.e.

$$\varepsilon' = \frac{C_b}{C_o} \tag{13.15}$$

where C_o is the vacuum capacitance of the cell (the same electrode arrangement but with a vacuum between the electrodes).

Care must be taken over the use of ideas and phrases such as 'dielectric constant' when referring to materials which are conductors because dielectrics and conductors have completely opposed electrical properties. With solid electrolytes, the dielectric constant is the value which would occur in the absence of long range migration of ions; it can usually be observed experimentally in a.c. measurements if the frequency is sufficiently high that the polarity of the applied voltage switches before the ions have time to move significantly; ε' and C_b then correspond to the polarization of atoms and electrons, such as that which occurs in normal dielectrics. Typically, ε' is in the range 5 to 20. If the cell constant is unity, $\varepsilon' = C_b/e_o$, and therefore $C_b \simeq 10^{-12} F$ (or 1 pF).

If the solid electrolyte has an electronic conductivity as well as an ionic conductivity, this is represented by a separate resistance, R_e, in parallel with the rest of the equivalent circuit. The electronic resistance could, if small enough in magnitude, short-out the rest of the circuit, including the double-layer capacitance at the electrode–electrolyte interface; in most solid electrolytes, $R_e \gg (R_{gb} + R_b)$ and R_e can effectively be ignored.

During the balancing of an admittance bridge (Fig. 13.21) at a certain frequency, single values of R and C are obtained at the null point. Since in practice the representation of solid electrolytes and cells may require a complex equivalent circuit such as shown Fig. 13.22, the bridge readings must correspond to composite R and C values of the cell. In general these bridge R and C values change with frequency. The crux of the problem in analysing a.c. data is, therefore, to (a) determine the appropriate equivalent circuit for the cell and (b) evaluate the various R and C components in the equivalent circuit. While space does not permit a full treatment of a.c. methods, a brief outline of the applications to solid electrolytes can be given.

The current, I, passing through a resistor, R, under an applied field, E, is given by Ohm's Law,

$$I = \frac{E}{R} \tag{13.16}$$

and is independent of frequency.

A pure capacitor does not allow the passage of a d.c. current but does allow a frequency dependent a.c. current given by

$$I = j\omega CE \tag{13.17}$$

where ω is the angular frequency ($\omega = 2\pi f$) and $j = \sqrt{-1}$. These equations for R

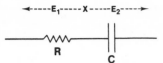

Fig. 13.23 Series combination of resistance and capacitance with respective voltage drops, E_1 and E_2, across them

and C may be written in the form

$$I = \frac{E}{Z} \tag{13.18}$$

where Z is the *impedance*. The impedance of the capacitance is imaginary since it contains the term $\sqrt{-1}$. This indicates that there is a difference in phase angle of $90°$ between the sinusoidal current and voltage; by convention, the current leads the e.m.f. by $90°$. In a circuit that contains a resistance and capacitance in series (Fig. 13.23), the total voltage drop across the circuit, E, is given by

$$E = E_1 + E_2$$

and, therefore, the total impedance of the circuit is given by

$$Z = R + \frac{1}{j\omega C} = R - \frac{j}{\omega C} \tag{13.19}$$

This impedance contains real and imaginary terms (i.e. R and $1/j\omega C$, respectively) and is therefore called the *complex impedance*, Z^*, where

$$Z^* = Z' - jZ''$$

and

$$Z' = R, \qquad Z'' = \frac{1}{\omega C} \tag{13.20}$$

For the circuit shown in Fig. 13.24, which has R and C in parallel, it is not possible to simply add the impedances as was done for the series circuit in Fig. 13.23. Instead, since resistances in parallel add as $1/R$ and capacitances in parallel add as $j\omega C$, we can write the reciprocal complex impedance for Fig. 13.24 as

$$A^* = \frac{1}{Z^*} = \frac{1}{R} + j\omega C \tag{13.21}$$

where A^* is the *complex admittance*. Again, the complex admittance separates into real and imaginary components, i.e.

$$A^* = A' + jA'' \tag{13.22}$$

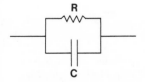

Fig. 13.24 Parallel combination of resistance and capacitance

where

$$A' = \frac{1}{R} \quad \text{and} \quad A'' = \omega C$$

The complex impedance of Fig. 13.24 may be evaluated by taking reciprocals; i.e.

$$Z^* = A^{*-1} = \left(\frac{1}{R} + j\omega C\right)^{-1}$$

$$= \frac{R}{1 + j\omega RC}$$

$$= \frac{R(1 - j\omega RC)}{(1 + j\omega RC)(1 - j\omega RC)}$$

$$= \frac{R}{1 + (\omega RC)^2} - R\frac{j\omega RC}{1 + (\omega RC)^2} \tag{13.23}$$

Therefore,

$$Z' = \frac{R}{1 + (\omega RC)^2} \quad \text{and} \quad Z'' = R\frac{\omega RC}{1 + (\omega RC)^2}$$

Equations can be written for the complex impedance and admittance of any circuit built of RC elements. However, the equations become rapidly more complex as the number of circuit elements increases. For the circuit in Fig. 13.25 which could represent a single crystal solid electrolyte between blocking electrodes, the impedance is given by

$$Z^* = \left(\frac{1}{R} + j\omega C_1\right)^{-1} + \frac{1}{j\omega C_2} \tag{13.24}$$

and the admittance by

$$A^* = \left[\left(\frac{1}{R} + j\omega C_1\right)^{-1} + \frac{1}{j\omega C_2}\right]^{-1} \tag{13.25}$$

For the circuit given in Fig. 13.22,

$$A^* = \frac{1}{R_e} + \left[\frac{1}{j\omega C_{dl}} + \left(\frac{1}{R_b} + j\omega C_b\right)^{-1} + \left(\frac{1}{R_{gb}} + j\omega C_{gb}\right)^{-1}\right]^{-1} \tag{13.26}$$

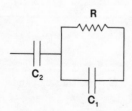

Fig. 13.25 Equivalent circuit for simple solid electrolyte without grain boundary resistance and with blocking electrodes

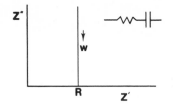

Fig. 13.26 Spike in a complex impedance plane plot obtained with a simple series RC element

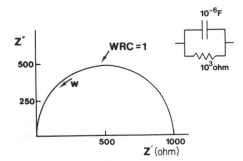

Fig. 13.27 Semi-circle in a complex impedance plane plot

When this equation is 'worked up' in order to separate out the real and imaginary components, it becomes very much longer. (Try it!)

Analysis of a.c. data is often carried out by *complex plane* methods which involve plotting the imaginary part of, say, Z^*, i.e. Z'', against the real part, Z'. When plotted on a linear scale, the data usually take the form of semi-circles and or spikes. For example, the series circuit shown in Fig. 13.23 gives a vertical spike in the Z^* complex plane because Z' is of fixed value, R, and Z'' decreases with increasing ω (Fig. 13.26). However, the equations for the parallel RC circuit, shown in Fig. 13.24, give rise to a semi-circle in the Z^* plane (Fig. 13.27). This is calculated for $R = 10^3$ ohm and $C = 10^{-6}$F by substituting into eq. 13.23. The semi-circle has intercepts on the Z' axis of zero and R; the maximum of the semi-circle equals $0.5\,R$ and occurs at sa frequency such that $\omega RC = 1$.

For more complex circuits, each parallel RC element gives rise to a semi-circle in the complex Z^* plane. For example, for Fig. 13.22, *two* semi-circles would be observed due to the parallel combinations R_b, C_b and R_{gb}, C_{gb}. For the circuit shown in Fig. 13.25, the elements R and C_1 give rise to a semi-circle and the series capacitance, C_2, to a spike, as shown in Fig. 13.28. The value of R may be determined from the intersection of either the spike or the semi-circle with the Z' axis. Bearing in mind that each point on the semi-circle and spike corresponds to a certain frequency, it should be clear why it is important to scan a range of frequency values. If a measurement is made at only one frequency, it is impossible to know if the data fall on a spike, in which case the measured R value may be the correct one, or on a semi-circle, in which case the measured R value (i.e. Z') is less than the correct one. Also, the equivalent circuit of a cell or solid electrolyte can

488

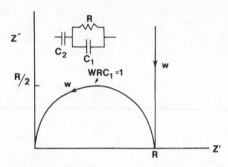

Fig. 13.28 Semi-circle and spike in a complex Z^* plot

be deduced only if a wide range of frequencies are studied. For example, the occurrence of a vertical spike at low frequencies in the complex Z^* plane indicates the presence of a large series capacitance in the circuit. This means that the electrodes are probably blocking and, therefore, that electronic conduction in the solid electrolyte is small or negligible compared to the magnitude of the ionic conductivity. Suppose that one's experimental results all fell on a semi-circle, as in Fig. 13.27; this could mean that either the equivalent circuit does not contain a large series capacitance or that such a capacitance is present but is observed in the Z^* plane, as in Fig. 13.28, only at lower frequencies.

Let us return for a moment to the actual experimental procedure involved in constructing a complex plane plot. Assuming that an impedance bridge is being used to make the measurements, the bridge readings obtained correspond to a series resistance, R_s, and a series capacitance, C_s. These measurements may be converted into impedances using the relations, $Z^* = R_s + 1/j\omega C_s$, $Z' = R_s$ and $Z'' = 1/\omega C_s$. A plot of Z'' against Z' is then made. No *a priori* assumptions need to be made regarding the equivalent circuit of the cell nor the equations for the impedance of the cell. The method works because, at each frequency, the null point is found by adjusting the variable R and C components: at balance, $Z'_{cell} = R_s$ and $Z''_{cell} = 1/\omega C_s$. In making a complex plane diagram, we are simply plotting one variable against another with the objective of finding an equation which fits the experimental data.

The above examples are all for complex impedance planes. There are a total of four basic formalisms that may be used to represent and analyse a.c. data. They are as follows:

$$\text{Complex impedance: } Z^* = R_s - \frac{j}{\omega C_s} \quad (s = series) \tag{13.27}$$

$$\text{Complex admittance: } A^* \text{ (or } Y^*) = (Z^*)^{-1}$$

$$= (R_p)^{-1} + j\omega C_p \quad (p = parallel) \tag{13.28}$$

$$\text{Complex permittivity: } \varepsilon^* = \frac{A^*}{j\omega C_0}$$

$$= \varepsilon' - j\varepsilon'' \tag{13.29}$$

Complex electric modulus: $M^* = (\varepsilon^*)^{-1}$

$$= j\omega C_0 Z^*$$

$$= M' + jM'' \qquad (13.30)$$

For any RC circuit, the different formalisms correspond to different ways of writing the equations for the circuit and basically all contain the same information. However, different formalisms highlight different features of a circuit and, for more complex circuits, it may be worth while to plot the data in more than one formalism in order to extract all possible information from the results. For example, the complex impedance formalism gives prominence to the most resistive elements in an equivalent circuit. Thus in a polycrystalline material that has relatively large grain boundary resistances and small crystal resistances, these grain boundary resistances dominate the a.c. response in the complex impedance plane and may even mask completely the effect of the crystal resistances. In the complex electric modulus formalism, however, prominence is given to those elements that have the smallest capacitance. In this case, the response of the crystals dominates the results and grain boundary effects may be effectively masked. By a comparison of the results analysed in these two different formalisms, it may therefore be possible to effectively separate grain boundary and bulk effects.

13.4 Other experimental techniques

Many other experimental techniques have been applied to solid electrolytes with varying degrees of success. Far infrared, Raman and microwave measurements have been used to cover the high frequency end of the a.c. conductivity spectrum (i.e. 10^9 to 10^{13} Hz). Broad peaks have been observed which are related to the vibrations of the mobile ions and this has given some information about mechanisms of conduction. For example, in mixed Na^+/K^+ β-alumina, a Raman peak at $69\,cm^{-1}$ has been interpreted as being due to the presence of K^+ split interstitial pairs; this has been proposed as further evidence for the interstitialcy mechanism of conduction (Section 13.2.1). It is often the case, however, that interpretation of spectroscopic data such as these is rather difficult and ambiguous. The use of thermodynamic measurements to study phase transitions in solid electrolytes is mentioned in Section 12.8.3; from entropy data the degree of disorder in the solid electrolytes may be determined. NMR has been used to study mechanisms of ion migration, examples being given in Section 3.2.3.3.

A useful technique that gives data complementary to conductivity data is the measurement of diffusion coefficients. In the tracer diffusion method, a thin film of a substance containing, say, radioactive Na^{+*} ions is painted on the surface of the material under study, say a crystal of sodium β-alumina. The crystal is then placed inside a furnace and the radioactive Na^{+*} ions begin to penetrate the crystal as a result of random diffusion processes. After a certain time, the crystal is removed from the furnace and cut into thin sections, parallel to the original crystal face. The radioactivity of each section is measured and a profile of the radioactive Na^{+*} ion concentration as a function of distance through the crystal

490

is obtained. From this, the diffusion coefficient D of Na^+ may be calculated and compared with the value of the conductivity, using the Nernst–Einstein relation, eq. 13.12. The method therefore provides an independent way of estimating conductivity. Alternatively, if both D and σ can be measured accurately, information about the conduction mechanism may be obtained from the value of the Haven ratio. The disadvantages of this method are that it is destructive and time-consuming. A new crystal is needed for each temperature that is studied and this may lead to problems of reproducibility of results.

An alternative method which is non-destructive is to immerse the crystal under study in a solution or melt containing, say, radioactive Na^{+*} ions. These radioactive ions diffuse into the crystal and ion-exchange with some of the non-radioactive ones which diffuse out. After a certain interval of time, the crystal is removed, washed and its total radioactivity counted. The crystal is then returned to the radioactive liquid and the process is repeated several times. Diffusion and conductivity data can be calculated from the results, as before.

13.5 Applications of solid electrolytes

13.5.1 Electrochemical cells—principles

Cells can be constructed that contain solid electrolytes instead of liquid electrolytes. A wide range of scientific and technological applications of such cells have been suggested, many of which are not possible with liquid electrolyte-containing cells; this is one reason for the interest in solid electrolytes. Consider the schematic cell shown in Fig. 13.29 which contains a solid electrolyte membrane separating two electrode compartments. The latter may contain solids, liquids or gases. There may be similar or dissimilar materials in both, e.g. oxygen gas at two different pressures or the two components of a cell, e.g. sodium and sulphur.

The e.m.f. of a cell reaction is given by the *Nernst equation*:

$$E = E_0 + \frac{RT}{nF} \log_e \frac{[O_x]}{[Red]} \tag{13.31}$$

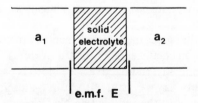

Fig. 13.29 Electrochemical cell containing solid electrolyte

Two such equations are usually needed, one for the reaction occurring at each electrode. Suppose that at the anode the following oxidation reaction occurs:

$$M \rightarrow M^+ + e$$

Then

$$E_1 = E_{0_{M/M^+}} + \frac{RT}{F} \log_e \frac{[M^+]}{[M]} \tag{13.32}$$

in which $E_{0_{M/M^+}}$ is the standard redox potential for that reaction. $[M^+]$ and $[M]$ are the concentrations of the two species and F is the Faraday unit, 96500 Coulombs. The M^+ ions that are generated at the anode diffuse through the solid electrolyte and react at the cathode with X^-, produced according to

$$X + e \rightarrow X^-$$

and for which

$$E_2 = E_{0_{X^-/X}} + \frac{RT}{F} \log_e \frac{[X]}{[X^-]}$$

Summation of E_1 and E_2 gives the overall e.m.f. for the reaction: $M + X \rightarrow MX$, and this is related to the free energy of formation of MX by

$$\Delta G = -nEF \tag{13.33}$$

Electrochemical cells embodying solid electrolytes may be used to obtain thermodynamic data about materials, in particular free energies of formation of compounds, e.g. the cell $Ag(s)/AgI(s)/Ag_2S(s)$, $S(l)$, $C(s)$ has been used to measure the free energy of formation of Ag_2S. The cell reaction is $2Ag + S \rightarrow Ag_2S$ and since the reactants, silver and sulphur, are in their standard states

$$\Delta G^0(Ag_2S) = -2EF \tag{13.34}$$

If the temperature dependence of the e.m.f. can be determined, one can go further and calculate entropies and enthalpies of reaction.

13.5.2 Batteries

Much of the impetus for research on solid electrolytes has come from their possible use in new types of battery. The *sodium–sulphur cell* utilizes Na^+ β-alumina solid electrolyte and is probably the most important one (Fig. 13.30). It is a high density secondary battery, i.e. it has a high energy/power to mass ratio and is undergoing extensive development and testing in several countries for use in, for example, electric cars and for power station load levelling. Basically, it consists of a molten sodium anode and a molten sulphur cathode separated by β-alumina solid electrolyte. Usually, the β-alumina is fabricated in the form of a tube closed at one end with the sodium inside and the sulphur outside (or vice versa). Since molten sulphur is a covalently bonded solid it is a non-conductor of

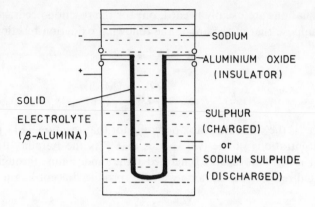

CELL REACTION: $2Na + 5S \rightleftharpoons Na_2S_5$ 2·08 V

Fig. 13.30 The sodium–sulphur cell

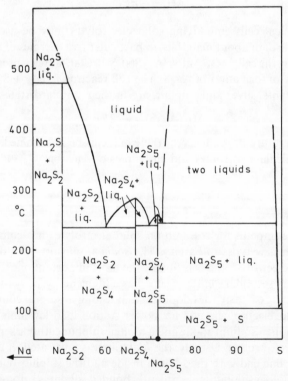

Fig. 13.31 Phase diagram for the sodium–sulphur system

electricity and the cathode material that is used consists of graphite felt impregnated with sulphur. The outer casing of the cell is made of stainless steel and serves as the current collector. The cell discharge reaction is:

$$2Na + xS \rightarrow Na_2S_x$$

where x depends on the level of charge in the cell. In the early stages of discharge, x is usually given as five, which corresponds to the formula of the most sulphur-rich sodium sulphide, Na_2S_5, but x decreases as discharge continues. The phase diagram for the sodium–sulphur system is shown in Fig. 13.31. The sodium–sulphur cell operates at 300 to 350 °C, which is the lowest temperature at which the discharge products are molten for a large range of compositions. From the phase diagram it can be seen that when discharge reaches the stage at which $x \lesssim 3$ (i.e. 60% S, 40% Na), the liquidus curve rises rapidly and crystalline Na_2S_2 begins to form; further discharge would lead to a gradual freezing of the cathode material. The open circuit voltage of the cell depends on both the level of charge and temperature. The maximum open circuit voltage for a fully charged cell is 2.08 V at 300 °C. This decreases on discharge to ~ 1.8 V for $x = 3$. Thermodynamic calculations give a theoretical energy storage in the cell of 750 Wh kg^{-1}. Experimental values of the current density are usually 100 to 200 Wh kg^{-1} and are unlikely to be improved upon because most batteries give only a small percentage of their theoretical maximum output.

Other types of cells that are finding applications are miniature primary cells which operate at room temperature and which have a long life rather than a high power output. They are used in electronic watches, heart pacemakers and in military applications. Various cells have been used satisfactorily, e.g. Ag/RbAg$_4$I$_5$/I$_2$ (0.65 V) and Li/LiI/I$_2$ (2.8 V). In both of these, iodine alone cannot be used as the cathode because it does not have sufficient electronic conductivity to sustain a discharge current; instead complexed iodides are used, e.g. $(CH_3)_4NI_5$ containing polyiodide anions and iodine poly-2 vinyl pyridine charge transfer complex, respectively. The Li/I$_2$ cell is widely used for pacemaker applications. When it is operated at 37 °C and gives current densities of 1 to 10 μA cm^{-2}, it is estimated that energy densities as high as 0.8 Wh cm^{-3} are possible and that the batteries should operate for at least ten years.

13.5.3 Oxygen concentration cells and sensors

Electrochemical cells containing solid electrolytes may be used for the measurement of partial pressures of gases or the concentrations of gases dissolved in liquids. An oxygen concentration cell that utilizes a stabilized zirconia solid electrolyte in the form of an open ended tube is shown schematically in Fig. 13.32. Inside the tube is a reference gas such as air. The tube is coated with porous metal electrodes to allow absorption and liberation of oxygen gas. If the partial pressure of oxygen that is to be measured, P'_{O_2}, is less than the reference pressure, P''_{O_2}, the electrode reactions shown in (b) take place and oxide ions migrate through the solid electrolyte from right to left. The Nernst

494

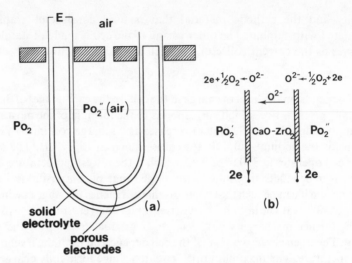

Fig. 13.32 Oxygen concentration cell with stabilized zirconia solid electrolyte

equations for the reactions at each electrode may be combined into a single equation that relates the difference in oxygen partial pressure to the cell voltage, i.e.

$$E = \frac{RT}{4F} \log_e \left(\frac{P''_{O_2}}{P'_{O_2}} \right) \qquad (13.35)$$

The cell operates at temperatures between ~ 500 and $1000\,^\circ$C (in order that transport of O^{2-} ions through the electrolyte occurs sufficiently rapidly) and can be used to measure oxygen partial pressures as low as 10^{-16} atm. At lower pressures, the zirconia becomes an electronic conductor and the cell short-circuits. Stabilized thoria ThO_2, solid electrolyte is ionically conducting over a wider range of oxygen partial pressures and may be used for $P_{O_2} < 10^{-16}$ atm. Oxygen concentration cells such as the zirconia probe have various uses, e.g. the analysis of exhaust gases and pollution, measurement of the consumption of oxygen gas during respiration and study of equilibria such as CO/CO_2, H_2/H_2O and metal/metal oxide. They may also be used to probe the oxygen activity in molten metals, e.g. steel, by dipping the probe into the melt. Oxygen concentrations are read from a calibrated meter and the response of the probe is usually very rapid.

13.5.4 Fuel cells

Stabilized zirconia is used in either tube or disc form as the solid electrolyte or separator in some types of high temperature fuel cells. The two electrode compartments contain (a) air or O_2 and (b) a fuel gas, e.g. H_2, CO. The zirconia is again coated with porous metal electrodes and the cell reaction is, for example, $H_2 + \frac{1}{2}O_2 \rightarrow H_2O$ or $CO + \frac{1}{2}O_2 \rightarrow CO_2$. The advantages of this type of fuel cell are that electrode polarization problems are minimal and high current densities, e.g. $0.5\,A\,cm^{-2}$ and $0.5\,W\,cm^{-2}$, can be achieved. However, constructional difficulties sometimes arise and the cell can be heated up only slowly to the working temperature.

The fuel cell can also be used as an electrolyser and as a means of storing energy by reversing the direction of current flow. Thus, steam can be decomposed and the products, H_2 and O_2, stored. It could also be used to regenerate oxygen from CO_2 in, for example, spaceships and to deoxidize liquid metals.

Questions

13.1 What effect if any, do you expect small amounts of the following impurities to have on the conductivity of NaCl crystals: (a) KCl; (b) NaBr; (c) $CaCl_2$; (d) AgCl; (e) Na_2O?

13.2 What effect, if any, do you expect small amounts of the following impurities to have on the conductivity of AgCl crystals: (a) AgBr; (b) $ZnCl_2$; (c) Ag_2O?

13.3 Using the conductivity data for NaCl given in Fig. 13.5, estimate (a) the enthalpy of cation vacancy migration, (b) the enthalpy of Schottky defect formation. Your results should be similar to those given in Table 13.2.

13.4 A particular solid has a conductivity of $10^{-5}\,ohm^{-1}\,cm^{-1}$ at room temperature. Suggest ways for determining the charge carrier responsible for conduction. How could you distinguish conduction by (a) electrons, (b) Na^+ ions, (c) O^{2-} ions?

13.5 Using the conductivity data for different ion exchanged β aluminas given in Fig. 13.14, calculate the pre-exponential factor A and the activation energy E for each.

13.6 Derive equations for (a) the complex impedance, (b) the complex admittance for the equivalent circuit shown in Fig. 13.25. Sketch the complex plane plots using the following values for the parameters: $R = 10^6\,ohm$, $C_1 = 10^{-12}F$, $C_2 = 10^{-6}F$. What would be the most useful frequency range to use to estimate the values of R, C_1 and C_2?

13.7 With the aid of Fig. 13.31, explain how the Na/S cell would differ in its mode of operation or usefulness if operated at (a) $200\,°C$, (b) $300\,°C$.

13.8 The Na/S cell is being developed for possible applications in electrically powered buses and cars. For such applications, what advantages and disadvantages do you think the Na/S cell would have in comparison with conventional Pb/acid batteries?

13.9 Suggest a design for a device containing a solid electrolyte that would be sensitive to fluorine gas.

13.10 Compare the relative merits of a.c. and d.c. methods for studying the ionic conductivity of polycrystalline ceramics.

496

References

L. W. Barr and A. B. Lidiard (1970). In *Physical Chemistry, An Advanced Treatise*, Vol. 10, *Solid State*, Academic Press.

C. E. Derrington and M. O'Keeffe (1973). Anion conductivity and disorder in lead fluoride, *Nature Phys. Sci.*, **246**, 44–45, Nov. 19.

S. Geller (Ed.) (1977). Solid electrolytes, in *Topics in Applied Physics*, Vol. 21, Springer Verlag.

J. B. Goodenough, H. Y-P. Hong and J. A. Kafalas (1976). Fast Na^+ ion Transport in Skeleton Structures, *Mat. Res. Bull.*, **11**, 203.

S. Hackwood and R. G. Linford (1981). Physical Techniques for the Study of Solid Electrolytes. *Chem. Revs.*, **81**, 327.

W. Hayes (1978). Superionic conductors, *Contemp. Phys.*, **19**(5), 469–486.

G. Holzäpfel and H. Rickert (1977). Solid state electrochemistry—new possibilities for research and industry, *Die Naturwiss.*, **64**(2), 53–58.

A. Hooper (1978). Fast ionic conductors, *Contemp. Phys.*, **19**, 147–168.

M. O'Keeffe and B. G. Hyde (1976). The solid electrolyte transition and melting in salts, *Phil. Mag.*, **B3**, 219–224.

H. Rickert (1978). Solid ionic conductors—principles and applications, *Angew. Chem. Int. Ed.*, **17**, 37–46.

B. C. H. Steele (1972). Electrical Conductivity in Ionic Solids, in *MTP International Reviews of Science. Solid State Chemistry* (Ed. L. E. J. Roberts), p. 117, Wiley.

T. Takahashi (1973). Solid silver ion conductors, *J. Appl. Electrochem.*, **3**, 79–90.

M. S. Whittingham and R. A. Huggins (1972). *Solid State Chemistry*, NBS special publication 364.

Chapter 14

Electronic Properties and Band Theory: Metals, Semiconductors, Inorganic Solids, Colour

14.1 Introduction—metals, insulators and semiconductors

Metals have long been known to man for their ability to conduct electricity. Following on from the discovery of semiconductors and transistor action in 1948 by Bardeen, Shockley and Brattain, there has been a tremendous upsurge of interest in electronic properties of materials, typified by, for example, applications in silicon chip devices. The purpose of this chapter is to give a non-mathematical treatment of electronic properties in terms of the band theory of solids. Most emphasis is given to metals and semiconductors but other inorganic solids are also discussed.

At a first glance, the main difference between metals, semiconductors and insulators is in the magnitude of their conductivities. Metals conduct electricity very easily, $\sigma \sim 10^4$ to $10^6 \, \text{ohm}^{-1} \, \text{cm}^{-1}$, insulators very poorly, or not at all, $\sigma \lesssim 10^{-15} \, \text{ohm}^{-1} \, \text{cm}^{-1}$, and semiconductors lie in between, $\sigma \sim 10^{-5}$ to $10^3 \, \text{ohm}^{-1} \, \text{cm}^{-1}$ (Table 13.1). The boundaries between the three sets of values are somewhat arbitrary and a certain amount of overlap occurs.

There is, however, a fundamental difference between the mechanism of conduction in metals, on the one hand, and semiconductors/insulators, on the other. Very simply, the conductivity of most semiconductors/insulators increases rapidly with increasing temperature whereas that of metals shows a slight but gradual decrease.

The conductivity, σ, is given by the equation

$$\sigma = n e \mu \qquad (13.1)$$

where n is the number, e the charge and μ the mobility of the charge carriers. The temperature dependence of σ in different materials can be understood by considering the temperature dependence of the terms n, e and μ. For all electronic conductors, e is constant and independent of temperature. The mobility term is similar in most materials in that usually it decreases slightly with increasing temperature due to collisions between the moving electrons and phonons, i.e. lattice vibrations. The main source of the different behaviour in metals, semiconductors and insulators therefore lies in the value of n and its temperature dependence:

(a) For metals, n is large and essentially unchanged with temperature. The only variable in σ is μ and since μ decreases slightly with temperature, σ also decreases.

(b) For semiconductors and insulators, n usually increases exponentially with temperature. The effect of this dramatic increase in n more than outweighs the effect of the small decrease in μ. Hence, σ increases rapidly with temperature. Insulators are extreme examples of semiconductors in that n is very small at normal temperatures. Thus some insulators become semiconducting at high temperatures where n becomes appreciable and, conversely, some semiconductors become more like insulators at low temperatures.

Semiconductors may be divided into two broad types:

(a) *Elemental semiconductors.* The elements silicon and germanium are the classic semiconductors. These elements are in Group IV of the Periodic Table. With increasing atomic weight in Group IV, the elements change from being insulators (diamond) to semiconductors (Si, Ge, grey Sn) to metals (white Sn, Pb). Apart from the two metals, all have the diamond structure in which each atom is tetrahedrally surrounded by four others. The tetrahedra link up by sharing corners to form a rigid three-dimensional framework structure that has cubic symmetry (Chapter 7). The diamond structure appears to be especially favourable to semiconductivity.

(b) *Compound semiconductors.* Many inorganic and some organic compounds are semiconductors. The most well-known inorganic compound semiconductors are the so-called III–V compounds. These compounds are 1:1 combinations of Group III and Group V elements, some of which are isoelectronic with an intermediate Group IV element. For example, GaAs and InSb are isoelectronic with germanium and tin, respectively. Other nonisoelectronic combinations such as GaP are also semiconductors. Most III–V compounds have the zinc blende structure (Section 7.2.1), which is closely related to that of diamond.

A wide range of other compounds—oxides, sulphides, etc.—with a variety of crystal structures are also semiconductors. Some of these are discussed later.

14.2 Electronic structure of solids—band theory

The electronic structures of metals, semiconductors and many other solids may be described in terms of band or zone theory. In a metal such as aluminium, the inner core electrons—1s, 2s and 2p—are localized in discrete atomic orbitals on the individual aluminium atoms. However, the 3s and 3p electrons that form the valence shell occupy energy levels that are delocalized over the whole of the metal crystal. These levels are like giant molecular orbitals, each of which can contain only two electrons. In practice, in a solid material, there must be an enormous number of such levels and they are separated from each other by very small energy differences. Thus in a crystal of aluminium that contains N atoms, each atom contributes one 3s orbital and the result is a band that contains N closely spaced energy levels. This band is called the 3s valence band. The 3p levels in aluminium are similarly present as a delocalized 3p band of energy levels.

The band structure of other materials may be regarded in a similar way. The differences between metals, semiconductors and insulators depend on:

(a) the band structure of each,
(b) whether the valence bands are full or only partly full,
(c) the magnitude of any energy gap between full and empty bands.

The band theory of solids is well supported by X-ray spectroscopic data and by two independent theoretical approaches. The 'chemical approach' to band theory is to take molecular orbital theory, as it is usually applied to small, finite-sized molecules and to extend the treatment to infinite, three-dimensional structures. In the molecular orbital theory of diatomic molecules, an atomic orbital from atom 1 overlaps with an atomic orbital on atom 2, resulting in the formation of two molecular orbitals that are delocalized over both atoms. One of the molecular orbitals is 'bonding' and has lower energy than that of the atomic orbitals. The other is 'antibonding' and is of higher energy (Fig. 14.1).

Extension of this approach to larger molecules leads to an increase in the number of molecular orbitals. For each atomic orbital that is put into the system, one molecular orbital is created. As the number of molecular orbitals increases, the average energy gap between adjacent molecular orbitals must decrease (Fig. 14.2). The gap between bonding and antibonding orbitals also decreases until the situation is reached in which there is essentially a continuum of energy levels.

Metals may be regarded as infinitely large 'molecules' in which an enormous

500

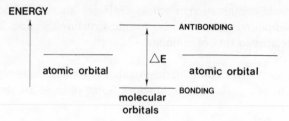

Fig. 14.1 Molecular orbitals in a diatomic molecule

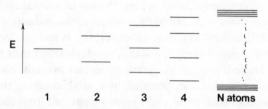

Fig. 14.2 Splitting of energy levels on molecular
orbital theory

number of energy levels or 'molecular' orbitals is present, $\sim 6 \times 10^{23}$ for one mole of metal. It is, however, no longer appropriate to refer to each level as a 'molecular' orbital since each is delocalized over all the atoms in the metal crystal. Instead, they are usually referred to as energy levels or energy states.

The band structure of metallic sodium, calculated using the 'tight binding approximation', is shown in Fig. 14.3. It can be seen that the width of a particular

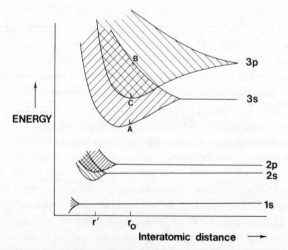

Fig. 14.3 Effect of interatomic spacing on atomic energy levels and bands for sodium, calculated using tight
binding theory

energy band is related to the interatomic separation and hence to the degree of overlap between orbitals on adjacent atoms. Thus, at the experimentally determined value for the interatomic separation, r_0, the 3s and 3p orbitals on adjacent atoms are calculated to overlap significantly to form broad 3s and 3p bands (shaded). The upper levels of the 3s band have similar energies to the lower levels of the 3p band. Hence, there is no discontinuity in energy between 3s and 3p bands. Overlap of bands is important in explaining the metallic properties of certain other elements such as the alkaline earths (see later).

At the interatomic distance, r_0, the 1s, 2s and 2p orbitals on adjacent sodium atoms do not overlap. Instead of forming bands they therefore remain as discrete atomic orbitals associated with the individual atoms. They are represented in Fig. 14.3 as thin lines. If it were possible to reduce the internuclear separation from r_0 to r' by compression of sodium under pressure, then the 2s and 2p orbitals would also overlap to form bands of energy levels (shaded). The 1s levels would, however, still be present as discrete levels at distance r'. It has been suggested that similar effects would occur in other elements subjected to high pressure. For instance, it has been calculated that hydrogen would become metallic at a pressure $\gtrsim 10^6$ atm.

Sodium has the electronic configuration $1s^2 \, 2s^2 \, 2p^6 \, 3s^1$. It therefore has one valence electron per atom. Since the 3s, 3p bands overlap (Fig. 14.3), the valence electrons are not confined to the 3s band but are distributed over the lower levels of both the 3s and 3p bands. This accounts for the occurrence of a $K\beta$ line in the X-ray emission spectrum of sodium metal. The $K\beta$ line corresponds to the transition $3p \rightarrow 1s$.

The 'physical approach' to band theory is to consider the energy and wavelength of electrons in a solid. In the early *free electron theory* of Sommerfeld, a metal is regarded as a potential well, inside which the more loosely held valence electrons are free to move. The energy levels that the electrons may occupy are quantized (analogy with the quantum mechanical problem of a particle in a box) and the levels are filled from the bottom of the well with two electrons per level. The highest filled level at absolute zero is known as the Fermi level. The corresponding energy is the Fermi energy, E_F (Fig. 14.4). The work function, ϕ, is the energy required to remove the uppermost valence electrons from the potential well. It is analogous to the ionization potential of an isolated atom.

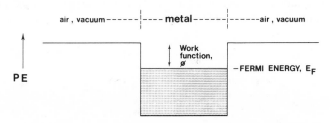

Fig. 14.4 Free electron theory of a metal; electrons in a potential well

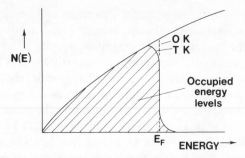

Fig. 14.5 Density of states plot on the free electron theory

A useful diagram is the density of states, $N(E)$, diagram which is a plot of the number of energy levels, $N(E)$, as a function of energy, E (Fig. 14.5). The number of available energy levels increases steadily with increasing energy in the Sommerfeld theory. Although the energy levels are quantized, there are so many of them and the energy difference between adjacent levels is so small that, effectively, a continuum occurs. At temperatures above absolute zero, some electrons in levels near to E_F have sufficient thermal energy to be promoted to empty levels above E_F. Hence at real temperatures, some states above E_F are occupied and others below E_F must be vacant. The average occupancy of the energy levels at some temperature T above zero is shown as shading in Fig. 14.5.

The high electrical conductivity of metals is due to the drift of those electrons that are in half-occupied states close to E_F. Electrons that are in doubly occupied states lower down in the valence band cannot undergo any net migration in a particular direction whereas those in singly occupied levels are free to move. The promotion of an electron from a full level below E_F to an empty one above E_F therefore gives rise, effectively, to two mobile electrons.

The free electron theory is a big oversimplification of the electronic structure of metals but it is a very useful starting model. In more refined theories, the potential inside the crystal or well is regarded as periodic (Fig. 14.6) and not constant as in Sommerfeld's theory. The positively charged atomic nuclei are arranged in a regularly repeating manner. The potential energy of the electrons passes through a minimum at the positions of the nuclei, due to coulombic attraction, and passes through a maximum midway between adjacent nuclei. Solution of the Schrödinger equation for a periodic potential function such as shown in Fig. 14.6 was carried out by Bloch using Fourier analysis. An important conclusion was

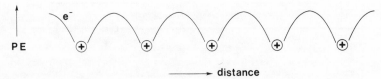

Fig. 14.6 Potential energy of electrons as a function of distance through a solid

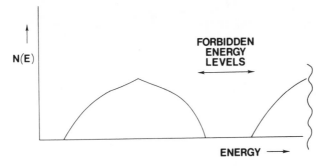

Fig. 14.7 Density of states on band theory

that an uninterrupted continuum of energy levels does not occur but that, instead, only certain bands or ranges of energies are permitted for the electrons. The forbidden energies correspond to electron wavelengths that satisfy Bragg's Law for diffraction in a particular direction in the crystal. This effect is discussed further in Section 14.3. Consequently, the density of states diagrams given by Bloch's results show discontinuities, as in Fig. 14.7.

Similar conclusions about the existence of energy bands in solids are obtained from both the molecular orbital and periodic potential approaches. From either theory, one obtains a model with bands of energy levels for the valence electrons. In some materials, overlap of different bands occurs. In other materials, a forbidden gap exists between energy bands.

Experimental evidence for the band structure of solids is obtained from spectroscopic studies. Electronic transitions between different energy levels may be observed using various spectroscopic techniques (Chapter 3). For solids, X-ray emission and absorption are the most useful techniques for gaining information about both inner core and outer valence electrons. A certain amount of information about outer valence electrons also comes from visible and UV spectroscopy.

X-ray emission spectra of solids usually contain peaks or bands of various width. Transitions between inner or core levels appear as sharp peaks, e.g. the $2p \rightarrow 1s$ transition in copper, Fig. 5.2. This indicates that these levels in copper metal are discrete atomic orbitals. Transitions involving valence shell electrons give broad spectral peaks, however, especially for metals. This indicates that valence electrons have a broad distribution of energies and are therefore located in energy bands.

The L emission spectrum of aluminium metal is shown in Fig. 14.8. It spans an energy range of 13 eV and contains the transitions from $n = 3$ to $n = 2$, i.e. $M \rightarrow L$ transitions. The cut-off at ~ 73 eV represents the transition of electrons that are in the $3p$ band and whose energy is close to E_F. The shape of the L emission spectrum (Fig. 14.8) is similar to the density of states diagram calculated for aluminium. It shows a broad band at lower energies that corresponds to transitions from the $3s$ band. This overlaps with another band at higher energies

504

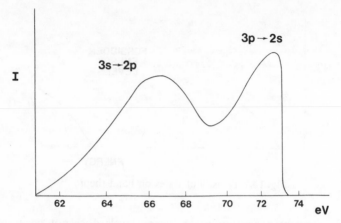

Fig. 14.8 X-ray L-emission spectrum of aluminium metal

that corresponds to transitions from the $3p$ band. Clearly, only the lower levels of the $3p$ band contain electrons.

X-ray emission spectra give information about energy levels below E_F. X-ray absorption spectroscopy can be used to study energy levels above E_F (Section 3.2.3.5).

14.3 Refinements to simple band theory — k space and Brillouin zones

Further insight into the origins of energy bands in solids comes from consideration of electron diffraction effects. The freely mobile valence electrons in a metal may, under certain circumstances, be diffracted internally by the periodic arrangement of nuclei in the crystal structure. The basic law of diffraction is Bragg's Law (Section 5.2.2.2). This relates the wavelength of the particles or waves, whether they be electrons (Section 3.2.1.4), X-rays (Sections 3.2.1 and 5.2) or neutrons (Section 3.2.1.5), to the interplanar spacing, d, and the diffraction angle, θ, i.e.

$$n\lambda = 2d \sin \theta \qquad (5.3)$$

In structural studies using diffraction methods, the radiation is, of course, provided by an external source. Usually it is monochromatic (fixed λ) although, sometimes, as in 'time of flight' neutron diffraction (Section 3.2.1.5) and EXAFS studies using synchrotron radiation (Section 3.2.3.5) the wavelength is variable.

In considering internal diffraction effects associated with the mobile valence electrons in a solid, it is found that the possible occurrence of electron diffraction places a limitation on the wavelength, energy and freedom of motion of the electrons. To be specific, it is forbidden that free electrons should at any stage in their motion be able to satisfy Bragg's law. The free electrons in a metal or semiconductor have variable energy and, therefore, variable wavelength. In evaluating possible electron diffraction effects, account must be taken of the wavelength, λ, diffraction angle, θ, and interplanar spacing, d.

The kinetic energy of a particle such as a free electron is given by

$$E = \frac{mv^2}{2} \qquad (14.1)$$

Momentum, mv, and wavelength, λ, are related by the wave-particle duality equation of de Broglie:

$$\lambda = \frac{h}{mv} \qquad (14.2)$$

It is useful to describe waves by a vector, $\mathbf{k}$, whose direction is parallel to the direction of propagation of the waves and whose magnitude is related to reciprocal wavelength by

$$k = \frac{2\pi}{\lambda} \qquad (14.3)$$

Combining equations (14.1), (14.2) and (14.3) gives

$$E = \frac{h^2 k^2}{8\pi^2 m} \qquad (14.4)$$

This indicates a parabolic relation between the energy and wave vector of an electron, as shown graphically in Fig. 14.9.

Returning to Bragg's Law (equation 5.3), substitution for λ (equation 14.3) gives

$$k = \frac{n\pi}{d \sin \theta} \qquad (14.5)$$

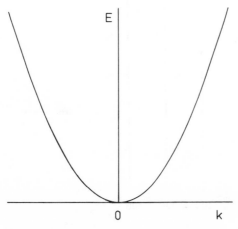

Fig. 14.9 Parabolic relation between energy, E, and wavenumber, k, of a free electron

Thus, we have equations that give the values of k and hence E for which the Bragg condition holds in a particular crystal structure.

Consider, now, application of equation (14.5) to the case of a primitive cubic substance with unit cell parameter, a. The longest d-spacing in such a structure is that for the (100) planes, for which $d_{100} = a$. Let us evaluate the conditions, in terms of k, for diffraction from the (100) planes. For first-order ($n = 1$) diffraction from (100) planes, k can have any value equal to or greater than π/a. For $k = \pi/a$, $\theta = 90°$ and $\sin\theta = 1$; this represents the limiting situation in which incidence of the electron wave is normal to the (100) planes. For $k > \pi/a$, $\sin\theta < 1$ and $\theta < 90°$; this represents the general case of diffraction at a Bragg angle $\theta < 90°$, as shown in Fig. 14.10(a). The vertical lines in (a) represent (100) planes in projection; the x axis of the unit cell is therefore horizontal.

The picture of diffraction shown in (a) is a real space description. It may be contrasted with the reciprocal space or k space description shown in (b). In this, the wave vector $\mathbf{k}$, of magnitude $\pi/(a\sin\theta)$ and angular direction $(90-\theta)°$ to the x axis, is represented by a line of the same magnitude (scaled) and direction, but drawn from an arbitrary point of reference or origin. The $\mathbf{k}$ vectors are shown in (b) for the special cases where $\theta = 45, 60$ and $90°$, as well as for a general value of θ. It can be shown, by consideration of equation (14.5) and by geometry, that the k vectors all terminate on the vertical line XY shown. In three dimensions, line XY becomes a plane. This plane forms part of the boundary of the (100) Brillouin zone. It represents in k space the conditions under which Bragg's Law holds for first-order ($n = 1$) diffraction from the (100) planes using radiation of variable wavelength and $\mathbf{k}$ vector.

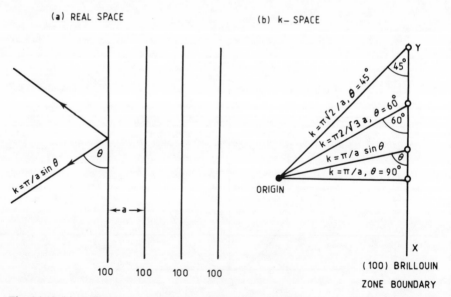

Fig. 14.10 (a) Diffraction of electrons from (100) planes in real space. (b) Origin of Brillouin zone boundary in k space

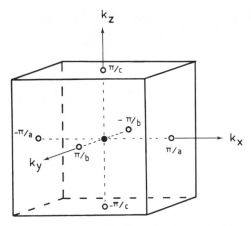

Fig. 14.11 First Brillouin zone for a primitive cubic lattice

In three dimensions, it is necessary to consider the set of planes {100} for our primitive cubic material. This set contains the planes (100), ($\bar{1}$00), (010), (0$\bar{1}$0), (001) and (00$\bar{1}$). Each of these contributes a face to the (100) Brillouin zone which, in three dimensions, takes the form of a cube (Fig. 14.11). This zone is known as the first Brillouin zone since it is the zone associated with the planes of longest d-spacing in the crystal structure. It is also the zone associated with the smallest k values or energies, since $k \propto d^{-1}$. Another point to note is that the surfaces of the Brillouin zone are parallel to the diffracting planes {100}. This also applies to higher Brillouin zones.

As discussed above, the first Brillouin zone in a primitive cubic structure corresponds to diffraction from the {100} planes. The second Brillouin zone is for diffraction from the {110} planes. There are twelve sets of planes in the set {110}. The Brillouin zone for this takes the shape of a regular dodecahedron (not shown) that envelops the cube-shaped, first Brillouin zone. A two-dimensional section through first and second Brillouin zones is shown in Fig. 14.12.

For structures that are body centred cubic, e.g. sodium, the first Brillouin zone is the same dodecahedron indicated above that corresponds to reflection from the {110} planes. This is because the 100 reflection is systematically absent in body centred lattices.

For structures that are face centred cubic, e.g. γ-Fe, the first Brillouin zone takes the shape of a truncated octahedron with fourteen faces. This is because both {111} and {200} planes are involved, with multiplicities of 8 and 6, respectively. A truncated octahedron may be regarded as an octahedron with each of its six corners sliced off, leaving a square butt.

Let us consider now the energies of valence electrons and their positions in k space. Electrons that lie deep down in the valence band of, for instance, a metal or semiconductor have relatively low energy and a low k value. In k space, they would therefore be well inside the first Brillouin zone since their **k** vectors would terminate short of the zone boundary. As the energy of the electrons increases, the

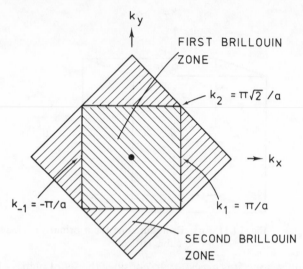

Fig. 14.12 Section through first and second Brillouin zones in a primitive cubic lattice

k vectors gradually lengthen to fill the first Brillouin zone. However, the situation in which k terminates on the zone boundary represents a forbidden condition. With still greater energy of the electrons, the **k** vectors pass through the zone boundary and enter the second zone.

The forbidden condition corresponding to **k** vectors that terminate on a zone boundary is, in fact, rather more complicated. This is because it is not so much the k value that is forbidden but rather the associated energy. Furthermore, in the region of a zone boundary, the energy of a moving electron is not given by equation (14.4)! Instead, for k values that are just inside a zone boundary, the associated energies are less than given by equation (14.4), whereas for k values just outside the boundary, they are greater. Hence, although a discontinuity in k does not occur at a zone boundary, a discontinuity in energy does occur. The parabolic relation between E and k is then modified, as shown in Fig. 14.13. The E/k curves turn over and become horizontal as the critical k values are approached. Consequently a gap of forbidden energy levels opens up.

The mathematical explanation of the occurrence of forbidden energy gaps is rather complex, but we can at least understand qualitatively why these energy gaps should occur at all. In a hypothetical crystal for which the potential inside is constant (Fig. 14.4), the electrons would be free to move, unconstrained by any possible diffraction effects. In a real crystal, however, the potential inside the crystal is periodic (Fig. 14.6). The shapes of the potential curves, and the depths and breadths of the minima, depend on the amount of charge associated with the nuclei. It is these positively charged nuclei that are responsible for the diffraction effects. Effectively, the potential minima act as secondary sources and re-emit spherical waves of electrons. If the nuclei are highly charged, as in multivalent

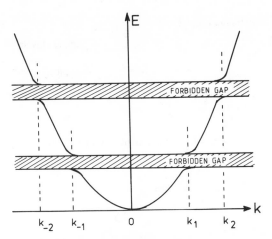

Fig. 14.13 The effect of diffraction of electrons on their energies, giving rise to gaps of forbidden energy levels. The boundaries of the first zone k_1, k_{-1} and second zone, k_2, k_{-2} are marked

atoms, the potential minima are deep and the electrons are diffracted strongly by them.

These secondary electrons interfere, not only with each other but also with the primary travelling wave that constitutes the free, valence electrons. At the edge of the zone boundary (where diffraction effects are significant), secondary electrons generated by adjacent layers of positively charged atomic nuclei are in phase and interfere constructively with each other. However, the nature of their interference with the travelling wave of the valence electrons depends on direction. For secondary electrons propagating in the same direction as the valence electrons, interference is constructive. Consequently, the net amplitude and energy of the valence electrons increases. For electrons propagating in the opposite direction to the valence electrons, partial destructive interference occurs. Consequently, the amplitude and energy of the valence electrons is diminished. The effect of secondary diffraction is therefore to cause an unavoidable modification of the energy of the valence electrons. Hence, an energy gap opens up. The magnitude of the gap depends on the degree of Bragg scattering that occurs, the crystal structure of the material and the element(s) that are present.

We are now in a position to consider the band structure of different types of material and the controlling influence that this has on electrical properties. We shall not consider Brillouin zones any further since a more simplified, schematic description of bands and band gaps is sufficient to explain the various effects.

14.4 Band structure of metals

Metals are characterized by a band structure in which the highest occupied band, the valence band, is only part full (Fig. 14.14). The occupied levels are shown schematically by the shading; some levels just below the Fermi level are

510

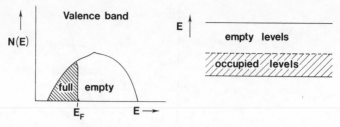

Fig. 14.14 Band structure of a metal

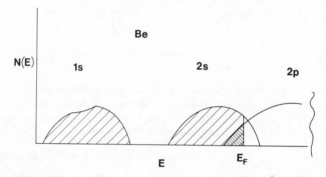

Fig. 14.15 Overlapping band structure of beryllium metal

vacant whereas some above E_F are occupied. Electrons in singly occupied states close to E_F are able to move and are responsible for the high conductivity of metals.

In some metals, such as sodium, energy bands overlap (Fig. 14.3). Consequently, both $3s$ and $3p$ bands in sodium contain electrons. Overlap of bands is responsible for the metallic properties of, for example, the alkaline earth metals. The band structure of beryllium is shown in Fig. 14.15. It has overlapping $2s$, $2p$ bands, both of which are only partly full. If the $2s$ and $2p$ bands did not overlap, then the $2s$ band would be full, the $2p$ band empty and beryllium would not be metallic. This is the situation that holds in insulators and semiconductors.

14.5 Band structure of insulators

The valence band in insulators is full. It is separated by a large, forbidden gap from the next energy band, which is empty (Fig. 14.16). Diamond is an excellent insulator with a band gap of $\sim 6\,\mathrm{eV}$. Very few electrons from the valence band have sufficient thermal energy to be promoted into the empty band above. Hence the conductivity is negligibly small. The origin of the band gap in diamond is discussed in Section 14.6; it is similar to that for silicon.

14.6 Band structure of semiconductors: silicon

Semiconductors have a similar band structure to insulators but the band gap is not very large; usually it is in the range 0.5 to 3.0 eV. At least a few electrons have

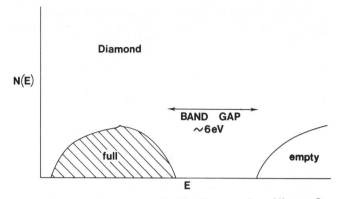

Fig. 14.16 Band structure of an insulator, carbon (diamond)

sufficient thermal energy to be promoted into the empty band.

Two types of conduction mechanism may be distinguished in semiconductors (Fig. 14.17). Any electrons that are promoted into an upper, empty band, called the *conduction band*, are regarded as negative charge carriers and would move towards a positive electrode under an applied potential. The vacant electron levels that are left behind in the valence band may be regarded as *positive holes*. Positive holes move when an electron enters them, leaving its own position vacant as a fresh positive hole. Effectively, positive holes therefore move in the opposite direction to electrons.

Semiconductors may be divided into two groups:

(a) *Intrinsic* semiconductors are pure materials whose band structure may be as shown in Fig. 14.17. For these, the number of electrons, n, in the conduction band is governed entirely by (i) the magnitude of the band gap and (ii) temperature. Pure silicon is an intrinsic semiconductor. Its band gap, and that of other Group IV elements, is given in Table 14.1.

The band structure of silicon and germanium is quite different to that which might be expected by comparison with sodium and magnesium. In these elements, the $3s, 3p$ levels overlap to give two broad bands, both of

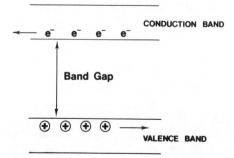

Fig. 14.17 Positive and negative charge carriers

Table 14.1 *Band gaps of Group IV elements*

Element	Band gap (eV)	Type of material
Diamond, C	6.0	Insulator
Si	1.1	Semiconductor
Ge	0.7	Semiconductor
Grey Sn (> 13 °C)	0.1	Semiconductor
White Sn (< 13 °C)	0	Metal
Pb	0	Metal

which are part full. If the trend continued, it might be expected that two similar bands would be present in silicon. In this case, the bands would, on average, be half full and silicon would be metallic. Clearly this is not the case and instead silicon contains two bands separated by a forbidden gap. Further, the lower band contains four electrons per silicon atom and is full. If the forbidden gap merely corresponded to separation of s and p bands, the s band could only contain two electrons per silicon. This cannot be the explanation, therefore.

The explanation of the band structure of silicon lies in quantum mechanics and the fact that the crystal structure of silicon is quite different to that of sodium: the structure of sodium is body centred cubic with a coordination number of eight whereas that of silicon is derived from face centred cubic with a coordination number of four. However, a simplified and non-mathematical explanation of the band structure of silicon (and germanium, diamond, etc.) may be given. It starts from the assumption that each silicon forms four equal bonds that are arranged tetrahedrally; these bonds or orbitals may be regarded as sp^3 hybridized. Each hybrid orbital overlaps with a similar orbital on an adjacent silicon to form a pair of molecular orbitals, one bonding, σ, and the other antibonding, σ^*. Each can contain two electrons, one from each silicon atom. The only step that remains is to allow the individual σ molecular orbitals to overlap to form a σ band; this becomes the valence band. The σ^* orbitals overlap similarly and become the conduction band. The σ band is full since it contains four electrons per silicon atom. The σ^* band is empty.

(b) *Extrinsic* semiconductors are materials whose conductivity is controlled by the addition of dopants. Silicon may be converted into an extrinsic semiconductor by doping with an element from either Group III or Group V of the Periodic Table. First, let us see the effect of doping silicon with a small amount, e.g. 10^{-2} atom %, of a trivalent element, e.g. gallium. Let the gallium atoms replace silicon atoms in the tetrahedral sites of the diamond structure to form a substitutional solid solution. In pure silicon, using covalent bonding theory, all of the Si—Si bonds may be regarded as normal, electron pair covalent bonds since silicon has four valence electrons and is coordinated to four other silicon atoms (Fig. 14.18). Gallium has only three valence

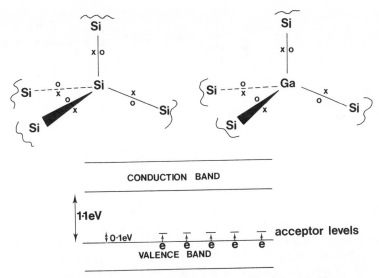

Fig. 14.18 p- type semiconductivity in gallium-doped silicon

electrons, however, and in galluim-doped silicon, one of the Ga—Si bonds must be deficient by one electron. Using band theory, it is found that the energy level associated with each single-electron Ga—Si bond does not form part of the valence band of silicon. Instead, it forms a discrete level or atomic orbital just above the top of the valence band. This level is known as an *acceptor level* because it is capable of accepting an electron. The gap between the acceptor levels and the top of the valence band is small, ~ 0.1 eV. Consequently, electrons from the valence band may have sufficient thermal energy to be promoted readily into the acceptor levels. The acceptor levels are discrete if the concentration of gallium atoms is small and it is therefore not possible for electrons in the acceptor levels to contribute to conduction. The positive holes that are left behind in the valence band are able to move and gallium-doped silicon is a positive hole or *p-type semiconductor*.

At normal temperatures, the number of positive holes created by the presence of gallium dopant atoms far exceeds the number created by the thermal promotion of electrons into the conduction band, i.e. the extrinsic, positive hole concentration far exceeds the intrinsic concentration of positive holes. Hence, the conductivity is controlled by the concentration of gallium atoms. With increasing temperature, the concentration of intrinsic carriers increases rapidly. At sufficiently high temperatures, the intrinsic carrier concentration may exceed the extrinsic value, in which case, a changeover to intrinsic behaviour would be observed (see later).

Let us consider now the effect of doping silicon with a pentavalent element such as arsenic. The arsenic atoms again substitute for silicon in the

514

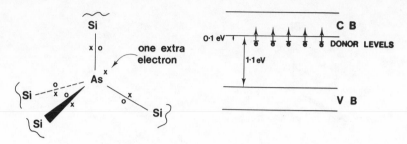

Fig. 14.19 n-type semiconductivity in arsenic-doped silicon

diamond-like structure, but now for each arsenic atom there is one electron more than needed to form four Si—As covalent bonds (Fig. 14.19). On the band description, this extra electron occupies a discrete level that is found to lie about 0.1 eV below the bottom of the conduction band. Again, the electrons in these levels cannot move directly as there is insufficient of them to form a continuous band. Instead these levels can act as *donor levels* because the electrons in them do have sufficient thermal energy to get up into the conduction band where they are free to move. Such a material is known as an *n-type semiconductor*.

It is worthwhile at this stage to summarize the differences between extrinsic and intrinsic semiconductors. These are principally:

(a) Extrinsic semiconductors have much higher conductivities than similar intrinsic ones at normal temperatures. For example, at 25 °C, pure silicon has an intrinsic conductivity of only $\sim 10^{-2}$ ohm^{-1} cm^{-1}. By appropriate doping, however, its conductivity is increased by several orders of magnitude.
(b) The conductivity of extrinsic semiconductors may be accurately controlled by controlling the concentration of dopant. Materials with desired values of conductivity may therefore be designed and manufactured. With intrinsic semiconductors, the conductivity is very dependent on temperature and the presence of stray impurities.

The temperature dependence of (a) carrier concentration and (b) conductivity for a semiconducting material that is extrinsic at low temperatures and intrinsic at high temperatures is shown schematically in Fig. 14.20. At low temperatures, A, the carrier concentration is temperature dependent since the relatively small gap, ~ 0.1 eV, between acceptor/donor states and the valence/conduction band is sufficiently large that only a limited number of electrons can make the transition. With increasing temperature 'saturation' or 'exhaustion' occurs, B, when the concentration of extrinsic carriers attains its maximum value. At this stage the carrier concentration is independent of temperature and the conductivity may show a gradual decrease with a further rise in temperature due to

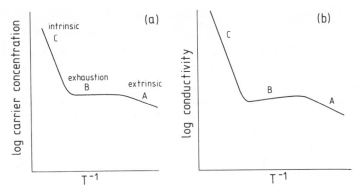

Fig. 14.20 Extrinsic, exhaustion and intrinsic regions in a semiconductor

the mobility effect. At still higher temperatures, the intrinsic concentration of carrier exceeds the extrinsic concentration and both carrier concentration and conductivity increase sharply, C.

For use in semiconductor devices, it is usually desirable to have materials that are in the exhaustion range since they are then relatively insensitive to temperature changes. The exhaustion range, B, can be made to extend over a wide range of temperatures by (a) choosing a material with a large intrinsic band gap and (b) dopants whose associated energy level is close to the relevant valence or conduction band. In this way region A is displaced to lower temperatures and region C to higher temperatures.

14.7 Controlled valency semiconductors

Certain transition metal compounds that have an element present in more than one oxidation state are electrical conductors. A good example of this effect is oxidized nickel oxide. Pure nickel oxide, NiO, is a pale green solid of low conductivity. It is an intrinsic semiconductor and its colour is due to internal d–d transitions within the octahedrally coordinated Ni^{2+} ions.

On oxidation of NiO, by heating in air at, for example, 1000 °C, a slight increase in weight due to uptake of oxygen occurs, the solid turns black and it becomes a moderately good electrical conductor. On oxidation, some of the Ni^{2+} ions are oxidized to Ni^{3+} and the composition of the solid product may be written:

$$Ni^{2+}_{1-3x}Ni^{3+}_{2x}V_xO$$

In this process, oxygen molecules absorb on to the surface, dissociate and undergo a redox reaction with some Ni^{2+} ions to form Ni^{3+}, O^{2-} ions. In order to give better local charge balance, Ni^{2+} ions diffuse out to the surface and leave cation vacancies inside the crystals. This black nickel oxide is a conductor because electrons can transfer from Ni^{2+} to Ni^{3+} ions:

$$Ni^{2+} \rightarrow Ni^{3+} + e$$

or

$$Ni^{3+} \rightarrow Ni^{2+} + p$$

Effectively, the Ni^{3+} ions are able to move by virtue of these electron transfer reactions and black nickel oxide is a p-type semiconductor. In contrast to gallium-doped silicon, which is also p-type, black NiO is best regarded as a *hopping semiconductor*. This is because the Ni^{2+}/Ni^{3+} exchange process is thermally activated and therefore highly temperature dependent. On the band model, this would correspond to poor overlap between the nickel d orbitals, giving rise to either a narrow d band or d levels that are still localized on the individual nickel ions. A further discussion of d levels is given in Section 14.9.

The disadvantage of using oxidized NiO as a semiconductor is that its conductivity is difficult to control; it depends on both temperature and the partial pressure of oxygen. To overcome this difficulty, Verwey (1948) introduced the idea of controlled valency semiconductors in which the concentration of, say, Ni^{3+} ions is dependent not on temperature but on the addition of a controlled amount of dopant.

Lithium oxide may be reacted with nickel oxide and oxygen to form solid solutions of formula:

$$Li_x Ni^{2+}_{1-2x} Ni^{3+}_x O$$

In these, the concentration of Ni^{3+} ions, and hence the conductivity, depends on the concentration of Li^+ ions. The magnitude of the conductivity varies enormously with x, from $\sim 10^{-10}$ ohm^{-1} cm^{-1} for $x = 0$ to $\sim 10^{-1}$ ohm^{-1} cm^{-1} for $x = 0.1$ at 25 °C.

14.8 Applications of semiconductors

The main use of semiconductors is in solid state devices such as transistors, silicon chips, photocells, etc. A simple example and the main one to be discussed here is the *pn junction*. This is the solid state equivalent of the diode rectifier valve. Suppose that a single crystal of, for example, silicon is doped in such a way that one half is n-type and the other half p-type. The band structure at the junction is approximately as shown in Fig. 14.21. The Fermi levels are at different heights in the two halves. Consequently, electrons are able to flow spontaneously from the n-type to the p-type regions across the junction. The Fermi energy of electrons is similar to their electrochemical potential: as long as a difference in potential exists, electrons flow from a region of high potential to one of low potential.

In the absence of an externally applied potential difference a few electrons move from left to right across the junction but the band structure then rapidly adjusts itself so that E_F becomes the same on either side. An alternative explanation is that a space charge layer rapidly develops at the junction as electrons flow across it. The space charge then acts as a barrier to further electron flow.

This situation changes, however, if an external potential difference is applied such that the p-type end is positive and the n-type end is negative. A continuous

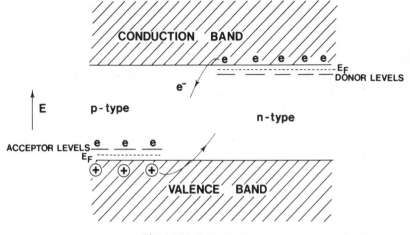

Fig. 14.21 A pn junction

current is able to flow through the circuit; electrons enter the crystal from the right-hand electrode. They flow through the conduction band of the n-type region, drop into the valence band of the p-type region at the pn junction and then flow through the valence band, via the positive holes, to leave at the left-hand electrode. A continuous current cannot flow in the opposite direction since, under normal circumstances, with a relatively low applied voltage, electrons cannot surmount the barrier necessary in order to pass from left to right across the junction. The pn junction is therefore a rectifier in that current may pass in one direction only. It may be used to convert a.c. into d.c. electricity. Silicon based pn junctions have now largely replaced diode valves.

A more complex arrangement is the pnp or npn junction. This acts as a current or voltage amplifier. It forms the basis of the transistor which has almost completely taken the place of triode valves.

Controlled valency semiconductors find application as *thermistors*, thermally sensitive resistors. In these, use is made of the large temperature dependence of the conductivity associated with the fact that these materials are hopping semiconductors. For example, the material $Li_{0.05}Ni_{0.95}O$ shows Arrhenius-type conductivity behaviour over a wide temperature range up to $\sim 200\,°C$. The activation energy is $\sim 0.15\,eV$. If the conductivity behaviour is reproducible, lithium nickel oxide can be used in devices to control and measure temperature. In order to achieve reproducibility, materials that are insensitive to impurities must be used, e.g. Fe_3O_4, Mn_2O_3, Co_2O_3 doped NiO and certain spinels.

Some semiconductors are *photoconductive*, i.e. their conductivity increases greatly on irradiation with light. Amorphous selenium is an excellent photoconductor and forms an essential component of the photocopying process (Chapter 18). Conventional band theory cannot be used to explain the properties of amorphous materials such as selenium since they lack any long range periodicity.

14.9 Band structure of inorganic solids

So far, we have concentrated on materials that are conductors of electricity. However, most inorganic solids can be treated profitably using band theory, whether or not they are electrical conductors. Band theory provides an additional insight into the structures, bonding and properties of inorganic solids that complements the insight obtained with the ionic/covalent models. Most inorganic materials have structures that are more complex than those of the metals and semiconducting elements. They have also received less theoretical attention of the type involving band structure calculations. Consequently, their band structures are usually known only approximately.

Above, we have considered Group IV elements, especially silicon and closely related III–V compounds, such as GaP. These latter compounds are isoelectronic with Group IV elements, at least as far as the numbers of electrons in the valence shells are concerned. Let us now take this one stage further and consider more extreme cases, with I–VII compounds such as NaCl and II–VI compounds such as MgO.

The bonding in these materials is predominantly ionic. They are white, insulating solids with negligibly small electronic conductivity. Addition of dopants tends to produce ionic rather than electronic conductivity (Chapter 13). On the assumption that NaCl is 100 per cent ionic, the ions have the configurations:

$$Na^+ : 1s^2 2s^2 2p^6$$
$$Cl^- : 1s^2 2s^2 2p^6 3s^2 3p^6$$

Hence, the $3s$, $3p$ valence shell of Cl^- is full and that of Na^+ is empty. Adjacent Cl^- ions are approximately in contact in NaCl and the $3p$ orbitals may overlap somewhat to form a narrow $3p$ valence band. This band is composed of anion orbitals only. The $3s$, $3p$ orbitals on Na^+ ions may also overlap to form a band, the conduction band. This band is formed from cation orbitals only. It is completely empty of electrons under normal conditions since the band gap is large, $\sim 7\,eV$. The band structure of NaCl is very like that of an insulator (Fig. 14.16), therefore, but with the additional detail that the valence band is composed of anion orbitals and the conduction band of cation orbitals. Any promotion of electrons from the valence band to the conduction band may therefore also be regarded as back transfer of charge from Cl^- to Na^+.

This conclusion leads us to expect some kind of correlation between the magnitude of the band gap and the difference in electronegativity between anion and cation. A large electronegativity difference favours ionic bonding. In such cases back transfer of charge from anion to cation is expected to be difficult and this correlates with the general observation that ionic solids have large band gaps. The band gaps of a variety of inorganic solids are given in Table 14.2. A quantitative relation between band gap and ionicity has been proposed by Phillips and van Vechten (equation 8.31). The band gap is regarded as being made of two components: part is due to the 'homopolar band gap' which is the

Table 14.2 *Band gaps (eV) of some inorganic solids*

I–VII compounds		II–VI compounds		III–V compounds	
LiF	11	ZnO	3.4	AlP	3.0
LiCl	9.5	ZnS	3.8	AlAs	2.3
NaF	11.5	ZnSe	2.8	AlSb	1.5
NaCl	8.5	ZnTe	2.4	GaP	2.3
NaBr	7.5	CdO	2.3	GaAs	1.4
KF	11	CdS	2.45	GaSb	0.7
KCl	8.5	CdSe	1.8	InP	1.3
KBr	7.5	CdTe	1.45	InAs	0.3
KI	5.8	PbS	0.37	InSb	0.2
		PbSe	0.27	β-SiC	2.2
		PbTe	0.33	α-SiC	3.1

Some of these values, especially for the alkali halides, are only approximate.

band gap that would be observed in the absence of any difference in electronegativity between the constituent elements; the other part is associated with the degree of ionic character in the bonds. A further discussion of this correlation is given in Section 8.4.

In compounds of transition metals, an additional factor that is of great significance is the presence of partly filled d orbitals on the metal ions. In some cases, these overlap to give a d band or bands and the material may have high conductivity. In other cases, d orbital overlap is very limited and the orbitals are effectively localized on the individual atoms. An example in which the latter occurs is stoichiometric NiO. Its pale green colour is due to internal d–d transitions within the individual Ni^{2+} ions. It has a very low conductivity, $\sim 10^{-14}$ ohm^{-1} cm^{-1} at 25 °C, and there is no evidence for any significant overlap of the d orbitals to form a partly filled d band. Examples at the other extreme are TiO and VO. These have the rock salt structure, as has NiO, but by contrast, d orbitals of the type d_{xy}, d_{xz}, d_{yz} (Section 8.6.1) on the M^{2+} ions overlap strongly (Fig. 14.22a) to form a broad t_{2g} band. This band is only partly filled by electrons. Consequently, TiO and VO have almost metallic conductivity, $\sim 10^3$ ohm^{-1} cm^{-1} at 25 °C.

An additional difference between TiO and NiO is that the t_{2g} band, capable of containing six electrons per metal atom, must be full in NiO. The two extra d electrons in Ni^{2+} are in e_g levels, $d_{z^2}, d_{x^2-y^2}$. These e_g orbitals point directly at the oxide ions (Fig. 14.22b). Because of the intervening oxide ions, the e_g orbitals on adjacent Ni^{2+} ions cannot overlap to form a band. Hence e_g orbitals remain localized on the individual Ni^{2+} ions.

Some general guidelines as to whether good overlap of d orbitals is likely to occur have been given by Phillips and Williams. In these guidelines, d band formation is likely to occur if:

(a) The formal charge on the cations is small.

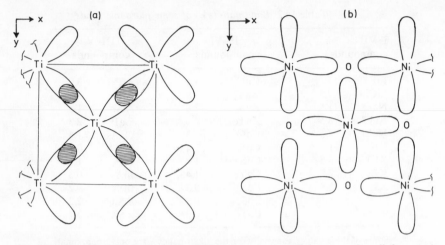

Fig. 14.22 (a) Section through the TiO structure, parallel to a unit cell face and showing Ti^{2+} positions only (see Fig. 5.8). Overlap of d_{xy} orbitals on adjacent Ti^{2+} ions, together with similar overlap of d_{xz} and d_{yz} orbitals, leads to a t_{2g} band. (b) Structure of NiO, showing $d_{x^2-y^2}$ orbitals pointing directly at oxide ions and, therefore, unable to overlap and form an e_g band

(b) The cation occurs early in the transition series.

(c) The cation is in the second or third transition series.

(d) The anion is reasonably electropositive.

The reasoning behind these guidelines is fairly straightforward and is based on similar arguments that are used in ligand field theory. Effects (a) to (c) are all concerned with keeping the d orbitals spread out as far as possible and reducing the amount of positive charge that they 'feel' from their parent transition metal ion nucleus. Effect (d) is associated with a reduction in ionicity and band gap, as discussed earlier in this section.

A variety of examples can be found to illustrate each of the guidelines:

Thus, for (a), TiO is metallic whereas TiO_2 is an insulator. Cu_2O and MoO_2 are semiconductors whereas CuO and MoO_3 are insulators.

For (b), TiO, VO are metallic whereas NiO and CuO are poor semiconductors.

For (c), Cr_2O_3 is a poor conductor whereas lower oxides of Mo, W are good conductors.

For (d), NiO is a poor conductor whereas NiS, NiSe, NiTe are good conductors.

The d electron structure of solid transition metal compounds is also sensitive to the crystal structure of the solid and to any variation in oxidation state of the transition metal. Some interesting examples are provided by complex oxides with the spinel structure:

(a) Both Fe_3O_4 and Mn_3O_4 have a spinel structure but whereas Mn_3O_4 is virtually an insulator, Fe_3O_4 has almost metallic conductivity. The structure of Fe_3O_4 may be written as:

$$[Fe^{3+}]^{tet}[Fe^{2+}, Fe^{3+}]^{oct}O_4; \text{ inverse spinel}$$

whereas the structure of Mn_3O_4 is:

$$[Mn^{2+}]^{tet}[Mn_2^{3+}]^{oct}O_4: \text{ normal spinel}$$

Since Fe_3O_4 is an inverse spinel, it contains Fe^{2+} and Fe^{3+} ions distributed over the octahedral sites. These octahedral sites are close together since they belong to edge-sharing octahedra. Consequently, positive holes can migrate easily from Fe^{2+} to Fe^{3+} ions and hence Fe_3O_4 is a good conductor.

In Mn_3O_4, the spinel structure is normal which means that the closely spaced octahedral sites contain only Mn^{3+} ions. The tetrahedral sites, containing Mn^{2+} ions, share corners only with the octahedral sites. The $Mn^{2+}-Mn^{3+}$ distance is greater, therefore, and electron exchange cannot take place easily.

(b) A related example, which is really an example of guideline (b) above, is provided by the lithium spinels, $LiMn_2O_4$ and LiV_2O_4. The structural formulae of these spinels are similar:

$$[Li^+]^{tet}[Mn^{3+} Mn^{4+}]^{oct}O_4$$
$$[Li^+]^{tet}[V^{3+} V^{4+}]^{oct}O_4$$

A mixture of $+3$ and $+4$ ions are present in the octahedral sites of both but since d orbital overlap is greater for vanadium than for manganese, this is reflected in the electrical properties: $LiMn_2O_4$ is a hopping semiconductor whereas LiV_2O_4 has metallic conductivity.

14.10 Colour in inorganic solids

Colour arises usually because a solid is in some way sensitive to visible light. In most cases, if a coloured solid is irradiated with white light, radiation from part of the visible spectrum is absorbed. The colour that we see then corresponds to the unabsorbed radiation and its range of associated wavelengths.

Colour is often, but not always, associated with transition metal ions. In molecular chemistry, colour can arise from two possible common causes. The d–d electronic transitions within transition metal ions give rise to many of the familiar colours of transition metal compounds, e.g. the various shades of blue and green associated with different copper (II) complexes. Charge transfer effects in which an electron is transferred between an anion and a cation are often responsible for intense colours as in, for example, permanganate (purple) and chromates (yellow). In solids, there is an additional source of colour; it involves the transition of electrons between energy bands.

Colour may be measured in a quantitative way by spectroscopic techniques (Chapter 3) in which the emission or absorption of radiation by a sample is recorded as a function of frequency or wavelength. These techniques can also

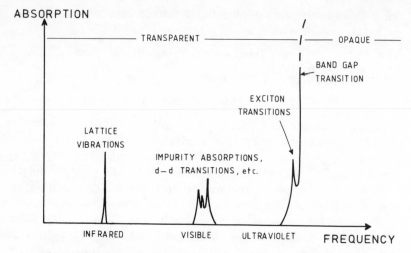

Fig. 14.23 Schematic absorption spectrum of a non-metallic solid

detect transitions or effects that lie outside the visible region, in either the infrared or ultraviolet. A schematic absorption spectrum that covers the IR, visible and UV regions is shown in Fig. 14.23. We note the occurrence of absorption peaks associated with lattice vibrations in the infrared. At higher frequencies, electronic transitions associated with d-level splitting, impurity ions, crystal defects, etc., are possible. Many of these occur in the visible region and are responsible for colour.

The position of the band gap transition is important. It obviously influences the conductivity of a solid, it may affect colour and it provides a frequency limit beyond which the solid is no longer transparent to radiation.

In the example shown, the band gap is quite large, several electronvolts and falls in the UV region. Such a solid would therefore be a poor electronic conductor. Also it would be colourless unless electronic transitions involving discrete energy levels occurred in the visible region. Examples are rutile, TiO_2 (band gap 3.2 eV), which is white, and Cr_2O_3 (band gap 3.4 eV), which is green due to $d-d$ transitions in the Cr^{3+} ions. Similarly, NiO (band gap 3.7 eV) is green due to $d-d$ transitions.

In some solids, the band gap falls in the visible region (1.7 to 3.0 eV) and is directly responsible for colour as in CdS (band gap 2.45 eV), which is bright yellow.

If the band gap is less than ~ 1.7 eV, the solid is inevitably dark coloured and absorbs visible light; e.g. PbS (band gap 0.37 eV) and CuO (band gap 0.6 eV) are black. Such materials are also appreciable or very good conductors of electricity.

The experimental determination of band gaps is carried out using spectra such as Fig. 14.23. Difficulties can arise, especially with large band gap solids, in that exciton transitions may occur at frequencies somewhat less than the band gap transition. These involve transitions to discrete energy levels that lie somewhat below the bottom of the conduction band and the distinction between exciton and band gap transitions may not be clear-cut.

Some of the above-mentioned solids are used in radiation detection devices, depending on where the onset of absorption associated with the band gap transition occurs. Thus, PbS, PbSe and PbTe with band gaps, $\sim 0.3\,eV$, in the infrared region are used in IR detectors. Corresponding cadmium chalcogenides have band gaps in the visible and are used in lightmeters.

14.11 Bands or bonds: a final comment

The three extreme types of bonding in solid materials are ionic, covalent and metallic. Within each category many good examples can be found. Most inorganic solids do not belong exclusively to one particular category, however. We shall not consider here the wide variety of bonding effects that do occur in solids. Some of these have already been discussed in Chapter 8, especially mixed ionic/covalent bonding. Instead, we shall confine ourselves to a few comments on when it is appropriate to use a band rather than a bond model.

The band model for bonding is clearly appropriate in systems where there are freely mobile electrons, as in metals and some semiconductors. Experimental measurements of mobility show that these electrons are highly mobile and are not associated with individual atoms. In other semiconductors, such as doped nickel oxide, the band model would not appear to be suitable for explaining the electrical behaviour. The evidence indicates that these materials are best regarded as hopping semiconductors in which the electrons do not have high mobility. Instead, it appears to be more appropriate to regard the d electrons as occupying discrete orbitals on the nickel ions. It is important to remember, however, that the question of conduction in nickel oxide refers to only one or two sets of energy levels. Nickel oxide, like other materials, has many sets of energy levels. The lower lying levels are fully occupied by electrons. These are discrete levels associated with the individual anions and cations. At higher energy there are various excited levels that are usually completely empty of electrons. However, these levels may well overlap to form energy bands. In asking whether or not a bond or band model is the most suitable, one has to be clear about the particular property or set of energy levels to which the question refers. Thus many ionically bonded solids may, under UV irradiation, show electronic conductivity that is best described in terms of band theory.

Questions

14.1 The electronic configuration of calcium is $1s^2\,2s^2\,2p^6\,3s^3\,3p^6\,4s^2$. Explain why calcium shows metallic conductivity.

14.2 Sketch the energy band structure of silicon with a band gap of $1.1\,eV$. What elements would you add to silicon to make it p-type? Sketch the resulting energy level structure for acceptor levels that are located $0.01\,eV$ above the top of the valence band. What fraction of the acceptor levels would be occupied at room temperature? Assume that the probability of excitation is proportional to $\exp(-E/kT)$. If the impurity concentration is 10^{-4} atom per cent, what is the carrier density due to the impurities? What would be the intrinsic carrier concentration at room temperature, in the absence of impurities? At what temperature is the intrinsic density equal to that caused by the impurities?

14.3 Repeat the above problem but for germanium with a band gap of 0.7 eV.

14.4 For use in practical semiconductor devices, why is it desirable to have materials with a large band gap between valence and conduction band but a small gap between valence/conduction band and impurity levels?

14.5 The potassium halides are all transparent to visible light. Calculate the wavelengths at which they become opaque; band gap data are given in Table 14.2.

References

There are many solid state books devoted to semiconductors and band theory. The less mathematically oriented ones include:

C. Barrett and T. B. Massalski (1982). *Structure of Metals*, Pergamon.

W. J. Moore (1967). *Seven Solid States*, W. A. Benjamin.

K. J. Pascoe (1973). *Properties of Materials for Electrical Engineers*, Wiley.

C. S. G. Philips and R. J. P. Williams (1965). *Inorganic Chemistry*, Oxford.

R. M. Rose, L. A. Shepard and J. Wulff (1966). *The Structure and Properties of Materials. IV. Electronic Properties*, Wiley.

C. A. Wert and R. M. Thomson (1970). *Physics of Solids*, McGraw-Hill.

D. A. Wright (1966). *Semiconductors*, Methuen.

Chapter 15

Other Electrical Properties

In addition to the electrical conductivity of materials, Chapters 13 and 14, a variety of other electrical effects may be observed. Some of the more important ones are summarized in this chapter.

15.1 Thermoelectric effects

At a junction between two dissimilar metals, an e.m.f. is automatically generated whose value depends on (a) the metals used and (b) temperature. If a temperature gradient exists along a metal rod, an e.m.f. is generated whose value depends on (a) the metal and (b) the temperature gradient. These effects are called thermoelectric effects and several types may be distinguished. Their main applications are in thermocouples, which are used for measuring temperature, in refrigeration devices and as sources of power.

15.1.1 The Thomson effect

A temperature gradient ΔT across a homogeneous conductor, e.g. a bar of metal, gives rise to a potential gradient, ΔV; this is known as the Thomson effect. Its origin may be understood by considering the mobile conduction electrons in the metal as particles, whose velocity and kinetic energy increase with

526

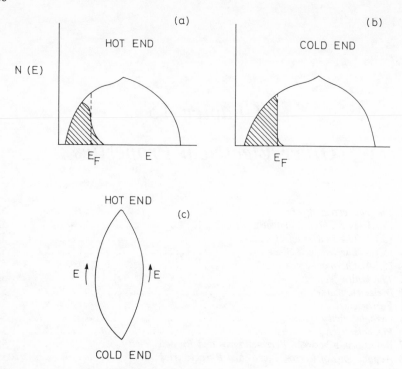

Fig. 15.1 The Thomson effect

temperature. When a temperature gradient is established, the electrons at the hot end have more thermal energy than those at the cold end and, statistically, more current carriers flow from hot to cold than vice versa. An excess of electrons therefore accumulates at the cold end, giving rise to a potential difference, with the cold end negative.

The Thomson effect may also be explained using band theory (Fig. 15.1). The conduction electrons at the hot end (a) are distributed over a range of energy levels to either side of E_F; a similar effect occurs at the cold end but to a lesser extent (b). Since the 'hottest' electrons (i.e. those of highest energy) occupy higher energy levels at the hot end than at the cold end, a net drift of electrons from hot to cold occurs whose magnitude depends on ΔT. The e.m.f., E, that develops is given by

$$E = \sigma \Delta T \tag{15.1}$$

where σ is the Thomson coefficient; E is usually of the order of a few millivolts.

Semiconductors also give a Thomson effect and the sign of the e.m.f. can be used to distinguish between n-type and p-type conduction mechanisms. If the semiconductor is n-type, the cold end becomes negative, as is the case with metals. If, however, the semiconductor is p-type and is operating in the extrinsic region (Fig. 14.20), the charge carriers are positive holes and the cold end becomes positively charged. This is because more electrons are promoted from the valence band to the acceptor levels at the hot end than at the cold end. An excess of positive holes builds up at the hot end and some of these flow to the cold end, making it positively charged.

It is difficult to use the Thomson effect, by itself, as a source of energy. It cannot, for example, be used to drive a current round a closed circuit, since, if two points on a circuit comprising a ring of metal are at different temperatures, equal and opposing e.m.f's are set up in the two paths from the hot end to the cold end (Fig. 15.1c). On the other hand, if different metals are used in the two paths, additional thermoelectric effects arise at the junctions of the two metals, as discussed next.

15.1.2 The Peltier effect

At a junction between two dissimilar conductors, e.g. iron and copper, heat is absorbed when a current flows in one direction and emitted when current flows in the reverse direction. Alternatively, when two dissimilar metals are simply brought into contact, some electrons cross the junction from one metal to the other until an electric field or 'space charge' is established of sufficient magnitude to prevent further electron flow. This process occurs because the Fermi levels in any two different metals are not generally the same. The metal junction therefore becomes a source of e.m.f., known as the Peltier e.m.f., π, whose magnitude depends on the two metals and the temperature of the junction.

Peltier e.m.f's are usually of the order of a few millivolts. They are largest with the metals tin and bismuth and with some semiconducting compounds. The band structure at a metal/n-type semiconductor junction is shown schematically in Fig. 15.2. At any such junction, the Fermi levels are at the same height on either side once equilibrium has been established. In order to reach equilibrium,

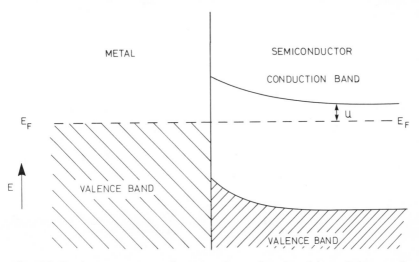

Fig. 15.2 Band structure at metal–semiconductor junction giving a Peltier e.m.f

however, some modification of the band structure at the semiconductor surface is necessary. (Similar effects are observed at semiconductor–air surfaces in which a space charge layer builds up inside the semiconductor surface and modifies the band structure.) The Fermi level is at a depth, U, below the bottom of the conduction band in an n-type semiconductor. In order for electrons to flow from left to right across the junction, an energy, U, is required in order to raise electrons from the valence band of the metal into the conduction band of the semiconductor, together with an extra amount of energy, $\frac{3}{2}kT$, in order for them to have the kinetic energy of free electrons. Each electron that accomplishes this transition extracts energy from the metal, thereby producing a cooling of the junction. Conversely, as an electron current, I, flows from right to left, heat, Q, is liberated at the junction, given by

$$Q = \pi I$$

$$= \frac{I}{e_0(U + \frac{3}{2}kT)} \tag{15.2}$$

15.1.3 The Seebeck effect

Consider a closed circuit formed by two different conductors, A and B (Fig. 15.3), with their junctions at temperatures T_1 and T_2. A temperature gradient occurs in both metals and a Thomson e.m.f. is set up in each. A Peltier e.m.f. occurs at each junction but these have unequal values since the

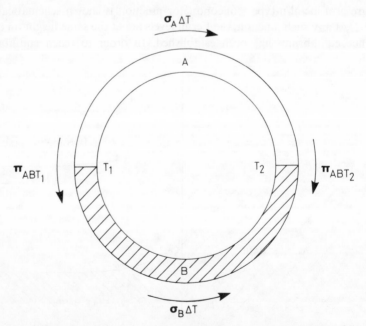

Fig. 15.3 Peltier and Thomson effects in a thermocouple

temperatures of the two junctions are different. The net e.m.f. in the circuit is the algebraic sum of two Thomson e.m.f's and two Peltier e.m.f's; usually, it is not zero. If the individual e.m.f's and their directions are as shown in Fig. 15.3, the net e.m.f. is given by

$$E = (\sigma_A - \sigma_B)\Delta T + (\pi_{ABT_2} - \pi_{ABT_1}) \tag{15.3}$$

A current flows in the circuit as long as the junctions are kept at different temperatures. This is known as the Seebeck effect. It is the basis on which thermocouples operate. The Seebeck coefficient or thermoelectric power, α, is defined by

$$\alpha = \frac{\pi}{T} \tag{15.4}$$

Typically, α has magnitudes of the order of microvolts per degree celsius, but values as high as $1 \, mV \, °C^{-1}$ are observed with some semiconductors.

15.1.4 Thermocouples

Thermocouples are invaluable for measuring temperature. They can be used for temperatures covering a very wide range, up to the melting point of the metals forming the couple which, for platinum-based alloys, is $\sim 1700 \, °C$. A thermocouple consists of wires of two dissimilar materials which are joined at their ends to form a closed loop. Somewhere in the circuit a millivoltmeter is placed. An e.m.f. develops if the two junctions of the wires are at different temperatures (Fig. 15.4). One junction, the reference junction, is kept at a fixed temperature, usually $0 \, °C$. The e.m.f. therefore depends on the temperature of the other junction, the probe junction. Tables are available for converting e.m.f. to temperature. Electromotive forces are usually of the order of millivolts and are measured with a high resistance voltmeter or potentiometer. In order that the voltage across the meter terminals is the same as that developed at the probe junction, it is important that no current be drawn during the measurements. This then avoids the possibility of 'I²R heating losses' in the circuit.

A typical thermocouple measuring circuit is shown in Fig. 15.4. The two metals in this case are platinum and Pt/13%Rh alloy. A third metal, usually copper, is added to the circuit such that the Pt–Cu and Pt/13%Rh–Cu junctions are both kept at $0 \, °C$. The e.m.f. is measured between the two copper leads. It is unaffected by the presence of copper in the circuit provided the two reference junctions are at the same temperature (Fig. 15.4b). The arrangement shown therefore gives the e.m.f. of the Pt–Pt/13%Rh couple whose junctions are at 0 and $T°C$.

The temperature dependence of the e.m.f. of a thermocouple can be expressed fairly accurately by

$$E = a_{AB}(T_2 - T_1) + \tfrac{1}{2}b_{AB}(T_2^2 - T_1^2) \tag{15.5}$$

where a and b are the thermoelectric coefficients characteristic of the metals A, B

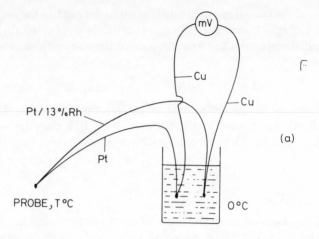

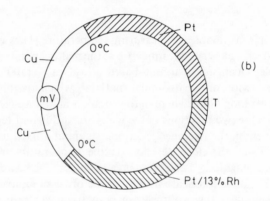

At 0 °C: $\Pi_{Pt,Cu} - \Pi_{Pt/13\% \, Rh,Cu} = \Pi_{Pt,Pt/13\%Rh}$

Fig. 15.4 A thermocouple measuring circuit

Table 15.1 *Thermoelectric coefficients*

Metal or alloy	a	b
Antimony	+ 35.6	+ 0.145
Iron	+ 16.7	− 0.0297
Copper	+ 2.71	+ 0.0079
Platinum	− 3.03	− 3.25
Nickel	− 19.1	− 3.02
Constantan (60% Cu, 40% Ni)	− 38.1	− 0.0888
Bismuth	− 74.4	+ 0.032

and T_2, T_1 are the temperatures of the two junctions. In order to have a comparative listing of a and b for different metals, a reference metal is needed for which a and b are set equal to zero. Conventionally, lead is taken as the reference metal. (In the same way, redox potentials are tabulated relative to a value of zero for the $\frac{1}{2}H_2/H^+$ electrode.) A selection of thermoelectric coefficients is given in Table 15.1. To calculate E for any couple AB, $a_{AB} = a_A - a_B$ and $b_{AB} = b_A - b_B$. A couple with high sensitivity is therefore one with a large a_{AB} value, e.g. iron–constantan, for which $a = 16.7 - (-38.1) = 54.8$. Platinum–based couples have small values of a but have the advantage of being usable to very high temperatures.

From equation (15.5), a graph of E versus T_2 is a parabola (Fig. 15.5). $E = 0$ when $T_2 = T_1$, E passes through a minimum when $T_2 = -a/b$ and E has a second value of zero when $T_2 = -2a/b - T_1$. For most practical thermocouple circuits, $T_2 \gg T_1$ and the relation between e.m.f. and temperature is essentially confined to the high temperature wing of the parabola.

Thermoelectric effects have various other applications. A thermopile is a bank of thermocouples connected in series; for n thermocouples, the e.m.f. generated is n times the e.m.f. of one thermocouple. Thermopiles are used as solar energy convertors and in astronomy to measure the heat radiated by stars.

Materials that have large Seebeck and Peltier coefficients may be used in thermoelectric power sources and thermoelectric coolers for refrigeration. The coefficients for semiconductors are usually much larger than for metals but the difficulty is that semiconductors with large α and π values usually have low conductivity, σ. Large Joule, I^2R, heating losses then occur during current flow

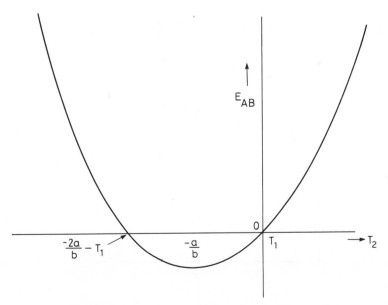

Fig. 15.5 Thermocouple e.m.f. against temperature

532

and the efficiency of the device is lowered. Optimization of performance depends on maximizing the quantity Z, known as the figure of merit and given by

$$Z = \frac{\alpha^2 \sigma}{K}$$

where K is the thermal conductivity. In practice, it is found that Z is largest for certain semiconducting compounds of high formula weight, e.g. Bi_2Te_3, Bi_2Se_3 and PbTe. Couples are used in which one wire is p-type and the other n-type. This then maximizes the difference in Fermi level at the junction and therefore maximizes π. Thermoelectric coolers based on Bi–Te–Se couples give cooling of at least 70 °C below room temperature and are used in small, silent refrigerant devices.

15.2 The Hall effect

Measurement of the Hall effect gives an important source of information about conduction mechanisms, especially in semiconductors. Conductivity is given by equation (13.1), $\sigma = ne\mu$; from conductivity measurements alone it is not possible to separate the number of current carriers, n, from their mobility, μ. Combined measurements of conductivity and Hall voltage do enable the separation of n and μ. The Hall effect is demonstrated in Fig. 15.6. If a current of I amperes flows through a solid in one direction and a magnetic field of H oersteds is applied at right angles to the direction of current flow, a potential difference is developed across the sample in the direction perpendicular to H and I. This is because the

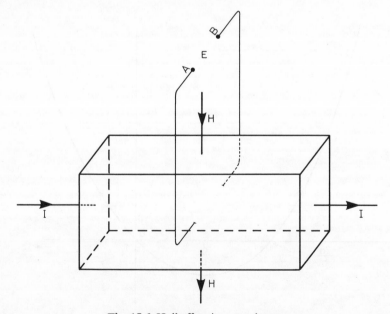

Fig. 15.6 Hall effect in a conductor

magnetic field causes the electrons to follow curved paths and therefore one side of the conductor becomes depleted in electrons relative to the other. The unequal electron concentration across the sample causes an electric field to build by until an equilibrium situation is reached where the tendency for electrons to deflect is balanced by the Hall potential acting in the opposite direction. The Hall coefficient, R, for a particular material is the potential gradient produced when I and H are both unity. The direction of R depends on the sign of the current carriers and is therefore different for positive holes and electrons. In Fig. 15.6, terminal A is positive for electrons flowing left to right.

If the current density flowing through the solid is σ and the applied field is H, the deflecting force, F_1, experienced by the moving charges, due to H, is given by

$$F_1 = He\mu \tag{15.6}$$

where μ is the mean drift velocity or mobility of the electrons and e is their charge. The Hall potential, E, applies an equal and opposing force, F_2, to the deflecting electrons, given by

$$F_2 = Ee = H\sigma Re$$
$$= Hne\mu Re \tag{15.7}$$

Equating F_1 and F_2 gives rise to

$$R = \frac{1}{ne} \tag{15.8}$$

From measurements of the Hall voltage and with known values of I and H, it is possible to calculate R and hence n. If σ is known, the mobility μ may then also be determined.

Most metals have negative Hall coefficients, which indicates that electrons are the current carriers. Some, e.g. Zn, Cd, are anomalous, however, and have positive R values. The qualitative explanation given by band theory is that conduction in zinc and cadmium is primarily by positive holes, or vacancies, at levels just below E_F (similar to p-type conduction in semiconductors). In an intrinsic semiconductor, the number of free electrons is equal to the number of holes; if both had the same mobility the net Hall voltage would be zero. In practice, electrons are usually more mobile than positive holes and so a net Hall voltage, indicating n-type carriers, is obtained in intrinsic semiconductors. In such cases, however, the value of n deduced from R has no significance unless the mobility ratio is sufficiently large that one type of carrier controls σ. The temperature dependence of R is different for n-type and p-type semiconductors in the region where the changeover to intrinsic behaviour occurs. Caution in the interpretation of the value of R for semiconductors is therefore needed.

Hall voltages are usually too small to be detected in ionically conducting materials because n and μ are both low. The effect has been observed, however, in some solid electrolytes, e.g. in $RbAg_4I_5$ which has high Ag^+ ion conductivity ($\sigma \sim 1\, ohm^{-1}\, cm^{-1}$ at $25\,°C$).

15.3 Dielectric materials

Dielectric materials are electrical insulators. They are used principally in capacitors and electrical insulators. They should possess the following properties in order to be of use in practical applications. They should possess high *dielectric strength*, i.e. they should be able to withstand high voltages without undergoing degradation and becoming electrically conducting. They should have low *dielectric loss*, i.e. in an alternating electric field, the loss of electrical energy, which appears as heat, should be minimized. Other dielectric related properties; pyro-, piezo- and ferroelectricity are discussed in subsequent sections. Let us consider first the behaviour of dielectric materials in an alternating electric field.

Application of a potential difference across a dielectric does lead to a polarization of charge within the material although long range motion of ions or electrons cannot occur. The polarization disappears when the voltage is removed. Ferroelectric materials are a special type of dielectric in that they retain a large, residual polarization of charge after the electric field has been removed (Section 15.4).

Dielectric properties may be defined by the behaviour of the material in a parallel plate capacitor. This is a pair of conducting plates, parallel to one another and separated by a distance, d, that is small compared with the linear dimensions of the plates (Fig. 15.7). With a vacuum between the plates, the capacitance C_0 is defined as

$$C_0 = \frac{e_0 A}{d} \tag{15.9}$$

where e_0 *is the permitivity of free space*, $8.854 \times 10^{-12}\,\mathrm{F\,m^{-1}}$, and A is the area of the plates. Since e_0 is constant, the capacitance depends only on the dimensions of the capacitor. On applying a potential difference, V, between the plates, a quantity of charge, Q_0, is stored on them, given by

$$Q_0 = C_0 V \tag{15.10}$$

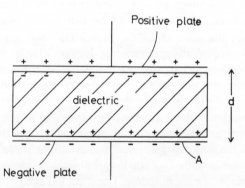

Fig. 15.7 Dielectric material between the plates
of a parallel plate capacitor

If a dielectric substance is now placed between the plates and the same potential difference applied, the amount of charge stored increases to Q_1 and the capacitance therefore increases to C_1. The dielectric constant or relative permittivity, ε', of the dielectric is related to this increase in capacitance by

$$\varepsilon' = \frac{C_1}{C_0} \tag{15.11}$$

The magnitude of ε' depends on the degree of polarization or charge displacement that can occur in the material. For air, $\varepsilon' \simeq 1$. For most ionic solids, $\varepsilon' = 5$ to 10. For ferroelectric materials such as $BaTiO_3$, $\varepsilon' = 10^3$ to 10^4.

The polarizability, α, of the dielectric is defined by

$$p = \alpha E \tag{15.12}$$

where p is the dipole moment induced by the local electric field, E. The polarizability has four possible components and is given by the summation:

$$\alpha = \alpha_e + \alpha_i + \alpha_d + \alpha_s \tag{15.13}$$

These four components are as follows:

(a) The *electronic polarizability*, α_e, is caused by a slight displacement of the negatively charged electron cloud in an atom relative to the positively charged nucleus. Electronic polarizability occurs in all solids and in some, such as diamond, it is the only contributor to the dielectric constant since ionic, dipolar and space charge polarizabilities are absent.

(b) The *ionic polarizability*, α_i, arises from a slight relative displacement or separation of anions and cations in a solid. It is the principal source of polarization in ionic crystals.

(c) *Dipolar polarizability*, α_d, arises in materials such as HCl or H_2O that contain permanent electric dipoles. These dipoles may change their orientation and they tend to align themselves with an applied electric field. The effect is usually very temperature dependent since the dipoles may be 'frozen in' at low temperatures.

(d) *Space charge polarizability*, α_s, occurs in materials that are not perfect dielectrics but in which some long range charge migration may occur. In NaCl, for instance, cations migrate preferentially towards the negative electrode by means of crystal defects such as cation vacancies; consequently, an electrical double layer builds up at the electrode–NaCl interface. When such effects are appreciable, the material is better regarded as a conductor or solid electrolyte than as a dielectric. Apparent dielectric constants as high as 10^6 to 10^7 may be measured (corresponding to double-layer capacitances of $\sim 10^{-6}$ F) but these values have no significance in the conventional dielectric sense.

The magnitude of α usually decreases in the order, $4 > 3 > 2 > 1$, although, clearly, not all materials show all types of polarization. Experimentally, the four

536

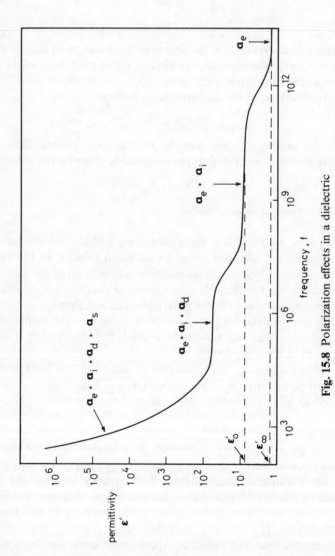

Fig. 15.8 Polarization effects in a dielectric

contributions to α and ε' may be separated by making measurements over a wide range of a.c. frequencies using a combination of capacitance bridge, microwave and optical measurements (Fig. 15.8). At low frequencies, e.g. audiofrequencies ($\sim 10^3$ Hz), all four (if present) may contribute to α. At radio frequencies, ($\sim 10^6$ Hz), space charge effects may not have time to build up in most ionically conducting materials and are effectively 'relaxed out'. At microwave frequencies ($\sim 10^9$ Hz) dipoles do not usually have time to reorient themselves and are effectively relaxed out. The timescale of ionic polarizations is such that they do not occur at frequencies higher than infrared ($\sim 10^{12}$ Hz). This then leaves the electronic polarization which is observable into the UV but is relaxed out at X-ray frequencies.

In good dielectric materials which do not contain contributions from α_s and α_d, the limiting low frequency permittivity, ε'_0, is composed mainly of α_i and α_e polarizations. This permittivity ε'_0 may be obtained from a.c. capacitance bridge measurements in which the value of the capacitance is determined with and without the dielectric substance placed between the plates of the capacitor or cell (equation 15.11). The value of ε'_∞ which contains α_e contributions only may be obtained from refractive index measurements (visible light frequencies) using the simple relation:

$$n^2 = \varepsilon'_\infty \tag{15.14}$$

Values of ε'_0 and ε'_∞ for NaCl, which are fairly typical of ionic crystals, are 5.62 and 2.32, respectively.

For detailed information about dielectrics, measurements are usually made over a range of frequencies covering the audiofrequency, radiofrequency and microwave regions. The results are plotted as Cole–Cole complex permittivity diagrams or as the dielectric loss factor, $\tan \delta$. Let us consider the response of a dielectric to an alternating electric field. If a low frequency sinusoidal e.m.f. is applied across a dielectric, the various polarizations can all occur before the field reverses. These polarization processes act to transmit energy through the dielectric and are equivalent to an alternating current. At low frequencies, the a.c. current leads the e.m.f. by exactly 90° (Fig. 15.9a and b). This is the behaviour desired of an ideal dielectric because energy losses due to the Joule heating are zero, i.e. the vector product $i \times v$ is zero for a phase difference of 90° between i and v. In such a case, energy is transmitted through the sample without *dielectric losses* occurring. As the frequency is raised, a stage is reached where the ionic polarizations (assuming that α_s, $\alpha_d = 0$) can no longer keep up with the alternating e.m.f. Consequently, the current leads the voltage by less than 90°, i.e. by an angle $(90 - \delta)$. The current has a component $i \sin \delta$ that is in phase with the voltage (Fig. 15.9c). This gives rise to dissipation of energy as heat, i.e. dielectric losses.

The frequencies at which δ is significant correspond to the region in Fig. 15.8 where ε' is not constant but is varying between ε'_0 and ε'_∞. In this region, it is useful to represent the permittivity as a complex number, ε^*, given by

$$\varepsilon^* = \varepsilon' - j\varepsilon'' \tag{15.15}$$

538

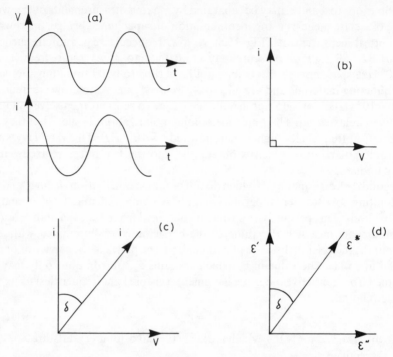

Fig. 15.9 (a), (b) Phase lag of 90° between i and v in a dielectric, (c) Dielectric losses when $\delta \neq 0$ and (d) $\tan \delta = \varepsilon''/\varepsilon'$

ε' is the real part of ε^* and is equivalent to the measured dielectric constant; ε'' is the loss factor and is a measure of the conductance or dielectric losses in the material; $\tan \delta$ is given by the ratio $\varepsilon''/\varepsilon'$ (Fig. 15.9d). The variation of ε' and ε'' with frequency is shown in Fig. 15.10; ε'' has a characteristic frequency dependence, passing through a maximum value at the frequency where ε' undergoes its maximum rate of change with frequency. The shape of the ε'' peak is called a Debye peak. It is characterized by the formula:

$$\varepsilon'' = (\varepsilon_0' - \varepsilon_\infty') \frac{\omega\tau}{1 + \omega^2\tau^2} \tag{15.16}$$

where $\omega = 2\pi f$ and τ is the characteristic relaxation time or decay time of the ionic polarization; the maximum of the ε'' peak occurs when $\omega\tau = 1$ (i.e. $\omega = \tau^{-1}$). It is given by $\varepsilon''_{max} = \frac{1}{2}(\varepsilon_0' - \varepsilon_\infty')$.

Results may also be presented as complex plane diagrams, known as Cole–Cole plots, in which ε'' is plotted against ε' (Fig. 15.11). Note that ε' is given by

$$\varepsilon' = \varepsilon_\infty' + \frac{\varepsilon_0' - \varepsilon_\infty'}{1 + \omega^2\tau^2}$$

The results fall on a semi-circle and the values of ε_0' and ε_∞' may be obtained by extrapolating the data to cut the ε' axis.

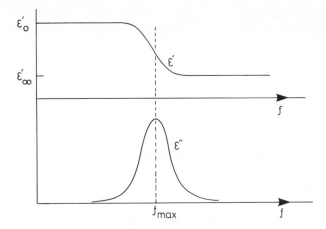

Fig. 15.10 Frequency dependence of ε' and ε''

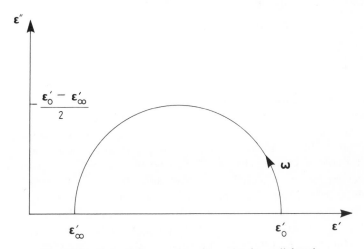

Fig. 15.11 Cole–Cole complex plane plot for a dielectric

In practice, the Cole–Cole plots are often not exactly semi-circular but are distorted to varying degrees. Similarly, the peaks in the dielectric loss, ε'', are not symmetric Debye peaks but are often broadened asymmetrically. The traditional approach to describing such distorted peaks is to regard them as the superposition of an appropriate number of individual Debye peaks, each, occurring at different frequency. This then introduces the concept of a distribution of relaxation times.

In recent years, this approach has been questioned, principally by Jonscher, who has formulated his so-called 'Law of Universal Dielectric Response'. Jonscher has derived a set of equations that fit experimental data remarkably well. One such equation which applies to dielectric loss in materials that are also

540

conductors of electricity is

$$\varepsilon'' \propto \frac{\sigma}{\omega e_0}$$

$$\propto \left(\frac{\omega}{\omega_p}\right)^{n_1-1} + \left(\frac{\omega}{\omega_p}\right)^{n_2-1} \tag{15.17}$$

where ω is the angular frequency $2\pi f$, ω_p is the hopping frequency of the carriers and n_1, n_2 are constants. Inherent to this equation is the notion that individual polarization events, whether they be the hopping of ions in conductors or the reorientation of dipoles in dielectrics, do not take place independently of each other but, rather, interact cooperatively. This means that if a dipole in a crystal reorients itself, this must influence neighbouring dipoles in the crystal. At this stage, however, it is not clear how Jonscher's Law can be applied quantitatively to the characterization of such phenomena. A further discussion of complex plane diagrams has been given in Chapter 13, in which principal attention is given to conductivity rather than dielectric behaviour.

15.4 Ferroelectricity

Ferroelectric materials are distinguished from ordinary dielectrics by (a) their extremely large permittivities and (b) the possibility of retaining some residual electrical polarization after an applied voltage has been switched off. As the potential difference applied across a dielectric substance is increased, a proportional increase in the induced polarization, P, or stored charge, Q, occurs (equation 15.10). With ferroelectrics, this simple linear relation between P and V does not hold, as shown in Fig. 15.12. Instead, more complicated behaviour with a hysteresis loop is observed. The polarization behaviour that is observed on increasing the voltage is not reproduced on subsequently decreasing the voltage. Ferroelectrics exhibit a saturation polarization, P_s, at high field strength (for $BaTiO_3$, $P_s = 0.26\,C\,m^{-2}$ at 23 °C) and a remanent polarization, P_R, which is the value retained as V is reduced to zero after saturation. In order to reduce the polarization to zero, a reverse field is required; this is the coercive field, E_c.

Some common ferroelectric materials are listed in Table 15.2. All are characterized by structures in which one type of cation present, e.g. Ti^{4+} in $BaTiO_3$, can undergo a significant displacement, e.g. 0.1 Å, relative to its anionic neighbours. These charge displacements give rise to dipoles and the high dielectric constants that are characteristic of ferroelectrics.

The unit cell of $SrTiO_3$, which has the same perovskite structure as $BaTiO_3$, is shown in Fig. 6.15. In the primitive cubic unit cell, titanium ions occupy corner positions, oxygen ions occupy the cube edge centres and strontium is at the cube body centre. Alternatively, the unit cell may be displaced so that its origin is at strontium. In this case, titanium is at the body centre and oxygen ions occupy the face centre positions. Either way, the structure is composed of (TiO_6) octahedra which link together by sharing corners to form a three-dimensional

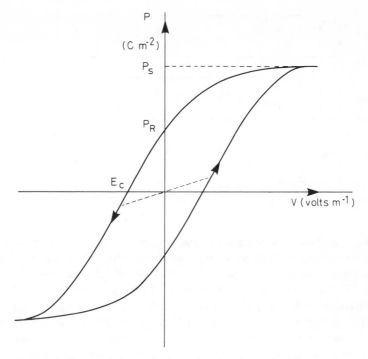

Fig. 15.12 Hysteresis loop of a ferroelectric. The dashed line passing through the origin represents the behaviour of normal dielectric materials

Table 15.2 *Some ferroelectric materials*

	$T_c(°C)$
Barium titanate, $BaTiO_3$	120
Rochelle salt, $KNaC_4H_4O_6.4H_2O$	Between -18 and $+24$
Potassium niobate, $KNbO_3$	434
Potassium dihydrogen phosphate, KDP, KH_2PO_4	-150
Lead titanate, $PbTiO_3$	490
Lithium niobate, $LiNbO_3$	1210
Bismuth titanate, $Bi_4Ti_3O_{12}$	675
Gadolinium molybdate, GMO, $Gd_2(MoO_4)_3$	159
Lead zirconate titanate, PZT, $Pb(Zr_xTi_{1-x})O_3$	Depends on x

framework; strontium ions occupy twelve coordinate cavities within this framework.

This ideal, cubic, perovskite structure, which is stable above 120 °C in $BaTiO_3$, does not possess a net dipole moment since the charges are symmetrically positioned. The material therefore behaves as a normal dielectric, albeit with a very high dielectric constant. Below 120 °C, a structural distortion occurs in

542

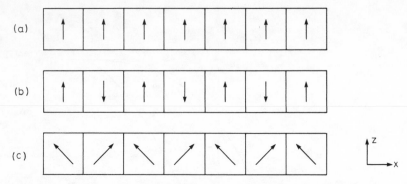

Fig. 15.13 Dipole orientation (schematic) in (a) a ferroelectric, (b) an antiferroelectric and (c) a ferrielectric

$BaTiO_3$. The TiO_6 octahedra are no longer regular because titanium is displaced
off its central position and in the direction of one of the apical oxygens. This gives
rise to a spontaneous polarization. If a similar parallel displacement occurs in all
of the TiO_6 octahedra, a net polarization of the solid results.

In ferroelectric $BaTiO_3$, the individual TiO_6 octahedra are polarized all of the
time; the effect of applying an electric field is to persuade the individual dipoles to
align themselves with the field. When complete alignment of all the dipoles occurs
the condition of saturation polarization is reached. From the observed magnitude of P_s, it has been estimated that titanium is displaced by ~ 0.1 Å off the
centre of its octahedron and in the direction of one of the oxygens. This has been
confirmed by X-ray crystallography. This distance of 0.1 Å or 10 pm is fairly small
when compared with the average Ti—O bond distance of ~ 1.95 Å in TiO_6
octahedra. Alignment of dipoles is shown schematically in Fig. 15.13(a); each
arrow represents, for example, one distorted TiO_6 octahedron and all are shown
with a common direction of distortion.

In ferroelectrics such as $BaTiO_3$, domain structures form because adjacent
TiO_6 dipoles tend to align themselves parallel to each other (Fig. 15.14). The
domains are of variable size but are usually quite large, tens or hundreds of
angstroms across. Within a single domain, the polarization of the dipoles has a
common crystallographic direction. The net polarization of a piece of ferroelectric material is the vector resultant of the polarizations of the individual
domains.

Application of an electric field across a ferroelectric leads to a change in the net
polarization. This can arise from several possible processes:

(a) The direction of polarization of the domains may change. This would happen
   if all the TiO_6 dipoles within a domain were to change their orientation, e.g. if
   all the dipoles in domain (ii) in Fig. 15.14 changed their orientation so as to be
   parallel to the dipoles in domain (i).
(b) The magnitude of P within each domain may increase, especially if some
   randomness in dipole orientation is present before the field is applied.

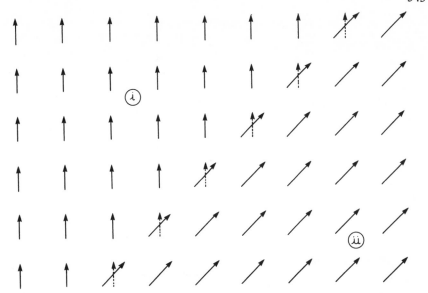

Fig. 15.14 Ferroelectric domains separated by a domain wall or boundary

(c) Domain wall migration may occur such that favourably oriented domains grow in size at the expense of unfavourably oriented ones. For example, domain (i) in Fig. 15.14 may grow by migration of the domain wall one step to the right. To effect this, the dipoles at the edge of domain (ii) change their orientation to the positions shown dashed.

The ferroelectric state is usually a low temperature condition since the effect of increasing thermal motions at high temperatures is sufficient to break down the common displacement in adjacent octahedra and destroy the domain structure. The temperature at which breakdown occurs is the ferroelectric Curie temperature, T_c (Table 15.2). Above T_c, the material is paraelectric (i.e. non-ferroelectric). High dielectric constants still occur above T_c (Fig. 15.15), but no residual polarization is retained in the absence of an applied field. Above T_c, ε' is usually given by the Curie–Weiss Law, $\varepsilon' = C/(T - \theta)$, where C is the Curie constant and θ the Curie–Weiss temperature. Usually, T_c and θ either coincide or differ by only a few degrees. The ferroelectric–paraelectric transition, at T_c, is an example of an order–disorder phase transition. However, unlike order–disorder phenomena in, say, brass, no long range diffusion of ions occurs. Rather, the ordering that occurs below T_c involves preferential distortion or tilting of polyhedra and is therefore an example of a displacive phase transition (Chapter 12). In the high temperature paraelectric phase, the distortions or tilts of the polyhedra, if they occur at all, are randomized.

A necessary condition for a crystal to exhibit spontaneous polarization and be ferroelectric is that its space group should be non-centrosymmetric (Chapter 6). Often the symmetry of the paraelectric phase stable above T_c is centrosymmetric

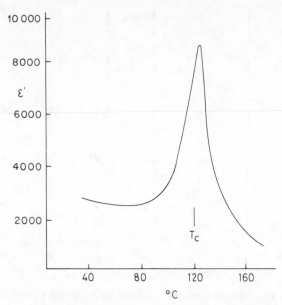

Fig. 15.15 Dielectric constant of $BaTiO_3$ ceramic

and the ordering transition that occurs on cooling simply involves a lowering of symmetry to that of a non-centric space group.

Several hundred ferroelectric materials are now known, including a large number of oxides that have distorted (non-cubic) perovskite structures. These contain cations that are happy in a distorted octahedral environment—Ti, Nb, Ta—and the asymmetric bonding within the MO_6 octahedron gives rise to spontaneous polarization and a dipole moment. Not all perovskites are ferroelectric, e.g. $BaTiO_3$ and $PbTiO_3$ are whereas $CaTiO_3$ is not, and this may be correlated with the ionic radii of the ions involved. It appears that the larger Ba^{2+} ion causes an expansion of the unit cell relative to Ca^{2+}; this results in longer Ti–O bonds in $BaTiO_3$ and allows the Ti^{4+} ions more flexibility to move within the TiO_6 octahedra. Other ferroelectric oxides contain cations that are asymmetrically bonded because of the presence of a lone pair of electrons in their outer valence shell. These are the cations of the heavy p-block elements that are in oxidation states two less than the group valency, e.g. Sn^{2+}, Pb^{2+}, Bi^{3+}, etc.

Ferroelectric oxides are used in capacitors because of their high dielectric constants, especially near to T_c (Fig. 15.15). In order to maximize ε', for practical applications, it is therefore necessary to displace the Curie point so that it is close to room temperature. The Curie point of $BaTiO_3$, 120 °C (Fig. 15.15), may be lowered and broadened when either Ba^{2+} or Ti^{4+} are partially replaced by other ions. The substitution $Ba^{2+} \rightleftharpoons Sr^{2+}$ causes a unit cell contraction and reduction in T_c; replacement of 'active' Ti^{4+} by 'non-active' tetravalent ions such as Zr^{4+} and Sn^{4+} causes a rapid decrease in T_c.

A related type of spontaneous polarization occurs in *antiferroelectric* ma-

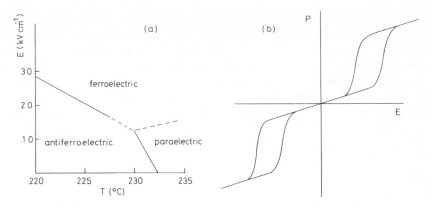

Fig. 15.16 (a) Antiferroelectric–ferroelectric transition in PbZrO$_3$ as a function of the applied field, E. (b) Polarization behaviour across this transition

terials. In this, individual dipoles again occur but they generally arrange themselves so as to be antiparallel to adjacent dipoles (Fig. 15.13b). As a result, the net spontaneous polarization is zero. Above the antiferroelectric Curie temperature, the materials revert to normal paraelectric behaviour. Examples of antiferroelectrics, with their Curie temperatures, are: lead zirconate, PbZrO$_3$, 233 °C; sodium niobate, NaNbO$_3$, 638 °C; and ammonium dihydrogen phosphate, NH$_4$H$_2$PO$_4$, -125 °C.

The electrical characteristics of antiferroelectrics are rather different to those of ferroelectrics. The antiferroelectric state is a non-polar one and no hysteresis loop occurs, although a large increase in permittivity may occur close to T_c (for PbZrO$_3$, $\varepsilon' \simeq 100$ at 200 °C and $\simeq 3000$ at 230 °C). Sometimes, the antiparallel arrangement of dipoles in the antiferroelectric state is only marginally more stable than the parallel arrangement in the ferroelectric state and a small change in conditions may lead to a phase transition. For example, application of an electric field to PbZrO$_3$ causes it to change from an antiferroelectric to a ferroelectric structure (Fig. 15.16a); the magnitude of the field required depends on temperature. The polarization behaviour is then as shown in Fig. 15.16(b). At low fields, no hysteresis occurs and PbZrO$_3$ is antiferroelectric; at high positive and negative fields, hysteresis loops occur and PbZrO$_3$ is ferroelectric.

A related type of polarization phenomenon in which the structure is antiferroelectric in certain direction(s) only is shown in Fig. 15.13(c); in the x direction, the net polarization is zero and the structure is antiferroelectric, but in the z direction, a net spontaneous polarization occurs. This type of structure is known as *ferrielectric*; it occurs in, for example, Bi$_4$Ti$_3$O$_{12}$ and lithium ammonium tartate monohydrate.

The importance of hydrogen bonding in certain ferroelectric and antiferroelectric materials is shown in Fig. 15.17. Ferroelectric KH$_2$PO$_4$ (Fig. 15.17a) and antiferroelectric NH$_4$H$_2$PO$_4$ (Fig. 15.17b) are both built of isolated PO$_4$ tetrahedra that are linked together by K$^+$ and NH$_4^+$ ions (not shown) and hydrogen bonds. These hydrogen bonds link up oxygens in adjacent

PO$_4$ tetrahedra. The two structures differ mainly in the positions of hydrogen in the hydrogen bonds; both have large orthorhombic unit cells, parts of which are shown as projections down c. The PO$_4$ tetrahedra are seen edge-on and are represented by squares; the diagonal line in each corresponds to the upper oxygen–oxygen edge of the tetrahedron. Oxygens are therefore at the corners of the squares. The relative c heights of the tetrahedra are given by the phosphorous positions in Fig. 15.17(a); they are the same in Fig. 15.17(b). The tetrahedra are staggered along c such that the upper edge of one tetrahedron, e.g. XX′, is at about the same c height as the lower edge, YY′, of two adjacent tetrahedra.

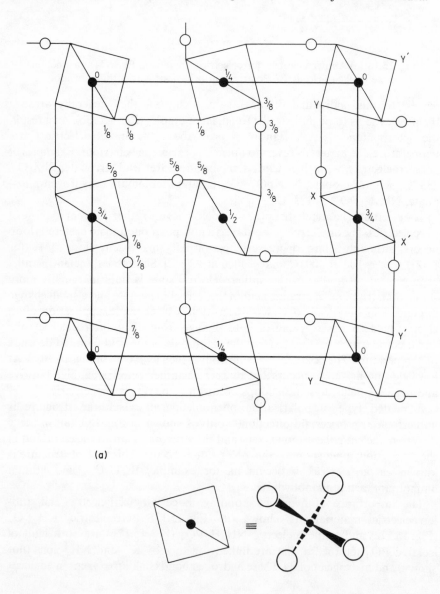

(a)

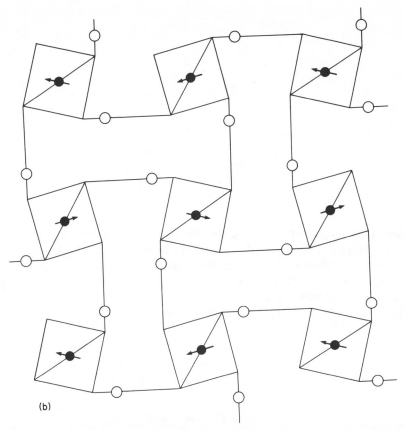

Fig. 15.17 (a) Ferroelectric KH_2PO_4 and (b) antiferroelectric $NH_4H_2PO_4$;
●P ○H, (001) projections

Each PO_4 tetrahedron forms four hydrogen bonds with adjacent PO_4 tetrahedra. In each of these hydrogen bonds, the hydrogens are displaced so as to be nearer to one oxygen or the other, i.e. in each hydrogen bond, the hydrogen atom has a choice of two positions, neither of which is midway along the bond. For each PO_4 tetrahedron, therefore, two hydrogens are close and two are somewhat further away. In the high temperature, paraelectric forms of KH_2PO_4 and $NH_4H_2PO_4$, the hydrogen positions are randomized over the two positions in each bond and a disordered structure obtains. In ferroelectric, low temperature, KH_2PO_4, the hydrogens order themselves so that both are associated with the upper edge of each PO_4 tetrahedron. The hydrogens are responsible indirectly for spontaneous polarization within the PO_4 tetrahedra since the phosphorus atoms are displaced downward away from the hydrogen atoms (Fig. 15.18). This generates dipoles whose directions are parallel to c. In order to reverse the directions of the dipoles it is not necessary to bodily invert the tetrahedra. Instead, a simple movement of hydrogen atoms within the H bond

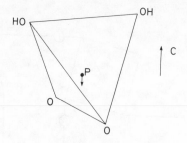

Fig. 15.18 Displacement of phosphorus within a $PO_2(OH)_2$ tetrahedron giving rise to spontaneous polarization

achieves the same effect. The two hydrogens associated with the upper oxygens in Fig. 15.17(a) move away laterally to associate themselves with the lower oxygens of adjacent tetrahedra. At the same time, two hydrogens move in to associate themselves with the lower oxygens. This motion of hydrogen atoms perpendicular to c leads to dipole reversal parallel to c.

In antiferroelectric $NH_4H_2PO_4$, the two hydrogens of each tetrahedron are associated with one upper and one lower oxygen (Fig. 15.17b); this creates dipoles in a direction perpendicular to c (i.e. in the (001) plane). The dipole directions are shown as small arrows in Fig. 15.17(b), from which it can be seen that the net polarization over the whole crystal is zero.

15.5 Pyroelectricity

Pyroelectric crystals are related to ferroelectric ones in that they are non-centrosymmetric and exhibit a net spontaneous polarization, P_s. Unlike ferroelectrics, however, the direction of P_s cannot be reversed by an applied electric field. P_s is usually temperature dependent:

$$\Delta P_s = \pi \Delta T \tag{15.18}$$

where π is the pyroelectric coefficient. This is mainly because the thermal expansion that occurs on heating changes the sizes (i.e. lengths) of the dipoles. A good example of a pyroelectric crystal is ZnO, which has the wurtzite structure. It contains a hexagonal close packed array of O^{2-} ions with Zn^{2+} ions located in one set of tetrahedral sites, say the T_+ sites (Chapter 7). The ZnO_4 tetrahedra all point in the same direction and since each tetrahedron possesses a dipole moment, the crystal has a net polarization. Opposite, (001) surfaces of a ZnO crystal must contain, respectively, Zn^{2+} and O^{2-} ions as the outermost layers of ions. Usually, however, polar impurity molecules are absorbed onto the crystal in order to neutralize the surface charges. Consequently, the pyroelectric effect in a crystal is often not detectable under constant temperature conditions but becomes apparent only when the crystal is heated, thereby changing P_s.

15.6 Piezoelectricity

Under the action of an applied mechanical stress, piezoelectric crystals polarize and develop electrical charges on opposite crystal faces. As is the case for ferro- and pyroelectricity, the crystal must belong to one of the non-centrosymmetric point groups. The occurrence of piezoelectricity depends on the crystal structure of the material and the direction of applied stress, e.g. quartz develops a polarization when subjected to a compressive stress along [100] but not when stressed along [001]. The polarization, P, and stress, σ, are related to the piezoelectric coefficient, d, by

$$P = d\sigma \qquad (15.19)$$

Many crystals that contain tetrahedral groups, e.g. ZnO, ZnS, are piezoelectric since application of a shearing stress distorts the tetrahedra. One of the most important piezoelectrics is PZT, lead zirconate titanate, which is a series of solid solutions between $PbZrO_3$ and $PbTiO_3$. These solid solutions are also antiferroelectric and ferroelectric at certain compositions, as shown by a partial phase diagram (Fig. 15.19). The best piezoelectric compositions occur at $x \simeq 0.5$.

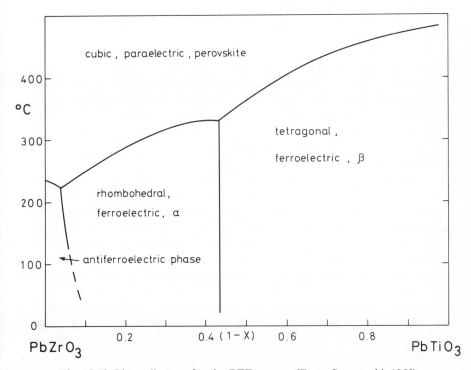

Fig. 15.19 Phase diagram for the PZT system. (From Sawaguchi, 1953)

15.7 Relationship between Ferro-, Piezo- and Pyroelectricity

All three of these properties are concerned with polar effects in crystals and, obviously, they have a great deal in common. They are related as follows. A great many materials may be included in the general class of dielectrics in which their properties, especially electrical properties, are influenced by electric fields. Piezoelectrics are a subclass of dielectrics. Piezoelectrics develop an electric charge when mechanically stressed; conversely, piezoelectrics generate mechanical stresses under the action of an applied electric field. A subclass of piezoelectrics is pyroelectrics, materials that are spontaneously polarized and therefore exhibit a net dipole moment. Some pyroelectric materials are also ferroelectric because the spontaneous polarization can be reversed under the action of an applied electric field. By definition, therefore, ferroelectric materials are also pyroelectric and piezoelectric; further, pyroelectric materials are also piezoelectric. The reverse does not hold, however; i.e. not all piezoelectric materials are pyroelectric, etc.

15.8 Applications of Ferro-, Piezo- and Pyroelectrics

Applications of these materials are considered together because they overlap considerably and, in some cases, more than one property may be involved in a particular application. Applications are very varied, especially of ferroelectrics, and only a brief summary of the less esoteric uses can be given. An excellent up-to-date account is given by Burfoot and Taylor (1979), to whom the interested reader is referred.

The main commercial application of ferroelectrics is in capacitors. Because ferroelectrics have a high permittivity or dielectric constant, ε', usually in the range 10^2 to 10^4, they can be used in the construction of large capacitors (equations 15.9 and 15.11). The main commercial materials are $BaTiO_3$ and PZT (lead zirconate titanate) which are used in the form of dense, polycrystalline ceramics. By contrast, conventional dielectrics such as TiO_2 or $MgTiO_3$ have ε' in the range of 10 to 100. Hence, for a given volume, a $BaTiO_3$ capacitor has 10 to 1000 times the capacitance of a dielectric capacitor.

An important use of certain ferroelectrics such as $BaTiO_3$ and $PbTiO_3$, which does not depend directly on their ferroelectricity, is in PTC thermistors (i.e. positive temperature coefficient thermally sensitive resistors). In most non-metallic materials, the electrical resistivity decreases with increasing temperature, i.e. the resistivity has a negative temperature coefficient (NTC). However, some ferroelectrics including $BaTiO_3$, show an anomalous and large increase in resistivity as the temperature approaches the ferroelectric–paraelectric transition temperature, T_c (Fig. 15.20). This increase in ρ matches the large increase in ε' close to T_c although the reasons for the increase in ρ appear not to be well understood. PTC thermistors are used as switches; when a current is passed through any resistive material, Joule heating losses, given by $I^2 R$, cause the material to heat up. With a $BaTiO_3$ thermistor, the resistivity increases

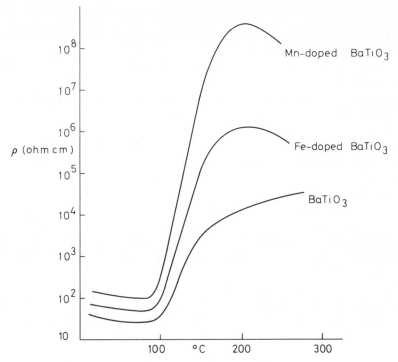

Fig. 15.20 Positive temperature coefficient resistivity in semiconducting BaTiO$_3$ ceramic with various dopants. (From Ueoka, 1974)

dramatically as it heats and consequently the current switches off. Applications include (a) thermal and current overload protection devices, in which the thermistor acts as a reusable fuse, and (b) time delay fuses. For a description of NTC thermistors based on, for example, MnO and NiO semiconductors, see Chapter 14.

Pyroelectric crystals are used mainly in infrared radiation detectors. If desired, they can be made spectrally sensitive by coating the probe surface of the crystal with appropriate absorbing material. For detectors, it is desirable to maximize the ratio π/ε', which means that ferroelectric materials with high dielectric constant are not suitable. The best detector material found to date is triglycine sulphate.

Piezoelectric crystals have been used for many years as transducers for converting mechanical to electrical energy, and vice versa. Applications are diverse, e.g. as bimorphs in microphones, earphones, loudspeakers and stereo pick-ups; as fuses, solenoid ignition systems and cigarette lighters, sonar generators and ultrasonic cleaners. More complex devices are used in transformers, filters and oscillators. Most of these applications use PZT ceramics, quartz, Rochelle salt or Li$_2$SO$_4 \cdot$H$_2$O.

Questions

15.1 Give two methods for distinguishing between conduction by positive holes and electrons in a semiconductor.

15.2 Construct a graph of e.m.f. against temperature, in the range -100 to $500\,^\circ\mathrm{C}$, for an iron–constantan thermocouple.

15.3 What are the differences between paraelectric (i.e. dielectric), ferroelectric, ferrielectric and antiferroelectric materials?

15.4 If the following substances were placed between the plates of a capacitor, approximately what values would you expect for their apparent dielectric constant: (a) Ar gas; (b) water; (c) ice; (d) pure single-crystal silicon; (e) pure single-crystal KBr; (f) single crystal of KBr doped with $CaBr_2$; (g) Na β-alumina; (h) $BaTiO_3$?

15.5 What differences might you expect between the temperature dependence of the conductivity of (a) $CaTiO_3$, (b) $PbTiO_3$?

15.6 Which, if any, of the following crystals might you expect to exhibit piezoelectricity: (a) $NaCl$; (b) CaF_2; (c) $CsCl$; (d) ZnS, wurtzite; (e) $NiAs$; (f) TiO_2, rutile?

15.7 Comment on the validity of the following statement. 'Pyroelectric substances are those that develop a net spontaneous polarization on heating.'

15.8 What are dielectric losses, what are they caused by and how might they be minimized in materials that are to be used as electrical insulators?

References

J. C. Burfoot and G. W. Taylor (1979). *Polar Dielectrics and Their Applications*, Macmillan.

A. K. Jonscher (1983). *Dielectric Relaxation in Solids*, Chelsea Dielectrics Press, London.

Mansel Davies *Molecular Properties*.

R. E. Newnham (1975). *Structure—Property Relations*, Springer–Verlag.

E. Sawaguchi (1953). *J. Phys. Soc. Japan*, **8**, 615.

H. Ueoka (1974). *Ferroelectrics*, 7(3), 351.

Chapter 16

Magnetic Properties

16.1 Introduction

Inorganic solids that exhibit magnetic effects other than diamagnetism, which is a property of all substances, are characterized by having unpaired electrons present. These are usually located on metal cations. Magnetic behaviour is thus restricted mainly to compounds of transition metals and lanthanides, many of which possess unpaired d and f electrons, respectively. Several magnetic effects are possible.

The unpaired electrons may be oriented at random on the different atoms, in which case the material is *paramagnetic*. They may be aligned so as to be parallel, in which case the material possesses an overall magnetic moment and is *ferromagnetic*. Alternatively, they may be aligned in antiparallel fashion, giving zero overall magnetic moment and *antiferromagnetic* behaviour. If the alignment of the spins is antiparallel but with unequal numbers in the two orientations, a net magnetic moment results and the behaviour is *ferrimagnetic*.

554

Clearly, there are strong analogies between these magnetic properties and corresponding electrical properties such as ferroelectricity (Chapter 15). One difference, of course, is the absence of the magnetic monopole. This would be the magnetic equivalent of an electrical charge, as possessed by an ion or an electron.

Magnetic oxides, especially ferrites such as $MgFe_2O_4$, are modern-day materials with uses in transformer cores, magnetic recording and information storage devices, etc. The theory of magnetic behaviour is, unfortunately, rather complicated. There is a plethora of terms, symbols and units that is bewildering to the uninitiated. To make matters worse, there appear to be two different and rather contradictory ways of evaluating the magnetic moments of ions containing unpaired electrons. Here we shall use a bare minimum of theory in order to be able to appreciate the different kinds of magnetic behaviour and how they are related to crystal structure.

16.2 Theory

16.2.1 Behaviour of substances in a magnetic field

First, let us see how different materials react when placed in a magnetic field. If a substance is placed in a magnetic field, H, then the density of lines of force in the sample, known as the *magnetic induction*, B, is given by H plus a contribution $4\pi I$ due to the sample itself:

$$B = H + 4\pi I \tag{16.1}$$

where I is the *magnetic moment* of the sample per unit volume. The *permeability*, P, and *susceptibility*, κ, are defined as

$$P = \frac{B}{H} = 1 + 4\pi\kappa \tag{16.2}$$

$$\kappa = \frac{I}{H} \tag{16.3}$$

The *molar susceptibility*, χ, is given by

$$\chi = \frac{\kappa F}{d} \tag{16.4}$$

where F is the formula weight and d the density of the sample.

The different kinds of magnetic behaviour may be distinguished by the values of P, κ, χ and by their temperature and field dependences (Table 16.1).

Diamagnetic substances are those for which $P < 1$ and κ, χ are small and slightly negative. For *paramagnetic* substances, the reverse is the case, $P > 1$ and κ, χ are positive. When placed in a magnetic field, the number of lines of force passing through a substance is greater if it is paramagnetic and slightly less if it is diamagnetic, than would pass through a vacuum (Fig. 16.1). Consequently, paramagnetic substances are attracted by a magnetic field whereas diamagnetic substances experience a slight repulsion.

Table 16.1 *Magnetic susceptibilities*

Behaviour	Typical χ value	Change of χ with increasing temperature	Field dependence?
Diamagnetism	-1×10^{-6}	None	No
Paramagnetism	0 to 10^{-2}	Decreases	No
Ferromagnetism	10^{-2} to 10^6	Decreases	Yes
Antiferromagnetism	0 to 10^{-2}	Increases	(Yes)

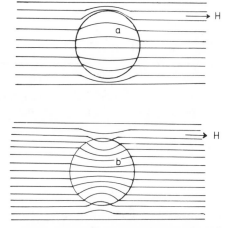

Fig. 16.1 Behaviour of (a) diamagnetic and (b) paramagnetic substances in a magnetic field

In *ferromagnetic* substances, $P \gg 1$ and large values of κ, χ are observed. Such materials are strongly attracted to a magnetic field. In *antiferromagnetic* substances, $P > 1$ and κ, χ are positive; values of κ, χ are comparable to or somewhat less than those for paramagnetic substances.

16.2.2 Effects of temperature: Curie and Curie–Weiss Laws

The susceptibilities of the different kinds of magnetic material are distinguished by their different temperature dependences as well as by their absolute magnitudes. Many paramagnetic substances obey the simple Curie Law, especially at high temperatures. This states that the magnetic susceptibility is inversely proportional to temperature:

$$\chi = \frac{C}{T} \tag{16.5}$$

where C is the Curie constant. Often, however, a better fit to the experimental

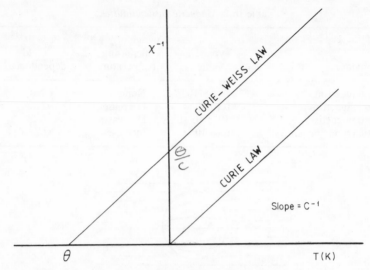

Fig. 16.2 Plot of reciprocal susceptibility against temperature showing
Curie and Curie–Weiss Law behaviour

data is provided by the Curie–Weiss Law:

$$\chi = \frac{C}{T + \theta}$$ (16.6)

where θ is the Weiss constant. These two types of behaviour are shown in
Fig. 16.2 in which χ^{-1} is plotted against T.

For ferro- and antiferromagnetic substances, the temperature dependence of χ
does not fit the simple Curie/Curie–Weiss Laws, as shown schematically in
Fig. 16.3. Ferromagnetic materials show a very large susceptibility at low
temperatures that decreases increasingly rapidly with rising temperature
(Fig. 16.3b). Above a certain temperature (the ferromagnetic Curie temperature,
T_C), the material is no longer ferromagnetic but reverts to paramagnetic, where
Curie–Weiss Law behaviour is usually observed. For antiferromagnetic materials (Fig. 16.3c), the value of χ actually increases with rising temperature up to
a critical temperature, known as the Néel point, T_N. Above T_N, the material again
reverts to paramagnetic behaviour.

The magnitude of χ in the different materials and its variation with
temperature may be explained as follows.

The paramagnetic χ values correspond to the situation where unpaired
electrons are present in the material and show some tendency to align themselves
in a magnetic field. In ferromagnetic materials, the electron spins are aligned
parallel due to cooperative interactions between spins on neighbouring ions in
the crystal structure. The large χ values represent this parallel alignment of a
large number of spins. In general, not all spins are parallel in a given material
unless (a) very high magnetic fields and (b) low temperatures are used. In

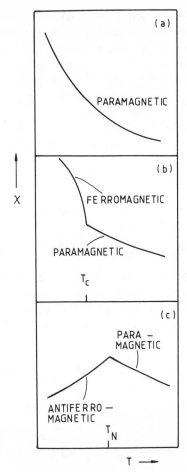

Fig. 16.3 Temperature dependence of the magnetic susceptibility for (a) paramagnetic, (b) ferromagnetic and (c) antiferromagnetic materials

antiferromagnetic materials, the electron spins are aligned antiparallel and have a cancelling effect on χ. Hence small χ values are expected. Residual χ values may be associated with disorder in the antiparallel spin arrangement.

For all materials, the effect of increasing temperature is to increase the thermal energy possessed by ions and electrons. There is, therefore, a natural tendency for increasing structural disorder with increasing temperature. For paramagnetic materials, the thermal energy of ions and electrons acts to partially cancel the ordering effect of the applied magnetic field. Indeed, as soon as the magnetic field is removed, the orientation of the electron spins becomes disordered. Hence, for paramagnetic materials, χ decreases with increasing temperature, in Curie/Curie–Weiss Law fashion.

For ferro- and antiferromagnetic materials, the effect of temperature is to introduce disorder into the otherwise perfectly ordered parallel/antiparallel arrangement of spins. For ferromagnetic materials, this leads to a rapid decrease in χ with increasing temperature. For antiferromagnetic materials, this leads to a decrease in the degree of antiparallel ordering, an increase in the number of 'disordered' electron spins and hence an increase in χ.

The magnetic properties of materials are often conveniently expressed in terms of the magnetic moment, μ, since this is a parameter that may be related directly to the number of unpaired electrons present. The relationship between χ and μ is

$$\chi = \frac{N\beta^2\mu^2}{3kT} \tag{16.7}$$

where N is Avogrado's number, β is the Bohr magneton and k is Boltzmann's constant. Substituting for N, β and k gives

$$\mu = 2.83\sqrt{\chi T} \tag{16.8}$$

Magnetic susceptibilities and moments are often determined experimentally using a Gouy balance. The sample is placed in the jaws of an electromagnet and the variation in sample mass is monitored as a function of applied field. For paramagnetic substances, unpaired electrons are attracted by a magnetic field and this is shown by an apparent increase in mass of the sample when the field is switched on. The measured susceptibility is corrected for various factors, including diamagnetism of the sample and sample holder.

16.2.3 Calculation of magnetic moments

The approach generally used by spectroscopists and chemists for the calculation of magnetic moments of ions that contain unpaired electrons is as follows.

The magnetic properties of unpaired electrons are regarded as arising from two causes, electron spin and electron orbital motion. Of most importance is the spin component. An electron may usefully be visualized as a bundle of negative charge spinning on its axis. The magnitude of the resulting spin moment, μ_s, is 1.73 Bohr magnetons (BM), where the Bohr magneton is defined as

$$1\,\text{BM} = \frac{eh}{4\pi mc} \tag{16.9}$$

where e = electron charge
h = Planck's constant
m = electron mass
c = velocity of light

The formula used for calculating μ_s for a single electron is

$$\mu_s = g\sqrt{s(s+1)} \tag{16.10}$$

Table 16.2 *Experimental and theoretical magnetic moments for some transition metal ions.*
(Data taken from Cotton and Wilkinson, 1966)

Ion	Number of unpaired electrons	$\mu_{s(calc)}$	$\mu_{S+L(calc)}$	$\mu_{(observed)}$
V^{4+}	1	1.73	3.00	~1.8
V^{3+}	2	2.83	4.47	~2.8
Cr^{3+}	3	3.87	5.20	~3.8
Mn^{2+}	5(high spin)	5.92	5.92	~5.9
Fe^{3+}	5(high spin)	5.92	5.92	~5.9
Fe^{2+}	4(high spin)	4.90	5.48	5.1–5.5
Co^{3+}	4(high spin)	4.90	5.48	~5.4
Co^{2+}	3(high spin)	3.87	5.20	4.1–5.2
Ni^{2+}	2	2.83	4.47	2.8–4.0
Cu^{2+}	1	1.73	3.00	1.7–2.2

where s is the spin quantum number, $\frac{1}{2}$, and g is the gyromagnetic ratio, ~ 2.00. Substituting for s and g gives $\mu_s = 1.73$ BM for a single electron.

For atoms or ions that contain > 1 unpaired electron, the overall spin moment is given by

$$\mu_s = g\sqrt{S(S+1)} \qquad (16.11)$$

where S is the sum of the spin quantum numbers of the individual unpaired electrons. Thus, for high spin Fe^{3+}, containing five unpaired $3d$ electrons, $S = \frac{5}{2}$ and $\mu_s = 5.92$ BM. Calculated values of μ_s for different numbers of unpaired electrons are given in Table 16.2.

The motion of an electron around the nucleus may, in some materials, give rise to an *orbital moment* which contributes to the overall magnetic moment. In cases where the orbital moment makes its full contribution,

$$\mu_{S+L} = \sqrt{4S(S+1) + L(L+1)} \qquad (16.12)$$

where L is the orbital angular momentum quantum number for the ion. Equations (16.10) to (16.12) are applicable to free atoms or ions. In practice, in solid materials, equation (16.12) does not hold because the orbital angular momentum is either wholly or partially *quenched*. This happens when the electric fields of the surrounding atoms or ions may restrict the orbital motion of the electrons.

Experimentally observed magnetic moments for a variety of ions are given in Table 16.2, together with the values calculated using equations (16.11) and (16.12). In most cases, the observed moments are similar to or somewhat larger than the calculated, spin-only values.

The methods outlined above for the calculation of magnetic moments have their origin in quantum mechanics. The details of the methods are actually quite involved, but, even so, agreement between theory and experiment is often not good (Table 16.2). An alternative, much simpler method is often used, especially by people working with properties such as ferro- and antiferromagnetism and their applications. In this, the magnetic moment of a single unpaired electron is set equal to one Bohr magneton. For an ion that possesses n unpaired electrons, the magnetic moment is given by nBM. Thus high spin Mn^{2+}, Fe^{3+} would both have a magnetic moment of 5 BM. This method may be quantified by the simple equation

$$\mu = gS \tag{16.13}$$

where $g \simeq 2.00$ and S, the spin state of the ion, equals $n/2$. The values obtained in this way underestimate the true values (compare columns 2 and 5 of Table 16.2); nevertheless, they provide a rough and useful guide to the magnitude of μ. A modification to equation (16.13) is to allow g to become an adjustable parameter. Equation (16.13) is regarded as a spin-only formula and by allowing g to exceed two, a contribution to μ arising from orbital momentum is effectively allowed for. Thus, for Ni^{2+}, a g value in the range 2.2 to 2.3 is often used. In discussing the magnetic properties of ferrite phases, (Section 16.3). we shall use equation (16.13) to evaluate μ because of its simplicity.

16.2.4 Mechanisms of ferro- and antiferromagnetic ordering, superexchange

In the paramagnetic state, the individual magnetic moments of the ions containing unpaired electrons are arranged at random. Alignment occurs only on application of a magnetic field. The energy of interaction between dipoles and a magnetic field may be calculated readily, although details are not given here. It is, however, generally greater than the thermal energy, kT, possessed by the ions or dipoles.

In the ferro- and antiferromagnetic states, alignment of magnetic dipoles occurs spontaneously. There must therefore be some positive energy of interaction between neighbouring spins that allows this to occur, either in parallel or antiparallel fashion. The origin of this coupling of spins or cooperative interaction is quantum mechanical. Qualitatively, the effect is understood, although a complete rationale for the behaviour of, for example, ferromagnetic iron or cobalt is still needed.

One process by which coupling of spins occurs to give rise to anti-ferromagnetism in, for example, NiO, is known as superexchange. It is shown schematically in Fig. 16.4. The Ni^{2+} ion has eight d electrons. In an octahedral

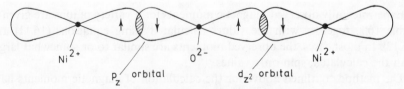

Fig. 16.4 Antiferromagnetic coupling of spins of d electrons on Ni^{2+} ions through p electrons of oxide ions

environment, two of these electrons singly occupy the e_g orbitals, d_{z^2} and $d_{x^2-y^2}$. These orbitals are oriented so as to be parallel to the axes of the unit cell and therefore point directly at adjacent oxide ions. The unpaired electrons in the e_g orbitals of Ni^{2+} ions are able to couple magnetically with electrons in the p orbitals of the O^{2-} ions. This coupling may well involve the formation of an excited state in which the electron transfers from the e_g orbital of the Ni^{2+} ion to the oxygen p orbital. The p orbitals of the O^{2-} ion contain two electrons each, which are also coupled antiparallel. Hence, provided Ni^{2+} and O^{2-} ions are sufficiently close that coupling of their electrons is possible, a chain coupling effect occurs which passes through the crystal structure (Fig. 16.4). The net effect of this is that neighbouring Ni^{2+} ions, separated by intervening O^{2-} ions, are coupled antiparallel.

16.2.5 Some more definitions

Ferromagnetic materials have a domain structure, similar to the domain structure of ferroelectric materials (Chapter 15). Within each domain all the spins are aligned parallel but unless the material is in the saturation condition, different domains have different spin orientations.

The response of ferromagnetic materials to an applied magnetic field is similar to that of ferroelectrics in an applied electric field (Section 15.4). A hysteresis loop occurs in the plot of magnetization, M, or induction, B, against the applied field, H. A similar loop (Fig. 15.12) is observed for the polarization of a ferroelectric plotted against voltage. At sufficiently high fields the saturation magnetization condition is reached when the spins of all the domains are parallel. During the processes of magnetization and demagnetization in an alternating magnetic field, energy is dissipated, usually as heat. During one complete cycle this amount of energy, the *hysteresis loss*, is proportional to the area inside the hysteresis loop. For certain applications, low loss materials are required, one essential for which is that the area encompassed by the hysteresis loop should be as small as possible.

Materials that are magnetically soft are those of low *coercivity*, H_c. The coercivity (Fig. 15.12) is the magnitude of the reverse field required to achieve demagnetization. Soft materials also have high *permeability* and hence a hysteresis loop that is 'narrow at the waist' and of small area. Materials that are magnetically hard are those with a high coercivity and a high *remanent magnetization*, M_r. This latter is the magnetization that remains after the field is switched off (Fig. 15.12). Hard materials are not easily demagnetized, therefore, and find uses as permanent magnets.

Ferromagnetic materials have a preferred or 'easy' direction of magnetization; in iron, this is parallel to the axes of the cubic unit cell (Fig. 16.5a, see later). The *magnetocrystalline anisotropy* is the energy required to rotate the magnetization out of this preferred direction.

An additional source of energy loss in an alternating magnetic field is associated with electrical currents called *eddy currents* that are induced in the material. The varying magnetic field induces a varying voltage and the eddy

562

current losses are given by I^2R or V^2/R. Eddy currents are therefore minimized in highly resistive materials. One advantage that most of the magnetic oxides have over metals is their much higher electrical resistance.

Most magnetic materials exhibit the property of *magnetostriction*, i.e. they change their shape on magnetization. For instance, nickel and cobalt both contract in the direction of magnetization but expand in the perpendicular directions. With iron the reverse happens at low fields. At high fields, iron behaves like nickel and cobalt. The dimensional changes involved are small. The coefficient of magnetostriction, λ_s, defined as $\lambda_s = \Delta l/l_0$, increases with H up to a maximum value in the range 1 to 60×10^{-5} for saturation magnetization. The effect is therefore comparable to changing the temperature of the material by a few degrees.

16.3 Selected examples of magnetic materials, their structures and properties

16.3.1 Metals and alloys

Five transition elements—Cr, Mn, Fe, Co, Ni—and most of the lanthanide elements exhibit either ferro- or antiferromagnetism. A large number of alloys and intermetallic compounds also show some kind of magnetic ordering.

Iron, cobalt and nickel are ferromagnetic, as shown in Fig. 16.5. In body centred cubic α-Fe, the spins point in a [100] direction, parallel to a cubic cell edge, whereas in face centred cubic nickel, they point in a [111] direction, parallel to a cube body diagonal. Cobalt has a hexagonal close packed structure and the spins are oriented parallel to the c axis of the unit cell. These examples demonstrate clearly that ferromagnetism is not associated with a particular type of crystal structure!

Chromium and manganese are both antiferromagnetic at low temperatures, $T_N = 95\,\text{K (Mn)}$, $313\,\text{K (Cr)}$. Mangenese has a complex crystal structure, but that of chromium is body centred cubic, similar to α-Fe. In chromium the spins are arranged in antiparallel fashion, parallel to one of the cube unit cell axes.

Some of the characteristics of ferromagnetic materials are given in Fig. 16.6. In

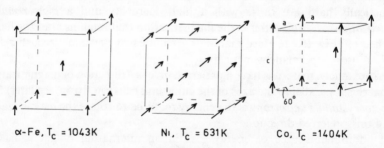

α-Fe, $T_c = 1043\,\text{K}$ Ni, $T_c = 631\,\text{K}$ Co, $T_c = 1404\,\text{K}$

Fig. 16.5 Ferromagnetic ordering in body centred cubic α-Fe, face centred cubic Ni and hexagonal close packed Co

(a) is shown the temperature dependence of the magnetic susceptibility or magnetic moment, although the axes are labelled somewhat differently to this. The vertical axis represents the saturation magnetization of iron, relative to its maximum possible value, which occurs at absolute zero. The horizontal axis has a 'reduced temperature' scale in which the ratio of the actual temperature to the Curie temperature is plotted. Therefore, at the Curie point, $T/T_C = 1$. Use of reduced axes such as these facilitates comparison between materials with different Curie points and different magnetic moments. When plotted in this way,

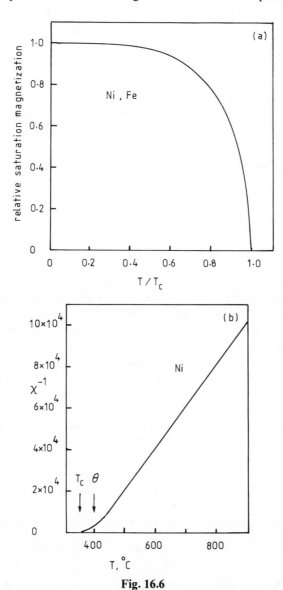

Fig. 16.6

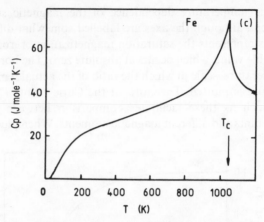

Fig. 16.6 Some properties of ferromagnetic materials: (a) saturation magnetization relative to value at absolute zero, as a function of reduced temperature. (b) Curie–Weiss Law plot for nickel of inverse susceptibility against temperature, showing a deviation close to T_C, (c) heat capacity of iron as a function of temperature. Data taken from Wert and Thomson; original references given therein

iron and nickel are found to behave very similarly: with increasing temperature above OK, the saturation magnetization stays almost constant at small T/T_C and then begins to decrease increasingly rapidly as T_C is approached.

Above the Curie point, iron, cobalt and nickel are paramagnetic. At temperatures well above T_C, Curie–Weiss Law behaviour is observed but deviations occur close to T_C (Fig. 16.6b). These deviations are attributed to the existence of short range order between the spins. The long range order of the ferromagnetic state is lost but residual, short range order remains just above T_C; hence the Weiss temperature, θ, is somewhat removed from T_C. Data are shown for nickel; iron and cobalt behave similarly.

The transition from ferromagnetic to paramagnetic behaviour at T_C has many of the characteristics of a second-order or lambda phase transition (Chapter 12). It is a classic example of an order–disorder phase transition. Order is perfectly attainable only at absolute zero; at all real temperatures, disorder is present and increases rapidly with increasing temperature. This is shown by the heat capacity which passes through a maximum at T_C (Fig. 16.6c).

At this stage, let us give an airing to an unsolved problem. One of the mysteries of ferromagnetism concerns its dependence on position in the Periodic Table and in particular the question of how many unpaired electrons there are available to contribute to ferromagnetism. The facts are as follows. Provide your own explanation!

The three ferromagnetic elements in the first transition series have the electronic configurations shown in Table 16.3. Column 2 gives the configuration of the free ion in its ground state; the $4s$ level is fully occupied in each case. In the

Table 16.3 *Electronic constitution of iron, cobalt and nickel*

Metal	Free ion configuration	Ferromagnetic state	
		Number of unpaired spins	Configuration
Fe	$d^6 s^2$	2.2	$d^{7.4} s^{0.6}$
Co	$d^7 s^2$	1.7	
Ni	$d^8 s^2$	0.6	

ferromagnetic state (column 4), the $4s$ band is not full but some of its electrons are transferred to the $3d$ band. Evidence for this comes from band theory calculations and the value of the saturation magnetization which is proportional to the number of unpaired spins (column 3). Thus, iron has a net moment of 2.2 BM per atom and hence has, on average, 2.2 unpaired d electrons per iron atom; i.e. of the 7.4 d electrons, 4.8 are of one sign and 2.6 of the other. So far, we are on familiar ground because chemists are used to the idea that electrons are transferrable between $4s$ and $3d$ levels, depending on circumstances. The sticking point concerns the maximum number of unpaired electrons that can contribute to ferromagnetism in elements or alloys of the $3d$ series. The answer, apparently, is 2.4 per atom but no satisfactory explanation of the significance of this number 2.4 is available at present. The effective number of unpaired electrons varies with total electron content across the $3d$ series. The maximum value of 2.4 is found for an alloy of composition $Fe_{0.8}Co_{0.2}$. With increasing total electron content, the number of unpaired spins decreases gradually, passing through cobalt and nickel before decreasing to zero in the alloy $Ni_{0.4}Cu_{0.6}$. Thus pure copper is paramagnetic. To the other side of the $Fe_{0.8}Co_{0.2}$ composition, the number of unpaired spins also drops systematically, passing through iron, manganese and chromium. Both manganese and chromium are antiferromagnetic at low temperatures.

One factor that appears to be important is the degree of overlap of the d orbitals and the breadth of the d band. This is directly related to the interatomic separation in the metals, which increases with atomic number across the $3d$ transition series. At short distances, overlap is strong and quantum mechanical exchange forces dictate an antiparallel coupling of spins, as found in anti-ferromagnetic Cr, Mn. With increasing interatomic distance, overlap is still strong but the coupling now gives a parallel arrangement, as in ferromagnetic iron, cobalt and nickel. At larger separation, coupling is weaker and para-magnetic behaviour is observed as in copper.

The lanthanide elements have magnetically ordered structures that are associated with unpaired $4f$ electrons. Exceptions are those elements for which the $4f$ shell is empty: La, $4f^0$; or full: Yb, $4f^{14}$; Lu, $4f^{14}$. Most of the lanthanides are antiferromagnetic below room temperature. Some, especially the later lanthanides, form both ferro- and antiferromagnetic structures at different

Table 16.4 *Néel (antiferromagnetic) and Curie (ferromagnetic) temperatures (K) in lanthanides.* (Data taken from Taylor, 1970).

Element	Neel temperature T_N	Curie temperature T_C
Ce	12.5	
Pr	25	
Nd	19	
Sm	14.8	
Eu	90	
Gd	—	293
Tb	229	222
Dy	179	85
Ho	131	20
Er	84	20
Tm	56	25

temperatures. For these, the sequence with decreasing temperature is always:

Paramagnetic → antiferromagnetic → ferromagnetic

Néel and Curie temperatures are given in Table 16.4. The exception appears to be gadolinium which does not form an antiferromagnetic structure.

Several antiferromagnetic lanthanides exhibit the additional property known as *metamagnetism*. In this, they can be induced to switch over to a ferromagnetic state by application of a suitably high magnetic field. For instance, dysprosium is ferromagnetic below 85 K but at higher temperatures changes to antiferromagnetic. By application of a magnetic field, the ferromagnetic state can be preserved above 85 K and up to the Néel temperature, 179 K.

16.3.2 Transition metal oxides

The oxides of the first transition series show very large, systematic changes in properties with atomic number and number of d electrons. The variation in electrical conductivity of the divalent oxides, MO, has been discussed in Chapter 14, with, for instance, the extremes of metallic conductivity in TiO and insulating behaviour in NiO. The oxides MO also show a large range of magnetic behaviour which parallels roughly the electrical behaviour. The oxides of the early elements, TiO, VO and CrO are diamagnetic. In these oxides, the d electrons are not localized on the individual M^{2+} ions but are delocalized over the entire structure in a partly filled t_{2g} band. There appear to be no magnetic interactions between these delocalized electrons; consequently, the materials are diamagnetic as well as being conductors of electricity. The oxides of the later elements, MnO, FeO, CoO and NiO are paramagnetic at high temperatures and exhibit ordered magnetic structures at low temperatures. In these oxides, the d electrons are localized on the individual M^{2+} ions. This localization of the unpaired electrons is responsible for both the observed magnetic properties and the virtual absence of electrical conductivity.

The oxides MnO, FeO, CoO and NiO are all antiferromagnetic at low temperatures and change to paramagnetic above the Néel temperature, T_N. Values of T_N, in degrees celsius, are MnO, -153; FeO, -75; CoO, -2; and NiO, $+250$. All have similar structures in both the antiferromagnetic and paramagnetic forms. Using nickel oxide as an example, at high temperatures it has the rock salt crystal structure (Fig. 5.9). The structure may be viewed and described in various ways. For our purposes, if the structure is viewed along any of the four equivalent [111] directions, parallel to the body diagonals of the face centred cubic unit cell, alternate layers of Ni^{2+} and O^{2-} ions are found.

Below 250 °C, the structure of NiO undergoes a rhombohedral distortion: a slight contraction of the structure occurs along one of the threefold axes parallel to a [111] direction. A similar contraction occurs in MnO but in FeO the structure is, instead, slightly elongated. The symmetry of the structure is lowered since all of the fourfold axes and three of the threefold axes are lost, leaving a single threefold axis. In practice, the degree of distortion of the structure from cubic symmetry is small and the effect is barely noticeable in, for example, the X-ray powder pattern of NiO.

The cause of the rhombohedral distortion in NiO is antiferromagnetic ordering of the Ni^{2+} ions. Within a given layer of Ni^{2+} ions, the spins of all the Ni^{2+} ions are aligned parallel but in adjacent layers they are all antiparallel (Fig. 16.7). Magnetic superstructures such as these may be studied very elegantly using neutron diffraction (Section 3.2.1.5). Two types of scattering contribute to the observed diffraction pattern: scattering by atomic nuclei and scattering by unpaired electrons. The former type gives a diffraction pattern similar to that observed using X-rays, although the intensities may be somewhat different. In antiferromagnetic structures, the second type of scattering gives rise to extra lines in the neutron powder diffraction pattern. This is because (a) cooperative

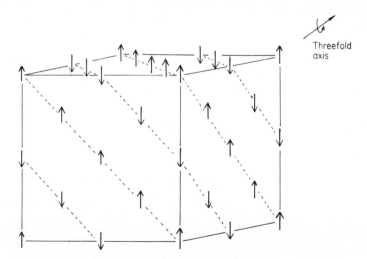

Fig. 16.7 Antiferromagnetic superstructure in MnO, FeO and NiO, showing pseudo-cubic unit cell for which: a (supercell) $= 2a$ (subcell). Oxygen positions are not shown

interactions between unpaired electrons may give rise to a superstructure and (b) neutrons are scattered strongly by unpaired electrons whereas X-rays are not.

Neutron powder diffraction patterns for MnO below and above the Neel temperature, together with a schematic X-ray powder pattern at room temperature, are shown in Fig. 16.8. Comparison of the two patterns above T_N (Fig. 16.8b and c) shows that lines appear in the same positions but are of very different intensities. In the rock salt structure, the condition for reflection is that h, k, l should be either all odd or all even. Hence, the first four lines to be expected in the powder pattern are 111, 200, 220 and 311. All four lines appear in both patterns but 200 and 220 are weak in the neutron pattern (b). The small intensity of 200 and 220 in (b) is largely because the neutron scattering powers of Mn^{2+} and O^{2-} are opposite in sign although slightly different in magnitude. Partial cancellation therefore occurs for the 200 and 220 reflections since Mn^{2+} and O^{2-} ions on the same planes scatter out of phase with each other. This is precisely the opposite of the case with scattering of X-rays, for which the scattering factors of all elements have the same sign and for the 200 and 220 reflections Mn^{2+} and O^{2-} scatter in phase with each other.

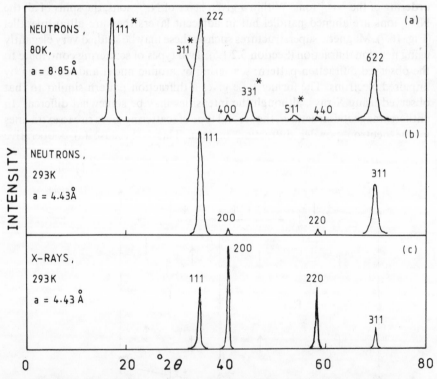

Fig. 16.8 Schematic neutron and X-ray powder diffraction patterns for MnO for λ = 1.542 Å. Peaks are assigned Miller indices for the cubic unit cells given. Neutron data are adapted from Shull, Strauser and Wollan (1951). X-ray data are from Powder Diffraction File, Card No. 7-230

Comparison of Fig. 16.8(a) and (b) shows that below T_N extra lines (asterisked) appear in the neutron diffraction pattern. These extra lines are associated with the antiferromagnetic superstructure. Although, as mentioned above, the true symmetry of the antiferromagnetic structure is rhombohedral, to a first approximation it can be treated as cubic with cell dimensions that are twice the value for the high temperature paramagnetic structure, i.e. $a = 8.85\,\text{Å}$ at $80\,\text{K}\,(\,<T_N)$ whereas at $293\,\text{K}\,(\,>T_N)$, $a = 4.43\,\text{Å}$ (Fig. 16.7). The volumes of the unit cells are therefore, in the ratio of 8:1. The extra lines in the powder pattern of the antiferromagnetic structure may be indexed as shown; observed reflections are those for which h, k and l are all odd.

16.3.3 Spinels

Several of the commercially important magnetic oxides have the spinel structure. The parent spinel is $MgAl_2O_4$. It has an essentially cubic close packed array of oxide ions with Mg^{2+}, Al^{3+} in tetrahedral and octahedral interstices, respectively. There are well over a hundred compounds with the spinel structure reported to date. Most are oxides. Some are sulphides, selenides and tellurides. A few are halides. Many different cations may be introduced into the spinel structure and several different charge combinations are possible, viz.:

2, 3	as in	$MgAl_2O_4$
2, 4	as in	Mg_2TiO_4
1, 3, 4	as in	$LiAlTiO_4$
1, 3	as in	$Li_{0.5}Al_{2.5}O_4$
1, 2, 5	as in	$LiNiVO_4$
1, 6	as in	Na_2WO_4

Similar cation combinations occur with sulphides, e.g. 2, 3: $ZnAl_2S_4$ and 2, 4: Cu_2SnS_4. With halide spinels, cations are limited to charges of 1 and 2, in order to give an overall cation: anion ratio of 3:4, e.g. Li_2NiF_4.

Crystallographic data for the spinel structure are given in Table 16.5. A projection of the unit cell of $MgAl_2O_4$ is shown in Fig. 16.9. Unit cell contents are eight formula units ($Z = 8$), corresponding to $Mg_8Al_{16}O_{32}$. The atoms are in three sets of special positions (Table 16.5). For instance, magnesium is in an eightfold set, label 8a; coordinates of two positions are given in the table: 0, 0, 0 and $\frac{1}{4}\frac{1}{4}\frac{1}{4}$. The remaining six are generated by the face centring operation, i.e. for an atom at position x, y, z, there are three other identical atoms at the equivalent positions, $x + \frac{1}{2}$, $y + \frac{1}{2}$, z; $x + \frac{1}{2}$, y, $z + \frac{1}{2}$ and x, $y + \frac{1}{2}$, $z + \frac{1}{2}$. The six other Mg^{2+} ions are therefore at $\frac{1}{2}\frac{1}{2}0$, $\frac{1}{2}0\frac{1}{2}$, $0\frac{1}{2}\frac{1}{2}$, $\frac{3}{4}\frac{3}{4}\frac{1}{4}$; $\frac{3}{4}\frac{1}{4}\frac{3}{4}$; $\frac{1}{4}\frac{3}{4}\frac{3}{4}$. The positions of the Al^{3+}, O^{2-} ions not included in Table 16.5 may be generated similarly.

The coordination environments of Mg^{2+}, Al^{3+} are shown in Fig. 16.9; Mg^{2+} is tetrahedral and Al^{3+} is octahedral. The arrangement of the oxide ions is ideally cubic close packed if the u parameter (Table 16.5) has the value $\frac{3}{8}$. To see this, the unit cell of Fig. 16.9 may be divided into octants (only quadrants appear in

Table 16.5 *The spinel structure: crystallographic data*

Space group: Fd3m, No. 227, face centred cubic
Atomic coordinates for $MgAl_2O_4$

Atom	Position	Fractional coordinates
O	32e	$uuu; \bar{u}\bar{u}u; u\bar{u}\bar{u}; \bar{u}u\bar{u};$
		$\frac{1}{4}-u, \frac{1}{4}-u, \frac{1}{4}-u; \frac{1}{4}+u, \frac{1}{4}-u, \frac{1}{4}+u;$
		$\frac{1}{4}-u, \frac{1}{4}+u, \frac{1}{4}+u; \frac{1}{4}+u, \frac{1}{4}+u, \frac{1}{4}-u$
		+ face centring
Al	16d	$\frac{555}{888}; \frac{577}{888}; \frac{757}{888}; \frac{775}{888};$
		+ face centring
Mg	8a	$000; \frac{111}{444}$ + face centring

Coordination numbers:
Mg,   8a,   tetrahedral, MgO_4
Al,   16d,   octahedral, AlO_6
O,   32e,   tetrahedral $OMgAl_3$

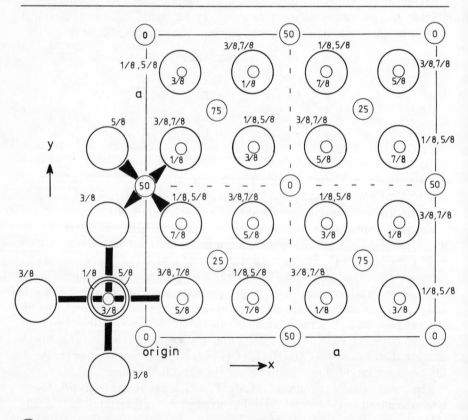

Fig. 16.9 Projection of the spinel structure

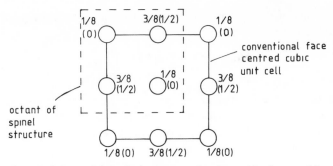

Fig. 16.10 Part of the spinel structure showing cubic close packing arrangement of oxide ions

projection). The origin is then displaced by one eighth of a body diagonal of the cube such that it coincides with an oxide ion, instead of with Al^{3+} (Fig. 16.10).

A complicating factor in the spinel structure is that the cation distribution over the 8a and 16d sites may vary. Two extreme types of behaviour may be distinguished. In *normal* spinels, the cations are in the sites that would be expected from the formula AB_2O_4, i.e. with A in tetrahedral 8a sites and B in octahedral 16d sites. Examples of normal spinels are $MgAl_2O_4$ and $MgTi_2O_4$.

In *inverse* spinels, half of the B ions are in tetrahedral 8a sites, leaving the remaining B ions and the A ions to occupy the 16d sites. Usually, the occupancy of these 16d sites is disordered. Examples of inverse spinels are $MgFe_2O_4$ and Mg_2TiO_4.

A representative selection of normal and inverse spinels, with their unit cell parameters and u values (where known), is listed in Table 16.6.

Table 16.6 *Crystallographic data for some spinels*

Crystal	Type	$a(\text{Å})$	u	Structure
$MgAl_2O_4$	2, 3	8.0800	0.387	Normal
$CoAl_2O_4$	2, 3	8.1068	0.39	Normal
$CuCr_2S_4$	2, 3	9.629	0.381	Normal
$CuCr_2Se_4$	2, 3	10.357	0.380	Normal
$CuCr_2Te_4$	2, 3	11.051	0.379	Normal
$MgTi_2O_4$	2, 3	8.474	—	Normal
Co_2GeO_4	2, 4	8.318	—	Normal
Fe_2GeO_4	2, 4	8.411	—	Normal
$MgFe_2O_4$	2, 3	8.389	0.382	Inverse
$NiFe_2O_4$	2, 3	8.3532	0.381	Inverse
$MgIn_2O_4$	2, 3	8.81	0.372	Inverse
$MgIn_2S_4$	2, 3	10.708	0.384	Inverse
Mg_2TiO_4	2, 4	8.44	0.39	Inverse
Zn_2SnO_4	2, 4	8.70	0.390	Inverse
Zn_2TiO_4	2, 4	8.467	0.380	Inverse
$LiAlTiO_4$	1, 3, 4	8.34	—	Li in 8a
$LiMnTiO_4$	1, 3, 4	8.30	—	Li in 8a
$LiZnSbO_4$	1, 2, 5	8.55	—	Li in 8a
$LiCoSbO_4$	1, 2, 5	8.56	—	Li in 8a

The formulae of spinels may be expanded in order to distinguish between normal and inverse:

Normal: $[A]^{tet}[B_2]^{oct}O_4$

Inverse: $[B]^{tet}[A, B]^{oct}O_4$

As well as the extremes of normal and inverse, intermediate cation distributions are possible. Sometimes, the cation distribution varies with temperature. The cation distribution may be quantified in a simple way by using a parameter, γ, which corresponds to the fraction of A ions on the octahedral sites:

Normal: $[A]^{tet}[B_2]^{oct}O_4,$ $\gamma = 0$

Inverse: $[B]^{tet}[A, B]^{oct}O_4,$ $\gamma = 1$

Random: $[B_{0.67}A_{0.33}]^{tet}[A_{0.67}B_{1.33}]^{oct}O_4$ $\gamma = 0.67$

The cation distribution in spinels and the degree of inversion, γ, have been studied in considerable detail. Several factors influence γ, including the site preferences of ions in terms of size, covalent bonding effects and crystal field stabilization energies. More details and examples are given in Section 8.6.1. The actual γ value in any particular spinel is given by the net effect of these various parameters taken together.

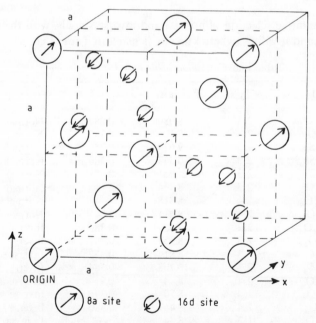

Fig. 16.11 Magnetic structure of antiferromagnetic and ferrimagnetic spinels

The commercially important magnetic spinels, known as ferrites, are those of the type MFe_2O_4, where M is a divalent ion such as Fe^{2+}, Ni^{2+}, Cu^{2+}, Mg^{2+}. They are all inverse, either partially or completely. This is probably because Fe^{3+}, being a d^5 ion, has no crystal field stabilization energy in an octahedral site; hence, the larger divalent ions go preferentially into octahedral sites and Fe^{3+} is distributed over both tetrahedral and octahedral sites.

These ferrite phases have interesting magnetic structures and are all either antiferromagnetic or ferrimagnetic. This is because the ions on the tetrahedral, 8a sites have magnetic spins that are antiparallel to those of the ions on the octahedral, 16d sites. This is shown in Fig. 16.11; the contents of four octants of the unit cell are shown and the origin coincides with an 8a ion as in Fig. 16.9. Oxide ions, at positions such as $\frac{1}{8}, \frac{1}{8}, \frac{3}{8}$, are not shown. The orientation of the unit cell is the same as in Fig. 16.9. When the unit cell is drawn in this way, it may be described as a face centred cubic array of 8a ions, at corner and face centre positions, with additional 8a ions in the centre of one set of alternate octants of the unit cell. This gives, overall, eight 8a ions per unit cell. The 16d ions are arranged tetrahedrally inside the other set of alternate octants, giving sixteen 16d ions per unit cell. The magnetic spins of 8a and 16d ions are antiparallel, as shown. Let us now calculate the magnetic moments of different spinels using equation (16.13). $\mathcal{M} = g \cdot S$

$ZnFe_2O_4$ is a normal spinel, $Zn^{tet}Fe_2^{oct}O_4$, but is antiferromagnetic at very low temperatures ($T_N = 9.5\,K$). This is because the Fe^{3+} ions on the octahedral 16d sites fall into two groups which couple antiferromagnetically but the coupling between octahedral sites is very weak and hence T_N is low. [Note, the coupling between tetrahedral and octahedral sites is much stronger and gives rise to much higher T_N, T_c values].

A similar result would be expected for $MgFe_2O_4$ but it has, in fact, a residual, overall moment and is ferrimagnetic. Two explanations are possible. Either the spinel is partially inverse and there are more Fe^{3+} ions on the octahedral sites than on the tetrahedral ones. This would give only partial cancellation of the spins. Or, the effective magnetic moment per Fe^{3+} ion is different for the two types of site. Experimental measurements indicate that the first explanation is correct. At high temperatures, $MgFe_2O_4$ gradually transforms to the normal spinel structure. The degree of inversion in a sample at room temperature depends very much on its thermal history, especially the rate at which it was cooled from high temperature. Thus, rapidly quenched samples have a lower degree of inversion and hence a higher magnetic moment than slowly cooled ones.

Manganese ferrite, $MnFe_2O_4$, is about 80 per cent normal, 20 per cent inverse, but since both cations, Mn^{2+} and Fe^{3+}, are d^5, the overall magnetic moment is insensitive to the degree of inversion and to heating/thermal history effects. $MnFe_2O_4$ is expected to be ferrimagnetic with an overall moment of ~ 5 BM, and this is found to be the case.

An interesting example of the importance of cation site occupancies and solid

574
$Zn^{2+} = d^{10}$

solution effects is provided by the mixed ferrites, $M_{1-x}Zn_xFe_2O_4$:M = Mg, Ni, Co, Fe, Mn. These ferrites are largely inverse for $x = 0$, i.e.

$$[Fe^{3+}]^{tet}[M^{2+}, Fe^{3+}]^{oct}O_4$$

Expected values of μ for purely inverse spinels are Mg:0, Ni:2, Co:3, Fe:4 and Mn:5. Experimental values are in fact somewhat larger, as shown by the left-hand axis of Fig. 16.12. The zinc ferrite, $ZnFe_2O_4$, $x = 1$ is, by contrast, almost entirely normal at room temperature. However, the spins of the Fe^{3+} ions on the octahedral sites of $ZnFe_2O_4$ are not aligned but are random. $ZnFe_2O_4$ is therefore paramagnetic and shows no saturation magnetization. On formation of the ferrite solid solutions by partial replacement of M^{2+} by Zn^{2+}, a gradual change from inverse to normal behaviour is found to occur. Introduction of Zn^{2+} into the tetrahedral sites causes Fe^{3+} ions to be displaced onto the octahedral sites, i.e.

$$[Fe^{3+}_{1-x}Zn^{2+}_x]^{tet}[M^{2+}_{1-x}Fe^{3+}_{1+x}]^{oct}O_4$$

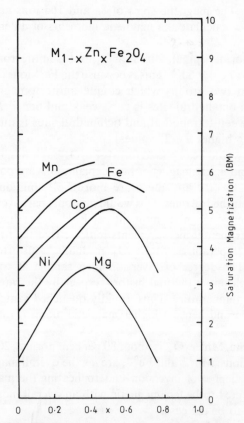

Fig. 16.12 Variation in saturation magnetization with composition for ferrite solid solutions. (From Gorter, 1950)

If the solid solutions retained the antiferromagnetic character of the MFe_2O_4 ferrites, $x = 0$, a linear increase in μ should occur and attain the value of 10 for $x = 1$, $ZnFe_2O_4$. Long before $x = 1$ is reached, however, the antiferromagnetic coupling between 16d and 8a sites is destroyed and the saturation magnetization values drop (Fig. 16.12). For small values of x, the experimental values of the saturation magnetization increase, consistent with retention of the antiferromagnetic/ferrimagnetic ordering, but pass through a maximum for $x = 0.4$ to 0.5.

In addition to the magnetic moment of a ferrite, other parameters are important in controlling the magnetic properties. These include the saturation magnetization, M_{sat}, the magnetostrictive constant, λ_s, the permeability, P, and the magnetocrystalline anisotropy constant, K_1. Suffice it to say here that the values of these parameters vary widely between the different ferrite phases. A particular ferrite may be chosen depending on the values of these properties and the application in mind. Further variation in the magnetic properties may be achieved by making mixed ferrites which are solid solutions of two or more pure ferrites. For instance, replacement of Mn^{2+} by Fe^{2+} in $MnFe_2O_4$, giving a solid solution $Mn_{1-x}^{2+} Fe_x^{2+} Fe_2^{3+} O_4$, reduces the magnetic anisotropy parameter to zero. This parameter measures the ease with which the orientation of the magnetic moment may be altered in an applied magnetic field. Reduction of the magnetic anisotropy yields an increased permeability which is usually desirable in commercial ferrites. An undesirable side-effect, however, is an increase in electrical conductivity with increasing Fe^{2+} content.

16.3.4 Garnets

The garnets are a large family of complex oxides, some of which are important ferrimagnetic materials. They have the general formula $A_3B_2X_3O_{12}$. A is a large ion with a radius of ~ 1 Å and has a coordination number of eight in a distorted cubic environment. B and X are smaller ions which occupy octahedral and tetrahedral sites, respectively. The garnets with interesting magnetic properties have A = Y or rare earth, Sm, Gd, Tb, Dy, Ho, Er, Tm, Yb, Lu; B, X = Fe^{3+}. One of the most important is yttrium iron garnet (YIG), $Y_3Fe_5O_{12}$. Many other A, B, X combinations are possible, e.g.:

	A	B	X	O
Grossular:	Ca_3	Al_2	Si_3	O_{12}
Uvarovite:	Ca_3	Cr_2	Si_3	O_{12}
Pyrope:	Mg_3	Al_2	Si_3	O_{12}
Andradite:	Ca_3	Fe_2	Si_3	O_{12}
	Ca_3	$CaZr$	Ge_3	O_{12}
	Ca_3	Te_2	Zn_3	O_{12}
	Na_2Ca	Ti_2	Ge_3	O_{12}
	$NaCa_2$	Zn_2	V_3	O_{12}

Table 16.7 *The garnet structure : crystallographic data*

Space group: la3d, No 230, body centred cubic
Atomic coordinates for $Y_3Fe_5O_{12}$(YIG)

Atom	Position	Fractional Coordinates
Y	24c	$\frac{1}{8}, 0, \frac{1}{4}$
Fe(1)	24d	$\frac{3}{8}, 0, \frac{1}{4}$
Fe(2)	16a	$0, 0, 0$
O	96h	$u, v, w : u = -0.0275,$ $v = 0.0572, w = 0.1495$

Coordination numbers:

Y	24c,	CN = 8,	distorted cube
Fe(1)	24d,	CN = 4,	tetrahedron
Fe(2)	16a,	CN = 6,	octahedron
O	96h,	CN = 4,	distorted tetrahedron, $2Y^{3+}$, $1Fe^{3+}(1)$, $1Fe^{3+}(2)$

Crystallographic data for YIG are given in Table 16.7. The body centred cubic cell is large, $a = 12.376$ Å, and contains eight formula units. No attempt is made to give a drawing of the structure here, but the structure may be regarded as a framework built of corner-sharing BO_6 octahedra and XO_4 tetrahedra. The larger A ions occupy eight coordinate cavities within this framework. In YIG and the rare earth garnets the B and X ions are the same, Fe^{3+}.

YIG and the rare earth garnets are all ferrimagnetic with a Curie temperature in the range 548 to 578 K. In order to evaluate the overall magnetic moment of these garnets, the three types of ion, on sites 24c, 24d and 16a have to be considered. It appears that the spins of the 24d ions are coupled so as to be antiparallel to those of both the 24c and 16a ions. If we consider the two types of Fe^{3+} ion first, their spins partially cancel, giving a net moment of one Fe^{3+} ion per formula unit $M_3Fe_5O_{12}$, i.e. 5 BM. Since Y^{3+} is a d° ion, it has no magnetic moment. Consequently, the expected moment of YIG is 5 BM, which is in excellent agreement with the measured value.

For the rare earth garnets, the overall moment is expected to be given by

$$(3\mu_M - 5) \quad \text{BM}$$

where μ_M is the moment of the 24c ion. For Gd^{3+}, which is an f^7 ion, $\mu_{Gd} = 7$ BM and the net moment for GdIG is calculated to be 16 BM, which also agrees with experiment. For Lu^{3+}, which is f^{14}, $\mu_{Lu} = 0$ and the net moment is therefore 5 BM, in agreement with experiment. For the other ions, Tb to Yb, it appears that the orbital moment is not completely quenched and the μ_M values are larger than given by the spin-only formula with $g = 2.00$. Comparison between experimental and calculated values is shown in Fig. 16.13. For the latter, two curves are given, corresponding to the spin-only formula and the spin + orbital moment formula. Experimental values generally fall between the two theoretical curves and show that the orbital moment is only partially quenched.

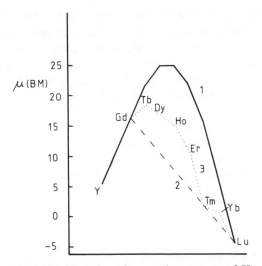

Fig. 16.13 Variation of magnetic moment at 0 K of garnets. Curve 1; calculated, spin + orbital formula. Curve 2: calculated, spin only formula. Curve 3: experimental. (Data from Standley, 1972)

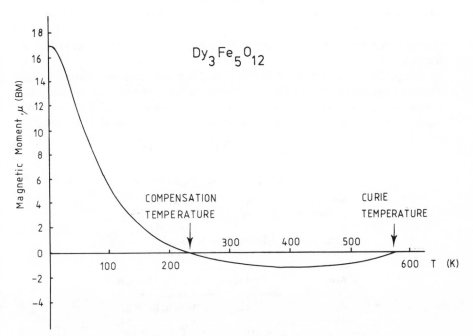

Fig. 16.14 Spontaneous magnetization in dysprosium iron garnet as a function of temperature

The magnetic moments of the rare earth garnets show an interesting and unusual temperature dependence. The spontaneous moments at absolute zero, shown in Fig. 16.13, decrease with rising temperature and fall to zero at the *compensation temperature*. They then rise again, but in the opposite direction, and fall to zero a second time at the Curie temperature, as shown for dysprosium iron garnet in Fig. 16.14. This effect occurs because the spins on one of the Fe^{3+} sublattices randomize more rapidly than those on the other sublattices.

Many ionic substitutions in the garnet structure are possible and the magnetic properties may be systematically varied. For instance, the large trivalent ion in the 24c site may be partially replaced by Ca^{2+} and to compensate for the charge mismatch, some of the Fe^{3+} ions on the tetrahedral sites may be replaced by V^{5+} according to the formula:

$$[Y^{3+}_{3-2x}Ca^{2+}_{2x}]\,Fe^{3+}_2\,[Fe^{3+}_{3-x}V^{5+}_x]O_{12}$$

16.3.5 Ilmenites and perovskites

Ilmenites are a group of phases of formula ABO_3:A = Fe, Co, Ni, Cd, Mg; B = Ti, Rh, Mn. Their structure is closely related to that of chromium sesquioxide, Cr_2O_3; haematite, α-Fe_2O_3; or corundum, α-Al_2O_3. The symmetry is rhombohedral but the structure can be drawn and visualized more easily using a larger, hexagonal unit cell (Table 16.8). The structure may be described as an approximately hexagonal close packed array of oxide ions with the cations in two-thirds of the octahedral sites. The cations are segregated so that, along the c axis, there are alternate layers of A and B cations. Another way of looking at the ilmenite structure is as a derivative of NiAs with one-third of the octahedral sites vacant.

The perovskite structure of $SrTiO_3$ has been described in Chapter 6. Some oxides containing Fe^{3+} and $Mn^{3+,4+}$ have a perovskite structure and interesting ferromagnetic properties. These are mixtures of La^{3+} Mn^{3+} O_3 and A^{2+} $Mn^{3+,4+}O_3$ which form double substitutional solid solutions of formula:

$$[La^{3+}_{1-x}A^{2+}_x][Mn^{3+}_{1-x}Mn^{4+}_x]O_3$$

Table 16.8 *Crystallographic data for ilmenite*

Space group: $R\bar{3}$ for $FeTiO_3$ ($R\bar{3}c$ for Cr_2O_3)
Rhombohedral cell dimensions, $a = 5.538\,\text{Å}$
$\qquad\qquad\qquad\qquad\qquad\quad \alpha = 54.41°$
Hexagonal cell dimensions, $a = 5.048\,\text{Å}$
$\qquad\qquad\qquad\qquad\qquad\quad c = 14.026\,\text{Å}$

Atom	Position	Coordinates
Fe	6c	$0, 0, u$: $u = 0.358$
Ti	6c	$0, 0, u$: $u = 0.142$
O	18f	$u\,v\,w$:$u = 0.305$, $v = 0.015$, $w = 0.25$

In these, the larger La^{3+}, A^{2+} ions occupy the twelve-coordinate sites and $Mn^{3+,4+}$ occupy the octahedral sites. Ions A can be Ca^{2+}, Sr^{2+}, Ba^{2+}, Cd^{2+}, Pb^{2+}. Closer study has shown that the crystal chemistry, phase diagrams and magnetic properties of these systems are, in fact, quuite complex.

16.3.6 Magnetoplumbites

Magnetoplumbite is the mineral $PbFe_{12}O_{19}$; its barium analogue $BaFe_{12}O_{19}$, known as BaM, is an important component of permanent magnets. The structure of magnetoplumbite is closely related to that of β-alumina, 'NaAl$_{11}$O$_{17}$' (Chapter 13). The latter is basically a close packed structure with five oxide layers in the subrepeat unit. Each layer contains four oxide ions per unit cell and the distinctive feature of the structure is that every fifth layer has three-quarters of the oxide ions missing. Hence there are $(4 \times 4) + 1 = 17$ oxide ions to each five layer subrepeat unit. In β-alumina, the empty spaces in the fifth layer are partially occupied by Na^+ ions. Magnetoplumbite has a similar five-layer subrepeat unit contains $(4 \times 4) + (1 \times 3) = 19$ oxide ions with one Ba^{2+}, Pb^{2+} alumina. The fifth layer contains three-quartérs of its quota of oxide ions with a large divalent ion—Pb^{2+}, Ba^{2+}—in the other oxide ion site. Hence, the subsequent unit contains $(4 \times 4) + (1 \times 3) = 19$ oxide ions with one Ba^{2+}, Pb^{2+} ion completing the fifth layer.

The magnetic structure of BaM is complex because there are Fe^{3+} ions in five different sets of crystallographic sites. However, the net effect is that in the formula unit $BaFe_{12}O_{19}$, eight of the Fe^{3+} ions have their spins oriented in one direction and the remaining four are antiparallel, giving a resultant of four Fe^{3+} ions with a moment of 20 BM.

16.4 Applications: structure/property relations

A large number of parameters affect the magnetic properties of materials. By careful control of composition and fabrication procedures, it is now possible to carry out 'crystal engineering' and deliberately prepare materials with a desired set of properties. In this section, we shall briefly summarize some applications of magnetic materials and the factors that influence the selection of a material for a particular application.

16.4.1 Transformer cores

A major application of ferro- and ferrimagnetic materials is in transformer and motor cores. Materials are required that are magnetically soft with large power handling capacity and low losses. Magnetically soft materials have a high permeability—they are magnetized easily at low applied fields—and a low coercive field. They also tend to have low hysteresis losses. All of these properties are favoured in materials that have a small magnetostriction coefficient, λ_s, and a low magnetocrystalline anisotropy coefficient, K_1. Soft magnetic materials are, mechanically, those in which the domain walls are able to migrate easily. These various parameters can all be optimized by attention to the details of composition and fabrication.

As well as hysteresis losses, eddy current losses are a serious problem at high frequencies and especially in materials of low resistivity. This is because eddy currents are proportional to $(\text{frequency})^2$. Eddy currents in metals such as iron can be reduced by alloying the iron with, for example, nickel or silicon. This is because alloys generally have a much higher resistivity than the pure component metals. One great advantage of the ferrimagnetic oxides such as Mn, Zn ferrite, Ni, Zn ferrite and YIG is that, provided they are prepared correctly, they have very high resistivity and negligibly small eddy currents. This is particularly so for the garnets such as YIG; these contain trivalent cations only and so there is no easy mechanism for electronic conduction to occur. The conductivity of YIG at room temperature is only 10^{-12} ohm^{-1} cm^{-1}. With the ferrites, it is necessary to ensure that all the iron present is in the $+3$ oxidation state; otherwise $Fe^{2+} \rightleftharpoons Fe^{3+} + e^-$ redox transfer may give rise to a high conductivity. For instance, magnetite, Fe_3O_4 or $Fe^{2+}Fe_2^{3+}O_4$, has a conductivity of 10^2 to 10^3 ohm^{-1} cm^{-1} at room temperature which is ~ 15 orders of magnitude higher than that of YIG. This high conductivity is associated with the mixture of valence states of iron.

16.4.2 Information storage

A critical requirement for magnetically based information storage components is that they should be soft with low eddy current losses and a certain type of hysteresis loop, either square or rectangular (Fig. 16.15). With this characteristic, a reverse field may be applied to a magnetized sample and it should undergo no

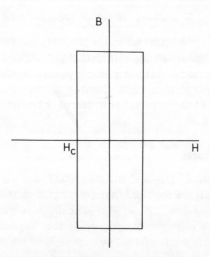

Fig. 16.15 Rectangular hysteresis loop required for information storage devices

change until the coercive field H_c, is exceeded, at which point a sudden switch in magnetization occurs. The two orientations of magnetization, $+$ and $-$, can be used to represent 0 and 1 in the binary digital system. Certain of the magnetic ferrites have the required characteristics for this application and, with switching times of 10^{-6} seconds or less, they are essential components to modern computer technology.

16.4.3 Magnetic bubble memory devices

An interesting recent development in information storage uses thin films of garnet, a few micrometres thick, which are deposited epitaxially onto a non-magnetic substrate (Chapter 2). The films are deposited at high temperature and, by careful choice of the composition of the garnet and, in particular, its lattice parameters, a slight difference in thermal contraction occurs on cooling the sample to room temperature. The stresses that are generated are sufficient to induce a preferred direction of magnetization in the garnet film which is perpendicular to the plane of the film. The resulting domain structure of the film has spins pointing up or down and these appear as bubbles when viewed in a polarizing microscope. As indicated in Section 16.4.2, therefore, these magnetic bubble materials can be used as memory components for binary digital computers.

16.4.4 Permanent magnets

For use in permanent magnets, the following properties are desired: high saturation magnetization, high energy product BH, high coercive field, high remanent magnetization, high Curie temperature, high magnetocrystalline anisotropy. Materials that are competing for applications as permanent magnets are likely to be based on either the metals Fe, Co, Ni or 'hard' oxides such as BaM, barium magnetoplumbite. Let us see how the performance of these two types of material can be optimized.

The hardness of magnets can be increased if ways can be found of either pinning or reducing the ease of motion of the domain walls. This may be achieved in steels by adding suitable dopants such as chromium or tungsten that cause either precipitation of, say, a carbide phase or a martensitic transformation on cooling. A novel feature of the Alnico group of magnets is that the ferromagnetic Co, Ni-based material is present as a large number of small crystalline regions embedded in an aluminium-based matrix. These small regions are all magnetized in the same direction and it is very difficult to demagnetize them or change their magnetic orientation.

The oxide magnets such as BaM are relatively light and cheap. Although their intrinsic magnetic properties are generally inferior to those of Alnico magnets, they can be improved if prepared with a magnetically aligned texture. To achieve this, the powdered starting materials are subjected to a magnetic field while they are being prepared into a shape and subsequently sintered at high temperature.

The effect of the applied field is to cause magnetic alignment of the grains and this increases the remanent magnetization of the material.

Questions

16.1 Indicate how you would distinguish between paramagnetic, ferromagnetic and antiferromagnetic behaviour using a Gouy balance.

16.2 Vanadium monoxide is diamagnetic and a good conductor of electricity whereas nickel oxide is paramagnetic/antiferromagnetic and a poor conductor of electricity. Account for these observations.

16.3 Account for the magnetic behaviour of the following spinels: (a) $ZnFe_2O_4$ is antiferromagnetic; (b) $MgFe_2O_4$ is ferrimagnetic and its magnetic moment increases with temperature; (c) $MnFe_2O_4$ is ferrimagnetic and its magnetic moment is independent of temperature.

16.4 Explain why magnetic materials that are to be used in information storage should have either square or rectangular shaped hysteresis loops.

16.5 Why cannot pure iron metal be used in transformer cores?

16.6 Show that the following magnetic susceptibility data for Ni fit the Curie–Weiss Law:

$T(K)$	800	900	1000	1100	1200
$\chi \times 10^{-5}$	3.3	2.1	1.55	1.2	1.0

Evaluate T_c or θ and C.

References

F. A. Cotton and G. Wilkinson (1966). *Advanced Inorganic Chemistry*, Wiley.

D. J. Craik (Ed.) (1975). *Magnetic Oxides*, Parts 1 and 2, Wiley.

A. Earnshaw (1968). *Introduction to Magnetochemistry*, Academic Press.

J. B. Goodenough (1963). *Magnetism and the Chemical Bond*, Wiley.

E. W. Gorter (1950). Nature (London), **165**, 798.

H. Hibst (1982). Hexagonal ferrites from melts and aqueous solutions, magnetic recording materials, *Angew. Chemie, Int. Ed. Engl.*, **21**, 270.

W. D. Kingery, H. K. Bowen and D. R. Uhlmann (1976). *Introduction to Ceramics*, Wiley.

M. M. Schieber (1967). *Experimental Magnetochemistry*, North Holland.

C. G. Shull, W. A. Strauser and E. O. Wollan (1951). Neutron diffraction by paramagnetic and antiferromagnetic substances, *Phys. Rev.*, **83**, 333.

K. J. Standley (1972). *Oxide Magnetic Materials*, Clarendon Press.

K. N. R. Taylor (1970). The rare-earth metals *Contemp. Phys.*, **11**, 423.

R. S. Tebble and D. J. Graik (1979). *Magnetic Materials*, Wiley.

J. H. Wernick (1973). Structure and composition in relation to properties, in *Treatise on Solid State Chemistry* (Ed. N. B. Hannay), Vol. 1, Plenum Press.

C. A. Wert and R. M. Thomson (1970). *Physics of Solids*, McGraw-Hill.

Chapter 17

Optical Properties : Luminescence, Lasers

Absorption of light and the origins of colour in solids have been discussed in Chapter 14. Here we are concerned with materials and applications that are associated with the emission of light.

17.1 Luminescence and phosphors

17.1.1 Definitions and general comments

Luminescence is the name generally given to the emission of light by a material as a consequence of it absorbing energy. Various types of excitation source may be used and are indicated as a prefix. *Photoluminescence* uses photons or light, often UV, for excitation. *Electroluminescence* uses an electrical energy input. *Cathodoluminescence* uses cathode rays or electrons to provide energy. Two types of photoluminescence may be distinguished. For a short time lapse, $\lesssim 10^{-8}$ sec, between excitation and emission, the process is known as *fluorescence*. Fluorescence effectively ceases as soon as the excitation source is removed. For much longer decay times, the process is known as *phosphorescence*. This may continue long after the source of excitation is removed.

Photoluminescent materials generally require a *host* crystal structure, ZnS, $CaWO_4$, Zn_2SiO_4, etc., which is doped with a small amount of an *activator*, a cation such as Mn^{2+}, Sn^{2+}, Pb^{2+}, Eu^{2+}. Sometimes, a second type of dopant is added to act as a *sensitizer*. The mode of operation of inorganic luminescent materials, known generally as *phosphors*, is shown schematically in Fig. 17.1. Note that the energy of the emitted light is generally less than that of the exciting

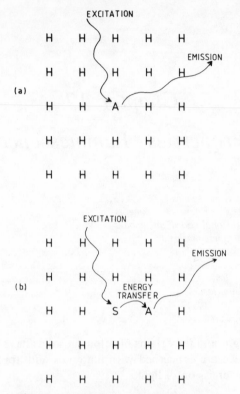

Fig. 17.1 Schematic representation of the luminescence process involving (a) an activator, A, in a host lattice, H, and (b) both a sensitizer, S, and an activator, A. (Adapted from DeLuca, 1980)

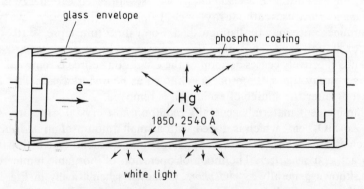

Fig. 17.2 Schematic design of a fluorescent lamp

radiation and is, therefore, of longer wavelength. This effective increase in wavelength is known as the *Stokes shift*. In fluorescent lamps, which provide the most important application of phosphors, the exciting radiation is UV light from a mercury discharge. Phosphor materials are required that absorb this UV radiation and emit 'white' light. The construction of a fluorescent lamp is shown schematically in Fig. 17.2. It consists of a glass tube lined on the inside with a coating of phosphor material and filled with a mixture of mercury vapour and argon. On passage of an electric current through the lamp, the atoms of mercury are bombarded by electrons and are excited into upper electronic energy states. They can then return to the ground state, accompanied by the emission of UV light of two characteristic wavelengths, 2540 and 1850 Å. This light irradiates the phosphor coating on the inner surface of the glass envelope which subsequently emits white light.

Examples of emission spectra of ZnS phosphors activated with a variety of cations are shown in Fig. 17.3. Each activator gives a characteristic spectrum and colour to ZnS. Various types of electronic transition occur in activator ions, e.g.:

Ion	Ground state	Excited state
Ag^+	$4d^{10}$	$4d^9 5p$
Sb^{3+}	$4d^{10} 5s^2$	$4d^{10} 5s 5p$
Eu^{2+}	$4f^7$	$4f^6 5d$

The host materials used for phosphors fall into two main categories:

(a) Ionically bonded, insulating materials, such as $Cd_2B_2O_5$, Zn_2SiO_4 and apatite, $3Ca_3(PO_4)_2 \cdot Ca(Cl, F)_2$. In these, a set of discrete energy levels is associated with the activator ion and these levels are modified by the local

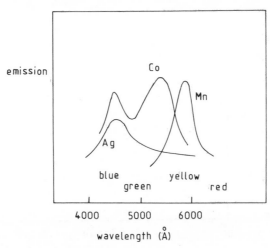

Fig. 17.3 Luminescence spectra of activated ZnS phosphors after irradiation with UV light

environment of the host crystal structure. For ionic phosphors, the *configurational coordinate model* provides a useful way of representing qualitatively the luminescence processes.

(b) Covalently bonded, semiconducting sulphides, such as ZnS. In these, the energy band structure of the host is modified by the addition of localized energy levels associated with the activator ions.

17.1.2 Configurational coordinate model

The potential energy of the ground and excited electronic states of the luminescent centre is plotted against a general coordinate which is often the internuclear distance. This is shown schematically for the ground state in Fig. 17.4. The solid curve shows qualitatively how the potential energy varies as a function of interatomic distance. It passes through a minimum at the equilibrium bond length, r_e. Within this electronic ground state, different quantized vibrational states of the ion are possible, as shown by the horizontal lines V_0, V_1, etc.

Each electronic state for the luminescent centre has a potential energy curve roughly similar to that in Fig. 17.4. Typical curves for a ground state and an excited state are shown in Fig. 17.5; using this diagram, many of the features of luminescence can be explained.

First, the process of excitation involves raising the active centre from its ground state, level A, into a higher vibrational level of the excited state, B. Second, some energy is then dissipated as the ion quickly relaxes to a lower level, C, in the excited state. This energy is lost to the host lattice and appears as heat. Third, the active centre returns to its ground state, level D or A, and, in so doing, emits light. Since the energy of excitation A → B is greater than that of emission C

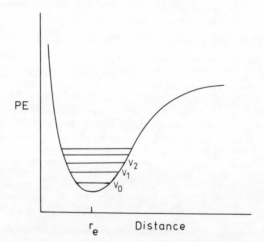

Fig. 17.4 Ground state potential energy diagram for a luminescent centre in an ionic host crystal

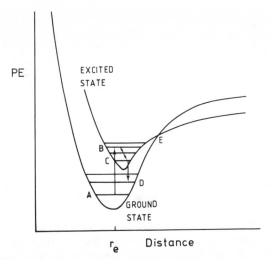

Fig. 17.5 Ground and excited state potential energy diagrams for a luminescent centre

→ D, the emitted radiation is of longer wavelength than the exciting radiation. This therefore accounts for the Stokes shift.

An effect known as *thermal quenching* in which the luminescence efficiency decreases markedly above a certain temperature can also be explained with the aid of Fig. 17.5. The potential energy curves for the ground and excited states cross over at point E. At this point, an ion in the excited state can transfer back to its ground state, at the same energy. It can then return to the lower vibrational levels of the ground state by means of a series of vibrational transitions. Point E represents a kind of spillover point, therefore. If an ion in the excited state can acquire sufficient vibrational energy to reach point E, it can spill over into the vibrational levels of the ground state. If this happens, all the energy is released as vibrational energy and no luminescence occurs. The energy of point E is obviously critical. In general, it is likely to be reached as a consequence of increasing the temperature since, with rising temperature, ions have increasing thermal energy and are able to move to progressively higher vibrational levels.

The type of transition described above to explain thermal quenching is an example of a *non-radiative transition*. In this the excited ion gets rid of some of its excess energy by imparting vibrational energy to the surrounding host lattice. In this way, the excited ion is able to return to a lower energy level but no electromagnetic radiation, i.e. light, is emitted.

Another type of non-radiative transition is involved in the operation of sensitized phosphors. This transition, known as *non-radiative energy transfer*, is shown schematically in Fig. 17.6. It depends on (a) there being similar energy levels in the excited states of both sensitizer and activator ions and (b) sensitizer and activator ions being relatively close together in the host crystal structure. In operation, the exciting radiation promotes sensitizer ions into an excited state.

588

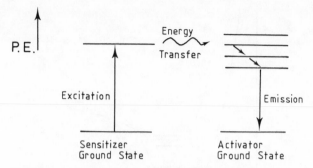

Fig. 17.6 Non-radiative energy transfer involved in operation of a sensitized phosphor

These then transfer energy to neighbouring activator ions, with little or no loss of energy during transfer, and at the same time the sensitizer ions return to their ground state. Finally, the activator ions return to their ground state by the emission of luminescent radiation.

Non-radiative energy transfer is also involved in the *poisoning* effect of certain impurities. In this, energy is transferred from either a sensitizer or an activator to a poison site at which the energy is lost to the host structure in the form of vibrational energy. Ions that have non-radiative transitions to the ground state and which must be avoided in the preparation of phosphors include Fe^{2+}, Co^{2+} and Ni^{2+}.

17.1.3 Some phosphor materials

An enormous number of host/activator combinations have been studied for luminescence, with a fair degree of success. However, future improvements and the development of new materials will probably depend on improved understanding of the relation between crystal structure and the energy levels of dopant ions.

A phosphor that is used extensively in fluorescent lamps is an apatite, doubly doped with Mn^{2+} and Sb^{3+}. Fluorapatite is $Ca_5(PO_4)_3F$. When doped with Sb^{3+} it fluoresces blue and with Mn^{2+} it fluoresces orange-yellow; the two together give a broad emission spectrum that approximates to white light. A modification of the wavelength distribution in the emission spectrum is possible by partly replacing the F^- ions in fluorapatite by Cl^- ions. The effect of this is to modify the energy levels of the activator ions and hence their emission wavelengths. By careful control of composition in this way, the colour of the fluorescent light may be optimized. A selection of other lamp phosphor materials is given in Table 17.1.

Trivalent europium is an important activator ion, especially for use in red phosphors for colour television screens. In $YVO_4:Eu^{3+}$, the vanadate group absorbs energy in the cathode ray tube but the emitter is Eu^{3+}. The mechanism of charge transfer between the vanadate group and Eu^{3+} appears to involve a non-

Table 17.1 *Some lamp phosphor materials.* (Data taken from Burrus, 1972)

Phosphor	Activator	Colour
Zn_2SiO_4, willemite	Mn^{2+}	Green
Y_2O_3	Eu^{3+}	Red
$CaMg(SiO_3)_2$, diopside	Ti	Blue
$CaSiO_3$, wollastonite	Pb, Mn	Yellow-orange
$(Sr, Zn)_3 (PO_4)_2$	Sn	Orange
$Ca_5(PO_4)_3(F, Cl)$ fluorapatite	Sb, Mn	'White'

radiative, superexchange process via the intervening oxide ions. The effect is related to the superexchange mechanism postulated to explain the antiferromagnetic ordering of Ni^{2+} ions in NiO (Chapter 16). In order to get efficient energy transfer by superexchange, it appears to be important that the metal–oxygen–metal bond should be approximately linear, so as to maximize the degree of orbital overlap. In YVO_4:Eu, the vanadium–oxygen–europium angle is 170° and the energy exchange takes place rapidly.

Transitions from and to several different f energy levels are possible, in principle, with Eu^{3+}. The ones that are actually observed, and hence the colour, depend on the host crystal structure and, in particular, on the symmetry of the site that it provides for Eu^{3+}. If the Eu^{3+} is located at a centre of symmetry, as it is when doped into $NaLuO_2$ and Ba_2GdNbO_6, the favoured transition is of the type $^5D_0 \rightarrow {}^7F_1$. The resulting colour is orange. When located on a non-centrosymmetric site, as in $NaGdO_2$:Eu^{3+}, the preferred transition is $^5D_0 \rightarrow {}^7F_2$ and the colour of emission is red. The crystal structure of the host material therefore has a major effect on the resulting colour.

Colour television screens require three primary cathodoluminescent colours. These are:

(a) red, mentioned above, for which YVO_4:Eu^{3+} is often used,
(b) blue—ZnS:Ag^+;
(c) green—ZnS:Cu^+.

For black and white screens, mixtures of the blue-emitting ZnS:Ag^+ and yellow-emitting $(Zn, Cd)S$:Ag^+ are used.

17.1.4 Anti-Stokes phosphors

A relatively new class of phosphors that has aroused considerable interest is the anti-Stokes phosphors. These exhibit the remarkable property of emitting light or photons of higher energy (shorter wavelength) than the incident exciting radiation. Using these, it is possible, for instance, to convert infrared radiation

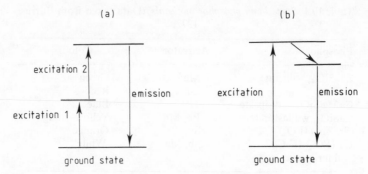

Fig. 17.7 Schematic representation of (a) anti-Stokes and (b) normal luminescence phenomena

into higher energy, visible light. There must be a catch in this somewhere, of course! The law of conservation of energy cannot be violated. Instead, the process of excitation takes place in two or more stages, as shown schematically in Fig. 17.7.

The best studied anti-Stokes phosphors to date are host structures such as YF_3, $NaLa(WO_4)_2$ and α-$NaYF_4$, which have been doubly doped with Yb^{3+} as a sensitizer and Er^{3+} as an activator. These materials can convert infrared radiation into green luminescence. During irradiation, Yb^{3+} ions transfer two photons of infrared radiation to nearby Er^{3+} ions which are then raised into a doubly excited state and decay by the emission of visible light.

17.2 Lasers

The solid state laser is basically a luminescent solid in which certain special requirements have been met. The name laser stands for light amplification by stimulated emission of radiation. The excitation process involves pumping the active centres into an excited state that has a reasonably long lifetime; a situation can then be reached in which a 'population inversion' occurs and there are more active centres in the excited state than in the ground state. During luminescence, the light emitted from one centre stimulates others to decay in phase with the radiation emitted from the first centre. In this way, an intense beam or pulse of coherent radiation is built up.

The first laser system, the ruby laser, was reported in 1960 by Maiman; from this has grown a large area of modern science and technology with applications in photography, surgery, communications and precise measurements, to name but a few. Many types of laser system have been discovered. The commercially available ones fall into three main categories: gas lasers, dye lasers with their wavelength tunability and solid state lasers. We are here concerned only with solid state lasers and the chemistry involved in their operation. It is not intended to give a general review of lasers.

17.2.1 The ruby laser

This was the first laser system to be discovered and more than twenty years later it is still an important one. The essential component to the ruby laser is a single crystal of Al_2O_3 doped with a small amount, 0.05 wt%, of Cr^{3+}. The Cr^{3+} ions substitute for Al^{3+} ions in the distorted octahedral sites of the corundum crystal structure (this structure is similar to that of ilmenite, $FeTiO_3$; Section 16.3.5). On addition of Cr_2O_3 to Al_2O_3, the colour changes from white, in Al_2O_3, to red, at low Cr^{3+} levels, to green, for larger Cr^{3+} contents; these solid solutions are discussed in general terms in Chapter 10.

The energy levels of the Cr^{3+} ion in ruby are shown schematically in Fig. 17.8. On shining intense visible light from, for example, a xenon flash lamp, onto a ruby crystal, d electrons on Cr^{3+} ions may be promoted from the $^4A_2$ ground state into $^4F_2$ and $^4F_1$ upper states. These then rapidly decay, by a non-radiative process, into the ^2E level. The lifetime of the ^2E excited state is fairly long, $\sim 5 \times 10^{-3}$ sec, which means that a considerable population inversion has time to build up. Laser action then occurs by transition from the ^2E level to the ground state. During this transition many ions are stimulated to decay, in phase with each other, giving an intense, coherent pulse of red light, of wavelength 6934Å.

The design of a ruby laser is shown schematically in Fig. 17.9. It contains a ruby crystal rod, several centimetres long and 1 to 2 cm in diameter. The flash lamp is shown as being wrapped around the ruby rod. Alternatively, it may be

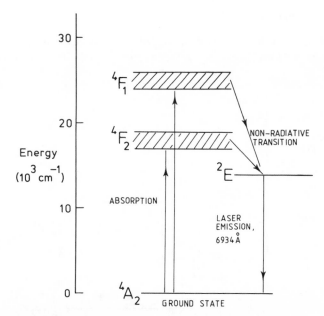

Fig. 17.8 Energy levels of Cr^{3+} ion in ruby crystal and laser emission

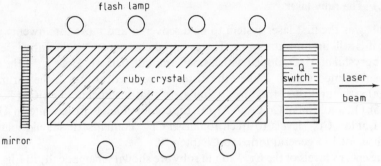

Fig. 17.9 Design of a ruby laser

placed alongside the rod; the two are then arranged inside a reflection cavity such that the rod is effectively irradiated from all sides. At one end of the rod is a mirror for reflecting the light pulse back through the rod. At the other end is a device known as a Q switch which may either allow the laser beam to pass out of the system or may reflect it back through the rod for another cycle. The Q switch may simply be a rotating mirror timed to allow out the laser beam when it has reached its optimum intensity: as the light pulse passes back and forth along the rod, it builds up in intensity as more of the active centres are stimulated to emit radiation that is coherent with the initial pulse.

17.2.2 Neodymium lasers

The host material for the laser active Nd^{3+} ion is either a glass or yttrium aluminium garnet (YAG), $Y_3Al_5O_{12}$ (see Section 16.3.4 for a discussion of YIG

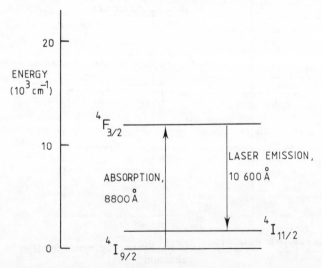

Fig. 17.10 Energy levels of the Nd^{3+} ion in neodymium lasers

and other garnets). The energy levels and transitions involved in the operation of neodymium lasers are shown in Fig. 17.10. During irradiation from a high energy lamp, several absorption transitions occur, although only one is shown. These excited states all decay non-radiatively to the $^4F_{3/2}$ level from which laser action occurs to the $^4I_{11/2}$ level with a wavelength of 10600 Å for Nd–glass and 10640 Å for Nd^{3+}–YAG. The $^4F_{3/2}$ state is long lived, $\sim 10^{-4}$ sec, but depends somewhat on the Nd^{3+} concentration. This again allows a large population inversion to build up and allows Nd^{3+} to be used in high power lasers.

Questions

17.1 What are the general requirements for a solid material that is to be used as a laser source?

17.2 Antistokes phosphors emit light of shorter wavelength than that used for excitation. Explain why the law of conservation of energy is not violated.

References

H. L. Burrus (1972). *Lamp Phosphors*, Mills and Boon.

J. A. Deluca (1980). An introduction to luminescence in inorganic solids, *J. Chem. Ed.*, **57**, 541.

Ted R. Evans (1982). *Applications of Lasers to Chemical Problems*, Wiley.

T. H. Maiman (1960). Stimulated optical radiation in ruby, *Nature*, **187**, 493.

C. B. Moore (1974). *Chemical and Biochemical Applications of Lasers*, Academic Press.

R. E. Newnham (1975). *Structure—Property Relations*, Springer-Verlag.

Chapter 18

Glass

Glass is one of the oldest synthetic materials used by man and knowledge of glass has been gradually acquired over many centuries. Scientific study of glass began with Faraday and others at the beginning of the nineteenth century and today it is still a rapidly developing subject, both in the development of new glassy materials with special properties and in the application of new scientific techniques to improve our understanding of the structure and behaviour of glass. There are two main definitions of a glass. One concentrates on the conventional method of preparation. This involves cooling from the liquid state, without crystallization occurring, until the material becomes a rigid solid through increase in its viscosity. The other definition focuses on the structure of a glass, i.e. a glass is an amorphous solid, without long range order or periodicity in the arrangement of atoms. Neither definition appears to be ideal. Thus, glasses can now be prepared by methods other than by cooling from the liquid state, for instance by the drying of aqueous gels or by vapour deposition. As an example, an amorphous solid material similar to silica glass may be prepared by hydrolysis of tetramethylorthosilicate, $Si(OCH_3)_4$, and subsequent heating of the gelatinous product to $\sim 550\,°C$. It is also debatable, therefore, whether all amorphous solids are glasses and the all-embracing term 'non-crystalline solids' has been proposed of which glasses are but one category.

A wide variety of materials form glasses. The most important commercially are certain inorganic silicates and borates but glasses can also be prepared from some aqueous solutions, molten salt mixtures, liquid metals, chalcogenides and organic macromolecules. This chapter deals mainly with the structures, properties and applications of oxide (usually silicate) glasses and with other new glasses that have interesting properties and applications.

18.1 Factors that influence glass formation

18.1.1 Oxide glasses—electronegativity and bond type

The main glass-forming oxides are SiO_2, B_2O_3, GeO_2 and P_2O_5, all of which come from a certain area of the Periodic Table (see Fig. 18.1). They are oxides of elements with intermediate electronegativity: these elements are not sufficiently electropositive to form ionic structures, such as MgO, NaCl, but also are not sufficiently electronegative to form covalently bonded, small molecular structures, such as CO_2. Instead, bonding is usually a mixture of ionic and covalent and the structures are best regarded as three-dimensional polymeric structures. Oxides of other elements around this group in the periodic table also show a tendency to glass formation. Some, such as As_2O_3 and Sb_2O_3, form glasses if cooled very rapidly. Others, such as Al_2O_3, Ga_2O_3, Bi_2O_3, SeO_2 and TeO_2, are *conditional glass-formers*, i.e. they do not form glasses alone, but may do so in the

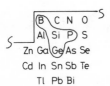

Fig. 18.1 Elements whose oxides form glasses readily

presence of certain other non-glass-forming oxides. For instance, a range of liquid compositions in the $CaO-Al_2O_3$ system forms glasses although CaO and Al_2O_3 do not, themselves, form glasses. The glass-forming oxides, SiO_2, B_2O_3, etc., can form glasses either alone or when mixed with considerable quantities of non-glass-forming oxides. For instance, glasses may be readily formed from SiO_2 or B_2O_3 to which up to 20 to 40 per cent of an alkali oxide has been added.

18.1.2 Viscosity

A factor that is undoubtedly important in glass formation is the viscosity of liquids above their melting points. The glass-forming oxides shown in Fig. 18.1 all form very viscous liquids; e.g. the viscosity of silica just above its melting point, 1715 °C, is 10^7 poise (Section 18.6). By contrast, most inorganic substances, which do not form glasses readily, are fluid in the liquid state, e.g. the viscosity of liquid H_2O at 0 °C or molten LiCl at 613 °C is $\sim 2 \times 10^{-2}$ poise, P. Viscosity is related to the structure and bonding that is present. Thus, molten silica may be regarded as an amorphous, polymeric structure with strong Si—O bonds and this is the reason for its high viscosity. In order for such a liquid to crystallize, many strong bonds must break and reform and considerable atomic reorganization is necessary. Clearly these processes take place with more difficulty in viscous, polymeric liquids than they do in fluid, ionic or molecular liquids. There are exceptions to this general relation between melt viscosity and glass-forming ability, however. Two examples of liquids that are fluid but readily yield glasses on cooling are (a) aqueous solutions of $ZnCl_2$ and (b) molten mixtures of $LiNO_3$ and $Ca(NO_3)_2$.

18.1.3 Structural effects—Zachariasen's rules (1932)

Zachariasen considered the relative glass-forming ability of simple oxides and concluded that the ideal condition for glass formation is that the material should be capable of forming an extended three-dimensional network structure without any long range order. To meet this requirement, he developed a set of rules which have had a great influence on subsequent studies of glass. Since the structures of glasses are not usually known, the rules are applied to the crystal structures of the relevant oxides from which the glasses may or may not be formed. This is a reasonable assumption since similar bonding, coordination polyhedra, etc., are likely to be present in both crystalline and liquid (glassy) states. In discussing these rules, let us distinguish between the oxygen atoms and the 'other' atoms in the binary oxide. Zachariasen's rules are as follows:

1. An oxygen atom is linked to, at most, two other atoms.
2. The coordination number of the other atoms is small.
3. The coordination polyhedra, formed by oxygen atoms around the other atoms, share corners and not edges or faces.
4. The polyhedra link up to form a three-dimensional network.

Rule 4 has already been discussed and the presence of a network structure, especially when coupled with a high liquid viscosity, acts in favour of glass formation.

Rules 1 and 3 make it possible for a glass or a liquid to have a three-dimensional network structure which lacks long range order. Thus, vitreous silica is built of corner-sharing SiO_4 tetrahedra in which each oxygen is bonded to only two silicons. This gives rise to a rather open structure in which it is possible for the Si–O–Si angle to vary without distorting the SiO_4 tetrahedra themselves; hence a three-dimensional network structure is possible that lacks long range order or periodicity. In crystalline silica, the Si–O–Si angle is constant in each polymorph (quartz, cristobalite, etc.) and this gives rise to a regular arrangement of SiO_4 tetrahedra. Crystalline and vitreous silica both contain network structures, therefore, but they differ in the structure of the network: one is regular and ordered, the other is irregular and disordered. For structures in which the oxygen has a high coordination number (rule 1), and especially if the polyhedra share edges or faces (rule 3), it is difficult to form a random network structure without grossly distorting the polyhedra.

Rule 2 is closely related to rule 1; for a given formula the coordination numbers of the different atoms are interdependent. Thus, in SiO_2, the silicon coordination number is four and hence the oxygen coordination number must be two (Chapter 8).

Let us see now how these rules may be applied to oxides of different Groups of the Periodic Table. Oxides of the alkali and alkaline earth elements, which do not form glasses, clearly do not obey Zachariasen's rules. For instance, the coordination number of oxygen is eight in Na_2O and six in MgO (rule 1) and the resulting polyhedra, (NaO_4), (MgO_6), share edges, not corners (rule 3).

Oxides of Group 3 satisfy the rules only if the other element is three-coordinate to oxygen: for a general formula M_2O_3, the other element must be three-coordinate in order for the oxygen to be two-coordinate (rule 1). B_2O_3 obeys the rules since boron is in triangular coordination to oxygen, but Al_2O_3, containing octahedrally coordinated aluminium, does not. This correlates nicely with the ability of B_2O_3 but not Al_2O_3 to form a glass.

Oxides of Groups 4 and 5 obey the rules only if the other element is tetrahedrally coordinated to oxygen. This then gives an oxygen coordination number of two for Group 4, or somewhat less than two, on average, for Group 5. Partial coordination numbers are, of course, not possible; in P_2O_5, for instance, some oxygens have CN = 2 whereas others have CN = 1.

Using these rules, Zachariasen predicted the ability of certain oxides to form glasses; subsequently his predictions were confirmed experimentally, albeit on small quantities of material. Examples were V_2O_5, Nb_2O_5, Ta_2O_5, Sb_2O_5, P_2O_3 and Sb_2O_3.

In recent years, it has been realized that Zachariasen's rules do have certain limitations. For instance, oxide glasses have been prepared which do not satisfy the rules, e.g. glasses which do not have three-dimensional network structures. These examples do not detract from the usefulness of Zachariasen's rules but do serve to illustrate the difficulty in finding a comprehensive structural theory of glass formation.

18.1.4 Criteria of Sun and Rawson

Other correlations have been proposed between structural features and the glass-forming tendency of simple oxides. Sun suggested that the other element–oxygen bond strength is important. He noted that glass-forming oxides have bond strengths $\gtrsim 330\,kJ\,mol^{-1}$, whereas modifier ions, which are not part of the network structure, have bond strengths to oxygen that are below this value.

Rawson modified Sun's criterion and related glass-forming tendency to the ratio of bond strength: melting temperature. This ratio accounts for both the bond strength and the thermal energy available to break the bonds, which depends on temperature. It is virtually impossible to crystallize B_2O_3 glass and this can be understood from Rawson's criterion, since B_2O_3 has a relatively low melting point, $\sim 400\,°C$. This criterion may also explain why, in binary systems, the glass-forming compositions are often located around the low melting eutectics. A good example is provided by the $CaO-Al_2O_3$ system: neither CaO nor Al_2O_3 is itself capable of forming a glass (except perhaps by ultra rapid quenching of thin films) but compositions between $CaAl_2O_4(CA)$ and $Ca_3Al_2O_6(C_3A)$ (see Fig. 19.8) readily form glasses. These compositions are in the region of low melting eutectics with liquidus temperatures in the range ~ 1400 to $1600\,°C$.

An additional factor in binary and more complex glass-forming systems is that the liquid or glass composition may be quite different to that of the related crystalline phases. In order for crystallization to occur on cooling of the liquid it may therefore be necessary for long range diffusion of atoms or ions to occur. Consequently, this leads to a reduction in the rate of crystallization and an increase in the ease of glass formation.

18.2 Thermodynamics of glass formation—behaviour of liquids on cooling

There are two main types of pathway that a liquid may follow on cooling. Either it may crystallize at or below the melting temperature, T_m, or it may undercool sufficiently, without crystallization, to form a glass. The volume–temperature characteristics for a liquid that follows either of these two pathways are shown in Fig. 18.2. The behaviour of most non-glass-forming liquids is similar to the changes represented by curves *abcd*. Crystallization occurs, *bc*, at temperature T_m, although for kinetic reasons, the liquid may undercool somewhat before freezing actually occurs. The difference in slope of regions *ab* and *cd* indicates that the coefficient of thermal expansion of liquids is usually greater than that of solids.

The behaviour of glass-forming liquids on cooling is similar to the changes represented by curves *abef* (or *abgh*). In region *be*, the liquid is undercooled but does not freeze. At each temperature in this region, the liquid rapidly reaches a state of internal equilibrium following a temperature change but is, nevertheless, thermodynamically metastable relative to the crystalline state. With decreasing temperature, however, the liquid viscosity gradually increases until a stage is reached at which the liquid can no longer maintain itself in internal equilibrium. The atomic arrangement that is present in this undercooled liquid then becomes effectively 'frozen in' and on further cooling, the material acquires the rigid, elastic properties of a crystalline solid but without the regular three-dimensional periodicity of a crystal structure. This change in properties or behaviour, from an undercooled liquid to a glass, takes place at a temperature or range of temperatures called the *glass transition temperature*, T_g.

For any given composition, it is generally possible to prepare glasses with different degrees of stabilization; i.e. with somewhat different T_g values. With a slow rate of cooling, and provided that crystallization does not occur, the undercooled liquid may be able to maintain itself in internal equilibrium until a somewhat lower temperature than it would if it had been rapidly cooled. Consequently, T_g is lowered somewhat (compare points e and g of Fig. 18.2). The glass transition effectively represents the crossover of two timescales: the timescale to measure some property such as viscosity (value $\sim 10^{13} \, P \lesssim T_g$) and the timescale for internal reorganization of the liquid to occur on changing the temperature. The latter timescale is of the order of a few minutes at T_g and, in

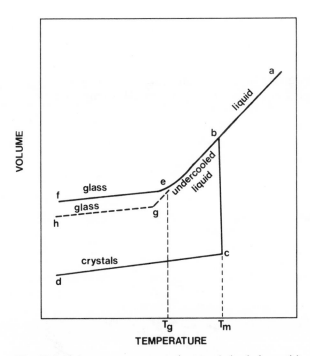

Fig. 18.2 Volume–temperature characteristics (schematic) for crystals, liquid and glass

practice, it is not possible to extend the dashed curve *eg* very far since progressively longer times are required for the liquid to contract and attain internal equilibrium.

Angell (1970) has shown that, thermodynamically, there is a theoretical lower temperature limit at which the glass transition can occur and which he has termed the *ideal glass transition temperature*, T_0. The explanation for this is found by considering the relative heat capacities and entropies of liquid and crystalline phases of the same composition. The entropy, S, of a substance is related to its heat capacity, C_p, by

$$S = \int_0^T \frac{C_p}{T} dT = \int_0^T C_p d(\ln T) \tag{18.1}$$

From a graph of C_p versus $\ln T$, the entropy at a particular temperature is given by the area under the curve up to that temperature, as shown in Fig. 18.3 for temperature T_1. In cases where a phase change has occurred below the temperature of interest, it is necessary to add the entropy of the transition to that derived from the heat capacity curve. For instance, in Fig. 18.3, at temperature T_m,

$$S(\text{crystals}) = \int_0^{T_M} C_p d(\ln T) \tag{18.2}$$

but

$$S(\text{liquid}) = \int_0^{T_M} C_p d(\ln T) + \Delta S(\text{melting}) \tag{18.3}$$

Let us now compare the relative entropies of an undercooled liquid and crystals, both of which are at the same temperature. At a temperature between T_g and T_m

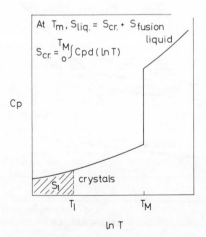

Fig. 18.3 Heat capacity and entropy as a function of temperature

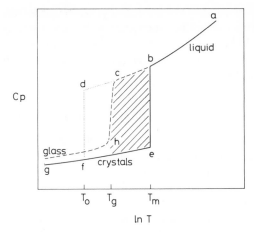

Fig. 18.4 Estimation of ideal glass transition
temperature, T_0

(Fig. 18.4), an undercooled liquid, curve *bc*, has a higher heat capacity than the corresponding crystalline phase, curve *eg*. Therefore, on cooling below T_m, a liquid loses entropy more rapidly than does the crystalline phase that is cooled by the same amount. At the glass transition temperature, T_g, the shaded area corresponds to the net loss of entropy of the undercooled liquid relative to the loss of entropy of the crystals. This net loss of entropy is, nevertheless, smaller than the entropy of fusion and so the undercooled liquid, or glass, has a higher entropy than the crystals. Suppose, now, that T_g is displaced to lower temperatures by using a slower cooling rate. A corresponding increase in the net loss of entropy of the liquid or glass results. The limiting situation arises at some lower temperature, T_0, at which the area bounded by curves *bdfe* is equal to the entropy of fusion of the crystals. At this temperature, the undercooled liquid or glass would have lost all of its excess entropy and would have the same entropy as the crystalline phase. Temperature, T_0, therefore represents the theoretical lower limiting temperature for the glass transition since, if it were possible to supercool a liquid to below T_0, the unlikely situation would exist in which an amorphous phase had lower entropy than the corresponding stable, ordered, crystalline phase.

Table 18.1 *Glass transition temperatures T_g (measured)
and T_0 (calculated)*

Glass	$T_g(°C)$	$T_0(°C)$
B_2O_3	250	60
Pyrex	550	350
Window glass	550	270
(Na_2O, CaO, SiO_2)		
Lead crystal (PbO, SiO_2)	440	150
$Ca(NO_3)_2 \cdot 4H_4O$	-50	-70

In practice, it is found that always $T_g > T_0$; i.e. the glass transition occurs, on cooling, well before the limiting temperature, T_0, is reached. Infinitely slow cooling rates would be needed for the glass transition to occur at temperatures approaching T_0. Glass transition temperatures T_g and T_0 are given for a selection of glasses in Table 18.1.

18.3 Kinetics of crystallization and glass formation

In order for a glass to form, the rate of crystallization of the undercooled liquid must be sufficiently slow that crystallization does not occur during cooling. It is possible, therefore, to treat glass formation in terms of kinetic criteria as well as the structural and thermodynamic criteria referred to above.

Crystallization of an undercooled liquid is a two-stage process that involves (a) the formation of crystal nuclei followed by (b) their subsequent growth. A kinetic condition for glass formation is that the rate of nucleation and/or the rate of crystal growth should be slow. In some undercooled liquids, nucleation is easy because there are plenty of nucleation sites available: foreign particles, container surfaces, etc. The rate of crystallization is then largely controlled by the rate of growth, which varies with temperature in a manner shown in Fig. 18.5. The rate is zero at the melting point, increases to a maximum at a certain degree of undercooling and then falls to zero again at still lower temperatures.

The general form of Fig. 18.5 can be observed experimentally in undercooled liquids which crystallize sufficiently slowly that their crystallization rates can be

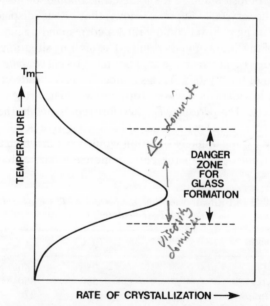

Fig. 18.5 Dependence of rate of crystallization of an undercooled liquid on temperature. (After Tammann)

measured. Most liquids freeze very rapidly at or just below T_m and their rate of crystallization cannot be measured. The explanation of the behaviour shown in Fig. 18.5 is as follows.

At temperatures close to the melting point T_m, crystals and liquid have similar free energy. There is no driving force for any changes to occur and, therefore, the net rate of crystallization of liquid is essentially zero.

At temperatures below T_m, the free energy of the crystals is less than that of the liquid. Assuming the entropy of fusion, ΔS_m, to be independent of temperature, the difference in free energy between liquid and crystals is given by

$$\Delta G = \Delta H - T\Delta S_m$$

However, at T_m:

$$\Delta G = \Delta H_m - T_m\Delta S_m = 0 \quad \text{and} \quad \Delta H_m = T_m\Delta S_m$$

Therefore,

$$\Delta G = \Delta S_m(T_m - T) \quad \text{for} \quad T < T_m \tag{18.4}$$

The increase in the rate of crystallization below T_m (Fig. 18.5) therefore corresponds to an increasing difference in free energy between crystals and liquid and hence is associated with a greater driving force for crystallization.

At lower temperatures, especially for glass-forming liquids, an additional factor, the viscosity of the undercooled liquid, becomes increasingly important. With increasing viscosity, the diffusion of atoms or ions through the liquid to the surface of the growing crystal becomes increasingly difficult and the rate of crystallization tends to decrease accordingly.

With decreasing temperature, there are therefore two competing effects. The increased difference in free energy between crystals and liquid favours crystallization whereas the increased viscosity of the undercooled liquid reduces the tendency to crystallization. The peak in the rate of crystallization (Fig. 18.5) corresponds to the situation where these two competing effects have equal weight. On the low temperature side of the peak, the viscosity effect dominates whereas on the high temperature side it is the difference in free energy between crystals and liquid that predominates.

In considering the crystallization of undercooled liquids (Fig. 18.5) and their ability to form a glass, there is a 'danger zone' for glass formation that corresponds to the maximum in the crystallization rates. If it is possible to undercool a liquid through this danger zone, it should be relatively safe from subsequent crystallization (or devitrification) and form a kinetically stable glass.

The above discussion refers to the kinetics of crystal growth and assumes that nucleation has either already taken place or is relatively easy. Two nucleation mechanisms—heterogeneous and homogeneous—may be distinguished. If heterogeneous nuclei, such as foreign particles, are not present in the undercooled liquid, spontaneous homogeneous nucleation may occur. This takes place throughout the bulk of the liquid, without the necessity of artificial nucleation sites. For example, water usually freezes at or just below 0 °C, because heterogeneous nucleation sites are present. If the water is very clean, however, it may undercool to $-20\,°C$ and crystallization may be delayed, kinetically, for a considerable time. On cooling to $-40\,°C$, however, no matter how stringent the precautions taken to keep the water clean, it always crystallizes very rapidly. This is because homogeneous nucleation occurs rapidly at this degree of undercooling.

The rate of homogeneous nucleation has a similar temperature dependence to the rate of growth shown in Fig. 18.5, but the whole curve is usually displaced to lower temperatures for nucleation. There are two reasons for this. First, nucleation is a three-dimensional process whereas growth is, at worst, a two-dimensional process (on the surface of existing crystals); hence the activation energy for nucleation is considerably greater than that for growth. Second, the surface energy of small nuclei, and hence the associated *increase* in free energy on nucleation due to the creation of new surfaces, has to be balanced against the *decrease* in free energy which occurs on crystallization, i.e. the volume free energy of the crystal nuclei (Chapter 12). For small degrees of undercooling, where the net decrease in free energy on crystallization is very small, nuclei are unstable and redissolve because of their relatively high surface energies. Just below the melting point, therefore, only infinitely large homogeneous nuclei with almost zero surface energy would be stable. (An alternative, but similar, viewpoint is that small nuclei have lower melting points than larger nuclei and hence only large nuclei can be stable at temperatures close to T_m.)

The *critical size* of a nucleus is determined by the relative size of the two terms mentioned above, i.e. the surface energy and bulk free energy. For a nucleus to be stable, the decrease in bulk free energy must be greater than the increase in energy due to the surface of the nucleus and the critical size represents the situation where these two energies are comparable. Hence, nuclei below the critical size are unstable and redissolve, whereas those larger than the critical size are stable and can grow larger. The critical size of nuclei is very temperature dependent. At and above T_m, the critical size is infinitely large but it decreases rapidly with increased undercooling below T_m. At a sufficient degree of undercooling, the critical size is sufficiently small that stable nuclei can readily form, due to random thermal or compositional fluctuations in the liquid. At this stage, spontaneous nucleation occurs.

Some liquids, therefore, have a range of temperatures below T_m over which, in the absence of heterogeneous nuclei, they are kinetically stable and can be maintained as homogeneous, undercooled liquids, without crystallization occurring. On decreasing the temperature further, nucleation and crystallization may occur spontaneously (e.g. H_2O).

Mathematical theories of crystallization, developed especially by Turnbull and coworkers, give general curves for nucleation and growth rates that are similar to the experimentally observed curves. Some of the variables are difficult to quantify, especially the surface energy of nuclei, and at present the correlation between theory and experiment is only qualitative.

Table 18.2 *Classification of glass-forming materials by the type of bonding.* (After Doremus, 1973)

Bond type	Examples
Covalent	Oxides (silicates, etc.), chalcogenides, organic high polymers
Ionic	Halides, nitrates, etc.
Hydrated ionic	Aqueous salt solutions
Molecular	Organic liquids
Metallic	Splat-cooled alloys

18.4 Structure of glasses

Glasses are not restricted to inorganic silicates but form in widely different types of material. As shown in Table 18.2, materials of every bond type form glasses. Each group usually has a different type of structure and the factors that influence glass formation vary somewhat from group to group. The structures of some of the more important oxide glasses are described in this section.

18.4.1 Vitreous silica

Vitreous silica is the simplest of the SiO_2-based glasses and the study of its structure and properties has been very important in understanding the chemically more complex silicate glasses. The generally accepted view of the structure of vitreous SiO_2 is largely the same as that proposed by Zachariasen (1932) and supported by the X-ray diffraction results of Warren (1936). The structure is built up of corner-sharing SiO_4 tetrahedra which link up to form a three-dimensional infinite network that lacks symmetry or long range order. In order to maintain electroneutrality, each corner oxygen is shared between only two tetrahedra and consequently the structure is rather open.

The X-ray 'powder' diffraction pattern of glasses is very diffuse, consisting of broad humps rather than sharp peaks; compare the X-ray powder diffraction patterns of vitreous and crystalline cristobalite (Fig. 18.6). Because there is no such thing as a unit cell in glass structure, the only kind of information that can be obtained from X-ray studies is a *radial distribution curve*. This is a curve giving the probability of finding a second atom as a function of distance from a chosen atom. A radial distribution curve for vitreous SiO_2 is given in Fig. 18.7 in which the probability of finding a second atom is represented on the ordinate by a pair distribution function. The straight line gives the results expected for a hypothetical material that consists of a random array of non-interacting point atoms. The large peak in the SiO_2 curve at 1.62 Å and the second, smaller peak at 2.65 Å correspond to silicon–oxygen and oxygen–oxygen distances in SiO_4 tetrahedra. These values are similar to those found in crystalline SiO_2 and in silicates. These first two peaks are reasonably narrow in Fig. 18.7 because the corresponding

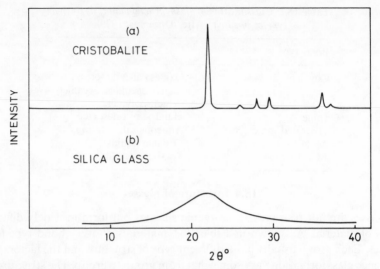

Fig. 18.6 X-ray powder diffraction pattern of (a) cristobalite and (b) glassy SiO_2, $CuK\alpha$ radiation.

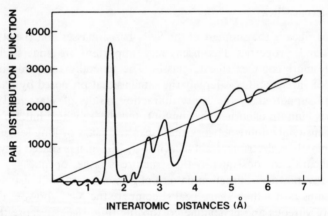

Fig. 18.7 X-ray diffraction results for SiO_2 glass. (From Mozzi and Warren, 1969)

distances are approximately constant in both glasses and in crystals, i.e. the SiO_4 tetrahedra are not distorted. The other peaks, however, are increasingly broadened since a spread of corresponding distances is present in the glass. The third peak at $\sim 3.12\,\text{Å}$ represents the nearest silicon–silicon distance, i.e. the distance between the centres of two SiO_4 tetrahedra. Since the Si—O—Si angle varies somewhat as the tetrahedra are twisted or rotated in various ways relative to each other, there is a spread in values of the silicon–silicon distance. Other peaks are assigned as follows: $\sim 4.15\,\text{Å}$—silicon to second oxygen: $\sim 5.1\,\text{Å}$—combined peak for oxygen to second oxygen and silicon to second silicon. No clear peaks are observed beyond 6 to 7 Å and this is consistent with a random network model for the structure of vitreous SiO_2.

In spite of the fairly clear-cut nature of Warren's results and the way in which they support a random network model of the glass structure, there has been considerable support over the years for microcrystallite models of the structure of vitreous SiO_2. In these models, very small ordered regions or microcrystallites are presumed to be present and are connected together by disordered regions. In order to account for the broadness of the experimental X-ray powder diffraction peaks (Fig. 18.6), the size of the microcrystallites must be very small, no larger than 8 to 10 Å, and it is questionable whether the term 'crystallite' has any meaning on this scale. Certainly an isolated crystallite of these dimensions would be unstable because of its relatively very high surface energy; in the structure of vitreous SiO_2, however, if we accept that it is a network structure of some sort, there need be no unsatisfied silicon or oxygen valencies at the interface between ordered and disordered regions and hence this high surface energy would not be present.

Evidence that is cited to support the microcrystallite model includes the observation that the first devitrification or crystallization product of a glass may reflect any structural features that are present in the glass (Ostwald's rule of successive reactions). A good example is provided by vitreous silica which may be prepared by melting either crystalline quartz or cristobalite; on subsequent crystallization of the glass, either quartz or cristobalite may appear, depending on the conditions. This is then cited as evidence for the presence of quartz-like or cristobalite-like microcrystallites in the glass. Molten SiO_2 is very viscous, however, even at 1800 °C, and it is difficult to ensure that the process of melting and homogenization on an atomic scale is thoroughly completed, even though the resulting glass may be amorphous to X-rays and optically homogeneous. The present status of the microcrystallite model is that, although it is unlikely to be correct, it cannot be discounted entirely.

18.4.2 Silicate glasses

The structure and properties of binary silicate glasses depend very much on the nature of the second oxide. With *network modifying oxides* such as alkali and alkaline earth oxides, the silica network is gradually broken up as more of the second oxide is added. This is shown by, for example, the lower viscosity of the melts as compared to that of fused SiO_2. As extra oxide ions are added to SiO_2, Si—O—Si bridging linkages are cleaved to give non-bridging oxide ions (Fig. 18.8a).

By the time that the ratio (second oxide to silica) has increased to 1:2 (e.g. as in Na_2O: $2SiO_2$ or $Na_2Si_2O_5$), the silicon to oxygen ratio has decreased to 1:2.5. For every SiO_4 tetrahedron, this means that, on average, one of the four corner oxygens must be a non-bridging oxygen (Section 7.3). In crystalline silicates of this formula, e.g. $Na_2Si_2O_5$, the silicate anions are usually infinite two-dimensional sheets; in the glasses, small pieces of sheet anion may also be

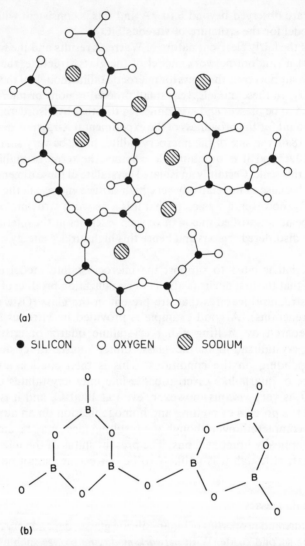

Fig. 18.8 (a) A sodium silicate glass structure (schematic).
Only three oxygens around each silicon are shown. (b) Two
boroxyl groups linked by a bridging oxygen

present but it is more likely that an open three-dimensional framework structure exists and the cations, Na$^+$, etc., occupy relatively large holes in the framework.

The distribution of the cations such as Na$^+$ is not entirely random as there is some evidence from X-ray diffraction studies that the cations may cluster together. The significance of this is not well understood. It could be simply that the arrangements of available holes in the glass network is not random; an alternative possibility is that there is some kind of attractive cation–cation interaction, presumably also involving oxide ions, which leads to clustering.

As the amount of second oxide is increased further, the silica network breaks up even more, the melts become more fluid and the tendency to devitrify on cooling increases. Usually by the time the ratio (second oxide: silica) is 1:1, it is difficult, if not impossible, to retain the liquids as glasses on cooling.

The effect of addition to silica of other glass-forming oxides or conditional glass-forming oxides is somewhat different to the effect of adding network-modifier oxides. The glass-forming oxides effectively substitute for silica and a random three-dimensional network structure is retained. Consequently, devitrification occurs with difficulty on cooling. Ranges of glass formation are usually extensive; e.g. lead silicate glasses may be prepared containing up to $\sim 80\,mol\%\,PbO(PbO$ is a conditional glass-forming oxide) and B_2O_3–SiO_2 glasses may be prepared over the entire range of compositions between SiO_2 and B_2O_3.

18.4.3 Vitreous B_2O_3 and borate glasses

Although borate glasses are of little commercial importance because they are water soluble, B_2O_3 is an important constituent of borosilicate glasses such as Pyrex. In contrast to SiO_2 and silicate glasses in which the silicon is present as SiO_4 tetrahedra, B_2O_3 glass contains BO_3 triangular units and borate glasses contain a mixture of BO_3 triangles and BO_4 tetrahedra, depending on composition. An important constituent of vitreous B_2O_3 is the boroxyl group (Fig. 18.8b). It is a planar, six-membered ring of alternate boron and oxygen atoms. These groups are linked into a three-dimensional network by bridging oxygens. However, with the planar coordination for boron, in comparison with the tetrahedral coordination of silicon in SiO_2, glassy B_2O_3 has a rather open structure; molten B_2O_3 is also more fluid than molten SiO_2.

The triangular coordination of boron in B_2O_3 glass was deduced from X-ray diffraction and various spectroscopic studies, especially B^{11} NMR spectroscopy. From the X-ray diffraction measurements, the radial distribution curve for B_2O_3 glass gives peaks at 1.37 and 2.40 Å; these correspond to the boron–oxygen and oxygen–oxygen distances of BO_3 triangles. These distances differ from those in crystalline borates that contain BO_4 tetrahedra and in which, for instance, the boron–oxygen distance is larger, 1.48 Å.

Addition of alkali oxide to vitreous B_2O_3 gives rather different results to those obtained in the corresponding alkali silicates and an effect known as the *boron oxide anomaly* is observed. In for example, the system Na_2O–B_2O_3, the viscosity of the melts increases with increasing soda content and passes through a maximum at $\sim 16\,mol\%$ Na_2O. The coefficient of thermal expansion of the glasses decreases with increasing Na_2O content and passes through a minimum at 16% Na_2O. Other properties also show either minima or maxima around this composition. By contrast, the alkali silicates become more fluid, their coefficient of thermal expansion increases steadily with increasing alkali content and no maxima or minima in properties are observed.

There appear to be several factors that contribute to the boron oxide anomaly and the maxima and minima in properties observed at $\sim 16\%$ Na_2O, but after many years of discussion and experiment the effect is still not well understood.

Bray has shown, using B^{11} NMR spectroscopy, that a gradual change in the coordination number of boron from three to four occurs as alkali oxide is added to B_2O_3. By the time that about 30 per cent. alkali oxide has been added, approximately 40 per cent of the borons have changed to tetrahedral coordination and this is independent of the nature of the alkali. In triangular coordination, the B^{11} nucleus shows strong quadrupole coupling with a broad resonance line whereas in tetrahedral coordination, the quadrupole coupling is weak and the resonance narrow. The NMR results show that, in terms of the boron coordination number, there is nothing special about the composition 16% Na_2O: the fraction of tetrahedrally coordinate borons increases linearly over the range 0 to 30% Na_2O.

A partial explanation of the boron oxide anomaly is that, with small amounts of added alkali oxide, some boron atoms change to tetrahedral coordination and these act to 'tie-in' the network, thereby increasing the viscosity. Thus, the boron to oxygen ratio, which is 1:1.5 in B_2O_3, increases towards the value of 1:2 that exists in vitreous SiO_2 as alkali oxide is added. A fully tetrahedral network could be achieved, in theory, at 50 per cent. alkali oxide (i.e. $Na_2B_2O_4$ composition), but it appears that long before this situation is reached, the viscosity has started to decrease again. Presumably this is because a network containing a large amount of tetrahedral boron would not have the strength or rigidity of the corresponding silicate network. Because of the lower formal charge on boron (+ 3) than on silicon (+ 4), tetrahedral B—O bonds are considerably weaker than tetrahedral Si—O bonds.

18.5 Liquid immiscibility and phase separation in glasses

For most uses, glasses are required to be homogeneous and single phase, i.e. on melting the constituents, a homogeneous melt should form which retains its single phase character on subsequent cooling. For certain uses, however, e.g. in the preparation of Vycor glass, the presence of fine scale liquid immiscibility in the glass is very important. In order to understand liquid immiscibility, it is necessary to consider the appropriate phase diagram (see Figs 11.9 and 11.10). On melting silica-rich compositions in, for example, the system $CaO–SiO_2$, the phase diagram of Fig. 11.8 shows that a region of liquid immiscibility exists above $\sim 1700\,^\circ C$ and it is not possible to prepare a homogeneous liquid of composition, say, 20% CaO, 80% SiO_2. Although, in theory, a single phase liquid of this composition may be prepared above $\sim 2000\,^\circ C$, it would enter the immiscibility dome on cooling and separate very rapidly into two liquids. The relatively small, high temperature immiscibility dome that appears on the $CaO–SiO_2$ phase diagram is, in fact, only the tip of a much larger, metastable, dome at lower temperatures, as shown schematically in Fig. 11.10. In order to observe this larger dome, it is essential that the liquids can undercool without crystallization occurring, i.e. the liquids must behave on cooling as though the phase diagram, with solidus and liquidus curves, did not exist.

In the $CaO-SiO_2$ system, then, the large metastable immiscibility dome which exists in undercooled liquids is capped by the stable immiscibility dome that appears on the phase diagram. In other systems, such as Li_2O-SiO_2 and Na_2O-SiO_2, a metastable immiscibility dome exists that is entirely metastable and does not protrude through the liquidus surface (Fig. 11.10b).

The appearance of a glass that contains an immiscibility texture depends on the composition and the temperature at which the liquid enters the immiscibility dome on cooling. If the latter occurs at a very high temperature, e.g. 1500 °C, the diffusion rates are high at these temperatures and even on rapid cooling a coarse liquid immiscibility texture results. The texture depends on the volume fraction of the two liquid phases; it may take the form of droplets of the second liquid phase immersed in a matrix of the first liquid or it may consist of interconnecting liquid networks if the two volume fractions are more similar. If the unmixing process commences at much lower temperatures, however, e.g. 800 °C, diffusion rates are much slower since (a) the atoms or ions have less thermal energy for migration and (b) the liquid is more viscous. This leads to much finer immiscibility textures.

The visual appearance of a glass that contains an immiscibility texture depends on the scale of the texture relative to the wavelengths of visible light. If the texture is much coarser than these wavelengths and, especially, if there is a large difference in refractive index between the two glassy phases, then the resulting glass may be opaque and look like enamel. This is because light is not transmitted through such a structure but is reflected internally at the boundaries between the two liquid phases. A liquid of composition, say, 30% CaO, 70% SiO_2 would be expected to give such an opaque glass on cooling.

If the immiscibility texture is on a scale that is comparable to the wavelengths of visible light and there is a fairly small difference in refractive index between the two liquid phases, then an opalescent glass may result. Lithium silicate glasses of approximate composition 25% Li_2O may be prepared which are opalescent, i.e. they appear blue in reflected light and orange in transmitted light. In this case, some of the visible light is scattered at the internal boundaries.

At the other extreme, if the immiscibility texture is much finer than the wavelength of light and, especially, if the two glassy phases have similar refractive indices, then the light may not be scattered at all by the internal boundaries; consequently, the glass appears optically to be transparent and homogeneous. In order to observe the immiscibility, it is then necessary to examine the glass at high magnification using an electron microscope. An example of a glass that would give such a structure is 20% Na_2O, 80% SiO_2.

18.5.1 Structural theories of liquid immiscibility

Liquid immiscibility is a feature of silica-rich glasses in, for example, the systems alkali oxide–SiO_2 and alkaline earth oxide–SiO_2. In some systems, an immiscibility dome appears on the phase diagram (Fig. 11.10a, with MgO, CaO, SrO); in others, the dome is entirely metastable (Fig. 11.10b, with BaO, alkali oxides). For these two groups of systems, the tendency to immiscibility is greatest (i.e. the dome is most extensive) for $MgO-SiO_2$ and least for Cs_2O-SiO_2. Warren and Pincus (1940) observed this trend and proposed a correlation between the tendency towards immiscibility and the increasing strength of the bond between the modifier cation and oxygen as given by the ratio cation charge, z: cation radius, r; clearly z/r is greatest for Mg^{2+} and least for Cs^+. Qualitatively, immiscibility may be regarded as resulting from a competition between the modifier cation and silicon for the oxygen atoms. Modifier cations prefer a coordination environment of non-bridging oxide ions and this tends to break up a melt into silica-rich and silica-deficient regions. As the ratio z/r of the modifier cation increases, so the tendency to immiscibility increases.

While these simple ideas on modifier cation strength and immiscibility appear to work well for some systems, there are a number of exceptions, and other factors must also affect immiscibility or, more particularly, the absence of immiscibility. One such factor is the coordination number of the second cation; if it is tetrahedral, then it may substitute for silicon in the glass network, in which case the tendency to immiscibility is much reduced. Hence oxides such as BeO, Al_2O_3, GeO_2 and P_2O_5 do not show stable immiscibility domes on their binary phase diagrams with SiO_2, even though their z/r ratios are large. This is because the cations in these oxides can enter tetrahedral sites and form part of the network. By contrast, ions such as Mg^{2+} have higher coordination numbers, perhaps six, and break up the silica network.

Levin and Block (1957) took coordination numbers into consideration and applied Pauling's rules for ionic crystals (Chapter 8) to glasses. The electrostatic bond strength (e.b.s.) is defined as

$$\text{E.b.s.} = \frac{\text{cation charge, } z}{\text{cation coordination number, CN}}$$

Levin and Block noted that the tendency to immiscibility in binary silicate and borate systems depends on the e.b.s. values of the second cation. For small values— $\frac{1}{8}$ for K^+, $CN = 8$; $\frac{1}{6}$ for Na^+, Li^+, $CN = 6$—only metastable immiscibility domes occur. Values of $\frac{1}{4}$, e.g. for Sr^{2+}, Ba^{2+}, $CN = 8$, are associated with the changeover from stable to metastable immiscibility domes and for values of $\frac{1}{3}$ or greater (e.g. Fe^{2+}, Zn^{2+}, $CN = 6$) stable immiscibility domes exist in both silicate and borate melts. The exceptions to the latter, Be^{2+}, Al^{3+}, are again systems in which the network structure of the melt is retained because the second cation is also tetrahedrally coordinated.

18.5.2 Thermodynamics of liquid immiscibility

Let us consider the thermodynamic factors that influence whether or not immiscibility occurs in a system of two liquids, A and B. The overall parameter is, of course, free energy and if a particular composition gives a mixture of two liquids it is because this assemblage has lower free energy than a single phase liquid of the same overall composition. In order to understand this further, it is necessary to break down the free energy into its two components, enthalpy, ΔH, and entropy, ΔS, according to

$$\Delta G = \Delta H - T\Delta S$$

On mixing two liquids, A and B, to form a single phase, homogeneous liquid mixture, an increase in entropy always occurs, i.e. $\Delta S > 0$. Whether or not the two liquids A and B actually mix together depends, therefore, on the enthalpy of mixing; this in turn depends on the relative magnitudes of the interactions A–B, A–A and B–B. Two possible situations arise:

(a) The net enthalpy change on mixing A and B is either negative or zero, i.e. the attractive interaction A–B is greater than or equal to the attractive interactions A–A and B–B and, therefore, $\Delta H \leqq 0$. Since $\Delta S > 0$, $\Delta G < 0$ and, therefore, the two liquids are miscible at all temperatures.
(b) The net enthalpy of mixing is positive, i.e. there is a net A–B repulsion relative to the A–A and B–B forces. In this case, there is a tendency for like atoms to cluster together. The positive ΔH value for mixing must be balanced against the negative $(T\Delta S)$ value, which clearly is temperature dependent. Therefore, if $\Delta H > T\Delta S$, immiscibility occurs since $\Delta G > 0$. The general tendency is for immiscibility to occur at low temperatures (when $\Delta H > T\Delta S$) but increasing miscibility occurs with increasing temperature and above the consolute temperature, T_c, complete miscibility occurs, when $\Delta H < T\Delta S$.

18.5.3 Mechanisms of phase separation

Phase separation is the name given to the process of separation of a homogeneous liquid into two liquid phases as it enters an immiscibility dome. Recently, there has been considerable interest in the unmixing process and two possible mechanisms have been identified: (a) nucleation and growth and (b) spinodal decomposition. Cahn and Hillert have shown that the two mechanisms differ in how the free energy of an initially homogeneous, single phase liquid changes with small fluctuations in the composition of the liquid.

Figure 18.9(a) shows the variation of free energy with composition for two temperatures T_1 and T_2 in a system that exhibits liquid immiscibility at T_2 but not at T_1 (Fig. 18.9b). At T_1, the liquid is single phase for all compositions and the curve of free energy against composition is convex downwards. Thus, on mixing A and B in any proportions at temperature T_1, a decrease in free energy occurs. Conversely, if compositional fluctuations occur in a liquid mixture, these fluctuations lead to an overall increase in free energy; hence the fluctuations are unstable and the liquid rehomogenizes. For example, if composition x experiences compositional fluctuations, to give regions of composition y and z, then a net increase in free energy, indicated by the arrow, occurs.

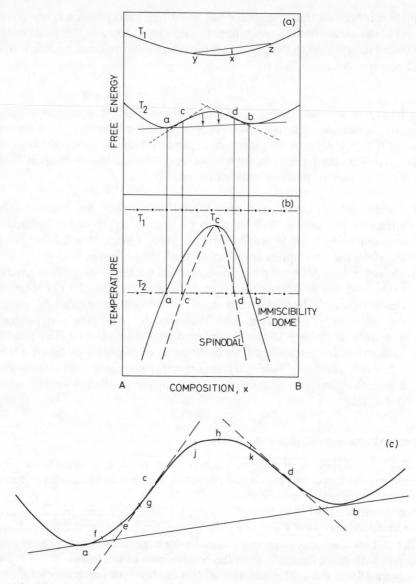

Fig. 18.9 Free energy–composition curves for systems that exhibit immiscibility phenomena

At temperature T_2, liquid immiscibility occurs between compositions a and b. Hence any composition between a and b is stable as a mixture of the two liquids of these compositions. The free energy curve at this temperature shows minima at both a and b. Between a and b, the free energy curve, which represents a metastable or unstable single phase liquid, rises and passes through a maximum; in this range, separation into two liquids of composition a and b is accompanied by a lowering of free energy, as shown by the arrows. The resulting two-phase mixtures have a free energy which lies on the straight line connecting a and b.

Consider now the effect of small compositional fluctuations in a single phase, homogeneous liquid at temperature T_2; compositions between c and d behave in a different way to the two ranges of compositions between a and c and between d and b. Small fluctuations in a liquid of composition e, between a and c (Fig. 18.9c), result in regions of composition f and g. These are associated with a net increase in free energy since point e lies below the line connecting points f and g; hence these small compositional fluctuations are unstable and redissolve. If, however, large compositional fluctuations occur then a lowering of free energy, to give compositions a and b, may result. Composition e is therefore stable to small compositional fluctuations but is unstable to large fluctuations. Composition e separates into two phases by a mechanism of *nucleation and growth*. The separation is not a spontaneous process and composition e may be regarded as metastable, i.e. there is a considerable barrier, associated with the necessity of large compositional fluctuations and hence long range diffusion of ions, to be surmounted in order for phase separation to occur. Kinetically, nucleation and growth processes often take place only slowly and there is an activation energy associated with nucleation.

The behaviour of compositions between c and d is quite different. For composition h, any small compositional fluctuation, e.g. to give regions of compositions j and k, would result in a lowering of free energy. Since fluctuations arise from the normal, random, thermal motions of atoms or ions in the liquid, composition h is unstable and spontaneously separates into two liquids. There is no barrier to unmixing, such as occurs with composition e. The process of spontaneous phase separation that occurs in composition h is called *spinodal decomposition*. It usually takes place extremely rapidly and it would be difficult to

Table 18.3 *Some differences between spinodal and nucleation and growth textures.* (From Cahn and Charles)

Nucleation and growth	Spinodal decomposition
1. Composition of second phase is unchanged with time (temperature constant).	Compositions of both phases vary with time until equilibrium is reached.
2. A sharp boundary always exists between the two phases	The boundary between phases is initially very diffuse but sharpens with time.
3. The second phase particles tend to be of random size and distribution.	The second phase is characterized by a regular distribution in size and position with a characteristic spacing.
4. The second phase particles tend to be spherical droplets or have low connectivity.	The second phase tends to form non-spherical particles with high connectivity.

quench a homogeneous liquid or glass of composition h to room temperature without phase separation occurring.

The spinodal decomposition process is the reverse of the normal trend to increasing randomization and increasing entropy which occurs on mixing liquids. Random diffusion processes normally act to even out compositional variations, but in spinodal decomposition the reverse occurs and diffusion effectively occurs in an uphill compositional gradient. The driving force for this decrease in entropy is the large decrease in enthalpy which occurs on phase separation and which leads to an overall decrease in free energy.

Spinodal decomposition occurs for compositions between c and d; the locus of c and d with changing temperature (the dashed curve in Fig. 18.9b) defines the *spinodal*. At points c and d (Fig. 18.9a) the tangent to the free energy-composition curve undergoes a change in sign, i.e. $\partial^2 G/\delta x^2$ is zero.

The two mechanisms of phase separation, spinodal decomposition and nucleation and growth, give rise to different characteristics in the resulting texture, at least in the early stages of decomposition. Some of these are summarized in Table 18.3.

18.6 Viscosity

Viscosity is a very important parameter in the manufacture and use of glass:

(a) It influences the choice of melting conditions, i.e. the temperature of melting and the time for which the mixture of raw materials is heated in order to obtain a homogeneous melt. If the melt is very viscous at a certain temperature, it may take a long time for all the crystalline starting materials to dissolve, or melt. In such cases, it may be preferable to melt at a higher temperature, where the liquid is more fluid, in order to obtain a homogeneous melt more rapidly.

(b) For a good quality product, it is necessary to 'fine' the melt, i.e. to remove all gas bubbles; the ease of fining increases with decreasing viscosity.

(c) After the melting and fining stages, the material is cooled to form a glass. At temperatures in the region of the glass transition, further heat treatment is usually necessary, for two reasons. First, the glass body is worked or shaped into its final form. Second, the glass is annealed so as to remove mechanical stresses which may have been generated during cooling. If a glass body has not been properly annealed it may shatter unexpectedly during subsequent use. The temperatures of working and annealing are governed largely by the viscosity of the undercooled melt.

(d) In use, the upper temperature at which a glass retains its characteristic properties is usually given by the onset of softening and/or devitrification (i.e. crystallization); viscosity is an important parameter in controlling the temperatures at which these processes occur.

Viscosity is defined as follows. If a shearing force, F, is applied to a liquid, the liquid flows. The viscosity, η, is given by the ratio of the applied force to the rate of flow. If the liquid is contained between two parallel plates, each of area A and distance d apart, and a shearing force is applied to the plates, then

$$\eta = \frac{Fd}{Av} \tag{18.5}$$

where v is the relative velocity of the two plates. The unit of viscosity is the poise (P), which has the dimensions of grams per centimetre per second (gm cm^{-1} sec^{-1}).

Commonly, a fluid liquid has a viscosity of $\sim 10^{-2}$P. The viscosity of a glass at various stages is:

Working point	10^4P
Softening point	$10^{7.6}$P
Annealing point	$10^{13.4}$P
Strain point	$10^{14.6}$P
Glass transition, T_g	$10^{14} - 10^{15}$P

The theories of viscous flow in glasses and undercooled liquids appear not to be very satisfactory. For simple liquids, the viscosity often obeys the Arrhenius equation over a considerable temperature range:

$$\eta = \eta_0 \exp\left(\frac{Q}{RT}\right) \tag{18.6}$$

where Q is the activation energy for viscous flow. However, deviations from straight line plots occur in graphs of $\log \eta$ versus T^{-1} and in multicomponent

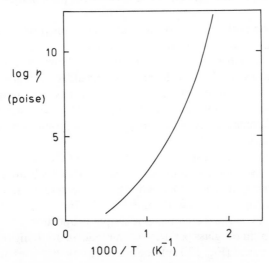

Fig. 18.10 Viscosity of B_2O_3 above the glass transition temperature

liquids especially, the Fulcher equation may provide a better fit to the data:

$$\eta = \eta_0 \exp\left(\frac{B}{T - T_0}\right) \qquad (18.7)$$

where B is a constant and T_0 is some ideal, lower limiting temperature at which the viscosity should become infinitely large. In practice, the Fulcher equation may fit the data for $T \gg T_0$, but deviations often occur as $T \to T_0$.

The viscosity of B_2O_3 in the undercooled liquid region above T_g is shown in Fig. 18.10 in the form of an Arrhenius plot of $\log \eta$ versus T^{-1}. The data fall on a curve and, clearly, do not fit the Arrhenius equation.

18.7 Electrical (ionic) conductivity of glass and the mixed alkali effect

Most borate and silicate glasses are essentially electrical insulators at room temperature with conductivities in the range 10^{-10} to 10^{-20} ohm^{-1} cm^{-1}. At elevated temperatures, however, alkali ions are mobile and alkali silicate glasses have conductivities of $\sim 10^{-5}$ ohm^{-1} cm^{-1} at T_g, 500 °C. An interesting and convincing demonstration that Na$^+$ ions are mobile at high temperatures in normal soda–lime–silica glass was described by Burt in 1925. An electric light bulb (of the vacuum type, not gas filled) was partly immersed in a molten salt bath containing Na$^+$ ions. When the current was switched on, electrons were emitted from the tungsten filament and neutralized Na$^+$ ions at the inner surface of the glass bulb. Under the action of an applied d.c. voltage between the melt and the filament, sodium ions were transported from the melt through the wall of the bulb. This resulted in the gradual formation of a film of metallic sodium on the inside of the bulb and showed that Na$^+$ ions were the principal current carriers in the glass.

The conduction pathways of, for example, Na$^+$ ions through a glass are not known since only the average, short range structure of glasses is known. In crystalline solids, conduction probably occurs by the hopping of ions over 'energy barriers' which are all of the same height (see Appendix 8). In glasses, however, it is likely that a distribution of barrier heights exists, due to the disordered structure of glass and the fact that not all sites are of the same size or shape. Thus, a migrating Na$^+$ ion may have some 'easy' jumps and some more difficult jumps, as shown schematically in Fig. 18.11(a).

The conductivity of glass is very composition dependent, as shown in Fig. 18.11(b) for the conduction of Na$^+$ ions in sodium silicate glass at 400 °C. Vitreous silica has a very low conductivity and the actual value is sensitive to impurities. On addition of soda, the conductivity increases by many orders of magnitude. However, the increase is not simply proportional to the number of alkali ions present in the glass per unit volume; e.g. on doubling the soda content from 10 to 20 per cent (Fig. 18.11b), the conductivity increases about ten times. The reasons for this are not well understood. Partly, the effect is due to the variation with composition of the activation energy for conduction: as the soda

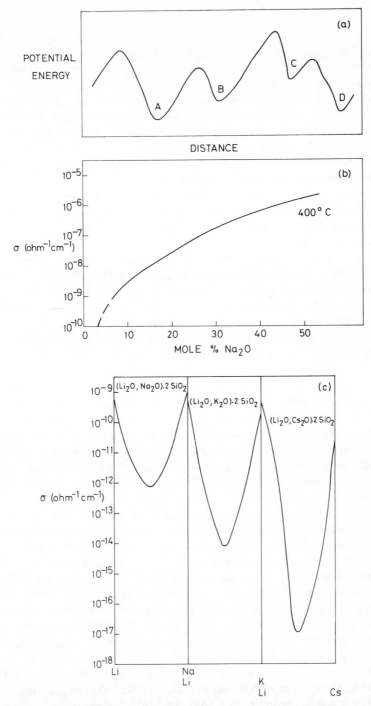

Fig. 18.11 (a) Schematic energy wells and barriers in a glass. (b) Variation of the conductivity of sodium silicate glass with composition at 400 °C. (c) The mixed alkali effect in the conductivity of disilicate glasses containing Li$^+$ and one other alkali ion

content of the glass increases, the activation energy decreases and it becomes easier for Na^+ ions to move. An attractive, recent theory that has been proposed is that glass is an example of a *weak electrolyte*: only a small proportion of the Na^+ ions in soda–silica glass are thought to be mobile and this proportion increases with increasing temperature. This theory may account for the non-linear conductivity Arrhenius plots of $\log \sigma T$ versus T^{-1} (see Appendix 8) since the pre-exponential factor contains the number of mobile ions, N, which would be temperature dependent. This may also account for the complex dependence on conductivity on composition.

An unusual, well-characterized but little understood phenomena in glasses is the so-called *mixed alkali effect*. For a given alkali silicate glass, e.g. 33.3% Li_2O, 66.7% SiO_2, a dramatic decrease in conductivity occurs as Li_2O is gradually replaced by a second alkali oxide, e.g. Na_2O. At intermediate compositions the conductivity passes through a minimum, as shown in Fig. 18.11(c) for three series of mixed alkali silicate glasses of composition 33.3% A_2O, 66.7% SiO_2 at 150 °C. The pure lithium silicate glass has a conductivity of $6 \times 10^{-10}\, ohm^{-1}\, cm^{-1}$ (due to Li^+ ions) and the conductivity of pure sodium silicate glass is about two times greater, $\sim 10^{-9}\, ohm^{-1}\, cm^{-1}$. In the mixed glasses, however, a drop in con-ductivity of up to three orders of magnitude occurs. A similar, but more pronounced effect is observed in Li–K, Li–Rb and Li–Cs silicate glasses. The effect is accentuated at lower temperatures (not shown) since the mixed alkali glasses have a higher activation energy than the end-member glasses. Measurements of diffusion coefficients have shown that in the mixed alkali glasses, the mobility of both alkali cations is decreased. Also, the decrease in conductivity that occurs in the mixed glasses is greater, the greater the difference in size between the two cations. In spite of many years of study, the mixed alkali effect is still only poorly understood. Its occurrence is not limited to conductivity values; other properties such as viscosity and volume also exhibit an anomaly, but rather less dramatic than the changes in conductivity that occur.

The most recent theory of the mixed alkali effect is due to Hendrickson and Bray (1972). Basically, it proposes that there is an energy of interaction between dissimilar cations located on adjacent cation sites. The interaction occurs because different ions have different natural vibrational frequencies, v. This theory would predict a mixed alkali effect in glasses that contain different isotopes of the same alkali and such an effect has indeed been observed.

Although most glasses have very low conductivity at room temperature, there has been considerable interest recently in glasses with high conductivity, both ionic and electronic. Chalcogenide glasses are electronic conductors and are discussed in Section 18.9. Glasses with high ionic conductivity, e.g. $10^{-2}\, ohm^{-1}\, cm^{-1}$ at 25 °C are mainly those containing mobile Ag^+ ions. They are made from mixtures of AgI and oxysalts of silver such as Ag_2MoO_4 and Ag_3AsO_4, e.g. a glass of composition $3AgI \cdot Ag_2MoO_4$ has a conductivity of 10^{-2} to $10^{-1}\, ohm^{-1}\, cm^{-1}$ at room temperature. These materials are of interest as novel solid electrolytes and have potential uses in solid state batteries. They are discussed further in Chapter 13.

18.8 Commercial silicate and borate glasses

Although pure silica glass has very desirable properties such as a high glass transition temperature, $\sim 1200\,^\circ\text{C}$, high softening point, resistance to devitrification and chemical attack and high transparency to visible and ultraviolet light, it is expensive to produce. The melting point of SiO_2 is $1713\,^\circ\text{C}$ and the resulting melt is very viscous. Hence, temperatures considerably in excess of $1700\,^\circ\text{C}$, together with long times, are needed in order to achieve a homogeneous melt. Therefore, pure silica glass is unsuitable for use in windows, bottles, lamps, etc., in which production on a massive scale is necessary. Instead, various oxides are added to silica which serve to reduce the melting temperature and maintain the resistance to chemical attack and devitrification.

Addition of Na_2O ($\sim 25\,\text{mol}\,\%$) to SiO_2 causes the liquidus (i.e. melting) temperature to drop from ~ 1700 to $\sim 800\,^\circ\text{C}$, (see Fig. 20.3); however, the resulting glass is water soluble and rather prone to devitrify. Other oxides— CaO, MgO, Al_2O_3—are therefore added to Na_2O–SiO_2 which act to maintain a fairly low liquidus temperature but also increase the resistance to devitrification and the chemical durability. Window glass is therefore a multicomponent material. Impurities such as transition metals must be avoided in order to produce transparent glasses; for brown bottles this is unimportant and impure sand containing iron oxide may be used as the source of silica.

During manufacture, it is very important that glasses are *annealed* correctly. The volume and refractive index of a glass depend somewhat on the cooling schedule that is adopted and this is of practical importance. The objective of annealing is to heat the glass for some time in the region of or just below the glass transition in order to ensure that the glass is stabilized and uniform and that the glass transition temperature has decreased to an approximately constant value. This is important in high quality, optical glass in which a uniform and constant refractive index is essential. Annealing also acts to relieve undesirable mechanical stresses; in extreme cases, unannealed glasses may shatter due to the internal stress that is present. While for most glasses annealing times of a few hours are sufficient, occasionally for special glasses or applications longer times are used. For instance, the large mirror used in the Mount Palomar telescope was annealed by cooling from 500 to $300\,^\circ\text{C}$ over a period of nine months!

Although high-silica glasses are expensive to make by normal means, the Vycor process uses an ingenious method to make glasses containing $\sim 96\%\ SiO_2$, but which circumvents the very high melting temperatures that are normally needed to prepare a glass of such a composition. The method involves first the preparation of a sodium borosilicate glass of approximate composition 10% Na_2O, $30\%\ B_2O_3$, $60\%\ SiO_2$. During manufacture, the undercooled liquid enters a metastable immiscibility dome and separates into two liquid phases, probably by a mechanism of spinodal decomposition. One of the resulting liquids has a composition close to pure silica; the other is rich in Na_2O and, especially, B_2O_3. Figure 18.12 shows the partial phase diagram for the system sodium tetraborate ($Na_2B_8O_{13}$)–silica, on which is superposed the immiscibility dome that is

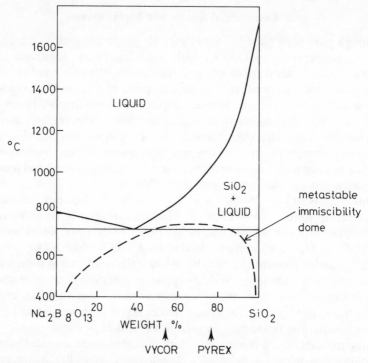

Fig. 18.12 Metastable immiscibility dome in the system sodium tetraborate–silica. (From Rockett and Foster, 1966)

observed in undercooled liquids; Vycor compositions lie on this join or close to it in the ternary system Na_2O–B_2O_3–SiO_2 and in the region of the centre of the immiscibility dome. The two liquids form an interconnected texture. The next stage in the process is to leach out the sodium borate-rich component by acid treatment, thus leaving behind a fragile, honeycomb glassy matrix of almost pure silica; this is subsequently heat treated at $\sim 1000\,^{\circ}C$ to cause the material to flow and form a non-porous, clear glass similar to fused SiO_2.

Pyrex glass is another glass which benefits from phase separation. It is also made from Na_2O, B_2O_3 and SiO_2 but the starting composition, e.g. $\sim 4\%$ Na_2O, 16% B_2O_3, 80% SiO_2, arrowed in Fig. 18.12, contains rather more SiO_2 than that of Vycor glass. The position of this composition relative to the immiscibility dome is such that the volume fraction of the sodium borate-rich phase is quite small (10 to 20 per cent) and it therefore forms as isolated droplets within the silica-rich matrix, in contrast to Vycor glasses. Since the droplets are trapped, they cannot be leached out; hence the chemical durability of the phase separated glass is controlled by the properties of the silica-rich component liquid. The chemical durability of Pyrex glass is therefore considerably enhanced by the occurrence of phase separation.

18.9 Chalcogenide and other semiconducting glasses

There is currently much interest in glassy or amorphous semiconductors made from the chalcogens–sulphur, selenium and tellurium—either alone (selenium is used in photocopying machines) or in combination with other elements. More traditional semiconductors, such as silicon and germanium, may also be made as amorphous thin films and it is now possible to dope them in the same way that crystalline silicon and germanium may be doped. Amorphous silicon and germanium have potential applications in large surface area devices such as solar energy converters.

18.9.1 Chalcogenide glasses

Sulphur and selenium give viscous liquids on melting which readily form glasses on cooling. The bonding is fairly covalent and the liquids contain rings and chains of sulphur, selenium atoms. This contrasts with liquid oxygen which has a very low melting point, is a molecular liquid and is not a glass former.

18.9.1.1 *Sulphur*

Sulphur is a complex substance. It exists in a number of crystalline polymorphs (orthorhombic, monoclinic, etc.), but, most unusually, its liquid structure is also very temperature dependent. On melting sulphur at 114°C, a fluid liquid of viscosity $\sim 10^{-2}$ P forms which consists largely of S_8 rings. With increasing temperature, the ring structure and fluidity is retained up to 160°C. Between 160

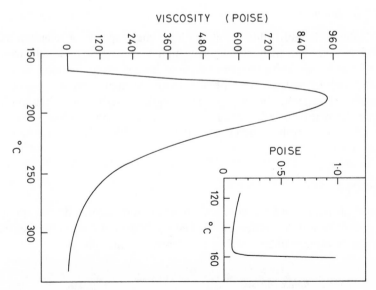

Fig. 18.13 Viscosity of molten sulphur as a function of temperature. (From Bacon and Fanelli, 1943)

and $\sim 180\,°C$, however, the viscosity *increases* by about five orders of magnitude; it then decreases again above $180\,°C$ (Fig. 18.13). The increase in viscosity that occurs above $160\,°C$ is due to polymerization of the sulphur. Many of the S_8 rings break open and join together, resulting in polymers with an average chain length of 10^5 to 10^6 atoms. With rising temperature, two competing effects occur; more of the S_8 rings open and polymerize but, also, the average chain length decreases rapidly. Above $180\,°C$, the latter effect dominates and hence the viscosity decreases.

The changes in the structure of liquid sulphur that occur with changing temperature are rapidly reversible, equilibration reactions of the type: $xS_8 \rightleftharpoons S_{8x}$; the reactions that occur on heating are reversed on the cooling cycle. It is possible, however, by rapid quenching from $> 160\,°C$ to retain a considerable fraction of the high molecular weight polymer to room temperature. The resulting material, *plastic sulphur*, is in fact a metastable, undercooled, viscous liquid which devitrifies on prolonged standing at room temperature; it turns into glass on cooling to below room temperature, where it has a glass transition at $-27\,°C$.

The structure of liquid sulphur is very sensitive to impurities. Small amounts of iodine lead to a large reduction in viscosity at, for example, $180\,°C$, probably by acting as a chain breaker:

$$\{-S-S-S-\} + I_2 \rightarrow \{-S-S-I + I-S-S-\}$$

Phosphorus, on the other hand, has the opposite effect, probably by acting as a crosslinking agent; it leads to an increase in viscosity.

18.9.1.2 *Selenium*

Crystalline selenium, which contains long chain spirals of selenium atoms, melts at $217\,°C$. The melt has a fairly high viscosity, 30 P, which decreases gradually with increasing temperature and there is no anomalous increase in viscosity, such as occurs with molten sulphur. Molten selenium also contains long chain polymers but the average chain length is less than in molten sulphur at the same temperature and hence the viscosity of selenium is less. On cooling, liquid selenium readily undercools and forms a glass at room temperature ($T_g = 31\,°C$).

18.9.1.3 *Tellurium*

Crystalline tellurium melts at $453\,°C$ and the liquid viscosity drops rapidly to a value typical of a fluid molten metal. Tellurium does not form a glass, except possibly as a rapidly quenched, vapour deposited thin film.

18.9.1.4 *More complex glasses*

The thermal stability of sulphur and selenium glasses may be increased by combining them with certain Group 4 and Group 5 elements, especially arsenic and germanium. For instance, selenium and sulphur may be replaced by up to 60 atom % As to form wide ranges of stable glasses. In the S–As system, the glass transition temperature increases from $-27\,°C$ in pure sulphur to $\sim 160\,°C$ in As_2S_3 glass. This is because arsenic acts as a cross-linking atom to increase the viscosity; the structural units present in As_2S_3 glass are puckered, covalently bonded, sheets. Because of its high T_g, As_2S_3 glass is stable to devitrification and is used commercially as an infrared transmitting material.

On addition of two other elements to the chalcogens, the range of glass-forming compositions may be increased considerably. For example, sulphur forms a limited range of glasses with both phosphorus and germanium but on addition of phosphorus and germanium together, an extensive range of ternary compositions forms glasses. A more dramatic example is provided by the system Si–As–Te; none of the three elements themselves form glasses readily nor do the three binary combinations, Si–As, Si–Te and As–Te; however, a range of tellurium-rich, ternary compositions do form stable glasses. These glasses, and those in the system Ge–P–S, are useful materials since they are stable to atmospheric attack, can be used up to $500\,°C$ and have good optical properties; as such they find applications in aircraft optical systems.

18.9.2 Electrical properties

Chalcogenide-based glasses are semiconductors and, usually, have electronic conductivities in the range 10^{-3} to 10^{-13} ohm^{-1} cm^{-1}. In the pure chalcogens, the conductivity increases with increasing atomic weight and, for example, molten tellurium is almost like a liquid metal in its electrical properties. There are characteristic differences between amorphous semiconductors, such as selenium, and crystalline ones, such as silicon and germanium, apart from their obvious differences in structure.

(a) Amorphous semiconductors may be made over wide composition ranges and are not confined to stoichiometric compositions, whereas in crystalline systems the composition of the crystal is usually stoichiometric or close to stoichiometric. This enables systematic variations in the composition and properties of amorphous semiconductors to be made.

(b) Amorphous semiconductors are versatile in the form that they take, e.g. the Xerox process uses a thin film of selenium. Single crystals, however, are much less versatile and it would be impracticable (and expensive!) to prepare single crystals in the form of thin films.

(c) Amorphous semiconductors are usually insensitive to dopants, in complete contrast to crystalline semiconductors. Recently, however, methods of doping amorphous germanium and silicon have been found.

To understand the effect of dopants on amorphous semiconductors, let us consider the structural differences between crystalline and amorphous germanium. Crystalline germanium has the diamond structure in which each atom

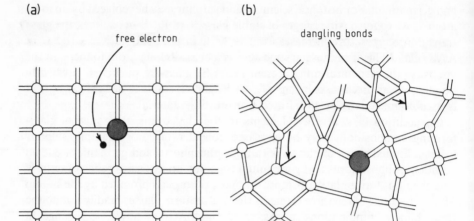

(a)   (b)

free electron   dangling bonds

○ germanium   ● arsenic

Fig. 18.14 Two-dimensional representation of the structure of (a) crystalline germanium and (b) amorphous germanium, both containing one arsenic atom. (After Davis, 1977.) In crystalline germanium, every atom including arsenic, is at a lattice site and is surrounded by four atoms

is tetrahedrally coordinated to four others, giving a three-dimensional network structure. On doping germanium with, for example, arsenic, a pentavalent arsenic atom is constrained to enter a tetrahedral site in place of a germanium atom; it therefore forms four covalent bonds to its germanium neighbours. Since arsenic contains five valence electrons and only four are used in bonding, the fifth electron is able to move through the structure and therefore to conduct electricity. (According to band theory, this electron occupies the conduction band; see Chapter 14). This is shown schematically in two dimensions in Fig. 18.14(a).

In amorphous germanium, a network structure again forms in which most germanium atoms are tetrahedrally coordinated but some germanium atoms are only three-coordinate and are left with a single electron in a 'dangling bond'; two of these are shown schematically in Fig. 18.14(b). In an amorphous network, the arsenic atoms in arsenic-doped germanium are less constrained and show a preference for three-coordinate sites. Three of the electrons of each arsenic atom are used in covalent bond formation and the remaining two form a non-bonding lone pair. The arsenic atoms are therefore electrically neutral and have no unpaired free electrons available for conduction. Hence, the efficiency of doping amorphous germanium with arsenic to give n-type conduction is much lower than with crystalline germanium.

There are two main reasons for the virtual absence of doping effects in amorphous silicon and germanium when they are prepared by conventional methods. First, the arsenic dopant atoms do not usually provide a free electron if they occupy three-coordinate sites. Second, if any free electrons could be introduced into the structure, they would soon become trapped at the dangling bonds associated with three-coordinate germanium and thereby be lost for conduction. One conventional method of preparing thin films is to use the sputtering technique in an argon atmosphere (Chapter 2). Argon ions are driven into the target material, e.g. crystalline germanium, and the germanium atoms that are ejected travel to the substrate where they deposit as a thin amorphous film.

Recently, a new method for preparing amorphous silicon and germanium has been developed that appears to avoid the creation of dangling bonds and consequently allows doping. Germane (GeH_4) gas containing small amounts of phosphine (PH_3) or diborane (B_2H_6) is decomposed by glow discharge in an RF-induced plasma. The atoms of Ge, Si, P and B that are produced form a thin film deposit which appears to be relatively free from defects such as dangling bonds. This is probably because any dangling bonds that would be present are able to form covalent bonds with atomic hydrogen; hence they cannot act as traps for free electrons. Although some of the phosphorus and boron atoms that are introduced occupy sites with coordination number three, some must enter tetrahedral sites and these give rise to conduction in the usual way, i.e. by n-type and p-type mechanisms.

18.9.3 The photocopying process

The photocopying process makes use of two properties of amorphous selenium. First, it may be prepared in the form of a semiconducting thin film. Second, it is a *photoconductor*, i.e. its electronic conductivity is enhanced enormously by exposure to light. The photocopying machine contains a cylindrical metal drum upon which has been deposited, by vacuum evaporation, a thin film of amorphous selenium. The steps involved in photocopying are as follows (see Fig. 18.15).

The surface of the selenium is charged positively by a corona discharge induced from a wire, held at a high potential, which is moved parallel to the surface of the selenium (Fig. 18.15a). The page to be copied is then imaged onto the screen by exposure to visible light; the light areas of the page reflect photons onto the selenium screen and generate electron-hole pairs in the selenium (Fig. 18.15b). Under the action of an electric field, these pairs dissociate; the holes move inwards towards the metal drum and the electrons drift out to the surface of the selenium where they neutralize some of the induced positive charge (Fig. 18.15c). Hence, the selenium film is left with an image, in the form of positive charges, of the dark areas on the paper that is being photocopied. The selenium screen is now dusted with a negatively charged carbon black ink; the ink adheres to those positively charged areas of the film that have not been discharged (Fig. 18.15d). The ink is transferred to clean paper with the aid of a second corona discharge (Fig. 18.15e); the paper is then removed and heated in order to make the image permanent.

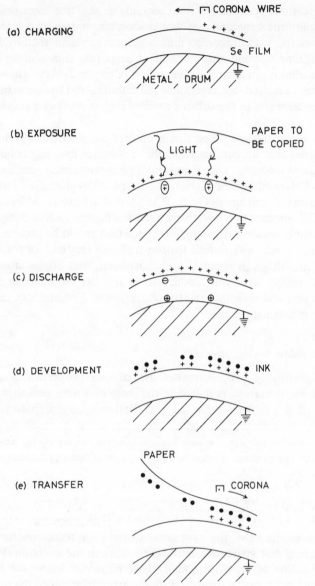

Fig. 18.15 Stages in photocopying. (After Davis, 1977)

18.10 Metallic glasses

Liquid metals do not usually supercool to form glasses, but recently certain alloy compositions have been shown to do so. In the structures of liquid and crystalline metals, the bonding forces between the atomic nuclei and the conduction electrons are non-directional. Thus, in order for a liquid metal to crystallize, it is not necessary to break any strong, covalent bonds. The effect of this can be seen in, for example, the chalcogens: liquid selenium contains covalent —Se—Se—linkages and selenium readily forms a glass, but liquid tellurium has much more metallic bonding and behaviour and does not usually form a glass. A similar situation arises in ionic materials: in molten NaCl, the ionic bonds between Na^+ and Cl^- ions are non-directional. This greatly reduces the energy barrier to atomic shuffles and hence to crystallization; NaCl, in common with many other simple molten salts, does not form a glass.

In order to prepare glassy metals, special ultra-rapid quenching techniques are necessary. In *splat quenching*, liquid droplets are propelled at high velocity (by a pressure gun) onto a cold surface and the liquid is effectively cooled within a small fraction of a second. In *roller quenching*, the melt is propelled onto a cooled rotating drum as shown in Fig. 18.16(a) and forms a thin ribbon of glass. Alternatively, the melt may be injected between two cooled, counter-rotating drums. Cooling rates of the order of 10^6 to 10^8 K sec^{-1} may be achieved by this method.

Only certain liquid metal compositions may be quenched to yield glasses and, usually, at least two elements must be present in the melt composition. One of these is a conventional metal, e.g. a transition element such as iron or palladium. The other is an element on the metal-insulator borderline, e.g. a semiconductor such as silicon or phosphorus. In each system, it is usually possible to prepare glasses over a range of compositions, e.g. for compositions containing 15 to 25% Si in Pd, Si mixtures. Glass-forming ability is generally associated with the presence of low melting eutectics on the equilibrium phase diagram (Section 18.1.4). The structures of metallic glasses approximate to a random close packed arrangement of spheres of two different sizes, such as would be obtained by shaking together two kinds of sphere inside a container.

Glassy metals have certain properties that are unusual and valuable when compared to those of conventional crystalline metals:

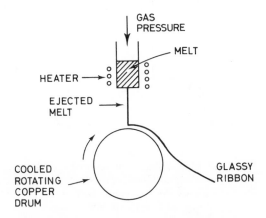

Fig. 18.16 Schematic roller quenching method

(a) Glassy metals are usually much stronger than crystalline metals and, in favourable cases, their strength may approach the theoretically attainable limit. Many pure metals are relatively soft due to the presence of dislocations (Chapter 9) which can move very easily under the action of an applied stress. Dislocations are able to move readily in a crystal partly because of the regular periodicity of the structure: the dislocation moves from one region of the structure to an identical adjacent region. It is a debatable point whether or not dislocations can exist in a non-periodic structure such as a glass, although one extreme point of view is that the glass structure may be considered as built entirely of dislocations. If dislocations do exist in glassy metals, they are unlikely to be able to move as easily as in crystalline materials. Glassy metals therefore have high strength, but they also exhibit a high plasticity, unlike strong and hardened metals such as cast iron which are usually brittle. Typically, glassy metals can withstand shear strains of 50 per cent or more before finally failing by ductile fracture mechanisms. It is likely that they will find applications in fibre reinforcement.

(b) Glassy metals are more resistant to chemical attack such as corrosion than are polycrystalline metals. Metals are usually most reactive at grain boundaries and at surface sites of high energy, e.g. at the sites of emergent dislocations or other defects. Since glasses do not contain grain boundaries or dislocations, in the usual sense of the word, they are less reactive chemically.

(c) Certain glassy metals have interesting magnetic properties. Cobalt and iron-containing glasses have low coercivity and may be easily magnetized and demagnetized (Fig. 18.17). This property is also attributed to the absence of grain boundaries in the glasses and may find application in memory devices in which rapid recording and erasure of information is important.

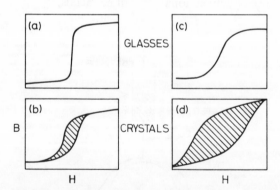

Fig. 18.17 Magnetization curves for $Fe_{75}P_{15}C_{10}$ (a, b) and $Co_{78}P_{22}$(c, d). Glassy curves (a, c) show smaller and more square hysteresis loops with lower coercive forces and higher permeabilities

18.11 Glass-ceramics

Glass-ceramics are crystalline substances made from glassy starting materials. They possess a valuable combination of the favourable properties of both glasses and ceramics. They maintain their mechanical strength to much higher temperatures than do most glasses, which soften above $\sim 500\,°C$, and by control of the composition it is possible to control properties such as the coefficient of thermal expansion. Glass-ceramics are prepared by subjecting selected glasses to a carefully regulated, heat-treatment schedule which results in the nucleation and growth of crystalline phases within the glass. In many cases, the crystallization process can be taken almost to completion, but a small proportion of residual glass phase is often present.

Devitrification or crystallization of a glass normally signifies the end of its useful life since the glass may lose its mechanical strength, transparency, workability, etc. However, in glass-ceramics, first developed by Stookey of Corning Glass in 1957, the glass is deliberately devitrified in a controlled manner to give a fine-grained ceramic with many valuable properties such as high mechanical strength. For controlled devitrification, it is first necessary to create a large number of crystal nuclei distributed throughout the bulk of the glass body, typically 10^{12} to 10^{15} nuclei cm^{-3}. Crystallization that begins only at a few surface nuclei must be avoided. Various ways have been used to generate these nuclei:

(a) By creating a dispersion of colloidal particles of metals such as Cu, Ag, Au and Pt in the melt. These metal particles probably do not dissolve completely and subsequently act as nucleation sites when the glass is annealed at lower temperatures (e.g. $500\,°C$). Precipitation of the metal nuclei is aided by irradiation with ultraviolet light if the glasses are also photosensitive.

(b) By adding oxides such as TiO_2, P_2O_5 and ZrO_2, which are soluble in the melt at high temperatures but which precipitate on annealing at lower temperatures, perhaps as a consequence of phase separation, and hence act as nucleation sites.

(c) By homogeneous nucleation of the glass, as discussed in Section 18.3. The glass is annealed at a certain temperature, usually in the vicinity of the glass transition, to form crystal nuclei throughout the volume of the glass.

The success of heterogeneous nucleation (methods a and b) depends on two main factors. First, the interfacial tension between the nucleation catalyst and the crystallizing phase must be low, i.e. the contact angle, α (see Fig. 20.2), between the nucleation substrate and the nucleated crystal phase should be low. A low contact angle leads to a reduction in the free energy of nucleation.

Second, there should be some similarity in crystal structure and, in particular, in low index d-spacings between the nucleation catalyst and the nucleating phase. This allows oriented or epitaxial growth to occur if the atomic repeat units in the nucleating phase can match those in the catalyst to within 10 to 15 per cent.

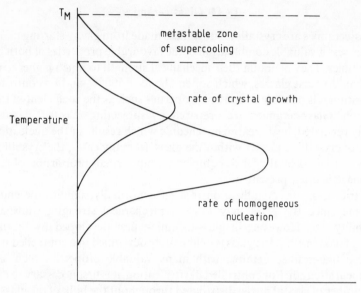

Fig. 18.18 Rates of homogeneous nucleation and growth in a viscous liquid. (After Tammann)

After nucleation—which is usually carried out close to the glass transition temperature where the viscosity is high, 10^{11} to 10^{12} P, and crystal growth rates are small—the glass is heated to a higher temperature, at which crystal growth occurs on the surface of the existing nuclei. Because there are a large number of nuclei present, distributed throughout the bulk of the glass, each one can grow only by a small amount until it impinges on neighbouring nuclei. Hence, the final crystal size in the glass-ceramic is small, typically in the range 0.1 to 1 μm (10^{-7} to

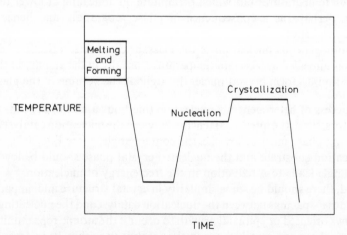

Fig. 18.19 Two-stage heating process used in the production of glass-ceramics

10^{-6} m) diameter. The temperature that is used for the crystal growth stage is higher than the temperature used for nucleation because of the way in which nucleation and growth rates vary with temperature (Fig. 18.18). Because nucleation is generally a more difficult process than growth, a large degree of undercooling is necessary in order to build up the necessary difference in free energy for the process of nucleation to occur.

The stages involved in the manufacture of a glass-ceramic article are, therefore, as follows. First, the glass is prepared, in the usual way, by melting the component oxides, together with any nucleating agents. The resulting melt is formed into the desired shape of the final product, cooled and subjected to a two-stage nucleation and growth process, as shown schematically in Fig. 18.19.

18.11.1 Some important glass-ceramic compositions

Glass-ceramic compositions are complex and contain several components, but the properties can usually be referred back to simpler systems. Perhaps the most important prototype glass-ceramic system is Li_2O–SiO_2, whose phase diagram is given in Fig. 18.20. Addition of ~ 30 mol % Li_2O to SiO_2 causes the liquidus temperature to drop rapidly from 1713 to 1030 °C. The resulting liquid forms a clear glass on cooling. (Liquids containing $\lesssim 25\%$ Li_2O give opalescent or opaque glasses on cooling, due to phase separation within a metastable immiscibility dome.) On crystallization between the glass transition temperature,

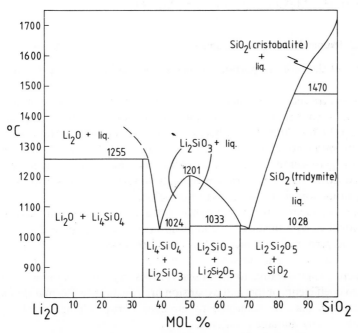

Fig. 18.20 Phase diagram for the system Li_2O–SiO_2

$\sim 500\,°C$, and the solidus temperature (i.e. melting temperature), $1030\,°C$, $Li_2Si_2O_5$ is the main product, together with small amounts of either SiO_2 and/or Li_2SiO_3. In order to produce a fine-grained lithium silicate glass-ceramic, the two-stage crystallization would typically involve nucleation at 450 to $500\,°C$ followed by growth of the $Li_2Si_2O_5$ crystals at 650 to $700\,°C$.

One of the most important commercial glass-ceramic systems is $Li_2O–Al_2O_3–SiO_2$. A nucleation catalyst is usually added. This may be a metal such as Au or an oxide such as TiO_2 or P_2O_5. Photosensitively nucleated glass-ceramics form in compositions with low Al_2O_3 content in which the main crystalline phase is $Li_2Si_2O_5$ and/or Li_2SiO_3. In more Al_2O_3-rich compositions, the main crystalline phase is β-spodumene, $LiAlSi_2O_6$ or β-quartz solid solutions and these have the unusual property of a very low or negative coefficient of thermal expansion ($\lesssim 1 \times 10^{-6}$).

Lithium silicate-based glass-ceramics with added MgO have applications because of their high coefficients of thermal expansion ($\sim 1.4 \times 10^{-5}$); various crystalline phases, including Li_2MgSiO_4, form. Similar materials but with added ZnO have high expansion coefficients and high mechanical strengths; various crystalline phases again form, including Li_2ZnSiO_4 and willemite, Zn_2SiO_4.

Glass-ceramics made from the system $MgO–Al_2O_3–SiO_2$, with added TiO_2 or P_2O_5 nucleation catalyst, are good electrical insulators since they are alkali free and have high mechanical strength at high temperatures. The main crystalline phase is α-cordierite, $2MgO\cdot2Al_2O_3\cdot5SiO_2(\equiv Mg_2Al_4Si_5O_{18})$, together with smaller amounts, depending on composition, of cristobalite, SiO_2, clinoenstatite, $MgSiO_3$, and forsterite, Mg_2SiO_4.

18.11.2 Properties of glass-ceramics

Some of the more important properties of glass-ceramics are as follows:

(a) Glass-ceramics are usually stronger than conventional glasses under strained load or impact. The crossbreaking strength of glass rods is usually in the range 3000 to 10 000 p.s.i. whereas the corresponding value for glass-ceramics is 40 000 to 60 000 p.s.i. This is probably because the crystals that are present in the glass-ceramic limit the size of the flaws that are possible and this acts to slow down the rate of crack propagation. Glass-ceramics also usually have superior abrasion resistance to normal glasses.

(b) The coefficient of thermal expansion of glass-ceramics (and glasses) may be controlled by adjusting the composition; values as high as 2×10^{-5} and as low as zero may be achieved. It is therefore possible to match the expansion characteristics of a glass-ceramic to that of, for example, a metal in cases where a glass-ceramic to metal seal is necessary. Glass-ceramics with low or zero coefficients of thermal expansion are resistant to thermal shock, i.e. the materials may be subjected to large or abrupt changes in temperature without any damage occurring.

(c) Glass-ceramics have much higher deformation temperatures than corresponding glasses of the same composition. For example, many oxide glasses have a T_g of ~ 450 °C and soften readily above 600 to 700 °C. A glass-ceramic of the *same* composition may, however, retain its mechanical strength and rigidity to higher temperatures, e.g. 1000 to 1200 °C.

(d) Glass-ceramics are usually good electrical insulators, especially if they do not contain alkali oxides.

(e) The visual appearance of glass-ceramics depends on the crystalline phases present. They may be transparent, translucent or opaque, depending on crystal size, crystal birefringence and the difference in refractive index between different crystals or between crystals and any residual glassy matrix.

(f) Glass-ceramics have zero porosity, unlike most ceramic bodies formed by conventional powder pressing techniques, which may be up to 10 per cent porous. This is because, during crystallization, the glass can flow to accommodate any changes in volume.

18.11.3 Applications of glass-ceramics

Glass-ceramics that are resistant to thermal shock are used in cooking ware (Pyrosil), both as containers and as oven tops with built-in electric heaters. They are used as bearings, where they have several times the wear resistance of steel, as metal coatings (similar to enamel) and in ceramic to metal seals. With their high temperature stability and thermal shock resistance they are used as heat-protective nose cones on rockets. 'Slagceram' is a glass-ceramic made from waste blast furnace slag and has potential applications as a building material. Slag is a complex, lime-rich material which may be converted into a glass by remelting with added SiO_2. This is then crystallized to give a glass-ceramic in which the main crystalline phases are wollastonite, $CaSiO_3$, diopside, $CaMgSi_2O_6$, and anorthite $CaAl_2Si_2O_8$.

'Chemically machined' glass-ceramics have uses as printed circuit boards in the electronics industry. In the chemical machining process, photosensitive glasses, such as $Li_2O–Al_2O_3–SiO_2$ glasses containing small amounts of copper, silver or gold are nucleated with UV light. On suitable heat treatment crystals of Li_2SiO_3 grow and these are removed by etching in hydrofluoric acid. The important factor is that the lithium silicate crystals are much more soluble in HF than is the surrounding glass. By masking the glass prior to irradiation, it is, therefore, possible to produce a photographic image of the mask inside the glass; after suitable heat treatment and etching, the resulting glass contains an intricate network of holes. If it is then desired to convert the residual glass substrate into a glass-ceramic, the normal nucleation and growth procedures are used.

Chemically machined glasses and glass-ceramics utilize light irradiation effects to cause crystal nucleation. A related mechanism occurs in *photochromic glasses*. In these, the base glass contains a dispersion of very small crystals of AgCl; the crystals are of submicrometre size. On irradiation of the glass with visible light, some of the Ag^+ ions are converted to silver atoms which coagulate to form small clusters of metallic silver. These are responsible for a darkening of the glass and their mechanism of formation is probably similar to that which occurs in the darkening of photographic film on exposure to light. In the photochromic glasses, however, the darkening process is reversible and on removal of the light source, the silver clusters gradually redissolve and the glass recovers its transparency. Photochromic glass is used as the lens material in automatic sunglasses; many other applications are possible.

Questions

18.1 Using Zachariasen's rules would you expect the following to form glasses: (a) ZrO_2; (b) BeF_2; (c) MgF_2?

18.2 Explain why the experimentally observed glass transition temperature, T_g, is always above the ideal value, T_o.

18.3 The glass transition is sometimes described as a second order phase transition. Do you think this is valid?

18.4 Why is it difficult to obtain structural information on glass? What kind of information on glass do you expect to be obtained using diffraction, spectroscopic and microscopic techniques?

18.5 The mixed alkali effect in glasses is a well-studied phenomenon but as yet, it has not been satisfactorily explained. Suggest experiments that might lead to a better understanding of the mixed alkali effect.

18.6 Repeat question 18.5 but address your answer to the boron oxide anomaly rather than to the mixed alkali effect.

18.7 Of the chalcogens, sulphur and selenium readily form glasses but oxygen and tellurium do not. Explain.

18.8 Glasses may be used for the following applications. What special properties are required and what compositions would be suitable. (a) A solid electrolyte to be used in a Ag/I_2 cell, (b) A window in a house, (c) A container to withstand high temperatures, e.g. 900 °C. (d) An electrically insulating glass-ceramic. (e) A possible replacement for steel in fibre reinforcement. (f) A semiconducting film whose conductivity is sensitive to visible light. (g) A magnetic memory device.

18.9 A hypothetical example of unstable equilibrium (Chapter 11) would be a homogeneous glass or liquid whose temperature and composition was such that it lay inside a spinodal. Explain.

References

C. A. Angell (1970). The data gap in solution chemistry. The ideal glass transition puzzle, *J. Chem. Educ.*, **47**, 583–587.

R. F. Bacon and R. Fanelli (1943). The Viscosity of Sulphur. *J. Amer. Chem. Soc.*, **65**, 639.

R. W. Cahn (1980). Metallic glasses, *Contemp. Physics*, **21**, 43–75.

E. A. Davis (1977). Non-crystalline materials, *Endeavour*, **1** (3/4), 103.

R. H. Doremus (1973). *Glass Science*, Wiley.

J. J. Gilman (1975). Metallic glasses, *Physics Today*, May **1975**, 46–53.

G. O. Jones (1971). *Glass*, 2nd ed., Chapman and Hall.

J. D. Mackenzie (1964). *Modern Aspects of the Vitreous State*, Butterworths.

P. W. McMillan (1979). *Glass-Ceramics*, 2nd ed., Academic Press.

G. W. Morey (1954). *The Properties of Glass*, Reinhold.

R. L. Mozzi and B. E. Warren (1969). The Structure of Vitreous Silica. *J. Appl. Cryst.*, **2**, 164.

A. Paul (1982). *Chemistry of Glasses*, Chapman and Hall.

R. Parthasarathy, K. J. Rao and C. N. R. Rao (1984). The Glass Transition: Salient Facts and Models, *Chem. Soc. Revs.*, **12**, 361.

H. Rawson (1967). *Inorganic Glass-Forming Systems*, Academic Press.

H. Rawson (1980). *Properties and Applications of Glass*, Vol. 3 of *Glass Science and Technology*, Elsevier.

T. J. Rockett and W. R. Foster (1966). *J. Amer. Ceram. Soc.*, **49**, 31.

D. R. Uhlmann (1976). Inorganic amorphous solids and glass-ceramic materials, in *Treatise on Solid State Chemistry* (Ed. N. B. Hannay), Vol. 3, p. 293, Plenum Press.

W. A. Weyl and E. C. Marboe (1962). *The Constitution of Glasses*, Wiley-Interscience.

Chapter 19

Cement and Concrete

A cement can be defined as a substance that sets and hardens due to chemical reactions that occur on mixing with water. Various types of cement have been in use since ancient times. The Egyptians used calcined impure gypsum, $CaSO_4 \cdot 2H_2O$; the Greeks and Romans used calcined limestone, $CaCO_3$, and later added aggregate—sand and crushed stone—to form concrete. These *lime mortars* reacted unevenly with water to give a product of poor quality. Improved quality was obtained by the Romans who added volcanic ash—a source of reactive silica and alumina—to form what is now called *pozzolanic cement*. Modern *Portland cement* originated in Britain in the nineteenth century when high temperatures were first used in the preparation of cements. The strongly cementitious calcium silicates, Ca_2SiO_4 and Ca_3SiO_5, were produced during the high temperature reactions. Portland cement was originally prepared by heating a mixture of clay and chalk. Nowadays, various raw materials are used: either chalk, limestone or gypsum as a source of lime, together with sand, clay and iron oxide.

In use, Portland cement is mixed with aggregate and water which sets to form concrete. The role of the aggregate is twofold. It reduces the cost of concrete production and it increases the fracture strength. In the absence of aggregate, set cement crumbles easily. Other additives, such as pozzolans, fuel ash from power stations and blast furnace slag may be mixed in with Portland cement to reduce costs. Some of these additives are beneficial in that they mop up $Ca(OH)_2$, liberated during hydration of cement, and thereby increase the resistance of the set cement to chemical attack.

19.1 Portland cement

19.1.1 Manufacture

Portland cement is the most important cement in everyday use; the stages in its production and use are summarized in Fig. 19.1. The name derives from the similarity in colour and appearance to Portland stone, found in Dorset, England.

In the commercial production of Portland cement, the raw materials are first ground and intimately mixed, either dry or as a slurry with water. This feed is introduced into the top, cool end of a long rotary kiln whose axis is inclined slightly to the horizontal. The bottom of the kiln is heated to 1300 to 1500 °C by coal, oil or natural gas. As the feed passes down the kiln, it gradually heats up, first losing H_2O and CO_2. Further on, reactions between the solids begin to occur and these are completed in the hottest zone where partial melting also occurs. The presence of the liquid phase greatly speeds up the reactions by acting as a medium for the transport of matter. Reactions between solids without the help of a liquid phase are characteristically slow (Chapter 2). An oxidizing atmosphere in the kiln ensures that any iron present is in the + III state. The partially fused black lumps, known as *clinker*, are discharged from the bottom of the kiln, cooled rapidly in an air blast and crushed. Usually, gypsum is added at this stage to

$$\text{RAW MATERIALS} \xrightarrow[\sim 1500\,°C]{\text{KILN}} \text{CLINKER} \xrightarrow[25\,°C]{H_2O} \text{SET CEMENT}$$

(CaO, Al_2O_3, SiO_2, Fe_2O_3, etc.)	(Ca_3SiO_5, $Ca_3Al_2O_6$, β-Ca_2SiO_4, etc.)	(calcium silicate hydrates)

Fig. 19.1 Stages in the manufacture and use of Portland cement

Table 19.1 *Oxide and phase compositions of a typical Portland cement*

Oxide composition (weight %)		Phase composition (weight %)	
CaO	63	$C_3A(Ca_3Al_2O_6)$	5–12
SiO_2	20	$C_3S(Ca_3SiO_5)$	50–70
Al_2O_3	6	β-$C_2S(Ca_2SiO_4)$	20–30
Fe_2O_3	3	$C_4AF(Ca_4Al_2Fe_2O_{10})$	5–12
SO_3	2		
MgO	2		
K_2O Na_2O	1		
Others	3		

640

prevent *flash set* from occurring, i.e. rapid setting on subsequent reaction with water. The resulting powder is the familiar Portland cement which, when mixed with water and appropriate aggregate, hydrates to form set cement or concrete.

The oxide composition of a typical Portland cement clinker is given in Table 19.1, together with the phase composition, i.e. the approximate amounts of the different phases present. The most important phases are beta dicalcium silicate, β-Ca_2SiO_4, and tricalcium silicate, Ca_3SiO_5. Also present are tricalcium aluminate, $Ca_3Al_2O_6$, and a phase known simply as 'ferrite phase', of approximate formula $Ca_4Al_2Fe_2O_{10}$. Cement scientists have devised their own notation for these phases. The oxide constituents are each represented by a letter: $CaO \equiv C$; $SiO_2 \equiv S$; $Al_2O_3 \equiv A$; $Fe_2O_3 \equiv F$; and $H_2O \equiv H$. The phases are then written in terms of their constituent oxides:

$$Ca_2SiO_4 \equiv 2CaO \cdot SiO_2 \equiv C_2S$$
$$Ca_3SiO_5 \equiv 3CaO \cdot SiO_2 \equiv C_3S$$
$$Ca_3Al_2O_6 \equiv 3CaO \cdot Al_2O_3 \equiv C_3A$$
$$Ca_4Al_2Fe_2O_{10} \equiv 4CaO \cdot Al_2O_3 \cdot Fe_2O_3 \equiv C_4AF$$

19.1.2 Phase diagram considerations

The phase composition of cement clinker and the reactions that occur in the kiln can be understood from a knowledge of the relevant phase diagrams (Chapter 11). A complete representation, including the effect of iron oxide and the minor components such as alkali oxides and MgO, would be very complicated, involving a system of six or seven component oxides. However, for

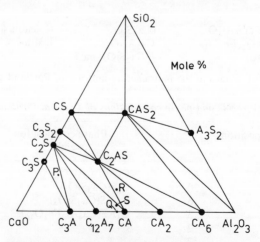

Fig. 19.2 Subsolidus equilibria in the system CaO–Al_2O_3–SiO_2. Typical compositions of Portland cement, P, and aluminous cement, Q, are marked. $C = CaO$, $A = Al_2O_3$, $S = SiO_2$: e.g. C_3A $= 3CaO \cdot Al_2O_3 = Ca_3Al_2O_6$

most practical purposes and to a reasonable approximation, it is necessary to consider only the ternary system CaO–Al$_2$O$_3$–SiO$_2$ (Figs 19.2 and 19.3). The point marked P shows the composition of a typical Portland cement clinker. Consideration of the phase diagram gives us the following information.

Point P lies within the triangle which has Ca$_2$SiO$_4$(C$_2$S), Ca$_3$SiO$_5$(C$_3$S) and Ca$_3$Al$_2$O$_6$(C$_3$A) as corners (Fig. 19.2). These three phases would, therefore, be the constituents of a cement clinker of composition P that had reached equilibrium at subsolidus temperatures. This is usually the case in practice and so a reasonable state of equilibrium is reached both in the kiln and during the cooling of the clinker. The relative amounts of the three phases could be determined from a lever rule calculation (Chapter 11).

The three-phase assemblage, C$_2$S, C$_3$S and C$_3$A, is stable only below 1455 °C, as indicated in Fig. 19.3. Above 1455 °C, a liquid phase is present and the absence of at least one of these three crystalline phases is required by the phase rule. This can be seen from the shape of the liquidus surface and the distribution of primary phase fields in Fig. 19.3. The primary phase fields of C$_2$S, C$_3$S and C$_3$A (i.e. the ranges of temperature and bulk composition over which these phases exist alone in the presence of liquid) meet at the peritectic point Y. Only at temperatures below this peritectic temperature, 1455 °C, can C$_2$S, C$_3$S and C$_3$A coexist in equilibrium. In the hottest part of the kiln, $\sim$ 1500 °C, composition P will have partially melted, giving a mixture of C$_2$S, C$_3$S and liquid of composition B; composition B is located at the intersection of the C$_2$S–C$_3$S univariant boundary curve and the 1500 °C isotherm. The bulk composition P lies within the three-

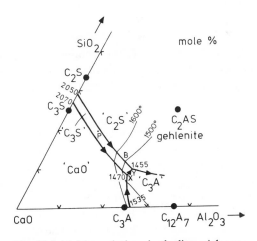

Fig. 19.3 Melting relations in the lime-rich corner of the system CaO–Al$_2$O$_3$–SiO$_2$, showing primary phase fields (e.g. CaO). Neighbouring primary phase fields meet at univariant curves and at invariant points. Point X is a peritectic invariant point that belongs to the compatibility triangle, C + C$_3$S + C$_3$A. Point Y is a peritectic that belongs to the triangle, C$_2$S + C$_3$S + C$_3$A

phase triangle: C_2S, C_3S and liquid B, at 1500 °C, as of course it must if it is to contain these three phases in equilibrium at this temperature. The relative amounts of these phases and hence the amount of melting could be determined from a lever rule calculation.

One consequence of the presence of the liquid phase in the kiln is that reaction of the starting materials to give C_2S and C_3S is greatly speeded up. Thus, the length of time that the materials need to spend in the hot zone of the kiln is no more than a few hours, whereas to obtain the same degree of reaction in the absence of a liquid (and therefore at somewhat lower temperature) would need days rather than hours.

Although the presence of C_3A appears to contribute little to the final strength of set cement and concrete, it is an important phase in the economics of cement manufacture because of the fluxing action of alumina. Thus, although a cement containing only C_3S and C_2S may be very desirable, melting temperatures in this region of the binary lime–silica system are well above 2000 °C (Figs 19.3 and 11.8). Addition of Al_2O_3 lowers the solidus temperature by ~ 600 °C, thus allowing kilns to operate efficiently at much lower temperatures. In the absence of alumina flux and partial melting, formation of C_2S and C_3S from the starting materials would require heating for several days at 1400 to 1500 °C.

The texture of cement clinker, i.e. the size and distribution of particles of the different phases present, can also be understood with the aid of Fig. 19.3. At 1500 °C, the phases present in equilibrium are C_2S, C_3S and liquid. The C_2S and C_3S crystals are relatively large, usually 10 to 50 μm diameter, because they have had time to grow in the presence of liquid. As the clinker moves out of the hot zone prior to discharge, it cools somewhat. More C_2S and C_3S precipitate, probably on the surface of the crystals already present, as the liquid composition moves from B to Y. At or below 1455 °C, the remaining liquid usually crystallizes to give a fine-grained matrix that surrounds the larger C_2S and C_3S grains. The texture and phase composition of this matrix varies, depending particularly on the cooling rate; in general it contains at least two crystalline phases, one of which is C_3A. Another possibility which is thought to occur to a limited extent is that some of the residual liquid fails to crystallize and forms a glass.

After CaO, Al_2O_3 and SiO_2, the next major component in Portland cement clinker is Fe_2O_3. The relevant phase diagrams show that at 1500 °C, Fe_2O_3 plays a similar role to Al_2O_3 and is present as part of the liquid phase. On discharge from the kiln, the iron oxide crystallizes in the 'ferrite' phase whose composition is often given as C_4AF but which, in reality, is a solid solution of variable composition between $Ca_2Fe_2O_5$, C_2F and the hypothetical phase, $Ca_2Al_2O_5$, C_2A (see Fig. 19.7 later).

19.1.3 Polymorphism of calcium silicates

Before considering the hydration of cement, some features of the polymorphism and crystal structures of the anhydrous phases merit attention.

Tricalcium silicate, C_3S, is stable from its incongruent melting point, 2070 °C,

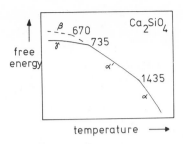

Fig. 19.4 Schematic free energy relations for the polymorphs of C_2S

down to $\sim 1250\,°C$ (Fig. 11.8). Below $1250\,°C$, C_3S is metastable relative to C_2S and lime; in practice its rate of decomposition below $1250\,°C$ is slow and there is usually no problem in preserving C_3S in cement clinker to room temperature. The crystal structure of C_3S is an orthosilicate in that it contains isolated SiO_4^{4-} tetrahedra. However, it also contains free oxide ions and so its formula is best written as $Ca_3(SiO_4)O$. Small amounts of other ions such as Al^{3+} and Mg^{2+} can be incorporated into the C_3S structure and *alite*, the name for the C_3S phase present in clinker, probably has a somewhat variable composition.

The polymorphism of dicalcium silicate, C_2S, is summarized in a schematic free energy diagram, Fig. 19.4. There are three stable polymorphs, which in order of decreasing temperature are labelled α, α' and γ. The $\alpha' \rightarrow \gamma$ transformation on cooling (at $735\,°C$) is sluggish and with rapid cooling α' undercools somewhat before it transforms at $670\,°C$ to a metastable polymorph, β. The polymorph β-C_2S is the phase normally present in cement clinker and is the C_2S polymorph with the most valuable cementing properties. Although pure β-C_2S is metastable at room temperature, relative to γ-C_2S, many different foreign ions are capable of entering the β-C_2S structure and some of them stabilize β-C_2S relative to γ-C_2S.

Additives which cause the $\alpha' \rightarrow \gamma$ or $\beta \rightarrow \gamma$ reaction to occur readily on cooling the clinker must be avoided since γ-C_2S has poor cementing properties. Also, the conversion of any unhydrated β-C_2S to γ-C_2S in hardened cement or concrete must be avoided as the increase in volume associated with the $\beta \rightarrow \gamma$ transformation may cause disintegration of the cement.

19.1.4 Hydration of Portland cement

The silicates and aluminates present in cement react with water to form products of hydration and, in time, these set to a hard mass. The different anhydrous phases have very different cementitious properties, as shown in Fig. 19.5; C_3S hydrates rapidly and develops high early strength whereas β-C_2S hardens more slowly. Hydration products of C_3A and C_4AF have very little strength. The hydration of commercial cement can be represented approximately by the summed hydration of the components. C_3S is the phase mainly responsible

644

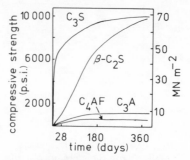

Fig. 19.5 Development of compressive strengths on hydration. (Data from Bogue and Lerch, 1934)

for the initial hardening; C_3S and β-C_2S give set cement and concrete its long time strength.

Hydration of cement is a complicated process and part of the difficulty in studying it is that the main products of hydration are either gelatinous or poorly crystalline, thus making conventional X-ray diffraction studies extremely difficult. The main product and the most important one for high strength is a poorly crystalline calcium silicate hydrate (C–S–H), sometimes called C–S–H gel or, incorrectly, tobermorite gel. The composition of this material is uncertain and is probably variable both in lime to silica and in silica to water ratios; it may also contain Al^{3+}, Fe^{3+} and SO_4^{2-} ions. In addition to C—S—H gel, hardened cement contains unreacted clinker, $Ca(OH)_2$, aluminate hydrates, aluminosulphate hydrates and water.

Many of the physical and mechanical properties of hardened cement and concrete seem to depend on the physical structure of the hydration products on a colloidal scale, rather than on their chemical composition. Advances in understanding the hydration processes are currently being made by using various techniques, including electron microscopy combined with microanalysis of the phases present. Hydration appears to involve at least two stages. First, a coating of C–S–H gel forms rapidly on the surface of the anhydrous cement particles. Second, this coating thickens by both growing outwards and eating into the anhydrous cement particles. The coatings subsequently begin to join up within a few hours and the product stiffens or sets.

The water to cement ratio affects the properties of cement. Once cement paste has set its apparent volume stays approximately constant; this final volume increases with increasing water to cement ratio in the original mix. Set cement is porous and contains both very small water-filled holes, ~ 10 to 20 Å across (gel pores), and much larger channels, $\sim 1\,\mu m$ across (capillary pores). Interconnected capillary pores are mainly responsible for the permeability of set cement and its vulnerability to frost damage. The absence of interconnected capillary pores is clearly desirable. It may be attained after a sufficiently long time of moist curing, i.e. hydration in a moist atmosphere and by using a sufficiently low water to cement ratio. Thus, for water to cement ratios of ~ 0.4, about three days are needed before the capillaries are no longer connected, whereas for a ratio of ~ 0.7 at least one year is needed.

The problem of *flash set* is caused by the very rapid reaction between C_3A and water. The C_3A appears to dissolve very rapidly, followed by the precipitation of calcium aluminate hydrates, and is accompanied by much evolution of heat. Although this reaction is rapid, the mechanical properties of a cement that has undergone flash set are very poor. In practice, flash set is avoided by adding 1 to 2% gypsum to cement clinker. In a complex reaction, gypsum, in the presence of $Ca(OH)_2$, acts to retard the hydration of C_3A. Instead, aluminosulphate phases, either ettringite, $Ca_6Al_2(OH)_{12}(SO_4)_3 \cdot 26H_2O$, or monosulphate, $Ca_4Al_2(OH)_{12}$-$SO_4 \cdot 6H_2O$, are formed, probably as a protective coating on the surface of the C_3A crystals.

There are several different anhydrous calcium silicates but only two, C_3S and β-C_2S, have cementitious properties suitable for use in hydraulic cements (i.e. they react with water to form an insoluble product which sets to form a rigid mass). The reasons for this are only poorly understood and there are probably several factors which affect cementing ability.

One feature that C_3S and β-C_2S have in common and which distinguishes them from other calcium silicates is their metastability. Below 670 °C, β-C_2S is metastable relative to γ-C_2S. At temperatures below 1250 °C, C_3S is metastable with respect to the combination $CaO + C_2S$. The latter is shown in Fig. 19.6, a schematic free energy diagram for the C_3S composition. The solid, lower lines, represent the equilibrium assemblage at any temperature. At room temperature, C_3S is shown as being metastable relative to β-C_2S + lime which, in turn, is metastable relative to γ-C_2S + lime. Because C_3S and β-C_2S are metastable, the

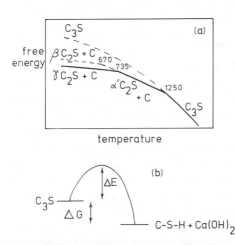

Fig. 19.6 (a) Schematic free energy diagram for the C_3S composition; polymorphic transformations in Ca_3SiO_5 are not shown. (b) Schematic energy changes on hydration of C_3S

decrease in free energy on their hydration should be greater than for hydration of the corresponding stable phases, i.e. $\gamma\text{-}C_2S + CaO$ and $\gamma\text{-}C_2S$, respectively. This, of course, assumes that the products of hydration of $\beta\text{-}C_2S$ and $\gamma\text{-}C_2S$ have the same free energy. However, although C_3S and $\beta\text{-}C_2S$ are thermodynamically metastable, it does not automatically follow that their hydration will be kinetically more rapid than that of the corresponding stable phases. In Fig. 19.6(b) are shown schematically the free energy changes which occur during hydration. There is an activation energy for hydration, ΔE, whose magnitude is probably governed by the ease of attack of water molecules on the anhydrous phase and is therefore related to the crystal structure of the anhydrous phase. It has been suggested that C_3S and, to a certain extent, $\beta\text{-}C_2S$ have rather open crystal structures which may facilitate penetration and subsequent attack by water molecules at the surface of the crystals, thereby reducing ΔE for hydration of these phases. Another factor which may reduce ΔE for hydration of C_3S is the presence in the crystal structure of CaO-like regions which may provide the initial points for attack by H_2O. The problem in comparing the cementing ability of two phases in terms of diagrams such as Fig. 19.6(b) is that it is difficult to evaluate the various factors which may influence the values of their hydration activation energies.

Particle size greatly affects hydration kinetics, because with a smaller particle size the surface area of the crystals is greater and so hydration, which is mainly a surface reaction, proceeds more rapidly. In practice, the economics of grinding the cement clinker to a finer size have to be balanced against the more rapid development of strength which is thereby obtained. For ordinary Portland cement in Britain, regulations require a minimum surface area of $225 \, m^2 \, kg^{-1}$, although, in practice, $300 \, m^2 \, kg^{-1}$ is a more typical value.

19.1.5 Types of Portland cement

By varying the composition of cement clinker or the amounts of various additives, the properties of cement can be varied. *Ordinary Portland cement* is the most common, general purpose cement but it is susceptible to sulphate attack and so is not used in contact with seawater, for example. There are two modes of sulphate attack. One is by reaction of sulphate with hydrated calcium aluminates to form less-dense calcium sulphoaluminates; the consequent expansion leads to disintegration of the cement. The other is by reaction of sulphate with $Ca(OH)_2$ to form gypsum which also causes disruption of the hardened cement. The remedy used in *sulphate-resisting cement* is to reduce the C_3A content. This is achieved by increasing the $Fe_2O_3:Al_2O_3$ ratio in the raw materials, thereby increasing the amount of the 'ferrite phase' C_4AF in cement clinker. For reasons that are not clear, C_4AF appears to make hydrated cement more sulphate resistant. *Rapid hardening Portland cement* is produced (a) by increasing the C_3S content relative to that of C_2S and (b) by grinding the clinker to a finer particle size. Because rapid hardening is associated with a high rate of heat development, this cement cannot be used in massive structures as cracking would occur. On the other hand, it may be of use in low temperature environments where the heat developed on hydration may serve to prevent frost damage in the early, critical stages of hydration. Cement which hardens even more rapidly can be produced by intergrinding the rapid hardening clinker with 1 to 2% $CaCl_2$. For use in massive concrete structures, *low heat Portland cement* is used in which the C_3A and C_3S contents are reduced; the development of strength is slowed down although the ultimate strength is unaffected. *Portland blast furnace cement* has rather similar properties to ordinary Portland cement and is produced by mixing Portland cement clinker with blast furnace slag. The slag is produced as a byproduct in the extraction of iron, by reaction of limestone with silica, alumina and other components of the ore. Alternatively, the slag composition and structure may be modified so that it can be used directly, together with limestone, as a raw material for conventional cement manufacture. This modification involves quenching the molten slag so that it contains a significant proportion of a glassy phase which reacts rapidly with water.

19.2 Aluminous cement and high alumina cement

Aluminous cement was developed in France at the beginning of this century as a result of a search for sulphate-resisting cement. The main constituents of normal aluminous cement are lime and alumina in roughly equal parts with smaller amounts of iron oxide, silica, magnesia, alkali oxides and titania. Raw materials are limestone or chalk and bauxite and these are melted completely in a kiln at 1500 to 1600 °C. The cement clinker produced on cooling is more expensive than Portland cement because of the cost of bauxite, the high melting temperatures and the hardness of the clinker, which makes it difficult to crush. The important properties of aluminous cement are its rapid hardening—high strengths are reached well within twenty-four hours—its sulphate resistance and its refractoriness at high temperatures.

High alumina cement has a much higher alumina content than normal aluminous cement, ~ 80 wt % Al_2O_3, and the amounts of iron oxide, silica and other impurities are kept very low; it may be used as *refractory concrete* up to ~ 1800 °C.

In normal aluminous cements, monocalcium aluminate, $CaAl_2O_4$ (CA), is the main phase and the one with the valuable cementitious properties. Another calcium aluminate, $C_{12}A_7$, forms but as it hydrates very rapidly, causing flash set, the bulk composition is chosen so as to limit the quantity of $C_{12}A_7$ present. This is achieved by the addition of Fe_2O_3 to the raw material. The role of Fe_2O_3 is seen by consideration of the relevant part of the $CaO–Al_2O_3–Fe_2O_3$ phase diagram (Fig. 19.7). Most of the phases in this system, including CA, $C_{12}A_7$ and 'ferrite', C_2F, form ranges of solid solution by $Al^{3+} \rightleftharpoons Fe^{3+}$ replacement. The compositions of typical, normal aluminous cements fall within the shaded area. These compositions lie in various compatibility triangles, all of which have CA s.s. as one corner with various calcium iron oxide phases as the other corners. In none of these triangles is $C_{12}A_7$ present and so, under conditions of equilibrium

648

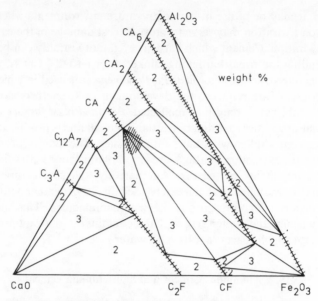

Fig. 19.7 Subsolidus phase diagram for the system CaO–Al_2O_3–Fe_2O_3. Compositions of aluminous cements are shaded. Hatched lines correspond to ranges of solid solutions. The numbers 2 and 3 refer to compatibility triangles which contain two and three phases in equilibrium, respectively

which are approximately maintained on cooling, $C_{12}A_7$ is absent from aluminous cement clinker.

The amount of silica in aluminous cement is kept low, to no more than a few per cent. This is because, with any more silica than this, a lime-rich aluminosilicate phase, gehlenite, $Ca_2Al_2SiO_7$ (C_2AS), forms. A relatively small amount of silica is sufficient to form a large amount of gehlenite (Fig. 19.2), as shown by drawing a line from point Q, which represents aluminous cement, to SiO_2. For instance, with $\sim 10\%$ SiO_2, point R, much of the CA has disappeared to be replaced by gehlenite. As gehlenite has poor hydraulic properties and forms at the expense of calcium aluminates, the silica content of aluminous cement is restricted to ~ 5 wt $\%$. Small amounts of silica are, in fact, quite acceptable because then β-C_2S forms in preference to gehlenite, e.g. composition S in Fig. 19.2. Other phases which occur in clinker in small quantities are glass, wustite (FeO) and a fibrous, pleochroic phase of uncertain composition.

Reaction of aluminous cement with water at $\lesssim 25\,°C$ gives, as the initial product CAH_{10}, a calcium aluminate hydrate phase, with small amounts of C_2AH_8 and alumina gel. CAH_{10} is the phase mainly responsible for the strength of set aluminous cement but it is a metastable phase. At higher temperatures, 30 to $40\,°C$, and especially under humid conditions, CAH_{10} (and C_2AH_8) converts to C_3AH_6, alumina gel (which gradually crystallizes to form gibbsite, $Al(OH)_3$) and free water. Because C_3AH_6 is more dense than CAH_{10} and C_2AH_8, this 'conversion' causes an increase in porosity and permeability and can lead to a severe loss in compressive strength. In Britain in 1973, several buildings collapsed due to the failure of concrete structural members made from aluminous cement. The most dramatic was the caving-in of the roof of a swimming pool in a London school, only minutes after the pool had been evacuated. Although conversion appears to be a generally occurring reaction in aluminous cements, it leads to a failure of strength *only* if the water to cement ratios which are used are too high. For complete conversion of CA to CAH_{10}, the calculated water to cement ratio is 0.5. For ratios less than 0.5 (0.35 is a common value and does not give problems of failure) there will be unhydrated CA present in set cement and it appears that water which is liberated by conversion of CAH_{10} reacts with this CA to fill the pores created by the preceding conversion. Aluminous cement is therefore a safe cement provided the water to cement ratio used is in the correct range. It is always a temptation to add more water because a mix made with a ratio of 0.35 appears to be too dry, especially in comparison with a typical Portland cement mix. Human errors such as this were probably the cause of failure in the buildings in Britain. Because of conversion, aluminous cement is banned in many countries from use in certain types of structure.

The sulphate-resistant properties of set aluminous cement arise because $Ca(OH)_2$ is absent, unlike in Portland cement, (Any $Ca(OH)_2$ that formed during hydration would react with alumina gel.) Also, the alumina gel that is produced forms a protective coating around the aluminate and aluminate hydrate crystals. On hydration any $C_{12}A_7$ present sets first, followed by CA. Therefore, for a more rapidly setting cement, the lime to alumina ratio is increased somewhat to increase the $C_{12}A_7$ content. On the other hand, the presence of glass in the clinker appears to slow down the hydration reactions. Because high strength is attained rapidly (~ 80 per cent of the final strength after twenty-four hours), there is a high rate of heat development and so aluminous cement cannot be used to build massive structures; it is therefore always used to make thin sections.

Concrete made from aluminous cement (especially high alumina cement) is an important refractory material with good mechanical strength and resistance to dry heat. The concrete is mixed and placed in the normal way and after twenty-four hours it is heated. The hydration products lose their water and the strength drops to a minimum value at 900 to 1100 °C. At higher temperatures, the dehydration products react with the aggregate—crushed firebrick, Al_2O_3 or SiC—and a strong ceramic bond replaces the hydraulic bond present in the original set concrete. Thus, the strength increases again. The most refractory concrete, suitable for use up to 1800 °C, is made from white aluminous cement and corundum, Al_2O_3, aggregate. The concrete contains almost entirely calcium aluminate phases and Al_2O_3 with no impurities such as Fe_2O_3 to lower melting temperatures. The CaO–Al_2O_3 phase diagram is shown in Fig. 19.8. Reaction of CA from the dehydrated cement (M.P. 1608 °C) with Al_2O_3 aggregate yields CA_2 and, more important, CA_6 (M.P. 1860 °C). This therefore leads to a considerable improvement in refractoriness. By comparison, Portland cement is of little use above about 500 °C.

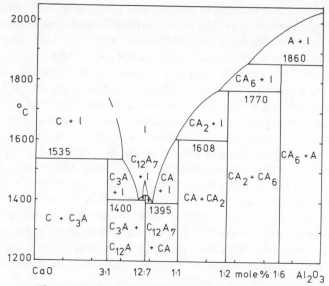

Fig. 19.8 Phase diagram for the system $CaO-Al_2O_3$

19.3 Pozzolans and pozzolanic cement

Pozzolan is a natural or synthetic form of reactive silica such as volcanic ash or pulverized fuel ash. The silica is reactive because it is metastable and/or finely divided. Crystalline quartz is usually unreactive, but materials such as poorly crystalline silica of small particle size, opaline (i.e. phase-separated) silica glass or silica gel react with H_2O and $Ca(OH)_2$ at room temperature to form cementitious products. The Romans used such lime–pozzolan mortars and some of their structures are still standing today. The lime–pozzolan reaction produces a calcium silicate hydrate gel similar to the C–S–H gel formed on hydration of Portland cement, but the whole process is carried out at room temperature. This contrasts with the use of Portland cement where high temperatures are first needed to make the cementitious calcium silicates, C_3S and β-C_2S.

Pozzolans can also be mixed with Portland cement to form *pozzolanic cement*. The Portland cement component reacts as usual and the liberated $Ca(OH)_2$ reacts with the pozzolan to form more C–S–H gel. Removal of the $Ca(OH)_2$ by subsequent reaction with the pozzolan has two advantages: first, pozzolanic cement has good resistance to chemical attack, especially by sulphate; second, the final strength may be higher than that of Portland cement alone, because the mechanically weak $Ca(OH)_2$ is replaced by more C–S–H gel. Pozzolanic cement hardens more slowly than does Portland cement and so, because of the slow rate of heat development, it can be used in massive structures.

19.4 Autoclaved products

Most forms of silica, excluding the reactive pozzolans, do not react appreciably with lime and water at room temperature, but do react with steam and lime at 175 to 200 °C to form autoclaved calcium silicate products. Typically, for the manufacture of calcium silicate bricks, a mixture of lime (4 to 12 per cent), quartz sand and water (4 to 7 per cent) is moulded at ~ 5000 p.s.i. and then autoclaved in high pressure steam at 120 to 200 p.s.i. and 175 to 200 °C for 12 to 15 hours. The sand begins to react with the lime and water to form C–S–H gel which initially has a high CaO : SiO$_2$ ratio (~ 1.75). When all the lime is used up, the C–S–H gel reacts with more quartz and its CaO : SiO$_2$ ratio falls to a lower limit of about 0.8. With longer times of autoclaving, the C–S–H gel partially changes to crystalline tobermorite, C$_5$S$_6$H$_5$, although it is still not clear whether the main source of strength is the C–S–H gel or tobermorite.

High pressure steam is also used to speed up the hardening of concrete made from Portland cement. To manufacture pre-cast concrete blocks, Portland cement is mixed with fine sand and aggregate; the amount of sand added is critical in determining the compressive strength of the concrete. During the autoclave process (8 to 15 hours at 180 °C and 150 p.s.i.), the cement and sand react to give C–S–H gel of low lime to silica ratio (~ 0.8) which may then recrystallize to give tobermorite. Characteristics of autoclaved concrete blocks are their high compressive strength and resistance to sulphate attack; the latter is caused by the absence of lime, as any Ca(OH)$_2$ present would react with sand under autoclave conditions.

19.5 Oxychloride (Sorel) cements

Magnesium oxychloride cements, Sorel cements, are used, especially in the United States, for flooring, for decorative internal plasters and external stuccos. They have good acoustic and elastic properties, suitable for flooring, with a resistance to the accumulation of static charge and an attractive marble-like appearance.

The cement is made by dissolving finely divided, reactive MgO in aqueous solutions of MgCl$_2$. This gives a homogeneous, thixotropic gel which crystallizes to a dense, hard, aggregate of oxychloride phases $5:1:8(5Mg(OH)_2 \cdot MgCl_2 \cdot 8H_2O)$ and $3:1:8(3Mg(OH)_2 \cdot MgCl_2 \cdot 8H_2O)$. It is essential for the MgO to be finely divided; otherwise it dissolves only slowly, water evaporates, compositional gradients occur and deliquescent MgCl$_2$ and undissolved MgO precipitate out. The resulting product has poor dimensional stability and lacks weather and corrosion resistance.

As with Portland cement, chemical reactions continue long after the cement has set initially. The attainment of weather resistance appears to be associated with the pick-up of atmospheric CO$_2$ which forms an insoluble carbonate phase, $Mg(OH)_2 \cdot 2MgCO_3 \cdot MgCl_2 \cdot 6H_2O$, at least as a surface layer. After a long period of time, chloride is leached out leaving hydromagnesite $5MgO \cdot 4CO_2 \cdot 5H_2O$.

19.6 Recent advances—MacroDefectFree (MDF) cement

Scientists at ICI Laboratories in Britain have recently described a new cement that has far superior mechanical properties to conventional Portland cement and

652

concrete. This new cement. known as MacroDefectFree (MDF) cement has potentially useful mechanical properties such as a moderate flexural strength ($\sim$ 70 MPa compared to 10 MPa in Portland cement) and a moderate toughness or fracture energy (1 kJ m^{-2} compared to 0.02 kJ m^{-2} in conventional cements). These values are more than an order of magnitude less than the corresponding values for good quality steels but they do nevertheless offer exciting prospects for future improvements and development.

Information on the nature of the MDF cement is still rather limited but it appears that there is one characteristic feature to this new cement. This is that the texture or microstructure of the solid product is controlled in such a way that large pores or voids are absent. In conventional cements, large holes, e.g. 1 mm across, are present and are at least partly responsible for the poor mechanical properties. In the preparation of MDF cement, improved mixing and moulding methods are used together with the addition of surface-active organic chemicals whose rheological properties enable the cement grains to achieve better compaction.

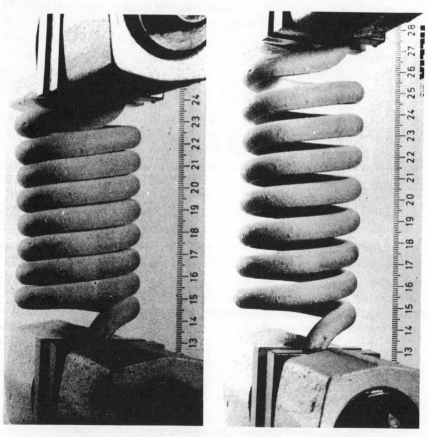

Fig. 19.9 A tension spring made from the new MDF cement.
Reproduced by permission of the Royal Society of Chemistry
from Birchall *et al.* (1982)

A convincing illustration of the potential of MDF cement is the cement spring shown in relaxed and extended form in Fig. 19.9,

Questions

19.1 Using Fig. 19.2, what phases do you expect to be present at equilibrium and at subsolidus temperatures in compositions P, Q, R and S? Using lever rule calculations (Chapter 11), estimate the relative amounts of the phases present in each.

19.2 Using Fig. 19.3, what phases do you expect to be present at $1500\,^{\circ}C$ in compositions (a) 80% CaO, 10% Al_2O_3, 10% SiO_2, (b) 60% CaO, 20% Al_2O_3, 20% SiO_2, (c) 70% CaO, 25% Al_2O_3, 5% SiO_2?

19.3 What kinds of cement/concrete could be used for the following: (a) for building a freshwater dam; (b) for building a pier; (c) for constructional work under arctic conditions; (d) for structural components that must subsequently withstand high temperatures, e.g. $1000\,^{\circ}C$?

19.4 The processes involved in the setting of cement and concrete are still only poorly understood. Why do you think this is? Suggest experiments that might yield more insight into the processes involved.

19.5 Using Fig. 19.8, what phase(s) do you expect to form on heating the following compositions: (a) 50% CaO, 50% Al_2O_3; (b) 70% CaO; 30% Al_2O_3; (c) 25% CaO, 75% Al_2O_3; (d) 10% CaO, 90% Al_2O_3? What is the temperature at which melting begins in each case?

References

J. D. Birchall, A. J. Howard and K. Kendall (1982). New cements—Inorganic plastics of the future, *Chem. Brit.*, **18**, 860.

R. H. Bogue and W. Lerch (1934). *Ind. Eng. Chem.*, **26**, 837.

F. M. Lea and C. H. Desch (1956). *The Chemistry of Cement and Concrete*, 2nd ed., Arnold, London.

S. Mindess and J. F. Young (1981). *Concrete*, Prentice Hall.

H. F. W. Taylor (Ed.) (1964). *The Chemistry of Cements*, 2 vols, Academic Press, London and New York.

H. F. W. Taylor (1966). *The Chemistry of Cements*, RIC Monograph No. 2, Royal Institute of Chemistry, London.

H. F. W. Taylor (1981). Modern chemistry of cements, *Chem. and Ind.*, **1981**, 620.

Chapter 20

Refractories

Refractories are materials which have high strength, mechanical stability and chemical inertness at temperatures of the order of 1400 °C and above. They play a key role in many industries such as the manufacture of iron and steel, glass, cement, etc., where they are used as furnace linings. Without this relatively inert lining, much of these industries could not exist, at least not in the form known to us.

Refractories are generally dismissed from consideration by chemists as being non-chemicals. By virtue of their characteristic properties of inertness and high melting points, they are thought to have nothing of interest to offer. In fact, of course, the reverse is true and a good understanding of topics such as chemical bonding, phase equilibria, kinetics and surface tension is necessary in order to appreciate the properties of refractories.

In this chapter, some general aspects of refractories are outlined, followed by a discussion of selected refractory materials.

20.1 Microstructure or texture

Refractories are polycrystalline bodies which contain one or more crystalline phase and often a liquid or glassy phase as well. Physical properties such as strength depend on the size and shape of the individual crystals, the nature of the bonding between crystals and the distribution of any liquid phase that is present. These characteristics are known collectively as the texture or microstructure of the refractory body.

The principal techniques for studying texture are reflected light microscopy and scanning electron microscopy (Section 3.2.2). For reflected light microscopy, a section is cut through the refractory piece and the cut surface is gradually polished down to give a smooth finish. The surface is then etched in a suitable reagent which either preferentially attacks certain of the phases present or attacks the boundary regions between phases. Either way, the texture is brought 'into relief' and features larger than $\sim 1\,\mu m$ (i.e. 10^4 Å) in size can readily be seen in the microscope. With experience, an analysis of the phases present can usually be made from a cursory visual inspection with the microscope and it is also an easy matter to measure particle sizes, volume fractions of the different phases present, porosity and effect of liquid on the texture.

Similar results may be obtained with scanning electron microscopy (SEM), but with the additional feature that much higher magnifications may be used. Also, with SEM, rough fracture surfaces may be examined, in which case the material is free from any possible artefacts or damage associated with polishing. The chemical composition of selected crystals or areas can be determined if the SEM instrument has a microanalysis facility. Alternatively, the polished section may be transferred to an electron microprobe analyser.

20.2 Grain size and grain growth

On heating a polycrystalline body at high temperatures, the average crystal size gradually increases. Small grains or crystals disappear and large ones consequently get bigger. The driving force for grain growth is provided by the reduction in relative surface area of a body composed of a few large grains in comparison with a body that has the same amount of material but present as a large number of small grains. *Grain growth* is an important process in reducing or eliminating the porosity of refractory bodies whereas *grain size* affects properties such as mechanical strength.

20.3 Sintering

Sintering is the general name for the process of densification of a polycrystalline body, with or without the presence of a liquid phase to aid the transport of matter. Starting with a finely powdered material, preferably in the form of a compact, sintering occurs on heating to temperatures close to and below the solidus. Sintering also occurs at temperatures above the solidus in which case partial melting also occurs; the presence of the liquid phase acts as a medium for the transport of matter from one grain to another. During the initial stages of sintering at subsolidus temperatures, an increase in the areas of interparticle contact occurs with time. 'Necks' form between grains which grow thicker and have the effect of pulling the crystals closer together, thereby increasing the density of the body. With increasing time or temperature, shrinkage of the body continues and the pores between particles become smaller and lose their connectivity. If these pores can shrink to zero size or be 'swept out' to the surface

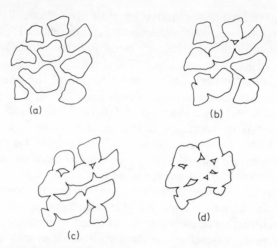

Fig. 20.1 Progressive stages of sintering, starting from (a) a loosely compacted powder to (b) the onset of contact between grains to (c) formation of a porous three-dimensional network of linked particles to (d) formation of a solid, non-porous piece with isolated pores. In the final stages (not shown), the isolated pores may be 'swept out' to the surface by grain growth

of the body by grain growth, then the bulk density of the body approaches the true or theoretical crystal density. Various stages in a sintering process are shown schematically in Fig. 20.1.

The presence of a small amount of liquid usually speeds up the sintering process greatly and the onset of sintering may occur at much lower temperatures than in the absence of a liquid. The presence of an excessive amount of liquid is obviously undesirable if the body is to maintain its shape and strength. It is also important in the manufacture and use of refractory bodies that the amount of liquid present does not increase too rapidly as the temperature is raised above the solidus; the *vitrification range*, defined as the temperature interval between the onset of densification due to the formation of liquid and the occurrence of slumping due to excessive amounts of liquid, should be as large as possible. The vitrification range depends critically on composition and is directly governed by the appropriate phase diagram. The worst case would be to have a composition that corresponded to a eutectic, as complete melting occurs over a range of a few degrees close to a eutectic composition. Conversely, then, the most suitable compositions for use as refractory bricks are those that are well away from eutectics.

20.4 Surface properties

The surface tension properties of crystals and liquids have an important bearing on sintering kinetics and on subsequent slag attack. Grain growth

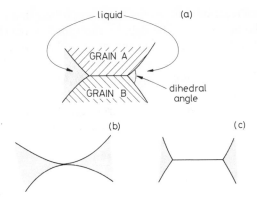

Fig. 20.2 The dihedral angle and its effect on the amount of grain-to-grain contact

kinetics and the final texture of the body depend on the *dihedral angle*, which is the angle in the liquid phase between two crystalline grains (Fig. 20.2). A low dihedral angle gives a small amount of grain-to-grain contact (Fig. 20.2b) and a large amount of liquid penetration between the grains. It is generally found that a small dihedral angle gives more rapid grain growth and a larger final grain size. The kinetics of grain growth also depend on the initial grain size, as a fine grained body sinters more rapidly than a coarse grained one. In practice, when it is desired to sinter a powdered compact, considerable attention must be paid to the preparation of fine grained starting materials, with a consequent high surface area. The kinetics of grain growth are temperature dependent and often the growth rate is given by

$$\frac{\mathrm{d}D}{\mathrm{d}t} = \frac{k}{D} \tag{20.1}$$

where D is the grain diameter and k a rate constant.

As well as influencing grain growth kinetics, the dihedral angle is important in determining the hot strength of refractories and the degree of slag attack. For strong solid–solid contacts, a large dihedral angle is required (Fig. 20.2c). If the angle is low or zero (Fig. 20.2b), a liquid slag may be able to enter a refractory body by penetrating between grains. This may cause 'washing out' of the crystals and disintegration of the body.

20.5 Slag attack

Before a liquid slag can attack a refractory lining, it must be able to *wet* the refractory, otherwise the slag forms isolated droplets on the refractory surface. Although the relevant surface tension theory is beyond the scope of this chapter, some general comments may be made. Wetting occurs if the surface tension of the liquid (liquid–air interface) is greater than the interfacial tension between the refractory and liquid slag. This is usually the case with slags but not so with liquid metals, which seldom wet refractory linings.

658

The amount of reaction that occurs between molten slag and refractory is related to the *fluxing action* of the slag, i.e. the ability of the slag to lower the melting point of the refractory. Lime and alumina have very different fluxing ability with silica bricks, as shown by the relevant parts of the phase diagrams of Figs 11.8 and 20.3. The addition of $\sim 4\,\text{mol}\%$ Al_2O_3 is sufficient to lower the melting point (i.e. liquidus) of SiO_2 from 1720 to 1595 °C (Fig. 20.3a), whereas $\gtrsim 30\%$ CaO is needed to achieve the same result (Fig. 11.8). The effect of added alkali, Na_2O or K_2O, is even more drastic (Fig. 20.3b), as the liquidus curve in these alkali silicate systems plummets to ~ 800 °C as alkali is added to silica.

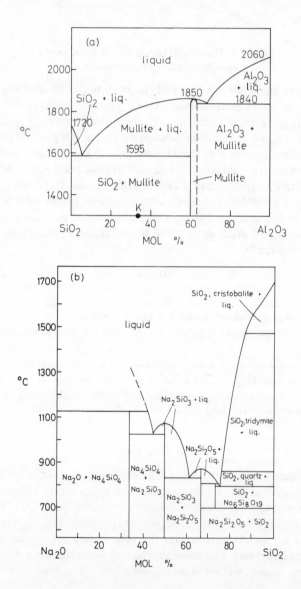

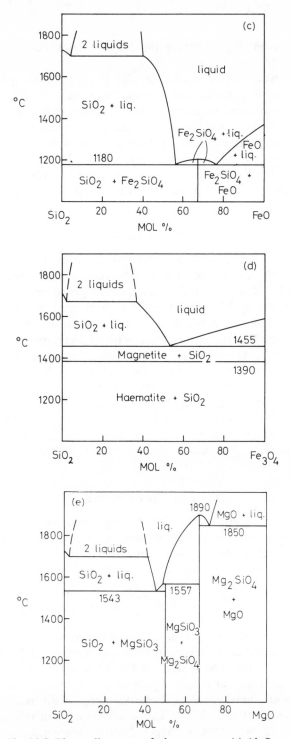

Fig. 20.3 Phase diagrams of the systems: (a) Al_2O_3–SiO_2, (b) Na_2O–SiO_2, (c) FeO–SiO_2, (d) Fe_3O_4–SiO_2, (e) MgO–SiO_2. (Data taken from *Phase Diagrams for Ceramists*, American Ceramic Society, 1964, 1969, 1975)

Data are shown in (Fig. 20.3b) for the $Na_2O–SiO_2$ system; the $K_2O–SiO_2$ diagram at the silica-rich end is similar.

For use in iron and steel blast furnaces, silica bricks must clearly be resistant to high concentrations of molten iron oxides. This is the case if the atmospheric conditions are reasonably oxidizing because the stable oxides of iron are Fe_3O_4 (magnetite) and Fe_2O_3 (haematite) and, for instance, the $Fe_2O_3–SiO_2$ system has high melting temperatures with an extensive region of liquid immiscibility (Fig. 20.3d). However, under strongly reducing conditions, FeO is the stable oxide of iron and in the $FeO–SiO_2$ system, SiO_2 and FeO react to form fayalite, Fe_2SiO_4, which melts at lower temperatures, $\sim 1200\,°C$ (Fig. 20.3c).

20.6 Strength

Refractories are usually brittle at room temperature, have little tensile strength and exhibit plastic deformation at high temperatures. Cold crushing strengths are usually high—up to several thousand pounds per square inch—unless the bodies are very porous. Microstructure has an important influence on strength, although the relationship between microstructure and strength is complex. A general guideline is that strength increases with (a) decreasing grain size and (b) decreasing porosity. Strength also depends on effects which arise during previous thermal cycling of the body and which cause a change in shape or volume of the individual grains. These volume changes are mainly due either to thermal expansion/contraction, which may be isotropic or anisotropic, or to polymorphic phase transitions. A good example of the latter is provided by pure ZrO_2 which is useless as a refractory material because of the disruptive effects of the monoclinic $\rightleftharpoons$ tetragonal transition which occurs at $\sim 1000\,°C$.

20.7 Refractory materials

Refractories are materials with high melting points and hence with strong interatomic bonds. Both ionic and covalent bonding occur in refractories. In ionically bonded refractories, it is an essential requirement that the lattice energy be high; from equation (8.19), the lattice energy, U, is given by

$$U \propto \frac{Z_+ Z_-}{r_e}$$

in which Z_+, Z_- are the charges on the ions and r_e is the interatomic distance between the anion and cation nuclei. By far the most important factor in U is the magnitude of the charge product $(Z_+ Z_-)$. For instance, on comparing an alkali halide and an alkaline earth oxide, both of which have the same rock salt structure and with similar r_e values, the lattice energy of the oxide is about four times greater than that of the halide. This is shown quantitatively by comparing melting points, e.g. NaCl, $800\,°C$: MgO, $2800\,°C$. A general guideline for the occurrence of high melting points in ionic materials is, therefore, that at least one, but preferably both ions, should be multivalent. Other examples are Al_2O_3, Cr_2O_3 and ZrO_2.

The influence of r_e on refractoriness may be seen by comparing the melting points of the alkaline earth oxides, all of which (except BeO) have the rock salt structure:

MgO	2800 °C	$r_e = 2.12$ Å
CaO	2580 °C	$r_e = 2.40$ Å
SrO	2430 °C	$r_e = 2.56$ Å
BaO	1923 °C	$r_e = 2.76$ Å

In this series, r_e is greatest and hence U least for BaO; this correlates with the fact that BaO has the lowest melting point.

Similar considerations apply to refractory materials in which the bonding is primarily covalent; strong bonds must form, giving a three-dimensional or network structure. Compounds of monovalent elements, such as the halogens, appear to be unsuitable because their compounds are either volatile, molecular structures or are low melting solids; in materials such as $AlCl_3$ (M.P. 190 °C), the monovalent chlorine atoms are bonded to six aluminium atoms and the individual Al—Cl bonds are weak. Thus, although these compounds may be solids, they are relatively low melting. Suitable refractory covalent compounds are those in which the bond strengths are large, perhaps as large as one. They therefore contain multivalent elements of fairly similar electronegativity, in which coordination numbers are small (usually four) and in which three-dimensional network structures form. Examples of potential refractory materials, with their melting points in degrees celsius are:

SiC	2700	HfN	3305
Si_3N_4	~ 1900	HfC	3890
BN	~ 3000	TaN	3360
B_4C	2350	TaC	3880
NbB_2	~ 2900	ZrC	3540

This list includes materials such as HfC which are among the highest melting materials known.

The above is a consideration of materials which are potential refractories. The main types of refractory material used on a large scale industrially are silica, chrome-magnesia basic, fired fireclay and high alumina. Other materials such as SiC and Si_3N_4 have more specialized applications.

Silica bricks are made from quartzite to which 2 to 3% CaO is added as an aid to sintering. Impurities such as the alkalies must be avoided because of their great fluxing action (Fig. 20.3b). One problem is that the bricks may partially disintegrate on heating due to volume changes that accompany the $\alpha \rightleftharpoons \beta$ quartz transition at 573 °C. This disintegration is known as spalling. Above 573 °C, the bricks are quite stable and maintain their strength up to melting temperatures, ~ 1700 °C. Because of the high viscosity of molten silica, a considerable amount of liquid in the bricks can be tolerated during use.

Certain types of clay have refractory properties after firing. Pure kaolin (also known as china clay) has the approximate formula $Al_2O_3 \cdot 2SiO_2 \cdot 2H_2O$ and its high temperature behaviour can be understood by reference to the Al_2O_3–SiO_2 phase diagram (Fig. 20.3a). After loss of water and various recrystallization reactions, the kaolin yields a mixture of mullite and silica with approximate composition K, which begins to melt above 1595 °C but is not completely molten until well over 1800 °C. On the other hand, montmorillonite clays such as bentonite have a much higher silica to alumina ratio ($\sim 4:1$) and a considerable amount of other cations, such as alkalies, present. Combination of these two effects gives much lower liquids temperatures, ~ 1300 °C, and hence much reduced refractoriness of bentonite.

High alumina refractory bricks usually contain $> 85 \, wt \%$ Al_2O_3 and have SiO_2 as the main impurity. They are made from diaspore or bauxite. The Al_2O_3–SiO_2 phase diagram (Fig. 20.3a) shows that these bricks contain a mixture of corundum, Al_2O_3, and mullite, $3Al_2O_3 \cdot 2SiO_2$, under equilibrium conditions at subsolidus temperatures, < 1840 °C. Above 1840 °C, partial melting to give Al_2O_3 + liquid occurs. The bricks are made by one of two processes. Fusion-cast alumina is made by melting in electric furnaces at ~ 2000 °C, the lining of which forms part of the charge. Sintered alumina blocks are made by firing at 1700 to 1800 °C. These lower temperatures are used for impure aluminas in which sintering occurs with the aid of a liquid phase; the product is non-porous and abrasion resistant. High purity aluminas containing $\gtrsim 99.8\%$ Al_2O_3 sinter by a solid state mechanism; usually $\sim 0.2\%$ MgO is added because, for reasons which are not clear, this leads to a transparent product with zero porosity. The preparation of refractory, high alumina concrete is described in Chapter 19.

Basic magnesia and chrome-magnesia bricks are widely used in metal extraction furnaces because of their resistance to molten slag. MgO melts at 2800 °C and is obtained from three main sources: brucite, $Mg(OH)_2$; magnesite, $MgCO_3$; and dolomite, $CaCO_3 \cdot MgCO_3$. The raw materials are calcined (i.e. heated) at 500 to 1000 °C to eliminate CO_2 and H_2O and leave a very fine grained 'active' MgO powder. Active MgO has a large surface area and is hygroscopic, forming $Mg(OH)_2$ which has a lower density than MgO. If hydration occurs in an MgO body during survice, the increase in volume causes the body to crack. It may be avoided by dead-burning the MgO, i.e. by heating it to a sufficiently high temperature to increase the average grain size and give a well-sintered, non-porous body. Temperatures of 1400 to 1700 °C, depending on impurity content, are usually necessary to achieve dead-burning.

Pure magnesia may be sintered to a non-porous, transparent compact by using LiF as a sintering aid. The LiF volatilizes during sintering, leaving behind pure MgO suitable for high temperature applications. Commercial MgO sources usually contain CaO, Al_2O_3, Fe_2O_3 and SiO_2 as impurities and so partial melting occurs during dead-burning. The greatly enhanced sintering that occurs at moderate temperatures in impure MgO appears to be due to (a) the presence of liquid and (b) the formation of periclase (MgO) solid solutions in which part of the magnesium is replaced by aluminium and iron.

Use of dolomite as a raw material gives, on calcination, an oxide mixture that is very hygroscopic due to the high lime content and it is difficult to eliminate the hydration tendency by dead-burning. Two methods are used to overcome this: either the calcined product is coated with hydroxyl-free tar which effectively seals off the free-lime surface or other materials are added that react with the free-lime during heating. For example, serpentine, $3MgO \cdot 2SiO_2 \cdot 2H_2O$, may be used which reacts with CaO to form Ca_3SiO_5 (and forsterite, Mg_2SiO_4).

Chrome-magnesia bricks are also used in blast furnaces. Chrome, Cr_2O_3, is a refractory oxide melting at 2275 °C, but chrome ores generally contain several other oxides. Usually the chromium is present in a spinel phase AB_2O_4 where A is magnesium and iron (II) and B includes aluminium and iron (III) as well as chromium. A second minor phase is usually present which has the approximate composition of serpentine. On heating chrome ore, the serpentine dehydrates and at ~ 1400 °C, any iron (II) present will have oxidized to Fe_2O_3. This may react with MgO from the serpentine to give more spinel, leaving a less refractory, silica-rich, $MgO-SiO_2$ liquid. One effect of added MgO in the chrome-magnesia bricks is that it reacts with this silica-rich liquid to form a much more refractory, forsterite phase, Mg_2SiO_4 (Fig. 20.3e).

In addition to the above general purpose refractories, various materials of more closely controlled impurity levels have special applications. Sintered alumina has already been mentioned and has many uses including spark plug cores, valve seats, wire-drawing dies, etc. Mullite, prepared by reaction of Al_2O_3 (bauxite or Bayer alumina) and SiO_2 (quartz or china clay) is used for crucibles, pyrometer tubes, etc. Forsterite melts at 1890 °C and is used as a foundry sand for steel moulds. It occurs as a mineral, usually with partial replacement of magnesium by iron (II), or can be synthesized by calcining MgO with talc, $3MgO \cdot 4SiO_2 \cdot H_2O$, or, more commonly, serpentine, $3MgO \cdot 2SiO_2 \cdot 2H_2O$. Steatite bodies are made from talc–clay mixtures and are used as low loss electrical ceramics. The amount of clay used is small and the main product after firing is clinoenstatite, one of the polymorphs of $MgSiO_3$. Another valuable electroceramic is cordierite, $2MgO \cdot 2Al_2O_3 \cdot 5SiO_2$, which has a very low coefficient of thermal expansion and is therefore resistant to thermal shock. It may also be prepared from talc–clay mixtures to which extra Al_2O_3 or sillimanite, Al_2SiO_5, has been added. Zirconia, ZrO_2, has considerable potential as a refractory, with a melting point of 2700 °C, but it usually shatters on heating because of the volume changes that accompany the monoclinic $\rightleftharpoons$ tetragonal transition. Monoclinic ZrO_2 has a density of $5.56\,g\,cm^{-3}$ and high temperature tetragonal ZrO_2, $6.10\,g\,cm^{-3}$. On heating, therefore, a 9 per cent contraction in volume accompanies the transition. The phase transition and the shattering may be avoided by adding 10 to 20 per cent of CaO, MgO or Y_2O_3. These oxides form solid solutions with the high temperature, cubic polymorph of ZrO_2 (stable only at > 2400 °C in pure ZrO_2) and these cubic solid solutions are stabilized to much lower temperatures. The formula of lime stabilized zirconia may be written as $Ca_xZr_{1-x}O_{2-x}$ and shows that replacement of Zr^{4+} by Ca^{2+} occurs together with the creation of oxide ion vacancies (Chapter 10). Stabilized zirconia has reversible expansion/contraction behaviour on thermal cycling without any disruptive phase transitions. It is finding applications as a solid electrolyte in high temperature fuel cells and oxygen selective electrodes because, at high temperatures, the oxide ion vacancies are mobile (Chapter 13).

Zircon, $ZrSiO_4$, has a low coefficient of thermal expansion and good thermal shock resistance and is used in the lining of some glass melting tanks and as a foundry sand. β-Alumina, of approximate formula $Na_2O \cdot (7-9)Al_2O_3$, forms in the refractory lining of glass melting furnaces by reaction of Na_2O from the glass with Al_2O_3 in the refractory. There is much interest currently in β-alumina because it has the unusual property of high Na^+ ion mobility and is being used as the solid electrolyte in novel types of solid state battery (Chapter 13).

Many non-oxide materials have very high melting points but their use as refractory materials is often limited by their poor oxidation resistance and the high cost of manufacture. One of the most promising is silicon nitride, Si_3N_4, which can be used in air up to $\sim 1400\,°C$. It has high hot strength and possible applications in turbine blades and high temperature bearings. Silicon carbide, SiC, is a well-known material and is stable in air up to $\sim 1700\,°C$ by virtue of forming a protective coating of SiO_2. It is used as an abrasive (carborundum), as electric furnace elements and in heavy refractories. SiC refractories are non-porous and have high thermal conductivity and excellent thermal shock resistance. Synthetic graphite, C, is an excellent refractory, resistant to slags and non-ferrous metals. It has high hot strength and good electrical and thermal conductivity, which allow its widespread use for electrodes in industrial furnaces. Molybdenum silicide, $MoSi_2$, forms a protective coating of SiO_2 in air and is used as an electric furnace heater element for temperatures up to $1600\,°C$.

20.8 Recent advances: sialons—nitrogen-based ceramics

Silicon nitride, Si_3N_4, is an inert material that can withstand higher temperatures than most metal alloys. It is therefore being seriously considered as a constructional material for use in high temperature, efficient, ceramic-based gas turbines. One difficulty is that, being a covalently bonded structure, it is difficult to sinter powdered Si_3N_4 into a dense polycrystalline body. During studies of the sintering of Si_3N_4 in the early 1970s. Jack and coworkers at Newcastle discovered a whole new family of phases which they termed *sialons* (see Jack, 1976). These are based on the four elements Si, Al, O and N–hence the name 'sialon'—although many other elements can also be incorporated into the structures. The sialons are oxynitride phases, built up of (Si, Al)(O, N)$_4$ tetrahedra that link up to form three-dimensional framework structures. Some sialon phases have superior mechanical properties and chemical resistance to Si_3N_4 at high temperatures and can be sintered more easily. It is anticipated that they may find commercial applications in the near future.

The sialons are interesting materials from a crystallochemical point of view as often they are isostructural with a known silicate phase. For example:

(a) $YSiO_2N$ is isostructural with wollastonite, $CaSiO_3$; its structure contains infinite chains, $(SiO_2N)^{3-}$, analogous to metasilicate chains, $(SiO_3)^{2-}$.

(b) $Y_2O_3 \cdot Si_3N_4$ or Y_2Si $[Si_2O_3N_4]$ is isostructural with akermanite, $Ca_2MgSi_2O_7$.

(c) $Y_5(SiO_4)_3N$ is isostructural with apatite, $Ca_5(PO_4)_3OH$.

(d) Li_2SiAlO_3N is a stuffed cristobalite-like phase in which interstitial Li^+ ions enter cavities in the cristobalite (SiO_2)-like framework.

Very recently, it has been found that certain compositions on the join $MgO–Si_3N_4$ may be prepared as glasses. Since addition of small amounts of nitride to oxide-based glasses leads to an increase in the glass transformation temperature, T_g, these new nitrogen-rich glasses may well have interesting properties, such as high temperature stability.

Questions

20.1 What are the requirements for refractory bricks that are to be used for lining iron and steel blast furnaces? Explain why silica bricks are suitable unless highly reducing conditions are present in the kiln.

20.2 Assess the importance of phase diagrams to the practical applications of refractories.

20.3 Explain why silica bricks would not be suitable as the refractory lining in furnaces used for making window glass.

20.4 Explain why useful refractory materials may be made by heating china clay (kaolin) but not by heating montmorillonite clay.

References

A. M. Alper (Ed.) (1971). *High Temperature Oxides*, Vols 1 to 4, Academic Press, New York.

J. Bell (1984). The ceramic age dawns, *New Scientist*, **1983**, Jan. 26, 10.

M. Chandler (1967). *Ceramics in the Modern World*, Aldus.

W. E. Ford (1967). *The Effect of Heat on Ceramics*, MacLaren and Sons.

K. H. Jack (1973). Nitrogen ceramics, *Trans J. Brit, Ceram. Soc.*, **19**, 376.

K. H. Jack (1976). Sialons and related nitrogen ceramics, *J. Materials Science*, **11**, 1135.

W. D. Kingery, H. K. Bowen and D. R. Uhlmann (1976). *Introduction to Ceramics*, Wiley.

I. J. McColm (1983). *Ceramic Science for Materials Technologists*, Chapman and Hall.

R. Pampuch (1976). *Ceramic Materials*, Elsevier.

E. Ryshkewitch (1960). *Oxide Ceramics*, Academic Press.

A. B. Searle (1940). *Refractory Materials*, Griffin and Co.

R. B. Sosman (1965). *The Phases of Silica*, Rutgers University Press.

J. White (1983). New perspective and aspirations of ceramic science and technology, *Trans. J. Brit. Ceram. Soc.*, **1983**, 108.

Chapter 21

Organic Solid State Chemistry

This book could not be concluded without at least some mention of organic materials. Usually, the fact that many organic substances are solids at room temperature has little bearing on their chemistry. There are exceptions, however. Two types of solid state effect are considered here.

First, there are organic reactions which give different products after reaction in the solid state as compared to more conventional reactions that take place in solution. This opens up avenues for novel synthetic pathways and for the production of new materials.

Second, there are organic solids that have interesting physical properties typified by the 'organic metals' such as doped polyacetylene.

21.1 Topochemical control of solid state organic reactions

In conventional organic chemistry, chemical reactivity and the products of chemical reactions are governed by the molecular structures of the compounds

666

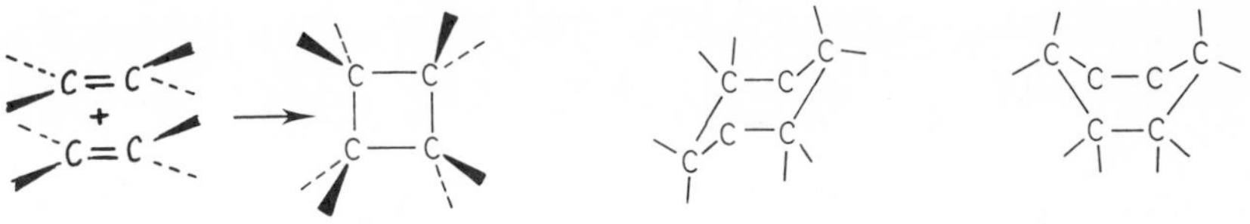

(a)   (b)   (c)

Fig. 21.1 (a) Addition across two double bonds to form a cyclobutane ring. (b) Chair and (c) boat configurations of cyclohexane

involved. In certain reactions that take place in the solid state, however, and which are subject to *topochemical control*, an additional factor comes into play. This is the crystal structure of the starting materials and, in particular, the way in which the molecules are packed together in the crystal. In most crystals, the molecules are packed in a highly ordered and regular manner and are present in only a very limited number of orientations. They are also unable to move about very much. Adjacent molecules are able to react together only if certain criteria are met. In particular, suitable reactive centres should be present on adjacent molecules, in the right orientation and sufficiently close together. For instance, polymerization reactions may occur by the addition together of olefinic groups $>C = C<$ in adjacent molecules. It is found that, for polymerization to occur in the solid state, the double bonds of the olefinic groups should be approximately parallel and no further than $\sim 4\,\text{Å}$ apart. Addition of two olefinic groups gives rise to a cyclobutane linking unit (Fig. 21.1a).

Reactions in the solid state do no usually occur spontaneously but need a catalyst such as UV light. The products of topochemically controlled reactions in the solid state are often quite different from those that are obtained in solution and this gives possibilities for synthetic routes to new polymers, asymmetrically pure compounds, etc. Much of the pioneering work in this area was carried out by Schmidt and coworkers in the 1960s (see Schmidt, 1971).

Stereochemical control of organic reactions is, of course, extremely important and familiar to organic chemists. We are concerned here only with the additional steric factors that are imposed by the particular arrangement of molecules in the crystal structure. Two types of effect may be distinguished: (a) intramolecular effects, where internal rearrangements, cyclizations, elimination reactions, etc., occur within a single molecule and (b) intermolecular effects, in which adjacent molecules react together. Some examples are now given, but mainly of type (b).

21.1.1 Intramolecular reactions: conformational effects

Most organic molecules in liquid and gaseous states are flexible and can change their shape. A simple example is cyclohexane which can exist in chair and boat shapes (Fig. 21.1b and c). The molecules can readily flip from one form to the other since no bonds are broken in the process; such effects are known as conformational effects. In the solid state, however, such flexibility of movement is not usually possible. Instead, one particular molecular shape or conformation is

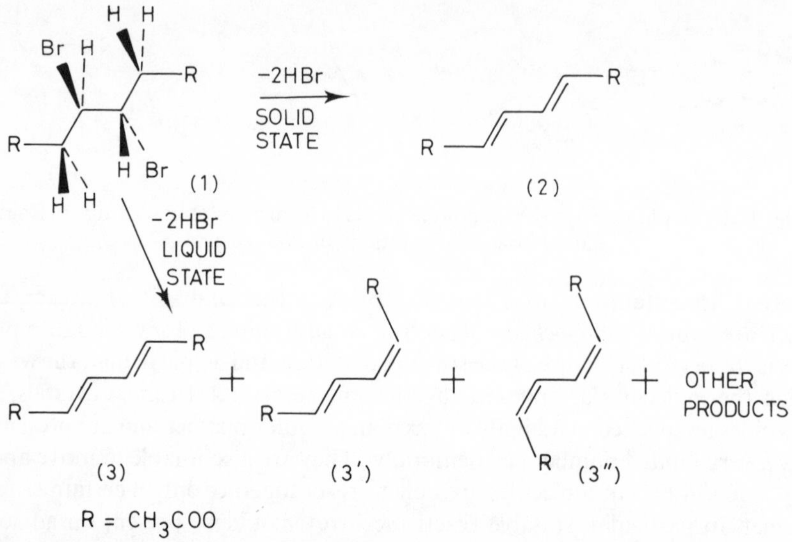

Fig. 21.2 Elimination reactions of dimethyl *meso-β, β'*-dibromoadipate

effectively frozen in. Subsequent reactions are then restricted to that particular conformation. An example is shown in Fig. 21.2. Molecule **1**, dimethyl *meso-β, β'*-dibromoadipate, adopts the conformation shown in the crystalline state. By reaction with gaseous NH_3, two molecules of HBr are eliminated from solid **(1)** to give solid dimethyl *trans, trans*-muconate **(2)**. Both reactant **(1)** and product **(2)** are centrosymmetric. If the same reaction is carried out in the liquid state, however, molecule **1** adopts a variety of conformations, each of which gives a different isomer **3, 3′, 3″** on elimination of HBr.

21.1.2 Intermolecular reactions: molecular packing effects

In a molecular liquid, the molecules are continually tumbling and moving about. Consequently, one particular molecule encounters others, of the same or different kind, in many different orientations. In the solid state, however, the molecules are in essentially fixed positions and are present in very few orientations. This has three consequences. It places a severe and specific limitation on the way in which such 'fixed' molecules can react together. It gives rise to stereochemically pure products. It also brings the molecules somewhat closer together than they would be in the liquid or gaseous phase and reactions may sometimes occur in the solid state which otherwise would not occur at all.

21.1.3 Photodimerization of o-ethoxy-trans-cinnamic acid

This substance has the molecular structure shown in **4** in Fig. 21.3. In the crystalline state it exists in three polymorphic forms, α, β and γ, which differ in the manner in which the molecules are packed together. On irradiation with UV light, the three forms behave quite differently.

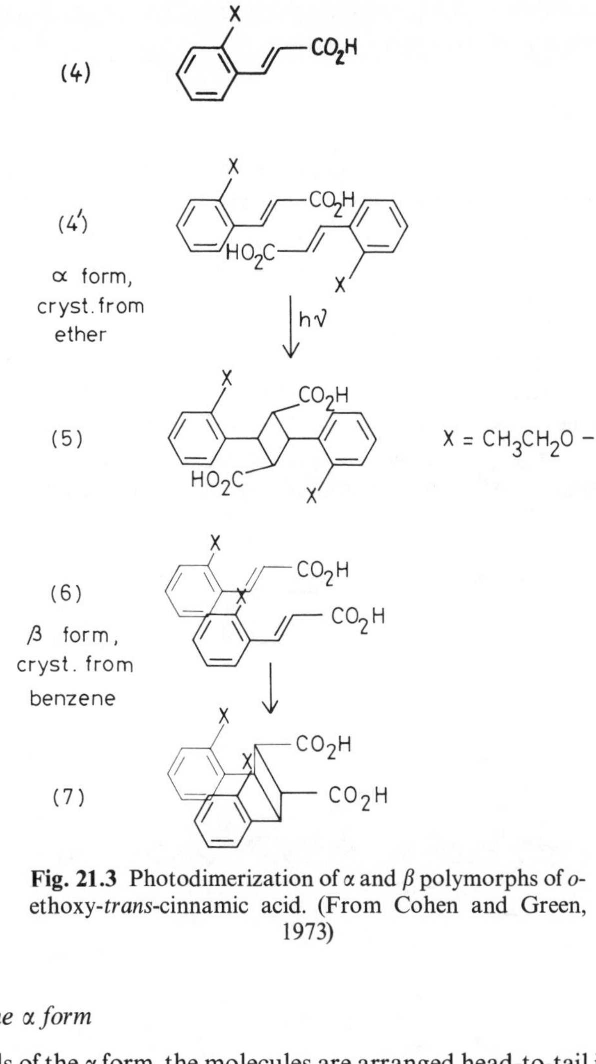

Fig. 21.3 Photodimerization of α and β polymorphs of *o*-ethoxy-*trans*-cinnamic acid. (From Cohen and Green, 1973)

21.1.3.1 *The α form*

In crystals of the α form, the molecules are arranged head-to-tail in the form of centrosymmetric pairs (**4′**). On irradiation with UV light, dimerization occurs to give truxillic acid (**5**) which is also centrosymmetric. The mechanism involves addition across the double bonds in adjacent molecules to give a cyclobutane linkage in **5**. Clearly the reaction can occur only because the two molecules and, especially, the two double bonds are favourably positioned. Very little bulk displacement of the molecules is necessary.

21.1.3.2 *The β form*

The molecules in crystals of the β form are arranged head-to-head and are related by a mirror plane of symmetry (**6**). The molecules are again able to dimerize, by addition across the two double bonds. The product is truxinic acid (7) in which the mirror symmetry is retained.

21.1.3.3 *The γ form*

The molecules in crystals of the γ form, which is crystallized from ethanol, are arranged in such a way that the double bonds on adjacent molecules are not close together. Consequently, no photodimerization reaction occurs. In the liquid state, also, no photodimerization occurs and the only reaction that does is cis–trans isomerization.

Many di-and polymerization reactions, such as the above, have been studied. It appears that the relative positions of the double bonds on adjacent molecules are critical. Distances of $\lesssim 4\text{Å}$ are favourable, but with larger distances, $\sim 5\text{Å}$ the molecules are usually too far apart for cycloaddition to occur. It appears also that the double bonds must be approximately parallel. This requirement, that the

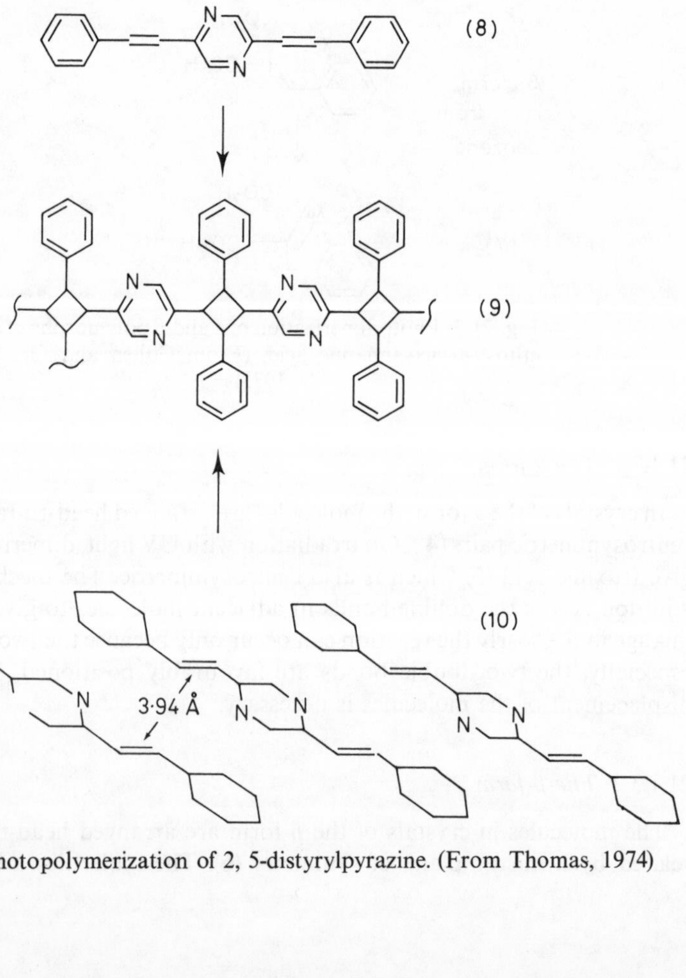

Fig. 21.4 Photopolymerization of 2, 5-distyrylpyrazine. (From Thomas, 1974)

double bonds be parallel and separated by no more than 4Å is known as Schmidt's criterion.

21.1.4 Photopolymerization of 2, 5-distyrylpyrazine

Solution grown crystals of 2, 5-distyrylpyrazine (8) polymerize on UV irradiation at − 60 °C to give a long chain polymer (9) (Fig. 21.4). This is also a topochemically controlled reaction since it depends very much on the relative positions of the molecules in the crystal. The molecules of 8 are stacked into rows, shown horizontally (10) in Fig. 21.4 and are oriented so that the rings on adjacent molecules face each other. Each molecule is displaced relative to its neighbours. Adjacent double bonds are only 3.94Å apart and readily undergo photo-cycloaddition. Since each molecule contains two double bonds, a stereoregular, long chain polymer (9) results. Polymerization of (8) also occurs in solution but the product is, instead, amorphous and of much smaller chain length.

21.1.5 Photopolymerization of diacetylenes

The acetylinic molecules shown in 11 (Fig. 21.5) contain two reactive centres (triple bonds) each. In the crystalline state, they are arranged (schematically) as shown, each being displaced laterally relative to its neighbours. By a suitable choice of substituent R, the triple bonds on adjacent molecules are close enough to add together. On UV irradiation, the stereoregular polymer (12) results.

21.1.6 Asymmetric syntheses

In the examples described so far, the symmetrical arrangement of the reactant molecules is usually preserved in the product. Asymmetric linkages are not formed, therefore. However, asymmetric linkages, with possible optical activity,

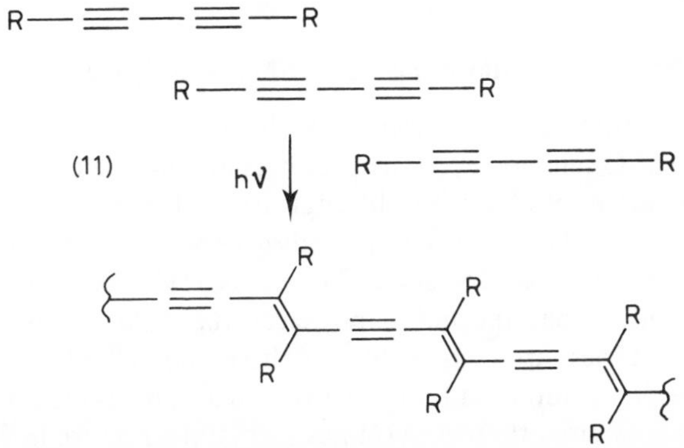

Fig. 21.5 Photopolymerization of a substituted diace-tylene

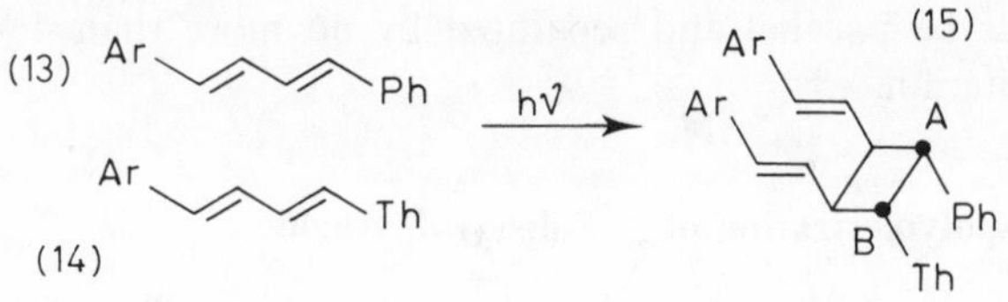

Ar = 2,6 dichlorophenyl, $C_6H_3Cl_2$

Ph = phenyl, C_6H_5 ; Th = thienyl, C_4H_3S

Fig. 21.6 Formation of molecules with optically active centres by photoaddition in solid solution systems

may be produced from solid solution starting materials. This is possible when adjacent molecules of different substances react together.

Inorganic and metallic solid solutions are described in Chapter 10. An *organic* solid solution contains molecules of two or more different compounds (in a binary solid solution) that are packed at random in the crystal. An example of two compounds that can form a solid solution is shown in Fig. 21.6. The two compounds (**13, 14**) are both substituted butadienes and differ in the nature of one of the substituents. Solid solution crystals can be prepared that contain molecules of both **13** and **14** packed at random. By photolysis under carefully controlled conditions, adjacent molecules of **13** and **14** may react to produce the addition product (**15**). It contains two asymmetric carbon atoms, A and B.

There is much scope for similar syntheses of other asymmetric molecules. This is also important as a way of getting fundamental information on the mechanisms of formation of optically active molecules, especially in those cases where the enantiomeric products form in unequal amounts. This is possible because the absolute geometries of reactant molecules and product may be determined by X-ray crystallography.

21.1.7 Dimerization of Anthracene—role of crystal defects

Substituted anthracenes undergo topochemically controlled photodimerization reactions. Often, however, the products are not the expected ones. Thus, some compounds photodimerize although from their packing geometry they would not be expected to; other compounds give products of different symmetry to that expected. Thomas, Moresi and Desvergne (1977) have shown, by elegant microscopic studies, that line and plane defects (i.e. dislocations, stacking faults, antiphase boundaries) are responsible for these unexpected reactions.

An example of a compound that dimerizes in an unexpected way is shown in Fig. 21.7. The 9-cyanoanthracene (**16**) has a crystal structure in which adjacent molecules are arranged head-to-head. Consequently the cis dimer (**17**) would be expected to form. In fact, the trans dimer (**18**) is obtained. This is because, although the vast majority of the molecules are arranged head-to-head in the

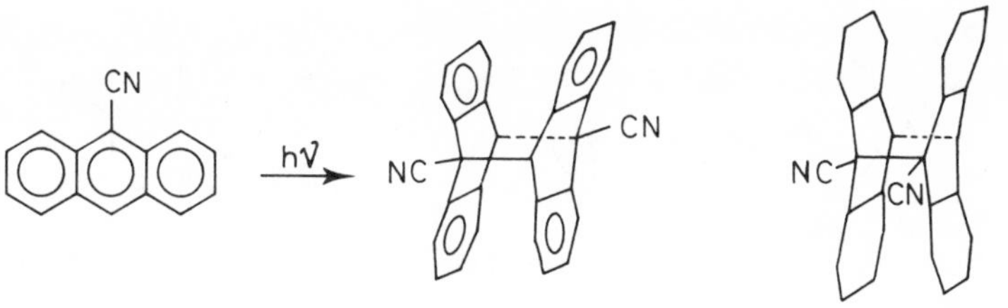

Fig. 21.7 Photodimerization of 9-cyanoanthracene

crystal of **16**, the crystals contain stacking faults in which one half of the crystal is effectively displaced laterally relative to the other half. This displacement brings molecules on either side of the fault into a head-to-tail overlap. On UV irradiation the molecules on the stacking fault appear to be more reactive than those in the bulk of the crystal and dimerize; consequently the trans product (**18**) is obtained. In some way, perhaps by generating new stacking faults, this then encourages the rest of the crystal to give the trans product. In a related reaction, the photodimerization of acenaphthylene, Thomas (1974) was actually able to observe directly in the microscope the formation of dimer on dislocation sites.

21.1.8 Control of molecular packing arrangements

The first topochemically controlled solid state reactions to be discovered depended on the chance suitability of the molecular packing arrangement in the crystals of the reactant compounds. Usually, the molecules in organic crystals are not in favourable orientations relative to each other and, in particular, the reactive centres on adjacent molecules are not in the right orientation and/or are not sufficiently close together. In order to make large scale and varied use of topochemical control in organic solid state reactions, it is therefore necessary to be able to control or modify the crystal structure of the reactant compounds involved. The possibility of achieving this, first recognized by Schmidt (1971) and termed 'crystal engineering', offers exciting prospects for the future.

Some progress in this direction has been made. For instance, Schmidt and Green (1971) found that by substituting halogens, especially chlorine atoms, into aromatic rings, the molecules tend to align in an essentially close packed, highly overlapped and parallel fashion, with an intermolecular separation of ~ 4 Å. The reasons why this should happen are not understood but possibly some kind of attractive force between halogen atoms on adjacent molecules is at least partly responsible. Using this discovery, a large number of chlorosubstituted compounds have been induced to crystallize in a similar close packed way and to subsequently undergo photoaddition reactions. A beautiful example, shown in Fig. 21.8, is an eight-centre dimerization; cycloaddition of two pairs of double bonds in crossconjugated dienones (**19**) yields the unusual tricyclicdiketone ring structure (**20**).

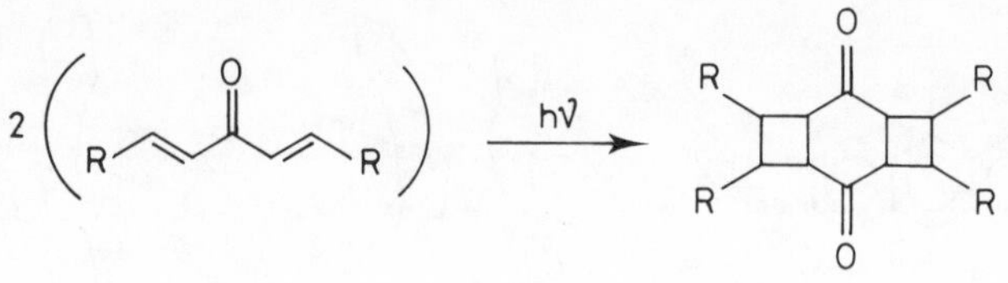

(19)   (20)

$$R = 3,4 - C_6H_3Cl_2$$

Fig. 21.8 Photodimerization of a dienone yielding a three ring structure

21.1.9 Organic reactions within inorganic host structures

The rationale behind topochemical control of organic solid state reactions is the bringing together of molecules in the correct orientation for them to be able to react together. A variation on this theme, developed by Thomas, Williams and others, has been to use inorganic layer compounds, such as sheet silicates and transition metal dichalcogenides, to act as hosts for organic molecules. These molecules intercalate the inorganic layers where they take up specific orientations and may be subsequently induced to undergo selective reactions. Using sheet silicates such as the montmorillonite clay minerals as hosts, it has been possible to synthesize peptides from amino acids and to carry out the benzidine rearrangement reaction. As yet this area has been little developed but it has considerable potential both for general organic syntheses and because of its applications in catalysis.

21.2 Electrically conducting organic solids: organic metals

The possibility of preparing electrically conducting polymers or 'organic metals' is a tantalizing one. Such materials would combine the mechanical properties of polymers—flexibility and ease of fabrication as thin films—with the high electrical conductivity normally reserved for metals. A great deal of research has been carried out in the last few years on such materials but because of problems of atmospheric instability and degradation, none have been commercially exploited as yet (1983). Some of the main materials that have been investigated are now given.

21.2.1 Conjugated systems

21.2.1.1 *Doped polyacetylene*

Organic solids are usually electrical insulators. Electrons cannot move freely within molecules or from one molecule to another in a crystal. Exceptions are conjugated systems that contain a skeleton of alternate double and single carbon

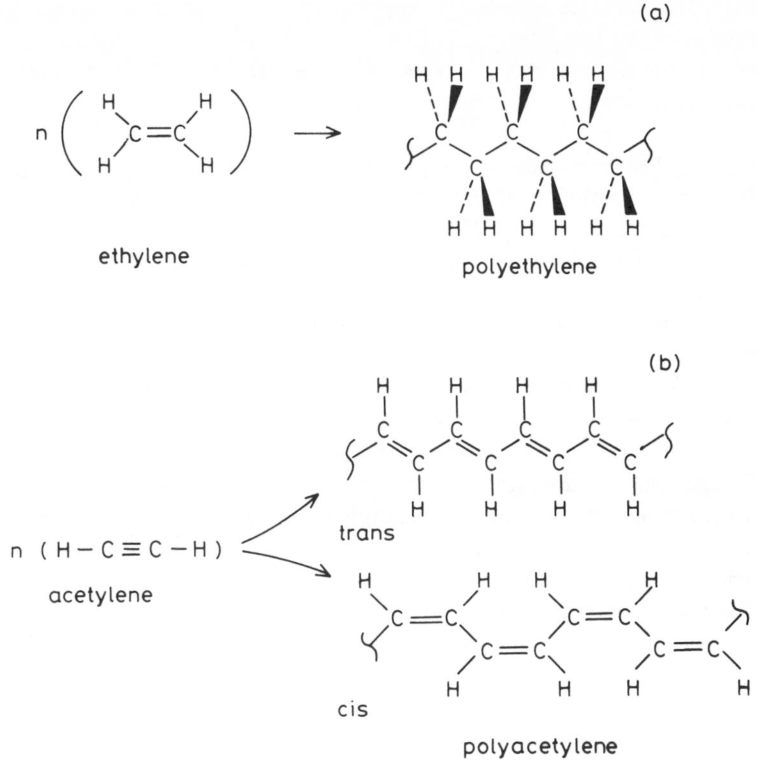

(a)

ethylene → polyethylene

(b)

n (H−C≡C−H)
acetylene

trans

cis

polyacetylene

Fig. 21.9 Formation of (a) polyethylene and (b) polyacetylene

to carbon bonds, as in graphite. Polymers such as polyethylene are insulators because although the polymer precursor, ethylene, contains a $>C=C<$ double bond, polyethylene itself is saturated and contains only single $—C—C—$ bonds (Fig. 21.9a). A conjugated long chain polymer with the potential for electrical conductivity is polyacetylene. The acetylene precursor contains a $—C≡C—$ triple bond whereas polyacetylene contains alternate single and double bonds (Fig. 21.9b). In fact, polyacetylene has a modest electrical conductivity, in the range 10^{-9} ohm^{-1} cm^{-1} (*cis* form) to 10^{-5} ohm^{-1} cm^{-1} (*trans* form), which is comparable to that of semiconductors such as silicon.

These conductivity values are quite low because, unlike graphite, the π electron system is not completely delocalized in polyacetylene. It has a band gap of 1.9 eV. A significant discovery, by MacDiarmid, Heeger and coworkers (1977) was that on doping polyacetylene with suitable inorganic compounds, its conductivity increases dramatically. With dopants such as (a) Br_2, SbF_5, WF_6 and H_2SO_4, all of which may act as electron acceptors, and (b) alkali metals, which may act as electron donors, conductivities as high as 10^3 ohm^{-1} cm^{-1} in *trans*-polyacetylene have been obtained. This is similar to the conductivity of many metals and indeed such materials have been termed 'synthetic metals'. The conductivity increases extremely rapidly as the dopant is added and a semiconductor to insulator transition occurs at about 1 to 5 mol % added dopant. As the dopant level is increased to about 10 per cent, a much more gradual further increase in conductivity occurs.

Preparation. Polyacetylene is prepared by the catalytic polymerization of acetylene in the absence of oxygen. A Ziegler–Natta catalyst may be used which is a mixture of $Al(CH_2CH_3)_3$ and $Ti(OC_4H_9)_4$. In one method acetylene is bubbled through a solution of the catalyst and a solid polyacetylene precipitate forms. In another method, acetylene gas is introduced into a glass tube whose inner surface is coated with a thin layer of catalyst; a layer of polyacetylene forms on the surface of the catalyst.

The relative amounts of cis and trans product depend on the reaction temperature. The trans form is more stable and forms at higher temperatures, $\sim 100\,°C$. The cis form is obtained preferentially at lower temperature, $\sim -80\,°C$. At room temperature, a mixture is obtained. It is usually desired to prepare the trans form because it has higher conductivity. It may be prepared directly, at $100\,°C$, or by heating the cis form or cis–trans mixtures, since the cis form changes rapidly to the trans form on heating to $\sim 150\,°C$.

Doping may be achieved simply be exposing polyacetylene to gaseous or liquid dopant. Similar reactions are well known with graphite (Chapter 2); dopant molecules or ions intercalate into the graphite layers and the conductivity is modified as electrons are added to or withdrawn from the conduction band of graphite. With polyacetylene, dopants such as bromine act as electron acceptors; bromine-doped polyacetylene may be regarded as $(CH)_n^{\delta+} \cdot Br^{\delta-}$. Partial or complete electron transfer probably takes place from the double bonds of polyacetylene to the bromine atoms. However, the electronic structure of polyacetylene films is still unclear, especially the mechanism of electron transfer between polyacetylene molecules. The films have a complex morphology in which chains of polyacetylene appear to fold up to form plates and the plates overlap to form fibres. More work is needed in order to better characterize and control the film texture and to relate specific textural features to the desirable, high electronic conductivity.

Applications. Conducting polyacetylene has a variety of possible applications although as yet (1985) it is not used in any commercial devices. By analogy with conventional semiconductors, pn diode junctions may be fabricated. In these, two films of polyacetylene are brought into contact; one is doped with an electron acceptor and is p-type; the other is doped with an electron donor and is n-type. Such devices are easy to make and, with their large surface areas, they have potential application in solar energy conversion. However, one problem that remains to be solved is the sensitivity of polyacetylene to oxygen. Perhaps substituted or modified polyacetylenes will be made in the future which retain the high conductivity but are not liable to atmospheric attack.

The above discussion is concerned with electronic conductivity. Recently,

MacDiarmid and coworkers have shown that ionic conductivity is also possible and that certain doped polyacetylenes may be used as reversible electrodes in new types of battery. The doping is carried out electrochemically. In one arrangement, a polyacetylene film is dipped into a liquid electrolyte composed of $LiClO_4$ dissolved in propylene carbonate. A lithium metal electrode also dips into the electrolyte. On charging the cell at 1.0 V, at room temperature, perchlorate ions from the electrolyte are deposited and react with the polyacetylene electrode to give $(CH)_y{}^+(ClO_4)_y{}^-$, where y is in the range 0 to 0.06. At the same time, Li^+ ions are deposited at the lithium electrode in order to preserve charge balance. The perchlorate ions enter the polyacetylene structure reversibly, i.e. they subsequently diffuse out of the polymer and back into the electrolyte on allowing the cell to discharge. The polyacetylene therefore behaves as a mixed ionic–electronic conductor.

The possibility of using polymeric electrodes in batteries, especially in solid state batteries, is very attractive for at least two reasons: (a) the polymers are very light (compared to lead in lead/acid batteries): (b) polymers have a flexible structure and this should reduce problems of contact resistances at the electrode–solid electrolyte interface.

21.2.1.2 *Polyparaphenylene*

Another long chain polymer with considerable potential is polyparaphenylene (Fig. 21.10a). It consists of a long chain of benzene rings and is also a semiconductor. It has been less extensively studied than polyacetylene but has been doped successfully to give much increased electronic conductivity. For instance, on doping with $FeCl_3$, a product of approximate composition $(C_6H_4(FeCl_3)_{0.16})_x$ has been obtained with a conductivity of 0.3 ohm^{-1} cm^{-1} at room temperature.

21.2.1.3 *Polypyrrole*

Pyrrole is a heterocyclic molecule with a five-membered ring C_4H_5N. It can be polymerized to give a long chain structure which effectively has alternate double

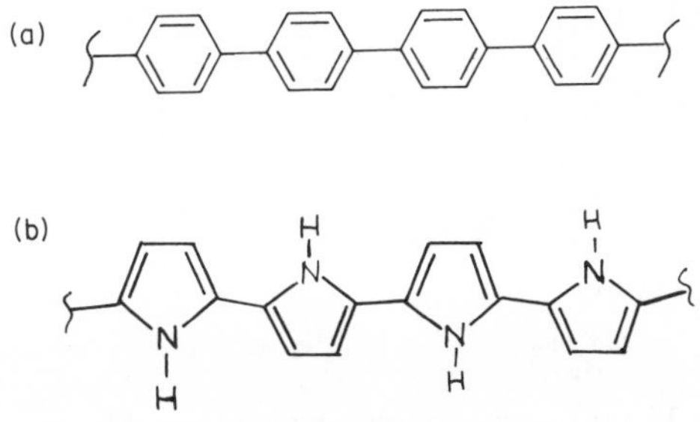

Fig. 21.10 (a) Polyparaphenylene and (b) polypyrrole

and single bonds, giving a delocalized π electron system (Fig. 21.10b). Poly-pyrrole itself has a low conductivity but it can be oxidized by perchlorate to give p-type conductivity as high as $10^2\,\text{ohm}^{-1}\,\text{cm}^{-1}$. It also has the advantages of being stable in air and able to withstand temperatures of up to $250\,^\circ\text{C}$.

21.2.2 Organic charge transfer complexes: new superconductors

There has been much interest for several years in two component organic systems in which one component is a π electron donor and the other an electron acceptor. Some of these behave as highly conducting synthetic metals and a few are superconducting at very low temperature. The tantalizing possibility of achieving superconductivity at a higher temperature, e.g. at 50 or 100 K, provides added impetus to research on related new materials.

Examples of strong π electron acceptors, shown in Fig. 21.11, are (a) tetracyanoquinodimethane (TCNQ) and (b) chloranil. Some strong π electron donors are (c) paraphenylenediamine (PD), tetramethylparaphenylenediamine (TMPD) and (d) tetrathiofulvalene (TTF). In crystalline complexes of π donors and acceptors, the molecules form stacks in which, often, donor and acceptor alternate. The π electron systems on alternate molecules overlap and permit transfer of charge in the stack direction. For example, TMPD and chloranil form mixed stacks which can be represented as:...$(\text{TMPD})^+$ $(\text{chloranil})^-$ $(\text{TMPD})^+$ $(\text{chloranil})^-$ $(\text{TMPD})^+$.... Complexes are also possible in which only one component is an aromatic system. For instance, TCNQ forms charge transfer

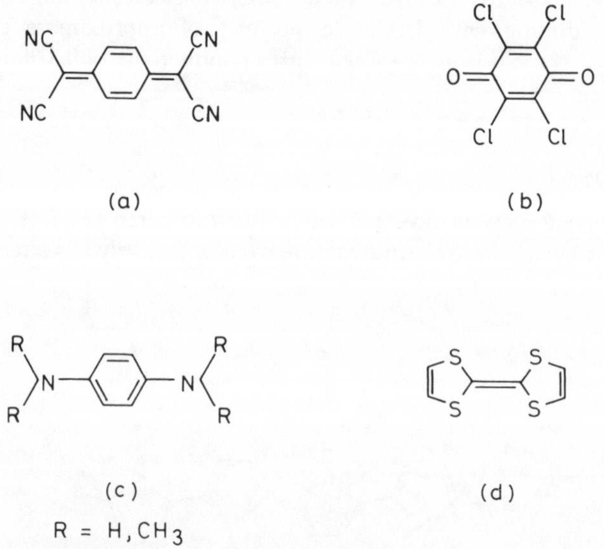

(a)

(b)

(c)

R = H, CH₃

(d)

Fig. 21.11 (a) Tetracyanoquinodimethane (TCNQ), (b) chloranil, (c) p-phenylenediamines and (d) tetra-thiofulvalene

complexes with alkali metals, e.g. $(TCNQ)_3Cs_2$, and TTF forms complexes with halogens, e.g. $(TTF)_2Br$.

Although many organic charge transfer complexes are semiconductors some, such as TTF–TCNQ and NMP–TCNQ (NMP = N-methyl phenazine), have very high conductivity, $\sim 10^3\,ohm^{-1}\,cm^{-1}$. The complex $(TTF)_2Br$ is superconducting at low temperature (below 4.2 K) and high pressure (at 25 kbar).

References

D. Bloor (1982). Plastics that conduct electricity, *New Scientist*, **1982**, 577–580.

M. D. Cohen (1975). The photochemistry of organic solids, *Angew. Chemie Int. Ed.*, **14**, 386–393.

M. D. Cohen and B. S. Green (1973). Organic chemistry in the solid state, *Chem. Brit.*, **9**, 490–497.

W. Jones and J. M. Thomas (1979). Applications of electron microscopy to organic solid state chemistry, *Progr. Solid St. Chem.*, **12**, 101–124.

P. J. Nigrey, D. MacInnes, D. P. Nairns and A. G. MacDiarmid (1981). Lightweight rechargable storage batteries using polyacetylene $(CH)_x$ as the cathode active material, *J. Electrochem. Soc.*, **128**, 1651–1654.

G. M. J. Schmidt (1971). Photodimerisation in the solid state, *Pure and Applied Chemistry*, **27**, 647–678.

R. B. Seymour (1981). *Conductive Polymers*, Plenum Press.

J. M. Thomas (1974). Topography and topology in solid state chemistry, *Phil. Trans. Roy Soc.*, **277**, 251–286.

J. M. Thomas, S. E. Moresi and J. P. Desvergne (1977). Topochemical phenomena in organic solid state chemistry, *Adv. Phys. Org. Chem.*, **15**, 64–151.

Appendices

Appendix A1

Geometrical Considerations in Crystal Chemistry

A1.1 Notes on the geometry of tetrahedra and octahedra

A1.1.1 Relation of a tetrahedron to a cube

In Fig. A1.1(a) is shown a cube with a tetrahedron inside; the centre of the tetrahedron, M, is at the cube body centre and the corners of the tetrahedron are at four alternate corners of the cube, X_1 to X_4. Using this relation it is relatively easy to make calculations on the geometry of tetrahedra.

A1.1.2 Relation between distances M–X and X–X in a tetrahedron

Let cube edge have length l. From Pythagoras, the distance X–X is the face diagonal of the cube, i.e. $X–X = \sqrt{2}\,l$. Distance M–X is half the body diagonal of the cube, i.e. $M–X = \sqrt{3}\,l/2$. Hence, the ratio $XX/MX = \sqrt{2}\,l/(\sqrt{3}\,l/2) = \sqrt{8/3} = 1.633$. In silica structures, the basic building blocks are SiO_4 tetrahedra. In these usually $Si–O \simeq 1.62\,\text{A}$. Therefore, the tetrahedron edges, $O–O \simeq 1.633\,(Si–O) \simeq 2.65\,\text{Å}$.

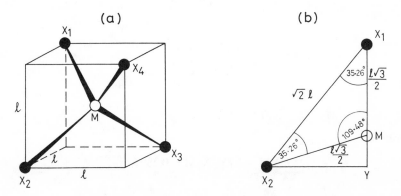

Fig. A1.1 Relation of a tetrahedron to a cube

681

A1.1.3 Angle XMX of a tetrahedron

Using the above result and the cosine equation (Fig. A1.1a and b):

$$(X_1-X_2)^2 = (M-X_1)^2 + (M-X_2)^2 - 2(M-X_1)(M-X_2)\cos\angle X_1MX_2$$

Therefore,

$$(1.633\ M-X_1)^2 = 2(M-X_1)^2 - 2(M-X_1)^2\cos\angle X_1MX_2$$

and so

$$\angle XMX = 109.48°$$

A1.1.4 Symmetry of a tetrahedron

A tetrahedron has a threefold rotation axis along each of the M–X directions, i.e. along each cube body diagonal. Both the tetrahedron and the cube therefore possess four threefold axes. The cube also possesses fourfold rotation axes passing through each pair of opposite cube faces. In the tetrahedron these axes are fourfold inversion axes, $\bar{4}$ (i.e. rotation by 90° followed by inversion through the centre of the tetrahedron).

A1.1.5 Centre of gravity of a tetrahedron

For this, we need to know the vertical height of position M above the triangular base of the tetrahedron. Consider the triangular section X_1X_2M (Fig. A1.1b). Point Y, the extension of the line X_1-M, lies in the centre of the base X_2, X_3, X_4 (Fig. A1.1a, not shown). Hence, distance M–Y gives the height of the centre of gravity above the base. Since

$$\angle X_1MX_2 = 109.48°$$

then

$$\angle MX_1X_2 = \angle MX_2X_1 = 35.26°$$

and so

$$\cos 35.26° = \frac{X_1Y}{X_1X_2} = \frac{X_1Y}{\sqrt{2}l}$$

i.e.

$$X_1Y = 1.155\,l$$

Therefore,

$$YM = X_1Y - X_1M = 0.289\,l$$

and

$$\frac{YM}{YX_1} = \frac{0.289\,l}{1.155\,l} = 0.25$$

i.e. the centre of gravity of a tetrahedron, given by position M, is at one quarter of the vertical height above the (or any) base and in the direction of the apex.

A1.1.6 Relation of an octahedron to a cube

In Fig. A1.2 is shown an octahedron centred at the body centre of a cube with the corners of the octahedron at the face centres of the cube. If the cube has edge l, the distance MX in the octahedron is $l/2$. From Pythagoras, distance XX is $l/\sqrt{2}$.

The octahedron has fourfold rotation axes (3) parallel to each XMX straight line. These axes coincide with the fourfold axes of the cube that pass through opposite cube faces. The cube also has threefold axes (4) which pass through opposite cube corners, along the body diagonal of the cube. In the octahedron these threefold axes pass through pairs of opposite faces

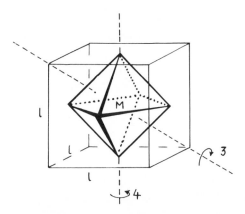

Fig. A1.2 Relation of an octahedron to a cube

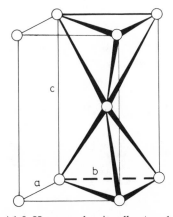

Fig. A1.3 Hexagonal unit cell; $c/a = 1.633$

A1.2 Hexagonal unit cell: proof that the axial ratio, c/a, ideally equals 1.633

A hexagonal close packed array of atoms contains atoms at the corners and inside the unit cell at $\frac{1}{3}, \frac{2}{3}, \frac{1}{2}$ (Fig. 7.6). Along the a and b edges, atoms are in contact; hence a is equal to the atom diameter. For the purpose of calculating c/a, tetrahedral arrays of atoms may be identified, as in Fig. A1.3, in which two tetrahedra share a common corner (i.e. the atom at $\frac{1}{3}, \frac{2}{3}, \frac{1}{2}$). Hence, the c dimension equals twice the vertical height of such a tetrahedron.

Therefore:

$$\frac{c}{a} = \frac{2X_1Y}{X_1X_2} = \frac{2 \times 1.155\,l}{\sqrt{2}l} = 1.633$$

Appendix A2
Model Building

The construction of two types of model, based on spheres and polyhedra, is outlined.

Equipment needed:

Polystyrene spheres (100 to 200), any diameter but 30 mm convenient
Solvent (5 to 10 ml) in bottle with dropper or paint brush; chloroform suitable
Card paper (2 to 3 m²), perhaps sheets of different colour Glue stick

A2.1 Sphere packing arrangements

In order to join polystyrene spheres, glue may be used but is often messy. An alternative is to use a solvent in which polystyrene dissolves. A small amount of solvent is spotted or brushed onto a sphere and a second sphere is placed in contact with the first in the region of the solvent. If too much solvent is used it may run and spoil the appearance of the sphere. The spheres usually stick immediately on contact but the bond hardens only after leaving for a few hours. It is therefore often better to construct larger three-dimensional models in stages.

It should be relatively easy to stick together spheres to form close packed layers (Fig. 7.1) and then build up layers in h.c.p. (Fig. 7.2) and c.c.p. (Fig. 7.3) stacking sequences; further details will not be given. Two useful additional exercises are the following.

A2.1.1 To show the relation between a c.c.p. structure and an f.c.c. unit cell

Stick together six spheres to form that part of a c.p. layer shown in Fig. A2.1(a). Place a single sphere in the layer above, as shown in (b). Repeat steps (a) and (b) so that two identical arrangements, (b), are obtained. Allow models to harden. Orient one model as in (c) and place the second model over the first as in (d). The resulting structure should be a face centred cubic unit cell with spheres in corner and face centre positions as in Fig. 7.5(a). The f.c.c. cell therefore contains c.p. layers parallel to the (111) planes of the unit cell. On standing the model on one corner so that the c.p. layers are horizontal, it should be apparent that the stacking sequence is ABC.

(a)

(b)

(c)

(d)

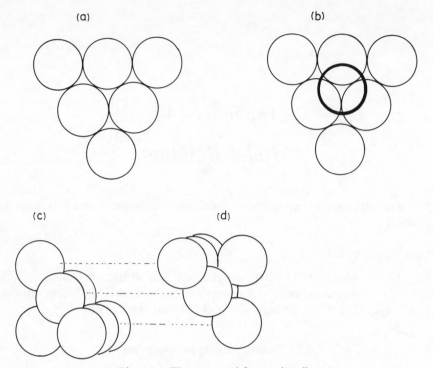

Fig. A2.1 The c.c.p. and f.c.c. unit cells

A2.1.2 To show the four orientations of the c.p. layers in an f.c.c. unit cell

One major difference between c.c.p. and h.c.p. is that in an h.c.p. array of spheres, only one set of c.p. layers occurs parallel to the basal plane of the hexagonal unit cell. In a c.c.p. array of spheres, c.p. layers occur in four orientations. This may be shown by constructing the pyramid shown in Fig. A2.2. First, construct a triangular base of six spheres, as in Fig. A2.1; then

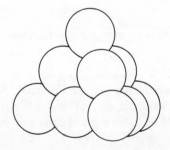

Fig. A2.2 The c.p. layers in a c.c.p. structure

add three spheres in the layer above and one sphere to form the apex to give the pyramid shown in Fig. A2.2. Check that the packing sequence is ABC, i.e. c.c.p. The pyramid may now be reoriented so that any of its other three faces forms the base and an identical structure results. The four equivalent orientation of the pyramid correspond, therefore, to the four orientations of c.p. layers in a c.c.p. structure.

A2.2 Polyhedral structures

Tetrahedra, octahedra and any other polyhedra may be made from card paper. Using templates such as in Fig. A2.3, the polyhedra may be copied, cut out, folded and glued at the tabs. A convenient size for making fairly rigid polyhedra is to make the polyhedron edge $\simeq$ 5 cm. The polyhedra may be linked up by glueing together corners, edges or faces. Glueing polyhedra by their corners only is a little tricky but can be done given time and patience. A selection of polyhedral linkages is given in Fig. A2.4.

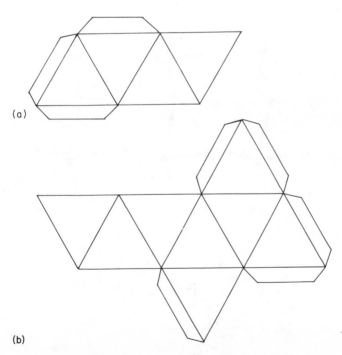

(a)

(b)

Fig. A2.3 Templates for making tetrahedra and octahedra

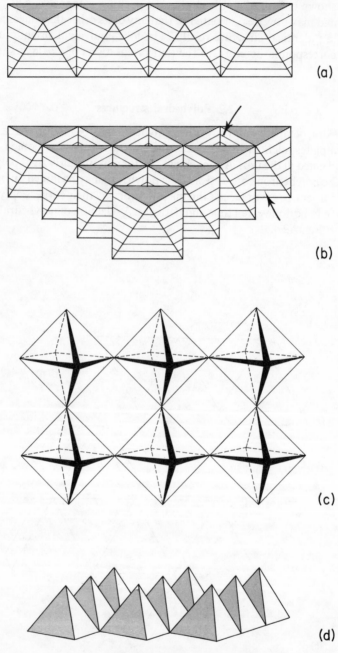

Fig. A2.4 Some polyhedral linkages

(a)

(b)

(c)

(d)

Appendix A3

How to Recognize Close Packed (Eutactic) Structures

On encountering a new or unfamiliar structure a common question is: Is it close packed? A definite answer can often not be given until a three-dimensional model of the structure has been examined. These notes are intended to provide guidelines for helping to decide whether or not a structure is c.p., although it is difficult to give hard and fast rules. The requirements may be summarized:

(a) For non-metallic, close packed (or eutactic) structures of general formula $A_x B_y$, either A atoms or B atoms, or both, should form a c.p. array.
(b) Coordination numbers, given by the number of B atoms surrounding an A atom, and vice versa, are often four and six but values between two and eight are possible.

Point (a) must hold in a eutactic c.p. structure but the difficulty is often to visualize a sufficiently large part of the structure (often > 1 unit cell is needed) that the packing arrangements of like atoms can be seen. The best test, I find, is to look for one close packed layer, i.e. to choose a central atom and look for six equidistant neighbours of the same type such that a coplanar, hexagonal arrangement exists, as in Fig. 7.1. Once such a layer has been identified, it is usually relatively easy to find six other atoms in the layers on either side that complete the c.p. requirement of twelve neighbour atoms. Often, different layers may be seen in a structure, some of which are c.p. and others non-c.p. As well as the above criterion about the number of neighbours of the same kind in a c.p. layer, the two types of layer are often distinguishable in that non-c.p. layers may contain a mixture of A and B atoms whereas c.p. layers have atoms of only one kind (unless mixed c.p. layers occur, as in perovskites). Thus, the unit cell faces in the rock salt structure may be regarded as layers of atoms but they do not constitute c.p. layers because (i) each Na^+ (or Cl^-) has only four Na^+ (or Cl^-) neighbours in the layer and (ii) the layers contain a mixture of Na^+ and Cl^- ions. By contrast, the layers parallel to the $\{111\}$ planes of the rock salt structure are c.p. because (i) each Cl^- (or Na^+) has six like neighbours and (ii) layers of ions of one kind may be identified which alternate with layers of the other kind.

Point (b) refers to the fact that in a c.p. array of A atoms, tetrahedral and octahedral interstitial sites exist for the B atoms. The coordination number of B is therefore four or six. The possibility of other coordination numbers arises when we consider the coordination number of A, as shown by the following examples:

(i) Na_2O (antifluorite) has c.c.p. oxide ions with tetrahedral coordination for Na^+. The coordination number of oxygen (by Na^+) is eight.

(ii) NiAs has h.c.p. As^{2-} ions with octahedrally coordinated Ni^{2+} ions. The coordination number of arsenic (by nickel) is six but is trigonal prismatic.

(iii) Cu_2O (cuprite) has c.c.p. Cu^+ ions with tetrahedral O^{2-} ions. The coordination number of Cu^+ (by O^{2-}) is two (linear).

In order to fully understand and classify a structure A_xB_y, therefore, it is necessary to examine (a) the packing arrangement of A atoms, (b) the packing arrangement of B atoms, (c) the coordination number of A and (d) the coordination of B.

Appendix A4

Positive and Negative Atomic Coordinates

In assigning fractional coordinates to the positions of atoms in a unit cell, it is customary to include in the cell those atoms whose coordinates lie between 0 and 0.999. Atoms with negative coordinates or with coordinates ≥ 1 then lie in adjacent unit cells. This is illustrated in Fig. A4.1, in which four unit cells are drawn. Let us consider the top, right-hand cell; its origin is given as the solid circle and positive directions of x, y and z are indicated by arrows. The diagram shows the coordinates of one pair of opposite face centre positions in each of the four cells. Relative to the origin of our chosen cell, all the face centre positions shown at the right-hand side have $x = \frac{1}{2}$ and those at the left have $x = -\frac{1}{2}$. All positions in the diagram have a positive y value of $\frac{1}{2}$. The z values are in three sets, with values of 1, 0 and -1. In considering the contents of our unit cell, the position at $\frac{1}{2}\frac{1}{2}0$ is regarded as belonging to the cell, but all other positions shown belong to neighbouring cells.

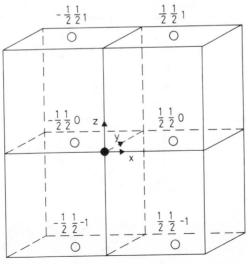

Fig. A4.1 Positive and negative atomic coordinates

Appendix A5
The Crystallographic Point Groups

The thirty-two crystallographic point groups are shown here as stereographic projections using the notation of *International Tables for Crystallography* and as discussed in Chapter 6. For the non-cubic point groups, the symmetry elements and the general equivalent positions are placed on the same diagram; for the five cubic point groups, different diagrams are used.

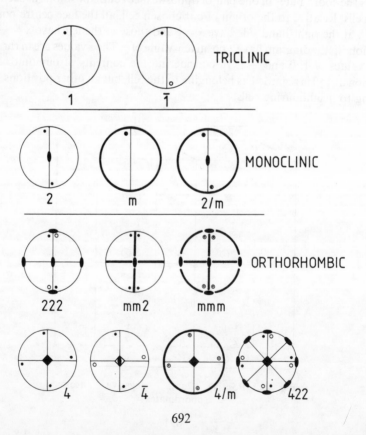

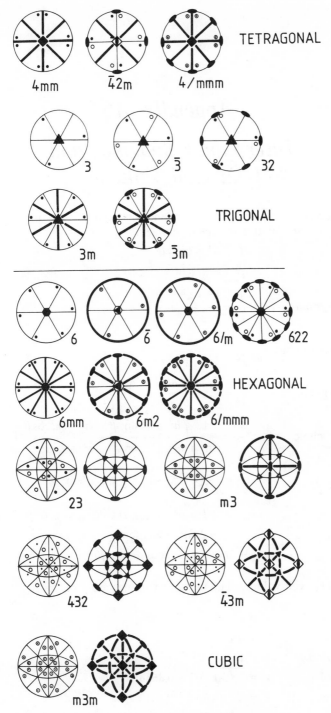

Fig. A5.1 The crystallographic point groups

Appendix A6

Interplanar Spacings and Unit Cell Volumes

The value of d, the perpendicular distance between adjacent planes in the set (hkl), may be calculated using the formulae:

Cubic
$$\frac{1}{d^2} = \frac{h^2 + k^2 + l^2}{a^2}$$

Tetragonal
$$\frac{1}{d^2} = \frac{h^2 + k^2}{a^2} + \frac{l^2}{c^2}$$

Orthorhombic
$$\frac{1}{d^2} = \frac{h^2}{a^2} + \frac{k^2}{b^2} + \frac{l^2}{c^2}$$

Hexagonal
$$\frac{1}{d^2} = \frac{4}{3}\left(\frac{h^2 + hk + k^2}{a^2}\right) + \frac{l^2}{c^2}$$

Monoclinic
$$\frac{1}{d^2} = \frac{1}{\sin^2\beta}\left(\frac{h^2}{a^2} + \frac{k^2\sin^2\beta}{b^2} + \frac{l^2}{c^2} - \frac{2hl\cos\beta}{ac}\right)$$

Triclinic
$$\frac{1}{d^2} = \frac{1}{V^2}[h^2b^2c^2\sin^2\alpha + k^2a^2c^2\sin^2\beta$$

$$+ l^2a^2b^2\sin^2\gamma + 2hkabc^2(\cos\alpha\cos\beta - \cos\gamma)$$
$$+ 2kla^2bc(\cos\beta\cos\gamma - \cos\alpha)$$
$$+ 2hlab^2c(\cos\alpha\cos\gamma - \cos\beta)]$$

where V is the cell volume. The unit cell volumes are given by:

Cubic $V = a^3$
Tetragonal $V = a^2c$
Orthorhombic $V = abc$
Hexagonal $V = (\sqrt{3}a^2c)/2 = 0.866a^2c$
Monoclinic $V = abc\sin\beta$
Triclinic $V = abc(1 - \cos^2\alpha - \cos^2\beta - \cos^2\gamma + 2\cos\alpha\cos\beta\cos\gamma)^{1/2}$

Appendix A7

The Reciprocal Lattice

In Chapter 5, diffraction of X-rays by crystalline materials is treated in terms of the unit cell, lattice type and symmetry of the crystals using Bragg's Law. The concept of lattice planes is introduced, the number, orientation and interplanar separation of which depends solely on the geometry of the unit cell. In deriving Bragg's Law, X-rays are regarded as being diffracted or reflected from these lattice planes. This provides us with a *real space* description of X-ray diffraction. The directions of the incident and diffracted beams are related to the interplanar spacing and X-ray wavelength by Bragg's Law.

For many purposes, it is advantageous and convenient to treat diffraction processes in terms of *reciprocal space*. Reciprocal space and the reciprocal lattice are concepts rather than a physical reality but are particularly useful for discussing single crystal diffraction phenomena. The purposes of this Appendix are to show (i) how the reciprocal lattice is derived from its real space counterpart, (ii) how different kinds of crystal system, lattice type and space symmetry elements manifest themselves in the reciprocal lattice, and (iii) how diffraction may be described in terms of the reciprocal lattice.

A.7.1 Real and reciprocal lattices

In order to show the relation between real and reciprocal lattices it is convenient to consider the example of a monoclinic unit cell (Fig. A7.1). In (a) is shown a projection of the unit cell down the b axis and onto the ac plane. Since b is the unique axis, by definition the angle $\beta \neq 90°$. The b axis is perpendicular to the plane of the paper since both the angles α and γ are 90°. The lattice planes (100) and (001) are shown, together with their corresponding d-spacings d_{100} and d_{001}. Note that, since $\beta \neq 90°$, $d_{100} \neq a$ and $d_{001} \neq c$. The origin of the unit cell is at point X.

In reciprocal space, the sets of lattice planes (100), (001), etc., are each represented by a single point, as shown in (b). These points are at a distance $1/d$ (for the planes in question) from the origin of the reciprocal lattice and are in the direction (relative to the origin) that is perpendicular to the particular set of planes. The directions perpendicular to the planes (100) and (001) define the axes a^* and c^* of the reciprocal lattice. Starring of axes and angles is used to

696

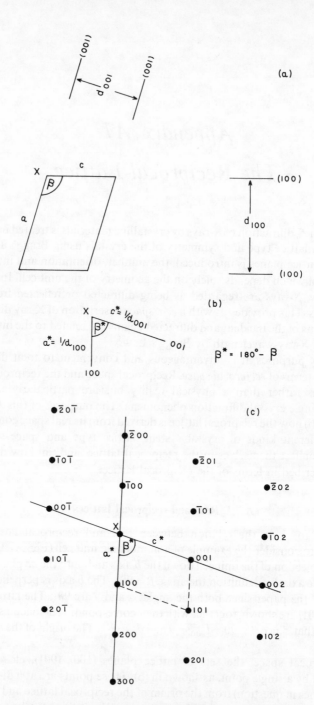

Fig. A7.1 (a) monoclinic unit cell in projection down b; (b) reciprocal lattice points associated with (a); (c) more extended 2D array of reciprocal lattice points.

distinguish reciprocal space parameters from the corresponding real space parameters. Note that, in this particular case, the reciprocal space axes a^* and c^* are not parallel to the real space axes a and c. This is because $\beta \neq 90°$. On the other hand, the axes b and b^* are parallel to each other (and perpendicular to the plane of the paper) because $\alpha = \gamma = 90°$.

Each set of lattice planes gives rise to a point in the reciprocal lattice and it can be shown that, when a large number of sets of planes is considered, a three-dimensional array of points builds up. A more extensive section through the reciprocal lattice is shown in (c). The reciprocal cell is outlined, dashed. It has the origin and the reciprocal lattice points 100, 001 and 101 as its four corners in the section shown. Along each of the reciprocal cell directions, equally spaced reciprocal lattice points occur. Hence 002 is at twice the distance of 001 from the origin X (since $d_{002} = \frac{1}{2}d_{001}$). Since negative reciprocal cell directions are present as well as positive directions, the origin of the reciprocal lattice lies at its centre.

Thus far, and implicit in (c), we have considered only those sets of lattice planes that are parallel to b, i.e. planes of the general type $(h0l)$. The reciprocal lattice points that correspond to these planes all lie on a coplanar section that passes through the reciprocal lattice, as in (c). In order to complete the reciprocal lattice, other sections, corresponding to non-zero values of k, must be added, to form a three-dimensional array or grid of points. The section shown in (c) may be regarded as the *zero layer* since for this, $k = 0$. The next layer above would have identical appearance in a primitive lattice but would have indices $h1l$; similarly the layer below would have indices $h\bar{1}l$. These layers are known as the *first layer*, and so on. The reciprocal lattice effectively extends to infinity in three directions; in practice it may be curtailed when for instance a certain minimum d-spacing (and maximum $1/d$ value) is reached.

A7.1.1 Proof that the reciprocal lattice is a three-dimensional array of points

For this proof, it is sufficient to show that all the sets of planes hkl with the same h value have reciprocal lattice points that form a layer perpendicular to a. For each h value, a different layer of points occurs. Since the same proof applies to sets of planes of constant k and l, the reciprocal lattice points must form a periodic 3-D array.

For the construction shown in Fig. A7.2, suppose that point X is taken as the origin of both the real and the reciprocal lattice. The x axis of the real cell is vertical; no assumptions need be made about the length of the a unit cell dimension. Two planes, 1 and 2, of the general set, hkl, are shown. Plane 1 passes through the origin. Plane 2 cuts the x axis at S. From the definition of Miller indices:

$$XS = a/h$$

Let point Z represent the reciprocal lattice point corresponding to the planes hkl shown. Therefore, line XZ is perpendicular to the planes hkl and is of length $1/d$:

$$XZ = 1/d$$

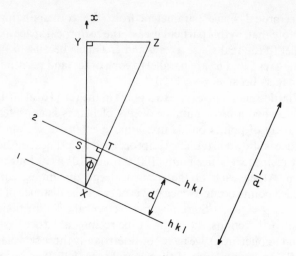

Fig. A7.2 Construction of reciprocal lattice

From triangle XYZ, in which angle $\hat{XYZ}$ is defined as 90°,

$$XY = XZ\cos\phi$$

$$= \frac{1}{d}\cos\phi$$

But, from triangle XST,

$$\cos\phi = XT/XS$$
$$= d/(a/h)$$

Therefore,

$$XY = \frac{1}{d}[d/(a/h)]$$

$$= \frac{h}{a}$$

Since both h and a are fixed for all hkl planes of constant h, the distance XY is fixed. Since YZ is perpendicular to XY, all hkl planes of the same h value have reciprocal lattice points, Z that form a layer perpendicular to x. For different h values, different layers of reciprocal lattice points form. Thus for $h = 0$, point S must coincide with X and hence the zero layer of the reciprocal lattice $0kl$ passes through the origin X. The first layer, $1kl$, cuts the x axis at $1/a$. The second layer, $2kl$, cuts x at $2/a$. etc.

Thus far, we have no information on the location of the reciprocal lattice points *within* a layer. However, this becomes defined as soon as we consider the sets of planes of constant k and constant l. Hence planes, $h0l$, $h1l$, $h2l$, etc., give layers of reciprocal lattice points perpendicular to y and spaced at intervals of $1/b$.

Similarly planes $hk0, hk1$, etc., form layers of points perpendicular to z and spaced at intervals of $1/c$. Each set of planes is effectively considered three times although only one reciprocal lattice point is produced for each. For instance, the point 201 is on the second layer for planes perpendicular to x, the zero layer for planes perpendicular to y and the first layer for planes perpendicular to z. Since the reciprocal lattice points occur only at the points of intersection of the different layers and since the layers are regularly spaced, the reciprocal lattice points must form a regular 3-D array.

A7.1.2 Relation between real and reciprocal lattice parameters

In the previous section, it is shown that reciprocal lattice points of the sets $0kl$, $1kl$, $2kl$, etc., occur in layers perpendicular to the real space axis x or real space cell dimension a. However, the reciprocal lattice directions b^* and c^* both lie in the $0kl$, $1kl$, etc., layers of reciprocal lattice points: this is seen in Fig. A7.1 (c) in which the $h0l$ section through the reciprocal lattice contains the a^* and c^* axes. Therefore,

> a is perpendicular to the b^*c^* plane and similarly
> b is perpendicular to the a^*c^* plane
> c is perpendicular to the a^*b^* plane

and vice versa

> a^* is perpendicular to the bc plane
> b^* is perpendicular to the ac plane
> c^* is perpendicular to the ab plane

In cases where all the angles of the real and reciprocal lattice are 90°, i.e. $\alpha = \beta = \gamma = \alpha^* = \beta^* = \gamma^* = 90°$, then corresponding pairs of real and reciprocal axes are parallel, i.e. $a \parallel a^*$, $b \parallel b^*$ and $c \parallel c^*$. The angles of the real and reciprocal lattice are always related in such a way that $\alpha + \alpha^* = 180°$, $\beta + \beta^* = 180°$, $\gamma + \gamma^* = 180°$.

A7.2 Systematically absent reflections and the reciprocal lattice

In Chapter 5, it is shown that centred (i.e. non-primitive) lattices give rise to sets of systematically absent reflections in the X-ray diffraction patterns. For each lattice type, there is a particular set of systematic absences. In addition, the presence of elements of space symmetry—glide planes and screw axes—gives rise to characteristic sets of systematic absences. Let us see how these systematic absences manifest themselves in the reciprocal lattice, but first with one simplification.

As shown in sections 3.2.1.3 and 5.4.2, single-crystal diffraction patterns give a direct representation or picture of the reciprocal lattice. However, the intensity of each reciprocal lattice point is proportional to the intensity of diffraction from that particular set of planes. Since intensities vary greatly between the different sets of planes, the intensities of the reciprocal lattice

•1̄1̄0 •1̄00 •1̄10 •1̄20   (a) P

X •01̄0 •000 •010 •020   b*→   • l=0,1,2....

•11̄0 •100 •110 •120

•21̄0 •200 •   •

•   •↓a*   •   •

•1̄1̄0 O1̄01 •1̄10 O   (b) I

X O01̄1 •000 O011 •020   b*→   hkl: h+k+l=2n

•11̄0 O101 •110 O121

O21̄1 •200 O211 •220   • l=0,2·····

•   O   •   O   O l=1,3·····

O ↓a* O   •

O1̄1̄1   O1̄11   (c) F

X •000   •020   b*→   hkl: h,k,l either all odd or all even

O11̄1   O111

  •200   •

O   O   • l=0,2···

  ↓a* •   O l=1,3···

O1̄1̄1 •1̄00 O1̄11 •1̄21   (d) A

X O01̄1 •000 O011 •020   b*→   hkl: k+l=2n

O11̄1 •100 O111 •120

O •200 O211 •220   • l=0,2···

O   •   O   •   O l=1,3···

O ↓a* O   •

(e) P

●T00 ●T̄20

●0T̄0 ●T̄10

●000 ●010 ●T̄20

γ*

●1T̄0 ●010

●100 ●020

b*

●2T̄0 ●110

●200

a*

γ* = 180 − γ = 60°

● l = 0,1,2···

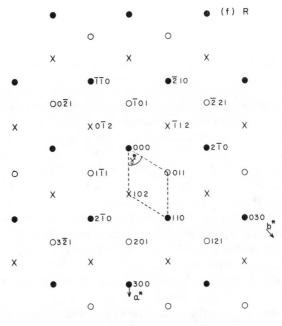

(f) R

●T̄T̄0 ●2̄10

○00 2̄1 ○0T̄01 ○02̄21

X 0T̄2 X T̄12

●000 ●2T̄0

○1T̄1 ○011

X102

●2T̄0 ●110 ●030 b*

○3 2̄1 ○201 ○121

●300

a*

hkl: −h+k+l = 3n

X : l = 2
○ : l = 1
● : l = 0

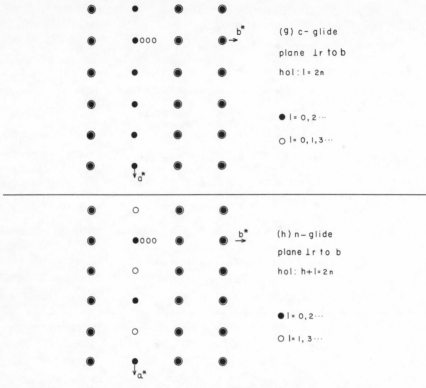

Fig. A7.3 Reciprocal lattices showing various types of systematic absence

points vary. If it is desired to represent intensities, then spots of different size may be used, as shown in Fig. 3.2. This representation is known as a *weighted reciprocal lattice*. Here we are concerned only with whether reflections are present, irrespective of their intensity, or are systematically absent. Allowed reflections are represented by equal-sized spots, therefore.

In the various reciprocal lattices shown in Fig. A7.3, it is desired to superpose different sections through the reciprocal lattice in order to show their three dimensional character. Small solid circles, large open circles and crosses are used, in the manner indicated, to represent different sections through the reciprocal lattice. For most of the examples, an orthorhombic lattice is used because orthorhombic crystals show the greatest diversity of lattice types, Table 5.2.

In Fig. A7.3(a) is shown a primitive lattice, P. There are no systematic absences; hence the zero layer appears to be the same, in projection as the first layer; upper layers are also the same. The origin is at X and may be given the reciprocal lattice coordinates, 000.

In (b) is shown a body-centred lattice, I, for which the conditions for reflection are that $h + k + l = 2n$. Hence, in the zero layer (solid circles), reflections such as 200, 020, 110 and 220 may be observed whereas ones such as 100, 010, 120 and

210 are systematically absent. In the first layer (open circles) the conditions are different. Reflections such as 011, 101, 121, 211 may be observed whereas ones such as 201, 021, 001, 111 and 221 are systematically absent. The second layer has the same appearance as the zero layer, the third layer is like the first layer, and so on.

It should be apparent, after inspection of diagram (b), that the pattern of allowed reflections forms a face-centred lattice: a face-centred cell may be imagined which has, as the corners: 000, 200, 020, 220, 002, 202, 022, and 222. The bottom and top face-centre positions are 110 and 112. The side face-centre positions are 011, 101, 121 and 211. This face-centred reciprocal cell has no real significance as such but it does allow us to understand the rather obscure but often repeated statement that: *A body-centred lattice in real space gives rise to a face centred lattice in reciprocal space.*

In (c) is shown a face-centred lattice, F, for which the conditions for reflection are that h, k, and l should be either all odd or all even for a particular reflection. In the zero layer, 020, 200 and 220 are present whereas 100, 010, 110, 120, 210, etc., are systematically absent. In the first layer, 111, 311, 131, etc., are present and 101, 011, 121, 211, etc., are absent. The pattern of spots in (c) forms a body-centred lattice. Hence, we can appreciate that: *A face-centred lattice in real space gives rise to a body-centred lattice in reciprocal space.*

In (d) is shown a lattice that is centred on only one pair of faces; the condition for A-centring is that for general reflections hkl, the sum $(k + l)$ should be even. This gives rise to the reciprocal lattice shown in (d) in which alternate rows of spots parallel to a^* are absent.

The only remaining lattice type to be considered is rhombohedral. This occurs with some hexagonal unit cells. In (e) is shown a primitive hexagonal reciprocal lattice. It differs from the primitive lattice shown in (a) only in that the angle γ^* is equal to 60°, whereas in (a) $\gamma^* = 90°$; c^* is again perpendicular to the plane of the paper.

In (f) is shown a rhombohedral lattice based on the same hexagonal axes a^* and b^* used in (e). Two-thirds of the reciprocal lattice points are systematically absent: the condition for reflections to be observed is that for $hkl: -h + k + l = 3n$. In order to represent fully a rhombohedral lattice, three sections through the reciprocal lattice are needed, for $l = 0, 1, 2$.

A rhombohedral reciprocal cell using rhombohedral and not hexagonal axes, may also be defined. It has, as its corners, the reciprocal lattice points:

zero layer: 000
first layer: 101, 0$\bar{1}$1, $\bar{1}$11
second layer: 012, 1$\bar{1}$2, $\bar{1}$02
third layer: 003

An example of the effect of screw axes on systematic absences has been given in Fig. 5.35(a). For a 2_1 screw axis parallel to the c axis, the only reflections that are affected are those of the type 00l. For these only those reflections with even values of l may be observed.

Reciprocal lattices containing two types of glide plane are shown in (g) and (h). In (g) is an axial glide plane, c, whose plane of reflection is perpendicular to b and whose translation step is $c/2$. The reflections that are affected by the presence of this glide plane are of the type $h0l$ and the condition for reflection is that l should be even. In the $h0l$ plane, therefore, alternate rows of spots parallel to a^* are systematically absent. c-glide planes may also occur in which the reflection plane is perpendicular to the a axis; in this case, the condition for reflection is $0kl : l = 2n$.

In (h) is shown the influence of a diagonal or n-glide plane. The reflection plane is perpendicular to b and the translation step is $(a/2 + c/2)$. The condition for reflection is $h0l : h + l = 2n$, i.e. for the $h0l$ plane, every alternate spot is systematically absent.

The third type of glide plane is a diamond or d-glide plane. This occurs rather infrequently and only in systems of orthorhombic, tetragonal or cubic symmetry. In orthorhombic systems, they give rise to systematic absences in reflections of the type, $h0l$ (and/or $hk0$, $0kl$). The condition for reflection is $h0l : h + l = 4n$. In tetragonal and cubic systems, d-glide planes influence reflections of the type hhl. For these, the condition for reflection is $hhl : 2h + l = 4n$. Diagrams of reciprocal lattices containing d-glide planes are not given but can be readily constructed following the methods given in the diagrams (a) to (h).

A7.3 Diffraction and the reciprocal lattice—Ewald sphere of reflection

Diffraction can be treated in terms of both the real and the reciprocal lattice. In Chapter 5, a real space description is used. Here a reciprocal space description is outlined.

In the construction given in Fig. A7.4(a), a set of planes hkl is shown with interplanar spacing, d. Let X be the origin of the reciprocal lattice and Z be the reciprocal lattice point associated with the planes hkl. Line YZ is drawn perpendicular to XZ and angle $X\hat{Y}Z = \theta$.

$$\sin \theta = \text{XZ}/\text{XY} = \frac{1}{d} \bigg/ \text{XY}$$

$$\therefore \text{XY} = \frac{1}{d} \sin \theta$$

But, from Bragg's Law, $1/d \sin \theta = 2/\lambda$

$$\therefore \text{XY} = 2/\lambda$$

Triangle XYZ may therefore be used to represent the conditions for diffraction and Bragg's Law. Since XYZ is a right angled triangle, points X, Y and Z lie on the circumference of a circle with XY as the diameter. This is shown further in (b). Suppose the sample is placed at the centre O, of the circle. Incident and diffracted beams are represented by YO and OZ, respectively: OZ is parallel to XY'. The origin of the reciprocal lattice is at X. It may seem odd that the origin of the reciprocal lattice is not arranged to coincide with the sample, but there is no reason why they should coincide.

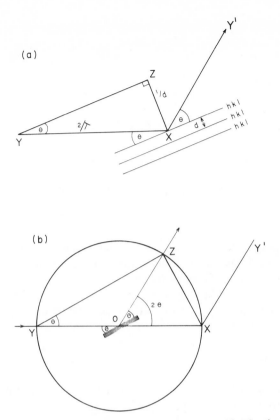

Fig. A7.4 Construction of the Ewald sphere of reflection

In the diffraction experiment, the sample rotates or crystals are present in different orientations. Hence the reciprocal lattice also rotates or is present in different orientations. Bragg's Law is satisfied and diffraction occurs only when the condition shown in (b) holds, i.e. the reciprocal lattice point falls on the circumference of the circle. In three dimensions, when diffraction in all directions is considered, the circle shown in (b) becomes a sphere. This sphere is known as the Ewald sphere of reflection. The objective of single crystal X-ray methods is therefore to orient or rotate the crystal so that the various reciprocal lattice points cut through the sphere of reflection at some stage. Each time a reciprocal lattice point cuts the sphere of reflection, a beam of diffracted radiation is produced in the appropriate direction OZ. By recording these various diffracted beams a picture of the reciprocal lattice may be built up. In some techniques, such as the Precession method, an undistorted picture of the reciprocal lattice is obtained directly. In others, such as the Weissenberg method, a distorted picture of the reciprocal lattice is obtained.

Appendix A8

The Arrhenius Equation for Ionic Conductivity

Ionic transport, whether in crystalline solids, glasses or melts, occurs by a process of activated hopping. In Fig. A8.1(a) is shown a schematic representation of the variation with position of the energy of, for example, a Na^+ ion in a crystalline Na^+ ion conducting solid as it moves along one direction in the crystal. The potential energy is lowest at sites A, B, which correspond to sites that may be normally occupied by Na^+ ions. Between A and B, the potential energy passes through a maximum which represents the saddle point for an ion that is jumping from one site to the next. The potential energy of the cation at sites A and B is identical.

All ions in a solid vibrate, due to their thermal energy, at a frequency, ν, that is around 10^{12} to 10^{13} Hz (i.e. cycles per second). In the absence of an applied electric field, the vibrating ions have a distribution of energies. At temperature TK, the probability that a Na^+ ion has vibrational energy greater than E and could therefore hop between sites A and B, or vice versa, is proportional to $\exp(-E/kT)$ where k is Boltzmann's constant. The number of jumps per second that each ion makes, on average, is proportional to $\nu \exp(-E/kT)$. Thus, provided vacant sites are available for ions to hop into, some cations are able to

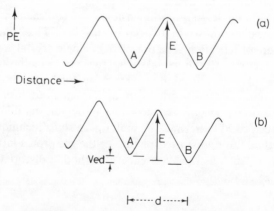

Fig. A8.1 Motion of an ion in a solid

706

hop, at random, even without an applied electric field. Since the hops occur in all directions, at random, there is no net current flow.

On application of an electric field across the solid, the potential energy diagram is slightly modified (Fig. A8.1b). For Na^+ ions that are jumping towards the negative electrode, the effective energy barrier is slightly reduced and for Na^+ ions jumping towards the positive electrode, the energy barrier is slightly greater. The magnitude of these energy differences may be quantified as follows.

For an applied field V, the work done by the field in moving an ion of charge e by a distance d in the direction of the field is equal to Ved. This work done corresponds to the difference in potential energy between sites A and B. The energy barrier to migration in the two directions $A \rightarrow B$ and $B \rightarrow A$ is given by $(E - \frac{1}{2}Ved)$ and $(E + \frac{1}{2}Ved)$, respectively, and hence it is more probable that an ion will jump from A and B than vice versa.

The net probability, P, of an ion jumping from A to B is given by

$$P \propto v \exp\left[\frac{-(E - \frac{1}{2}Ved)}{kT}\right] - v \exp\left[\frac{-(E + \frac{1}{2}Ved)}{kT}\right]$$

$$\propto v \exp\left(\frac{-E}{kT}\right)\left[\exp\left(\frac{eVd}{2kT}\right) - \exp\left(\frac{-eVd}{2kT}\right)\right]$$

The term in square brackets may be expanded as a power series and it may be shown that, for $eVd \ll kT$, it reduces to eVd/kT, i.e.

$$P \propto \frac{eVdv}{kT}\exp\left(\frac{-E}{kT}\right)$$

The mean velocity of the ions, u, is given by

$$u = Pd$$

The mobility, μ, of the ions is given by

$$\mu = \frac{u}{V}$$

and the conductivity, σ, is given by

$$\sigma = Ne\mu$$

where N is the number of mobile ions. Therefore,

$$\sigma = \frac{Ne^2d^2v}{gkT}\exp\left(\frac{\Delta S}{k}\right)\exp\left(\frac{-E}{kT}\right)$$

where $\exp(\Delta S/k)$ is a constant of proportionality, ΔS is the entropy of activation and g is a geometrical factor that includes the possibility that, for a given ion, several jump directions are possible. The equation is often rewritten as

$$\sigma T = \frac{Ne^2d^2v}{gk}\exp\left(\frac{\Delta S}{k}\right)\exp\left(\frac{-E}{kT}\right)$$

On taking logs and plotting $\log \sigma T$ against T^{-1}, a straight line should be obtained of slope equal to $-E/k$ and intercept of $(Ne^2d^2v/gk)\exp(\Delta S/k)$. Sometimes the concentration, N, of mobile ions is also thermally activated with a temperature dependence given by

$$N = N_0 \exp\left(\frac{\Delta S_f}{k}\right)\exp\left(\frac{-E_f}{kT}\right)$$

where N_0 is the total number of potentially mobile ions in the structure and E_f, ΔS_f are the activation energy and entropy for the creation of mobile ions.

Combining the last two equations:

$$\sigma T = \frac{e^2d^2v}{gk}N_0 \exp\left(\frac{\Delta S + \Delta S_f}{k}\right)\exp\left[\frac{-(E + E_f)}{kT}\right]$$

Appendix A9

The Elements and Some of Their Properties

Element	Symbol	Atomic weight[a]	Electronic configuration[b]	Atomic number	Main oxidation state(s)[c]	Some typical bond lengths to oxygen Å and coordination numbers[d]	Melting point (°C)	Crystallographic data: unit cell or structure; lattice parameters (Å); temperature[e]
Actinium	Ac	(227)	$(Rn)6d^17s^2$	89	3		1050	f.c.c., 5.311 (R.T.)
Aluminium	Al	26.98	$(Ne)3s^23p^1$	13	3	Al—O(4)1.79; Al—O(6)1.93	660	f.c.c.; 4.0495, (25 °C)
Americium	Am	(243)	$(Rn)5f^77s^2$	95	3, 4	Am(III)—O(6)2.40; Am(IV)—O(8)2.35	850	hex, ABAC; 3.642, 11.76 (R.T.)
Antimony	Sb	121.75	$(Kr)4d^{10}5s^25p^3$	51	3, 5		630	R3m; 4.5067, 57° 6.5' (25 °C)
Argon	A	39.95	$(Ne)3s^23p^6$	18			−189	f.c.c., 5.42(−233 °C)
Arsenic	As	74.92	$(Ar)3d^{10}4s^24p^3$	33	3, 5	As(V)—O(4)1.74; As(V)—O(6)1.90	814	R3m; 4.131, 54° 10'(25 °C)
Astatine	At	(210)	$(Xe)4f^{14}5d^{10}6s^26p^5$	85	1			
Barium	Ba	137.34	$(Xe)6s^2$	56	2	Ba—O(6)2.76; Ba—O(12)3.00	710	b.c.c.; 5.019 (R.T.)
Berkelium	Bk	(247)	$(Rn)5f^86d^17s^2$	97	3(4)	Bk(III)—O(6)2.36; Bk(IV)—O(8)2.33		
Beryllium	Be	9.01	$(He)2s^2$	4	2	Be—O(3)1.57; Be—O(4)1.67	1280	h.c.p.; 2.2856, 3.5832 (25 °C)
Bismuth	Bi	208.98	$(Xe)4f^{14}5d^{10}6s^26p^3$	83	1, 3(5)	Bi(III)—O(6)2.42; Bi(III)—O(8)2.51	271	R3m; 4.7457, 57° 14.2'(31 °C)
Boron	B	10.81	$(He)2s^22p^1$	5	3	B—O(3)1.42; B—O(4)1.52	2300	tet; 8.73, 5.03 (R.T.)
Bromine	Br	79.91	$(Ar)3d^{10}4s^24p^5$	35	1(3, 5, 7)	Br(VII)—O(4)1.66	−7	orth; 4.48, 6.67, 8.72(−150 °C)
Cadmium	Cd	112.40	$(Kr)4d^{10}5s^2$	48	2	Cd—O(4)2.19; Cd—O(6)2.35	321	h.c.p.; 2.9788, 5.6167 (21 °C)
Caesium	Cs	132.91	$(Xe)6s^1$	55	1	Cs—O(8)3.22; Cs—O(12)3.30	29	b.c.c.; 6.14(−10 °C)
Calcium	Ca	40.08	$(Ar)4s^2$	20	2	Ca—O(6)2.40; Ca—O(8)2.47	850	f.c.c.; 5.582 (18 °C)
Californium	Cf	(249)	$(Rn)5f^{10}7s^2$	98	3	Cf—O(6)2.35		
Carbon	C	12.01	$(He)2s^22p^2$	6	4	C—O(3)1.32	3500	hex, graphite; 2.4612, 6.7079(R.T.)
Cerium	Ce	140.12	$(Xe)4f^26s^2$	58	3, 4	Ce(III)—O(9)2.55; Ce(IV)—O(8)2.33	804	f.c.c.; 5.1604(20 °C)
Chlorine	Cl	35.46	$(Ne)3s^23p^5$	17	1(3, 5, 7)	Cl(V)—O(3)1.52; Cl(VII)—O(4)1.60	−101	tet; 8.56, 6.12(−185 °C)
Chromium	Cr	52.00	$(Ar)3d^54s^1$	24	3, 6	Cr(III)—O(6)2.02; Cr(VI)—O(4)1.70	1900	b.c.c.; 2.8846(20 °C)
Cobalt	Co	58.93	$(Ar)3d^74s^2$	27	2, 3	Co(II)—O(6)2.05 to 2.14; Co(III)—O(6)1.93 to 2.01	1492	h.c.p.; 2.507, 4.069 (R.T.)
Copper	Cu	63.54	$(Ar)3d^{10}4s^1$	29	1, 2	Cu(I)—O(2)1.86; Cu(II)—O(6)1.97 to 2.66	1083	f.c.c., 3.6147 (20 °C)
Curium	Cm	(247)	$(Rn)5f^76d^17s^2$	96	3(4)	Cm(III)—O(6)2.38; Cm(IV)—O(8)2.35		
Dysprosium	Dy	162.50	$(Xe)4f^{10}6s^2$	66	3(4)	Dy(III)—O(6)2.31	1500	h.c.p.; 3.5923, 5.6545 (20 °C)
Einsteinium	Es	(254)	$(Rn)5f^{11}7s^2$	99	3			
Erbium	Er	167.26	$(Xe)4f^{12}6s^2$	68	3	Er(III)—O(6)2.29	1525	h.c.p.; 3.559, 5.592 (20 °C)
Europium	Eu	151.96	$(Xe)4f^76s^2$	63	2, 3	Eu(III)—O(6)2.35; Eu(III)—O(7)2.43	900	b.c.c.; 4.578 (20 °C)

Element	Symbol	At. wt.	Electron configuration	At. no.	Oxidation states	M—O distances	m.p. (°C)	Crystal structure
Fermium	Fm	(253)	$(Rn)5f^{12}7s^2$	100	3			
Fluorine	F	19.00	$(He)2s^22p^5$	9	1		−220	
Francium	Fr	223	$(Rn)7s^1$	87	1			
Gadolinium	Gd	157.25	$(Xe)4f^75d^16s^2$	64	3	Gd—O(7)2.44	1320	h.c.p.; 3.6315, 5.777 (20 °C)
Gallium	Ga	69.72	$(Ar)3d^{10}4s^24p^1$	31	(1)3	Ga—O(4)1.87; Ga—O(6)2.00	30	orth; 4.520, 7.661, 4.526 (20 °C)
Germanium	Ge	72.59	$(Ar)3d^{10}4s^24p^2$	32	(2)4	Ge—O(4)1.79; Ge—O(6)1.94	958	diam; 5.6575 (20 °C)
Gold	Au	196.97	$(Xe)4f^{14}5d^{10}6s^1$	79	(1)(3)	Au(III)—O(4)2.10	1063	f.c.c.; 4.0783 (25 °C)
Hafnium	Hf	178.49	$(Xe)4f^{14}5d^26s^2$	72	4	Hf—O(6)2.11; Hf—O(8)2.23	2000	h.c.p.; 3.1946, 5.0511 (24 °C)
Helium	He	4.00	$1s^2$	2	—		−270	
Holmium	Ho	164.93	$(Xe)4f^{11}6s^2$	67	3	Ho—O(6)2.30; Ho—O(8)2.42	1500	h.c.p.; 3.5761, 5.6174 (20 °C)
Hydrogen	H	1.01	$1s^1$	1	1	H—O 1.02 to 1.22	−259	hex; 3.75, 6.12(−271 °C)
Indium	In	114.82	$(Kr)4d^{10}5s^25p^1$	49	(1)3	In—O(6)2.18; In—O(8)2.32	156	tet; 3.2512, 4.9467 (20 °C)
Iodine	I	126.90	$(Kr)4d^{10}5s^25p^5$	53	1(3,5,7)	I(V)—O(3)1.83	114	orth; 4.792, 7.271, 9.773 (R.T.)
Iridium	Ir	192.22	$(Xe)4f^{14}5d^76s^2$	77	3,4(6)	Ir(III)—O(6)2.13; Ir(IV)—O(6)2.03	2443	f.c.c.; 3.8389(R.T.)
Iron	Fe	55.85	$(Ar)3d^64s^2$	26	2,3(4,6)	Fe(II)—O(6)2.18; Fe(III)—O(4)1.86	1539	b.c.c.; 2.8664 (20 °C)
Krypton	Kr	83.80	$(Ar)3d^{10}4s^24p^6$	36			−157	f.c.c.; 5.68(−191 °C)
Lanthanum	La	138.91	$(Xe)5d^16s^2$	57	3	La—O(6)2.46; La—O(10)2.68	920	hex; ABAC; 3.770, 12.131 (20 °C)
Lawrencium	Lr	(257)	$(Rn)5f^{14}6d^17s^2$	103	3			
Lead	Pb	207.19	$(Xe)4f^{14}5d^{10}6s^26p^2$	82	2,4	Pb(II)—O(4)2.34; Pb(IV)—O(6)2.18	327	f.c.c.; 4.9502 (25 °C)
Lithium	Li	6.94	$(He)2s^1$	3	1	Li—O(4)1.99; Li—O(6)2.14	180	b.c.c.; 3.5092 (20 °C)
Lutetium	Lu	174.97	$(Xe)4f^{14}5d^16s^2$	71	3	Lu—O(6)2.25; Lu—O(8)2.37	1700	h.c.p.; 3.5050, 5.5486 (20 °C)
Magnesium	Mg	24.31	$(Ne)3s^2$	12	2	Mg—O(4)1.89; Mg—O(6)2.12	650	h.c.p.; 3.2094, 5.2105 (25 °C)
Manganese	Mn	54.94	$(Ar)3d^54s^2$	25	2,3,4,7	Mn(II)—O(6)∼2.10; Mn(VII)—O(4)1.66·	1250	cub. 8.914 (25 °C)
Mendelevium	Md	(256)	$(Rn)5f^{13}7s^2$	101	3			
Mercury	Hg	200.59	$(Xe)4f^{14}5d^{10}6s^2$	80	1,2	Hg(I)—O(3)2.37; Hg(II)—O(2)2.09	−39	R̄3m; 3.005, 70°32′(−46 °C)
Molybdenum	Mo	95.94	$(Kr)4d^55s^1$	42	(3,4,5)6	Mo(VI)—O(4)1.82; Mo(VI)—O(6)2.00	2620	b.c.c.; 3.1469 (20 °C)
Neodymium	Nd	144.24	$(Xe)4f^46s^2$	60	3(4)	Nd—O(6)2.40; Nd—O(8)2.52	1024	hex; ABAC; 3.6582, 11.802 (20 °C)
Neon	Ne	20.18	$(He)2s^22p^6$	10			−249	f.c.c.; 4.52(−268 °C)
Neptunium	Np	(237)	$(Rn)5f^57s^2$	93	(2,3)4(6,7)	Np(II)—O(6)2.50; Np(IV)—O(8)2.38	640	orth; 4.723, 4.887, 6.663 (20 °C)
Nickel	Ni	58.71	$(Ar)3d^84s^2$	28	2(3)	Ni(II)—O(6)2.10; Ni(III)—O(6)1.98	1453	f.c.c.; 3.524 (18 °C)
Niobium	Nb	92.91	$(Kr)4d^45s^1$	41	(4)5	Nb(V)—O(6)2.04; Nb(V)—O(7)2.06	2420	b.c.c.; 3.006 (20 °C)
Nitrogen	N	14.01	$(He)2s^22p^3$	7	(2,3,4)5	N(V)—O(3)1.28	−210	hex; 4.03, 6.59 (−234 °C)
Nobelium	No	(253)	$(Rn)5f^{14}7s^2$	102	3			

Element	Symbol	Atomic weight[a]	Electronic configuration[b]	Atomic number	Main oxidation state(s)[c]	Some typical bond lengths to oxygen Å and coordination numbers[d]	Melting point (°C)	Crystallographic data: unit cell or structure; lattice parameters (Å); temperature[e]
Osmium	Os	190.20	$(Xe)4f^{14}5d^66s^2$	76	4(6)8	Os(IV)—O(6)2.03	2700	h.c.p.; 2.7353, 4.3191 (20 °C)
Oxygen	O	16.00	$(He)2s^22p^4$	8	(1)2		−219	cub; 6.83 (−225 °C)
Palladium	Pd	106.40	$(Kr)4d^{10}$	46	2(4)	Pd(II)—O(4)2.04; Pd(IV)—O(6)2.02	1552	f.c.c.; 3.8907 (22 °C)
Phosphorus	P	30.97	$(Ne)3s^23p^3$	15	3, 5	P(V)—O(4)1.57	44	orth; 3.32, 10.52, 4.39 (black) (R.T.)
Platinum	Pt	195.09	$(Xe)4f^{14}5d^96s^1$	78	4(6)	Pt(IV)—O(6)2.03	1769	f.c.c.; 3.9239 (20 °C)
Plutonium	Pu	(242)	$(Rn)5f^67s^2$	94	3, 4, 6	Pu(III)—O(6)2.40; Pu(IV)—O(8)2.36		monocl. 6.18, 4.82, 10.97, 101.81°(21 °C)
Polonium	Po	(210)	$(Xe)4f^{14}5d^{10}6s^26p^4$	84	2, 4	Po(IV)—O(8)2.50	254	cub. 3.345 (10 °C)
Potassium	K	39.10	$(Ar)4s^1$	19	1	K—O(6)2.78; K—O(12)3.00	63	b.c.c.; 5.32 (20 °C)
Praseodymium	Pr	140.91	$(Xe)4f^36s^2$	59	3(4)	Pr(III)—O(6)2.41; Pr(IV)—O(8)2.39	935	hex, ABAC; 3.6702, 11.828 (20 °C)
Promethium	Pm	(147)	$(Xe)4f^56s^2$	61	3	Pm(III)—O(6)2.38		
Protactinium	Pa	(231)	$(Rn)5f^26d^17s^2$	91	4, 5	Pa(IV)—O(8)2.41; Pa(V)—O(9)2.35	3000	tet; 3.935, 3.238 (R.T.)
Radium	Ra	(226)	$(Rn)7s^2$	88	2		700	
Radon	Rn	(222)	$(Xe)4f^{14}5d^{10}6s^26p^6$	86	—		−71	
Rhenium	Re	186.23	$(Xe)4f^{14}5d^56s^2$	75	3, 4, 5, 7	Re(IV)—O(6)2.03; Re(VII)—O(6)1.97	3170	h.c.p.; 2.760, 4.458 (R.T.)
Rhodium	Rh	102.91	$(Kr)4d^85s^1$	45	3(4, 6)	Rh(III)—O(6)2.07; Rh(IV)—O(6)2.02	1960	f.c.c.; 3.8044 (20 °C)
Rubidium	Rb	85.47	$(Kr)5s^1$	37	1	Rb—O(6)2.89; Rb—O(12)3.13	39	b.c.c.; 5.70 (20 °C)
Ruthenium	Ru	101.07	$(Kr)4d^75s^1$	44	4(6)8	Ru(III)—O(6)2.08; Ru(IV)—O(6)2.02	2400	h.c.p.; 2.7058, 4.2816 (25 °C)
Samarium	Sm	150.35	$(Xe)4f^66s^2$	62	(2)3	Sm(III)—O(6)2.36; Sm(III)—O(8)2.49	1052	R; 8.996, 23° 13'(20 °C)
Scandium	Sc	44.96	$(Ar)3d^14s^2$	21	3	Sc—O(6)2.13; Sc—O(8)2.27	1400	h.c.p.; 3.3080, 5.2653 (20 °C)
Selenium	Se	78.96	$(Ar)3d^{10}4s^24p^4$	34	(2)4, 6	Se(VI)—O(4)1.69	217	hex; 4.3656, 4.9590 (25 °C)
Silicon	Si	28.09	$(Ne)3s^23p^2$	14	4	Si—O(4)1.66; Si—O(6)1.80	1410	diam; 5.4305 (R.T.)
Silver	Ag	107.87	$(Kr)4d^{10}5s^1$	47	1(2)	Ag(I)—O(2)2.07; Ag(I)—O(8)2.70	961	f.c.c.; 4.0857 (20 °C)
Sodium	Na	22.99	$(Ne)3s^1$	11	1	Na—O(4)2.39; Na—O(9)2.72	98	b.c.c.; 4.2906 (20 °C)
Strontium	Sr	87.62	$(Kr)5s^2$	38	2	Sr—O(6)2.56; Sr—O(12)2.84	770	f.c.c.; 6.0849 (25 °C)
Sulphur	S	32.06	$(Ne)3s^23p^4$	16	2, 4, 6	S(VI)—O(4)1.52	119	orth; 10.414, 10.845, 24.369 (R.T.)
Tantalum	Ta	180.95	$(Xe)4f^{14}5d^36s^2$	73	(3, 4)5	Ta(III)—O(6)2.07; Ta(V)—O(6)2.04	3000	b.c.c.; 3.3026 (20 °C)
Technetium	Tc	98.91	$(Kr)4d^65s^1$	43	(3)4(5, 6)7	Tc(IV)—O(6)2.04	2700	h.c.p.; 2.735, 4.388 (R.T.)
Tellurium	Te	127.60	$(Kr)4d^{10}5s^25p^4$	52	(2)4(6)	Te—O(3)1.92	450	hex; 4.4566, 5.9268 (25 °C)
Terbium	Tb	158.93	$(Xe)4f^96s^2$	65	3, 4	Tb(III)—O(6)2.32; Tb(IV)—O(8)2.28	1450	h.c.p.; 3.599, 5.696 (20 °C)
Thallium	Tl	204.37	$(Xe)4f^{14}5d^{10}6s^26p^1$	81	1(3)	Tl(I)—O(6)2.90; Tl(I)—O(12)3.16	304	h.c.p.; 3.4566, 5.5248 (18 °C)

Element	Symbol	Atomic weight	Ground state config	Z	Oxidation states	Bond distances	m.p. (°C)	Crystal structure
Thorium	Th	232.04	$(Rn)6d^27s^2$	90	(3)4	Th—O(6)2.40; Th—O(9)2.49	1700	f.c.c.; 5.0843 (R.T.)
Thulium	Tm	168.93	$(Xe)4f^{13}6s^2$	69	(2)3	Tm(III)—O(6)2.27; Tm(III)—O(8)2.39	1600	h.c.p.; 3.5372, 5.5619 (20°C)
Tin	Sn	118.69	$(Kr)4d^{10}5s^25p^2$	50	2, 4	Sn(II)—O(8)2.62; Sn(IV)—O(6)2.09	232	tet; 5.8315, 3.1814 (25°C)
Titanium	Ti	47.90	$(Ar)3d^24s^2$	22	(2)4	Ti(II)—O(6)2.26; Ti(IV)—O(6)2.01	1680	h.c.p.; 2.9506, 4.6788 (25°C)
Tungsten	W	183.85	$(Xe)4f^{14}5d^46s^2$	74	(4)6	W(VI)—O(4)1.81; W(VI)—O(6)1.98	3380	b.c.c.; 3.1650 (25°C)
Uranium	U	238.03	$(Rn)5f^36d^17s^2$	92	(3)4(5)6	U(IV)—O(9)2.45; U(VI)—O(4)1.88	1133	orth; 2.854, 5.869, 4.955 (27°C)
Vanadium	V	50.94	$(Ar)3d^34s^2$	23	2-5	V(II)—O(6)2.19; V(V)—O(4)1.76	1920	b.c.c.; 3.028 (30°C)
Xenon	Xe	131.30	$(Kr)4d^{10}5s^25p^6$	54	2, 4, 6		−112	f.c.c.; 6.24 (−185°C)
Ytterbium	Yb	173.04	$(Xe)4f^{14}6s^2$	70	2, 3	Yb(III)—O(6)2.26; Yb(III)—O(9)2.38	824	f.c.c.; 5.481 (20°C)
Yttrium	Y	88.91	$(Kr)4d^15s^2$	39	3	Y—O(6)2.29; Y—O(9)2.50	1500	h.c.p.; 3.6451, 5.7305 (20°C)
Zinc	Zn	65.37	$(Ar)3d^{10}4s^2$	30	2	Zn—O(4)2.00; Zn—O(6)2.15	419	h.c.p.; 2.6649, 4.9468 (25°C)
Zirconium	Zr	91.22	$(Kr)4d^25s^2$	40	4	Zr—O(6)2.12; Zr—O(8)2.24	1850	h.c.p.; 3.2312, 5.1477 (25°C)

a. Based on the relative atomic mass of $^{12}C = 12.00$. These atomic weights are for natural isotopic abundances except for those given in brackets which refer to short-lived elements without a natural abundance. For these, the mass of the isotope with longest half-life is given.

b. These are the ground state configurations of the elements. In some cases with transition elements, lanthanides and actinides, the difference in energy between the ground state and the first excited state is small.

c. Most elements show a variety of unstable oxidation states in addition to the stable one(s). In compiling this table, principal note is taken of the oxidation state(s) found in crystalline oxides and halides.

d. Values are taken from Shannon and Prewitt, *Acta Cryst*, **B25** 925 (1969); **B26** 1046 (1970). Note that bond distances generally increase with cation coordination number and decrease with increasing cation oxidation state. Bond distances to fluorine are usually 0.05 to 0.10 Å shorter than distances to oxygen; bond distances to chlorine are usually 0.20 to 0.40 Å larger.

e. Data taken mainly from *International Tables for X-ray Crystallography*, Vol. III, p. 278. f.c.c. = face centred cubic (cubic close packed); hex. = hexagonal; h.c.p. = hexagonal close packed; R3m, R3̄m = rhombohedral; b.c.c. = body centred cubic; tet = tetragonal; orth = orthorhombic; cub = cubic; monocl = monoclinic; diam = diamond structure.

Formula Index

718

Subject and Author Index

720

726

734

Ward, R., 262
Warren, B. E., 186, 605, 607, 612, 637
Warren particle size formula, 175
Weak electrolyte theory, 620
Weighted reciprocal lattice, 702
Weissenberg method, 143, 147, 149 ff, 705
Welfel, E., 269, 316
Wells, A. F., 212, 226, 229, 262, 317
Wendlandt, W. W., 114
Wernick, J. H., 582
Wert, C. A., 357, 524, 564, 582
West, A. R., 211, 222, 225, 262, 367, 370,
    373, 440, 444, 451
Weyl, W. A., 637
Wheatstone bridge, 482
Whipple, A., 17, 46
White, E. W., 87, 101
White, J., 665
White radiation, 117, 177
Whittaker, E. J. W., 186
Whittingham, M. S., 46, 471, 496
Wilkinson, G., 559, 582
Willemite, 359, 397, 589, 634
Williams, R. J. P., 519, 524
Williams, J. O., 675
Wilson, A. J. C., 186
Window glass, 601, 621
Witte, H., 269, 316
Wold, A., 17, 32, 46
Wölfel, E., 269, 316
Wollan, E. O., 568, 582
Wollastonite, 589, 635, 665
Wolten, G. M., 448, 451
Woolfson, M. M., 186
Wooster, W. A., 186
Work hardening of metals, 345
Wormald, J., 186
Wright, D. A., 524
Wulff, J., 524
Wustite, 330, 362, 376, 648
Wurtzite structure, 44, 221, 222, 241,
    242 ff, 281, 302, 369
Wyckoff, R. W. G., 211, 212, 247, 249,
    255, 256, 257, 262

Xerox process, 625

Xonotlite, 42, 79
XPS, 48, 80, 92 ff
X-ray, absorption coefficient, 88
    absorption spectra, 85 ff
    diffraction, 49 ff, 57, 58, 59, 62, 69,
        115 ff, 333, 335, 355, 609, 699
    emission spectra, 85 ff, 117
    fingerprint characterization, 147
    fluorescence, 16, 48, 85, 96
    line broadening, 51, 52, 173
    line profiles, 169, 171
    powder diffraction, 15, 47, 49 ff, 72, 75,
        144, 166 ff, 366 ff, 607
    powder diffractometer, 166
    powder method, 16, 47, 49 ff, 166 ff
    reflections, intensities of, 153 ff
    scattering power, 58, 154
    spectroscopy, 48, 58, 85 ff, 499, 503
    tube, 118
    wavelengths, tables of, 118
X-rays, generation of, 116
XRF, 16, 48, 85, 96

Yao, Y. F., 30
Yield point, 349
Yttrium aluminium garnet (YAG), 592
Yttrium iron garnet (YIG), 575

Zachariasen's rules, 596 ff, 605
Zeeman splitting, 98
Zeolites, 19 ff, 25
    conductivity of, 479
    synthesis, 18 ff
Zero point energy, 283
Ziegler–Natta catalyst, 676
Zinc blende structure, 221, 224, 230, 240,
    281, 302, 499
Zinc tungstate, 24
Zircon, 664
Zirconia, 363, 365, 466, 663
Zone melting, 37, 39
Zone refining, 37
ZrO_2, conductivity of, 466
Zussman, J., 72, 101